NEW RISKS
Issues and Management

ADVANCES IN RISK ANALYSIS

This series is edited by the Society for Risk Analysis.

Volume 1 **THE ANALYSIS OF ACTUAL VERSUS PERCEIVED RISKS**
Edited by Vincent T. Covello, W. Gary Flamm, Joseph V. Rodricks, and Robert G. Tardiff

Volume 2 **LOW-PROBABILITY/HIGH-CONSEQUENCE RISK ANALYSIS**
Issues, Methods, and Case Studies
Edited by Ray A. Waller and Vincent T. Covello

Volume 3 **RISK ANALYSIS IN THE PRIVATE SECTOR**
Edited by Chris Whipple and Vincent T. Covello

Volume 4 **UNCERTAINTY IN RISK ASSESSMENT, RISK MANAGEMENT, AND DECISION MAKING**
Edited by Vincent T. Covello, Lester B. Lave, Alan Moghissi, and V. R. R. Uppuluri

Volume 5 **RISK ASSESSMENT AND MANAGEMENT**
Edited by Lester B. Lave

Volume 6 **NEW RISKS**
Issues and Management
Edited by Louis A. Cox, Jr. and Paolo F. Ricci

Volume 7 **RISK ASSESSMENT IN SETTING NATIONAL PRIORITIES**
Edited by James J. Bonin and Donald E. Stevenson

A Continuation Order Plan is available for this series. A continuation order will bring delivery of each new volume immediately upon publication. Volumes are billed only upon actual shipment. For further information please contact the publisher.

NEW RISKS
Issues and Management

Edited by

Louis A. Cox, Jr.
US West Advanced Technologies
Englewood, Colorado

and

Paolo F. Ricci
University of California, Berkeley
Berkeley, California

PLENUM PRESS • NEW YORK AND LONDON

Library of Congress Cataloging-in-Publication Data

New risks : issues and management / edited by Louis A. Cox, Jr. and Paolo F. Ricci.
p. cm. -- (Advances in risk analysis ; v. 6)
"Proceedings of the annual meeting of the Society for Risk Analysis, held November 9-12, 1986, in Boston, Massachusetts"--T.p. verso.
Includes bibliographical references.
ISBN 0-306-43537-3
1. Technology--Risk assessment--Congresses. I. Cox, Louis A. II. Ricci, Paolo F. III. Society for Risk Analysis. IV. Series.
T174.5.N48 1990
363.1--dc20 90-7162
CIP

Proceedings of the annual meeting of the Society for Risk Analysis,
held November 9-12, 1986, in Boston, Massachusetts

A Division of Plenum Publishing Corporation
233 Spring Street, New York, N.Y. 10013

Printed in the United States of America

Preface

This volume contains the proceedings of the 1986 annual meeting and conference of the Society for Risk Analysis. It provides a detailed view of both mature disciplines and emerging areas within the fields of health, safety, and environmental risk analysis as they existed in 1986. In selecting and organizing topics for this conference, we sought both (i) to identify and include new ideas and application areas that would be of lasting interest to risk analysts and to users of risk analysis results, and (ii) to include innovative methods and applications in established areas of risk analysis. In the three years since the conference, many of the topics presented there for the first time to a broad risk analysis audience have become well developed—and sometimes hotly debated—areas of applied risk research. Several, such as the public health hazards from indoor air pollutants, radon in the home, high-voltage electric fields, and the AIDS epidemic, have been the subjects of headlines since 1986. Older areas, such as hazardous waste site ranking and remediation, air emissions dispersion modeling and exposure assessment, transportation safety, seismic and nuclear risk assessment, and occupational safety in the chemical industry, have continued to receive new treatments and to benefit from advances in quantitative risk assessment methods, as documented in the theoretical and methodological papers in this volume.

A theme of the meeting was the importance of new technologies and the new and uncertain risks that they create. Thus, several comparatively new applications for quantitative risk assessment are addressed here—risks from biotechnology, genetic engineering, and microorganisms; from innovative technologies, projects, and practices in the telecommunications, transportation and energy industries; and from natural and man-made threats to computer systems. Methodological themes and issues in cancer risk analysis that are now much debated also appear here. These include (i) the use of "biologically motivated" stochastic models of carcinogenesis as opposed to more traditional statistical models; (ii) the relation (if any) between acute toxicity and carcinogenicity; and (iii) the conservatism of the one-hit statistical model for carcinogenesis.

Particularly noteworthy at the conference was a strong recognition, reflected in these papers, of the roles of societal risk management institutions and policy instruments as indispensable parts of practical risk management—especially tort law, regulation, insurance, and negotiated compensation and conflict resolution. Where previous SRA annual meetings had perhaps more strongly emphasized engineering determinations of "acceptable risk" levels and safety factors, the 1986 meeting saw risk acceptability issues being treated squarely from a social and policy perspective. Many papers in this volume are concerned with the interaction between corporate environmental and risk management decision making and the liability, insurance, and governmental risk management systems that shape a company's economic and societal environment. There is increased awareness of the importance of both effective risk communication and decision-making partnerships and agreements between industry and the public, between industry and government decision makers, and between management and labor for occupational risks. These themes are starting to draw industrial health and safety risk management practices away from reliance on their engineering roots and single-decision-maker orientation, and towards the social sciences and an emphasis on participatory and group decision making about risks. The recently emerged sub-field of risk communication has confirmed the importance and practical value of this shift in perspective for real-world corporate risk managers.

The public-sector side of health and safety risk management decision making also receives innovative treatments in this volume. Decision-analytic and cost-benefit approaches to public risk and environmental resource management are well represented. Several papers discuss public perceptions of uncertain risks and the roles of uncertainty quantification, uncertainty management, and uncertainty communication in shaping public- and private-sector responses to potential hazards. The involvement and participation of the public in risk management decision processes is an increasingly important part of practical risk-management for the public as well as the private sector. This trend, and the resulting need for public risk managers to communicate risk information effectively to the public and media, rather than only dealing competently with technical risk and decision analysis issues, have become increasingly recognized by analysts and decision makers and are well discussed in the papers collected here.

In summary, the 1986 meeting of the Society for Risk Analysis raised many of the new risk assessment and risk management issues that have filled the literature since. The papers in this volume introduce and explain many of the technical and research directions in which the field of risk analysis is now moving. The new technology areas and applications described here also raise a number of research challenges that have yet to be adequately addressed. Thus, this collection both provides a thorough introduction to currently active areas in health, safety, and environmental risk analysis and presents a set of challenging problems for new researchers and old hands in the field. We look forward to future responses to these challenges.

ACKNOWLEDGMENT

Many people deserve thanks for the high quality of the conference presentations and the papers: the authors, first and foremost; the New England Chapter of the Society for Risk Analysis, which organized and hosted the meeting; and the expert organizers and facilitators at the national level of the Society. Professors Roger Kasperson (Conference Chair) and Paolo Ricci (Program Chair) thank the many individuals at the local and national levels without whom the meeting would not have been possible. Arthur D. Little, Inc., supported Drs. Cox and Ricci in organizing the program. On behalf of the Society, we also give special thanks to Lorraine Abbott and Katie Ingersoll of Tec-Com, Inc., whose patience, persistence, and dedication to quality have made the production of this book a reality.

Louis Anthony Cox, Jr.
U S WEST Advanced Technologies

Paolo F. Ricci
UCLA, School of Public Health and
UC Berkeley (Bolt Hall) School of Law

November, 1989

Contents

Risk Management for Controlling Pollution Liability in a Corporate Environment

Marvin Feldman
Resource Decisions
San Francisco, CA

Leland D. Attaway
Law Environmental
San Rafael, CA

Robert F. Hahn
Pacific Energy Research Institute
Berkeley, CA

ABSTRACT

Even in well-managed corporations, deficiencies exist within the information gathering and incentive system with respect to pollution liability. This paper examines the legal basis for pollution liability and the typical corporate structures for managing this liability. A program is described for diagnosing the mismatch between responsibility and authority and reducing pollution liability in a cost-effective manner.

KEYWORDS: Environment, liability, pollution, compliance, hazardous substances, corporations

INTRODUCTION

Corporations today are faced with increasing regulations regarding handling, use, discharge, and release of chemicals and environmental contaminants. Their liability for pollution, especially that resulting from hazardous and toxic substances, has increased, as has the cost of cleanup. The courts and regulatory agencies are increasing settlement amounts and penalties for parties found responsible in some way for pollution, or for noncompliance with regulations. Associated with this is the drastic increase in the cost of environmental impairment insurance, if such insurance can even be obtained. In short, corporate pollution liability has become a major concern of many companies and of those responsible for controlling corporate risks.

Corporate pollution liability arises from a wide range of risks, some physical or biochemical, some legal and regulatory. This paper attempts to examine the universe of liability-generating risks and to compare these risks to the corporate structure responsible for mitigating them. Such a comparison leads to the inescapable conclusion that corporations are poorly structured to mitigate pollution liability. The basic problem is

New Risks
Edited by L. A. Cox, Jr., and P. F. Ricci
Plenum Press, New York, 1990

- Suboptimization, and
- The mismatch between responsibility and authority.

Suboptimization, the optimization of subsystems as opposed to the larger system, is endemic to almost all large organizations. To a certain extent, the effects of suboptimization can be ameliorated by periodically focusing on a problem-shed as opposed to the normal mode of focusing on functional responsibilities. Mismatch between responsibility at the operating level and authority at the corporate level must be addressed by analyzing the main elements of the pollution liability problem, considering the company's preferences, and suitably amending the corporate structure.

The remainder of this paper will examine the risk conditions which give rise to liability and relate these to the structure of typical manufacturing corporations. Procedures are suggested for improving corporate liability posture by diagnosing the dominant liability producing risks. Finally, a method is suggested for prioritizing a company's liability implementation plans. The aim of this plan is to attempt to enable corporations to develop a program that can be used in successfully defending suits associated with environmental impairment.

LIABILITY CONCEPTS

The events which give rise to legal liability include injury and noncompliance. Injury here includes harm done to employees, to the public, or to property through exposure to pollution. Being found out of compliance with regulations also gives rise to liability.

The field of pollution liability is fraught with lurid headlines, from Bhopal to your local Superfund site. However, one story which the authors have heard from many companies rarely makes headlines. It concerns a manufacturing firm which, faced with high disposal costs and limited options, sells or gives a solvent recycler its waste solvents or pays the recycler to accept them. The firm has complied with all applicable regulations such as labeling and manifesting its waste. Yet, when the recycler is found guilty of illegal disposal practices or becomes financially insolvent, the waste generator finds that it is party to a costly toxic pollution legal action.

This story illustrates a common fallacy—that regulatory compliance is sufficient to avoid liability. Regulatory compliance is, of course, a necessary condition for reducing pollution liability but it is by no means sufficient. Indeed, there is no practical way for an operating company to completely close what the lawyers term "the window of liability." The risks which give rise to most categories of liability are not totally controllable by the corporations. In fact, it is only with respect to regulatory liability that a firm can eliminate the underlying risks. Risks subjecting the firm to other types of liability—i.e., common law liability, employee liability, contractor liability, and product liability—can still be present and are frequently unrecognized.

Common Law Liability

Common law liability refers to the ability of a party to sue a company for real, perceived or fictitious damages resulting from a firm's activities. The last category (fictitious damages) illustrates why it is not possible to totally close the window of liability. No amount of risk prevention can avoid the possibility that a firm will be faced with the need to defend against a costly, albeit meretricious, claim. It is important to note that complete regulatory compliance does not eliminate common law liability, although the failure to comply will substantially weaken defense against a suit.

In general, common law liability rests on establishing negligence. The finding of negligence in turn rests on the presence of the following risk conditions:

(a) An injury occurred.
(b) Pollution caused the injury.
(c) The polluter breached the applicable duty (standard) of care.

Closing the window of liability involves reducing the probability of the risk conditions occurring. With respect to (a), (b), and (c), the following risk mitigation measures are available:

Injury. Reducing the risk of injury is the first line of defense against common law liability. Safety measures intended to reduce injury risk probability may take many forms. They may consist of equipment or facilities, training programs, monitoring, and process changes.

Causation. Liability involves establishing that an injury was not caused by the corporation's actions (if this is indeed true). This may only require record-keeping measures or environmental monitoring.

Duty of Care. The duty of care refers to the standards of good practice are prevalent among prudent operators in the same industry. Risk mitigation requires that the firm (1) be aware of the industry's duty of care, (2) adhere to standards at least as stringent as the rest of the industry, (3) has the ability to prove that it did, in fact, exercise the duty of care when the injury occurred, and (4) is aware of academic/scientific opinion.

If the company is engaged in an ultrahazardous activity, strict liability may apply. Under this concept, there can be legal liability without breach of the duty of care. Risk mitigation can be achieved by isolating substances and activities which are subject to strict liability so as to more effectively control public and employee exposure.

Another instance in which adherence to the duty of care is not a risk reduction measure is when a statute has been violated. In that instance, the statute may substitute for the industry-defined duty of care. This is termed negligence per se. Defense or risk mitigation here involves proof of compliance with the relevant regulations.

More severe common law liability situations include gross negligence and intentional tort. To establish gross negligence one must demonstrate that the firm persisted in negligent or flagrant activities even after it had been forewarned, or that injury was reasonably foreseeable. Treble damages are typically assessed for gross negligence. Intentional tort, which can result in criminal prosecution, involves actions with the intent of causing injury.

Employee Liability

Employee liability refers to the claims which an employee may make against an employer. In most jurisdictions the Workers Compensation system applies to these claims. Normally, Workers Compensation claims need only establish that the injury occurred in the scope of employment. The same risk mitigation measures discussed above apply. Risk probability reduction again could involve injury prevention. Record keeping, in this case to establish that the injury was not caused by the employer, might also be effective.

The size of the awards available under Workers Compensation are considerably less than those typically available under common law. Therefore, there is a growing tendency for employees to get outside of the Workers Compensation system by establishing that an employer was acting in a dual role (as direct provider of a service to the employee, or by establishing that the employer fraudulently concealed risks). Reducing the risks of workers getting outside the Workers Compensation system is an important source of increased liability.

Contractual Liability

Contractual liability (technically "vicarious liability") occurs as a result of associating with other businesses. When a firm has a contractual relationship with another firm, which then is guilty of negligent behavior, both firms may be penalized, even if the first firm had no direct part in the negligent activity. However, judgments are usually apportioned to reflect relative degrees of participation. Risk probability reduction can be achieved by careful screening of contractual partners, and to some extent by contractual language, such as hold harmless clauses and mutual indemnification. Another mitigation measure is to clearly define the contractual relationship so that it is possible to isolate liability.

In certain activities defined by statute (for example, the Federal CERCLA-Superfund), "joint and several" liability may apply. Under this concept, the degree of participation does not matter. Any corporation which participates in the activity may be held for the entire amount of the damages. Risk reduction strategies may be isolationism, or, conversely, seeking safety in numbers, particularly in seeking partners with deep pockets.

Product Liability

Product liability: Depending on state statutes and common law rulings, corporations may be liable for damages caused by their waste—or byproducts if they are defined as products. Product liability can also arise as a result of violations of the duty of care with regard to packaging, labeling or transporting hazardous products. Risk probability reduction can be achieved by adhering to the duty of care as well as by regulatory compliance.

CORPORATE STRUCTURE AND REPORTING LINES

Corporate officers who have direct responsibility for managing environmental pollution liability vary from company to company. These corporate officers include environmental and health and safety managers, risk and insurance managers, plant managers, financial managers, and corporate attorneys. The following discussion characterizes the central tendencies of many firms with respect to these functions.

Environmental and Health and Safety Officers

Environmental and health and safety officers are assigned to one or more locations and are responsible for environmental conditions affecting or affected by production. Health and safety officers are responsible for employee conditions and safety training, whereas environmental managers are responsible for public health and safety. The authority granted to these positions varies from tokenism to the authority to fire employees on the spot or to shut down plant operations if an unsafe condition occurs. Typically these positions are limited to ensuring compliance with health and safety environmental regulations. Since budgets for these activities are usually derived from operating revenues, there is a tendency to spend as little as possible, particularly in unfavorable economic conditions. With the recent proliferation of regulations at all levels of government, these individuals are usually overwhelmed with keeping up with the regulations.

Risk and Insurance Managers

Risk and insurance managers are typically located in the parent company headquarters rather than at operating plants. In general, the function of risk managers has been restricted to placing insurance coverage for the entire range of corporate risks, including those associated with pollution liability. Although their function may include what are termed loss control or "loss prevention" activities, these activities do not typically relate to pollution liability activities. There is usually little contact between the risk manager and the environmental managers. Prior to the 1980s, it was possible to insure companies against environmental pollution liability, as a part of the firm's comprehensive general liability (CGL) policy. Then later, when pollution liability was excluded from CGL policies, separate environmental impairment liability insurance was offered. Presently this type of coverage is very scarce and expensive. In many corporations, coverage for pollution liability is negligible and exposures are excessive. Yet this has not typically translated into more active programs for risk analysis and reduction by risk managers.

Plant Managers

Plant managers are responsible for seeing that their facilities remain productive and profitable. They are also concerned with worker safety and to some extent with public safety, but this concern is often in conflict with productivity and profitability. Naturally, the plant manager is likely to be satisfied concerning safety if the plant operation is in compliance with applicable regulations. There is little incentive for plant managers to be concerned with low-probability catastrophic loss events because such losses are not typically paid out of the operating plant's budget.

Finance Managers

Finance managers are typically corporate vice presidents or members of the board of directors (comptrollers). These individuals are responsible for furthering the financial goals of the entire corporation. Although their financial responsibilities encompass the tradeoffs between risk probability reduction measures and the expected value of the savings which these measures might produce, their attention is normally focused on productive efficiency. Only when faced with large pollution-related losses do these managers typically become aware of environmental liability issues.

Corporate Attorneys

Corporate attorneys like financial managers, do not typically concern themselves with pollution liability issues until losses occur. The legal department usually focuses upon matters of product liability and contracts and tends to be reactive rather than proactive.

MISMATCH BETWEEN RESPONSIBILITY AND AUTHORITY

From the above description of the typical corporate structure, it is apparent that a fairly wide gap exists between responsibility for action on environmental liability and authority to act. The structure does not ensure that pollution liabilities are adequately addressed. The organizational structure of corporations in general introduces two major problems with regard to pollution liability management:

- Inappropriate allocation of responsibility and authority, and
- Lack of requisite information transfer between the lowest corporate levels and the highest levels.

These corporate structure anomalies seriously inhibit the identification and cost-effective mitigation of pollution-related liabilities. Responsibilities fall between the slats of the corporate structure. Incentives for action by the relevant corporate officers are not designed to cope with the current environmental situation of complex and rapidly evolving changes in regulatory and common law liability and increasingly high costs associated with the consequence of low-probability events. In short, the corporate structure which has evolved over the years is inadequate to deal with pollution liability.

Corporate policy concerning pollution liability recognition and mitigation must be set at the highest levels. This policy must address both routine liability management processes and the response to crisis situations, and create accountability for loss control performance at all levels of management. The environmental officer usually focuses upon compliance matters only. The risk manager typically sees only insurance and claims and has minimal cognizance of environmental and health problems. And the financial vice president is usually unaware of the possible tradeoffs between environmental preparedness and corporate financial success.

At the operating level, it is the plant manager who has direct control of the corporate resources: raw materials; finished products; time; and manpower. He often gives low priority to proactive liability management because such measures might impose overhead costs which unnecessarily reduce his profitability and therefore his personal success. Further, the plant manager's budget frequently does not include resources for the preventive mitigation of environmental/health liabilities, and he does not have the authority to reprogram funds to such ends. In this case, these resources must be so allocated from higher levels within the company. Finally, the plant manager has control over the basic information necessary to identify and describe the potential liabilities needing mitigation.

As a result of these corporate structural phenomena, we see the authority for setting and overseeing corporate policy and for allocating necessary resources located in the centralized offices of the corporation, but the control of requisite information and mitigation elements is located at the middle management levels and below. The necessary information flow for managing the actions of these two levels is usually seriously lacking.

LIABILITY MANAGEMENT PROGRAM

Diagnosing and correcting pollution liability problems is difficult to accomplish from within the existing corporate structure. A programmatic approach is necessary to facilitate the flow of information between the levels of management. Representing a departure from normal corporate functions, this approach will require a facilitator, who may be someone within the company or an outside consultant. In either case it will be necessary to involve all of the corporate functionaries described above in an active and creative role. Specialized expertise in legal and regulatory matters, health risk assessment and decision analysis is also needed to complete the team.

A program which attempts to create action agenda for pollution liability reduction has been devised by the authors. The program aims to reduce the exposure to and cost of pollution liability through the following steps:

- Identifying corporate goals and objectives;
- Identifying existing problems and concerns;
- Determining critical failure modes and costs;
- Predicting the costs of reactive risk management;
- Performing regulatory and legal research to define potential problems;
- Identifying a range of possible risk-mitigation measures and their probable consequences; and

- Compiling an environmental liability implementation plan that takes into account costs and benefits of potential mitigation measures.

These steps are discussed in more detail below.

Goals and Objectives

The first step in developing an implementation plan is identifying and, to the extent possible, quantifying the goals and objectives of the corporation with respect to pollution liability. This can be accomplished through focused discussions with key corporate managers. This statement is corroborated by review of established environmental management procedures and compliance documentation. Corporate goals include regulatory compliance, productivity, worker and public safety, and long-term viability and growth of the company. Ideally, these multiple and invariably conflicting objectives could be reconciled through group assessments using simplified multiattribute utility assessment techniques. The assessment is simplified by first assuming a certainty of outcomes. Then risk preferences and acceptable levels of risk are qualitatively assessed. Constraints with regard to timing and availability of staff and budget for these activities are also assessed at this time, although it is important to retain the flexibility to direct resources to action items in proportion to their probable payoffs.

Existing Problem Identification

The team meets with operating level managers (plant managers and health and safety and environmental managers) in small groups to gather information on operating conditions and perceived problems. This is followed by a carefully planned site walk-through in which some or all of the following are reviewed first-hand:

- Environmental/health in-house records, especially regulatory compliance;
- Hazardous materials tracking;
- Hazardous materials handling;
- Controls on hazardous materials brought in by outside contractors, suppliers and vendors;
- Regulatory compliance records of contractors, suppliers, and vendors;
- On-site hazardous materials storage;
- Off-site hazardous materials disposal;
- General conditions of plant(s)—potential for environmental impact;
- General working conditions;
- Training; and
- Contingency preparedness.

This site walk-through is followed by one-on-one meetings with key legal and engineering personnel. Once the team has collected, collated, and digested adequate information on the above facets of corporate operations, it meets with corporate management and key operations personnel to review its findings and amend the information as appropriate.

Critical Failure Modes and Costs

If corporate policy for cost-effective environmental liability mitigation is to be effectively established and overviewed at corporate level, and successfully implemented at the operating level, then information describing the environmental/health aspects of the operating plants must be produced at the lower levels and transferred on a regular basis to corporate management. Two kinds of information are needed: descriptions of actual and potential critical failure incidents; and related costs of reacting to their occurrence (including all significant hidden costs).

To accomplish effective environmental/health liability management corporate-wide, it is necessary to apply a systematic approach to identifying safety, environmental, and operational problem areas. The objectives of this approach are to produce a list identifying all critical areas of chemical usage, storage or discharge/release, to evaluate each identified area for critical failure effect, and to prioritize each of these critical failure effects. Given this information, the corporate liability managers will then know the problems to be managed and their relative priority. However, in most corporations such a systematic listing/description of critical failure modes is not available at either the plant or corporate management levels. This lack of knowledge is one of the two most important reasons for a firm's failure to recognize and manage environmental liabilities.

Costing Profile

The other important reason for inadequate management of environmental/health liabilities is the lack of recognition of the excessive costs of reactive risk management as opposed to preventive risk management. In order to design cost-effective risk management policies, knowledge of the costs of reactive management is required. Indeed, these costs are the principal means of convincing management at all levels that preventive risk management is preferred to strictly reactive measures. These costs are usually overlooked or simply included in the overhead as the cost of doing business. These real costs of environmental/health incidents can be measured and controlled, but only if information of the type discussed here is produced at the local plant level and transferred to the corporate offices responsible for liability management. As is the case for critical failure identification, the information necessary to estimate the real (hidden) costs of critical failures is available neither at the plant nor at the top corporate levels. It must be produced at the plant level and transferred to the corporate risk managers if environmental/health liabilities are to be managed in a cost effective manner.

Regulatory and Legal Research

Based on the site visit, the team identifies agencies and compliance issues most critical to the firm. Guidelines and timetables are developed for communicating with the key agencies to sustain a positive compliance record. Workers Compensation appeals' decisions in related industries are reviewed in order to determine the types of claims which were brought and the rules of evidence which prevail in the firm's jurisdiction(s). Legal literature is reviewed to determine the industry standards of care as defined by the case law in the appropriate jurisdictions. The objective of this analysis is to identify the most probable causes of action against the company.

Risk Mitigation Measures

This entire data collection process is designed to acquire all the information necessary to identify existing and potential liability issues, and to design and recommend potential action recommendations. Based on analysis of the results of the site inspection, the corporate structure, and the legal and regulatory research, a wide range of potentially relevant risk-mitigation measures are developed. An example of some of the measures which might be considered include the following:

Operational:
- Train employees in health and safety.
- Modify materials-handling procedures.

Process:
- Modify plant facilities.
- Change product line.
- Modify processes.

Waste Disposal:
- Change disposal method or contractor.

- Use on-site disposal.
- Incorporate waste-minimization strategies.

Legal:
- Improve compliance assurance procedures.
- Improve documentation of handling and disposal.
- Develop contingency plans for emergencies.
- Merge or separate manufacturing divisions.

Other:
- Insure residual risks.
- Improve management information flows.

Note that few of these measures are required for regulatory compliance.

Implementation Plan

It would not be possible nor perhaps even desirable to implement all of the potentially applicable risk-mitigation alternatives. In the next stage of the program, the team uses the objectives and constraints identified at the beginning of the program to develop a draft liability management implementation plan. This plan consists of a prioritized list of recommended liability reduction tactics, the cost to implement those tactics, and a measure of the extent to which liability will be reduced. The expected value of the benefit of each liability reduction tactic is estimated. The economic benefit of each tactic can be estimated using statistics on probability of suits being filed, probability of adverse judgment occurring with and without such records, and the average awards in these types of suits. When it is too costly to obtain accurate benefit data, informed expert opinion is an acceptable substitute.

The probable benefit of each liability mitigation tactic is compared with its implementation cost. Cost-effective measures are then prioritized with reference to the goals, objectives and constraints developed at the beginning of the process. At the present state of knowledge, it is not possible to adequately quantify the expected value of the outcomes with sufficient rigor to optimize a specified multiattribute utility function. However, the function can provide guidance in the judgmental process of prioritizing the implementation plan. Specific plans for implementing the program are then developed. Finally, the implementation plan should be periodically reviewed and amended as appropriate to take into account the effectiveness of the plan and legal and regulatory changes which may have occurred.

CONCLUSIONS

The program described here provides corporate management with the information needed to embark on a realistic and cost-effective liability management plan. The need for a programmatic approach to the liability problem stems from the inherent mismatch between responsibilities and authority which exist in even well-run corporations. The key to this process is discovering the weaknesses in the information-gathering flows and incentives in the existing corporate structure and making appropriate accommodations. Awareness of the large costs implied by a passive as opposed to a proactive liability management plan should convince management of the need for action.

Corporate Management of Liability Risk

Judith E. Selvidge
Strategic Decisions Group
Menlo Park, CA

ABSTRACT

Top level corporate managers responded to the 1985-1986 "crisis" in the area of liability risk by making strategic decisions that have impacted the types of business they are in and the way their firms do business. With the drastic changes in the availability and cost of insurance, which is the primary mechanism for transferring liability risk, managers are increasingly requiring explicit analyses of these risks in order to improve their understanding of the firm's options and to support the managers' decisions. These assessments identify the sources and magnitudes of the risks and evaluate the cost/benefit trade-offs among different risk reduction activities, such as, making changes in production process, divestiture and promoting tort reform.

KEYWORDS: Corporate risk management, liability risk, risk transfer, risk reduction

INTRODUCTION

To manage their liability risks, corporations go through a three step process: (1) assessing risks, (2) reducing risks, and (3) transferring risks. For a given firm, this process results in a level of retained risk intended to be compatible with the organization's overall ability to bear risk. The natural order of this process is shown in Fig. 1; however, the actual order and degree of emphasis on the different steps varies considerably from firm to firm.

The risk assessment step consists of the identification and quantification of sources of liability risks: for example, product liability and environmental impairment risks for manufacturers, professional liability risk for service providers such as consultants, lawyers and accountants, and the liability of the directors and officers of a firm in the execution of their duties. To quantify these risks, the firm assesses both the probabilities of occurrence and the consequences (given the occurrence) on a number of monetary and non-monetary dimensions.

Risk reduction refers to all the actions, voluntary and mandated, that are intended to reduce the probabilities, the consequences, or both. Testing new products, consumer education, proper disposal of wastes, peer review of professional work, and "due diligence" in presenting investment opportunities are all examples of activities intended to reduce risk.

In the area of risk transfer, the most common mechanism for shifting risk to another party is traditional insurance. The party laying off the risk (i.e., the purchaser of the

New Risks
Edited by L. A. Cox, Jr., and P. F. Ricci
Plenum Press, New York, 1990

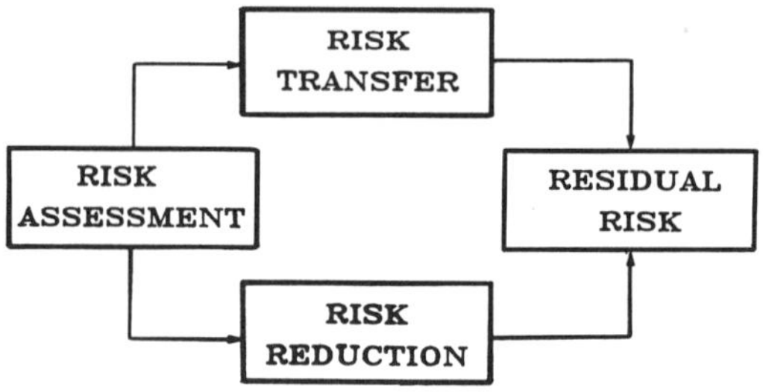

Fig. 1. Corporate risk management

insurance) is making the implicit judgment that the insurance premium is less than or equal to the firm's "certain equivalent"[a] for the gamble represented by bearing the risk. Other ways of transferring risk include contractual transfer of risk (e.g., a hold-harmless agreement between an investment banking firm and the clients it advises), divestiture (e.g., the sale and lease back of a facility), and legislative protection (e.g., the Price Anderson Act limiting the liability of nuclear power plant operators in the event of an accident).

ADAPTING TO CHANGES IN RISK TRANSFER OPTIONS

The objective of most firms is to operate at a fairly consistent level of risk. During the period 1985-1986, the availability and the cost of insurance as a transfer mechanism changed drastically. These changes were particularly noticeable in the liability risk area. Some types of coverage were withdrawn (e.g., liability for gradually occurring pollution); for many others the limit, that is, the maximum amount of loss covered, was reduced—sometimes by as much as 50% to 80%, and for all liability insurance the premiums charged went up—often by more than 500%.

Industry responded to the difference in terms of traditional insurance for managing liability risk in several ways, both internal and external to the firm.

Internal changes include changes in the ways of doing business, for example:

- altering the production process to eliminate the use of toxic chemicals during manufacturing,
- stricter screening of prospective clients by accounting and law firms,
- declaring bankruptcy to limit the extent of liability claims against the firm's assets (e.g., Johns Manville's response to asbestos liability claims).

Actions external to the firm and designed to increase the options for corporate risk management include the formation of new insurance-type mechanisms; for example,

- "captive" companies established by a particular firm or industry group,

---

a. For a discussion of certain equivalents, see, for example, R. L. Keeney and H. Raiffa, *Decisions with Multiple Objectives: Preferences and Value Trade-Offs*, John Wiley and Sons, New York, 1976.

- new shareholder-owned insurance companies to meet specific needs (e.g., ACE Insurance Company which provides $100 million in liability coverage in excess of the first $100 million of loss).

In these new formulations, the participants were able to tailor the features of the coverage to their particular needs, since the participants themselves embody the objectives of both the provider and the purchaser of the insurance.

Elimination of a product or divestiture of a business unit are actions that can be called the ultimate in risk transfer. Examples of such liability transfers in recent years include decisions by many pharmaceutical firms to cease production of vaccines and the selling by an electric utility of its coal mining subsidiary. Of course, when liability transfer is a factor in a firm's divestiture decision, the assumption of liability by the purchaser is reflected in the price paid.

And, finally, the firms have promoted legislative risk transfer by supporting initiatives for tort reform aimed at limiting or making more predictable the risks faced.

CONCLUSIONS

In designing, evaluating and implementing these new responses to risk, there is a role for formal risk assessment. It is mainly in the context of evaluating alternative risk transfer options and, to a lesser degree, making cost benefit trade-offs between risk reduction versus risk transfer. Corporations lean toward risk transfer for managing risk for a number of reasons. Among these are:

- its is simple and has an immediate impact;
- it often takes the form of a tax deductible expenditure; and
- it achieves an identifiable reduction in risk for a fixed cost.

The latter point is in contrast to risk reduction activities which often can only demonstrate a reduction in the "expected value"[b] of the loss. Risk reduction is most often practiced by firms with strong safety consciousness as a part of their corporate culture or by firms responding to regulatory requirements.

---

b. The mathematical expection of loss, i.e., the sum of the possible losses, each weighted by its probability of occurrence.

The Independent Auditors' Assessment and Conveyance of Risk (Preliminary Results)

Thomas A. Buchman
University of Colorado
Boulder, CO

ABSTRACT

The independent auditors' primary objective is to insure the reader (user) of a corporation's financial statements that those financial statements present "fairly" the information that they purport to present. The auditor does that by issuing an audit report to that effect; the audit report then accompanies the financial statements. By necessity, financial statements contain many estimates and judgments (such as the degree of collectibility of the firm's receivables, the amount of money to be realized from the sale of a company's inventory, the exposure associated with a lawsuit for which the company is a defendant, etc.). The auditor must evaluate the degree and type of uncertainty associated with the financial statements. If a firm is faced with material unusual circumstances, auditors are presently required to "qualify" their audit report by mentioning the uncertainty in the report. This research will examine how successful auditors are in determining and conveying whether a firm is faced with "material unusual circumstances."

KEYWORDS: Financial statements, audit report, business risk, qualified opinion, litigation

INTRODUCTION

A business faces several forms of uncertainty or risk (the terms uncertainty and risk will be used interchangeably in this paper), including various forms of business risk (e.g., lawsuits, bankruptcy, etc.), physical risks (earthquake, etc.), political risk (e.g., unfavorable legislative action, etc.). The management of a company is usually in a position to be aware of the probabilities and magnitudes associated with many of the risks. However, other persons, external to the firm, (e.g., stockholders and bondholders) having direct interests in the welfare of a company are not usually in a position to obtain the necessary information to make those assessments. Those external persons rely, primarily, on the information that is supplied by the company's management. The major source of that information is the company's annual report.

The independent auditors' primary objective is to insure the reader of a corporation's financial statements that those financial statements present "fairly" the information that they purport to present. The auditor does that by issuing an audit report that states whether or not the financial statements present fairly the financial information of the company; the audit report then accompanies the financial statements in the annual report. By necessity, financial statements contain many estimates and judgments (such as the degree of collectibility of the firm's receivables, the amount of money to be realized from the sale of a company's inventory, the exposure associated with a lawsuit for which the company is a defendant, etc.). The auditor must evaluate the degree and type of uncertainty associated with the financial statements. If a firm is faced with material unusual circumstances, auditors are presently required to "qualify" their audit report by mentioning the uncertainty in the report. An audit report that mentions an unusual uncertainty is called a "subject to" qualified report.

At the time the practice of issuing a qualified opinion was developed, it was apparently thought that the auditor, because of their expertise, their intimate knowledge of management's plans, etc., should appropriately communicate their assessment of the probability of the resolution of an uncertainty if it could have a material effect on the financial statements being audited [Banks and Kinney, 1982, p. 243]. The profession's guidance was the following:

> In cases where the probable effects of a matter are not reasonably determinable at the time of the opinion, such as the case of certain lawsuits, tax matters, and other contingencies which may have a material effect upon the financial statements, and the final outcome is dependent upon the decision of parties other than management, the independent auditor should appropriately qualify his opinion [AICPA, 1963].

Statement on Auditing Procedures No. 33 gives additional guidance. When the events which will resolve a contingency are:

> probable (likely to occur), but the amount of the loss cannot be estimated,...the auditor should modify his opinion when he believes that the resolution of the uncertainty will have a material effect on the financial statements he is examining [AICPA, 1963, para. 9509.14].

Also, when the probability of a loss is:

> reasonably possible (chance of occurrence is more than remote but less than likely) ...the auditor might modify [his opinion] if he concludes that the potential effect of the resolution of the uncertainty is material [AICPA, 1963, para. 9509.14].

However, in 1978, the Commission on Auditors' Responsibility (hereafter CAR or the Cohen Commission) felt there were several deficiencies in the auditing profession. Of relevance to this paper was the CAR's view of the reporting on uncertainties by independent auditors:

1. The auditor normally evaluates whether financial information presented by management conforms with appropriate standards; but for uncertainties the auditor is required to be a "reporter and interpreter" of financial information as well [CAR, 1978, p. 25].

2. The "subject to" phrase is ambiguous to users because they don't know if the auditor intends to highlight information more fully disclosed or to indicate a deficiency in the financial statements [CAR, 1978, p. 25].

3. Since auditors qualify their opinion for some material uncertainties, the absence of a qualification may lead users to believe that a company faces no uncertainties that could materially affect its financial condition or operating results [CAR, 1978, p. 26]. In fact, all companies face a wide variety of economic risks.

For the reasons above, the Cohen Commission recommended that the "subject to " qualification be eliminated [CAR, 1978, p. 29].

Very shortly after the CAR conclusions were published, AudSEC (then a committee of the American Institute of Certified Public Accountants—the standard setting body for the auditing profession) recommended, in an exposure draft, that the subject to qualified opinion be eliminated and that adequate disclosure in the footnotes be made as to the nature of the contingency which would have otherwise led to a subject to opinion [Auditing Standards Executive Committee, 1978]. AudSEC subsequently withdrew the exposure draft when it created a great deal of dissension among AICPA members. Later, the Auditing Standards Board (ASB), which superseded the AudSEC, voted a draft Statement on Auditing Standards which would eliminate the use of the subject to qualified opinion if the client company has reasonably estimated and disclosed any significant uncertainty [*Journal of Accountancy*, 1982, p. 3]. The ASB also withdrew its tentative proposal after failing to receive sufficient support for it.

Hence, the profession has not abandoned the "subject to" qualified opinion, but as a result of actions by the accounting profession, researchers began questioning if users do use the information contained in a "subject to" qualified opinion. The studies have used several approaches: security price research methods (SPRM), questionnaire studies asking subjects to make the types of decisions that financial statement users might make, and questionnaire studies gathering information about subjects' perceptions of audit reports.

The major thrust of that accounting research has been to see if investors act as if the "subject to" opinion conveys some information about added uncertainty. Little research has investigated whether users should attribute more uncertainty to financial statements containing a "subject to" audit opinion.

While, as might be expected, the results of the studies are somewhat inconclusive, it appears, in general, that users do not act as if the subject to qualified opinion conveys much, if any, information. Most of the SPRM studies found no price reaction when the "subject to" qualified opinion was made public, but there does seen to be a reaction when the event causing the qualification (lawsuit, etc.) becomes public knowledge [Banks and Kinney, 1982; Chow and Rice, 1982; Davis, 1982; Ball *et al.*, 1979; Firth, 1978; Alderman, 1977; Shank *et al.*, 1979].

The laboratory and questionnaire studies that attempt to see if a "subject to" qualified opinion made a difference in a subject's decision have also reached similar conclusions [Buchman, 1984; Friedman and Neumann, 1981; Estes and Reimer, 1979; Libby, 1979].

RESEARCH DESIGN

This research looked at firms receiving "subject to" qualifications from an inadequately explored perspective. It is argued that, in attempting to see if there is information content in a "subject to" qualification, SPRM researchers have assumed that the opinion contains information if investors change their perception about the fair value or risk associated with the firm's securities. Other research methodologies have made similar assumptions, such as, an opinion contains information if a bank loan officer requires different terms for a loan to a firm receiving a"subject to" qualified opinion than for a loan to a firm receiving an unqualified opinion (but having footnote disclosure of the uncertainty).

However, it is further argued, what should be of equal or greater interest to the auditing profession is the accuracy of the auditor's assessment of the uncertainty, i.e., if an auditor calls attention to some uncertainty, can a financial statement user expect that there is a greater probability of that uncertainty than if the opinion was unqualified?

METHODOLOGY

Firms receiving audit opinions mentioning "subject to" in the audit reports accompanying their fiscal 1977 financial statements were obtained from the NAARS data base. Of the 610 companies identified by NAARS:

319 were qualified for one reason,

144 were qualified for more than one reason,

119 audit reports mentioned "subject to" because of a removal of a qualified opinion from a prior year,

19 were not qualified opinions,

8 were subsidiaries of other companies and not publicly held, and

1 was a company not using Generally Accepted Accounting Principals.

610

The 319 firms receiving a "subject to" qualification for one reason formed the experimental group for the study. The audit reports for these firms were read and the audit opinions classified into one of seven reasons for the qualification:

1. Litigation,
2. Valuation/realization of assets,
3. Discontinued operations,
4. Going concern,
5. Tax or renegotiation liability,
6. Contract claim, and
7. Utility rate increase approval pending.

For firms in the litigation and going concern groups, the subsequent financial press, annual reports, and 10-K's were examined to determine the outcome of the uncertainty. This search process gave us a measure of the auditors' "hit" rate—or percentage of times auditors qualified opinions were associated with the actual event for which the auditor's opinion was qualified.

As a basis for comparison a control group of an equal number of firms receiving unqualified opinions in the same year were also selected. (The control group firms will be selected from the first year the matched subject to firms' qualification appeared. That is, if a firm receiving a subject to qualification in 1977 is found to have received a qualified opinion for the same reason in 1974, then 1974 will be the year used in selecting the control firm.) The control group will be matched on the variables that research indicates may be relevant (such as firm size, industry, exchange, etc.). For the control group, these same documents will be examined to see if there occurred an event which caused the qualification in the experimental group (i.e., bankruptcy, loss of a lawsuit, etc.).

RESULTS

A very small number of experimental and control firms have been looked at to date, so no statistical analysis has been performed and the conclusions must be viewed as tentative.

Going Concern Qualification. Approximately 73% of the firms receiving a going concern qualification did declare bankruptcy or file for reorganization. The remaining firms either merged or subsequently had the qualified opinion removed. Control firms were matched on industry and a measure of profitability (net income/total asset). Seventy percent of the control firms later experienced bankruptcy or reorganization. This leads us to conclude that auditors are fairly good at predicting the event of bankruptcy/reorganization; but given that the firms in the control group experienced almost an identical "failure" rate, we must conclude that a qualified opinion provides little information to a user.

Litigation. Of the data analyzed to date related to litigation, the auditor had a much lower "hit rate." Firms receiving litigation qualified opinions reported a loss only 22% of the time; those losses averaged 60% of income. Only 4% of the control firms (matched on industry, sales, and footnote disclosure of a lawsuit) reported a lawsuit loss; the losses averaged 15% of income. The litigation results are much more interesting than those of the going concern qualifications. They show that a firm receiving a litigation lawsuit is much more likely to lose a lawsuit (and more money) than a company not receiving such a qualification. However, the auditor has a fairly large error rate—78% of the time there is not a material loss to a firm receiving a qualified opinion for litigation.

REFERENCES

Alderman, C. W., 1977, The Role of Uncertainty Qualifications: Evidence to Support the Tentative Conclusions of the Cohen Commission, *Journal of Accountancy*, pp. 97-100 (November).

American Institute of Certified Public Accountants, 1963, Statement on Auditing Procedures No. 33.

Auditing Standards Executive Committee, 1978, Auditors' Reports When There Are Contingencies, Exposure Draft, AICPA, New York.

Ball, R., Walker, R. G., and Whittred, G. P., 1979, Audit Qualifications and Share Prices, *Abacus*, pp. 23-24 (June).

Banks, D. W., and Kinney, Jr., W. R., 1982, Loss Contingency Reports and Stock Prices: An Empirical Study, *Journal of Accounting Research*, pp. 240-254 (Spring).

Buchman, T. A., 1984, An Effect of Hindsight on Predicting Bankruptcy with Accounting Information, *Accounting, Organizations and Society* **10(3)**:267-285.

Chow, C. W., and Rice, S. J., 1982, Qualified Audit Opinions and Share Prices—An Investigation, *Auditing: A Journal of Practice and Theory*, pp. 35-53 (Winter).

Commission on Auditors' Responsibilities, 1978, Report, Conclusions, and Recommendations, AICPA.

Davis, R. R., 1982, An Empirical Evaluation of Auditors' 'Subject To' Opinions, *Auditing: A Journal of Practice and Theory*, pp. 13-32 (Fall).

Estes, R., and Reimer, M., 1979, A Study of the Effect of Qualified Auditors' Opinions on Bankers' Lending Decisions, *Accounting and Business Research*, pp. 250-259 (Autumn).

Firth, M., 1978, Qualified Audit Reports: Their Impact on Investment Decisions, *The Accounting Review*, pp. 642-650 (July).

Friedman, L. A., and Neumann, B. R., 1981, A Test of the Effect of Qualified Audit Opinions on Investment Decisions, Working Paper, University of Colorado.

Journal of Accountancy, 1982, ASB Votes On 'Subject To' Opinions, p. 3 (April).

Libby, R., 1979, The Impact of Uncertainty Reporting on the Loan Decision, *Journal of Accounting Research*, Supplement, pp. 35-57.

NAARS (National Automated Accounting Research System), LEXIS, Mead Data Central, Inc., New York.

Shank, J. K., Dillard, J. F., and Murdock, R. J., 1979, How Financial Executives Regard 'Subject To' Opinions, *Financial Executive*, pp. 824-35 (November).

Strict Liability and Insurance Under Loss Misestimation[a]

Rudy Yaksick
University of Pennsylvania
Philadelphia, PA

ABSTRACT

The objective of this paper is to examine the interaction between the strict liability law and third-party liability insurance with regard to controlling and allocating accident risks when injurers misestimate the severity of victims' losses. Our concern is that liability insurance reduces the incentives for risk reduction provided by liability law when the standard assumption of perfect risk information is invalid. We address the following question: How does an injurer's over- or underestimation of victims' losses affect (a) the efficiency of the injurer's accident prevention and insurance purchasing behavior and (b) the efficiency of the competitive insurance market?

Using Shavell's hybrid Accident-Contingent Claims Model, we obtain one counterintuitive result. When risk-averse injurers overestimate losses and moral hazard is present, they may *ex ante* purchase partial coverage that may equal the victims' realized *ex post* losses. In other words, an injurer's error in estimating the severity of risk leads to an efficient allocation of accident prevention resources and accident risk in the presence of moral hazard. In contrast, moral hazard normally causes inefficient resource allocation when insurers cannot observe an injurer's protection activity.

Our results have two major implications: (1) Liability insurance acts as a substitute for liability law with respect to providing incentives for risk reduction; and (2) given the uncertainty about risk severity, difficulties are encountered in designing optimal incentive schemes. Consequently, policy makers may be unable to balance the conflicting policy objectives of fair and timely victim compensation, accident risk reduction, and the preservation of socially necessary but hazardous economic activities.

---

a. Only the abstract of this paper is published here, since the full paper was published in *Insurance and Risk Theory*, M. Goovaerts, F. de Vylder, and J. Haezendonck, eds., pp. 465-479, Vol. 171, NATO Advanced Science Institutes Series C: Mathematical and Physical Sciences, D. Reidel Publishing Co., Dordrecht (1986).

New Risks
Edited by L. A. Cox, Jr., and P. F. Ricci
Plenum Press, New York, 1990

Comparison of EPA, ILO, and World Bank Guidelines for Risk Identification Assessment, and Management of Chemical Emergencies[a]

S. C. Morris, P. D. Moskowitz, V. M. Fthenakis and L. D. Hamilton
Brookhaven National Laboratory
Upton, NY

ABSTRACT

In the wake of the Bhopal accident, the U.S. Environmental Protection Agency, the International Labour Organisation, and the World Bank have each issued guidelines for identification, assessment, and management of risks from major hazards at industrial facilities. These different approaches are compared in both theoretical and practical perspectives.

KEYWORDS: Risk assessment, risk management, chemical emergencies, chemical hazards, industrial accident hazards, hazard assessment

INTRODUCTION

The tragic accident in Bhopal, India, in December, 1984, resulted in increased international concern for the risk of major hazards at industrial facilities. This has included reevaluation of such risks by professional organizations, corporations, and all levels of government. Three different guidelines for assessment of these risks were issued in the year following Bhopal by the International Labour Organisation (ILO), the U.S. Environmental Protection Agency (EPA), and the World Bank. Concern with major industrial accident hazards did not begin, however, with Bhopal. Guidelines were established by the European Community (EEC) in 1982 following major accidents in the 1970s in the U.K. (Flixborough, 1974), Holland (Beek, 1975), and Italy (Seveso and Manfredonia, 1976).[1] Another important set of guidelines were those developed by the U.K. Health and Safety Executive.[2] The EPA, ILO and World Bank guidelines drew heavily on this past work.

The nature of guidelines for evaluation of risks of major hazards at industrial facilities may differ depending on the intended users, the specific problems addressed, and the way they are addressed. Guidelines for local community groups, for example, will

---

a. A modified version of this paper appears in *Environment International* **23**:305-310 (1987).

Table 1. Criteria for Inclusion in EPA List of Acutely Toxic Chemicals[a]

Route	Acute Toxicity Measure	Value[b]
Dermal	Median lethal dose (LD_{50})	≤50 mg/kg
Oral	LD_{50}	≤25 mg/kg
Inhalation	Median lethal concentration (LC_{50})	≤0.5 mg/l[c]

[a]From EPA, 1985.

[b]Criteria matched against all mammalian test species evaluated for all chemicals.

[c]Time of exposure up to 8 hours.

differ from guidelines intended for consulting engineers. Concern may focus on the probabilities or consequences of events, on risks to the public or to workers or to both, on assessment of the hazard alone or more broadly on assessment, management, and emergency preparedness. In this context, this report compares the guidelines published by EPA, ILO, and the World Bank.

SUMMARY OF THE THREE APPROACHES

EPA Guidance on Chemical Emergency Preparedness

In June, 1985, EPA announced a strategy to deal with the problem of air toxics, including accidental releases of acutely toxic chemicals. The goals were to increase community awareness of chemical hazards and to develop state and local emergency response plans for chemical accidents. An interim guidance document was issued in November, 1985.[3] It drew in part on earlier guidance developed jointly by EPA and the Federal Emergency Management Agency.[4]

EPA's guidelines aimed at local communities. They provide general qualitative guidance on how to organize a community working group, develop community awareness, gather data, identify facilities which should be investigated, determine what control measures exist at those facilities, and evaluate the facilities' and community's emergency response capabilities.

Aside from this general discussion, the EPA guidance has two distinctive features. The first is a list of "acutely toxic chemicals." Presence of a listed chemical serves only to identify a need for further action. It does not in itself indicate a potential serious hazard and has no regulatory ramifications. Criteria for listing are given in Table 1. Consideration is given only to acute toxicity; substances which present other acute hazards such as explosives, flammables, and corrosives are not considered (the guidance refers to Department of Transportation guidance).[5] Chronic hazards such as carcinogenicity are not considered.

The second feature of the EPA guidelines is a quantitative tool to rank chemicals by degree of concern. This reduces the results of an atmospheric dispersion model to a nomograph. Entering this nomograph with a toxicity index (EPA recommends the National Institute for Occupational Safety and Health Immediately Dangerous to Life and Health or

Table 2. Factors for Which Only Qualitative Discussion Is Provided in EPA Guidelines

- Physical state of the chemical
- Volatility
- Potential for interaction with other chemicals
- Specific safety precautions currently used
- How the chemical is handled
- Is the chemical under pressure?
- Is the chemical at ambient or elevated temperature?
- Is the chemical involved in a reactive process?
- What is the nature of the area surrounding the site?

IDLH level be used)[6] and the distance to the population at risk, one determines the quantity of the chemical which represents a concern. Assumptions behind this calculation are a large rupture or spill resulting in complete dispersal of all the material in a 1-minute puff release (which gives the highest air concentration per unit release) at ambient temperatures (no plume rise) from a 3 feet high release point. It applies to catastrophic release of a gas, volatile liquid or liquefied gas which under high temperature or pressure can become quickly airborne. The atmospheric dispersion calculation is an upper bound (highest 30 minute concentration between 100 feet and 10 miles downwind under the worst meteorological conditions). Furthermore, the atmospheric dispersion model used is not a heavy gas model, and thus in general overestimates downwind concentration for heavy gases.

The EPA method overestimates concentration and underestimates the threshold quantity of concern for situations where the chemical has low volatility or is stored under ambient conditions. Qualitative guidance is provided on such factors and others which could affect exposure (see Table 2). No help, however, is given on how to modify the quantitative levels of concern to take these factors into account.

ILO

The ILO guidelines are embodied in a working paper produced by a Tripartite (employers, labor, government) Ad Hoc Meeting of Special Consultants on Methods of Prevention of Major Hazards in Industry.[7,8] These guidelines include methods of analysis, hazard management, and emergency plans for workers and the public for preventing major hazards in industry. They encourage nations to establish a regulatory program and provide the framework for such a program. Much of the approach drew on the earlier EEC guidelines.[9] Specifically, the EEC list of hazardous substances with threshold quantities above which facilities must notify regulatory authorities was adopted. Identification of a hazard is based on classes of materials considered hazardous (Table 3).

The ILO guidelines place the fundamental responsibility for ensuring safe operation on the employer and indicate that the operating organization conduct its own hazard surveys in addition to those conducted by the regulatory authority. Hazard management is achieved through design standards, protective devices, maintenance, inspecting and testing, emergency planning, and training. Considerable administrative guidance is provided, but not technical guidance. For example, guidelines for the integrity of pressurized systems include recommendations that inspection, preventive maintenance and standby equipment for pumps are necessary only if failure could create a hazardous situation. Determination of whether pump failure could lead to a hazardous situation and the design of preventive

Table 3. ILO Classes of Hazardous Materials

1. Toxic gases (releases in "tonnage" quantities, lethal or harmful at considerable distances from the point of release).
2. Extremely toxic materials (releases in "kg" quantities).
3. Flammable liquids or gases (releases in vapor form in "tonnage" quantities).
4. Unstable or highly reactive materials with explosion potential.

maintenance programs and standby systems is left to professionals on the staff of the employer with oversight by the regulatory agency.

The EEC and ILO guidelines place two levels of responsibility on manufacturers. The first is that they must "take all measures necessary to prevent major accidents and to limit their consequences..." and "to prove to the competent authority at any time ... that they have identified existing major-accident hazards, adopted the appropriate safety measures, and provided the persons working on the site with information, training, and equipment in order to assure their safety." Guidelines similar to EPA's (Table 1) are provided for identification of very toxic, toxic, flammable, and explosive substances for this purpose. Although specific for toxicity and flammability, they allow considerable interpretation regarding the chemical and physical conditions which might result in risk of a major accident.

The second responsibility requires the manufacturer to notify competent authorities if amounts of specific chemicals over those specified in a list included in the guidelines are present in storage, processing, manufacture, or produced as residues or by-products. ILO, in adopting this approach, indicates that "... notification to the relevant health and safety enforcement authority is absolutely fundamental to any system of control." This requirement constitutes the quantitative aspects of the ILO guidelines. In order to avoid overwhelming authorities in countries where no previous system existed, ILO recommends increasing all EEC threshold levels between 1 and 100 kg to 1 tonne in those countries. This permits them to concentrate on larger installations with the greater risk.

The World Bank

In supporting industrial programs in developing countries, the World Bank evaluates the adequacy and effectiveness of measures to control major hazards affecting people and the environment outside the plant boundary. In September, 1985, the World Bank issued guidelines based heavily on the earlier EEC and UK guidelines.[10] The following month, this was supplemented with a technical manual.[11] Guidelines and manual aim at developers of projects in developing countries. The World Bank *Manual of Industrial Hazard Assessment Techniques* provides a framework for the structured identification of such hazards. The basic procedures included in this manual guide engineers and safety analysts through a series of steps to identify basic failure modes, estimate the quantities of material released, and calculate the impacts of these materials on plant equipment and personnel or the surrounding population and environment. Attention is also focused on approaches for estimating event frequencies and for examining the need for remedial measures, though these are not integrated with the quantitative methods described earlier.

Table 4. Example Compounds

Substance	Annual feedstock (kg)	Average weekly stock (kg)	Threshold quantity (kg) ILO	EPA	WB
Hydrochloric acid	3,850	80	NA	80	NA
Hydrogen sulfide	,600	12	50,000	230	1,000
Silane	1,500	51	200	NA	15,000

The first step is to determine if a hazard assessment is needed based on the guidelines. The manual provides the framework for the assessment if needed. It focuses first on the "Release Case." Representative failures and associated releases for many possible components (i.e., pipes, flexible connections, filters, valves, pressure/process vessels, pumps, compressors, storage tanks, storage vessels, and flare vents) are given. The second step is consequence analysis. Based on properties of the discharged material (i.e., phase, pressure, temperature, flammability, and toxicity), tree diagrams guide the analyst to a selected set of transport and effects models.

CASE APPLICATIONS

In order to evaluate these guidelines from a practical perspective, they were applied to three example compounds: hydrochloric acid, hydrogen sulfide, and silane. There were drawn from work at Brookhaven on potential health implications of the developing photovoltaics industry.[12,13] Data on quantities of material corresponding to a photovoltaic cell manufacturing plant with an annual output of cells capable of producing 10 MWp, along with corresponding threshold quantities from the EPA, EEC/ILO, and World Bank guidelines for the three compounds are discussed below and summarized in Table 4. The distance to the population at risk was assumed to be 500 meters.

Hydrochloric Acid

Hydrochloric acid is used an etchant in the photovoltaics industry. Material is placed in an HC1 bath and spent etchant is piped to a waste treatment system prior to discharge to the environment. The potential hazard is from leaks in the piping system or a storage tank operating under atmospheric conditions. The EEC/ILO list and the World Bank guidelines include only liquefied hydrogen chloride gas, not the aqueous acid. EPA guidelines include hydrochloric acid, but the accompanying data sheet gives an IDLH of 100 ppm. Only by referring to the original NIOSH reference is it clear that this applies only to anhydrous HC1. A different section in the EPA data sheet, however, indicates that hydrochloric acid mist is considered less harmful than anhydrous HC1. Applying the inappropriate IDLH to the EPA method yields a threshold quantity of 80 kg at 500 meters, just equal to the average weekly storage.

The World Bank manual recommends that hazard analyses assume openings of 20-100% of the pipe diameter for piping and assume total loss for storage at atmospheric pressure. Release would result in a spreading liquid for which pollution of water and soil must be considered; evaporation of the material is necessary, however, for significant concentrations to become airborne. The manual advises that this does not generally occur if the boiling point of the chemical is above ambient temperature (hydrochloric acid boils at 110°C). In this example case, one would conclude from the manual that "potential impacts are not likely to cause off-site fatalities and detailed consequence modeling is not needed."

This example illustrates the difficulty with a broad general approach in both the EPA and the World Bank guidelines. The EPA method overestimates the inhalation hazard by misapplying the IDLH for anhydrous HCl to hydrochloric acid. The World Bank may underestimate the hazard by assuming little becomes airborne since the boiling point is higher than ambient temperature, ignoring hydrochloric acid's property of fuming in air at ambient temperature. Department of Transportation guidelines do not provide an evacuation distance for a "small" spill (single drum) of HCl solution, but recommend 190 feet in all directions for a "small" spill of anhydrous hydrochloric acid.[14] It would seem that, while a spill of 80 kg may not affect the public 500 meters or more distant, it may pose some hazard to workers.

Hydrogen Sulfide

Hydrogen sulfide gas is used as a feedstock in the preparation of cadmium sulfide photovoltaic cells. Leaks may occur from gas cylinders or from gas distribution systems. The EEC/ILO list gives a threshold quantity of 50,000 kg. The EPA method yields a threshold quantity of 230 kg for a population 500 meters distant. Although hydrogen sulfide is listed in the World Bank guidance under "Other toxic substances" for which a "major hazard assessment is necessary if quantities exceed the threshold," no specific threshold quantity is given so it falls under the general criterion of 1 tonne. The average weekly stock of 12 kg is well below the quantitative threshold levels of all three guidelines. Were a hazard analysis indicated, the World Bank manual recommends that hazard analyses for pressurized storage and piping system components assume openings ranging from 20% of the piping diameter to total rupture and complete loss of inventory. Size and duration of the predicted release is calculated with gas flow equations for subcritical and critical flows. Once discharge rates are calculated, the manual recommends that a dense-gas dispersion model be used to evaluate cloud characteristics and pollutant dispersion following release. The "dose-response" portion of this consequence analysis is based on a probit function. Familiarity with engineering and risk assessment methods, and access to computer models is assumed.

Silane

Silane is used as a feedstock material in the production of amorphous silicon photovoltaic cells. It is pyrophoric and highly explosive. Leakage and subsequent fire or explosion may occur in storage, distribution, process, or effluent systems. Annual use of silane in amorphous silicon glow discharge system is estimated to be 1500 kg with an average weekly stock of 51 kg. In a chemical vapor deposition system, a competing process, the corresponding quantities are 600 kg and 12 kg. EPA guidelines deal quantitatively only with toxic materials and do not include flammables or explosives. EEC/ILO guidelines include explosive substances as a criterion for inclusion at the first level, that is, requiring employers to identify the hazard and take appropriate safety measures. Silane is not on the EEC/ILO list of substances with threshold quantities, however, and the guidelines do not provide help in determining an appropriate threshold level. The closest quantitative threshold category is flammable gases, with a threshold quantity of 200 tonnes; clearly this is inadequate. ILO guidelines, however, also recommend establishment of a 500 meter "consultation zone" around any site with more than 15 tonnes of flammable gas. The lowest threshold level of a specifically named substance on the consultation zone list is 2 tonnes for hydrogen. World Bank guidelines do not specifically list silane, but it would fall under the indicative criteria for highly reactive substances. The World Bank manual describes models for unconfined vapor explosion to

Table 5. Points of Comparison

Item	EPA	ILO	World Bank
Intended users	Public	Firms	Professionals
Population at risk	Public	Public	Public
Media evaluated	Air	Air	Air
Release mechanisms	No	No	Yes
Toxic pollutants	Yes	Yes	Yes
Exposure/flammables	No	Yes	Yes
Corrosives/oxidizers	No	Yes	Yes
Dispersion modeling	No	No	Yes
Event probabilities	No	No	No
Materials with guidance	403	178	29
End-point evaluated	Quantity	Quantity	Effect

calculate damage radii as a function of the amount of explosive material in the cloud and the theoretical combustion energy of the material.

While silane is not specifically treated in any of the guidelines, the quantities in this example fall under the threshold quantities of the closest general or exemplary categories for ILO and World Bank. While the quantities in this example do not pose a threat to populations 500 meters or more distant, they do represent a potential hazard to workers at the facility. Moreover, quantities well under the guidelines' threshold levels could pose a public risk.

COMPARISONS OF THE THREE DIFFERENT APPROACHES

Similarities and differences among the three different approaches for several criteria are presented in Table 5. In general, these approaches are best distinguished by their differences. These occur principally in the audience to whom the reports are addressed, the types of contaminants considered, sophistication of the release scenario and environmental transport, and the end-points to be estimated. All treat, or fail to treat, the remaining points in a comparable way.

The EPA guidelines are designed for local community groups. They provide help to these groups to organize, gather data on industrial hazards and emergency preparedness, and provide a quantitative method to establish priorities. They may take such a local community group as far as it can go without professional consultants. For the quantitative approach in particular, however, this tends not to take into account factors which affect release and exposure such as the physical state of the material, its role (or lack of role) in a chemical process, and the extent of existing containment or other safety features. This may create a burdensome number of false positives, that is, identification of potential hazards which turn out upon further professional analysis not to be hazardous.

The EEC/ILO approach is aimed at manufacturing firms and government regulatory agencies which are presumed to have professional staff or consultants competent to carry out the analysis. Thus, the guidelines provide little more than a checklist of what must be done in an assessment and analysis of potential major hazards. They require plant management to assess hazards and take necessary steps to prevent major accidents and to limit their consequences. Further, they provide a list of 178 chemicals with threshold quantities for each. Any plant with more than the threshold quantity must notify the appropriate regulatory authority.

The World Bank guidelines aim at project developers, who also are assumed to have access to professionals. The manual accompanying the guidelines is more like an engineering handbook than is the case with the ILO working paper, but it is still a handbook for use by trained professionals, not a citizens group. The World Bank guidelines include a list with threshold quantities of materials above which a major hazard assessment is required similar to the EEC/ILO list. It contains, however, fewer substances with thresholds (these are generally lower than the EEC/ILO thresholds) and has a standard 1 tonne threshold for non-listed hazardous substances.

EEC/ILO and World Bank guidelines include toxic, corrosive, flammable, and explosive substances. EPA guidelines treat only toxic substances quantitatively. Thus, in the example of silane, only the first two sets of guidelines were applicable.

CONCLUSION

On the most simplistic level, the EEC/ILO and World Bank guidelines provide a list of compounds with a threshold quantity for each. The EPA approach is only slightly more complex, requiring the user to look up a "level of concern" such as the IDLH and enter a nomograph with the distance to the population at risk to determine the threshold quantity of the material. Going beyond this, the World Bank manual provides engineering data and formulae, but these are aimed at trained professionals. EPA provides only general qualitative discussion for the layman, but does not reach the depth of analysis that a professional could provide. The examples illustrate the problems that can arise in use of broad general screening methods.

REFERENCES

1. Council Directive on the Major-Accident Hazards of Certain Industrial Activities, *Official Journal of the European Communities* **25**:L230/1-18 (1982).
2. Health and Safety Executive, *A Guide to Control of Industrial Major Accident Hazards Regulations 1984*, UK Health and Safety Booklet HS (R)-21, HMSO, London (1985).
3. U.S. Environmental Protection Agency, *Chemical Emergency Preparedness Program Interim Guidance*, Washington, D.C. (1985).
4. Federal Emergency Management Agency and U.S. Environmental Protection Agency, *Planning Guide and Checklist for Hazardous Materials Contingency Plans*, Washington, D.C. (1981).
5. Department of Transportation, *1984 Emergency Response Guidebook*, U.S. Department of Transportation, Washington, D.C. (1984).

6. *NIOSH/OSHA Pocket Guide to Chemical Hazards*, U.S. Department of Health and Human Services, Washington, D.C. (1981).
7. International Labour Organisation, Tripartite ad hoc Meeting of Special Consultants on Methods of Prevention of Major Hazards in Industry (MHC/1985/6 Rev.), Geneva (1985).
8. International Labour Organisation, Working paper on Control of Major Hazards in Industry and Prevention of Major Accidents (MHC/1985/1), International Labour Office, Geneva (1985).
9. Council Directive on the Major-Accident Hazards of Certain Industrial Activities, *Official Journal of the European Communities* **25**:L230/1-18 (1982).
10. Office of Environment and Scientific Affairs, *World Bank Guidelines for Identifying, Analyzing and Controlling Major Hazard Installations in Developing Countries*, World Bank, Washington, D.C. (1985).
11. P. J. Kayes, (Ed.), *Manual of Industrial Hazard Assessment Techniques*, The World Bank, Washington, D.C. (1985).
12. P. D. Moskowitz, V. M. Fthenakis, L. D. Hamilton, and J. C. Lee, "Public Health Issues in Photovoltaic Energy Systems: An Overview of Concerns," presented at the SERI Photovoltaics Safety Conference, Lakewood, CO (1986).
13. V. M. Fthenakis, P. D. Moskowitz, and J. C. Lee, Manufacture of Amorphous Silicon and GaAs Thin Film Solar Cells: An Identification of Potential Health and Safety Hazards, *Solar Cells* **13**:43-58 (1984).
14. Department of Transportation, *1984 Emergency Response Guidebook*, U.S. Department of Transportation, Washington, D.C. (1984).

Decision Analysis in Environmental Risk Management: Applications to Acid Deposition and Air Toxics

D. Warner North
Decision Focus Incorporated
Los Altos, CA

ABSTRACT

Uncertainty is a major difficulty in environmental risk management. While it is desirable to base environmental management decisions on the best scientific knowledge available, scientists often cannot give precise estimates of environmental risks that are based on scientific data and models that are validated from data. In situations where data are lacking, the methods of decision analysis are useful as a means of assessing and representing expert judgments about the uncertainties in a form that is easily communicated to those with management responsibility and to concerned members of the public.

Two applications are described. The first is an analysis of acid deposition policy for Wisconsin in which a panel of scientific experts provided probabilistic assessments of the impact of changes in sulfur dioxide emissions on potential damage to lakes and forests in Wisconsin. The second application is to air toxics that are thought to be human carcinogens based on animal bioassays. Perchloroethylene, a widely used dry cleaning solvent, is an illustrative example.

KEYWORDS: Decision analysis, acid deposition, perchloroethylene, risk assessment, risk management

INTRODUCTION: THE NEED TO MAKE ENVIRONMENTAL RISK MANAGEMENT DECISIONS IN THE FACE OF UNCERTAINTY

Environmental risk management would be far easier and much less controversial if the consequences of risk management options could be accurately predicted. For many areas of risk management accurate prediction is not possible with the existing level of scientific understanding. We do not understand the causes of cancer and we know relatively little about the cancer risk that chemicals in the air, water, food, or workplace may pose. We may know even less about chemicals that could cause birth defects or subtle changes in the human nervous system. We can make only rough predictions about the relationships between air emissions and ambient or deposition levels of pollutants many miles away. Our knowledge is even weaker with respect to the transport and fate of contaminants from solid waste disposal and pesticide usage. Yet regulatory agencies, industries, and individual citizens are constantly called on to make decisions on the

New Risks
Edited by L. A. Cox, Jr., and P. F. Ricci
Plenum Press, New York, 1990

management of environmental risks. In the face of such vast uncertainties, how can analysis be useful?

One response to this question is that analysis can be most useful by summarizing the extent of our knowledge and uncertainties so that those with decision responsibility can be as well informed as possible. When we are facing difficult decisions in the face of uncertainty, most of us find that we would like to rely on judgment as well as hard data. When we face a difficult legal decision, we seek the advice of legal experts. Faced with a decision on whether to undergo medical treatment, we seek recommendations from specialists in the appropriate branches of medical science. It seems obvious that just as difficult decisions in law, medicine, and business are based on judgment as well as data, decisions on environmental risk management should reflect the best judgment available from the applicable fields of environmental and health science.

How can judgment about uncertainty be brought into the analysis of complex decisions? Decision analysis provides a formal theory for choosing among alternatives whose consequences are uncertain. It has been widely taught and practiced in the business community for several decades, and it is being increasingly used in law and medicine.[1-5] There have been a few applications of decision analysis to environmental risk management decisions,[6-7] including some recent efforts on the health effects of ambient lead[8] and the consequences of radioactive waste disposal.[9]

The key idea in decision analysis is the use of judgmental probability as a way to quantify uncertainty. By judgmental probability we mean the use of probability to summarize judgment on the occurrence of an event or the accuracy of a scientific hypothesis. People commonly refer to the probability (or equivalently, the odds) of rain tomorrow, of winning an election or a sporting event, or of success in a business venture. In these examples, the probability numbers serve to summarize judgments on a multitude of complex factors. The judgments of weather forecasters, sports or political experts, business consultants, or scientists may be good or they may be poor. What probability provides is a way to describe the likelihood of uncertain events quantitatively, permitting explicit judgments about uncertainty to be incorporated into the decision making process. Where there are disagreements in the judgments, the implications for decisions of these differences can be examined. There is extensive literature on the assessment of judgments about uncertainty in the form of probabilities.[10-12]

APPLICATION OF DECISION ANALYSIS TO ACID DEPOSITION

Acid deposition has become one of the leading environmental controversies of the 1980s, and it is an excellent example of an environmental risk management issue plagued with uncertainties. Several years ago the Electric Power Research Institute sponsored the development of a decision framework based on principles of decision analysis to aid decision makers with responsibility for policy choices and research management on acid deposition. The approach and its implementation as a computer model have been extensively documented elsewhere.[13-16]

A major purpose of the Acid Deposition Decision Framework is to facilitate the assessment of expert judgment on the scientific issues related to acid deposition. Three stages can be distinguished in the relationship between control alternatives and impacts, as shown in Fig. 1. First, there is the effect that control strategies will have on emissions. Then changes in emissions must be related to changes in acid deposition. Finally, changes in acid deposition must be related to changes in the various impacts that can be identified, such as decreased forest productivity and the loss of fish in lakes and streams. There is scientific uncertainty about each of these stages. Relatively little is known about how specific changes in acid deposition will affect changes in impacts. The estimates given by

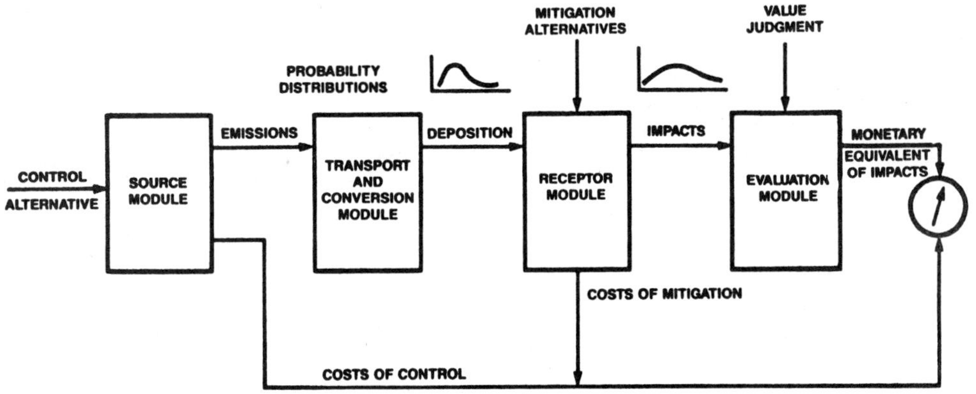

Fig. 1. Overview of decision framework.

respected scientists vary over a wide range. There is somewhat less uncertainty regarding how changes in emissions will affect changes in deposition; however, the range of uncertainty is still quite large due primarily to the complex nature of the chemical transformations that occur in the atmosphere. There is comparatively little uncertainty about how control strategies would affect changes in emissions. Accordingly, in implementing the framework, the importance of uncertainty in the other two stages, emissions to deposition and deposition to impacts, has been stressed.

At present, the scientific evidence regarding the effects of emissions is contradictory and subject to different interpretations by various experts. The decision framework allows an investigation of the implications of the differing assessments and evaluates the importance of the disagreements in terms of their effects on the choice of a control or mitigation strategy. Many experts who disagree about the interpretation of the current state of knowledge agree that in five to ten years many of those disagreements could be settled. Thus, in the decision framework, the choice is characterized as one in which we may act now and accept the possibility that emissions reduction will have little beneficial impact. Alternatively, we can wait five to ten years to act on better information that may be available and accept the possibility that damage may occur during that period. In each case, there is a possibility that the decision will turn out to have been incorrect. From our current state of knowledge, we cannot be sure.

Application to Wisconsin

An application to Wisconsin carried out by the author and colleagues in 1985 demonstrates how the Acid Deposition Decision Framework may be applied to evaluating alternative sulfur oxide emission control policies defined at the state and regional level.[17,18] The application relied heavily on the use of a panel of scientific experts to provide judgments on various aspects of acid deposition in Wisconsin: three atmospheric scientists involved in air pollution trajectory studies for Wisconsin, two experts on aquatic impacts, and two experts on forest impacts. These experts reviewed the assumptions and relationships in the modules of the decision framework in the area of their specialty. Lake and forest impacts within the state of Wisconsin were examined in this analysis, but this discussion will address only lake impacts.

Wisconsin's most sensitive lakes are seepage lakes with no stream inputs and no input of acid-neutralizing chemicals from groundwater. The experts on the panel reviewed the available information on the number of such lakes and their sensitivity to acid inputs from precipitation, using information from models of lake chemistry, field experiments, and measurements of alkalinity in many Wisconsin lakes. After initially differing in their assessments, the two experts on aquatic effects agreed on the potential extent of acidification over the next fifty years and the probability that such acidification would occur. Their judgments are summarized in Table 1. The two experts, Anders Andren and Douglas Knauer, differed in their judgment about the distribution of chemical retention times in the sensitive lakes, which was initially thought to be important for the analysis. The sensitivity analysis showed that this difference had a very small effect on the pattern of future lake acidification predicted by the model, as shown in Figs. 2 and 3.

Table 1. Judgment on Lake Impacts

Expert	Extent of Potential Future Acidification (1,000 acres)	Fraction of Lakes with Various Chemical Retention Times (years)			Probability of Acidification
		0-15	15-30	30 or more	
Anders Andren	15	0.75-0.8	0.19-0.24	0.01	0.1
Douglas Knauer	15	0.6	0.3	0.1	0.1

Results of the Wisconsin Application

The scientific information now available does not permit precise prediction of the impacts from acid deposition on Wisconsin's lakes and forests. Scientists are uncertain about whether such impacts will occur, and if they do occur, what the extent and severity of these impacts will be. While these statements are accurate as a summary of the state of scientific knowledge, they are not very helpful as a basis for a decision on whether to spend hundreds of millions of dollars on emissions control now or whether to defer the decision on imposing expensive controls until further research can be completed.

To provide a better basis for decision, we assessed judgments from the expert panel in the form of probabilities, and we used our acid deposition model to calculate how the extent of damage would depend on the choice of control alternative, given that adverse ecological impacts did occur. The scientists on the panel judged that it is unlikely that additional acidification of lakes or damage to forests will occur in the next fifty years, even in the absence of additional controls on emissions sources affecting Wisconsin. If such impacts do occur, their judgment was that the impacts would be limited to a small fraction of the resources in Wisconsin. The judgments on the probability of adverse impacts were stated in quantitative terms: a 10 percent probability of acidification of up to 15,000 acres of extremely sensitive lakes (2 percent of Wisconsin's lake acreage) within the next fifty years, and a one to two percent chance of reduced productivity affecting up to 10 percent of Wisconsin's forests. The analysis indicated that additional controls imposed regionally on the U.S. and Canadian sources affecting Wisconsin could be effective in avoiding potential lake acidification in Wisconsin and in greatly reducing the potential damage to Wisconsin forests. The risk of potential resource loss from waiting up to ten years before imposing the controls appeared to be very small for lakes and small for forests.

The results of any analysis depend on the assumptions, data, and judgments that go into the analysis, and the analysis process may be even more important than the analytical methodology. In carrying out this application, we held many meetings with representatives of state agencies in Wisconsin and concerned citizen groups in addition to representatives

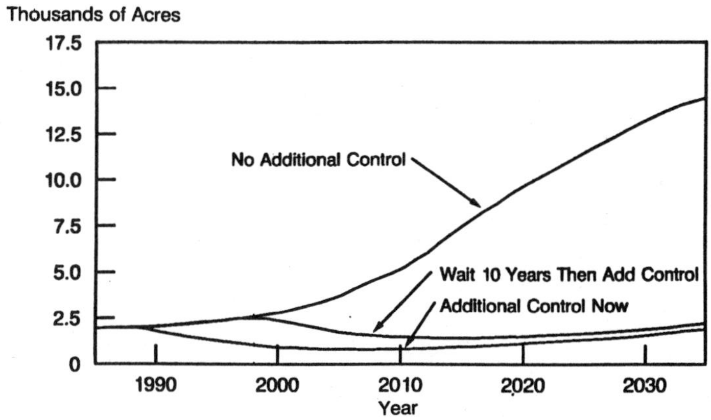

Fig. 2. Effect of no additional control, 50% near-term regional SO_2 control, and delayed regional control on the extent of lake acidification: Douglas Knauer's judgment used for distribution of chemical residence times.

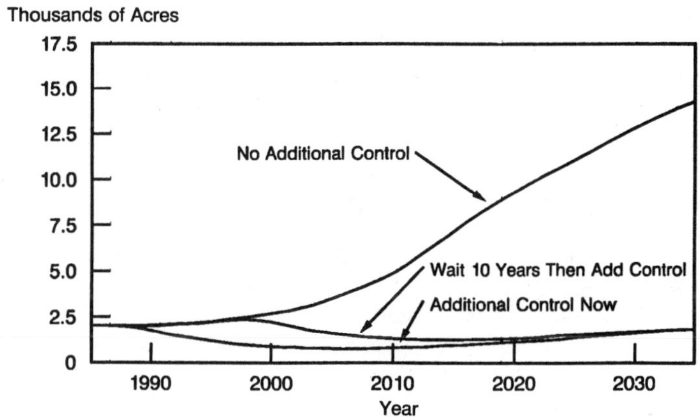

Fig. 3. Effect of no additional control, 50% near-term regional SO_2 control, and delayed regional control on the extent of lake acidification: Anders Andren's judgment used for distribution of chemical residence times.

of the electric utilities that sponsored our work. We presented the analysis and the expert judgments in preliminary form at these meetings to solicit criticism and encourage presentation of alternative viewpoints. At the conclusion of the analysis, the methodology and results were the subject of a day-long public workshop. At this workshop three of the four experts on aquatic and forest impacts made presentations on the scientific basis for their judgments. All of the panel members agreed to have probability assessments attributed to them specifically by name. While there were differences in their judgments, the differences appeared to be minor in terms of the conclusions from the analysis. The probability assessments from the panel members were prominently featured in a headline article in one of Wisconsin's leading newspapers.[19]

APPLICATION TO PERCHLOROETHYLENE

The second example did not involve actual assessments of expert judgment in the form of probabilities, but the same process as used in the Wisconsin acid deposition application might also provide a way to improve the basis for decision making on a complex environmental risk issue.

Toxicological research has shown that a number of widely-used chemicals can cause tumors in laboratory animals, and these chemicals may pose a risk of cancer in humans. How should such environmental risks be managed?

In the early 1980s the author and colleagues carried out a methodology development study for the Office of Toxic Substances of the U.S. Environmental Protection Agency.[20,21] We selected perchloroethylene as representative of the chemical substances over which EPA has regulatory jurisdiction under the Toxic Substances Control Act and Section 112 of the Clean Air Act. Our case study on perchloroethylene illustrates how decision analysis may be used for risk management decision making.

Perchloroethylene (also known as tetrachloroethylene and abbreviated as PCE) is a solvent that is widely used for dry-cleaning fabrics. In 1980 more than 300,000 metric tons of PCE were consumed in the U.S.; about 42% of this amount was for dry-cleaning uses.[22]

A bioassay carried out by the National Cancer Institute indicated that PCE exposure by gavage induced hepatocellular carcinomas in B6C3F1 mice. EPA's Carcinogen Assessment Group (CAG) had prepared a risk assessment on the basis of the bioassay data,[23] and a subcommittee of the EPA Science Advisory Board (SAB) had reviewed this risk assessment at a public meeting.[24]

The CAG risk assessment included a unit risk estimate of the dose response relationship made following CAG's standard procedures.[25] A review of the SAB transcript[24] indicated that alternative assumptions were viewed as plausible by the members of the SAB subcommittee. CAG had fitted its usual multistage model to the data using a 95% upper confidence limit, a procedure which leads to linear low-dose extrapolation. Yet the SAB scientists noted evidence that PCE does not act directly on DNA but indirectly through cellular toxicity. Given an epigenetic mechanism, a nonlinear dose response relation might plausibly be expected. Similarly, while CAG had used the B6C3F1 mouse data as the basis for its extrapolation, scientists at the SAB meeting argued that the rat was more representative of the human metabolism. Finally, while CAG had extrapolated dose level from animal to human using relative surface area, many scientific groups have recommended using daily dose per unit of body weight as an appropriate scaling procedure.

Three instances were thus identified in which there was uncertainty about whether CAG's assumption was right or whether an alternative assumption might be more appropriate. The sets of assumptions are summarized in Table 2. If one assumes for simplicity that for each of the three issues either the CAG assumption or the alternative assumption is correct, then we have eight possible combinations or cases, only one of which represents the correct dose response relationship. Using the methods of decision analysis, we might assign judgmental probabilities to the eight cases. Such probabilities were used in our report,[20] although the numbers are strictly illustrative. The probability numbers are less important than the concept of using a variety of cases based on alternative plausible assumptions. The change in estimated cancer incidence from these changes in the dose response assumptions is nearly five orders of magnitude.

Since perchloroethylene is used as a dry-cleaning solvent and PCE vapor is easily monitored, estimates of PCE exposure are relatively simple to make from existing data in the literature. Table 2 summarizes the results. Based on NIOSH data,[26] machine operators

Table 2. Dose Response Assumptions for PCE Case Study

Issue	CAG Assumption	Alternative Assumption
Choice of species	B6C3F1 mouse most sensitive species	Rat better represents human metabolism
Scaling of dose from animal to human	Ratio of surface area	Ratio of body weight
Low dose extrapolation	Multistage model (equivalent to linear with use of upper confidence limit)	Nonlinear response because of epigenetic mechanism (quadatric relation used as representative)

(Eight Combinations of Assumptions Possible)

are exposed to an average of about 30 ppm of PCE vapor during the working day, equivalent to a continuous exposure of 45,000 μg/m^3. Other workers in commercial and industrial dry cleaners are exposed at a lower level, 10,000 μg/m^3, and workers in coin operated laundromat dry-cleaning facilities have an estimated exposure level of 6,000 μg/m^3. The number of workers exposed is based on projections from industry and census data.

Users of dry-cleaning services are exposed when they visit the dry-cleaning facilities and to a lesser extent from cleaned clothing. As shown in Table 3, the exposure levels are far lower than for workers. The highest level exposures are for customers using coin-operated dry-cleaning machines. Urban residents are exposed to low ambient levels of PCE; the resulting exposure is less than for users of commercial dry-cleaning services.

The estimates for exposure and human cancer incidence given exposure can be combined into a risk assessment. A summary of the results is given in Table 4. If the CAG assumptions for the dose response relationship are used, the projected cancer incidence is about 350 cases per year, the majority of which occur in workers, with most of the remainder in service users. The lifetime probability of cancer for a machine operator is 23%, a high enough number that one would expect to see strong epidemiological evidence if these assumptions were correct.

If instead of the CAG assumptions we use the alternative assumptions on the right of Table 2, a very different picture emerges. The expected cancer incidence in that event is only one case per hundred years, a change of nearly five orders of magnitude. The lifetime probability of cancer estimate for a machine operator at 3×10^{-5} still is not negligible, because of the high level of the occupational exposure. The incidence and lifetime probability of cancer for service users and urban residents become negligible under the alternative assumptions.

In our 1982 analysis[20] we examined a set of control alternatives for reducing PCE exposure, and we illustrated how judgments of the probabilities of the various cases shown in Table 2 could be combined with a range of value judgments on the worth of avoiding a case of cancer to provide a basis for choosing among the control alternatives. We also showed that in the context of the control decision, resolution of the uncertainty about the risk assessment assumptions in Table 2 has a high value.

Table 3. Exposure Estimates for PCE Vapor

Classes of People	Number Exposed	Annual Average Exposure $\mu g/m^3$
Workers		
Machine Operators	17,000	45,000
Other Workers	130,000	10,000
Workers in Coin-Operated Facilities	33,000	6,000
Services Users		
Commercial	50 million	5
Coin-Op Cleaners	25 million	10
Coin-Op Laundry	37 million	38
Urban Residents	95 million	0.2-4

Table 4. Summary of PCE Risk Assessment Based on 1982 Study[20]

	CAG Assumptions (Present Exposure)	Alternative Assumptions (Present Exposure)
Expected Number of Annual Cancers:	347	0.01
Workers	181	0.01
Service Users	163	10^{-5}
Urban Residents	3	10^{-8}
Incremental Lifetime Probability of Cancer From PCE:		
Machine Operator	0.23	3×10^{-5}
Coin-Op Laundry User	2×10^{-4}	10^{-11}
Nearby Urban Resident	2×10^{-5}	10^{-13}

There are a great many assumptions that go into the risk assessment of toxic substances. The 1983 National Academy Report on risk assessment[27] lists about 50 of them. If one were to try to deal with all of these uncertainties explicitly using alternative scenarios and probabilities, it would be an overwhelming task. Therefore, the CAG calculation of a plausible upper bound for the dose response relationship at low doses serves as an excellent *first step* in the risk assessment process. If the numbers for cancer incidence are very low, then one might reasonably conclude that additional effort for risk management was not needed. However, if the numbers resulting from the CAG plausible upper bound are relatively high, then it is appropriate to explore the uncertainty in the underlying assumptions. The guidelines for regulatory impact analysis from Executive Order 12498 ask for the best estimate of the reduction in risk attributable to a proposed environmental standard. EPA has acknowledged in its Guidelines for Carcinogen Risk

Assessment[28] that it does not yet have an established procedure for making most likely or best estimates of cancer risk, but only estimates of a plausible upper bound. Our PCE illustrative calculations using judgmental probabilities (such as we assessed from scientific experts in the acid deposition example) provide a basis for best estimates in the sense of expected values over the range of the uncertainties.

Since 1982 the practice of risk assessment at EPA has evolved. The 1985 Health Assessment Document (HAD) on PCE[22] was one of the first uses by EPA of a risk assessment based on metabolic and pharmacokinetic information, that is, using the dose of reactive metabolites rather than the dose of PCE administered to the animal. The result is that the upper bound estimate of potency, or unit risk, from inhalation of PCE in air is reduced by a factor of about ten. EPA has made estimates of average lifetime PCE exposure to workers based on the same exposure data as we used; their estimates differ in that exposure to the workers is assumed for one third of their lifetime, and geometric rather than arithmetic averages of the occupational exposure data are used. Table 5 shows the results for expected cancer incidence and individual cancer risk if EPA numbers from the 1985 HAD[22] are used, replacing the values used in the 1982 analysis.

The 1985 HAD for PCE also explores through sensitivity analysis some of the key areas of uncertainty described in Table 2. With the pharmacokinetic data it makes little difference whether the mouse gavage bioassay or a negative inhalation study in rats is used as the basis for calculating the unit risk for PCE in air. The extrapolation of metabolized dose from animals to humans using body weight instead of surface area gives a reduction of about a factor of 10 in the risk estimates. The HAD does not include a discussion of nonlinear models, other than a brief mention that fitting Weibull and probit models to the data did not give meaningful results. A more appropriate comparison might be the use of a pure quadratic model for low-dose extrapolation instead of a linear model for low-dose extrapolation. A quadratic model would imply a reduction in risk to the machine operators of about a factor of one hundred compared to the results with the linear model, and larger reductions for other groups. The 1985 HAD also includes a discussion of predicted risk in the occupational population using the plausible upper bound unit risk estimate. This discussion indicates that with the low individual incremental lifetime cancer risks shown in Table 5, it is unlikely that a cohort study could be done with a sample size large enough to show any elevated risk of cancer.

Table 5. Revised Summary of PCE Risk Assessment, Based on 1985 HAD[22]

	Plausible Upper Bound Estimates
Expected Number of Annual Cancers:	16.4
Workers	3.1
Service Users	13.0
Urban Residents	0.3
Incremental Lifetime Probability of Cancer From PCE:	
Machine Operator	5.6×10^{-3}
Coin-Op Operator User	1.8×10^{-5}
Nearby Urban Resident	2×10^{-6}

The SAB is currently reviewing an Addendum[29] to the 1985 PCE HAD, which includes discussion of new inhalation bioassay data in rats and mice. The calculation of the unit risk from the new inhalation study in the mice corresponds closely to the previous result from the older gavage study.

Despite the apparent progress from better data and improved risk assessment methods, much controversy still attends the interpretation of the scientific information on the carcinogenicity of PCE and many similar chemical compounds. For many of these chemicals, a calculation of cancer risk using EPA's plausible upper bound estimate of unit risk or potency may be sufficient as a basis for risk management decisions, especially if the resulting numbers are small. But if the numbers are large, we may not wish to suppress uncertainty by using only this conservative estimate as the basis for risk management decision making.

Representing uncertainty explicitly by means of judgmental probabilities can give a much richer description of what the scientific specialists can tell us about the risks posed by toxic chemicals and other threats to the environment which call for risk management actions. These methods have recently been used by EPA's Office of Air Quality Planning and Standards for assessment of the health effects of lead in the review of the current National Ambient Air Quality Standard.[8] Similar efforts are currently underway for health and vegetation impacts of ambient ozone. The approach may prove useful for a wide spectrum of environmental risk management areas that involve significant uncertainties.

REFERENCES

1. H. Raiffa, *Decision Analysis: Introductory Lectures on Choices Under Uncertainty*, Addison-Wesley, Reading, MA (1968).
2. C. Holloway, *Decision Making Under Uncertainty: Models and Choices*, Prentice-Hall, Englewood Cliffs, NJ (1979).
3. R. Keeney, Decision Analysis: An Overview, *Operations Research* **30(5)**:803-838 (1982).
4. Program and Abstracts, The Society for Medical Decision Making Seventh Annual Meeting, Pacific Grove, CA, October 21-23, 1985.
5. S. E. Bodily, When Should You Go to Court? *Harvard Business Review* **59(3)**:103-113 (1981).
6. R. A. Howard *et al.*, The Decision to Seed Hurricanes, *Science* **176**:1191-1202 (1972).
7. *Air Quality and Stationary Source Emissions Control*, Commission on Natural Resources, National Academy of Sciences, U.S. Government Printing Office, Washington, DC (1975).
8. T. S. Wallsten and R. G. Whitfield, Assessing the Risks to Young Children of Three Effects Associated with Elevated Blood-Lead Levels, Report prepared for the Office of Air Quality Planning and Standards, U.S. Environmental Protection Agency by Argonne National Laboratory, Chicago, IL, ANL/AA-32, December 1986.
9. Office of Civilian Radioactive Waste Management, A Multi-Attribute Utility Analysis of Sites Nominated For the First Radioactive Waste Repository — A Decision-Aiding Methodology, U.S. Department of Energy, DOE/RW-0074, May 1986.
10. C. S. Spetzler and C. A. S. von Holstein, Probability Encoding in Decision Analysis, *Management Science* **22**:340-358 (1975).
11. D. Kahneman *et al.*, eds., *Judgment Under Uncertainty: Heuristics and Biases*, Cambridge University Press, Cambridge, MA (1982).
12. T. S. Wallsten and D. V. Budescu, Encoding Subjective Probabilities: A Psychological and Psychometric Review, *Management Science* **29**:151-172 (1983).

13. D. W. North and W. E. Balson, Risk Assessment and Acid Rain Policy: A Decision Framework that Includes Uncertainty, in *Acid Rain: Economic Assessment*, pp. 163-181, Plenum Press, New York (1985).
14. W. E. Balson *et al.*, Acid Deposition: Decision Framework, Vol. 1: Description of Conceptual Framework and Decision Tree Models, Electric Power Research Institute, Palo Alto, CA, EA-2540 (1982).
15. W. E. Balson *et al.*, Acid Deposition: Decision Framework, Vol. 5: ADEPT 2.0 Users Guide, Report prepared by Decision Focus Incorporated for the Electric Power Research Institute (1985).
16. Microcomputer Software Reviews, *J. Air Pollution Control Association* **35(12)**:1312-1313 (1985).
17. D. W. North, W. E. Balson, and G. R. Colville, Analysis of Sulfur Dioxide Control Strategies Related to Acid Deposition in Wisconsin, Vol. 1: Application of Decision Analysis, Report prepared by Decision Focus Incorporated for the Wisconsin Utilities Acid Deposition Task Force (1985).
18. D. W. North *et al.*, Representing Uncertainty Explicitly in Environmental Risk Management: An Example from Acid Deposition Policy Analysis, in *Environmental Risk Management: Is Analysis Useful?* APCA Specialty Conference Proceedings, Air Pollution Control Association, Pittsburgh, PA (1986).
19. P. G. Hayes, Wait, See Strategy on Acid Rain Advised, *The Milwaukee Journal*, p. 1, Tuesday, March 19, 1986.
20. G. L. Campbell, D. Cohan, and D. W. North, The Application of Decision Analysis to Toxic Substances: Proposed Methodology and Case Studies, Final report prepared by Decision Focus Incorporated for the Office of Toxic Substances, U.S. Environmental Protection Agency (1982).
21. D. W. North, Quantitative Analysis as a Basis for Decisions Under TSCA, in *TSCA's Impact on Society and Chemical Industry*, G. Ingle, ed., ACS Symposium Series 213, American Chemical Society, Washington, DC (1983).
22. Office of Health and Environmental Assessment, Health Assessment Document for Tetrachloroethylene (Perchloroethylene), Final report, U.S. Environmental Protection Agency, EPA/600/8-82/005F, July 1985.
23. R. E. Albert *et al.*, The Carcinogen Assessment Group's Carcinogenic Assessment of Tetrachloroethylene (Perchloroethylene), U.S. Environmental Protection Agency, (1980).
24. U.S. Environmental Protection Agency, Science Advisory Board, Subcommittee on Airborne Carcinogens, Public meeting held in Washington, D.C. on September 4 and 5, 1980, Transcript produced by Neal R. Gross, court reporters and transcribers, Washington, DC.
25. E. L. Anderson and the Carcinogen Assessment Group of the U.S. Environmental Protection Agency, Quantitative Approaches in Use to Assess Cancer Risk, *Risk Analysis* **3(4)**:277-295 (1983).
26. H. Ludwig *et al.*, Workers' Exposure to Perchloroethylene in the Commercial Dry-Cleaning Industry, *Am. Industrial Hygiene J.* **44(8)**:600-605 (1983).
27. *Risk Assessment in the Federal Government: Managing the Process*, Committee on the Institutional Means for Assessment of Risks to Public Health, Commission on Life Sciences, National Research Council, National Academy Press, Washington, DC (1983).
28. Guidelines for Carcinogen Risk Assessment, U.S. Environmental Protection Agency, *Federal Register* **51**:33992-34003 (1986).
29. Office of Health and Environmental Assessment, Addendum to the Health Assessment Document for Tetrachloroethylene (Perchloroethylene), External Review Draft, U.S. Environmental Protection Agency, EPA/600/8-82/005FA, April 1986.

Causality in Toxicologic Risk Assessment and Risk Management

Jerry LR Chandler
McLean, VA[a]

ABSTRACT

Analysis of the logical basis of chemical toxicity classes from four philosophical perspectives of causation is presented. Historical views of scientific "causality" are contrasted with the experimental basis of forming chemical toxicity classes. An analogy between a toxicity experiment and an Aristotelian causality is drawn. It is concluded that the causal prediction of human health risks is not directly related to either the connotation or denotation of a chemical toxicity class. An alternative strategy for risk management is proposed:

"Treat like chemicals alike,
treat different chemicals differently
in **proportion** to their relevant differences."

The scientific advantages of this alternative strategy of risk management are enumerated.

KEYWORDS: Causality, chemical toxicity classification, systems theory, risk management model, risk communication

INTRODUCTION

The notion of causality is intrinsic to risk assessment and risk management. Views of causation influence the perception, measurement, communication, and mitigation of risk. Despite the central role, the concept of causation plays in risk analysis and risk management, the different disciplines doing risk analysis employ the concept in fundamentally different ways. For example, in toxicologic taxonomy, classification of chemicals may be based on a differential model of causality while the prediction of risk may be based on a non-differential model of causality. Differing perceptions of risk may be generated by differing semantic usage of the term. [A visual representation of relationships among the four terms (causality, risk, risk analysis, and risk management) is presented in Fig. 1.]

An objective of this paper is to look at three of the historical origins of scientific usage of the term "causation" (Mill, Galileo, and Aristotle) and contrast them to the modern system's perspective. The influence of the assumed causal model on toxicity classifications and regulatory policy will be discussed. Chemical toxicity classes will be used to illustrate the roots of the problem. Finally, I will propose a referential basis for the management of chemical risks, intuitively associated with the syntax of systems theory, which does not presuppose any one semantic model of causation.

---

a. Full address: 837 Canal Drive, McLean, VA 22102.

New Risks
Edited by L. A. Cox, Jr., and P. F. Ricci
Plenum Press, New York, 1990

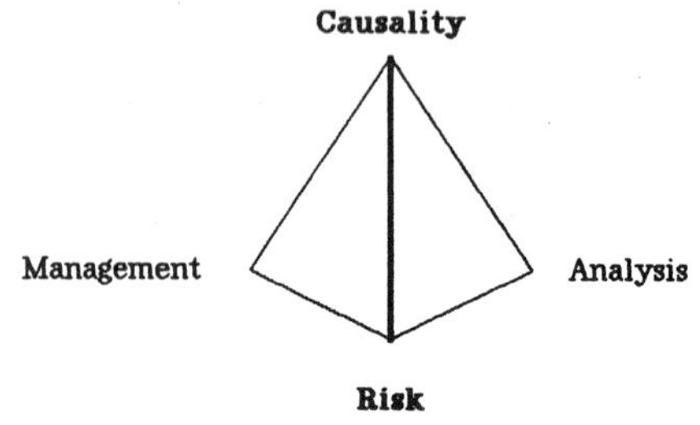

Fig. 1. Quaternary interactions in the design of public policy for toxic chemicals. Public policy is influenced by four domains; the central axis between the perceived cause and the perceived risk often motivate public efforts to analyze the past in order to change the direction into the future.

SEMANTIC MODELS OF CAUSATION AND MULTIDISCIPLINARY OBJECTIVES

Risk assessment and risk management often occur in a multidisciplinary environment.[5,6,13] Communication of risk among the professionals from different areas often difficult and can become a major impediment to policy development. Depending on the formal training of the participants, individuals may (knowingly or unknowingly) use different semantic models of causation. In order to limit the discussion (and extremely over-simplify the extensive literature on causation), I will restrict the consideration to models proposed by J. S. Mill, G. Galileo, and Aristotle.[a,b]

Fundamental to the structure of the following discussion is the concept of a "system," a concept not dependent on any one knowledge domain.[2,8,9,13] The term "system" is used here in the sense of the unit of discourse. It bounds the region of concern and is an essential starting point for fruitful analysis and prediction. Figure 2 illustrates one representation of a system. One attribute of a system is the idea of a boundary which distinguishes it from it's environment. Other attributes concern the inputs and outputs of a system and the processes which generate relationships among them. (As will be discussed later, changes in the attributes of a system are closely associated with the models of causation.)

LOGICAL BASIS OF CHEMICAL TOXICITY CLASSES

Health risks are often classified into various disease categories.[10,15] A disease category may refer to the state of an individual, of an organ system, or even of a specific organ. An example of chemically induced disease of an organ is blindness by methyl alcohol. An example of a chemically induced disease to organ system is the neurotoxicity induced by parathion. From observations with series of chemicals on different living systems, it is logically possible to construct toxicity classes of chemicals. It is critical to note the logical mechanism for assignment of any particular chemical to a toxicity class. The process of chemical toxicity class formation consists of *both* a forward chain of logic and a backward chain of logic. A simple representation of the forward chain is:

---

a. For detailed discussions of causation as it relates to the philosophy of science and to dynamical systems, the reader is referred to work of Bunge, *Causality*, Harvard Press, 1959, and to Patten *et al.* "Propagation of Cause in Ecology," in *Systems Analysis and Simulation in Ecosystems*, Academic Press, 1978.

b. While numerous other philosophers have discussed the notion of causation, these three views appear to be more widely used in the scientific community.

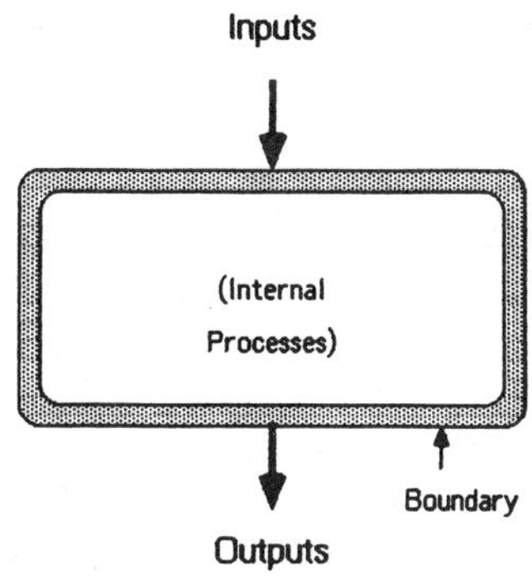

Fig. 2. A system's view of causality in toxicology as related to biological (internal) processes in the organism. The dynamics of internal biochemical process are a function of *both* the deisgn of the species' *genetic system* and the external *environment* of the organism.

> **If** a chemical is given to a biological system and
> **if**, after some time, a disease is observed in the organism,
> **then** the chemical is the "cause" of the disease.

A simple representation of the backward chain of logic is:

> **If** a disease is "caused" by exposure to chemicals,
> **then** a class of chemicals exists which "cause" the disease.

It is critical to note that a forward logical chain requires the participation of a specific (biological) system in order to establish the causal linkage. In the backward chain, the reference system is omitted from the classification.

The logical process for defining a chemical toxicity class is highly unusual since it depends on attributes which are not intrinic to the chemical structure. However, since experiments to demonstrate disease induction in animals are direct, controllable and reproducible, this paradigm has been widely used in risk analysis and risk management.

Careful scrutiny of the logic mechanism for assignment to toxicity classes reveals unstated assumptions which are masked by this usage of "cause." The term conceals the profound scientific differences between the hindsight of specific observation of disease states in a specific biological organism (system) and the prediction of the future health states in other (human?) biological systems. Before it is possible to discuss the semantic content of the classification, it is necessary to develop briefly the historical relationships between classification and "cause and effect" in the biological and physical sciences.

CLASSIFICATION

The idea of classification is quite general and transcends individual academic disciplines, as illustrated in the following examples.[6]

- Logically, the idea of a class is closely associated with set theory. A set can be defined in terms of a class of elements which are bound together by one or more attributes.

- A chemical class is bound together by one or more common structural features (attributes). Classification of molecules is often based on (static) structural theory. A molecule with a specified spatial relationship among different atoms (functional groups) is assigned to a class. Because of the very large number of different attributes (spatial relations) among the combinations of atoms, names are assigned in hierarchical preference.

- Biologically, living organisms are classified into groups based on similar function, on form, and on structure. A hierarchical classification of related living organisms is bound together by one or more common attributes at each level in the hierarchy.

- Higher order classifications at the group, social, and cultural levels are formed in analogous manner. Individual members are bound into the class by virtue of having one or more common attributes with other members of the class.

Thus, it appears that classification is a mechanism for assigning individual "systems" into groups which have common attributes. Classification is always done for a purpose; the purpose may be explicitly stated or may be unique to the values of the individual doing the classification or the purpose may be defined by the values of a group. *A priori*, a classification does not imply a "cause and effect" relationship, although the taxonomy for the classification could be so defined.

Despite the apparent logical simplicity of classification practical realities intrude. If all possible members of the class can be enumerated and everyone agrees with this denotation, then classification, as a scientific problem, is solved. More often than not in the real world, only attributes or criteria connoting class membership can be defined. The higher the level of complexity of the system and the more intertwined the relationships among the components and the attributes, the more difficult it becomes to assign each system to a class in a mutually exclusive and joint exhaustive way.

Connotation and denotation of chemical toxicity classes will be addressed later following a discussion of causation.[c]

ROOTS OF THE SCIENTIFIC PERSPECTIVE OF CAUSATION

Various notions about causation have been explicitly detailed by Bunge and by Patten *et al.*[3,9] While causality can be addressed broadly from the perspective of a doctrine, a principle or a category, our interest is to focus on specific models used in risk assessment.[d] Of particular interest to the scientific community is the notion of a "casual bond" which links the input to the output of a system.[4]

J. S. Mill proposed five general "canons" for inferring casual relationships.[e] From a systems perspective, the Millian canons ascribe "cause" in terms of differences between "phenomena." For example, the second canon states: "If an instance in which a phenomenon under investigation occurs, and an instance in which it does not occur, have every circumstance in common save one, that occurring in the former, the circumstance in which alone the two instances differ, is the effect, or cause, or a necessary part of the cause,

---

c. For a discussion of the logical attributes of classification, see *Logic*, by W. E. Johnson.

d. The philosophical, logical and scientific classical perspectives are described by Bunge, Causality, 1959.

e. Millian Cannon One: If two or more instances of the phenomena under investigations have only one circumstance in common, the circumstance in which all the instances agree is the cause (or effect) of the given phenomena.

of the phenomenon." At one extreme, the "phenomena" may differ only by "one circumstance," while at the other extreme, the "phenomena" may "have only one circumstance in common." The notion of a causal bond or linkage does not seem to play a substantial role in Millian canons, nor do ordering, reproducibility, or predictability. Rather, the canons focus on the differentia between phenomena. Hence, semantic descriptions of causation based on Millian canons will be referred to as differential causation. Clearly, the Millian semantic description of causation is sufficiently vague to permit the association of classes of phenomena without considering the nature of bonding, or linking, or bounding of the system.

Susser, in *Casual Thinking in the Health Sciences* notes that "To the eye of the epidemiologist more than a century later, Mill's canons seem appropriate to the construction and interpretation of research designs used in epidemiology today."[11] Correspondingly, the design and conduct of controlled toxicology experiments ("have ever circumstance in common save one") closely adheres to the Millian perspective. It is concluded that Millian thinking has a profound impact on current biological research and on health risk assessment. Neither classification nor prediction follow directly from the Millian canons.

In contrast to the Millian differential definition of causality, Galileo proposed criteria for causation based on the notion of "necessary and sufficient" conditions. Design and conduct of laboratory experiments are often in the Galileoian sense of causation. Scientific criteria for ordering, reproducibility, and consistency are closely allied to the concepts of "necessary and sufficient" conditions. From a systems perspective, "necessary and sufficient" conditions serve as one of the effective criteria for analysis and synthesis of systems. "Necessary" components of systems can be identified, isolated, and re-assembled with other components. The methodology related to "necessary and sufficient" conditions is widely used in the physical, chemical, engineering, and molecular biological sciences. The idea of "necessary and sufficient" conditions is closely associated with Aristotlian concept of "efficient" cause.

Aristotle's view of causation recognized the nature of process and thus is closer to the modern "systems" view than either Mill or Galileo. Aristotle recognized four kinds of causes: formal, efficient, material, and final.

> "Of these we hold that one is the essence or essential nature of the thing...;
> another is the matter or substrate;
> the third is the source of motion;
> and the fourth is the cause which is the opposite to this;
> namely the purpose...;

Aristotle's four kinds of causes are surprisingly analogous to current system's theory. The formal cause being analogous to the complete system under study; the material cause being analogous to the mechanism which associates inputs with outputs; the efficient cause being associated with the input or set of inputs to the system; the final cause being associated with the evolutionary law (differential equations describing the system) and the scientific laws used to justify the differential equations.

The results of toxicology experiments can be viewed from the perspective of Mill, Galileo, or Aristotle.

From the perspective of Mill, the differing circumstance between the exposed and control group becomes the "cause."

From the perspective of Galileo, the "necessary and sufficient" conditions for the "system" under study establish the "cause." That is, the population of animals, and the total set of experimental conditions that ensure the reproducibility of the result. Note that the "necessary and sufficient" condition for a response can not be met by the mere existence of the chemical. A toxicology experiment can also be viewed in terms of the four kinds of causes of Aristotle.

1. The essential nature of the experiment is established in a formal manner. The type, kind, health state, and other attributes of the experimental animals and the exposure conditions are selected in a purposeful manner.

2. The material cause which generates the response, can be associated with the biological (biochemical) processes in the body of the animal.

3. The dose can be associated with the notion of the efficient cause — "the source of the motion."[f]

4. The final cause (the toxic effect) can be associated with the evolutionary law of the genetic system. Backchaining this logic, the evolutionary law of the system can, in turn, be associated with the natural laws appropriate to the formal system, thus closing the circle.[g]

While the analogy between the Millian, Galileoian, and Aristotlian views of causation and the syntax of modern system's theory is limited, it provides a historical framework for examining the semantic relationships surrounding the concept of "cause."

Three conclusions may be drawn from this extremely brief survey on "cause." First, different academic disciplines have developed different perspectives of causation which can be associated with different historical roots. Second, the historical discussions on causation are not tightly coupled to the logical concepts of class connotation and denotation. Third, neither the historical concepts of causation nor the concept of classification relate directly to the problem of making quantitative predictions of the future, a fundamental scientific problem of risk analysis.[5] In order to associate the class (a non-dynamic concept) with a quantitative prediction for a dynamical system, the taxonomy for the class would have to be defined in terms of the evolutionary law of the system.

MODERN VIEW OF CAUSATION

In contrast to the historical semantic descriptions of causation, the current scientific view of causation is closely related to the evolutionary law of the system in terms of symbolic representations.[2,7,8,12]

Maruyama, expressing causal relations in terms of symbolic forms, proposed four different types of casual models at the meta level.[7] In contrast to the historical concepts, Maruyama focuses on evolutionary relationships which "generate differentiation and patterns, as well as analysis of interaction between heterogeneous elements." The change of the system with time may be described in terms of processes. These processes create the causal link between the input and output of the system. Thus, the evolutionary behavior of the system (which may be symbolically described in terms of mathematical dynamics/thermodynamics) can be associated with processes which link the input to the output.[4,8,9]

---

f. The widely quoted statement of Paracelus, "The dose makes the poison," reflects this perspective of causation.

g. Another view of the "final cause," one based on the meaning of symbols, is expressed by Bertrand Russell's statement: "Scientific laws can only be expressed in differential equations."

The four metatypes proposed by Maruyama are in terms of increasing complexity of the evolutionary law of the system:

1. "Nonreciprocal causal models: Causal relations may be either probabilistic or deterministic, but it is assumed that they can not form loops. They obey the transitive law."

2. "Independent event models: It is assumed that the most probable states of the universe or an isolated system are states of random distribution of independent events, each having its own probability. Nonindependent and nonramdom structures exist, but they are considered less probable and therefore tend to decay to more random, unstructured, homogeneous states"

3. "Homeostatic causal loop models": Causal relations may be probabilistic or deterministic and can form homeostatic causal loops."

4. "Morphogenetic causal loop models: Probabilistic or deterministic causal loops can increase heterogeneity, generate new patterns and symbiosis, and raise the level of sophistication of the system."

Maruyama notes that the "causal types may be mixed" and that subtypes exist. For example, "homeostatic causal loop models may either asymptotically converge to an equilibrium or periodically go through a path in state space, retracing the same path regularly." Thus, implicit to current views of causation are the dynamics of the system and feedback and feedforward mechanism which generate form and patterns. Specification of the nature of the causal bond generates a predictive form for the syntactical description of the system. The role of classification in the Maruyama scheme of causal metatypes is unclear. However, it is clear that the emphasis has switched from differentiating among phenomena to the underlying laws of genesis which create the causal bond that generates specific new attributes which are observable.

Quantitative risk estimates for chemicals are derived from models which can be related to Maruyama's metatypes. Reversible toxic effects can be expressed in terms of receptor theory, a non-reciprocal causal linkage. Probit and logit models can also be associated with the first Maruyama metatype. Hit and multi-stage theories of mutation and cancer can be associated with the first causal metatype with some overlap in the second type. The biochemical mechanistic model proposed by Chandler[4] can be associated with either the third or fourth causal metatype, depending on the mathematical linkages of the probability functions with steps in the mechanism.

HEALTH RISK PREDICTION

Wulff defines a disease process in terms of pathological changes with the body living in an environment.[15] Evaluation of the concepts of health and disease is far beyond the scope of this article; however, it is germane to point out that disease is closely associated with dysfunction. Disease often involves substantial change in biological form and function.[10,15] The morphogenesis of such changes may be associated with (biochemical) feedback and feedforward regulatory mechanisms within the body (which serve to maintain homeostasis.) Thus, concept of "morphogenic causal loops" can be a rational representation of a scientific description of disease induction at the molecular level of the system.

Public health policy should be based on "the best estimate of the true human risk."[1,4,5] In order to generate the best estimate of the risk, it is critical to start with the best possible model. For if the initial model is unrealistic, then the analysis may be unrealistic and the prediction unrealistic.

Estimates of human risk, derived from occupational, therapeutic and epidemiological studies, are available for a substantial number of chemicals. These estimates, when combined with pharmacokinetic and pharmacodynamic studies, provide the most plausible starting point for a systematic approach to risk analysis. These networks of observations from human beings under real world conditions, intrinsically include causal mechanisms projected by the human genome. The biological response (i.e., the risk function) is generated by the evolutionary law of the individual human system.

RISK COMMUNICATION

Risk management should take advantage of the existing knowledge base on the effects of chemicals on human systems in so far as practical. In order to effectively utilize the existing human information, I propose the starting point for risk management:[4]

> "Treat like chemicals alike and treat different chemicals differently,
> in proportion to their relevant differences."

This approach to risk management offers several advantages.

1. The essence of the proposal is to establish a referential structure and a referential metric for regulating chemical risks. No particular causal bond or logical classification is assumed.

2. Relative risks of exposure would be ordered on the relative weight of scientific evidence. This should substantially improve the scientific consistency of risk assessments.

3. Natural families of chemical structures, when quantitatively compared as "efficient causes" within a family of human systems (from an Aristotlian perspective, a family of "material causes") would establish a network of related dynamic parameters which would gradually accumulate into a body of self-consistent evidence. Such a body of self-consistent evidence would provide a systematic approach to bound the uncertainty of risk estimates.

4. As the various lines of evidence become integrated into a coherent body of knowledge, the predictive power of the referential network could be systematically improved by developing the relationships between the observed parameters in human populations with the parameters predicted on the basis of quantitative descriptions of the "efficient" and "material" causes.

This paper provides a brief introduction into role of causal models in risk assessment and risk management. Fuller elaboration of these relationships depends, in part, on a detailed analysis of the role of structure and function in biology and disease. Such an analysis is beyond the scope of the present work.

This paper elaborates on and enhances the proposal for a new principle of risk management for chemicals in terms of a network of relationships among chemicals and systems. Further elaboration of this concept is underway in terms of nomic predictions for systems.

REFERENCES

1. M. A. Alavanja, J. C. Brown, and J. LR Chandler, Biochemical Epidemiology: Potential Contributions to Cancer Risk Assessment," *J. National Cancer Inst.* (in press, 12/86).
2. L. von Bertalanffy, *General Systems Theory*, George Braziller, NY, (1968).
3. M. Bunge, *Causality: The Place of the Causal Principle in Modern Science*, Harvard Univ. Press, Cambridge (1959).
4. J. LR Chandler, New Mechanistic Models for Risk Assessment, *Fundamental and Applied Technology*, **5**:634-652 (1985).
5. J. LR Chandler, Environmental Risk Assessment, *Bioassays* **5**:176-180 (1986).
6. W. E. Johnson, *Logic*, pp. 99-101, Dover Publ., NY (1964).
7. M. Maruyama, Four Different Causal Metatypes in the Biological and Social Sciences, in *Self Organization in Non-Equilibrium Systems*, pp. 355-361, Univ. Texas Press, Austin (1984).
8. M. D. Mesarovic and Y. Takahara, *General Systems Theory: Mathematical Foundations*, Academic Press, NY (1975).
9. B. C. Patten, R. W. Bosserman, J. T. Finn, and W. G. Cale, Propagation of Cause in Ecosystems, in *Systems Analysis and Simulation in Ecology*, Vol. 3, pp. 457-579, Academic Press, NY (1978).
10. H. Stachowiak, Medicine and the Paradigm of Neo-Pragmatism: A Contribution to Medical Decision Theory, *Theory and Decision* **21**:189-208 (1986).
11. M. Susser, *Causal Thinking in the Health Sciences: Concepts and Strategies in Epidemiology*, Oxford Press (1973).
12. R. Thom, *Structural Stability and Morphogenesis*, W. A. Benjamin, Inc., Reading, MA (1975).
13. US HHS, *Determining Risks to Health*, Washington, DC (1986).
14. F. J. Varela, *Principles of Biological Autonomy*, North-Holland, NY (1979).
15. H. R. Wulff, *Rational Diagnosis and Treatment*, Blackwell Scientific, London (1976).

Environmentally Induced Cancer: A Model for Estimating Excess Cancer Risks Based on a Two-Stage Model of Carcinogenesis

Josephine Mauskopf
Research Triangle Institute
Research Triangle Park, NC

ABSTRACT

When chemicals that are known carcinogens are released to the environment, persons subsequently exposed experience excess risks of cancer. In this paper, an estimation model is presented that generates estimates such as excess cancer risks for each year after onset of exposure using a lifetable approach. This model was developed from the Moolgavkar-Knudson two-stage model for human carcinogenesis. Lifetime exposure duration is not assumed. Using this model, estimates are generated for a hypothetical carcinogen under two assumptions: the carcinogen increases the cell mutation rate, and the carcinogen promotes cell growth.

KEYWORDS: Two-stage model of carcinogenesis, exposure duration, excess cancer risk

INTRODUCTION

When chemicals that are known carcinogens are released to the environment, persons subsequently exposed experience excess risks of cancer. The U.S. Environmental Protection Agency (EPA) is currently attempting to reduce such risks to human health by regulating the use and disposal of such chemicals. However, before promulgating major new regulations, EPA is required by Executive Order (E.O.) 12291 to perform a benefit-cost analysis. The estimation of the health benefits expected as a result of the new regulation is an important part of such a benefit-cost analysis.

E.O. 12291 requires that the health and other benefits attributable to new regulations be estimated in such a way as to allow them to be compared directly to estimates of regulatory costs. When estimating regulatory costs, a time period of use or disposal is typically specified. Therefore, to make the estimates of benefits comparable to the costs, they should be estimated for the same time period of use or disposal of the carcinogen. Furthermore, the costs are typically estimated for each year after promulgation. Similar estimates of timing should also be obtained for the health benefits.

In this paper I present a model for estimating the excess cancer cases or deaths attributable to environmental exposure to a carcinogenic chemical. Using this model, cancer risk estimates for the exposed population can be obtained for any time period of use

New Risks
Edited by L. A. Cox, Jr., and P. F. Ricci
Plenum Press, New York, 1990

or disposal to match that used for the cost estimates. These cancer risk estimates are generated for each year after onset of exposure. The expected age of the cancer victims can also be estimated. The differences between such cancer risk estimates for exposures with and without the new regulations give estimates of the health benefits of the new regulation that can be directly compared to the estimated costs.

THE ESTIMATION MODEL

The estimation model is based on the assumptions that the carcinogenic chemical increases the incidence of an already occurring cancer, and that the age-specific incidence rates for this baseline cancer can be explained by the Moolgavkar-Knudson two-stage model of carcinogenesis. In the Moolgavkar-Knudson model for carcinogenesis, cancer is postulated to be the end result of two discrete and heritable events at the cellular level. The model can be illustrated as follows:[1]

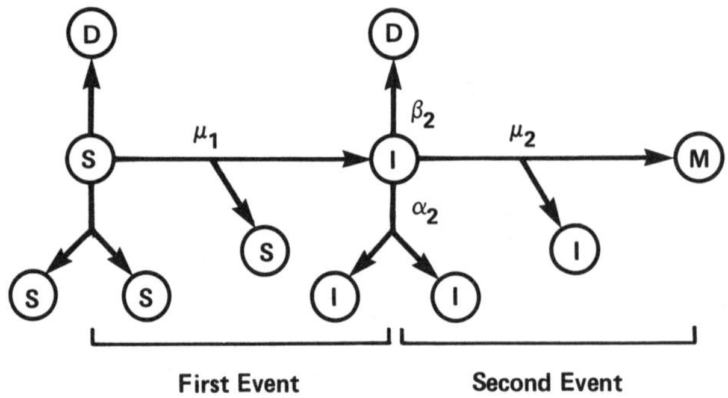

where

S = a normal susceptible cell
D = a dead or differentiated cell
I = an intermediate cell
M = a malignant cell
μ_1 = the mutation rate from normal to intermediate cells
μ_2 = the mutation rate from intermediate to malignant cells
α_2 = the rate of division of the intermediate cells
β_2 = the rate of differentiation or death of the intermediate cells.

Mathematically, the Moolgavkar-Knudson two-stage model postulates that, if the mutation rates are small and the mean time lag between cell malignancy and clinical diagnosis is five years, age specific cancer incidence rates can be generated from the expression:[1]

$$I(t) \approx \int_0^{(t-5)} X(s)\mu_1\mu_2 \exp\Big[(\alpha_2 - \beta_2)(t - 5 - S)\Big]\, ds \ , \qquad (1)$$

where

$I(t)$ = the age specific incidence rate
$X(s)$ = the number of susceptible cells in the affected organ at age S years
μ_1 = the mutation rate from normal to intermediate cells
μ_2 = the mutation rate from intermediate to malignant cells
$\alpha_2 - \beta_2$ = the growth rate of the intermediate cells (G_0)
t = the current age of the individual.

Assuming for simplicity that the number of normal susceptible cells remains constant over time, as in Thorslund and Charnley,[2] Eq. (1) can be rewritten:

$$I(t) \approx X\mu_1\mu_2 \int_0^{t-5} \exp\Big[G_0(t-5-s)\Big]\,ds$$
$$\approx X\mu_1\mu_2 \left[\frac{e^{G_0(t-5)}-1}{G_0}\right] \tag{2}$$

Using Eq. (2) and the observed age-specific incidence rates for the baseline cancer of interest, estimates can be obtained for two parameters, $(X\mu_1\mu_2)$ and G_0.

Two possible impacts of the carcinogen on this baseline carcinogenic process are postulated in this paper: either the carcinogen increases the mutation rate from normal to intermediate cells or the carcinogen increases the growth rate of the intermediate cells. The derivation of age-specific incidence rates for these two cases is described below.

Carcinogen Increases Cell Mutation Rate

If the mutation rate for normal cells is increased by the carcinogen such that $\mu_1 = \mu_{10} + \mu_{11}d(t)$,[3] from Eq. (2) the age-specific cancer incidence rate with a constant lifetime dose, d, of the carcinogen is given by:

$$I'(t) \approx X(\mu_{10}+\mu_{11}d)\mu_2 \left[\frac{e^{G_0(t-5)}-1}{G_0}\right] ,$$

where

$I'(t)$ = the age specific incidence rate for the baseline cancer in the presence of the carcinogen
μ_{10} = the baseline mutation rate from normal to intermediate cells
μ_{11} = the corresponding mutation rate constant for the carcinogen
d = the lifetime dose of the carcinogen in mg/kg/day.

Therefore, the age-specific excess cancer incidence rate for a cell mutagen can be written as:

$$EI(t) \approx X(\mu_{11}d)\mu_2 \left[\frac{e^{G_0(t-5)}-1}{G_0}\right]$$
$$\approx \gamma\cdot d\cdot I(t) , \tag{3}$$

where

$$\gamma = \mu_{11}/\mu_{10} .$$

If the lifetime dose, of d mg/kg/day, that results in an excess cancer risk of 1×10^{-5} has been estimated by EPA for the cell mutagen, using Eq. (3) the following relationship can be derived:

$$\sum_{t=5}^{T} Pr(A_t)\cdot EI(t) = 1\times10^{-5}$$
$$\approx \sum_{t=5}^{T} Pr(A_t)\gamma\cdot d\cdot I(t) = 1\times10^{-5} \tag{4}$$

where

$Pr(A_t)$ = the probability of being alive at age t
5 = the average time lag in years from cell malignancy to clinical diagnosis
T = the maximum lifespan
$\gamma=\mu_{11}/\mu_{10}$ = the mutation constant for the carcinogen relative to baseline
d = the lifetime constant dose that results in an excess lifetime risk of 1×10^{-5} in mg/kg/day (virtually safe dose = VSD)
$I(t)$ = the baseline age-specific cancer incidence rate.

Equation (4) can be solved for γ using age-specific incidence data for the baseline cancer.

The actual expected carcinogen exposure patterns with and without the new regulations can now be used to develop explicit expressions for the age-specific excess cancer incidence rates for any duration of use or disposal. In this paper, for an example, assume that the new regulation will prevent exposure altogether and that without the regulation, the people at risk of exposure will be exposed at a dose d_2 for a 45-year period some time during their life. The ages at first exposure depend on the age distribution of the population at risk. Thus, if the carcinogen increases the normal cell mutation rate, from Eq. (2) age-specific excess cancer risk for the exposed population without the new regulation is given by

$$EI(t) = 0 \qquad \text{for } t \le 5 + a$$

$$EI(t) = \gamma \cdot d_2 \left[\frac{e^{G_0(t-a-5)} - 1}{e^{G_0(t-5)} - 1} \right] \qquad \text{for } 5 + a < t \le 5 + a + ED$$

$$EI(t) = \gamma \cdot d_2 \left[\frac{e^{G_0(t-a-5)} - e^{G_0(t-a-5-ED)}}{e^{G_0(t-5)} - 1} \right] \cdot I(t) \qquad \text{for } t > 5 + a + ED, \quad (5)$$

where

t = the current age of the individual
a = the age at onset of exposure
ED = the duration of exposure
G_0 = the baseline growth rate of the intermediate cells, estimated from observed incidence data for the baseline cancer
5 = the average lag time between cell malignancy and clinical diagnosis
γ = the carcinogen cell mutation rate constant relative to the baseline cell mutation rate, estimated using EPA estimates of the VSD and observed incidence data
d_2 = the level of exposure in mg/kg/day.

Using the estimated values for γ and G_0, these age-specific excess cancer incidence rates can be estimated for the population at risk.

Carcinogen Promotes Intermediate Cell Growth

If the intermediate cell growth rate is increased by the carcinogen such that $G = G_0 + G_1 d(t)$,[1] from Eq. (2) the age-specific cancer incidence rate with a constant lifetime dose, d, of the carcinogen is given by

$$I'(t) \approx X\mu_1\mu_2 \left[\frac{e^{(G_0+G_1 d)(t-5)} - 1}{G_0 + G_1 d} \right],$$

where

$I'(t)$ = the age-specific incidence rate for the baseline cancer in the presence of the carcinogen
G_0 = the baseline growth rate
G_1 = the growth rate constant for the carcinogen
d = the lifetime dose of the carcinogen in mg/kg/day.

Therefore, the age-specific excess cancer incidence rate for a cell growth promoter can be written as:

$$EI(t) \approx \frac{G_0}{e^{G_0(t-5)}-1}\left[\frac{e^{(G_0+G_1d)(t-5)}-1}{G_0+G_1d} - \frac{e^{G_0(t-5)}-1}{G_0}\right]\cdot I(t)\ . \tag{6}$$

If the carcinogen promotes growth of the intermediate cells, EPA estimates of the lifetime dose, d, that results in an excess cancer risk of 1×10^{-5}, and Eq. (6) can be used to derive the following relationship:

$$\sum_{t=5}^{T} Pr(A_t)\cdot EI(t) = 1\times 10^{-5}$$

$$\approx \sum_{t=5}^{T} Pr(A_t)\cdot \left[\frac{e^{(G_0+G_1d)(t-5)}-1}{G_0+G_1d}\cdot\frac{G_0}{e^{G_0(t-5)}-1} - 1\right]\cdot I(t) = 1\times 10^{-5}\ , \tag{7}$$

where

$Pr(A_t)$ = the probability of being alive at age t
5 = the average time lag in years from cell malignancy to clinical diagnosis
T = the maximum lifespan
G_0 = the baseline growth rate
G_1 = the growth rate constant for the carcinogen
d = the lifetime constant dose that results in an excess lifetime risk of 1×10^{-5} in mg/kg/day (VSD)
$I(t)$ = the baseline age specific cancer incidence rate.

Equation (7) can be solved for G_1 using age-specific incidence data for the baseline cancer and an interative procedure. Assuming only 45 years exposure at level d_2, as for the cell mutagen, from Eq. (2) age-specific excess risk of cancer for the exposed population without the new regulation is given by

$$EI(t) = 0 \qquad \text{for } t \le 5+a$$

$$EI(t) = \frac{G_0}{e^{G_0(t-5)}-1}\cdot\left[\frac{1-e^{G_0(t-a-5)}}{G_0} + \frac{e^{(G_0+G_1d_2)(t-a-5)}-1}{G_0+G_1d_2}\right]\cdot I(t)$$

$$\text{for } 5+a < t \le 5+a+ED \tag{8}$$

$$EI(t) = \frac{G_0}{e^{G_0(t-5)}-1}\cdot\left[\frac{e^{G_0(t-a-5-ED)}-e^{G_0(t-a-5)}}{G_0}\right.$$

$$\left.+\frac{e^{(G_0+G_0d_2)(t-a-5)}-e^{(G_0+G_1d_2)(t-a-5-ED)}}{G_0+G_1d_2}\right]\cdot I(t) \quad \text{for } t > 5+a+ED$$

where

t = the current age of the individual
a = the age at onset of exposure
ED = the duration of exposure
G_0 = the baseline growth rate of the intermediate cells, estimated from observed incidence data for the baseline cancer
G_1 = the growth rate constant for the carcinogen, estimated using EPA estimates of the VSD and observed incidence data
d_2 = the level of exposure in mg/kg/day.

Using the estimated values for G_0 and G_1, these age-specific excess cancer incidence rates can be estimated for the population at risk.

USING THE ESTIMATION MODELS

The use of the model is illustrated using a hypothetical chemical that, in the absence of new regulations, will contaminate two community wells for 45 years at a concentration of 2.21 mg/liter. Assuming a daily intake of 2 liters of well water, this concentration is equivalent to a daily dose of 0.063 mg/kg/day for a 70 kg person. Two thousand persons per year are assumed to be at risk of exposure from each contaminated well. For this hypothetical chemical we assume that a lifetime dose of 0.000174 mg/kg/day is known to result in a lifetime excess risk of liver cancer of 1×10^{-5} no matter which mathematical model of carcinogenesis is assumed when estimating this dose. In fact, when estimating VSDs from animal data, the model of carcinogenesis and the mode of action of the carcinogen assumed will affect the estimated value of the VSD.

A lifetable model is used to convert the excess incidence of liver cancer at each age after onset of exposure [Eq. (5) or (8)], into lifetime excess cancer risks for the two cases described. This model assumes a nonstationary Markov process containing four possible states for an individual during each year of the rest of his life after the onset of exposure to a carcinogen. These four states are as follows:

1. Alive, no cancer attributable to exposure to the carcinogen;

2. Alive, cured of cancer attributable to exposure to the carcinogen;

3. Dead of a cancer attributable to exposure to the carcinogen; and

4. Dead of any other cause.

At the start of exposure the individual has a probability of 1 of being in state 1, and at age 90, a probability of 1 of being dead, i.e., in states 3 and 4. An individual is assumed to be able to contract only one cancer attributable to exposure to the carcinogen. If the chemical related cancer is fatal, the patient is assumed to die in the year that the cancer is detected.

The exposed population is assumed to be identical to the U.S. population in terms of sex, race, smoking habits, and age distribution.[4] To estimate the health effects of exposure to the carcinogen, exposed populations are divided into 10-year age groups and the health effects of exposure to the carcinogen estimated for each age cohort separately. To estimate the excess cancers for those born after the start of the contamination, the lifetable model is extended to allow for a new cohort of exposed individuals to be added each decade, at a mean age of 5 years and equal in size to the youngest cohort in the original population. Age specific death probabilities from all causes are estimated based on 1978 U.S. life tables. Baseline incidence rates for all the cancers thought to be related to chemical exposure are

Table 1. Excess Liver Cancer Cases Without New Regulations

Time since onset of exposure	Moolgavkar-Knudson first stage mutation	Moolgavkar-Knudson growth rate increase	Multistage 4-stages first stage mutation
0 - 10	0.0	0.0	0.0
10 - 20	0.05	0.01	0.01
20 - 30	0.15	0.05	0.05
30 - 40	0.30	0.19	0.15
40 - 50	0.50	0.51	0.34
50 - 60	0.70	1.12	0.59
60 - 70	0.79	1.97	0.81
70 - 80	0.78	2.79	0.90
80 - 90	0.68	3.19	0.82
90 - 100	0.50	2.90	0.62
100 - 110	0.30	2.13	0.37
110 - 120	0.12	1.12	0.15
120 - 130	0.02	0.25	0.02
Totals	4.89	16.23	4.83

taken from the Surveillance, Epidemiology and End Results (SEER) study.[5] The cure rates for the cancers of interest are estimated using data from Axtell, Asire, and Meyers.[6]

RESULTS

The excess cases of liver cancer expected without the new regulations, estimated for the hypothetical chemical using the models described, are presented in Table 1. Also presented are estimates of excess liver cancer cases expected without the regulation using an estimation model based on the multistage model for carcinogenesis.[3,7] All these estimation models generate estimates of excess cancer cases by time since onset of exposure. The mathematical model of carcinogenesis and the specific action of the carcinogen influence both the timing and the magnitude of the estimated cancer cases. When the carcinogen increases the mutation rate for normal cells to intermediate cells, the two-stage Moolgavkar-Knudson model and a four-stage multistage model give estimates that are approximately equal in size but vary a little in timing. For the two-stage model most of the excess cases are projected to occur 50 to 80 years after onset of exposure, while for the four-stage model most of the excess cases are projected to occur 60 to 90 years after onset of exposure. If the carcinogen increases the growth rate of the intermediate cells, a greater number of excess cancer cases are estimated, mostly occurring between 60 and 110 years after onset of exposure. Exposure levels above the VSDs will thus have greater adverse health effects for those carcinogens that act by promoting growth of intermediate cells than for those that act by increasing the mutation rates of normal to intermediate cells.

Since it is assumed that there will be no human exposure to the carcinogen with the new regulations, the estimates presented in Table 1 can also be interpreted as equal to the cancer cases that will be avoided if the new regulations are promulgated. These avoided cancers, therefore, are a measure of the expected health benefits of the new regulations.

CONCLUSIONS

In this paper, I have presented an estimation method, based on the Moolgavkar-Knudson two-stage model of carcinogenesis, that can be used to estimate the cancer cases avoided as a result of new regulations that reduce exposures to carcinogens for varying periods of time and at varying ages. Such an estimation method allows for greater flexibility than the more common method of estimating reductions in lifetime risks for reductions in constant lifetime exposures only. The proposed estimation method has the added advantage that it generates the estimates of avoided cancer cases for each year after reduction of exposure and can also estimate the ages of the victims of these cancers and thus the person-years of life gained due to the reduced exposure to the carcinogen.

The data requirements of the proposed estimation model are somewhat more demanding than those for the lifetime exposure estimates. However, both estimates depend on having estimates of the dose-response relationship for the carcinogen at low doses. These estimates are usually derived from animal data, at exposure levels far above those usually encountered in the environment. Thus, reliance on these estimates is a weak link for both estimation methods. The estimation method proposed in this paper requires, in addition, that we make assumptions about the type of cancer that the carcinogen is likely to cause in humans, and that we estimate the mutation and growth parameters that describe the behavior of this baseline cancer.

The lifetime exposure method, based on a linearized multistage model, assumes a linear relationship between dose and excess risk of cancer. The estimates presented in this paper have shown that if the carcinogen acts on the mutation rate of normal cells to intermediate cells, the relationship between dose and excess risk will be linear. If, however, the carcinogen acts to promote the growth of the intermediate cells, the effect of increasing dose levels of the carcinogen will not be linear. The model proposed can develop realistic estimates for both types of carcinogens.

As the federal government goes ahead with regulations to prevent releases to the environment of carcinogenic chemicals, it is important that the benefit-cost analyses that are required for major new regulations should include realistic estimates of the likely numbers and timing of avoided cases and deaths from cancer (the health benefits) if the regulation is promulgated. The estimation method proposed in this paper can be used to develop health benefits estimates that can be compared directly to the estimated costs of the regulation.

REFERENCES

1. S. H. Moolgavkar and A. G. Knudson, Mutation and Cancer: A Model for Human Carcinogenesis, *JNCI* **66**:1037-1052 (1981).
2. T. W. Thorslund and G. Charnley, Use of the Two-Stage Case of the Multistage Model for Cancer Risk Assessment, Presented at the Annual Meeting of the Society for Risk Analysis, Boston, MA, November 10, 1986.
3. J. Mauskopf and S. Curtis-Powell, Disposal of Hazardous Wastes in Unregulated Landfills: A Health Risk Assessment, in *Proceedings of the National Conference on Hazardous Wastes and Environmental Emergencies*, Cincinnati, OH, Hazardous Materials Control Research Institute, Silver Spring, MD, May 14-16, 1985.

4. U.S. Department of Commerce, *Statistical Abstract of the United States*, Bureau of the Census, Washington, DC (1980).
5. J. Young, C. Percy, A. Asire, J. Berg, M. Cusano, L. Gloeckler, J. Horm, W. Lourie, E. Pollack, and E. Shambaugh, *Cancer Incidence and Mortality in the United States, 1973-77*, Surveillance, Epidemiology and End Results Program EPA Discussion Paper, Monograph No. 57, National Cancer Institute, Washington, DC (1980).
6. L. M. Axtell, A. J. Asire, and M. H. Meyers, *Cancer Patient Survival, Report No. 5*, U.S. Government Printing Officer, Washington, DC (1976).
7. J. Mauskopf, "Estimating Excess Cancer Risks from Well Water Contaminated by Industrial Wastes, Manuscript in Preparation.

Is the One-Hit Model of Carcinogenesis Conservative? Evidence from the National Toxicology Program's Bioassay Database[a]

Donna Spiegelman and John C. Bailar, III
Harvard School of Public Health
Boston, MA

Edmund A. C. Crouch
Harvard University
Cambridge, MA

Rashid Shaikh
Health Effects Institute
Cambridge, MA

ABSTRACT

Linear (one-hit) formulas of carcinogenesis are widely believed to be "conservative," in the sense that they will rarely underestimate risks of cancer at low exposures. Such models are generally applied to the lifetime incidence of cancer at a specific site, with risks estimated from animal data at zero dose (control) and substantially higher doses that are appreciable fractions of the estimated maximum tolerated dose.

We know of no empirical study of whether the one-hit model is conservative in the sense described. The Carcinogenesis Bioassay Database System contains data on 1212 separate bioassays of 308 chemical substances tested at exactly 3 evaluable doses. These provided sufficient data to examine 8432 specific combinations of cancer site with sex, species, and chemical. For each of these, we fitted a one-hit model to the zero and maximum dose data points, then examined the relation of the fitted curve to the observation at the mid-dose with and without an adjustment for intercurrent mortality.

Both underestimates and overestimates of risk occurred more often than expected by chance. We cannot tell whether underestimates would occur at lower doses, but we offer six possible biologic reasons why underestimates might be expected.

In a high percentage of animal bioassays, the one-hit formula is not conservative when applied in the usual way to animal data. It remains possible that the one-hit formula may indeed be conservative at sufficiently low doses (below the observable range), but the usual procedure, applied to the usual dose range, can be non-conservative in estimating the slope of the formula at such low doses. Risk assessments for regulation of carcinogens should incorporate some measure of additional uncertainty.

---

a. A similar version of this paper appeared in *Risk Analysis* **8**:485-497, 1988.

New Risks
Edited by L. A. Cox, Jr., and P. F. Ricci
Plenum Press, New York, 1990

KEYWORDS: One-hit model, risk assessment, low dose extrapolation, carcinogenesis

INTRODUCTION

The one-hit or linear formula is often fitted to animal data at relatively high doses and used to estimate carcinogenic risk at considerably lower doses. This approach is considered "conservative" because the one-hit formula usually gives the highest risk estimates at low doses among the various models in common use for this purpose.

We examined the adequacy of fit of this formula to a large set of bioassay data, with the goal of assessing whether the one-hit formula is indeed "conservative" in the sense described. We cannot make this assessment in the low-dose region of greatest interest for human risk analysis; we can only assess the adequacy of fit at the observable dose points of animal studies, doses that are either effectively zero or very high. However, the finding that the one-hit formula is often not "conservative" in its estimate of risk at the middle experimental dose disproves the assumption that the "conservatism" of the one-hit formula spans the range from zero dose to the lowest testable doses.

Methods for fitting formulae to dose-response data usually exclude, by mathematical restrictions, any possibility of a supralinear curve. To the extent that nature produces supralinear dose-response curves, this is a grave defect. It is well known that this type of dose-response relationship occurs over a certain exposure range for vinyl chloride; we found numerous other examples consistent with such curves in several other carcinogen bioassays (Table 1). The data in this table have been adjusted for intercurrent mortality. The p-values are for the one-tailed exact binomial test of the hypothesis that the one-hit formula as fitted to the control and high dose groups describes the true state of nature, using the observed data at the middle-dose point. The procedures for adjustment for intercurrent mortality and performing the hypothesis test are described in detail in the Methods section below.

The examples presented in Table 1 could simply have occurred by chance. After examining hundreds of independent bioassays at numerous anatomic tumor sites, we would expect a certain number of apparently supralinear dose-response curves, simply due to random sampling variation. This report shows that the number of supralinear dose-response curves substantially exceeds that expected by chance, when a large number of independent rodent tumorogenicity experiments are evaluated.

METHODS

The Data

The National Toxicology Program (NTP) (Dr. M. Rowley, private communication) kindly provided a copy of the data in the Carcinogenesis Bioassay Database System (CBDS) (Linhart, 1974) as current at June, 1985. We chose from these data all bioassays (defined as an experiment performed over a continuous period of time at one laboratory, testing a unique substance by dosing at different levels animals of the same sex and species), that met the following criteria:

1. The bioassay tests a substance identifiable by a unique code under the National Cancer Institute system. This criterion was intended to exclude most mixtures, although some such as the caffeine, phenacetin and aspirin mixture were included. 1956 bioassays remained available for analysis after these exclusions.

Table 1. Examples of Supralinearity in Animal Carcinogenesis Data

Chemical	Sex	Species/ Strain[a]	Tumor Type	Route	Adjusted Lifetime Cumulative Incidence (Low)	(Mid)	(High)	p-Value[b]
4,4'-methylene-dianiline.2HC1	M	b	Hepatocellular carcinoma	water	0.22	0.66	0.70	0.025
1,2-dibromoethane	F	o	oral, GI squamous carcinoma	gavage	0.00	0.80	0.80	0.0002
1,4-dioxane	F	o	fibroadenoma	water	0.05	0.46	0.47	0.024
1,3 butadiene	F	b	hemangiosarcoma	inhalation	0.00	0.38	0.41	<0.00001
carbon tetrachhloride	M	b	hepatocellular carcinoma	gavage	0.07	0.98	0.98	0.008
dimethylvinyl chloride	M	b	oral, GI squamous carcinoma	gavage	0.00	0.86	0.90	0.004
cytembena	M	f	mesothelioma, osteosarcoma	peritoneal injection	0.05	0.74	0.68	0.00003
1,5-naphthalene-diamine	F	f	endometrial stromal polyp	food	0.07	0.28	0.27	0.04
iodinated glycerol	M	f	plasmacytic leukemia	gavage	0.31	0.72	0.76	0.04
2-methyl-aziridine	F	c	skin, breast	gavage	0.00	0.81	0.83	0.02

[a]b = B6C3F1 mouse; o = Osborne Mendel rat; f = Fischer 344 rat; c = Charles River rat.

[b]One-tail exact test of mid-dose to maximum likelihood one-hit model (non-decreasing) fitted to low and high dose points; see text.

2. Exactly three different doses were tested in the bioassay, including at least one control group. If both a vehicle control group and an untreated control group were available, we ignored the untreated control group. All remaining control groups were combined and treated as one. 679 of the 1956 bioassays did not have exactly three dose groups; most of these had two dose groups or no data at all.

3. The dosing information was interpretable by our computer software, and the dose points of the assay were sufficiently widely spaced (differing by more than 20% of each other). An additional 65 bioassays did not meet this criterion.

These exclusions left 1212 bioassays available for study.

Our analysis used the relative sizes of the doses for the two dosed groups. All doses were normalized to the maximum dose for each experiment, in order to obtain doses relative to this maximum. We always used the initial dose rates, although changes occurred in a few cases subsequent to the start of the experiment. Because the changes were, in each case, proportional to the initial dose rates, the initial dose rates should be an adequate measure of *relative* dosing. There were 13 different methods of dose administration observed in these data, including oral food, oral gavage, oral water, total body respiratory exposure, intraperitoneal injection, surface delivery, and intramuscular injection. Each of these had various ways of measuring dose.

Published reports of the data in the CBDS, such as the Technical Report series of the NTP/NCI and several journal articles, have used a wide range of schemes for categorizing tumors for analysis. Because of our need for a uniform classification to deal with so many bioassays, we developed our own tumor classification system from the most detailed data available. These were the 8-digit Systematized Nomenclature of Pathology (SNOP) codes, of which four digits describe histologic type, and four digits describe anatomic site. Our goal in developing this classification system was to obtain groups of tumor categories with numbers likely to be sufficiently large within individual bioassays to obtain meaningful results, yet to maintain within these groups, a reasonable degree of biologic uniformity. We used CBDS data to cross-classify the frequency of histologic types with anatomic sites without regard to substance, dose, species, sex or shape of the dose-response curve. Using this cross-classification, we produced a system of 102 tumor classes, 66 for primary malignant neoplasms and 36 for primary benign neoplasms. We excluded all metastatic sites and tumors of uncertain origin. Although our classification system does not reflect legitimate study-to-study differences in biologic justification for grouping neoplasms, it should be free of bias, because the development of the classification system was "blind" to both chemical substance, dose, and shape of the dose-response curves.

ANALYSIS

We analyzed each tumor class in which at least four animals were affected in the three dose groups combined. The purpose of the restriction was to assure that the tumor class would be frequent enough within the bioassay to provide minimal information about the shape of the dose-response curve. Analysis was directed at describing the deviation of the observed middle dose lifetime tumor probability from that expected from the null hypothesis of a one-hit dose response formula. Deviations were measured by "p-values." Each bioassay could produce up to 102 p-values. The p-values are not independent. In fact, 8432 tumor classes within the 1212 bioassays met the above restriction, an average of about seven p-values per bioassay.

The analysis was performed both without adjustment and with an adjustment for intercurrent mortality and other censorship. In both cases, we fit the linear formula to the observed lifetime probability tumors (adjusted or unadjusted) at the zero (control) dose and high dose points using the maximum likelihood method for the probability model

$$\text{Pr(neoplasm)} = 1 - \exp(-a-b^*\text{dose});\ a, b \geq 0, \tag{1}$$

where neoplasm means a tumor of one of the 102 tumor classes, and Pr refers to the lifetime tumor probability, to obtain the estimates $\hat{a}$ and $\hat{b}$ of a and b, respectively. When the observed lifetime incidence was higher at the high dose than in the control group, the fitted curve passes directly through the two probabilities. Otherwise, the procedure fits a horizontal line between the two points, with the location of the line determined as $(R_L+R_H)/(N_L+N_H)$ in the unadjusted case, and an equivalent weighted mean of the high and low-dose probabilities in the adjusted case, where R_L refers to the observed number of tumors in the class under consideration in the zero dose group, N_L refers to the observed number of animals in the zero dose group, and likewise for R_H, N_H in the high dose group.

We then defined our p-value as described above through a one-tailed exact binomial test:

$$\hat{p} = \sum_{i=R_M}^{N_M} \binom{N_M}{i} \hat{P}_M^{\,i} (1-\hat{P}_M)^{N_M-i}, \tag{2}$$

where

R_M = the observed number of animals with a primary tumor of the class under consideration in the middle dose group,

N_M = the total number of animals in the middle dose group,

$$\hat{P}_M = \begin{cases} 1-\exp(-\hat{a}-\hat{b}^*\text{dose}) \text{ if } R_H/N_H > R_L/N_L, & \text{the estimated lifetime tumor probability at the middle dose point under the one-hit formula.} \\ (R_L+R_H)/N_L+N_H) \text{ otherwise.} \end{cases}$$

This procedure would correspond to a one-tailed exact binomial test of the linear formula at the middle dose, if $\hat{p}$ was a true value rather than an estimate from the data. The p-value obtained will be close to 1 when the observed number of tumors at the middle dose point is well below that expected under the fitted curve, and the p-value will be close to 0 when the middle dose point is well above it.

The *adjusted* cumulative tumor probabilities at the zero and high dose group were computed as follows. We estimated *total* mortality as if these dose groups had the same mortality, from all other causes except cause j, as the middle dose group by the following formula:

$$\text{TPR}\hat{\text{O}}\text{B}_{ij} = \sum_{k=1}^{S_{ij}} (\hat{h}_{ij}(t_{ijk})) \prod_{m=1}^{k-1} (1-\hat{h}_{ij}(t_{ijm}))(1-\hat{h}_{\text{OM}}(t_{ijm})),\ i = L,H \text{ (the control, high dose groups)},\ j = 1,\ldots,102 \text{ (the tumor classes)}, \quad (3)$$

where

S_{ij} = the number of primary tumors of class j in dose group i

t_{ijk} = the ordered time to death (or time of terminal sacrifice) from tumor class j in dose group i; $k = 1,\ldots,S_{ij}$,

$\hat{h}_{ij}(t_{ijk}) = r_{ijk}/n_{ijk}$, the instantaneous probability of death from tumor class j at time t_{ijk}, given survival to time t_{ijk},

$\hat{h}_{\text{OM}}(t_{ijk}) = \sum_{j'=\text{all causes except } j} \frac{r_{ij'k}}{n_{ij'k}}$, the instantaneous probability of death from any neoplastic or non-neoplastic cause except j at the time t_{ijk}, given survival to time t_{ijk}, in the middle dose group,

r_{ijk} = number of events of type j, microscopically confirmed at autopsy, in dose group i at time t_{ijk}

n_{ijk} = number of animals at risk for event j in dose group i at time t_{ijk}.

This method of adjustment is a modification of the product-limit method of estimation of Kaplan and Meier (1958). The computations produced lifetime tumor probabilities at the three dose levels, middle (observed) and low and high (estimated) in which age-specific probabilities of death from causes other than the neoplasm of interest were identical; we then computed a p-value for the middle dose group, R_N/N_M, as described above.

Throughout our analysis, we censored any events that occurred either a) before the first neoplasm occurred in any of the three dose groups or b) after any dose group had no surviving members. Animals satisfying either of these conditions are counted in neither the unadjusted nor the adjusted analyses. The second constraint was enforced by noting the times of death from all causes, including sacrifice in every dose group, and omitting all events after any dose group had no surviving members. In practice, we relaxed this constraint slightly to include events at the next event time as though they occurred at the previous event time. This was done to insure that sacrifices of remaining animals in all dose groups performed within a few days always were included in the analysis, as there should be little biological significance in a few days difference in scheduled terminal sacrifice. Strict enforcement of constraint b) resulted in the possible loss of a substantial proportion of experimental data otherwise available for analysis, while just 1086 of the 8432 analyses were affected by the relaxed constraint.

Associated with each analysis was an unadjusted and adjusted p-value. If the one-hit formula holds for each bioassay and tumor class, and if the tumor rates in both the control and high doses groups were known exactly, we might expect that random variation would cause these p-values to be uniformly distributed between 0 and 1. If some of the observed sets of bioassays or tumor classes tend to produce supralinear dose-response curves, there would be an excess of low p-values; conversely, the presence of sublinear dose-response curves would produce an excess of high p-values.

However, the expected distribution will not be uniform, even if the one-hit formula is correct. The 8432 p-values are not independent, since subsets of them are derived from the same bioassays. The data have a discrete form, so that not all p-values are possible for any given sample size, and the experimental designs were often quite similar. There is the possibility that our various exclusion criteria, discussed above, introduced some bias. In addition, and perhaps most importantly, the binomial distribution is not appropriate for assessing the probability distribution of the data at the middle dose point, under the one-hit model. This is because probabilities $\hat{P}_M$ of equation (1) are estimates based on the data, with their own random sampling error, rather than fixed hypothesized values as required. Using our analysis procedure, we conducted a series of simulation experiments to generate the shape of the expected distribution of p-values.

The simulation datasets were intended to be as similar as possible to the actual data, except that cancer incidence conformed exactly to a one-hit model. We used the actual bioassay data by extracting, for each bioassay, the relative dose values, size of each dose group, and lifetimes of each animal in each dose group. If the bioassay contributed more than one set of p-values to the data because there was more than one tumor class with a total of four tumors in the three dose groups combined, we included that bioassay as many times as it produced p-values in the set of 8432. We assigned to each of these bioassays a parameter (a,b) randomly chosen with replacement from the distribution of pairs of estimated parameters obtained when performing the adjusted analysis. For each animal in each dose group, a probability of tumor was calculated assuming a one-hit formula:

$$\Pr(\text{neoplasm}) = 1 - \exp(-(a+b^*\text{dose})^*(t/t_o)^k). \qquad (4)$$

The extra term in age, $(t/t_o)^k$, was included to allow for the age dependence of tumor rates usually seen in animal experiments. The value of k, the power law for age dependence, was varied in different simulations, to check for sensitivity of the shape of the p-value distribution to this age dependence. Each animal was assigned a tumor randomly with this probability.

After each animal in each simulated bioassay had been assigned an outcome under the one-hit model, we analyzed these simulated data exactly as described above in the procedures for the adjusted and unadjusted analyses, and obtained the simulated adjusted and unadjusted p-value distributions expected under the one-hit formula. We then compared these to the actual p-value distributions.

RESULTS

Figure 1 shows the results of the unadjusted and adjusted analyses performed on the actual data. The peakiness of the unadjusted histogram is due to the discreteness of the input data and similar design of most experiments. There are relatively rare numbers of p-values at both extremes of very small and very large. For comparison, Fig. 2 shows the results of the simulations described above, with the parameter k set to 6. This value represents an extreme estimate for k. Again, there are peaks at low and high values of p, but the peak at small values of p is only about half the height (relative to the flat part of the distribution) as the peak occurring in the actual data. As can be seen by comparing the histograms for adjusted and unadjusted analyses, making the adjustments for intercurrent mortality reduces the size of the peaks in both the real data and the simulations, but this adjustment cannot completely account for the observed excess of low p-values in the real data.

The implication of the peak at low p-values is that there is an excess of experimental results in which a linear formula will underestimate risks at intermediate doses in the observable range of excess risk. The simulations show that the excess is real, and the adjusted analysis shows that it cannot be accounted for by dose-related intercurrent deaths. One can estimate the frequency of supralinearity and of sublinearity by comparing Fig. 1 with Fig. 2. Results are shown in Table 2. The simulated distributions were normalized to the observed distribution by demanding equality of their mean values over the range $0.2 < p < 0.8$, the flat part of the distributions. The observed distributions (adjusted and unadjusted) appear to exceed the simulated ones in the ranges $0.0 < p < 0.02$ and $0.99 < p < 1.0$. The values in Table 2 show that the one-hit model was "conservative" in the sense specified in about 4.5-7% of the bioassay and tumor classes we examined, and that the reverse was true ("anti-conservative") in about 3-5.5% of these data sets. The percentages vary within these ranges depending on the value of the parameter k used in the simulation, with the largest values of k (k=6) suggesting the lowest percentage of "conservative" results. Thus, a substantial majority of results are adequately fitted in the observable range by the one-hit model, and the remainder are about equally divided between sublinear and supralinear responses.

We also tested the effect of restricting the analyses to those experiments where there was clear evidence of carcinogenicity, by requiring that the zero and high dose unadjusted outcomes be significantly different at the $p < .05$ level, using a one-sided Fishers' exact test. This substantially reduced the number of experiments included (from 683 to 8432 endpoints), and led to higher estimates of the proportion of experiments in the low p-value peak — between 6% (adjusted analysis) and 11% (unadjusted analysis) as compared with simulations. This suggests that the demonstrated lack of conservatism of the one-hit model is an important consideration in assessing the risk of substances which are likely to be regarded as probable carcinogens.

We further classified the clear carcinogens by potency (shape) and found that substances with the largest potency estimates, $\hat{b}$, tended to produce p-value distributions with the largest excesses of low values as compared with the appropriate simulated distributions, although in all observed ranges of values of $\hat{b}$ at least small excesses occurred.

DISCUSSION

We have shown that the one-hit formula, applied in the usual way, often underestimates lifetime cancer risks in the experimentally observable range. Further, the extent of underestimation may increase with increased carcinogenic potency. Adjustment for intercurrent deaths from dose-related or other causes does not alter this conclusion. Again, we cannot assess how far down into the low dose-low risk range these results apply.

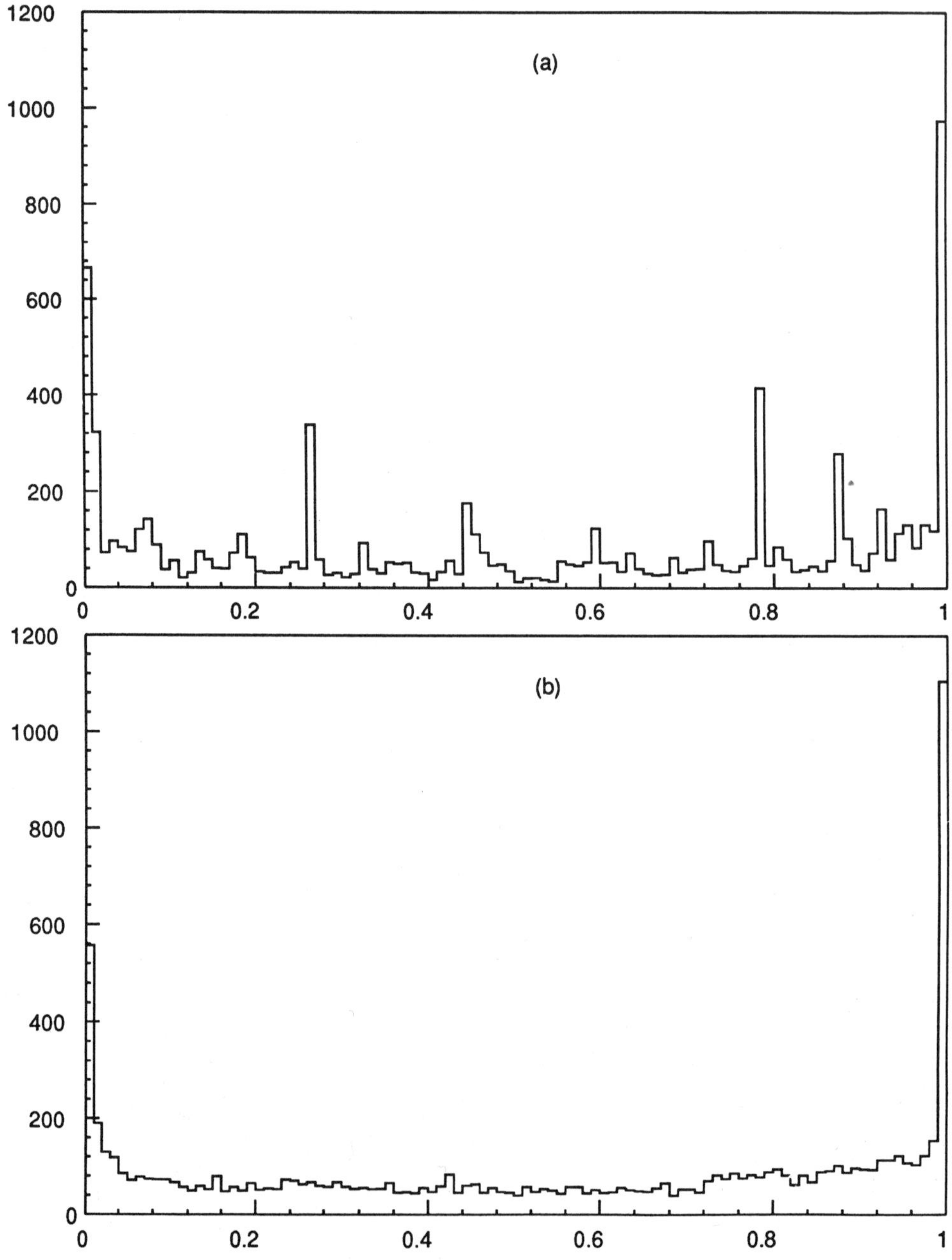

Fig. 1. (a) Unadjusted analysis of observed data mean over range 0.2 to 0.8 is 57.85. (b) Adjusted analysis of observed data mean over range 0.2 to 0.8 is 57.10.

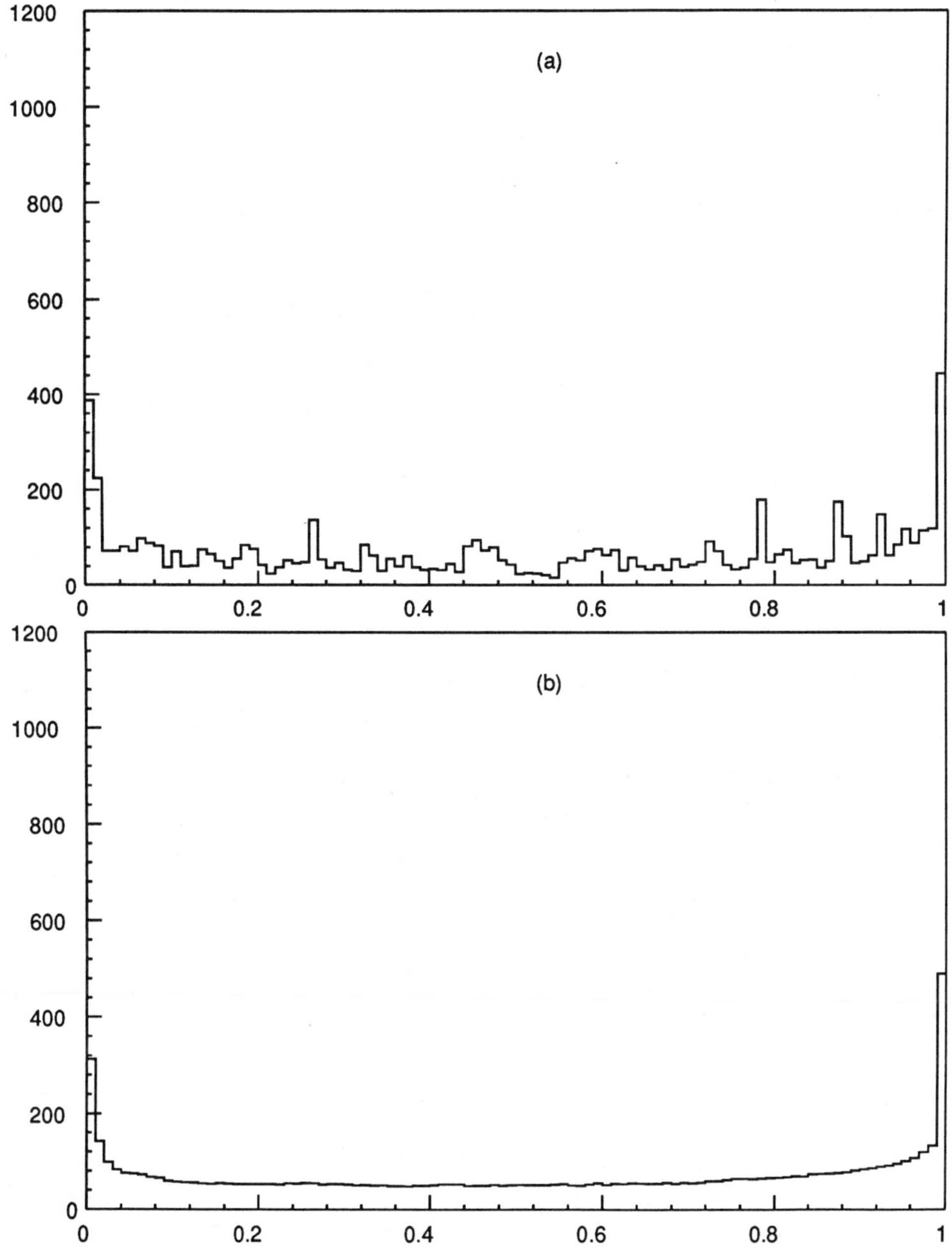

Fig. 2. (a) Unadjusted analysis [simulation (k=6.0)] mean over range 0.2 to 0.8 is 51.75. (b) Adjusted analysis [simulation (k=6.0)] mean over range 0.2 to 0.8 is 53.39.

There are numerous plausible biologic reasons why cancer risks at low doses may exceed those estimated by applying the one-hit formula, an artificial mathematical construct, to data at higher doses. One possibility is the well-known flattening of the dose-response curve due to dose-dependent intercurrent mortality. Our analysis controlled for this phenomenon in the adjusted analysis and the conclusions were unaltered. These adjustments are usually omitted from formal risk assessments, or are limited to the less

Table 2. Comparison of the Extreme Ends of the Frequency Distributions for p Observed and Obtained through Simulation

(a) Unadjusted Analysis

Unadjusted Analysis		$k = 1$ simulation		$k = 6$ simulation	
Range	Observed	Expected*	Excess	Expected*	Excess
$0 \le p < .01$	666	320.4	345.6	433.8	232.2
$.01 \le p < .02$	323	241.5	81.5	250.3	72.7
$.02 \le p < .03$	73	66.1	6.9	81.2	−8.2
$.03 \le p < .04$	97	70.8	26.2	81.4	15.6
$.96 \le p < .97$	84	96.2	−12.2	98.8	−14.8
$.97 \le p < .98$	132	132.1	−0.1	128.2	3.8
$.98 \le p < .99$	119	131.1	−12.1	133.1	−14.1
$.99 \le p < 1.0$	974	544.4	429.6	495.9	478.1

*Expected from the simulation normalized so that the mean values in the range $0.2 \le p < 0.8$ are equal in the observed and simulated distributions.

Total number of experiments 8432.

(b) Adjusted Analysis

Adjusted Analysis		$k = 1$ simulation		$k = 6$ simulation	
Range	Observed	Expected*	Excess	Expected*	Excess
$0 \le p < .01$	577	311.5	245.5	334.7	222.3
$.01 \le p < .02$	190	159.1	30.9	151.8	38.2
$.02 \le p < .03$	129	100.8	28.2	105.0	24.0
$.03 \le p < .04$	118	82.7	35.3	88.6	29.4
$.96 \le p < .97$	103	125.1	−22.1	114.2	−11.2
$.97 \le p < .98$	122	138.5	−16.5	127.7	−5.7
$.98 \le p < .99$	154	159.0	−5.0	142.1	11.9
$.99 \le p < 1.0$	1106	669.6	436.4	525.2	580.8

*Expected from the simulation normalized so that the mean values in the range $0.2 \le p < 0.8$ are equal in the observed and simulated distributions.

Total number of experiments 8432.

effective method of censoring all animals with events occurring before the first animal in any dose group developed the tumor of interest, as we did in our "unadjusted" analysis. Table 2 shows that the one-hit model is even less likely to be conservative when we use unadjusted data than when we used adjusted data.

A second possibility involves the apparent protective effect that limited weight gain has on carcinogenesis (Roe, 1983; Morris, 1982). If high dose animals fail to gain weight because of non-lethal toxicity or a decline in food-intake due to unpalatability, the dose-response curve may be flattened. Many animals in the high dose groups in the CBDS data had body weights well below those of the corresponding control groups.

A third possible biologic reason for supralinearity is saturation of the enzyme system which converts the administrated chemical to the proximate carcinogen. One such example is vinyl chloride (Watanabe, 1976; Gehring, 1978), where the dose-response curve rises approximately linearly at low doses but then flattens out at cancer rates still well below 100% (Maltoni, 1979). The metabolism of few carcinogens is understood in as a great detail as it is for vinyl chloride; thus, it is currently unknown how common this phenomenon is, and how much curve-flattening may occur. A break in the rise of the curve at some point below about 5% cumulative incidence would be virtually undetectable by standard bioassay designs, and risk assessments based on the higher dose and response region of the curve could underestimate risks considerably.

Fourth, we might also think about the effect of heterogeneity in susceptibility in individuals, animal or human, to carcinogenic agents. For example, although xeroderma pigmentosa in homo- and heterozygotes may account for a significant proportion of non-melanoma skin cancer in the human population (Swift, 1979), a dose-response curve for skin cancer versus exposure to sunlight at high doses, for the whole human population, would not reflect the sharp rise in risk at very low exposures due to this sensitive subpopulation. The animal bioassays use inbred strains with small genetic variability, as well as other homogeneous conditions, for other, unrelated reasons. However, we have no way of knowing whether the dose-response curve in a homogeneous animal experiment reflects the response, or even the general shape, of the dose-response curve in a heterogeneous human population.

The induction of intervening biologic mechanisms that reduce carcinogenicity, for example, the cell-killing or cell-sterilization phenomenon observed in radiation-induced breast cancer (National Academy of Sciences, 1972, 1980), is a fifth possible explanation for observed supralinear dose-response curves. Thus, the true dose-response curve may rise more rapidly at low doses than would be expected from its shape in the high-dose region. We do not know to what extent the induction of intervening mechanisms occurs in chemical carcinogenesis, or at what ranges of dose.

Finally, some therapeutic agents used in cancer suppression have also been found to induce cancer. For example, 7.6% of a group of ovarian cancer patients successfully treated with dihydroxybusulfan developed leukemia within several years, although comparable patients treated in other ways did not appear to be experience this elevated leukemia risk (Pedersen-Bjergaard, 1980). A simple shift of viewpoint suggests that some human carcinogens may be cancer suppressants at doses near the human maximum tolerated dose. Thus, we may expect supralinearity due to this phenomenon with some carcinogens.

Our purpose in discussing these six possible mechanisms is not to promote any combination of them as explaining the findings in Fig. 2 (other mechanisms and entirely different explanations are possible), but to show that quantitative risk assessment should be recognized as a short-hand way of representing complex biological phenomena. The rote fitting of mathematical formulae may be convenient, but as the data presented here show, is not (yet) a precise science. In particular, the one-hit formula — commonly considered "conservative," and itself usually the most conservative of all models in general use — sometimes substantially understates true risks at low exposures.

We fully recognize that the usual analysis of a carcinogen bioassay is likely to be modified whenever the analyst recognizes that there is a problem in fitting the usual dose-response models. However, our data suggest that the frequency of such problems may be substantially higher than is generally appreciated, and we suspect that it is substantially higher than the frequency of modified analyses. In fact, the one-hit model underestimates risks at 1/2 MTD almost as often as it overestimates risks, and there is no reason to think that this is any less frequent at doses much closer to zero. The one-hit model is not "conservative." This funding has direct and immediate application to the regulatory activities of governmental agencies that base their risk assessments (and regulations) on the assumption that the one-hit formula as currently used is conservative. In the United States, such agencies include the Environmental Protection Agency, the Food and Drug Administration, the Occupational Safety and Health Administration, the Consumer Product Safety Commission, and others, as well as numerous state agencies.

REFERENCES

Gehring, P. J., Watanabe, P. G., and Park, O. N., 1978, Resolution of Dose-Response Toxicity Data for Chemicals Requiring Metabolic Activation: Example — Vinyl Chloride, *Toxicol. Appl. Pharmacol.* **44**:581-591.

Kaplan, E. L., and Meier, P., 1958, Nonparametric Estimation from Incomplete Observations, *J. Am. Statist. Assoc.* **53**:457-481.

Linhart, M. S., Cooper, J. A., Martin, R. L., Page, N. P., and Peters, J. A., 1974, Carcinogenesis Bioassay Database System, *Computers in Biomedical Research* **7**:230-248.

Maltoni, C., Lefemine, G., Giliberti, A., Cotti, G., and Carretti, D., 1979, Vinyl Chloride Carcinogenicity Bioassays (BT Project) as an Experimental Model for Risk Identification and Assessment in Environmental and Occupational Carcinogenesis, in *Epidemiologie Animale et Epidemiologie Humane: le Cas du Chlorure de Vinyle Monomare, Proceedings of the 20th Meeting of Le Club Cancerogenese Chimiqure*, Paris, November 10.

Morris, H. R., Lustbader, E. D., and Bras, G., 1982, Dietary Practices of Early Life and Spontaneous Tumors of the Rat, *Nutrition and Cancer* **3**:150-167.

National Academy of Sciences (NAS), 1972, "The Effects on Populations of Exposure to Low Levels of Ionizing Radiation," Report of the Advisory Committee on the Biological Effects of Ionizing Radiation, National Academy of Sciences, Washington, D.C.

National Academy of Sciences (NAS), National Research Council, Committee on the Biological Effects of Ionizing Radiations (BEIR), 1980, *The Effects in Populations of Exposure to Low Levels of Ionizing Radiations*, National Academy Press, Washington, D.C.

Pedersen-Bjergaard, J., Nissen, N., Sorensen, H. M., Hou-Jensen, K., Larsen, M. S., Hernst, P., Ersbol, J., Knudtzon, S., and Rose, C., 1980, Acute Non-Lymphocytic Leukemia in Patients with Ovarian Cancer Following Long-Term Treatment with Tresulfan (=dihydroxybusulfan), *Cancer* **45**:19-29.

Roe, F. T. C., 1983, Testing for Carcinogenecity and the Problem of Pseudocarcinogenicity, *Nature* **303**:657-658.

Swift, M., and Chase, C., 1979, Cancer in Families with Xeroderman Pigmentosum, *J. Natl. Cancer Inst.* **62**:1415-1421.

Watanabe, P. G., McGowman, G. R., Madrid, E. O., and Gehring, P. H., 1976, Fate of 14C-Vinyl Chloride Following Exposure in Rats, *Toxicol. Appl. Pharmacol.* **37**:49-59.

Carcinogenicity Versus Acute Toxicity: Is There a Relationship?

Bernhard Metzger,[a] **Edmund Crouch, and Richard Wilson**
Harvard University
Cambridge, MA

ABSTRACT

Carcinogenic potency is compared in rodents with acute toxicity for a group of chemicals (155) which were tested independently of and mostly before the NCI/NTP program. For the entire data set and several subsets, we find partially biased statistically significant linear relationships between potency and the inverse of LD50. On average, the chemicals studied outside the NCI/NTP program are more carcinogenic compared to their acute toxicity than the NCI/NTP chemicals. Analysis shows a clear unbiased upper bound for carcinogenic potency. The correlation between potency and acute toxicity is robust with respect to species and route of administration. We find good agreement between oral and inhalation experiments, for toxicities and carcinogenicities.

KEYWORDS: Carcinogenicity, toxicity, correlation

INTRODUCTION

Zeise (1984) studied 250 chemicals included in the National Cancer Institute/National Toxicology Program (NCI/NTP) bioassays and found a statistically significant, albeit biased, correlation between potency and toxicity. This result is surprising since acute toxicity and potency represent two different biological end points.

Since these correlations could have been entirely an artifact of the data set used (Bernstein *et al.*, 1985) we examine data on 540 other chemicals studied outside the NCI/NTP program. These 540 chemicals were presumably studied because they were suspected of being the most carcinogenic. Accordingly, they might be *more carcinogenic* than suggested by the correlation by Zeise (1984). We ask in this paper:

1. Does there exist a statistically significant correlation between carcinogenic potency and acute toxicity for chemicals which are not included in the NCI/NTP series? If so, does it differ from that found earlier?

2. What is the effect on such a correlation of using bioassay data for different species and different routes of administration?

---

a. Now with Arthur D. Little, Inc., Acorn Park, Cambridge, MA 02140.

New Risks
Edited by L. A. Cox, Jr., and P. F. Ricci
Plenum Press, New York, 1990

SUMMARY OF PREVIOUS WORK

Parodi *et al.* (1982) compared carcinogenicity with biological activity in various test systems and with acute toxicity and found the correlation to be highest with acute toxicity for their particular sample of 21 chemicals. Zeise (1984) and Zeise and Crouch (1986) found that for the chemicals tested by the NCI/NTP program the correlation is just as strong as found by Parodi *et al.* (1982). Carcinogenic potency was correlated with chronic toxicity (measured by weight depression) and acute toxicity (LD50). It was recognized (Zeise and Crouch, 1985, 1986) that factors related to experimental design of the NCI/NTP bioassays introduced some bias in this correlation.

DATA

In this study we analyze two independent sets of data: one set comprises chemicals for which carcinogenesis data exist but which are not included in the NCI/NTP series; the other set is the NCI/NTP sample but re-examined using the same analysis procedure for both sets. We use potency and toxicity data as reported in two large readily available databases:

a. for carcinogenic potency we use the lowest reported, statistically significant ($p \leq 0.05$) estimate of TD50 cited by Gold *et al.* (1984);

b. for acute toxicity we use values of acute LD50 reported in the Registry of Toxic Effects of Chemical Substances (RTECS) (NIOSH, 1980; NIOSH, 1984).

TD50 is that dose which will cause an excess lifetime tumor probability of 50%. Acute LD50 is that dose which will kill 50% of the animals in a short period. A carcinogenesis experiment is included in the analysis if

a. it yields a tumor incidence (TD50) which is significant at $p \leq 0.05$ (H_0: TD50=∞);

b. it is conducted on either rats or mice using any means of oral administration or inhalation;

c. the tested chemical has a reported LD50 value in RTECS.

Of the 770 chemicals reported in Gold *et al.* (1984), 344 pass the screening procedure. Of those 344, 264, including 109 from the NCI/NTP series, are for oral administration. For chemicals with more than one bioassay passing the above selection criteria, we take as the estimate of the TD50 for a given route the geometric mean of the reported values. This approach introduces relatively small extra uncertainty. Application of the significance criterion (a) for the TD50 induces bias.

ANALYSIS

Regression analyses are performed (a) to test the equivalence between rat and mouse response and between inhalation and oral administration; and (b) to determine the correlation between LD50 and TD50, for different routes of administration and different species. Frequency analyses are performed for an index, *K*, a measure of comparison of the carcinogenic potency of a material with its acute toxicity. We define this index to be

$$K = \log_{10}(\mathrm{LD50}/\mathrm{TD50}). \quad (1)$$

Table 1. Interspecies Comparisons of Acute Toxicity (LD50) and Carcinogenic Potency (TD50) Between Mice and Rats, and Comparisons of LD50 and TD50 Between Inhalation and Oral Route of Administration

Regression Number	Data Set	y	x	n	C (se)	$\text{Log}_{10}(D)$ (se)	σ_r	R^2
1	all	Mouse TD50_{orl}	Rat TD50_{orl}	98	0.80 (0.06)	–0.50 (0.14)	0.81	0.62
2	all	Mouse LD50_{orl}	Rat LD50_{orl}	140	0.87 (0.04)	–0.40 (0.11)	0.44	0.79
3	all	R/M TD50_{inh}	R/M TD50_{orl}	5	1.02 (0.25)	0.01 (0.40)	0.53	0.86
4	all	R/M LD50_{inh}	R/M LD50_{orl}	18	0.99 (0.17)	0.15 (0.38)	0.71	0.68

Explanations:

$y = Cx + \log_{10}(D)$ (Eq. 2)

n: sample size
σ_r: standard error of regression
se: standard deviation
R^2: coefficient of determination
all: combined data sets (344 chemicals)
R/M: rats and mice combined
orl: all forms of oral administration
inh: inhalation

Regression Analysis

For all regressions we used ordinary least squares on a simple linear model in log-log space:

$$y = C{\cdot}x + \log_{10}(D) + \varepsilon, \tag{2}$$

where, e.g., $y = -\log_{10}(\text{TD50})$, $x = -\log_{10}(\text{LD50})$, and where ε is assumed to be normally distributed with zero mean.

The first two regressions in Table 1 demonstrate for the chemicals and experiments selected the practical equivalence between mouse and rat response. For mouse data regressed on rat data the slope parameter for both potency and acute toxicity is close to unity (C=0.8 for TD50; C=0.87 for LD50). Hence, data from both species can be combined in TD50-LD50 regression analyses without introducing large errors.

The third and fourth regressions in Table 1 demonstrate for the chemicals and experiments selected practical equivalence between inhalation and oral experiments. The slope parameters are unity within less than one standard deviation.

Regressing potency on acute toxicity (Table 2 and Figs. 1 and 2) shows that for large sample sizes the slope parameter is approximately unity and virtually invariant for different sets of chemicals, for different species (rats, mice), and for different routes of administration. The intercept parameter varies depending on the set of chemicals. Also, the regression parameters are only slightly sensitive to the presence of the two highly potent and toxic compounds, TCDD and HCDD (compare regressions number 7 and 10, Table 2). The correlation coefficients are generally higher and the intercept and the standard error smaller for the NCI/NTP chemicals.

Using the regression model introduced earlier [Eq. (2)] and rewritten in the form

$$\mathrm{TD50} = 1/D \cdot \mathrm{LD50}^{C} \cdot 10^{-\varepsilon}, \tag{3}$$

we find for the entire database the parameters $C \cong 1.0$ and $D \cong 13.4$ (regression number 7, Table 2). The standard error of regression is approximately one (0.8 for NCI/NTP series, regression number 9) corresponding to a factor of about ten.

Frequency Distribution of *K*

The index K is defined in Eq. (1). Chemicals for which $K=0$ (i.e., TD50=LD50) fall along the 45° line through the origin on plots such as Fig. 1. Chemicals with negative K-values fall below and chemicals with $K>0$ plot above this line. Histograms of K corresponding to Figs. 1 and 2 are plotted in Fig. 3 and appear symmetric and practically indistinguishable from normal.

The difference between the means of K for the NCI/NTP series (Fig. 3c) and the non-NCI/NTP data (Fig. 3b) is approximately $\Delta\mu_K \cong 0.7$, so that on average, chemicals in the latter set are more potent (by a factor of $10^{0.7} \cong 5$) when compared with their acute toxicity than those in the NCI/NTP series. The NCI/NTP data set contains no chemicals with $K>3$ while 12 chemicals (8% of the total) of the non-NCI/NTP data set fall in the range $3 \leq K \leq 4$.

DISCUSSION

Data Censoring Mechanisms

Zeise and Crouch (1985, 1986) demonstrated that a scatterplot such as Fig. 1 can be divided into three more or less distinct domains. Findings of this research suggest that there may be additional boundaries shaping the contours of a TD50-LD50 plot (Fig. 4).

Plots such as Figs. 1 and 2 have noticeable upper bounds for TD50 and LD50 (bottom and left). These limits (Fig. 4) arise because the test chemical should not constitute more than 5-10% of the diet in a long bioassay, and there are physical limits on how much can be administered in a single dose in LD50 tests.

Zeise and Crouch (1986) showed that, for a particular protocol, there is a lowest measurable potency below which complete censoring due to insignificant results occurs (Region B, Fig. 4). This is what we see in plots such as Figs. 1 and 3.

Furthermore, we find no chemicals with $K>4$. Twelve chemicals are in the range $3 \leq K \leq 4$ and they plot reasonably well along a line of unit slope (e.g., Fig. 1). Zeise and Crouch (1986) showed that the highest measurable potency is also a function of protocol. It occurs when all but one of the dosed animals are tumorous. Since we do not find near 100% tumor incidence in all dose groups of any experiment included in our analysis, the bounding exhibited by the $K=4$ contour is not an experimental artifact, but imposed by nature.

Table 2. Correlations Between Carcinogenic Potency (TD50) and Acute Toxicity (LD50) for Different Data Sets and Species and Routes of Administration

Regression Number	Data Set	y	x	n	C (se)	$\mathrm{Log}_{10}(D)$ (se)	σ_r	R^2
1	all	Rat $TD50_{eat}$	Rat $TD50_{orl}$	104	0.94 (0.11)	0.91 (0.31)	1.00	0.42
2	non-NCI/NTP	Rat $TD50_{eat}$	Rat $LD50_{orl}$	41	0.97 (0.22)	1.42 (0.66)	1.09	0.33
3	NCI/NTP	Rat $TD50_{eat}$	Rat $LD50_{orl}$	63	1.03 (0.11)	0.85 (0.30)	0.84	0.58
4	all	Mouse $TD50_{eat}$	Mouse $LD50_{orl}$	51	1.02 (0.14)	0.60 (0.41)	0.89	0.53
5	non-NCI/NTP	Mouse $TD50_{eat}$	Mouse $LD50_{orl}$	16	0.99 (0.32)	0.77 (1.04)	1.11	0.40
6	NCI/NTP	Mouse $TD50_{eat}$	Mouse $LD50_{orl}$	35	1.10 (0.15)	0.71 (0.41)	0.77	0.63
7	all	R/M $TD50_{orl}$	R/M $LD50_{orl}$	264	0.97 (0.07)	1.13 (0.20)	0.95	0.40
8*	non-NCI/NTP	R/M $TD50_{orl}$	R/M $LD50_{orl}$	155	0.97 (0.11)	1.40 (0.32)	0.96	0.34
9*	NCI/NTP	R/M $TD50_{orl}$	R/M $LD50_{orl}$	109	1.04 (0.09)	0.93 (0.25)	0.79	0.58
10**	all	R/M $TD50_{orl}$	R/M $LD50_{orl}$	262	0.89 (0.08)	0.89 (0.23)	0.94	0.32

Explanations:

$y = Cx + \log_{10}(D)$ (Eq. 2)

n: sample size
σ_r: standard error of regression
se: standard deviation
R^2: coefficient of determination
all: combined data sets (344 chemicals)
R/M: rats and mice combined
orl: all forms of oral administration
eat: normal diet
*: see Figs. 1 and 2
**: same as regression number 7 but without TCDD and HCDD

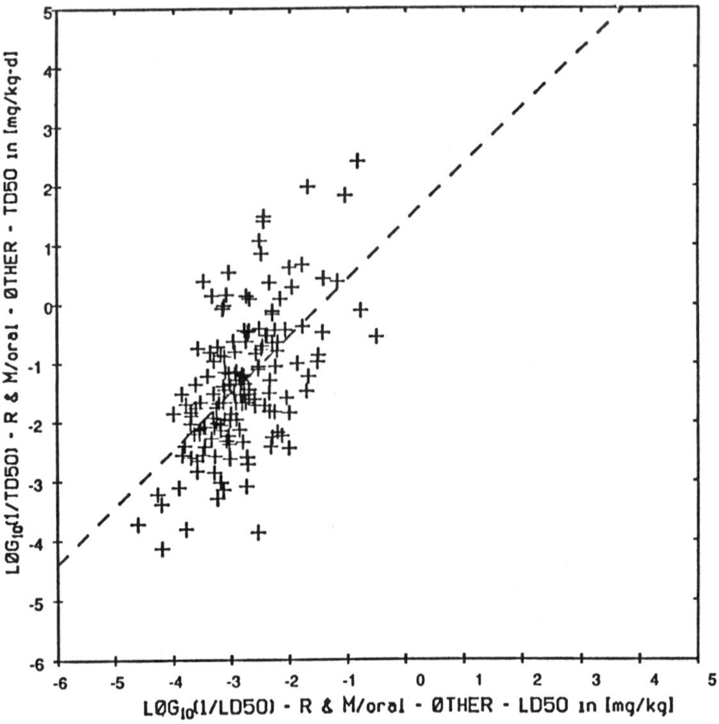

Fig. 1. TD50-LD50 correlation for non-NCI/NTP data (Table 2, regression number 8).

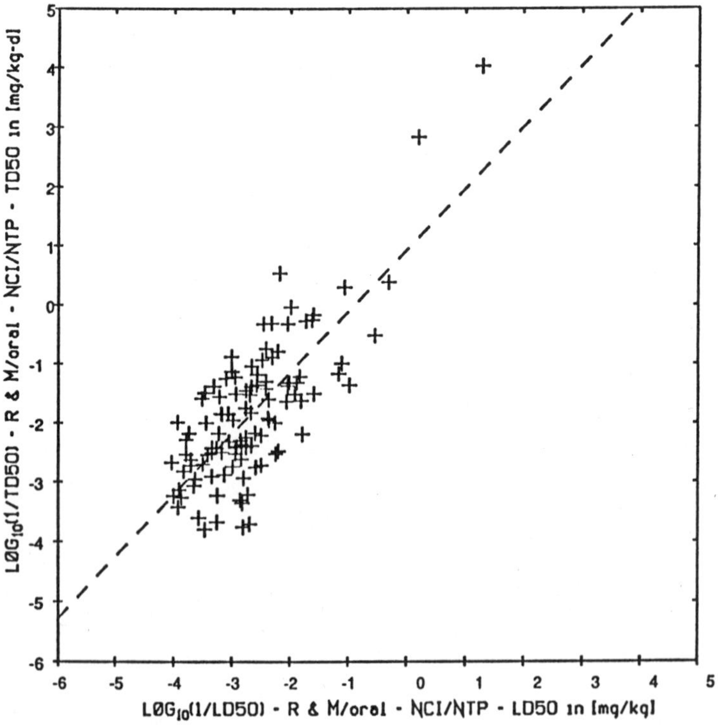

Fig. 2. TD50-LD50 correlation for NCI/NTP data (Table 2, regression number 9).

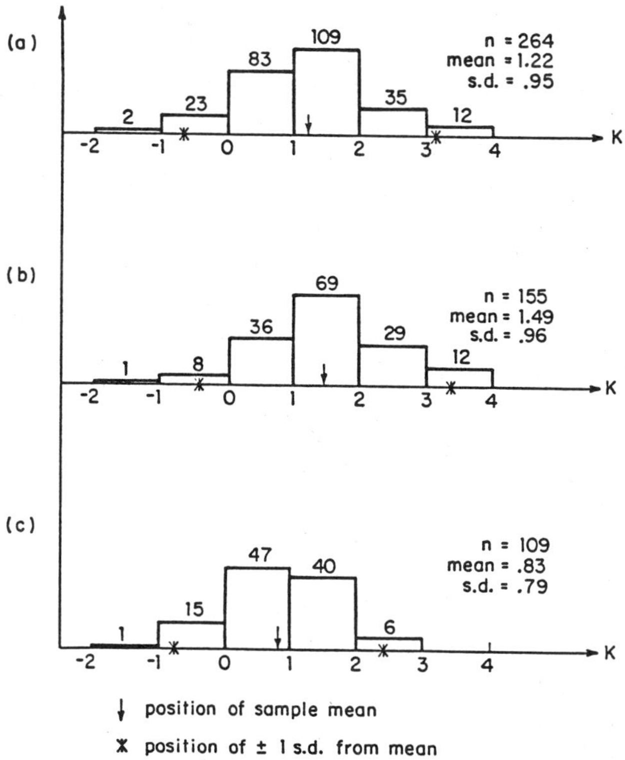

Fig. 3. Histogram of K for different sets of chemicals: (a) all, (b) non-NCI/NTP, and (c) NCI/NTP.

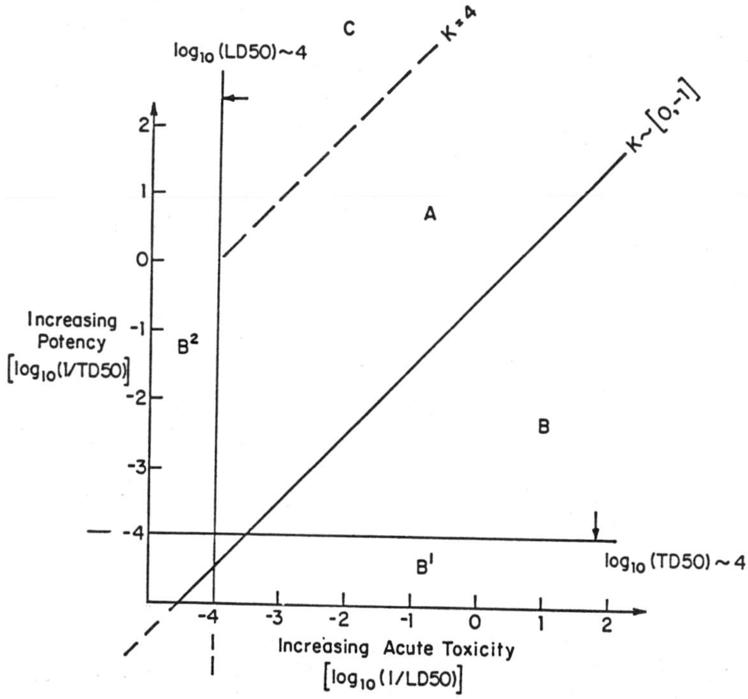

Fig. 4. TD50 vs LD50 scatterplot. Censoring mechanisms act in different regions: Region A—significant test results; Region B—insignificant test results due to early mortality and diet limitations (B^1, B^2); and Region C—100% tumor incidence.

Observed Correlations

The results suggest that little bias is introduced when mixing species (rats, mice) in TD50-LD50 regressions. Similarly, they indicate that oral data may be used as surrogates for inhalation data and vice versa (Zeise *et al.*, 1984; Tancrède *et al.*, 1986).

For large samples (n>50), a very nearly linear relationship (unit slope in log-log space) is found between carcinogenic potency and acute toxicity. The factor of proportionality, 1/D, depends on the particular data set used. Bias arises due to the fact that neither the NCI/NTP data nor the non-NCI/NTP data represent a random sample of a population of chemicals, but contain chemicals whose potency and acute toxicity are probably far above the median values of the population of, say, all chemicals in RTECS. Any inference drawn on the basis of the statistics of these samples can thus not simply be extrapolated to the universe of chemicals.

Due to experimental design, all chemicals which would fall in Region B (including B^1, B^2) of Fig. 4 are truncated. Of the 770 chemicals in Gold *et al.* (1984), about 30% are censored this way from further analysis (non-significant results). Since most of these chemicals have less than median LD50 values, the censoring is not believed to affect the regression statistics much. Nevertheless, the least squares parameter estimates are upwardly biased.

CONCLUSIONS

1. The correlation between carcinogenic potency and acute toxicity found by Zeise (1984) holds also for chemicals which are not part of the NCI/NTP series. However their carcinogenic potency is on average larger by a factor of $10^{0.7}$ for a given acute toxicity. Also, the standard error is larger and the correlation coefficient smaller ($R^2 \cong 30\%$-40% vs 60% for NCI/NTP). A clear, unbiased upper bound on potency is found for given values of acute toxicity in the range of $4 \geq \log_{10}(\text{LD50}) \geq -2$.

2. The correlation between potency and acute toxicity appears largely independent of species and route of administration.

3. The expected correlation for acute toxicity between rats and mice is found. Also, we find that carcinogenic potencies in mice and rats are approximately equal in this database.

REFERENCES

Bernstein, L., Gold, L. S., Ames, B. N., Pike, M. C., and Hoel, D. G., 1985, Some Tautologous Aspects of the Comparison of Carcinogenic Potency in Rats and Mice, *Fund. Appl. Toxicol.* 5:79-86.

Gold, L. S., Sawyer, C. B., Magaw, R., Backman, G. M., de Vecina, M., Levinson, R., Hooper, N. K., Havender, W. R., Bernstein, L., Peto, R., Pike, M. C., and Ames, B. N., 1984, A Carcinogenic Potency Database of the Standardized Results of Animal Bioassays, *Env. Health Perspectives* **58**:9-319.

NIOSH, 1984, *Registry of Toxic Effects of Chemical Substances*, U.S.DHHS, 1983 Supplement to the 1981-82 edition, Cincinnati, OH.

NIOSH, 1980, *Registry of Toxic Effects of Chemical Substances*, U.S.DHHS, 1979 edition, Cincinnati, OH.

Parodi, S., Taningher, M., Baero, P., and Santi, L., 1982, Quantitative Correlations Amongst Alkaline DNA Fragmentation, DNA Covalent Binding, Mutagenicity in the Ames Test, and Carcinogenicity for 21 Compounds, *Mutation Res.* **93**:1-24.

Tancrède, M., Wilson, R., Zeise, L., and Crouch, E. A. C., 1986, The Carcinogenic Risk of Organic Vapors Indoors: A Survey, Energy and Environmental Policy Center, E-86-06, Harvard University, Cambridge, MA.

Zeise, L., 1984, Surrogate Measures of Human Cancer Risk, Ph.D. dissertation, Harvard University, Cambridge, MA.

Zeise, L., Crouch, E. A. C., and Wilson, R., 1984, Use of Acute Toxicity to Estimate Carcinogenic Risk, *Risk Analysis* **4**:187-199.

Zeise, L., and Crouch, E. A. C., 1986, A Possible Relationship Between Toxicity and Carcinogenicity, *J. Am. Coll. Toxicol.* **5(2)**:137-151.

Zeise, L., and Crouch, E. A. C., 1985, Reply to Comments: On the Relationship of Toxicity and Carcinogenicity, Letter to the Editor, *Risk Analysis* **5(4)**:265-270.

Incorporating More Science Into Cancer Dose-Response Extrapolations

Robert L. Sielken, Jr.
Sielken, Inc.
Bryan, TX

ABSTRACT

Dose-response models can be an important part of the quantitative cancer risk assessments supporting risk management and regulatory decision making. Current dose-response models have limited biological rationale and tend to greatly over-simplify the carcinogenic process. While complete understanding of carcinogenic mechanisms and all-encompassing models are still beyond our grasp, new approaches to dose-response modeling can provide opportunities for these models to incorporate more scientific information. A new family of dose-response models called the Individualized Response Models is introduced and shown to provide such opportunities. The models allow for the use of more biologically relevant dose scales such as those provided by physiologically-based pharmacokinetic models of the absorption, delivery, metabolism, and elimination of a chemical as well as research on cell turnover rates, repair mechanisms, and immune system responses even if these phenomena are age-dependent. Since different individuals may have different physiological behavior and different background exposures, allowances are made for modeling populations with distributions of individual susceptibilities and background doses. The inclusion of time in the Individualized Response Models allows them to incorporate dose levels, susceptibilities, and background exposures which are not constant over time and also reflect age-dependent changes in the number of target cells and the effects of cell proliferation on the number of intermediate cells in a multistage process. Since the improved dose scales and other modeling extensions are related to physical phenomena, their nature, interrelationship, and functional dependence on dose, time, and age can frequently be investigated through independent scientific research prior to the fitting of the chronic dose-response data.

KEYWORDS: Individual variation, susceptibility, dose-response modeling, biologically effective dose, physiologically-based pharmacokinetic models, biological models

INTRODUCTION

Quantitative dose-response modeling can be an important input to cancer risk assessment which, in turn, can be a major ingredient in cancer risk management and the regulatory processes. However, the dose-response models currently used in the regulatory processes are generally overly simple probabilistic representations of highly complex biological phenomena; these models are not biological models. New dose-response

New Risks
Edited by L. A. Cox, Jr., and P. F. Ricci
Plenum Press, New York, 1990

modeling techniques which attempt to utilize more of the available scientific knowledge and expertise involve (i) representations of the exposure in terms of dose scales based on cell turnover rates, repair processes, immune system responses, and physiological and pharmacokinetic models of the absorption, delivery, metabolism, and elimination of chemicals, (ii) distributions of individual background exposure levels and individual susceptibilities to low levels of a chemical, (iii) age-dependent changes in the number of target cells and the time-dependent effects of cell proliferation on the number of intermediate cells in a multistage process, and (iv) dose levels, susceptibilities, and background exposures which are not necessarily constant over time.

MODELING ON MORE RELEVANT DOSE SCALES

The dose level should be expressed on a biologically relevant scale. For example, the dose level corresponding to an inhalation exposure can be expressed in terms of (i) the concentration of the chemical in the air inhaled, (ii) the total amount of the chemical inhaled in a particular period, (iii) the amount of the chemical or its active metabolite reaching the target cells in a particular tissue, (iv) the amount of interaction with DNA, (v) the amount of cell damage that escapes repair, and so on. That is, the dose scale can be defined in a progressively better manner as either (a) the administered dose, (b) an intermediate dose, (c) the dose delivered to the target site, or (d) the biologically effective dose (reflecting not only the amount delivered to the target site but also the net amount of cancer related activity generated which includes cell turnover, DNA repair, immune system responses, etc.). Using a more biologically reflective dose instead of the administered dose (1) allows the modeler to incorporate known biology, physiology, pharmacokinetics, and other factors, (2) reduces the uncertainty and the number of unknowns in the dose-response model, and (3) aids extrapolations from high doses to low doses, from species to species, from route to route, and from time frame to time frame.

If there is insufficient scientific knowledge to establish the complete transformation from administered dose to a biologically effective dose, then using what is known and even a partial transformation or best approximation should lead to a better risk assessment than using only the administered dose. When there is scientific uncertainty about the ultimate biologically effective dose, dose-response modeling should be evaluated using alternative expressions for the effective dose.

MODELING WITH DISTRIBUTIONS OF INDIVIDUAL SUSCEPTIBILITIES AND BACKGROUND EXPOSURES

Individual responses to a particular exposure to a potentially hazardous chemical vary. Individual variation is influenced by many factors including variation in background exposures, genetic traits, preexisting diseases, behavioral traits, etc. (e.g., Harris (1983)). The remainder of this section describes an extension of the linear multistage model to reflect individual variations in background exposure and susceptibility to low levels of a chemical. A further extension emphasizing the multistage carcinogenic process, age-dependence, time-dependence, and cell proliferation is described in the next section.

For a population of individuals the background exposure D_0 is a random variable. Each individual has his own particular value d_0 for the background dose D_0. An individual animal's background exposure is dependent upon such factors as environment and diet. A human's d_0 can also depend upon occupation, lifestyle, hygiene, etc.

The delivered dose corresponding to the transition from substantially lowered carcinogenic effectiveness to the dose region where the mechanisms resisting or suppressing carcinogenesis are overwhelmed is the "susceptibility frontier." For a population of individuals the susceptibility frontier F is a random variable. Each individual has his own particular value f for the susceptibility frontier F. Genetic differences in the metabolic detoxification, tissue sensitivity, repair efficiency, and immune system responses contribute to the individual variation in the value of the susceptibility frontier [e.g., Cleaver (1984) and Omenn (1984)].

A susceptibility frontier is different from a tolerance. If the delivered dose becomes greater than an individual's tolerance, then the individual's cancer probability jumps to 1.0; however, if the delivered dose becomes greater than the individual's susceptibility frontier, then the individual's cancer probability begins increasing more rapidly but doesn't immediately become 1.0.

The concept of a susceptibility frontier is also different from that of a threshold. A threshold denotes a dose level below which absolutely no carcinogenesis is possible whereas a susceptibility frontier indicates only a transition from the normal functioning of the body's carcinogenic defense mechanisms to dose levels overwhelming those mechanisms. Also, a threshold is the same for every member of the population whereas the susceptibility frontiers for different individuals are different.

The concept of a susceptibility frontier can be illustrated in simple terms as follows: Suppose that for a particular individual a chemical exposure causes a particular number of specific molecules to reach and invade a target cell. Also, suppose that the target cells have, for simplicity, a single type of cancer defense mechanism relative to these invaders, say a detoxification process. However, as an invading molecule moves within a target cell there may be multiple opportunities for the detoxification process to take place. In particular, suppose that the cell has up to 20 opportunities to carry out the detoxification of an invading molecule. However, at each opportunity the probability of detoxifying an invader is only 0.9 (not 1.0), and a successful detoxification reduces the number of detoxification opportunities the next invader must face by one. In such circumstances the expected number of invaders which avoid the random detoxification process can be shown to be approximately

$1 \times 10^{-(21-X)}$ if there are X invaders and $X \le 20$

and

$X-20$ if there are X invaders and $X > 20$.

If the probability of a carcinogenic response is proportional to the number of invaders which avoid detoxification, then the probability of a carcinogenic response as a function of dose generally increases approximately linearly with a very shallow slope for very small doses, increases exponentially with moderate slopes for medium doses, and increases linearly with a relatively steep slope once the dose generates more invaders than the defense mechanisms can handle. In this example the delivered dose level corresponding to the generation of 20 invaders would be called the individual's susceptibility frontier f. In general, if an individual has a nonzero susceptibility frontier, then it occurs where, as the dose increases, the effective dose shifts from a slowly increasing function of the delivered dose to a more rapidly increasing linear function of the delivered dose.

The Individualized Response Model is the result of the merger of the random background dose D_0 and the random susceptibility frontier F with the concept of a multistage carcinogenic process and its extensions based on time and cell proliferation. In the Individualized Response Model the proportion of the entire population at risk which is expected to develop a specified carcinogenic response by time t for an administered dose d is

$$P(t;d) = \int_0^{\infty} \int_0^{\infty} P[t; g(d, d_0, f)] \times p_{D_0}(d_0) \times p_F(f) \, dd_0 df$$

where

$P[t;g(d,d_0,f)]$ = the probability of the specified carcinogenic response occurring by time t at dose d when the individual's susceptibility frontier is f and his background dose is d_0,

$p_{D_0}(d_0)$ = the probability density function of D_0 evaluated at d_0 (i.e., the likelihood of $D_0 = d_0$),

$p_F(f)$ = the probability density function of F evaluated at f (i.e., the likelihood of $F = f$), and

$g(d,d_0,f)$ = the biologically effective dose.

Roughly speaking, $P(t;d)$ is the weighted average of $P[t;g(d,d_0,f)]$ where the weights reflect the proportion of the population that have a susceptibility frontier F equal to f and a background dose D_0 equal to d_0.

A reasonable approximation for the distribution of the susceptibility frontier F may often be the lognormal distribution. Such a distribution agrees with $F \geq 0$. In addition, lognormal distributions can be very skewed to match chemical carcinogens for which F is usually very near zero. On the other hand, lognormal distributions can be nearly symmetrical and more like the normal distribution if F is frequently moderately large. Similarly, the distribution of the background dose D_0 may be reasonably approximated by a lognormal distribution.

The Individualized Response Model corresponding to a multistage model which is linear in the biologically effective dose but without extensions based on time and cell proliferation has

$$P[t;g(d, d_0, f)] = 1 - \exp[-\alpha_0 - \alpha_1 \times g(d, d_0, f)].$$

The functional form of the biologically effective dose would be determined, preferably, prior to the fitting of the chronic dose-response data or estimated using a general form such as linear in the delivered dose when the delivered dose exceeds the susceptibility frontier and as an exponential (with associated low-dose linearity) when the delivered dose is less than the susceptibility frontier.

MODELING OVER TIME

In the past dose-response models have usually been constructed to represent lifetime probabilities without regard to when the carcinogenic response might occur during a lifetime. Time-to-response models were the first dose-response models to explicitly include the time of the carcinogenic response. The inclusion of time allows the model to better reflect (i) data where not all subjects are observed at the same time, (ii) biological processes which differ over time, and (iii) birth and death processes for cells. The inclusion of time also allows risk to be characterized in terms which reflect the way in which the probability of a carcinogenic response changes over time.

In the multistage theory of carcinogenesis a carcinogenic response occurs when k successive events stemming from a single cell have occurred [Armitage and Doll (1961)]. In this context, for a given individual's susceptibility frontier f and background dose d_0, the probability that the carcinogenic response occurs by time t is approximately

$$P[t; g(d, d_0, f)] = 1 - \exp\left\{ - \int_0^{t-\tau} \lambda_k(t_k) \times N_{k-1}(t_k) dt_k \right\} \qquad 4.1$$

where

$$N_{k-1}(t_k) = \int_0^{t_k} \lambda_{k-1}(t_{k-1}) \times N_{k-2}(t_{k-1})$$

$$\times \exp\left\{ \int_{t_{k-1}}^{t_k} [\beta_{k-1}(v) - \delta_{k-1}(v)] dv \right\} dt_{k-1},$$

.

.

.

$$N_1(t_2) = \int_0^{t_2} \lambda_1(t_1) \times N_0(t_1) \times \exp\left\{ \int_{t_1}^{t_2} [\beta_1(v) - \delta_1(v)] dv \right\} dt_1.$$

Here, τ represents the time for a malignant cell to become the specified carcinogenic response (a detectable tumor, death from tumor, etc.), N_0 (t_1) represents the number of normal stem cells (stage 0) at age t_1, $N_i(t_{i+1})$ represents the number of cells which are in stage i at a particular time t_{i+1},$\lambda_i(t_i)$ represents the transition rate from stage i to stage i+1 at time t_i, and $\beta_i(v)$ and $\delta_i(v)$ represent the birth and death rates, respectively, of a cell in stage i at time v. Any or all of the parameters (λ's, β's, and δ's) can depend on the age of the individual as well as the biologically effective dose $g(d,d_0,f)$. The lag time τ is usually treated as a constant but can be more generally represented as a random variable. The above form for $P[t;g(d,d_0,f)]$ assumes that each cell at a particular stage has the same probabilistic behavior and that all cells behave independently.

The above relationship for $P[t;g(d,d_0,f)]$ can be motivated in fairly rough terms as follows: Assume that cancer is a k-stage process. Then cancer occurs after at least one cell reaches the k-th stage. The number of cells which reach the k-th stage at time t_k depends on (a) the transition rate from the (k–1)-th stage to the k-th stage at time t_k (namely, λ_k (t_k)) and (b) the number of cells in the (k–1)-th stage at time t_k (namely, N_{k-1} (t_k)). The number of cells in the (k–1)-th stage at time t_k depends on

a. the number of cells which make the transition from the (k–2)-th stage to the (k–1)-th stage at time t_{k-1} (namely, $\lambda_{k-1}(t_{k-1}) \times N_{k-2}(t_{k-1})$) and

b. the growth in the number of (k–1)-th stage cells from the time (t_{k-1}) when the parent cell makes the transition from the (k–2)-th stage to the (k–1)-th stage until the time t_k (namely, the multiplicative growth factor

$$\exp\left\{ \int_{t_{k-1}}^{t_k} [\beta_{k-1}(v) - \delta_{k-1}(v)] dv \right\}$$

which incorporates the birth rate $\beta_{k-1}(v)$ and death rate $\delta_{k-1}(v)$ for (k–1)-th stage cells for the intervening times v between t_{k-1} and t_k).

The interrelationships continue backward to the number of cells in the first stage which are available for possible transition to the second stage at time t_2 (namely, $N_1(t_2)$) which depends on the number of normal cells which made the transition to a first stage cell at time t_1 (namely, $\lambda_1(t_1) \times N_0(t_1)$) multiplied by the growth rate

$$\exp\left\{ \int_{t_1}^{t_2} [\beta_1(v) - \delta_1(v)] dv \right\}$$

for first stage cells between times t_1 and t_2. The number of normal stem cells $N_0(\bullet)$ can be different at different times if necessary to reflect changes during early childhood, puberty, menopause, etc. If there are only two stages, then k=2, and (4.1) becomes

$$P[t; g(d, d_0, f)] = 1 - \exp\left\{ -\int_0^{t-\tau} \lambda_2(t_2) \times \int_0^{t_2} \lambda_1(t_1) \times N_0(t_1) \right.$$
$$\left. \times \exp\left\{ \int_{t_1}^{t_2} [\beta_1(v) - \delta_1(v)]dv \right\} dt_1 dt_2 \right\}. \tag{4.2}$$

Here λ_1, λ_2, β_1, and δ_1 may also depend on the biologically effective dose $g(d,d_0,f)$. Furthermore, the administered dose d may not be the same value for all times but rather may depend on time. In order to emphasize these possible dependencies on age and dose and to emphasize that dose may be a function of time, (4.2) can be rewritten in expanded form as

$$P[t; g(d, d_0, f)] = 1 - \exp\left\{ -\int_0^{t-\tau} \lambda_2\left\{t_2, g[d(t_2), d_0, f]\right\} \right.$$
$$\times \int_0^{t_2} \lambda_1\left\{t_1, g[d(t_1), d_0, f]\right\} \times N_0(t_1)$$
$$\left. \times \exp\left[\int_{t_1}^{t_2} \left\{ \beta_1\{v, g[d(v), d_0, f]\} - \delta_1\{v, g[d(v), d_0, f]\} \right\} dv \right] dt_1 dt_2 \right\}. \tag{4.3}$$

This expanded form shows explicitly how to incorporate a dose which is not constant over time.

The models described in Moolgavkar and Venzon (1979), Moolgavkar and Knudson (1981), and Thorslund *et al.* (1987) are special cases of the form (4.3) for $P[t;g(d,d_0,f)]$. These models assume that D_0 has the same value for every individual and that F is zero for every individual. In Moolgavkar and Knudson (1981) the λ's, β's, and δ's are assumed to be independent of time, and no parametric forms for the possible dependence of λ_1, λ_2, β_1, δ_1 on the biologically effective dose $g(d,d_0,f)$ are suggested.

The susceptibility frontier and background dose can both be extended to allow for age-dependence. For example, (4.3) can be generalized to

$$P[t; g(d, d_0, f)] = 1 - \exp\left\{ -\int_0^{t-\tau} \lambda_2\left\{t_2, g[d(t_2), d_0(t_2), f(t_2)]\right\} \right.$$
$$\times \int_0^{t_2} \lambda_1\left\{t_1, g[d(t_1), d_0(t_1), f(t_1)]\right\} \times N_0(t_1)$$
$$\left. \times \exp\left[\int_{t_1}^{t_2} \left\{ \beta_1\{v, g[d(v), d_0(v), f(v)]\} - \delta_1\{v, g[d(v), d_0(v), f(v)]\} \right\} dv \right] dt_1 dt_2 \right\}. \tag{4.4}$$

In (4.4) the random susceptibility frontier F taking on a particular value f for an individual means that, when that individual reaches age T, his susceptibility frontier (denoted by $f(T)$) can be calculated from the number f and the age T. (A further generalization would be to allow F to be a stochastic process with the $f(T)$'s for different T's being different (possibly dependent) random variables.) The interpretation of an age-dependent background dose $d_0(T)$ is analogous to that for the age-dependent susceptibility frontier $f(T)$.

SUPPLEMENTARY RESEARCH

The susceptibility frontier, the background dose, the biologically effective dose, the delivered dose, and the cell birth, death, and transition rates are all related to physical phenomena and hence subject to scientific investigation. Physiological and pharmacokinetic research should establish the delivered dose, and the careful analysis of a chemical's mechanistic behavior may establish the nature of its susceptibility frontier. Supplementary biological research may also indicate which phenomena are dose-dependent and which are age-dependent and even indicate the functional forms of the dependences. Further research should then identify the functional form of the relationship between the biologically effective dose and its components (the delivered dose and the susceptibility frontier). The results of such investigations should determine the particular form (or alternative forms) for the probability, $P[t;g(d,d_0,f)]$, of the specified response occurring by time t for a given dose, d, a given background dose, d_0, and a given susceptibility frontier, f. Estimates for most of the unknowns in $P[t;g(d,d_0,f)]$ should have been determined *before* the corresponding model is fit to the dose-response data to estimate the remaining parameter values. A careful study of the populations involved may provide good approximations for the distributions of the individual susceptibility frontiers and the individual background doses.

CONCLUSIONS

Quantitative cancer dose-response modeling is the subject of intensive research. New dose-response models can incorporate research on more biologically relevant dose scales, age-dependent and time-dependent changes, cell proliferation, and distributions of individual background exposure levels and individual susceptibilities to low levels of a chemical. The new models still involve some simplifying assumptions and are not exact biological models; however, the extensions described make it possible to incorporate *more* of the current scientific understanding of carcinogenesis and *more* biological reality. The new capabilities for dose-response modeling and an expanding data base provide the potential for substantial improvements in quantitative cancer risk assessment.

REFERENCES

Armitage, P., and Doll, R., 1961, Stochastic Models for Carcinogenesis, in *Proceedings of the Fourth Berkeley Symposium on Mathematical Statistics and Probability*, J. Neyman, Ed., Vol. 4, p. 19, University of California Press, Berkeley and Los Angeles.

Cleaver, J. E., 1984, Chairman's Comments: Variations in Individual Susceptibility to Genotoxic Agents, in *Banbury Report 16: Genetic Variability in Responses to Chemical Exposure*, G. S. Omenn and H. V. Gelboin, Eds., p. 225, Cold Spring Harbor Laboratory.

Harris, C. C., 1983, Concluding Remarks: Role of Carcinogens, Cocarcinogens, and Host Factors in Cancer Risk, in *Human Carcinogenesis*, C. C. Harris and H. N. Autrup, Eds., Ch. 38, p. 941, Academic Press.

Moolgavkar, S. H., and Venzon, D. J., 1979, Two-Event Models for Carcinogenesis: Incidence Curves for Childhood and Adult Tumors, *Math. Biosci.* **47**:55.

Moolgavkar, S. H., and Knudson, A. G., 1981, Mutation and Cancer: A Model for Human Carcinogenesis, *JNCI* **66**:1037.

Omenn, G. S., 1984, Risk Assessment, Pharmacogenetics, and Ecogenetics, in *Banbury Report 16: Genetic Variability in Responses to Chemical Exposure,* G. S. Omenn and H. V. Gelboin, Eds., p. 3., Cold Spring Harbor Laboratory.

Thorslund, T. W., Brown, C. C., and Charnley, G., 1987, Biologically Motivated Cancer Risk Models, *Risk Analysis* **7**:109.

Parameters Involved in Risk Assessment for Environmental Release of Biotechnology Products

Pamela J. Bridgen and Elizabeth Anderson
Clement Associates, Inc.
Fairfax, VA

ABSTRACT

Over the past ten years there has been a considerable decrease in the perception of risk associated with experiments involving recombinant DNA, and containment levels have been lowered. However, the industrialization of biotechnology has raised new concerns about the release of genetically modified microorganisms into the environment. The potential risks posed by such releases must be understood better, so that they can be weighed against the benefits in order to protect the public and the environment without unnecessarily inhibiting progress in this rapidly-developing field.

Risk assessment is the process of gathering and evaluating scientific data to determine the likelihood of the existence of risk under defined circumstances, and to predict its potential magnitude. Risk assessment has been used extensively to estimate risk associated with many types of chemicals and toxic pollutants. However, new questions arise when risk assessment is applied to the release of genetically engineered microorganisms. The parameters that need to be considered include whether these microorganisms are as genetically stable as the wild-type, whether they can survive, replicate, and spread in the environment, and whether they can be engineered so that they can easily be destroyed selectively. This paper focuses upon these and other factors related to the risk assessment and risk management of the products of biotechnology.

KEYWORDS: Risk assessment, biotechnology, genetically engineered microorganisms, environmental release

INTRODUCTION

Risk assessment is a process for assembling and evaluating scientific information to determine the likelihood that risk may exist under defined circumstances, and to predict the potential magnitude of this risk. The process of risk assessment can be divided into four steps:

1. Assessment of toxicity data to determine possible effects of the agent.
2. Characterization of the dose-response relationships.
3. Assessment of exposures.
4. Quantitative evaluation of the potential harm to human health and the environment.

New Risks
Edited by L. A. Cox, Jr., and P. F. Ricci
Plenum Press, New York, 1990

These methods are used routinely to estimate risk associated with a variety of concerns such as radiation exposures and exposures to toxic chemical pollutants. Similar methods can be applied to the products of biotechnology, but additional considerations of possible growth and multiplication need to be included, and laboratory methods must be developed to obtain the data necessary for inclusion in the risk assessment process.

Over the past ten years there has been a considerable decrease in the perception of risk associated with laboratory experiments involving recombinant DNA; consequently, containment requirements have been lowered. However, the industrialization of biotechnology has raised new concerns about the release of genetically modified microorganisms into the environment. The general public perceives the risks of biotechnology to be high; perhaps the fear of the unknown in this case is unnecessary. Even though molecular and cellular genetic manipulations have been scrutinized closely for almost ten years with no evidence of any significant risks associated with this technology, the risk perception involved with genetic engineering has resulted in delays of several years in field testing of genetically engineered organisms. It is important, therefore, to develop procedures for the risk assessment of genetically engineered microorganisms (GEMS) so that the risks can be assessed realistically and weighed against the social and economic benefits in order to protect the public and the environment without unnecessarily inhibiting progress in this important field.

THE RISK ASSESSMENT PROCESS

Before we can talk about assessing the risks from environmental release of biotechnology products, we need to define what this means. "Environmental release" and the related term "containment" have been difficult to define from a regulatory standpoint, and in the summer of 1986 the White House science office, the Office of Science and Technology Policy, asked for comments on how to determine when a biotechnology product should be considered "released into the environment" and what kind of "containment" is necessary for such products. Currently, environmental release has somewhat varying definitions for the regulatory and research review of the different governmental agencies. There may be minor differences between agricultural and non-agricultural approaches and between macro- and microorganisms. For the purposes of this paper we consider environmental release to be the transfer of a product from confined laboratory conditions to the outside world. Clearly, different uses of products result in different release potential—a pesticide sprayed on plants is likely to result in much greater exposure potential for the general population than internal use of an animal drug.

For the purposes of risk assessment, biotechnology products can be divided into two classes (see Fig. 1): chemicals that are produced by biotechnological processes and are not living organisms; and microorganisms that may be genetically engineered which have the ability to grow and replicate. In general, the major concern about risks from biotechnology products centers around the genetically engineered organisms produced as a result of recombinant DNA technology.

Chemicals that are produced by the processes of biotechnology can be evaluated in much the same way as toxic chemicals, with the only additional concern being purity. If it can be shown that a product contains no impurities as a result of its method of production, then its risk assessment should be a routine matter. If impurities are present, then their toxicity, persistence, and environmental mobility must be incorporated into the risk assessment process. There is particular concern that impurities present in the products of genetic engineering may contain part of the viral genome that was used as the vector or carrier for the DNA.

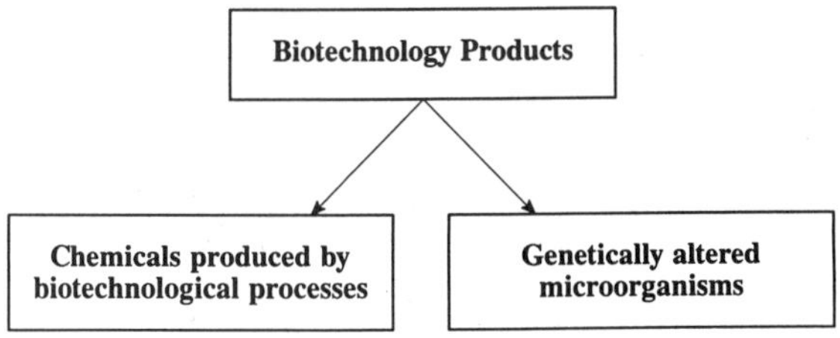

Fig. 1. Classes of biotechnology products.

Genetically engineered microorganisms (GEMS), however, require a more complex risk assessment because of their potential ability to survive, grow, and replicate in the environment. The range of possible organisms, applications, and environments indicates that no standard set of procedures is likely to be appropriate in all circumstances, but certain parameters are likely to be applicable to risk assessments for many of these genetically engineered microorganisms. Evaluation of microbial pesticides (i.e., microorganisms used to control pests) is not new. The first naturally-occurring microbial pesticide was registered in 1948 and there are currently more than 100 commercial products that contain microbial pest control agents. These agents are generally very insect specific and demonstrate low toxicity to humans and other mammals. Microbial pesticides have been used without known adverse effects in spite of the fact that they are living entities capable of survival, reproduction, and both pathogenicity and infectivity to non-target hosts. Thus, these parameters do not necessarily indicate a problem, but they must be evaluated in a determination of potential risks.

The organisms of particular concern in the field of biotechnology are intergeneric organisms (new organisms) and pathogens. Intergeneric organisms are those that deliberately formed to contain an intergeneric combination of material, excluding organisms that result from the addition of intergeneric material that is well characterized and contains only non-coding regulatory regions such as promoters (regions of DNA that interact strongly with RNA polymerase before transcription, i.e., formation of an RNA chain, can occur), operators (sites between the promoter and the region of DNA that codes for a protein to which a repressor protein can bind and prevent transcription), terminators (base sequences of DNA that signal termination of transcription), and ribosome binding regions (ribosomes are the nucleoprotein particles on which protein synthesis occurs). Pathogens are microorganisms (including viruses) that have an ability to cause disease in other living organisms. For genetically engineered organisms there is reason for concern if the microorganism belongs to a potentially pathogenic species, or if it has been derived from a pathogen or contains genetic material (other than a well characterized non-coding regulatory region) from a pathogenic donor.

Thus, the first parameter that should be considered in evaluating risks from release of biotechnology products is the nature of the GEM. Relevant factors include:

a. Characteristics of the non-modified parental organism, e.g., identification, reproductive cycle, capacity for genetic transfer;

b. Characteristics of the organism donating the DNA—identification, taxonomy, source, and strain;

c. Source and function of the modifying DNA sequences inserted into the organisms that have been altered.

d. Method of introduction of the DNA and procedure for selection of the modified organism—the amount and nature of any vector and/or DNA remaining in the GEM;

e. Results and interpretation of tests designed to measure genetic stability and expression of the introduced DNA in the modified organism;

f. Results of tests measuring the pathogenicity or toxicity of the GEM; and

g. Effect of the donation process on the organism—changes in physical/physiological properties, pathogenicity, etc.

Pathogenicity tests must be carried out under appropriate conditions of temperature, humidity, wind conditions, levels of inoculum, etc. One way to ensure that conditions are appropriate is to use the pathogenic parental strain as a positive control in conducting the experiments.

Once the GEM has been characterized fully, attention should be turned to the environmental aspects of the release. There is little information currently available on survival rate, growth potential, transformation potential, or environmental fate and transport of novel microorganisms. The Subcommittee on Risk Assessment of the Biotechnology Science Coordinating Committee is attempting to compile an extensive database on the effect that biotechnology has had on the environment, which will be useful in helping to assess the risks from new products. It is important that time and resources be spent on the development of generally accepted tests and protocols for determining survival, multiplication, dispersion, and possible ecological effects of GEMs. Protocols for release, containment and destruction procedures after release, and microcosm models in which to evaluate and screen GEMs for possible adverse effects in the environment all need to be developed.

Certain kinds of information are necessary to assess the environmental effects of release of a GEM in addition to those required for chemical agents. To help assess the survival potential of the GEM in the environment, relevant ecological characteristics of the parent organism and the GEM should be considered. The important factors are those in the GEM that are different from those in the parent organism. For instance, if the GEM requires additional nutrients it is less likely than the parent to survive if released into the environment. The physiological and ecological roles of the donated genetic sequences will help to determine the possible differences between the parent organism and the GEM.

The host range (variety of species in which the organism can survive and replicate) is an important factor in the spread of the organism—the more limited the host range, the less likely that the organism will spread out of control.

The likelihood of exchange of genetic information between the GEM and other organisms in nature is another factor that should be considered in assessing the potential for possible adverse effects in the environment. Even if the GEM is contained within a defined area, exchange of genetic information with other organisms and spread of these organisms could cause undesirable effects in the environment at large.

The ability to track the effects of the GEM or to curb the microbe's growth is a factor that would lessen the potential risk from environmental release. If the organism can be labelled in some way, for instance by genetic markers or immunologic procedures, it is possible to determine the survival rate and spread of the organism in the environment. Established natural populations of bacteria may retard or prevent the genetically altered bacteria from colonizing or becoming established in areas outside the test plots even if the GEM is able to survive and move off the test site in insects or via other vectors, and the migration and colonization can be monitored if the organism is labelled. If the microorganism can be engineered to be self-destructive once it has fulfilled its intended function, or if its growth or survival can be curbed by simple procedures, the risks from the GEM would be minimized.

The most complex factors that influence the survival, spread, and/or growth of the organisms in the environment are the interactions with and effects on other organisms. To some extent these have been covered in the preceding discussion, but they will be documented here. The relevant information for agricultural testing includes:

a. Ability to migrate outside the soil or target organism;

b. Effect on organisms other than the target;

c. Deposition rates;

d. Concentrations in the air;

e. Levels in soil and plants both inside and outside the test plot;

f. Levels in insects captured in the test plot; and

g. Existence of similar organisms in nature.

SUMMARY OF THE RISK ASSESSMENT PROCESS

A comparison of the risk assessment processes for chemical agents and genetically engineered microorganisms is shown in Table 1. The major differences in assessing the risk from GEMs are related to their potential for growth and multiplication in the environment. Laboratory tests must therefore include tests for pathogenicity and infectivity of the organism in addition to the tests generally required for chemical agents. Because epidemiology data are unlikely to be available for the GEM, epidemiology data on related organisms should be considered as part of the risk assessment. The dose-response relationships for altered organisms will be unknown and will add further uncertainty to the risk assessment. In addition to the factors considered in risk assessment of chemical agents, survival, multiplication, and growth of the organisms should be considered. If the organism is unable to become established in the environment, the environmental risk is likely to be minimal. The last step of the risk assessment process—determining quantitative risk—is similar whether the agent is a chemical or a microorganism, although, given our present state of knowledge, the uncertainties are likely to be greater in the estimate of the level of risk from genetically modified organisms.

AREAS REQUIRING FURTHER WORK

Because of the uncertainty involved in risk assessments of genetically engineered biological agents and their products, it is important that scientists involved in these developing techniques focus particular attention on developing the methods and data necessary to answer the risk assessment questions:

1. How likely are these agents to be threats to human health and the environment?

2. If they might be, what is the magnitude of the possible impact?

Table 1. A Comparison of Risk Assessment Procedures for Chemicals and Genetically Engineered Microorganisms

Risk Assessment	Chemicals	GEMS
Steps in Assessment		
Protocol for laboratory tests	Whole animal life study tests Acute studies Chronic studies In vitro tests, e.g., such as cell transformation Health effects tests, e.g., mutagenicity, carcinogencity	Whole animal tests Test for pathogenicity Test for infectivity Short-term tests for mutagenicity/cell transformation Chronic tests health effects, e.g., cancer
Hazard assessment (qualitative assessment)	Epidemiology Evaluation of data Identification of potential hazards associated with suspect agents Comparison with activity of other similar but better characterized chemicals	Epidemiology data, related organisms Assessment of data on health effects
Dose-response	Mechanisms of action Low dose and threshold effects	Unknown
Exposure	Evaluation of parameters, such as fate, transport, transformation, persistence, bioaccumulation, degradation, and route of exposure	Evaluation of parameters, such as survival, spread, fate, transport, transformation, persistence, accumulation, degradation, route of exposure, and metabolic variables
Quantitative risk	Evaluation of the potential of harm to human health and the environment Estimation of the magnitude of risk	Evaluation of the potential of harm to human health and the environment Estimation of the magnitude of risk

The specific parameters required to answer these questions have been discussed in this paper. As reliable methods for obtaining the data necessary for risk assessment and sound safety measures are developed, the controversy over the health and social-ethical issues surrounding biotechnology will subside and society will be able to benefit from the tremendous potential of biotechnology.

BIBLIOGRAPHY

Anderson, E. L., Albert, R. E., and Kamely, D., 1985, in *Genetically Altered Viruses and the Environment*, Banbury Report 22, pp. 33-46, Cold Spring Harbor Laboratory.

Environmental Protection Agency (EPA), 1986, Guidelines for Estimating Exposures, *Federal Register* **51**:34042 (September 24).

Environmental Protection Agency (EPA), 1986, Guidelines for Carcinogen Risk Assessment, *Federal Register* **51**:33992 (September 24).

Environmental Protection Agency (EPA), 1986, Guidelines for Mutagenicity Risk Assessment, *Federal Register* **51**:34006 (September 24).

Environmental Protection Agency (EPA), 1986, Guidelines for the Health Assessment of Suspect Developmental Toxicants, *Federal Register* **51**:34028 (September 24).

Environmental Protection Agency (EPA), 1986, Guidelines for the Health Risk Assessment of Chemical Mixtures, *Federal Register* **51**:34014 (September 24).

Lin, E. C. C., Goldstein, R., and Syvanen, M., 1984, *Bacteria, Plasmids, and Phages — An Introduction to Molecular Biology*, Harvard University Press.

National Research Council, 1983, *Risk Assessment in the Federal Government: Managing the Process*, prepared by the Committee on the Institutional Means for Assessment of Risk to Public Health, Commission on Life Sciences, National Academy Press, Washington, D.C.

Office of Science and Technology Policy, 1986, Coordinated Framework for the Regulation of Biotechnology, *Federal Register* **51**:23302 (June 26).

Do Genetically Engineered Microorganisms Pose Risks to Ecosystems?

P. W. Flanagan
Central Michigan University
Mount Pleasant, MI

ABSTRACT

Debate and controversy continues about proposals to release genetically engineered microorganisms (GEMS) into the environment. The concerns of ecologists and the public will have to be addressed by a program designed to build public trust about both risks and benefits of GEMS. At present there is no clear, coherent agenda of regulation in research for the public to relate to, and they know that the methods for assessing chemical or physical risks do not work with living GEMS. In this case the difference between basic research and applied research dissolves because any data in these GEM cases can be useful and applied immediately. This is because of the enormity of our ignorance about microbial ecology in general and in particular about the ecology of genes in soil and water. Molecular geneticists and ecologists must form a new discipline — Gene Ecology — in order to assess risks to the environment from horizontal gene transfer among GEMS and natural organisms in ecosystems.

KEYWORDS: Genetically engineered microorganisms (GEMS), ecosystem, microbial ecology, genes

INTRODUCTION

Releasing genetically altered organisms into the environment raises new problems in evaluating potential adverse ecological effects. Even if molecular geneticists know exactly which genes and processes they alter, at present no one can predict what will happen if genetically engineered microorganisms (GEMS) are released into the environment. That is to say, both the risks and the magnitude of the consequences are unknown. This stems from a dearth of relevant empirical data and from the complexity of ecological, molecular, and genetic disciplines involved. Inevitably, debate on the consequences of releasing GEMS raises the question of risks in proceeding before crucial data are available.

We have no field data on the probability of horizontal gene transfer (i.e. movement of recombinant genes out of host into non-target organisms of the same or another taxon and/or generation) or of the effects of recombinant genes on heredity and evolution in individuals, populations, or communities. It is not even known if GEMS and natural microbial populations actually engage in horizontal gene transfer under field conditions. At the ecosystem level, for example, we don't know if genes inserted into microorganisms for

New Risks
Edited by L. A. Cox, Jr., and P. F. Ricci
Plenum Press, New York, 1990

degrading xenobiotic molecules, persisting in the environment, might not degrade natural humus phenols and undermine the organic matrix of ecosystems.

We have some strong indirect evidence of horizontal gene exchange under field conditions. We are beginning to appreciate that antibiotics killed countless nonpathogenic bacterial genera. Surviving populations have responded by using an intrinsic ability to engineer their own genes and develop resistance to antibiotics so that many strains of bacteria on earth today are able to withstand antibiotic attacks that would once have killed them. Transfer to other species, genera, populations and communities follows rapidly. In the half century since man first used antibiotics, resistance has spread amongst many bacteria and some are resistant to dozens of antibiotics (O'Brien *et al.*, 1985). The spread of antibiotic resistance amongst bacteria is a sobering reminder that we must proceed with caution as GEMS are considered for use in nature. These data also show us how microbial populations develop characteristics when impelled by a selection pressure. With GEMS, a whole new variety of selection pressures could arise in the microbial world. Ecologists ask; how will GEMS alter ecological interrelationships and evolution of indigenous microorganisms when introduced to natural environments? Would GEM populations change the natural energy and mineral cycles at the basis of global ecosystems?

A first step towards the goal of minimum risk with GEMS may be to separate the elements which constitute genetic engineering (GE) from those which result from GE. The products of GE may only be as safe as the caution employed in their synthesis. The employment of GEMS in natural or managed environments raises questions separate from those needing attention during the GE process, but potentially negative consequences of accidental or/and deliberate release may be greatly reduced if the GE process is carefully controlled.

First concerns will be with the *taxonomy* of the gene donor and recipient organisms. Microbial taxonomy is not nearly as well understood as it might be. At the start of GE, the *gene* involved must be known in minute detail as must the gene *vector*. Characteristics of the *host* should be known at several levels including details of the structure and function of the genome and extra-chromosomal DNA which includes the plasmids. An examination of each of these concerns for a soon-to-be GEM follows.

TAXONOMY OF MICROORGANISMS

What kind of risks may be related to a microbe's taxonomy? It is generous to assume that we can isolate and cultivate even 10% of the microorganisms in nature. Microbial ecologists claim that ninety percent of soil microorganisms are unknown to science. If this is the case then microbial isolation and identification is a major component of risk prevention. It is impossible today to find funding for this kind of "descriptive research," but we can not minimize risks with GEMS without it.

In the laboratory many microorganisms can be characterized at the species level, but subspecific strains arise continually as species change morphologically and/or physiologically while being cultivated and stored. In fact, microorganisms which maintain their physiological integrity after prolonged periods *in vitro* are exceptions. Taxonomic factors and traits used in determining species are often borne on plasmids which are transmissible. Examples are found in the Rhizobiaceae and in certain hemolytic Streptococci (Berringer, 1982; Jacob *et al.*, 1975). Even chromosomal factors if close to repeated sequences may be transferred by transposable elements (Bukhari *et al.*, 1977, and Hopwood *et al.*, 1984). We do not know how many microbial species have similarly transferable taxonomic factors. To my knowledge, no taxonomic method is generally acceptable to the majority of microbiologists. If we cannot identify with precision the source of and target for gene transfer in GEMS, we may face a risk of unknown dimensions.

However, we may have the technology today to make very precise identifications. This new technology includes: serotyping and DNA homology and RNA sequence and structure. Serotyping is the identification of microorganisms based upon immunological responses of a host organism to surface antigens on microbial cells. Responses are very specific, but there are problems in that antigens may be encoded on plasmids which are transferable. However, antibodies specific for genome encoded cell surface glycosyl moities or cell surface proteins, are available (Gibbs *et al.*, 1972) and could be very important in the first steps of producing GEMS.

More specific DNA sequence homology determinations are possible and researchers are just starting to employ RNA base sequence analyses to explore microbial relationships (Woese *et al.*, 1983 and 1984; Stackbrandt and Woese, 1984; MacDonell and Colwell, 1984; and MacDonell *et al.*, 1986). An additional advantage of this new approach is that it could be used for identification and classification of viruses.

Microorganisms should not be included in genetic engineering experiments until they can be identified at the nucleic acid level. The technology to do this exists now and investigators in the forefront of research in this new approach should be encouraged to accelerate its development in consultation with ecologists.

Subject Genes. These are the exogenous DNA sections, which are inserted to function in a genome where they never before occurred naturally. Genetic engineering experiments may involve changing a gene to enhance its expression *in situ* or a gene may be transferred to another organism alone or with other genes or regulatory sequences.

It is known that genes may express themselves in different ways under different circumstances. Epistatic effects occur when a gene or its products suppress another gene or gene effect distant from it on the chromosome. Furthermore, some genes have complex multiple and combinational expressions and influences called pleiotropic effects. Other variations of gene amplifications have been reviewed by Anderson and Roth (1977) and Clarke (1984). Therefore, unforeseen properties of genes may result from putting them in places and associations heretofore alien to them. This is an area in need of immediate research.

As an example of gene amplification, it has been shown that certain microorganisms though incapable of metabolizing chlorophenols, have genes which, when mixed with a heterogeneous microbial gene pool, also incapable of utilizing chlorophenols, alter the gene content of the entire population, resulting in a new microbial gene complex able to deal with hitherto toxic substrates and eliminate the original population, i.e., those going without this new gene complex (Schmidt *et al.*, 1983). Such a phenomenon in a natural ecological community of decomposers could drastically alter microbial-substrate relationships and lead to new energy and mineral cycling pathways out of synchrony with plant root activities.

Very little is known about specific microenvironment effects on microorganisms under most field conditions. Nutritional aspects are deemed very important for microbial function (Alexander, 1985), but we are almost totally ignorant on questions of gene and environment interactions, including the microenvironment of intracellular space and all it contains as well as the physical conditions of microbial microhabitats in targeted soil systems. Doubtless, broader environmental variables, temperature, moisture PH, redox potential, etc., influence gene stability and transfer rates as well as expression. Data on relationships between genes and their macro and microenvironments in the laboratory and field are needed immediately.

VECTORS

There are at least four ways that microbial genes move from one cell to another; transformation, transduction, conjugation, and/or mobile plasmids. A more detailed discussion of these mechanisms may be found elsewhere (Stotsky and Babich, 1984). For both plasmid and viral vectors, we should know: (1) The host range of the vector. (2) The frequency of, and the influencing factors (environmental), transfer of the vector under laboratory and field conditions. (3) The exact gene content of the vector, e.g., a plasmid may carry genes encoding sulfa drug, heavy metal and/or antibiotic resistances, as well as any that can be attached. The former could give heightened survival traits to a host organism. Plasmids often carry transposons and insertion sequences which indicates that interactions amongst plasmids and with the chromosomal DNA may be common. This allows a potential for widespread dissemination of critical genes throughout a wide range of genera (Calos and Miller, 1980; Cohen, 1976; Kleckner *et al.*, 1977; Heffron *et al.*, 1975; James *et al.*, 1983).

Assuming a GEM persists in the soil fulfilling the role it was designed for and otherwise posing no risk, how do we know that "wild" genes from the soil microbial community will not enter its genome and alter it in ways which would cause it to become an ecological threat? Again we are constrained to approach risk assessment on a case by case basis. A solution would include research with the GEM in the presence of likely "suspects" amongst the natural microbial community about to receive it. Suspect organisms would have to be screened for potential conjugation plasmids and transferable genes which could enhance the GEMS survival function and pathogenicity. We need to know about microbial populations of sites and how specific microbes interact in their natural habitats. Also we need data on which of these mechanisms prevail at sites for transferal of genes.

These aspects of gene complexing and complex gene expression (in laboratory and natural habitats) are in need of extensive research and they continually emerge in considering different aspects of risk assessment.

The host organisms natural history in the target site should be understood: i.e. what other microorganisms are found in association (see gene section), what is its substrate(s), its predators, and with which other microorganisms might it exchange DNA?

RELEASING GEMS INTO THE ENVIRONMENT

Apart from considerations of survival, establishment, growth and reproduction of the introduced GEM which will be addressed below, there are other factors which should be accounted for earlier. These include the method of application, the numbers of potential microbial recipients of the "new" genes, the general characteristics of the site (drainage, slope, animal vector traffic, etc.), time (i.e., whether or not to apply before or after plowing), and concentration of organisms added.

The method of application may be important especially if opportunities exist for aerosoling the GEM to reach high atmospheric, transporting air masses or flowing water (Odum, 1985; Flanagan, 1986). The number of potential recipients relates to the seasonal microbial population dynamics of the site. Microbial populations change quantitatively and qualitatively with seasons in most natural systems. This factor also relates to the concentration of GEMS to be added. Considering site characteristics such as water movement and drainage allows a sampling regime to assess movement of the GEM in soil water to rivers and or lakes and to capture animal traffic (birds, invertebrates, etc.) to see if the organism is being moved, and if so, by what means and agent to where. Thus, research needs include some basic microbial ecological studies.

Genetically engineered organisms carrying the nitrogen fixation (NIF) genes were released in the field as early as 1974 at the Department of Microbiology in Galway, Ireland. The microorganism did well in the laboratory, fixing N2 with efficiency. It was spread "by the gallon" on soil plots but not a single living cell was detectable two weeks later. It could not survive in the field (Smith, Pers. Comm.). Similar rapid disappearances have been reported by Gillett *et al.* (1985). However, cells may survive in very low concentrations of less than ten per gram soil, i.e., the limit of our ability to detect them at present. We can not be sure that numbers decline to zero, but some opinions hold that most alien organisms in soil systems decline rapidly. However, Gillett *et al.* (1985) have reported that when added to soils, some organisms initially suffer exponential decline, but become stabilized at 10-100 cells/gram soil and persist indefinitely at this concentration. Conceivably, given the variability of conditions in most soils, persistent organisms, even those present at concentrations of less than one cell/gram soil, could encounter conditions favoring their growth and reproduction. The result could be a "take over" of microbial niches and the alteration, perhaps permanently, of the existing microbial community.

We must establish for each GEM the threshold level for the inoculum sizes for survival, persistence and multiplication in the soil types that receive it. The basic technique needed here is rudimentary microbial research but it must be refined now to allow very precise detection of cells down to concentrations of 10^{-6} cells/gram soil. This research is a must in attaining the goal of minimizing risks with GEMS. DNA hybridization/probe research on DNA extracts of soil (NSF sponsored research is beginning on this aspect) may be the most promising approach.

Microorganisms do not necessarily stay where they are placed. Gillett *et al.* 1985, have evaluated in detailed the mechanisms and risks of microorganisms dispersal following release. Linderman *et al.* (1982) discussed at length the release of microorganisms into the atmosphere as wind aerosols. This work should be studied in detail by those wishing to introduce GEMS into the environment. These authors show that even in calm weather the vertical flux of bacteria from plant surfaces alone could be $0.5 - 5.0 \times 10^6$ colony forming units (CFU)/ha^{-2}. However, risk of dispersal is an empty threat if an organism cannot survive transportation and establish itself in a site to which it is transferred.

Microorganisms can be globally dispersed in air flow (Flanagan, 1986) so there is little hope of evaluating the effects of a GEM organism on all environments and microorganisms which could be encountered. However, we must evaluate the mechanisms and extent of microbial transport systems and microbial survival within them to minimize the likelihood that GEMs may become widely dispersed and established distant from their site of introduction. This problem arises in considering pollen and or spores from genetically engineered plants.

ORGANIZING RESEARCH ON GEMS

Molecular geneticists manipulate genes with increasing precision but we know little about microbial survival and genetic exchange in the environment. What we need to do, to minimize risks to the environment, is to amalgamate the disciplines of microbial ecology and molecular biology and build the theoretical foundations for the resulting discipline.

The spectrum of research topics outlined above is not all-inclusive, yet it is very broad, and each facet must be considered simultaneously by those organizing research to produce, and release GEMS with minimal risk. The need for a team effort is obvious, but conservative philosophies and the logistics and cost of mustering and focusing so diverse a group may preclude its formation unless support is forthcoming from a number of sources including federal and local government, universities and industry.

Industry's involvement has to be developed and addressed in the open. This is not a subject where proprietary concerns rule. Nor does industry have sole responsibility for research needs since society stands to benefit from developments within biotechnology. Relying solely on government sponsored research may not be the answer either. What appears to be needed are new mechanisms of research support involving industry-university cooperatives with government support as well. All participants stand to benefit and so too will society as a whole. Problems of environmental risks of biotechnology will be solved only by cooperative,open approaches. Creating and nurturing this technical infrastructure presents a major challenge.

On examining a number of alternative suggestions for organizing appropriate team research, one formula more than others seems feasible, less costly, and more assured of success. It is as follows: three or four university groups could be established in a center to conduct interdisciplinary research across a range of topics like those discussed earlier. These could be supported primarily with federal funding following a national competitive proposal submission phase. Also, many states are aggressively seeking new industry and could prove to be a major source of funding. But funds should also come from industry. Research performed at the centers could, at once, help regulatory agencies fulfill their responsibilities and provide industry with data needed to meet regulatory agency requirements. Facilities and research protocols could be monitored by selected members of the university, federal government and local authorities, thus ensuring a healthy atmosphere in public perception and relations.

Perhaps there is no need to establish centers *per se* for this research. What is needed are focused groups of researchers integrated on campuses to address and work on questions about GEMs. As their usefulness wanes, they could be dissolved painlessly, the researchers moving on to other research topics *in situ*. Some labs may need special renovations, but new buildings may be unnecessary. These groups of integrated molecular biologists, geneticists, microbial, and other ecologists could apply for a government grant rather as some departments do already. The university, state and interested industry could provide matching funds and/or facilities. Local government must also be involved. This is a crucial ingredient.

Everybody is aware of the importance of federal and university support, and GEMS are clearly of high interest to industry. To evaluate the importance of local government involvement in these areas of research, consider the rejection of the ice-minus program in California by the Monterey authorities, despite EPA's permission for field trials and the University of California's announcement of a public meeting to discuss incipient field trials with the local community. The public perception of research in any area of genetic engineering can be extraordinarily influential in determining the inception, organization and success; and so it should be. From the outset, local representatives of the public in areas where research will be conducted must be included in the design and organization of the research facilities, goals, and agendas. Specific efforts should be made by researchers and their institutions to improve public understanding of various facets of recombinant DNA technology and genetic engineering in general.

To date, we in the U.S. have progressed little in designing and developing this infrastructure though we lead the world in genetic engineering. We have been in the forefront also in describing the problems which may arise, but we have done little to provide solutions for problems.

While the development of competitive biotechnology with GEMS has been slower in other nations, the pace of competition is accelerating and the extent to which GEMS in other countries will be subject to regulatory screening remains unknown. An interesting article appeared in the May 19, 1986 issue of the *London Financial Times*. It reported on a consortium of fifteen academic and industrial organizations (some from the U.S.) launching a "precompetitive" research programme to design a plant gene "toolkit" and users manual permitting the user to change the genetic makeup of Europe's most valuable crops. The initiative is government sponsored and originated in the laboratory of The Government Chemist in London. Moneys provided by industry are matched by government. Approximately $4.56 million has been provided for the next three years of research. It will involve about 70% of the U.K. expertise in this area. The article went on to describe participating institutions, individuals and goals but environmental consequences or risks were not even mentioned. It was as though they did not exist. Having fallen behind in electronics, automobiles, and optics, are we also to loose the competitive edge with genetically engineered organisms?

Microbial ecological, molecular and genetic tools are available to collect empirical data which at very least will narrow uncertainties about the consequences of releasing GEMS. However, the difference in scale and the complexity of ecosystems obviates, or greatly lessens, the value of data from microcosms. Our approach should consist of intragratable research from microcosms, mesocosms (sealed greenhouses?) and macrocosms (ecosystems). Scientists involved will have to brave some nebulous epistimes when trying to relate the results of laboratory or mesocosm studies of GEMS to field studies of the results of GEM processes at the ecosystem/landscape level. But we must proceed in this research and to provide the facilities and personnel to minimize risks with genetically engineered organisms. We should heed Francis Bacon here who counselled that progress comes faster from taking action, even mistakenly, than from stagnation. In the meanwhile, we are poorly informed about the consequences of releasing GEMS and so, as is always the case when we don't know, confidence is lowest and fear of risks is highest.

REFERENCES

Anderson, R. P., and Roth, J. R., 1977, Tandem Genetic Duplications in Phage and Bacteria, *Ann. Rev. Microbial.* **31**:473-505.

Alexander, M., 1985, Part I. Survival and Growth of Bacteria, Ecosystem Research Center Report, EPC-075, Cornell University.

Berringer, J. E., 1982, The Genetic Determination of Host Range in Rhizobacteria, *ISR. J. Bot.* **31**:89-93.

Bukhari, A. I., Shapiro, J. A., and Adhya, S. L. (eds), 1977, DNA Insertion Elements, Plasmids and Episomes, Cold Spring Harbor Laboratory, Cold Spring Harbor, N.Y.

Calos, M. P., and Miller, J. H., 1980, Transposable Elements, *Cell* **20**:579-595.

Clark, P. H., 1984, Evolution of New Phenotypes, in *Current Perspectives in Microbial Ecology*, Klug, M. J., and Reddy, C. A., eds, p. 71-78, ASM, Washington, D.C.

Cohen, S. N., 1976, Transposable Genetic Elements and Plasmid Evolution, *Nature* **263**:731-738 (London).

Flanagan, P. W., 1986, Genetically Engineered Organisms and Ecology, *Bull. Am. Ecol. Soc.* **67**:26-31.

Gillett, J. W., Levin, S. A., Harwell, N. A., Alexander M., Andow, D. A., and Stern, A. M., 1985, Potential Impacts of Environmental Release of Biotechnology Products: Assessment, Regulation, and Research Needs, Ecosystem Research Center Report, EPC-075, Cornell University.

Gibbs, P. A., Patterson, J. T., and Murray, J. G., 1972, The Flourescent Antibody Technique for the Detection of Salmonella in Routine Use, II, *J. Appl. Bacteriol.* **35**:415.

Heffron, F., Sublett, R., Hedges, R. W., Jacob, A., and Falkow, S., 1975, Origin of the TWM Betalactamase Gene bound on Plasmids, *J. Bacteriol.* **122**:250-256.

Hopwood, D. A., Hinterman, G., Kieser, T., and Wright, H., 1984, Integrated DNA Sequences in Three Streptomycetes from Related Autonomus Plasmids After Transfer to *Streptomyces lividans*, *Plasmid.* **11**:1-16.

Jacob, A. E., Douglas, G. J., and Hobbs, S. J., 1975, Self Transferable Plasmids Determining the Hemolysin and Bacteriocin of *Streptococcus faecolis* var. *zymogenes*, *J. Bacteriol.* **121**:863-872.

James, A., Morrison, P. T., and Kolodner, R., 1983, Isolation of Genetic Elements that Inverse Frequency of Plasmid Recombinants, *Nature* **303**:256-259 (London).

Kleckner, N., Roth, J., and Botstein, D., 1977, Genetic Engineering *in vitro* Using Translocatable Drug Resistance Elements, New Methods in Bacterial Genetics, *J. Mol. Biol.* **116**:125-139.

Linderman, J., Constantinidou, H., Barchet, W., and Upper, C., 1982, Plants as Sources of Airborne Bacteria, Including Ice Nucleation — Active Bacteria, *App. Env. Microbiology* **44**:1059-1063.

MacDonell, M. T., and Colwell, R. R., 1984, Identical 5S rRNA Nucleotide Sequences of *Vibiro cholerae* Strins Representing Temporal, Geographical and Ecological Diversity, *Appl. Environ. Microbial.* **48**:119-121.

MacDonell, M. T., Swartz, D. G., Ortz Conde, B. A., Last, G. A., and Colwell, R. R., 1986, Ribosomal RNA Phylogenies for the Vibrio-Enteric Group of Eubacteria, *Microbiological Sciences* **3**:172-178.

O'Brien, T. F., del Pilar Pla, M., and Mayer, K. H., 1985, Intercontinental Spread of a New Antibiotic Resistance Gene on an Epidemic Plasmid, *Science* **230**:87-88.

Odum, E. P., 1985, Biotechnology and the Biosphere, *Science* **229**:1138.

Schmidt, E., Hellwig, M., and Knackmuss, H. J., 1983, Degradation of Chlorophenols by a Defined Mixed Microbial Community, *Appl. Environ. Microbial* **46**:1038-1044.

Smith, P. R., 1980, Department of Microbiology, University College Galway, Galway, Ireland (Personal Communication).

Stackbrandt, E., and Woese, C. R., 1984, The Phylogeny of Prokaryotes, *Microbiological Sciences* **1**:117-122.

Stotsky, G., and Babich, H., 1984, Fate of Genetically Engineered Microorganisms in Natural Environments, *Recomb. DNA Tech. Bull.* **7**:163-188.

Woese, C. R., Gutell, R., Gupta, R., and Noller, H. F., Dec. 1983, Detailed Analysis of the Higher-Order Structure of 16S-Like Ribosomal Ribonucleic Acids, *Microbiological Reviews* 621-669.

Woese, C. R., Blanz, P., and Hahn, C. M., 1984, What Isn't a Pseudomonad?: The Importance of Nomenclature in Bacterial Classification, *System. Appl. Microbiol.* **5**:179-195.

Control of Microbial Pathogens in Poultry by Irradiation: Issues Related to Risks and Benefits

Ronald Krystynak
Agriculture Canada
Ottawa, Ontario, Canada

ABSTRACT

The use of food irradiation has been proposed to reduce the incidence of food-borne diseases, particularly salmonellosis and campylobacteriosis associated with poultry. The widespread adoption of irradiation for the control of food-borne hazards may be delayed by the divergence of actual and perceived risk of irradiated food and the divergence of social and private benefits evaluations of food irradiation. Food irradiation is, on balance, a risk reducing technology, yet risk perceptions may be to the contrary. Consumer private benefit assessments may differ from society benefits. These issues raise the policy implication of whether public health authorities should promote use of this technology to control microbial pathogens in poultry.

KEYWORDS: Food irradiation, poultry, food-borne diseases, benefit-cost ratio

INTRODUCTION

Food irradiation is a process that may be useful in augmenting conventional food processes. One of the more promising applications of food irradiation is to improve the safety of poultry and fresh meat. With the increase of some food-borne infectious diseases in many countries, irradiation has been proposed as a procedure which could eliminate microbial pathogen contamination of poultry and meat products at the processor level, and thus allow the delivery to the consumer of a safer product. Treatment with ionizing radiation at doses up to approximately 5 kGy (termed radicidation) is adequate for the elimination of nonsporing pathogenic micro-organisms in fresh foods.

Despite the fact that irradiation technology is well established and research on food applications of this technology began in the early 1950's, only a few countries use it on a limited commercial basis. To date no country is using irradiation on a commercial basis to treat poultry. Some of the factors which led to the lack of progress in the acceptance of food irradiation, particularly with respect to poultry, will be examined here. The case of poultry will be examined in detail because poultry is the suspected cause of a significant portion of food-borne illness and irradiation can readily be applied to the product.

New Risks
Edited by L. A. Cox, Jr., and P. F. Ricci
Plenum Press, New York, 1990

This paper will examine the actual and perceived risks both of food irradiation and the consumption of poultry. These risks will be contrasted with the potential benefits to be derived from irradiating poultry.

THE TECHNOLOGY OF FOOD IRRADIATION

Electromagnetic radiation suitable for food irradiation may be either electrons, X-rays or gamma radiation. Gamma radiation has more penetrating power than the other two forms and consequently is the most common form used for food irradiation. Gamma rays are produced by a radioactive source, Cobalt-60 or Cesium-137. Irradiation works by killing or inactivating food spoilage and disease causing organisms. The biological damage is caused by the release of energy at sufficiently high intensity to disrupt chemical bonds. The value of ionizing radiation is that it does this without at the same time splitting so many of the molecules of food that the foods are damaged.

TOXICOLOGICAL AND SAFETY CONSIDERATIONS

Contrary to some public misapprehension, irradiation of food does not make it radioactive.[1] A Joint Expert Committee on Irradiated Foods of the Food Agriculture Organization of the United Nations (FAO), the International Atomic Energy Agency (IAEA) and the World Health Organization (WHO) recommended 10 million electron volts as the maximum permissible energy for electron generators and 5 MeV for X-rays but recognized that gamma rays from the two approved sources (cobalt-60 and cesium-137) do not induce radioactivity in food at any dose.[2]

Toxicology

Ionizing radiation splits a few of the molecules of the food itself, producing stable molecules referred to as radiolytic products. Since the introduction of food irradiation, reassurance has been sought by national and international authorities concerned with food safety that any changes in food composition caused by irradiation did not render the food unwholesome. Although the chemical changes in food components resulting from irradiation are quantitatively very small, they could conceivably result in the possible appearance of harmful substances in foods so treated.

A large number of short and long term toxicity, carcinogenicity, mutagenicity, embryotoxicity, and teratogenicity studies have been performed over the past 35 years on at least one member of each existing class of foods processed by irradiation. Conclusions based on the available evidence are that there is no evidence of toxicologically significant compounds arising in irradiated food.[2-4] While there have been a few articles reporting various unfavorable effects of foods treated with ionizing energy, a critical evaluation of several of these reports showed the studies to be inadequately designed and thus unreliable.[4] In fact the Joint FAO/IAEA/WHO Expert Committee on Food Irradiation concluded that food irradiated up to an overall average dose of 10 kGy presented no toxicological hazard and that further toxicological testing of foods so treated was no longer required to demonstrate safety.[2] Subsequently, the Codex Alimentarius Commission developed a General Standard for Irradiated Foods and a Recommended International Code of Practice for Operating Irradiation Facilities Used for Treating Food.[5]

The evidence is that there is a commonality and a predictability of the formation of radiolytic products.[6] It has been conjectured that "unique" radiolytic products might exist which could affect safety, unique radiolytic products being defined as "substances not known to the present in nonirradiated food."[7] On the basis of such a definition no unique radiolytic compounds have been found in 30 years of research. Compounds produced in specific foods by ionizing energy have always been found in the same foods when processed by other accepted methods or in other foods.[4]

Toxicological evaluation of food irradiation, however, is difficult for a number of reasons. Food is a mixture of chemicals, and once they are treated by irradiation, it is impossible to identify and determine quantitatively every chemical compound that might conceivably give rise to health hazards in higher mammals and man. The standard toxicological approach to evaluate the toxicity of food treated with ionizing energy is inappropriate for determining whether the radiolytic products are toxic and for evaluating their safety in human diets. Neither the food processed with ionizing energy nor the radiolytic products can be added in large excess. Consequently, it is not possible to achieve adequate exposure of possible toxic components.

Nutritional Quality

Experiments on the response of experimental animals and humans have been conducted to investigate the nutritional quality of food treated with ionizing energy under conditions that could be used commercially.[4,8] Tests to determine the utilization of the nutrients and clinical tests of the subjects disclosed no unfavorable effects by foods processed with ionizing energy compared to similar foods processed by conventional means.[4,9]

Microbiological Safety

The microbiological safety of irradiated food is fully comparable to that of food handled with conventional and traditional treatments.[10] No evidence has been reported of an irradiation-induced enhanced pathogenicity of food-borne micro-organisms. According to the Codex Alimentarius Commission there are no microbiological safety problems with most foods, such as fresh meats, poultry and fish, that have been treated with medium doses (up to 10 kGy) of ionizing energy, as long as these foods are stored and distributed in a chilled state (2 to 5° C) according to good manufacturing practice.[11]

A problem can arise with *Clostridium botulinum*, which can survive high irradiation doses, and in particular *Clostridium botulinum* type E, which can grow under refrigeration. This may be of concern in fish, but should fresh poultry be contaminated with *Clostridium botulinum* type E, surviving microflora would be able to produce spoilage odors within 8 days, whereas *Clostridium botulinum* survivors could not produce toxin within 14 days.[12] This spoilage would ward against consumption of the product.

Overall Assessment

There is an impressive body of scientific evidence to the effect that food exposed to ionizing energy under the conditions proposed for commercial application is wholesome, that is, safe to eat. The Joint FAO/IAEA/WHO Expert Committee noted that "the question of wholesomeness has been raised in relation to the irradiation of foods with gamma rays, yet the wholesomeness of foods treated with radiation of longer wavelengths (heat or microwaves) has not been questioned to the same extent."[13] Nevertheless, the inability to carry out standard toxicological tests on irradiated foods allows critics to question the adequacy of the safety evidence.[14,15]

MICROBIAL HAZARDS ASSOCIATED WITH POULTRY

The prevalence of food-borne infectious diseases is on the increase in developed countries. Poultry is a vehicle for several intestinal diseases, salmonellosis and campylobacteriosis being the most prominent. The number of reported cases of *Salmonella* infections has doubled over the last decade in Canada and nearly so in the United States.[16,17] The number of infections with *Campylobacter* appears to approach or exceed those of *Salmonella*.[18,19] The evidence is that the number of reported cases represents only a small number of actual cases,[20] perhaps only 1-3 percent.[16] In Canada it is estimated that there are over 400,000 cases of salmonellosis[16] annually, while for the United States the estimate is 2 million, with a similar number of campylobacteriosis cases.[21] The incidence of *Salmonella* in poultry has shown a steady increase. A Canadian survey on the incidence of *Salmonella* in chicken carcasses has shown a steady increase from 15% in 1970 to 63% in 1982.[22,23] Comparable levels were noted in turkey carcasses.[23] The incidence of *Campylobacter* in poultry meat is reported to be high, although sample results vary from 30% to 100%.[24-27] *Staphylococcus aureus* and *Clostridium perfringens* are also import etiological agents responsible for poultry-associated disease outbreaks.[28]

Food-transmitted infections usually result from a dual failure: contamination of food followed by improper final processing which allows the contaminant to be present at contamination levels exceeding the minimal infectious or minimal toxic dose for the most sensitive consumer. The direct risk of microbial infection is eliminated if poultry is properly cooked. However, the role of cross-contamination must also be considered. Epidemiological studies have shown that secondary contamination during production, processing and culinary preparation of foods may play an even greater role than primary contamination. Numerous outbreaks have been reported from which it was clear that contaminated surfaces, kitchen utensils and human hands play a significant role in cross-contamination.[29]

RISK-COST-BENEFIT ASSESSMENT

The scientific evidence is that irradiated food poses minimal risk to the consumer. Poultry consumption, whether through direct contamination or cross-contamination, leads to increased risk of contracting intestinal diseases. The only practicable solution to the health problem has to rely on a final processing. Therefore radicidation of poultry would lead to a reduction in risk.

The economic cost to society of various food-borne diseases has been estimated. It was estimated, for Canada, that the annual cost of salmonellosis (in 1982 dollars), ranges from $59 million to $158 million, depending on the assumptions made concerning the number of unreported cases, with the mean being $84 million.[16] Of all cases of salmonellosis in Canada, it was estimated that 25% could be attributed to poultry; consequently 100% elimination of *Salmonella* in poultry would yield about $21 million in benefits. To irradiate all poultry produced in Canada would cost about $14 million annually, leading to a benefit-cost ratio for salmonellosis control in poultry of 1.5.[16] This analysis did not consider the costs of campylobacteriosis and other poultry-associated diseases which, if included, would have yielded a higher benefit-cost ratio. Roberts estimated the costs in the U.S. due to poultry contamination with *Salmonella* and *Campylobacter* and found a benefit-cost ratio of between 2.1 and 4.1 for irradiating poultry.[21]

RISK PERCEPTIONS

An important issue concerning the potential of irradiating poultry is how consumers perceive both the risks of food irradiation and the risks associated with poultry consumption. We know that people do not perceive risks in the ways predicted by models of rational choice and behavior.[30]

Food irradiation is still not well known or understood by consumers. In a Dutch study it was found that the spontaneous reaction to the term "irradiated food" was primarily negative.[31] The concept of "irradiated food" is unknown, people have no idea what it is, yet their first reaction is one of rejection. This negative attitude can be attributed to a number of aspects. The consumer wants to avoid being confused and alarmed, and rejects the confrontation with unknown problems. Consumers associate a radioactive source with unpleasant events like death and disease. Furthermore, there is a fear that radioactive residues are in the food and thus harmful to consumers. Finally, there is the perception that food is alive and that somehow irradiated food has been killed and devitalized by irradiation.

A Canadian survey of consumer attitudes toward irradiated fish products found a 3 to 1 positive response to the irradiation process.[32] However, respondents were not told the link between the irradiation process and nuclear energy, which might have biased the results. A U.S. study of consumer attitudes found that when consumers were given some educational information about food irradiation, most were only slightly concerned about the safety of irradiated food.[33] An interesting aspect of this study concerned ecologically sensitive consumers. This group showed a higher level of initial concern about the safety of irradiated food than did other consumers. Following educational effort the concern attitude of ecologically sensitive consumers, if anything, increased.

Willingness to buy revealed on a questionnaire and actual buying may be different. South Africa has been marketing irradiated foods for a number of years. Irradiated potatoes, mangoes, papayas and strawberries were first test marketed in 1978-1979. All products were marked with the RADURA emblem, and information was given at point of sale for the first few weeks. A 90% acceptance level was recorded.[34] This success led to further commercialization of the technology in South Africa. In September, 1986, irradiated mangoes from Puerto Rico were successfully test marketed in a grocery store located in North Miami Beach, Florida.[35] The evidence suggests that most consumers will accept irradiated food if benefits are apparent, in the same way that microwave ovens have been accepted. There will continue to be a small group opposed to irradiated food.

On the question of irradiating poultry the issue is whether consumers will recognize the benefits of radiation pasteurized (radicidated) poultry. A Gallup survey of women in the United States revealed that 74% were not familiar with *Salmonella*, 66% did not know how to minimize the spread of *Salmonella* and 34% thought that raw meat and poultry are regularly inspected for the absence of *Salmonella* by federal/state inspectors.[36] In a Canadian study of campylobacteriosis cases, the source of most infections was difficult to determine and only 30% of patients hazarded an opinion on possible contacts.[19] The majority of patients offering an opinion cited a sick relative as the possible contact. The survey demonstrated the difficulty in finding the original source of infection in sporadic cases, and it was observed that a patient's response to the question might be influenced by a particular dislike a pet, food or nationality.

The evidence demonstrates that consumers are not well informed about microbial hazards associated with food. Furthermore, in those instances where they contract a food-borne disease, it is unlikely they will correctly determine the source of their infection. They may even confuse their illness with the 'flu bug.' Because food-borne diseases are rarely life threatening, they are treated as trivial and as unavoidable as the common cold.

As noted, irradiation of poultry offers positive social benefits.[16,21] Consumers would face a direct additional cost (3-5¢ per kg)[37] for irradiated poultry. Improved safety benefits, on the other hand, are not quantifiable by the individual. Even a bout of gastroenteritis may not result in direct additional costs to the individual because of medical insurance and paid sick leave.

DISCUSSION AND CONCLUSIONS

Irradiation applied to poultry can lead to positive benefits to society through a reduction in certain intestinal diseases. This analysis leads to the conclusion that irradiation of poultry should be vigorously promoted. It has even been proposed that health authorities regulate the use of irradiation to protect human health, in the same way that pasteurization of milk was regulated.[38] Such a conclusion ignores the important part that perceptions, particularly of risks but also of benefits, play in the decision process. The divergence between actual and perceived risks creates a gap that may preempt commercialization of poultry irradiation.

If the analysis is correct and society will receive net benefits from irradiation of poultry, there are a number of possible options which could be pursued. Each option is not without its pitfalls.

1. *Public Promotion of irradiation of poultry.* Through a widespread publicity and educational campaign the public could be made aware of the potential health risks associated with poultry and informed that irradiation offers a safe method of improving the safety of poultry. There is a major difficulty with this proposal. The evidence is that any discussion of potential risks associated with a food product will lead to fear on the part of the public and a reduction in the consumption of the product.[39] A decline in consumption would adversely affect the poultry industry, and it is understandable that the industry would oppose such an option. Furthermore, there is no guarantee that consumers, given the choice, will buy the slightly more expensive irradiated poultry as opposed to the non-irradiated product.

2. *Regulation.* Government health agencies could regulate the maximum allowable microbial contamination level for poultry. The only method by which poultry processors could now consistently meet such a level would be to irradiate the product. There are also a number of difficulties with such a proposal. Again, it would raise in the consumers' mind the fear of consuming poultry. More importantly, if all poultry is irradiated, consumers will lose the option of consuming non-irradiated poultry. There would certainly be a group strongly opposed to irradiation of poultry. They could argue that, with no option to irradiated poultry, they have no freedom of choice.

3. *Adaptive approach.* This approach is to allow consumers to get used to the idea of food irradiation through promotion of the technology for potentially less controversial food products, for example, in disinfestation of such tropical fruits as mangoes and papayas. Once greater consumer awareness and acceptance of food irradiation develops, it may be possible to introduce irradiated poultry. This means foregoing potential health benefits for a number of years.

This examination of options suggests that radicidation of poultry may be some way off. Pasteurization of milk was first advocated in 1886.[40] Commercial application of the technology began around the turn of the century, while regulations requiring pasteurization followed much later. A similar time lapse may apply for radicidation of poultry. This case study indicates that in the evaluation of the risks and benefits of a new technology or new use of an existing technology, risk perceptions rather than scientific fact may drive the decision process. Over time, perceptions and fact will likely converge; in the meantime, however, early adoption of a potentially beneficial technology will be forestalled.

REFERENCES

1. J. Shubert, Toxicological Studies on Irradiated Foods and Food Constituents, *Food Preservation Irradiation*, Vol. II, Proc. of an International Symposium on Food Irradiation, IAEA/FAO/WHO, Nov. 1977, International Atomic Energy Agency, Vienna 13-24 (1978).
2. WHO, Wholesomeness of Irradiated Food, Report of a Joint FAO/IAEA/WHO Expert Committee, Technical Report Series 659, World Health Organization, Geneva, Switzerland (1981).
3. P. S. Elias, Toxicological Aspects of Food Irradiation, in *Recent Advances in Food Irradiation*, P. S. Elias and A. T. Cohen, eds., pp. 235-245, Elsevier Biomedical Press, Amsterdam, The Netherlands (1983).
4. CAST, Ionizing Energy in Food Processing and Pest Control: 1. Wholesomeness of Food Treated with Ionizing Energy, Report No. 109, Council for Agricultural Science and Technology (July 1986).
5. Codex Alimentarius Commission, *Codex General Standard for Irradiated Foods and Recommended International Code of Practice of the Operation of Radiation Facilities Used for the Treatment of Food*, CAC/V04XV-Ed.1, CAC, Rome (1984).
6. C. Merritt, Jr. and I. A. Taub, Commonality and Predictability of Radiolytic Products in Irradiated Meats, in *Recent Advanced in Food Irradiation*, P. S. Elias and A. J. Cohen, eds., pp. 27-57, Elsevier Biomedical Press, Amsterdam, The Netherlands (1983).
7. U.S. Food and Drug Administration, Irradiation in the Production, Processing and Handling of Food, Final Rule, *Federal Register* **51(75)**:13375-13399 (April 18, 1986).
8. T. K. Murray, Nutritional Aspects of Food Irradiation, in *Recent Advances in Food Irradiation*, P. S. Elias and A. J. Cohen, eds., pp. 203-216, Elsevier Biomedical Press, Amsterdam, The Netherlands (1983).
9. M. L. Beyers, L. den Drijver, C. W. Holzapfel, J. G. Niemand, I. Pretorius, and H. J. Van der Linde, Chemical Consequences of Irradiation of Sub-Tropical Fruits, in *Recent Advances in Food Irradiation*, P. S. Elias and A. J. Cohen, eds., pp. 171-188, Elsevier Biomedical Press, Amsterdam, The Netherlands (1983).
10. P. Teufel, Microbiological Aspects of Food Irradiation, in *Recent Advances in Food Irradiation*, P. S. Elias and A. J. Cohen, eds., pp. 217-233, Elsevier Biomedical Press, Amsterdam, The Netherlands (1983).
11. Codex Alimentarius Commission, *The Microbial Safety of Irradiated Food*, Codex Alimentarius Commission, CX/FH83/9, Rome, Italy (1983).
12. R. Firstenberg-Eden, D. B. Rowley, and G. E. Shattuck, Factors Affecting Growth and Toxin Production by *Clostridium botulinum* Type E on Irradiated (0.3 Mrad) Chicken Skin, *Journal of Food Science*, **47**:867-870 (1982).
13. World Health Organization, Wholesomeness of Irradiated Food, Report of Joint FAO/IAEA/WHO Expert Committee, Technical Report Series 604, World Health Organization, Geneva, Switzerland (1977).
14. L. R. Pim, *Gamma Irradiation as a Means of Food Preservation in Canada*, Pollution Probe, Toronto (1983).
15. W. Grady, Immortal Chicken, *Harrowsmith* **68**:25-33 (July/August 1986).
16. L. Curtin, Economic Study of *Salmonella* Poisoning and Control Measures in Canada, Working Paper 11/84, Agriculture Canada (1984).
17. L. Sweet, Recent Trends in Human Salmonellosis; Canada and United States, in *Human Salmonellosis Principles of Investigation and Control*, F. White and L. Sweet, eds., pp. 11-16, Dalhousie University, Halifax, Nova Scotia (1985).
18. M. J. Blaser, *Campylobacter jejuni*, and Food, *Food Technology* **36(3)**:89-92 (March 1982).
19. B. M. Coles, K. Tanner, P. McMyne, T. Matheson, W. A. Black, *Campylobacter* Enteritis in British Columbia — A 30 Month Study, *Canadian Journal of Public Health* **76**:343-346 (1985).

20. A. H. W. Hauschild and F. L. Bryan, Estimate of Cases of Food- and Water-Borne Illness in Canada and the United States, *Journal of Food Protection* **43**:435-440 (1980).
21. T. Roberts, Microbial Pathogens in Raw Pork, Chicken and Beef: Benefit Estimates for Control Using Irradiation, *American Journal of Agricultural Economics*, 957-965 (1985).
22. H. Pivnick, S. Handzel, and H. Lior, Product Contamination with Salmonella, in *Proc. of the International Symposium on Salmonella and Prospects for Control*, D. A. Barnum, ed., pp. 139-155, University of Quelph, Canada,
23. J. Y. D'Aoust, U. T. Purvis, R. M. Coulter, and K. Weiss, *Salmonella* in Fresh Pork Liver and Chicken Liver in Canada: A 1979-80 Survey, *Canadian Institute for Food Science and Technology Journal* **18**:323-325 (1985).
24. B. J. Juven and M. Rogol, Incidence of *Campylobacter jejuni* and *Campylobacter coli* Serogroups in Chicken Processing Factory, *Journal of Food Protection* **49**:290-292 (1986).
25. C. E. Park, Z. K. Stankiewicz, J. Lovett, and J. Hunt, Incidence of *Campylobacter jejuni* in Fresh Eviscerated Whole Market Chickens, *Canadian Journal of Microbiology* **27**:841-842 (1981).
26. J. M. Wempe, C. A. Genigeorgis, T. B. Farver, and H. I. Yusufu, Prevalence of *Campylobacter jejuni* in Two California Chicken Processing Plants, *Applied Environmental Microbiology* **45**:355-359 (1983).
27. N. J. Stern, M. P. Hernandez, L. Blankenship, K. E. Deibel, S. Doores, M. P. Doyle, H. Ng, M. D. Pierson, J. N. Sofos, W. H. Sveum, and D. C. Westhoff, Prevalence and Distribution of *Campylobacter jejuni* and *Campylobacter coli* in Retail Meats, *Journal of Food Protection* **48**:595-599 (1985).
28. E. C. D. Todd, Poultry-Associated Food-Borne Disease—Its Occurrence, Source and Prevention, *Journal of Food Protection* **43**:129-139 (1980).
29. E. H. Kampelmacher, Irradiation for Control of *Salmonella* and Other Pathogens in Poultry and Fresh Meats, *Food Technology* **37(4)**:117-119 (April 1983).
30. A. Whyte and I. Burton, Perception of Risks in Canada, in *Living with Risk: Environmental Risk Management in Canada*, I. Burton, C. D. Fowle, and R. S. McCullough, eds., pp. 39-70, Institute for Environmental Studies, Toronto (1982).
31. F. Defesche, How Does the Consumer React to Irradiated Food, in *Food Irradiation Now*, pp. 98-106, Martinus Nijhoff/Dr. W. Junk Publishers, The Netherlands (1982).
32. Gallup, Marketability Testing of Irradiated Fish and Seafood, Final Report, Canadian Gallup Poll Ltd., Toronto (1984).
33. C. M. Bruhn, H. G. Schutz, and R. Sommer, Attitude Change Toward Irradiation Among Conventional and Alternative Consumers, *Food Technology* **40(1)**:86-90 (January 1986).
34. H. J. Van der Linde, Commercial Experience with Irradiated Foods in South Africa, in *Proc. of the National Symposium on Food Irradiation*, O. J. McCarthy and C. G. Robertson, eds., pp. 173-180, Massey University (1984).
35. D. Phillips, Irradiated Produce Hits Miami Market, *The Packer*, p. 1A (September 13, 1986).
36. J. H. Silliker, Status of *Salmonella* — Ten Years Later, *Journal of Food Protection* **43**:307 (1980).
37. R. M. Morrison, Economies of Scale in Single-Purpose Food Irradiators, in *Proc. of a joint IAEA/FAO Symposium on Food Irradiation Processing*, Washington, D.C., pp. 407-428, IAEA, Vienna (1985).
38. D. A. A. Mossel, Irradiation: An Effective Mode of Processing Food for Safety, in *Proc. of an International Symposium on Food Irradiation Processing*, pp. 251-279, International Atomic Energy Agency, Vienna (1985).
39. E. Van Ravenswaay and M. Smith, Food Contamination: Consumer Reactions and Producer Losses, *National Food Review* **33**:14-16 (Spring 1986).
40. W. G. Savage, *Milk and the Public Health*, McMillan and Co. Ltd., London (1912).

Exposure Assessment for Sterilized Medical Devices: Implications for Microbiological Risk Analysis

Herbert C. Hammond
Office of Health Policy
Washington, DC

William C. Dierksheide
Food & Drug Administration
Rockville, MD

ABSTRACT

For many decades the conventional wisdom was that a "sterile" medical device necessarily posed no risk of infection. However, in recent years practitioners in infection control have recognized that a "sterile" device could serve as a "conduit" for microorganisms from within clinical environments. The NSF risk assessment framework applied to microbiological risks promises an effective method for advancing risk management decisions with respect to medical devices, because the framework requires all sources of an identified risk to be considered. The microbiological risk associated with medical devices discussed in this paper is nosocomial infection. Risk sources are medical devices as "conduits" of microorganisms and as "fomites," i.e., contaminated with microorganisms outside the clinical environment. The latter risk is preventable by sterilization, the former is not. The adequacy of sterilization to minimize fomite risk can be assessed by estimating the probability of surviving microorganisms. The probability of survivors is equivalent to the exposure to fomite risk using the NSF framework. Exposure to fomite risk can be estimated by two methods: the kinetics of the sterilization process and post hoc bioassay for viable microorganisms. Kinetic estimates are extrapolations predicated on the assumption that actual sterilization parameters exceed test conditions. Survivor probabilities of less than 10^{-6} are common. Post hoc bioassay results are confounded by the phenomenon of adventitious contamination estimated to occur between 1 in every 250 and 1 in every 9000 test manipulations. Experience with commercially sterilized medical devices revealed very low fomite exposure. Laboratory bioassay of 82 samples of a variety of 29 commercially sterilized medical devices identified only three non-sterile lots. One implication of the risk assessment framework is that research elucidating the conduit phenomenon is necessary to minimize infection risk. The risk of infection appears to be more sensitive to conduit exposure than to fomite exposure.

KEYWORDS: Nosocomial infection, sterilization, microbiological risk, infection risk

New Risks
Edited by L. A. Cox, Jr., and P. F. Ricci
Plenum Press, New York, 1990

INTRODUCTION

Post World War II technology has provided materials and sterilization methods that make possible the mass marketing of sterilized disposable medical devices. With this technology it is now possible to provide a wide variety of medical devices that enter the clinical environment virtually free of microorganisms. Nevertheless, nosocomial infection continues to be a serious problem. Data assembled by the Centers for Disease Control indicate that 3.4 nosocomial infections occur for every 100 patients discharged. Furthermore, there was no significant change in either infection rate or associated mortality in a study covering 1980-1982.[1] The problem of nosocomial infection is therefore large, between one and two million infections per year, and difficult to eradicate. The five stage National Academy of Science (NAS) framework[2] provides an analytic model for the problem of nosocomial infection. The authors describe how microbiological data, originally developed to minimize the microbial exposure associated with medical devices, fits into the NAS framework. Furthermore, the authors discuss these results in the context of the NAS framework as an explicit analytic model for further research related to managing the risk of nosocomial infection.

BACKGROUND

The principal microbiological risk associated with medical devices is the risk of infection. Devices can be the source of infection risk in two distinct ways. First, a device may transport an infectious microbe from the fabrication environment to a patient. This source of infection will be termed "fomite risk" in this paper. The data used in this paper are related primarily to fomite risk. Second, even though the device may be free of microbes when first brought in contact with the patient, the device subsequently may act as a conduit for microorganism from the clinical environment or a patient's own microflora. This infection phenomenon will be labeled "conduit risk" in this paper. Conduit risk is a function of the properties of the device in the clinical environment and the physiological response of tissues exposed to the device.

THE NAS FRAMEWORK

Given the preceding background, the five stages of the NAS framework are described below for the microbiological risks of medical devices:

- *Risk Identification*: Nosocomial infection.
- *Risk-Source Characterization*: Medical devices as conduits and/or fomites.
- *Exposure Assessment*: For low populations of microorganisms (approximately 1.0 per device) exposure is the Poisson probability of one or more organisms. For higher populations, expected exposure is the geometric mean (assuming a log normal distribution).
- *Dose-Response Assessment*: The probability of infection given exposure is based on the tissue exposed, the expected population, and class of organisms, e.g., saprophytic spores, non-sporeforming saprophytic microflora, and opportunistic microflora associated with clinical environments.
- *Risk Estimation*: The synthesis of exposure with dose-response of the form:

  $P(\text{infection}) = p(\text{fomite infection}) + p(\text{conduit infection}) - (p(\text{conduit infection}) * p(\text{fomite infection}))$

where

p(fomite infection) = p(fomite exposure) * p(infection | expected fomite exposure), and

p(conduit infection) = p(conduit exposure) * p(infection | expected conduit exposure)

MICROBIOLOGICAL TESTING AND THE NAS FRAMEWORK

Microbiological testing consists of methods used to estimate the population of viable microorganisms surviving or likely to survive after a specified antimicrobial treatment. The data obtained by these methods are estimates of exposure since the microorganisms may or may not be capable of causing infections.

THE CONCEPT OF STERILITY

A truly sterile device presents an exposure of zero viable organisms and no fomite risk. However, the concept of sterility is problematic. It is possible to demonstrate viable microorganisms in the laboratory, and therefore, non-sterility. However, failure to demonstrate organisms does not prove they are not present.

Devices can be sterilized to some statistically inferred degree of confidence. There are two methods for estimating this degree of confidence. One method is using the kinetics of microbial inactivation; the second method is the post hoc bioassay of a sample of the sterilized lot. The attributes and implications of both methods will be discussed.

THE CONCEPT OF BIOBURDEN

The probability of microorganisms occurring on devices is a function of the populations that occur in the fabrication environment. There are, as Russell[3] observed, ubiquitous and varied microorganisms that can populate devices. She described four major sources of microbial populations for medical devices. First, people bring bacteria, viruses, and other microbes into the production areas. Second, airborne dust particles contaminated with soil organisms including bacteria, yeasts and molds, enter the production area. Third, water contains both soil and water organisms, which include bacteria, viruses, and algae. Fourth, products that are derived in part from plant or animal materials will carry microorganisms associated with those plants and animals. Russell adds that there are two crucial factors related to the level of microbial population: first, the ability of the populating organisms to survive the production process, and second, the ability of these organisms to survive and multiply on the product.

The microbial population acquired during the manufacture of a device is often referred to as the "bioburden." In addition to providing data that enables hospitals and manufacturers to optimize sterilization, bioburden can be used for risk analysis as well. For devices composed of materials that do not allow microbial populations to multiply, the bioburden would be an upper estimate of patient exposure to fomite risk. Since many devices are composed of synthetic materials that will not support microbial proliferation, bioburden testing could become an integral component of exposure assessments.

The authors field tested the concept of bioburden assays.[4] A sample of an injection molded plastic device was collected after normal production and packaging but before terminal sterilization. The results were a Most Probable Number (MPN) of 1.049

organisms per device for 20 units tested in an aerobic growth medium, and an MPN of 0.1054 organisms per device for 20 units tested in an anaerobic growth medium (see note 1).

THE CONCEPT OF KINETIC EXTRAPOLATION

Based on a known or assumed bioburden, the probability of surviving microorganisms can be estimated by extrapolation. This extrapolation is based on the well established principle that microbial death whether caused by heat, radiation, or chemicals follows a first order kinetic model. For a uniform population of cells the expected number of survivors N is given by

$$N = N_0 \, e^{-kd}$$

where N_0 is the bioburden (assuming a bioburden of homogeneous resistance), d is the dose of sterilizing agent, and k is a constant determined experimentally. Values of less than one are estimates of the probability of survivors. Thus when $N = 1.0 * 10^{-6}$, the probability of a survivor is one in a million. This level, 10^{-6}, has been used widely.

A common methodology to estimate k is to begin with devices with a known total microbial population and expose samples to several sub-terminal doses of sterilizing agent. However, this tends to underestimate the amount of time required to achieve 10^{-6} expected survivors because microbial populations are not homogeneously resistant to a given sterilizing agent. The authors examined published data on radiation sterilization to produce the simulated curve shown in Fig. 1.[5] The total population method of analysis (total population versus dose to zero survivors) produces an estimate of k = 21. Ignoring the initial population and using the MPN of survivors versus time by regression analysis produced an estimate of k = 12.06. The simulation in Fig. 1 was produced from the following:

$$N = 2{,}000 \, e^{-21d} + 15 \, e^{-12d}$$

Instead of a perfect log linear decay, the results appear nonlinear. However, the results are simply the composite of two log linear decays: one, a relative high population of moderate resistance, and two, a small population of high resistance.

This simulation supports the validity of using regression to estimate expected survivor probabilities. However, these results also suggest that microbial resistance can be more important than initial population in determining survivor probabilities. However, the hardiest spores are produced by saprophytic soil bacteria that are not recognized as opportunistic pathogens. Thus an estimated survivor probability using highly resistant spores of soil organisms is a conservative estimate of the probability of exposure to organisms capable of causing infection.

POST HOC SAMPLING

The second method is to try to demonstrate sterility post hoc. For example, a forty unit sample can be split into two 20 unit groups and tested in two microbial growth media: a medium such as fluid thioglycollate that supports anaerobic organisms, and another medium such as soybean casein digest that will support aerobic microorganisms. Some bacteria will grow in both media. For those organisms capable of growing in both media, the test has an effective sample size of 40 units. However, many surviving organisms will grow in only one or the other medium and thus the effective sample size would be 20 units.

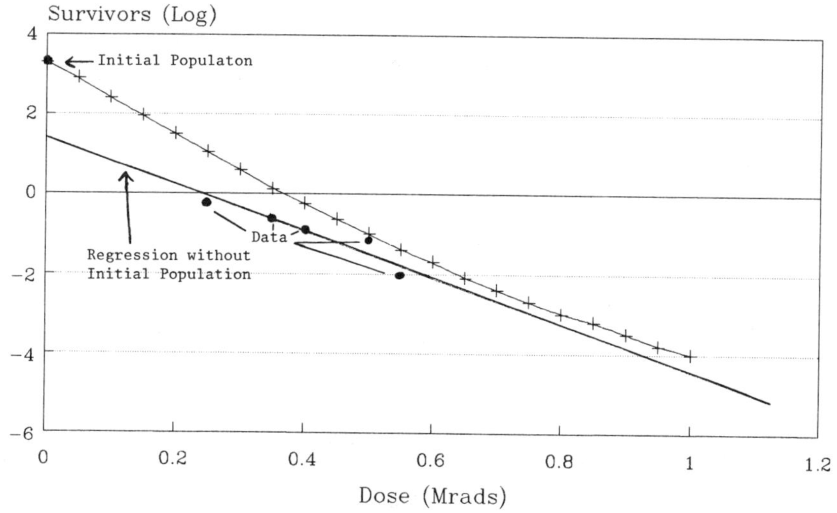

Fig. 1. Log linear decay of combined populations.

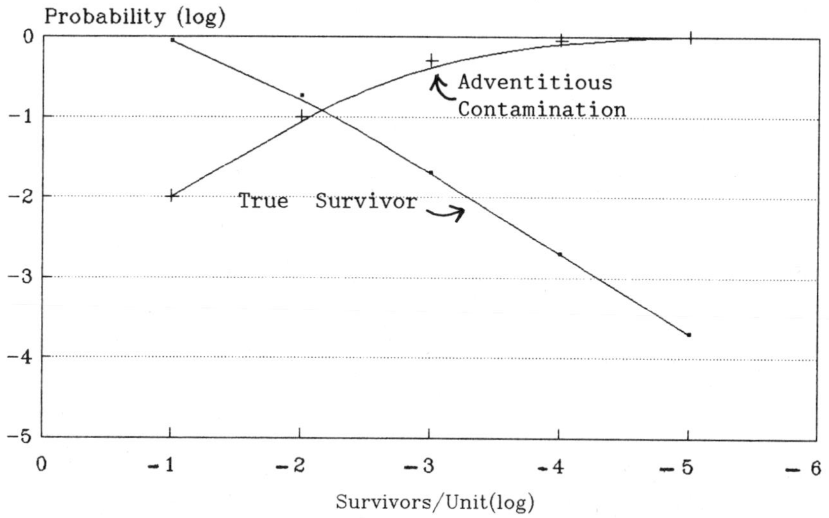

Fig. 2. Probability of survivors in a 20-unit sample, probability that growth in test medium is adventitious contamination, both as a function of survivor rate.

Given that the desired survivor level is zero and that there is a practical need for a modest sample size, post hoc sampling is inherently prone to a type I sampling error. The probability that a 20 unit sample will contain survivors is shown in Fig. 2 as a function of survivor rates. While there can be no true type II sampling error, microorganisms from the testing environment can turn up as growth in the test medium. This type of growth is commonly referred to as adventitious contamination. Since adventitious growth/contamination has the same implication as a type II error, i.e., rejecting a satisfactory lot, the authors have elected to use the term "biological" type II error to describe this phenomenon.

If the probability of adventitious contamination for a given test system can be estimated, and if, as already suggested, the probability of survivors occurring in a given sterilization procedure can be estimated, the probability of a "biological" type II error given growth in a sample can be calculated from the following equation:

$$P_{II} = (p(\text{cntm})/p(\text{srv}) + p(\text{cntm}) - p(\text{both}))^n$$

where

$p(\text{cntm})$ = probability of adventitious contamination

$p(\text{srv})$ = probability of survivors

$p(\text{both}) = p(\text{srv}) \times p(\text{cntm})$

n = number of units with observed growth

An authoritative estimate of the probability of adventitious contamination is 10^{-3}.[6] This estimate is used to calculate the probability of "biological" type II error as a function of the probability of survivors assuming 1 and only 1 unit shows growth. These results are also shown in Fig. 2. As the survivor rate declines, the probability that a sample might actually contain true survivors declines, and the probability that any growth was caused by adventitious contaminants increases.

As well as being limited by sample size, the utility of bioassay is also a function of the rate of adventitious contamination for a test system. It is, therefore, highly desirable to examine this issue quantitatively. The survey of intravenous catheters[7] provides data that can be used to generate a quantitative estimate of this phenomenon.

The catheter survey consisted of the bioassay for sterility of 1,916 intravenous catheters. To remove the catheters from the packaging involved a similar amount of manipulation. Comparably trained expert microbiologists performed the tests using the same methods in three specially equipped Food and Drug Administration laboratories. No organisms were found in the 958 catheters tested in soybean casein medium, and one catheter out of 958 tested produced growth in fluid thioglycollate medium. The available evidence was consistent with the explanation that the observed growth resulted from adventitious contamination rather than true survivors of sterilization. Assuming that one adventitious contamination was observed in 958 manipulations under more or less constant conditions, it is possible to apply inferential statistics to test hypotheses about the frequency of adventitious contamination.

The estimate of maximum likelihood for adventitious contamination is 1/958, or approximately 10^{-3}. The question is, however, whether this experience could be the result of sampling error. If a decision maker were to assert that laboratories were really very much worse, e.g., 1/500, would these results of 1/958 be a basis for rejecting this worse condition hypothesis? The issue also arises with a better condition hypothesis.

If the decision maker were convinced that the true rate was high, e.g., 1/500 what is the probability of getting as few as 1 growth in the 958 observed trials? The probability of as few as one can be calculated from the binomial distribution:

$$P = \sum_{r=0}^{1} \frac{n!\, p^r}{r!(n-r)!} q^{n-r},$$

where n=958 and p=0.002.

The solution, $P(r = 1 , r = 0)$, is $P = 0.43$. A decision maker should not reject a hypothesis of 1/500 at $P = 0.43$. A further question is how high the hypothesized contamination rate must be to be rejected at $P = 0.05$ or $P = 0.1$. The solution was found by the Newton method. The results are:

$P(1 , 0) = 0.05$ when p= 4.942E-03 (a rate of 1/202), and
$P(1 , 0) = 0.1$ when p= 4.054E-03 (a rate of 1/247).

The decision maker might also have been convinced that laboratories perform at consistently better levels than one contamination per one thousand manipulations. Then, how low must a hypothesized contamination rate be to be rejected? The results are:

$P(> 1) = 0.05$ when p= 5.305E-05 (a rate of 1/18,851), and
$P(> 1) = 0.1$ when p= 1.100E-04 (a rate of 1/9093).

In summary, the most likely rate of adventitious contamination that was present during the catheter study was roughly 1/1000. However, a worse case hypothesis of roughly 1/250 could not be rejected even at the $P = 0.1$ level and a better case hypothesis of 1/9000 could not be rejected either. To the extent that the catheter study represents post hoc sterility testing, these levels of 1/250 or 1/9000 can be viewed as worse case or better case estimates of the frequency of adventitious contamination.

The utility of bioassay data is limited where kinetic data are available. Table 1 contains a set of discrete calculations of "biological" type II error for adventitious contamination rates of 1/250 and 1/9000, three survivor rates from 10^{-4} to 10^{-6}, and as many as 5 units positive for growth. If a decision maker has evidence that the probability of true survivors is 10^{-6}; e.g., a conservative estimate based on linear extrapolation assuming both high bioburden and high resistance, Table 1 shows that as many as 5 units could show growth in a bioassay and the probability of a "biological" type II error would exceed 0.95 even using the better case rate of 1/9000 for adventitious contamination.

Table 1. Biological Type II: Error with Growth in 5 Units

Survivor Probability	*/250	*1/9000
0.0001	0.884	0.04
0.00001	0.988	0.650
0.000001	0.999	0.956

*Adventitious contamination.

EXPERIENCE WITH STERILIZED DEVICES

The results for 82 samples of commercially sterilized medical devices have been previously collated.[4] These samples included 29 types of devices tested over an 18 month period in Food and Drug Administration Laboratories. The devices covered included such items as surgeon's and exam gloves, needles, drapes, sutures, catheters, and syringes. Seventy-one of these samples produced no growth. Four samples exhibited one to four units with growth; however, four follow-up samples to these presumed "biological" type II errors produced no growth. Only three samples produced growth attributed to survivors. The results for the three samples with survivors is shown in Table 2 as a probability of

Table 2. Exposure from Inadequate Sterilization

Device	Number of Organisms* 0	1	2	3	4
Compress Bandage	.78	.19	.02	.00	.00
Muslin Sheet	.28	.36	.23	.01	.03
Muslin Sheet	.47	.36	.13	.03	.01

*Poisson probability.

patient exposure for levels of exposure assuming the survivor population follows a Poisson distribution.

The experience with devices reported in this paper suggests very low fomite exposure from commercially sterilized devices. The highest incidence of survivors observed in Table 2 gives a probability of exposure to one or more organisms of 0.722. If the probability of infection given exposure to the low number of saprophytic spores surviving on the sheet were as high as 0.01, the fomite risk of infection would be 0.00722. For devices where the probability of survivors can be inferred from kinetic data, e.g., 10^{-6}, the fomite risk of infection would be of the order of 10^{-8}.

IMPLICATIONS FOR FUTURE RESEARCH

The conventional wisdom is that a sterilization process should be chosen on the basis of its lethality to microorganisms; i.e., the lower the probability of survivors the lower the probability of infection. The profound advantage of the discipline imposed on risk analysis by the systematic NAS framework is that the assumptions that lead to conventional wisdom can be examined. The following analysis tests whether the assumption that increasing the lethality of sterilization necessarily reduces microbiological risk, given that sterilization itself affects devices in ways that may increase conduit risk (see note 2).

Table 3 contains a discrete analysis of total infection risk for two hypothetical sterilization processes. In this analysis the probability of fomite exposure is assumed to be either 10^{-6}, which can be established by kinetics, or 0.9, which is the upper limit based on the bioburden data reported earlier. The probability (infection | fomite exposure) is assumed to be lower than the probability (infection | conduit exposure), i.e., saprophytic spores surviving fabrication are less virulent than organisms in the clinical environment including the patient's own microflora. One order of magnitude of difference is assumed. Conduit exposure is assumed to be 0.1 given an initial assumption that the probability (infection | conduit exposure) is also 0.1. The combined result is a conservative estimate of conduit infection risk being equal to 0.01. If the probability (infection | conduit exposure) is raised to 0.2, a less conservative risk of 0.02 results.

In this hypothetical analysis, the sterilization process for trial II is almost six orders of magnitude more lethal than trial I. Nevertheless, the reduction in fomite risk does not result in reduced infection risk. This is the logical conclusion using the risk assessment framework. However, the microbiological testing discussed in this paper does not elucidate the conduit phenomenon; the data only speak to fomite risks. The need for research to

Table 3. Infection Risk (P)

P(Conduit, Fomite)	(Conduit)	(Fomite)
Trial I:	If p=0.1	If p=0.9
	& q=0.1	& q=0.01
P=0.0189	P=0.01	P=0.009
Trial II:	If p=0.1	If $p=10^{-6}$
	& q=0.2	& q=0.01
P=0.0200	P=0.02	$P=10^{-8}$
where p= exposure probability	and q= infection if exposure	

understand the parameters of the conduit phenomenon is clear. Even though conduit exposure and infection are low-frequency stochastic phenomena, the authors have shown with the case of sterilized devices that such phenomena are not intractable. The models applied in this paper to quantify bioburden, survivor probability, and the "biological" type II error suggest strongly that conduit phenomena can also be quantified.

The proper assessment of risk clearly needs to be multi-dimensional. The risk assessment framework can handle multi-dimensional biological risks and therefore extends the scientific understanding of the microbiological risks.

NOTE 1

Microorganisms are found in nature in pairs, short chains or even clumps. The authors do not use the term "organism" to necessarily mean a single cell. "Organism" in this paper is the minimum quantitative unit (one or more connected cells) that can produce growth.

The Most Probable Number (MPN) can be estimated in tests where the entire device is placed in a growth medium, provided the population of organisms is low enough that some individual units show no growth. The number showing no growth divided by the total number of units tested provides an estimate of the probability (P) of the zero term (r=0) of the Poisson distribution of the form:

$$P(r,p) = \frac{p^r}{r!} e^{-p}$$

The MPN is the expected value (p).

NOTE 2

There are several plausible reasons to assume this phenomenon occurs. For example, where ethylene oxide (ETO) is used as a sterilant the possibility exists that the residual ETO will be a sufficient irritant to mucosal tissues to increase the probability of infection.

Another example is the effect of radiation on plastic when used at sterilizing doses. Not only does the possibility exist of peroxides or other irritants, but the tendency of plastic to become brittle could significantly increase a device's physical properties as a conduit.

REFERENCES

1. Food and Drug Administration, Statement before the Joint Economic Committee, U.S. House of Representatives, September 25, 1986.
2. National Science Foundation, Risk Assessment and Risk Assessment Methods: the State of the Art, Washington, D.C., April 1985.
3. R. L. Russell, Environmental Microbiology, *Proceedings of the Educational Conference: Sterile Products*, Biomedical Division, American Society for Quality Control, pp. 16-18, 1980.
4. H. C. Hammond, The Device Sterility Program: Assessment of the Current State of the Art, *Proceedings of the Educational Conference: Sterile Products*, Biomedical Division, American Society for Quality Control, pp. 5-9, 1980.
5. J. D. M. White, *Biological Control of Industrial Gamma Radiation Sterilization, Industrial Sterilization*, pp. 101-116, Duke University Press, Durham, North Carolina, 1972.
6. I. J. Pflug, *Heat Sterilization, Industrial Sterilization*, pp. 239-282, Duke University Press, Durham, North Carolina, 1972.
7. H. C. Hammond and W. C. Dierksheide, Evaluation of Sterility Test Data from a Medical Device Product Survey, American Society for Microbiology Annual Meeting (Abstract), 1976.

Risks in New Technologies — Controlling Toxic and Hazardous Gas Releases During Photovoltaic Cell Manufacture

P. D. Moskowitz, V. M. Fthenakis, L. D. Hamilton
P. Kalb, and J. C. Lee
Brookhaven National Laboratory
Upton, NY

ABSTRACT

A large variety of materials may be used in the manufacture of photovoltaic devices; some may be released as by-products of normal or abnormal plant operations. These may adversely affect occupational and public health. This paper describes a logical and consistent framework for reviewing potential hazards associated with such releases and for their management. The most important hazards relate to accidental releases of toxic or flammable gases used in large quantity in thin-film photovoltaic cell manufacture. This study suggests that the most efficient strategies for reducing risks are to choose material and process options which inherently possess small risks (e.g., choice of processes which do not use hazardous gases) and to prevent accident initiating events (e.g., employee training). If accidents do occur, then strategies to prevent catastrophic releases (e.g., flow restricting valves), to control releases (e.g., flare towers and wet scrubbers), and to prevent public exposures to toxic agents (e.g., water curtains, siting and emergency evacuation) become important. Prevention and mitigation of consequences (e.g., medical treatment), form the final defensive barriers. Evaluations of these options prior to plant construction can permit plant engineers to design systems which are technologically competitive and present limited risks to health.

KEYWORDS: Risks, photovoltaic cells, toxic gases, hazardous gases

INTRODUCTION

Photovoltaic cells are solid state devices that convert sunlight directly into electricity. These devices, unlike conventional energy systems do not produce pollutants during their operation. Nevertheless, they are not without risk because a large variety of toxic or hazardous materials may be used in their manufacture; some may be released as by-products of normal or abnormal plant operations. These may adversely affect occupational or public health. In recognition of these potential hazards, efforts to define potentially important hazards and to devise logical and consistent frameworks for their management are being developed at Brookhaven National Laboratory (BNL) for the U.S. Department of Energy National Photovoltaics Program. This paper summarizes some lessons from these efforts.

WHAT ARE THE POTENTIAL HAZARDS?

Many different cell materials and production processes are used to fabricate photovoltaic cells in commercial production facilities and in research laboratories. These activities require a great variety of chemicals used mainly as process feedstocks, dopants, solvents, or transport agents. Because of relatively low limits of process efficiencies (e.g., 30% deposition efficiency) and the often complex stoichiometry associated with these processes, significant quantities of various chemicals are also produced as by-products. In Table 1, some gases commonly present in storage, distribution, and effluent systems in photovoltaic cell manufacture are listed. Most of these gases have been identified by the U.S. Environmental Protection Agency (EPA) as "Acutely Toxic Chemicals" (U.S. EPA 1985). These materials are capable of producing serious health effects that are manifested either immediately or shortly after an acute exposure from an accidental air release. Table 1 also presents comparisons between projected gas inventories at hypothetical production facilities and guidelines developed by the EPA/Chemical Emergency Preparedness Program and the United Nations International Labour Office (1985) which represent estimated minimum quantities capable of causing extensive damage to property or to persons within or outside the plant boundary. As shown, onsite inventories of some hazardous or toxic gases exceed the listed threshold quantities.

In general, routine release of small quantities of these materials should not present acute or unacceptable risks to health (Table 2). Accidental release of these same materials, in larger quantity, however, can present significant risks to health (Table 3). Consequently, strategies to prevent such events are of great importance.

WHAT ARE THE MANAGEMENT OPTIONS?

Toxic or hazardous gases accidentally released by photovoltaic manufacturing facilities can adversely affect public health. Since there are no comprehensive federal or state regulations to manage this type of problem, the burden falls on industry to identify, evaluate, and implement strategies to reduce or eliminate such hazards from abnormal plant operations. In Fig. 1, a simplified hazard management approach for such problems is presented (Bowonder *et al.*, 1985; Moskowitz *et al.*, 1986). It consists of six major hazard management strategies: Choice of Inherently Safer Technology, Prevent Events, Prevent Releases, Prevent Exposures, Prevent Consequences, and Mitigate Consequences.

It is obvious that the most efficient strategy to reduce hazards is simply to choose technologies which do not require use of large quantities of toxic or hazardous gases. In the photovoltaics industry, process engineers might elect to choose a-Si alternatives, instead of GaAs or Zn_3P_2, because of the inherent toxicity of arsenic or phosphorus compounds and because of the large volumes of feedstock gas required. The benefits of this tradeoff are not clear, however, because the substitution of a highly toxic material (e.g., phosphine) by a highly explosive material (i.e., silane) sustains the occupational risks, but reduces the public risks. Similarly, processes that use solids (e.g., closed space vapor transport) rather then gases (e.g., metallorganic chemical vapor deposition for Zn_3P_2 thin-film deposition), reduce potential hazards from accidents. This option, however, is not always available. Manufacturers of a-Si cells, for example, must use a silane compound. Alternative silicon-based materials which are not pyrophoric are not available. In these situations, the next series of management options must be considered.

Once a specific process and material alternative is chosen, safety managers efforts must focus on strategies to prevent accident initiating events and failure of control systems. In this step, there are again obvious measures which can be implemented: worker training, operating procedures and guidelines, maintenance, inspection and testing. Other measures (e.g., system design and integrity, safeguards against process deviations, etc.), however,

Table 1. A Comparison of On-Site Inventories with EPA and ILO Guidelines

Gas	Annual feedstock (kg)	Average weekly stock (kg)	ILO threshold quantity (kg)[a]	EPA threshold quantity (kg)[b]	Ratio (weekly stock/ threshold quantity)[c]
Arsine	23660	480	10	10	48
	6520	130	10	10	13
Chlorosilanes	4754	95	NA	5000	<1
Diborane	<1	0.02	NA	30	<1
Hydrochloric acid	3850	80	250000	80	1
Hydrogen	10000	200	50000	50000	<<1
Hydrogen selenide	1600	32	10	4	8
Hydrogen sulphide	600	12	50000	230	<<1
Phosphine	4000	80	100	150	0.8
	2	0.04	100	150	<<1
Silane	1500	51	NA	-	-
	600	12	NA	-	-

[a]Quantity beyond which major-accident hazard may be created as defined by ILO.

[b]Quantity identified by the USEPA which could present an acutely toxic exposure beyond 500m.

[c]Ratio presented is the maximum of either the ILO or EPA guidelines.

[d]Smaller quantities of hydrogen may be required for other process alternatives, for example, a-Si glow discharge, CdTe-CSVT, CdTe-CVD, CdTe-ED, and CVD.

may be more difficult to identify. Furthermore, their effectiveness in abnormal operating situations may not be clearly known. Fault and event tree analyses, used in the nuclear and aerospace industries, provide important theoretical tools for the identification and quantification of accidental events and potential safety system failures. Recently, we applied such techniques to a proposed manufacturing facility to examine the potential effectiveness of different safety systems in preventing some top events of interest (e.g., a silane explosion). This effort demonstrated the viability of this approach while at the same time demonstrating some of its weaknesses. Of critical significance is the lack of failure mode and rate data for components used in the proposed facility. Because many of these processes are still in experimental design and have not been scaled to commercial-sized facilities, exact specifications of the equipment, support systems and controls are often not clearly defined. As a result, application of this technique requires the extrapolation of failure mode and rate data from other systems. These extrapolations introduce uncertainty into the hazard analysis.

Table 2. Estimated Ground-Level Concentrations of Atmospheric Pollutants During Normal Operation of a 10-MWp a-Si Photovoltaic Cell Plant

Pollutant	TLV mg/m^3	"Safe" % of TLV	Emission Rate (lb/hr)[a]	Maximum Concentration (mg/m^3)[b]	Maximum Concentration % of TLV
SiH_4	7.	2.	7×10^{-3}	3×10^{-5}	4×10^{-4}
SiF_4	2.5[c]	0.33	2×10^{-3}	1×10^{-4}	4×10^{-3}
B_2H_6	0.1	0.33	1×10^{-4}	4×10^{-7}	4×10^{-4}
PH_3	0.4	0.33	2×10^{-4}	8×10^{-7}	2×10^{-4}

[a]Assuming a scrubber's efficiency of 98% for SiH_4 and SiF_4, 50% for B_2H_6 and PH_3.

[b]Estimated by the formula C = (emission rate) * 4.22/(emission height)$^{2.16}$; where emission height is assumed to be 25 ft above ground.

[c]No TLV is established for SiF_4, although the gas is considered highly toxic and it can form HF in the air. Here, the TLV for HF is assumed.

Table 3. Estimated Maximum Ground-Level Concentrations of Arsine and Phosphine from Accidental Instantaneous Releases of Weekly On-Site Gas Inventories at 10 MWp Photovoltaic Cell Plant

Pollutant/ Source	Quantity Released (kg)	Weather Pattern	Wind Speed (m/s)	Distance Downwind (km)[a] IDLH[b] (1 sec.)	Distance Downwind (km)[a] IDLH[b] (30 min.)
		Arsine			
GaAs Halide CVD	480	Unstable	8.5	1.8	0.19
		Stable	2.5	44	13
GaAs MOCVD	130	Unstable	8.5	1.1	0.085
		Stable	2.5	26	5.2
		Phosphine			
Zn_3P_2 MOCVD	80	Unstable	8.5	0.36	0.014
		Stable	2.5	7.6	0.73

[a]Ground-level concentrations are based on a dispersion algorithm developed by the U.S. Environmental Protection Agency. As discussed in the text, these estimates are very conservative. The column labeled "1 sec." represents the maximum distance downwind from the release at which an instantaneous maximum concentration equal to the IDLH level (see footnote b) would occur. Conversely, the column labeled "30 min." provides a measure of the downwind distance at which an individual would be exposed to a concentration equal to the IDLH level for 30 minutes.

[b]The "Immediately Dangerous to Life or Health" IDLH levels for arsine and phosphine are 0.024 g/m^3 and 0.267 g/m^3, respectively. The IDLH concentration represents the maximum level from which one could escape within 30 minutes without an escape-impairing symptoms or any irreversible health effects.

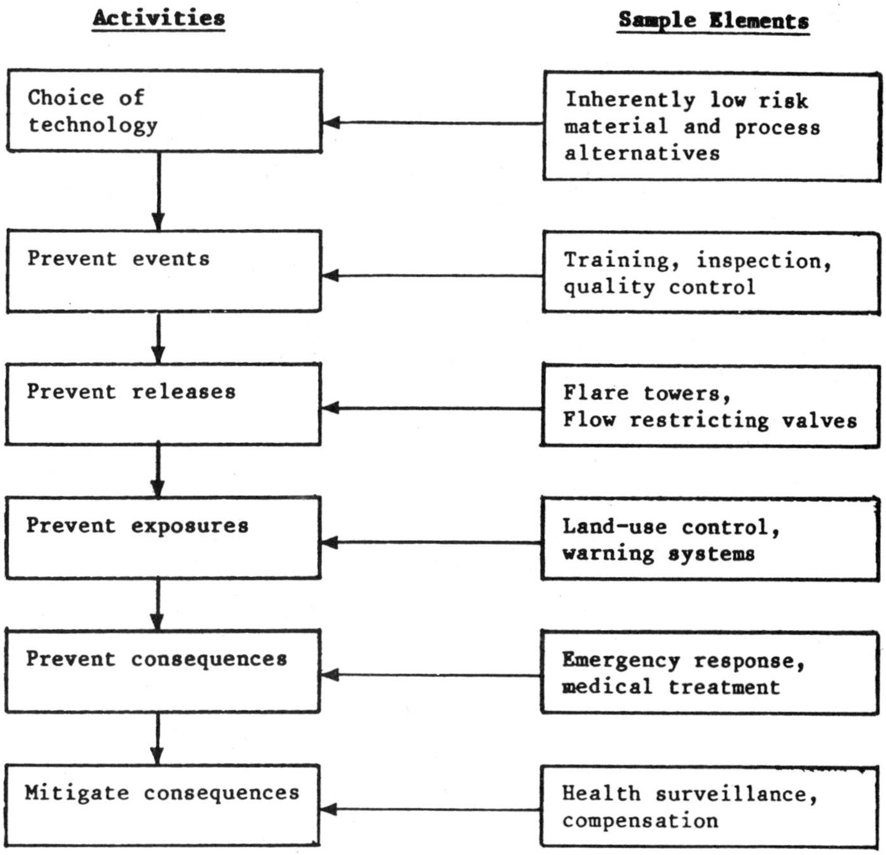

Fig. 1. A hierarchical approach for managing accidental gas releases from photovoltaic manufacturing facilities.

The next logical step in hazard management is to prevent catastrophic releases to the environment if accidents do occur. Again, there are simple administrative (e.g., limited inventory, emergency response training) and engineering (e.g., flow restricting valves) control options. More efficient control strategies (e.g., containment and neutralization), however, have not received great attention. We recently completed a study of methods for containing hydrogen selenide releases in the event of an accident (Fthenakis *et al.*, 1988). Two different options were examined: scrubbing and carbon adsorption, and containment-scrubbing and carbon adsorption. These options appear to present viable approaches for controlling an accidental release, but, because of the paucity of empirical data, they were based on many assumptions and extrapolations. Laboratory and pilot-scale experiments are needed to determine how well these systems would actually work in the event of an accident.

If there are releases to the environment, in spite of all previous strategies, plant safety managers must next consider strategies to prevent exposures to the public. This includes many "soft" options ranging from the remote siting of gas storage systems, to the application of exclusion zones adjacent to plant boundaries, to the implementation of early warning systems and emergency response plans. In theory, these are simply accomplished, but in practice there are many difficulties with this approach. Successful application of these strategies will require planning and liaison with outside public authorities, and establishing emergency control centers, site action plans and rescue systems and plant shut-down procedures.

Prevention of consequences forms the final defensive barrier. Plant managers should provide local health agencies and hospitals with information about materials which could be released to off-site environments and appropriate medical treatments required to mitigate such exposures. Subsequently, appropriate medical personnel (e.g., paramedics and hospital emergency room staff) should be trained to respond to such incidents; specialized medical equipment and antidotes may be required. In real emergencies, experience has shown that there is often much confusion. Hence, it is essential that such plans be regularly rehearsed and practiced under simulated emergency conditions in order to test the response of personnel, increase their base of experience, and evaluate the potential effectiveness of equipment and other response measures.

In the aftermath of an incident, industry may also need to mitigate consequences by implementing relief and health surveillance programs and possibly by providing compensation to aggrieved individuals.

DISCUSSION AND CONCLUSIONS

Gases used in the manufacture of photovoltaic cells may present hazards to health and welfare. This is particularly true for accidents at these facilities. In general, the most efficient approach for managing such hazards is to increase the inherent safety of the facilities by choosing manufacturing processes or cell types which use less hazardous chemicals. Once a particular process or chemical is selected, the best strategy to reduce hazards is to prevent accidental events through administrative and engineering control options. Should such an event occur, although control options to reduce the created hazard exist, their effectiveness is uncertain due to lack of testing and the human factors involved. Since many of these options are not yet directly required by any public agency, most burdens to protect health and welfare fall directly on industry.

REFERENCES

Bowonder, B., Kasperson, J. X., and Kasperson, R. E., 1985, Avoiding Future Bhopals, *Environment* **27(7)**:6-37.

Fthenakis, V. M., Moskowitz, P. D., and Sproull, R. D., 1988, Control of Accidental Releases of Hydrogen Selenide and Hydrogen Sulphide in the Manufacture of Photovoltaic Cells: A Feasibility Study, *J. Loss Prev. Process Ind.* **1**:206-212.

Moskowitz, P. D., Fthenakis, V. M., Hamilton, L. D., and Lee, J. C., Public Health Issues in Photovoltaic Energy Systems: An Overview of Concerns, *Solar Cell*, in press.

United Nations International Labour Office, 1985, Working Paper on Control of Major Hazards in Industry and Prevention of Major Accidents, Rep. ILO MHC/1985/1, Geneva, Switzerland.

U.S. Environmental Protection Agency, 1985, EPA Interim List of 403 Acutely Toxic Chemicals to be Included in with Emergency Planning Guidance Document, *Environment Reporter* **16(31)**:1492-1485.

Application of Systems Analysis Techniques to Vulnerability Studies of Complex Installations

T. F. Bott and D. F. Cameron
Los Alamos National Laboratory
Los Alamos, NM

ABSTRACT

Application of reliability and risk analysis methods (particularly fault trees) to failure analysis of systems in hostile manmade environments is discussed. Methods for constructing, analyzing, and interpreting the fault trees, graphs, and diagrams are described. The application of fault-trees, directed graphs, and functional diagrams to problems in sabotage, survivability, and information protection is discussed in general terms. Special problems encountered in this type of analysis are discussed, and future research directions are considered.

KEYWORDS: Systems analysis, sabotage, survivability, vulnerability

INTRODUCTION

Engineering analysis has long been concerned with reducing failures in complex systems. Often this analysis has focused on failures with randomly occurring causes, but recent analyses at the Los Alamos National Laboratory have included failure induced by malevolent or hostile actions of human adversaries. Los Alamos has worked since 1977 on applying reliability and safety analysis techniques to system vulnerability studies involving sabotage, security compromise, and combat damage.[1,2] In this paper, we will describe our experiences in this field, compare them with more traditional assessments, and discuss the special problems encountered in vulnerability assessments.

The details of most vulnerability analyses are classified because of their potential usefulness to an adversary. For this reason, the specific results of the analyses cannot be presented in an open forum. However, the techniques used and general examples of the results obtained from these studies are not classified.

METHODOLOGY

The purpose of vulnerability analysis is to provide an assessment of the likelihood of system failure in a manmade hostile environment. This environment may include stresses that are far beyond the system design capabilities and that are applied purposefully to the

New Risks
Edited by L. A. Cox, Jr., and P. F. Ricci
Plenum Press, New York, 1990

system by an adversary intent on causing system failure. Thus, the analysis includes elements of secondary cause, common cause, and human performance but from a unique viewpoint. Secondary causes that lead to failures of many components are the rule rather than the exception. Human actions may include acts of commission as a matter of course. The likelihood of the occurrence of a system failure cut set is a function of its attractiveness as a sabotage scenario, not a function of the basic reliability of the components in the cut set. In contrast to most reliability analyses, multiple initiating events and multiple mitigating system failures are likely, which makes the problem very complex from a systems standpoint. A major problem is just organizing the multidisciplinary information in a usable form. To approach this problem in a logical manner, Los Alamos has turned to the tools of reliability and risk analysis.

Deductive analysis methods are well suited to many of the vulnerability analyses. Undesired events are easier to determine than possible undesired sequences. For example, the unwanted results of sabotage at a nuclear reactor are limited: (1) radionuclide release and (2) economic loss to the plant. From these top undesired events, detailed sabotage scenarios may be deduced using fault trees.

The fault-tree approach has proven very useful in enforcing the discipline of systematic reasoning on the analysis. Many obscure, but significant, failure modes have surfaced as a result of the step-by-step logic involved in fault-tree construction. System interfaces and interrelationships are illuminated, and unanalyzed sequences of events often are discovered. The fault-tree development procedure guides the analyst in obtaining and correlating diverse system knowledge. Fault trees are graphic representations of abstract ideas and are easily understood by technical people, so a good review of the analysis by specialists on the system is possible without losing the independence of thinking of the vulnerability analysts. For these reasons, fault trees typically are chosen as the logic framework for the system modeling.

FAILURE MODELS

The failure of a system to perform a required mission is the undesired event in vulnerability analysis. This undesired event may be the result of a saboteur's carefully laid plans or the outcome of the more indiscriminate destructive capability of a hostile weapon. In either case, the first problem confronting the analyst is establishing the failure criteria for his system. This is not always easy because systems may degrade gradually in many environments. The point where system performance is no longer acceptable is often a matter of judgment. Multimission systems provide particular difficulties. Our experience has indicated the necessity for developing the top events of the tree carefully to insure a bounded problem and results that address the truly critical issues. This is accomplished by having detailed discussions with the systems experts for the facility or equipment under study.

After the failure criteria are established, a fault tree is constructed whose cut sets represent the sets of components whose failure will lead to mission failure. We have found that the vulnerability trees fall naturally into three principal sections. The top of the tree includes the undesired events (the goals of the adversary). Beneath the top of the tree are gates that describe system-level events. These gates typically have as inputs initiating events (adversary actions to stress the system) and mitigating system failures. Mitigating systems are designed to stop the progression of an undesired event (for example, a fire suppression system or a security alarm system). The adversary must either overwhelm or disable the mitigating systems to achieve his goals. Unambiguous system success criteria are critical to constructing this middle section of the tree, and a lack of understanding is very noticeable. In fact, sabotage analysis uncovered gaps in nuclear reactor safety knowledge before these issues were recognized generally as reactor safety issues.[3,4]

The final level of the tree is the subsystem and component level. This level is often the most straightforward part of the model to construct. In some cases, detailed failure analysis at the subsystem level already has been performed for safety or reliability purposes. However, emphasis on single-failure assumptions often reduces the usefulness of this work when adversary actions are considered.

The fault-tree cut sets are determined using the SETS computer code.[5] The fault tree is entered into the computer using a symbolic language that is converted automatically to Boolean equations for the various gates. The cut sets are solved by successive substitution from the top event to basic events, with simplification occurring along the way. The cut sets of the fault tree (in symbolic form) are then available as terms in a disjunctive normal form of the fault-tree Boolean equation,

$$F_T = \sum_{j=1}^{K} \prod_{i=1}^{N_j} B_{ij} \ , \tag{1}$$

where

Σ = a repeated cut set disjunction,
π = is a repeated event conjunction,
F_T = the top event,
j = the jth cut set,
B_{ij} = the ith event of the jth cut set,
N_j = the number of events in the jth in the cut set, and
K = the number of cut sets.

The SETS code is currently operable on the Cray-XMP computers at Los Alamos, and very large trees (~10^3 gates) may be reduced to cut sets in relatively short run times (~10 min).

BOOLEAN EQUATION TRANSFORMATIONS

The event cut sets generated from the system failure model fault tree are useful, but much additional information can be gained by transforming these equations into new Boolean equations representing other attributes of the system. For example, the locations from which a saboteur may induce a failure are of interest in sabotage analysis. This can be a major bookkeeping problem for a large facility with many remotely operated components. Nevertheless, the problem is made manageable by transforming variables.

Sets of components that a saboteur could attack to gain his objective are already available from the system failure model cut sets. For each failure, a Boolean equation is constructed that symbolically represents the locations from which that component may be affected. For example, failure F_A could be caused by visiting location B or both locations C and F in sequence. This is represented by the Boolean location equation

$$F_A = B + (C \times F) \ , \tag{2}$$

where + and × represent Boolean disjunction and conjunction operations. To convert a Boolean equation with event variables into its location equivalent, the Boolean equations for each failure event (such as F_A above) are substituted into the event equation and the equation is simplified. The resulting equation has locations as variables, with the terms of the equation representing sets of locations that allow the saboteur access to enough equipment to achieve his goals.

The transformation of variables is used frequently in vulnerability analysis. Transformations to locations, hazards, and threats are all useful for analysis of survivability in combat. For security, susceptibility to a surveillance method is a useful transformation. In these analyses, a large amount of information is sorted automatically into new forms for different perspectives on the problem.

NUCLEAR REACTOR SABOTAGE

The United States Nuclear Regulatory Commission (NRC) is charged with protecting the public from exposure to radiological hazards from nuclear power plants. This charter includes protecting nuclear power plants from sabotage that could cause the release of radionuclides to the environment. The NRC assesses the adequacy of security at nuclear plants, in part by ensuring that areas containing vital equipment are protected by physical and administrative safeguards. Los Alamos has provided technical assistance to the NRC by performing vulnerability analyses on commercial nuclear power plants to determine what equipment is vital to preventing radionuclide releases from the plant and the locations of this equipment.

This problem is addressed by constructing a fault tree with radiological release as a top event. To develop the fault-tree logic and to identify locations of equipment, the plant is visited by Los Alamos analysts because site-specific information is essential. A senior operator or engineer at the plant is involved in detailed discussions to determine system requirements and procedures. The physical layout of the plant is examined, and separable areas are designated as locations. A location is basically any area that can reasonably be enclosed to provide physical separation from the rest of the plant. The fault tree is developed based on reactor safety analysis and system performance specifications. The fault-tree model is realistic, but a conservative assumption is made when uncertainty exists. The fault-tree events are of two general classes: (1) initiating events that cause a potentially dangerous situation (for example, a pipe break) and (2) events that cause loss of components designed to mitigate the results of the initiators. The completed fault tree is transformed to location space by the methods described earlier and then solved for its cut sets, with the cut sets being in terms of locations a saboteur must visit to cause radionuclide release. The nuclear reactor fault trees are transformed to locations before solving for cut sets because of the great reduction in the number of location cut sets compared with event cut sets. This is because there are far fewer locations than component failure basic events. The cut sets typically consist of a number of single location sets and other sets that contain two or more locations. The Boolean equation for the cut sets becomes

$$\begin{aligned} \text{TOP-EVENT} = {} & \text{LOC1} + \text{LOC2}\ldots\ldots \\ & + (\text{LOCN} \times \text{LOCM}) + \ldots\ldots , \end{aligned} \tag{3}$$

where TOP-EVENT is the sabotage goal and LOCX is a plant location. LOC1 and LOC2 are single locations where enough equipment is accessible for the saboteur to cause a radiological release without visiting another area, and multiple location cut sets like LOCN×LOCM represent sets of areas that must all be visited to perform successful sabotage. The areas identified in the analysis are compared with the areas protected at the plant to insure that protection is given to vital equipment. To date, this analysis has been performed for over 80 nuclear power reactors in the U.S.

An outgrowth of nuclear plant vital area analysis is communicated threat credibility assessment. When a threat is made against a nuclear plant in the U.S., the NRC mobilizes a team of experts to assess its credibility. Los Alamos draws on its extensive accumulation of technical material to determine if the threatened act is a credible method for achieving the saboteur's objectives.

SPACE TRANSPORTATION SYSTEM VULNERABILITY

Security personnel for the Space Transportation System (STS) (commonly known as the Space Shuttle) at Kennedy Space Center (KSC) requested help from Los Alamos in determining the most likely STS sabotage targets and potential countermeasures. Los Alamos was chosen because of our experience with nuclear power plants and because we offered an independent, unprejudiced view of the problem.

This project was approached in a systematic manner by constructing a fault tree to describe the STS failure. The first problem encountered in the analysis was determining failure criteria for the STS failure. Thus, the systematic approach indicated areas of incomplete understanding for the STS security system from the beginning. To gather the information needed for constructing the fault tree, detailed discussions were held with technical experts in specialized areas at KSC. The fault tree was developed by identifying hazards, components vulnerable to the hazards, and actions defeating safeguards. The cut sets of the fault tree were calculated and represented sabotage scenarios (sets of actions leading to saboteur objectives). These scenarios then were evaluated in matrix form to determine their attractiveness as potential routes to achieving sabotage. The criteria for evaluation included accessibility, danger to the saboteur, likelihood of detection, likelihood of success, specialized knowledge required, and the time required to complete the actions. The scenarios fell into three main categories: (1) high-risk scenarios that were very destructive and achievable; (2) scenarios that were for some reason difficult, dangerous, or uncertain of success; and (3) scenarios that were malicious but not highly destructive. Countermeasures to reduce the risk from the five high-risk scenarios were suggested.

This study confirmed the ability of systems analysts to contribute to an area outside their specific fields of expertise by using systematic techniques to collect and interpret information collected from experts on diverse subjects. The results of the analysis helped focus the attention of security personnel in areas of the greatest security concern. In addition, the enumeration of possible scenarios gave security personnel considerable insight into what evaluations or equipment should be regarded with suspicion or challenged.

COMBAT SYSTEM VULNERABILITY

Survivability may be defined as the likelihood of a military system performing a mission successfully in a manmade hostile environment. Survivability is often considered in two phases. Susceptibility, the first phase, is the probability that a system interacts with the threats. Susceptibility reduction includes measures to avoid being acquired, targeted, tracked, or hit by a threat. Vulnerability, the second phase of survivability, deals with the probability of a system failing given that it interacts with a threat.

At Los Alamos, determining the vulnerability of combat systems to enemy threats is approached in a systematic manner by first constructing a failure model of the system to be analyzed. The failure model is, as in previous examples, a Boolean representation of system fault-tree cut sets. The basic events of the cut sets are component failures. Component failures are related to hazards that cause failure by Boolean equations of the form

$$\text{(LDRANT)} = \text{(/KE)} + \text{(/FIRE)} + \text{(/X-RAY)} + \text{(/EMP)} \ . \tag{4}$$

This equation means that a lookdown radar antenna for an aircraft may fail because of interaction with kinetic energy (/KE), fire (/FIRE), x-ray flux (/X-RAY) or electro-magnetic pulse energy (/EMP) from a nuclear detonation. (The "/" is a tag for convenience in working with the SETS code). Cut sets of events are transformed into cut sets with both events and hazards. The hazards are transformed to threats by equations like

$$\begin{aligned} \text{FIRE} &= /\text{HE} + /\text{INCENDIARY} \ , \text{and} \\ \text{EMP} &= /\text{JAMMER} + /\text{NUDET} \ , \end{aligned} \tag{5}$$

which means that fire can result from high explosive (HE) or incendiary devices and electro-magnetic pulse (EMP) can result from jamming or nuclear weapons. The transformed equations will have cut sets that indicate both the vulnerable components and the threats to which they are vulnerable. For example the "event loss of fire control radar" consists of loss of both lookdown (LDRADAR) and side-looking radar (SLRADAR),

$$\text{LOSS-FCRADAR} = \text{LDRADAR} \times \text{SLRADAR} \ = \text{LDRANT} \times \text{SLRANT} + ... \tag{6}$$

Transformed into a threat/event equation, it becomes

$$\text{LOSS-FCRADAR} = (/\text{HE} + /\text{JAMMER}) \times (\text{LCRANT} \times \text{SLRANT}) +..., \tag{7}$$

which indicates that either a high explosive or a jamming device can cause loss of the two-radar-antenna cut set LCRANT × SLRANT, causing loss of fire control radar and mission failure. The weapon system critical components and threats to them can be identified for a very large system by using the bookkeeping capabilities of the fault-tree Boolean equations and their transformations. This approach gives a logical foundation to vulnerability studies. The approach is being quantified by determining the hardness (for example, failure stress distribution) of each critical component to each hazard. The hardness may be represented by a plot of the expected component degradation as a function of stress. The probability of exceeding a given stress level for a component may be estimated by including spatial stress distributions, shielding, and critical-component presented areas. Work is just beginning in this area, but already the fault-tree approach has indicated areas of concern for focusing detailed analysis.

INFORMATION SECURITY

In many situations, a critical item of security is protecting information. An entire security system may be developed to prevent proprietary or classified documents, data, or equipment from being compromised. Protecting information about a military system that may be used by an adversary also can be considered from an information security viewpoint. It is important to assess the effectiveness of such a system. Los Alamos has applied system analysis techniques similar to those used in sabotage analysis to the assessment of information security effectiveness.

Information security attempts to control knowledge of a facility's existence, its mission or design details, or the relevant procedures. The controlled information can be compromised by "leakage" of observables through information flow paths to collectors of information. The attributes of the facility then can be inferred by the collector based on his context or "frame." To control this leakage of information, safeguards are set up to interrupt the information flow paths.

The first step in our analysis is to compile a list of observables. This is done by using logic diagrams that first divide the activity into phases, for example, the design, construction, and operation phases. The activities in each phase then are expanded to finer and finer detail until observables can be identified. To bound the problem, some measure of importance must be attached to observables during this process. This is done by expert judgments.

The observables are represented by source nodes in acyclic directed graphs (hereafter referred to as information flow graphs). The graphs consist of nodes that represent steps in a leak path (for example, copying a classified document) and safeguards against information leakage. The sinks of the graph are collectors of information. The graph is represented by Boolean equations for each node. The node equations indicate the adjacent nodes by Boolean equations such as

$$N_1 = e_{12} N_2 + e_{13} N_3 \ , \tag{8}$$

where the N's are nodes or vertices and the e's are the edges of the graph. This equation indicates that node N_1 is connected to nodes N_2 and N_3 by edges e_{12} and e_{13}, respectively.

Paths from observable sources to collector sinks are constructed by successive substitution into the graph equations. The nodes of the graph that represent safeguards failures are developed as fault trees, and the Boolean equations are substituted into the information flow graphs equations, giving a sequence of events leading to disclosure of an observable.

This approach has proven very useful for identifying qualitatively areas where information security can be upgraded. For example, steps in a procurement procedure that introduce many new leak paths can be circumvented.

Based on the above system model, an analysis of the relative likelihood of information disclosure through human intelligence was estimated quantitatively. The purpose of the analysis was to provide guidance on the relative cost effectiveness of various strategies for protecting sensitive information concerning a project.

The quantitative results were based on data collected from government sources with both time-variable or attribute data. For example, the rate at which classified document repositories were left open and unattended was estimated from security infraction historical data. The "occurrence rate" was estimated by

$$\chi = \frac{\text{number of violations}}{\text{number of custodian man years}} \ .$$

This rate was assumed to be sample independent so the rate for a population M of custodians was

$$y = M_\chi \ .$$

The occurrences were assumed to be Poisson distributed with χ, and the cumulative probability of one or more repository compromises as a function of time was calculated. Using such simple models, insight into the relative efficiency of various security strategies, such as compartmenting information, segmenting information, and human reliability programs, was obtained.

Los Alamos also has applied these techniques to reducing the susceptibility of military systems. In this case, the goal is to reduce the likelihood of an adversary acquiring, identifying, tracking, targeting, or hitting a military asset. For example, the list of observables for a satellite may be developed by working from satellite missions to required functions. The functions necessitate systems, which are made of components. The components generate observable emanations such as infrared radiation or optical emanations. Thus, a signature of the system can be developed with locations, intensities, and time dependencies included for each class of observable emanation. The reduction of observables can be addressed by shielding of large-signature components, by anti-simulation, or by other deceptive techniques.

SUMMARY AND CONCLUSIONS

Analysis tools developed for reliability and risk assessment of complex systems have been applied to the systematic analysis of sabotage, combat, and security vulnerability. The most useful techniques have been fault-tree analyses with variable transformations, functional diagrams, and directed graphs. Some tentative attempts at quantification of the analysis have begun using human factors and trust violation data, but much more effort is needed. The principal result of the work has been an increased qualitative understanding of system vulnerabilities, the effectiveness of countermeasures, and the ranking of scenarios by their attractiveness to sabotuers or other adversaries. The techniques have been applied to problems in nuclear reactors, the STS, defense, and information security.

REFERENCES

1. T. F. Bott, Sabotage Analysis and Risk Contributors to STS, in *Proc. of Seventh Annual System Safety Conference*, San Jose, California, July 25-29, 1985, Vol. I, pp. 31-41.
2. D. F. Cameron, Vital Areas at Nuclear Power Plants, in *Proc. of Seventh Annual Safety Conference*, San Jose, California, July 25-29, 1985, Vol. III, pp. 49-62.
3. J. W. Bolstad and R. A. Haarman, "Summary of Thermal-Hydraulic Calculations for a Pressurized Water Reactor," Los Alamos National Laboratory report LA-8361-MS (May 1980).
4. E. W. Barts and T. F. Bott, TRAC-PF1 Study of Loss of Pump Seals in A Westinghouse PWR, *Trans. Am. Nucl. Soc.* **46**:472-473 (June 1984).
5. D. W. Stack and M. S. Hill, "A SETS User's Manual for Vital Area Analysis," U.S. Nuclear Regulatory Commission Report NUREG/CR-3134 (April 1984).

Seismic Hazard in the Eastern United States: Bounding the Uncertainty

R. K. McGuire
Risk Engineering, Inc.
Golden, CO

J. C. Stepp
Electric Power Research Institute
Palo Alto, CA

G. R. Toro
Risk Engineering, Inc.
Golden, CO

ABSTRACT

Multiple alternative explanations of large earthquakes in the eastern US imply uncertainty in the calculated seismic hazard. A study conducted by the Electric Power Research Institute has developed and applied a technique to quantify and document alternative scientific interpretations of the causes and characteristics of large earthquakes, using a Bayesian framework. The method documents uncertainties in the sources of earthquakes, in the parameters of probability distributions describing their characteristics, and in the ground shaking that will accompany such events. The evaluation of hazard uncertainty follows a logic tree approach, with end branches representing specific alternative interpretations. Efficient computational procedures allow many end branches (~50,000) to be handled easily. An important part of the analysis is the interpretation of uncertainty in hazard represented by the 50,000 end branches; conditional distributions allow us to isolate effects of parameter uncertainties and to identify the major contributors to the total uncertainty in hazard. One of these is the range in expert interpretations of the causes of earthquakes. The method allows explicit documentation of expert uncertainties in scientific interpretations, and communicates the resulting uncertainties in earthquake hazard in a convenient form for a risk-based decision process.

KEYWORDS: Earthquakes, seismic risk, uncertainty

INTRODUCTION

There are multiple hypotheses to explain earthquake occurrences in the eastern U.S., and one or more of these explanations may be correct in various parts of the region. This state of understanding implies large uncertainties in the seismic hazard (probability of occurrence of earthquake shaking) for sites in the eastern US. In 1983 the Electric Power Research Institute (EPRI) and the Seismicity Owner Group (a group of electric utilities)

New Risks
Edited by L. A. Cox, Jr., and P. F. Ricci
Plenum Press, New York, 1990

began a study to develop a methodology to document and quantify earth science interpretations of the causes and effects of earthquakes in the U.S. east of the Rocky Mountains. Reported herein are test results obtained from application of this methodology.

The methodology that resulted from this effort has the following general steps:

1. Identification of hypotheses governing earthquake occurrences,

2. Interpretation of the contemporary tectonic stress regime,

3. Designation of candidate tectonic features that might cause moderate-to-large earthquakes,

4. Formulation of quantitative criteria to assess the tectonic capability of those features, explicitly incorporating both scientific uncertainty (in the causes of earthquake occurrences) and informational uncertainty (in the geophysical characteristics exhibited by any specific feature),

5. Application of the criteria to assess the probability that candidate features and sets of features are tectonically active,

6. Analysis of historical earthquakes in the region to achieve a seismicity data base with uniform statistical characteristics,

7. Development of methods to estimate seismicity parameters (rates of occurrence and characteristics of distributions, including maximum magnitude) for the tectonic features and seismic sources, from the historical data base, and

8. Development of efficient methods to calculate earthquake shaking hazard at any site, incorporating uncertainties in the activity of tectonic features and in characteristics of the earthquake distributions, given that a feature is active, and deriving methods to determine the sensitivity of uncertainty in results to uncertainty in the assumptions.

Application of this methodology in a consistent manner requires careful distinction among sources of variability. *Randomness* is probabilistic variability that results from the unpredictable nature of future events, given a state of nature consistent with current scientific knowledge. For an accurate set of active tectonic features with known probability distributions, the size, location, and time of occurrence of the next earthquake is random. *Uncertainty* is statistical or modeling (e.g., professional) variability that expresses lack of knowledge about the "true" state of nature. In concept, uncertainty is reducible with the collection of additional data or with the development of more refined theories.

Two underlying factors were important in developing the EPRI methodology. First, documentation of the scientific bases for the tectonic interpretations was critical. Also important was the calculation of the uncertainty in seismic hazard (and the causes for that uncertainty) as an expression of current scientific understanding of earthquake mechanisms and effects in the region. Details of the methodology are given in References 1-3; the remainder of this paper describes some of the major insights and results gained from the application.

INTERPRETATIONS OF TECTONIC MECHANISMS AND SEISMIC SOURCES

Identification and assessment of possible sources of future earthquakes began with the compilation of a data base of geological, seismological, and geophysical data for the

U.S. east of the Rocky Mountains. Six Earth Science Teams were formed (with about 30 earth scientists familiar with the eastern US) to identify tectonic features that might cause future earthquakes, evaluate the orientation and magnitude of the crustal stress field, and determine criteria that could be used to assess the potential activity of any particular feature, given its geophysical characteristics. The six teams made separate interpretations and developed their own criteria, but there was substantial interaction among the teams at workshops to exchange information and resolve misunderstandings of data. The final interpretations of each team consist of five items.

1. A set of tectonic features and seismic sources that might generate future moderate-to-large earthquakes.

2. A "matrix of physical characteristics" that quantifies the probability that any feature with a defined set of physical characteristics is active. Typical characteristics included the spatial association with historical seismicity, the orientation of the feature with respect to the crustal stress regime, and the depth of the feature in the crust.

3. A "feature assessment" that evaluates the probability that any feature exhibits any given set of physical characteristics.

4. A calculation of P^a, the probability that a given feature is active, from the matrix of physical characteristics and exhibits.

5. A specification of any dependencies of activity among features and sources caused, for example, by a common orientation in the crustal stress field.

Steps 2 and 3 allow a separation of scientific uncertainty from informational uncertainty. Specifically, the matrix of physical characteristics describes the assessment of the probability of activity given a set of associated characteristics; it describes (with subjective probability) the diagnostic power of characteristics. The feature assessments describe the degree to which each of the characteristics can be associated with each feature; it documents the *informational* uncertainty. The summation of element products of the two gives the probability of activity for the feature. The procedure was developed not to supplant scientific intuition and judgment, but rather to document it in a structure for review and evaluation by others.

The teams' interpretations were used to construct sets of sources that might be simultaneously active in the vicinity of a site. A probability for each set was calculated from the P^a's for each source and specifications of dependencies of activity of sources in each set. These sets were interpreted to represent the range and credibility associated with different tectonic interpretations for each Earth Science Team.

The method allows great flexibility for input interpretations. Seismic sources can be represented in any degree of complexity required to capture an Earth Science team's alternative interpretations of earthquake causes. An example seismic source configuration is shown in Fig. 1.

ANALYSIS OF EARTHQUAKE REOCCURRENCE

For seismic hazard computation one needs a probability distribution for the annual number and magnitude of earthquakes that may occur in each source. To estimate this distribution and its parameters, a historical data base was assembled; an unbiased magnitude estimate was calculated for each event, and clusters of earthquakes were eliminated. Issues of incompleteness were addressed by deriving statistical procedures to

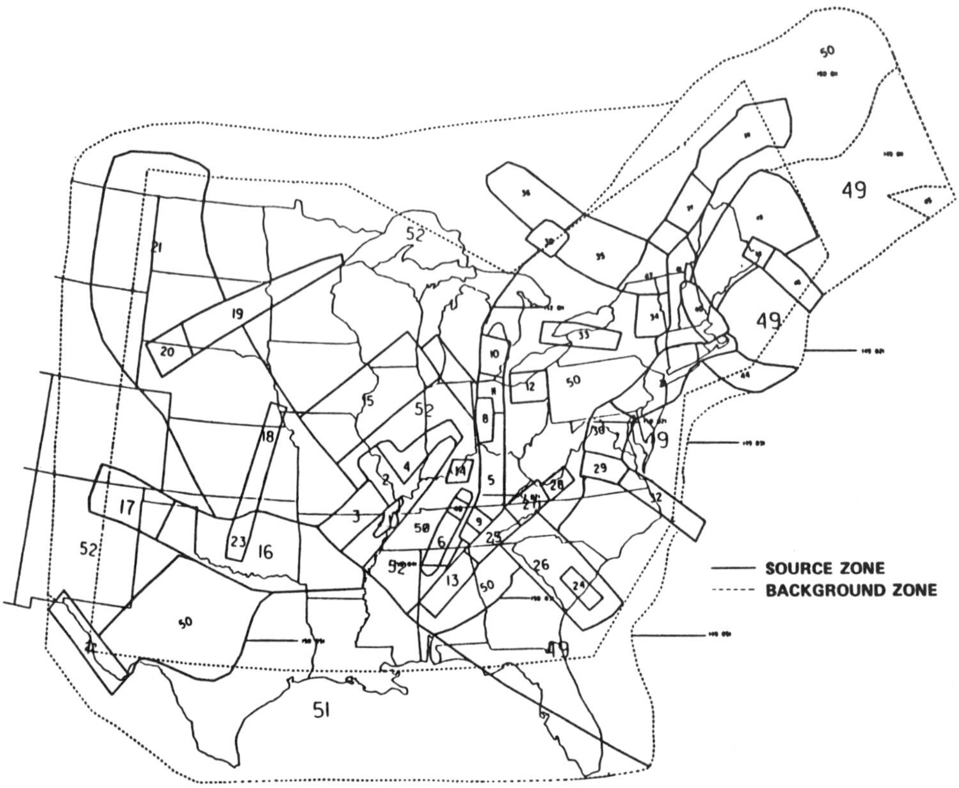

Fig. 1. Typical seismic source map for Eastern U.S.

quantify and account for the degree of incompleteness. Finally, procedures were developed to estimate parameters (the earthquake occurrence rate and the slope of an exponential magnitude distribution) for each source, allowing the parameters to vary in space within each source. External specification of smoothing functions allowed the parameters to vary smoothly or abruptly with each source, according to preferences expressed by the Earth Science Teams (see Ref. 3).

Additionally, a specification of the maximum magnitude earthquake in each source was required. Each Earth Science Team, using historical data, analogies from other regions, and inference from physical models, provided these specifications. A subjective distribution for maximum magnitude (discrete values and subjective probabilities) characterizes uncertainty in this parameter.

SEISMIC HAZARD, RISK, AND UNCERTAINTY

A seismic hazard computation integrates over randomness in earthquake occurrences and in the associated ground motion, the first type of variability described in the Introduction, and yields annual probabilities of exceedance for various ground motion amplitudes. The uncertainty in this hazard representation is governed by the uncertainty in the models used and assumptions made, the second type of variability. The uncertainties in tectonic interpretations and in seismicity parameters, given a specific interpretation, were documented and quantified by the Earth Science Teams as described above.

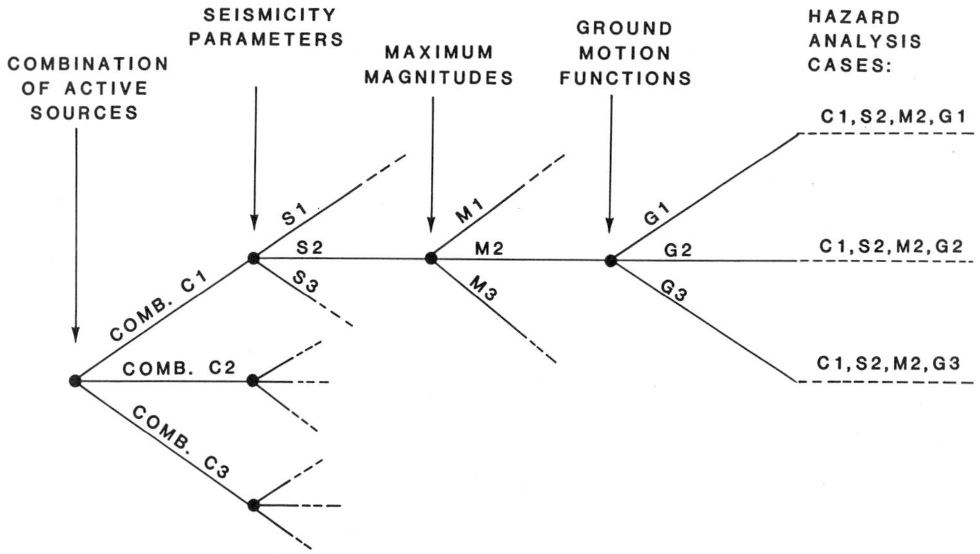

Fig. 2. Example of logic tree format used to represent uncertainty in hazard analysis input for one earth science team.

The format used to represent uncertainties in input to the hazard analysis is the logic tree, an example of which is shown in Fig. 2. Each node in the tree represents an unknown state of nature, and each branch represents an alternative discrete possibility for that state of nature. The set of branches emanating from one node is meant to represent the range of possible choices, with assigned probabilities (assessed subjectively) summing to unity. Each end branch thus describes a set of parameter values or model interpretations, for which a hazard analysis can be made. Because of the alternative seismic sources that are permitted by the data, a large number of combinations of sources and parameter values resulted. Typically the number of end branches for hazard analysis at a site was 50,000. This required that the analysis algorithm be efficient and that special procedures be derived to conduct sensitivity studies.

Application of the EPRI methodology using input from the six Earth Science Teams indicates that uncertainties in calculated seismic hazard for the eastern US are typically rather large. The uncertainty in annual probability of exceedance from all inputs, expressed as a range between the 15% and 85% fractiles, may span two orders of magnitude or more. These large uncertainties result from major uncertainties in input interpretations (What are the active tectonic features that may produce earthquakes in the future, and what earthquakes might they generate?), and in estimates of average levels of ground motion associated with these mid-plate earthquakes.

For nuclear power plants, there is an important implication from these large uncertainties. It is that uncertainties in seismic risk (the probability of monetary loss or human casualties) are dominated by the uncertainties in hazard. By comparison, the uncertainties in structural response and damage are relatively small (they do not span two orders of magnitude in probability, for a given ground motion). Therefore, understanding (and possibly reducing) the uncertainty in seismic risk involves addressing the uncertainties in seismic hazard, as illustrated by the EPRI study reported here.

To isolate the effects of input uncertainties on hazard, two procedures were adopted. The simpler of the two is useful when alternate, discrete choices (generally on the left side of the logic tree) affect all interpretations. Examples are differences among the six Earth Science Teams, and differences caused by alternative ground motion estimates. Alternatively, some uncertainties (generally on the right side of the logic tree) may be more interpretation-specific (such as uncertainties in maximum magnitude, which depend on the tectonic sources thought to be active). For these cases we use a "distribution of medians" approach wherein all possible sets of values are constructed, the probability of each set is determined, the conditional median hazard over all other uncertainties is calculated for each set, and a distribution is formed from these conditional median hazards and the probability of that set of values. Graphical demonstration of the sensitivity of median hazard to uncertainty in that input parameter is then possible.

ANALYSIS OF UNCERTAINTIES

The methodology described above was applied to nine test sites, using the interpretations of the six teams. Because of the large numbers of alternative hypotheses considered by the teams and the complex ways in which various parameters interact to affect seismic hazard, it is not possible to draw simple conclusions that hold for all sites. For example, it is not possible to divide the total uncertainty into pieces representing various contributions of the inputs. However, we can examine sensitivities of seismic hazard to uncertainty in input assumptions, using the methods described above. The following are general conclusions deduced by examining the results and sensitivity studies obtained at the nine test sites.

One important contributor to uncertainty in hazard is the range of alternative tectonic interpretations provided by the Earth Science teams, particularly on the east coast. Figures 3a and 3b show median seismic hazard curves for each of the six Earth Science teams for one site on the east coast and one in the Midwest, respectively. (The hazard curves are shown as annual probability of exceedance versus response spectral velocity at 10 Hz for 5% damping.) For the east coast site, the median among the six teams varies by a factor of about 30 in annual probability, and this is typical for the east coast. Generally, eastward of the western boundary of the Appalachian Mountain System, interpretations of the dominant source of earthquakes vary markedly among teams. The seismic hazard is quite sensitive to these differences. In the Midwest, the median hazard among teams varies by a factor of about 4, except for one team that is low with respect to the other five. This results because, in the Midwest, large background sources of seismicity, or sources with dimensions of several hundred km, typically dominate each team's interpretation, and the derived hazard is often similar among the teams.

A reasonable explanation of the east coast results is that they reflect the more complex tectonic history of the Appalachian System. The area has been frequently studied to determine the tectonic processes and active seismic sources causing earthquakes. These studies have resulted in increased understanding of the earthquake generating process, but have also led to multiple theories and interpretations that are given various credibilities by the Earth Science teams. The different credibilities reflect the different backgrounds and perspectives of the teams. These differences were sought and encouraged during the tectonic evaluation process. By comparison, the Midwest has undergone a more stable tectonic evolution and, except for specific sites, has not been as extensively studied. Thus there are fewer alternative interpretations available, and the expert teams tended to provide similar input.

The uncertainty in hazard given an Earth Science Team is generally small, often a factor of 3 between the 15th and 85th fractile. This is illustrated in Fig. 4, which shows typical results from an east coast site and a Midwest site, for a typical Earth Science Team. Each team was relatively confident about its interpretations; the difference among teams

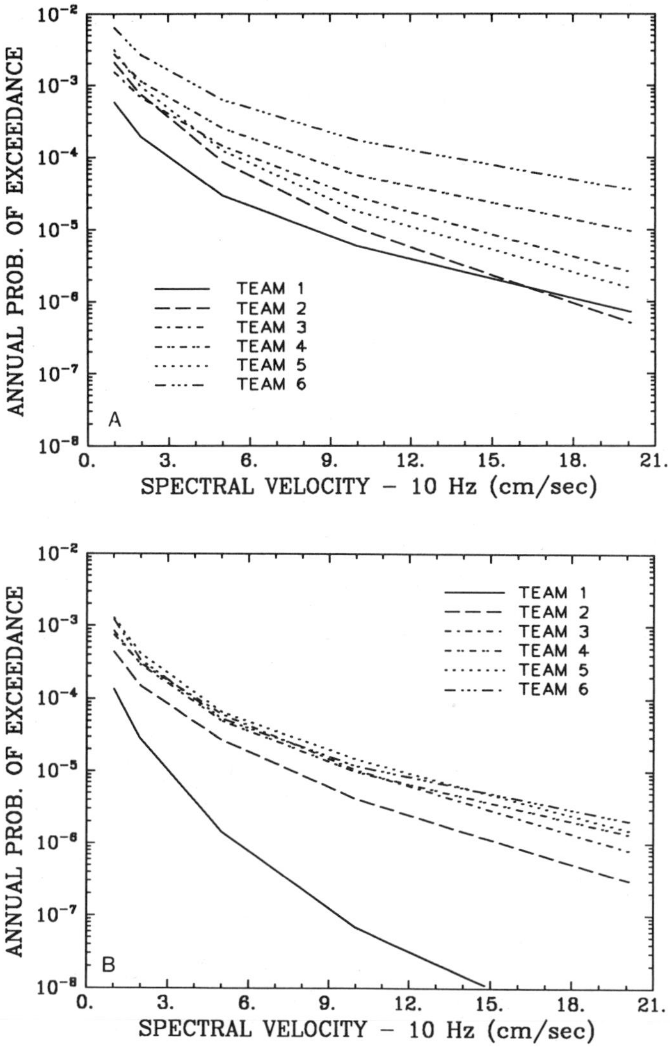

Fig. 3. Sensitivity of seismic hazard to earth Science Team.

was relatively large, however, illustrating a common result from studies where subjective interpretations are required to define uncertainties.

DISCUSSION AND CONCLUSIONS

The EPRI methodology allows explicit documentation of expert opinion in the form of a general procedure for describing the probability of activity of features and sources that might cause future earthquakes. The procedure allows multiple hypotheses and interpretations to be handled in a logical, quantitative format. The response of the expert teams was that, after an initial educational process, they found the format easy to apply and helpful in structuring and documenting group interpretations.

A common result was obtained during testing of the procedure to calculate seismic hazard at nine test sites in the eastern US. The difference in calculated hazard among the teams was often large (a factor of 30 range in median hazards for east coast sites), but the

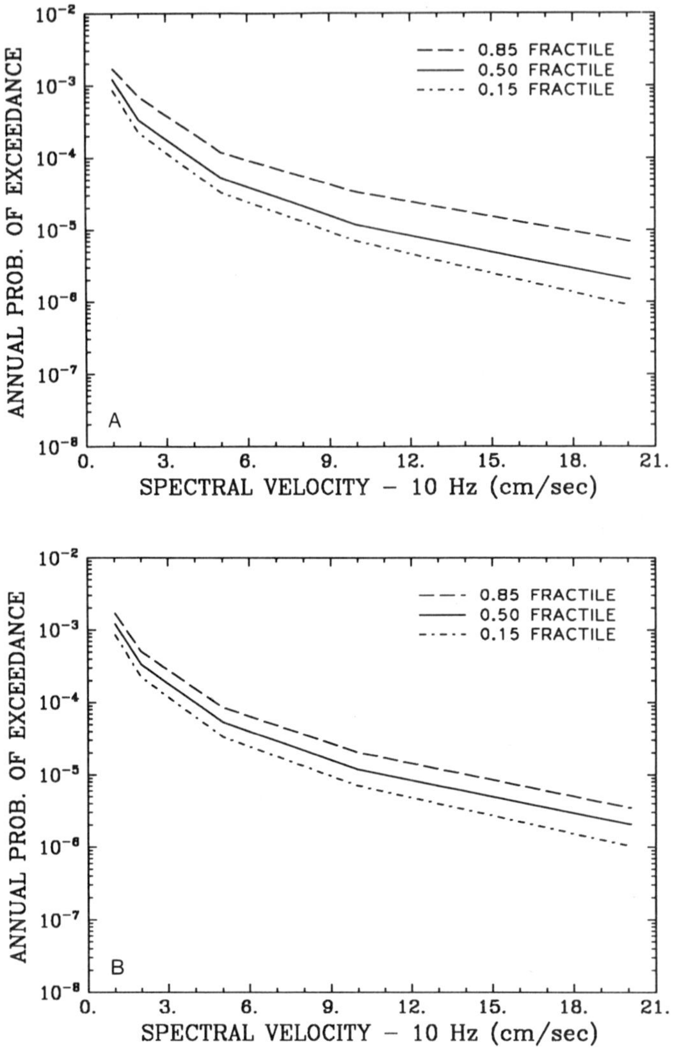

Fig. 4. Fractiles of seismic hazard illustrating within team variability at Midwest and East Coast site.

uncertainty expressed by a single team was smaller (often a factor of 3 between the 15th and 85th fractile). This indicates the necessity of employing multiple expert interpretations when conducting hazard analyses explicitly to quantify uncertainty in hazard.

With respect to specific seismic hazard results for the eastern US, we conclude that there are general differences between seismic hazard sensitivities at east coast sites and Midwest sites. For the former, a complex tectonic history of the Appalachian System has led to more varied interpretations of tectonic features and processes considered by the Earth Science teams, resulting in multiple seismic sources and source combinations. In the Midwest interpretations of seismic sources among teams are quite similar, reflecting the relatively less complex tectonic evolution of the region and leading to less uncertainty in calculated seismic hazard.

REFERENCES

1. Electric Power Research Institute/Seismicity Owners Group, *Seismic Hazard Methodology for Nuclear Facilities in the Eastern United States, Vol.1: Development and Application of Methodology*, EPRI/SOG Draft 85-1 (1985).
2. R. K. McGuire, R. K., Seismic Hazard Computational Methodology, in *Seismic Hazard Methodology for Nuclear Facilities in the Eastern United States, Vol. 1*, EPRI/SOG Draft 85-1 (1985).
3. R. K. McGuire *et al.*, *Seismic Hazard Methodology for the Central and Eastern United States, Vol. 1: Methodology*, EPRI NP-4726 (1986)

Thinking the Unthinkable: Preparing for Global Disaster

Robin Cantor and Steve Rayner
Oak Ridge National Laboratory
Oak Ridge, TN

ABSTRACT

The Federal Emergency Management Agency (FEMA) is responsible for national planning and management functions concerning emergency preparedness, natural disasters, civil defense, and major industrial accidents. These activities place the agency in the role of a risk manager and, in some cases, make it responsible for planning where too little is known about the likelihood or consequences of the risk. One area where risk analysis and management is particularly difficult is preparing for global disasters, for example, on the scale of nuclear war. Among the research areas related to this risk are issues pertaining to the restoration of the mechanisms for exchange and distribution of goods and services in the wake of global disaster. Investigations of this type present two major challenges to the risk analyst. First, the range of uncertainties involved implies that the number of possible scenarios to be examined must be capable of providing sufficient insight to be useful for planning purposes, while sufficiently limited in number to be feasibly researched. Second, the unique nature of this problem implies that the theoretical and empirical evidence must be extracted from a number of disciplines, including anthropology, economics, and sociology. Thus, the investigation requires the construction of a novel approach merging very diverse bodies of existing research. This paper discusses the design of an analytical framework used to examine the exchange and distribution problem, focusing on the challenges to be met and the implications for similar risk management problems.

KEYWORDS: Global disaster, research design

INTRODUCTION

Studying Global Disaster

When we consider the policy context, the problem of global disaster on the scale of nuclear war embodies many of the important features of risk management problems involving low probability, high consequences, and policy options which cannot be completely tested. In other words, we are faced with the task of trying to answer a question where too little is known about the consequences while the imagined responses are diverse. However, research is required for planning options. This paper discusses the design of an analytical framework used to examine one of the many problems following global disaster, focusing on the relevance of risk management guidelines for planning purposes. We discuss a number of guidelines for the design of research approaches used to study global

New Risks
Edited by L. A. Cox, Jr., and P. F. Ricci
Plenum Press, New York, 1990

disaster problems. The guidelines are illustrated by the example of a global disaster planning problem involving the restoration of free-market trade in an economy with various levels of institutional and resource damage.

The Research Question

Our research question concerned the socioeconomic conditions that would have to pertain for rational mechanisms of distribution and exchange to function in the wake of a major societal disaster. Specifically, what kinds of markets would be viable following the destruction or severe impairment of existing infrastructures which regulate and enforce contracts, and resources which define wealth? By infrastructure we mean the system of institutions whose functions facilitate demand and supply transactions. Infrastructure therefore includes stock and commodity exchanges, courts, legislatures, and regulatory agencies. In the case of a major societal catastrophe, the survival of a broad economic and legal infrastructure governing market activities cannot be assumed, even where there are surplus goods and services and willingness to participate in exchanges. Hence, novel adaptations of familiar market arrangements may arise according to local circumstances, in order to maintain trust between agents or to substitute confidence in an institution that will enforce contracts where that trust is missing. It should be pointed out that, at this stage, we are largely interested in the uncoerced exchange underlying the concept of a free market.

GUIDELINES FROM DISASTER-PLANNING RESEARCH

One reasonable approach to the study of market problems following a major societal disaster is to investigate the guidelines offered by disaster planning research. However, before these guidelines can be used in the problem, we must question to what extent there are parallels between natural disasters and global disasters and whether or not these guidelines are transferable.

Parallels Between Global Disaster and Natural Disasters

As we have structured the problem, we regard FEMA as a risk manager responsible for developing policy actions in the face of global disaster. This problem parallels the risk management problems traditionally examined in disaster planning, but with some major exceptions. In one very important aspect the two problems are identical in that "the ultimate goal in such planning is to enable an effective and efficient start towards the restoration of normal routines" (Dynes, Quarantelli, and Kreps, 1981).

The problem is similar to natural disasters in that FEMA has very limited influence on the occurrence of the disaster event. To some extent, FEMA is constrained to accept the position that planning is necessary to reduce the uncertainty of the event following global catastrophe, rather than preventing the catastrophe from happening. Thus, we are not debating the question of how to prevent a nuclear war, although some aspects of this argument will be important in the ethical problems considered in the conclusion.

Are Disaster Planning Guidelines Transferable?

The parallel problems suggest that planning for global disaster may benefit from existing guidelines for natural disaster planning. However, a closer examination of major specific guidelines reveals that some guidelines are very applicable while other are inappropriate and, if applied, could be misleading for global disaster planners. Below we consider four of the important disaster planning guidelines from the literature on natural disasters (Dynes, Quarantelli, and Kreps, 1981).

Emphasize the appropriate over the fast response. This guideline is particularly important for the market problem since other response concerns are likely to dominate in the period immediately following the disaster. In fact, the restoration of market activities does not reflect generally the crisis characteristics of what we consider emergencies. Thus, determining the appropriate response is likely to depend on allowing a sufficient amount of time to elapse so that potential trading agents are less concerned with emergency activities and begin reconstruction of social relationships.

Plan for most likely probability. This guideline is appropriate in the natural disaster context, where data exist from which probability information can be drawn. The lack of actual data even remotely applicable to our disaster event implies that we would have great difficulty in specifying most likely cases. Thus, considering a limited number of extreme cases is more suited to global disaster planning. This follows from the need to keep the planning process tractable, while recognizing that the unusual may come to pass. Lack of a direct experience base from which to define likely scenarios is one of the more difficult problems of planning for global disaster.

Focus on principles. Given that uncertainty is prevalent in the analysis, we believe that this guideline is crucial to the planning research design. We structured our analysis to emphasize the rules by which social interactions and exchange are governed. By doing so, we were able to specify market structures as packages of rules governing demand, supply and transactions. This specification allows direct comparison among the different types of market structures. More importantly, it facilitates the identification of social or technical conditions that give rise to particular rules and suggests principles which can be followed which account for these conditions.

Only exercised plans are realistic. Although sensible in concept, this guideline is not very practical for the global disaster problem. However, we agree that some exercise of the potential policy actions is desirable. For the market problem, such an exercise implies some application of the various policy actions designed to facilitate market exchange at different levels of surviving infrastructure and resources. One possibility that we are considering is to use experimental economic methods to define market structures under which participants may exchange.

ADDITIONAL GUIDELINES FOR THINKING THE UNTHINKABLE

An additional set of guidelines emerged from our research that we believe can be useful to similar planning questions. These guidelines are more concerned with the disaster planning research design, an area not well explored in the field of natural disaster planning. This is not an oversight of the natural disaster literature, since such research is strongly based in actual disaster response experiences. Without the benefit of actual data, the global disaster planner must rely on a well-developed research design to suggest the possible circumstances that will necessitate a response. In other words, possible scenarios must be limited in a meaningful way. Furthermore, the research design must be capable of suggesting the efficacy of possible disaster responses, again without the benefit of real-world applications. Thus, the analyst will be forced to combine a number of relevant theoretical and empirical studies in innovative ways to create a conceptual environment in which important hypotheses may be tested.

Limit the possibilities in a meaningful way. This guideline encourages the development of a defensible framework for determining the logical possibilities. Furthermore, given the number of scenarios that may be generated from incremental changes in the important variables, some reasoning is also necessary to constrain the analysis. In our study, we specified the range of initial scenarios within which institutional adaptations could arise in the wake of large-scale material and institutional devastation.

Any such catastrophe may be presumed to affect existing markets in two important ways: destruction of resources (skills, goods, and currency to exchange) and destruction of infrastructure. Depending on the extent of the catastrophe and the location of the affected market, either or both of these factors may sustain light or heavy damage. The possible combinations of these factors generate four distinct scenarios and determine the initial conditions for the development of possible post disaster market arrangements:

a. Business as usual: where infrastructure and resources survive largely intact. For example, one may consider a disaster on the scale of a natural disaster affecting a limited area, where resources and infrastructure are obtained from surrounding areas. Although the directly affected area may not be restorable, we expect normal market activities to continue elsewhere.

b. Bureaucratic nightmare: where resources are restricted but infrastructure remains intact. This scenario might occur in an undamaged urban area which is unable to obtain resources like food from suppliers elsewhere. Heavily regulated exchange, such as rationing, is a likely reaction on the part of the surviving infrastructure.

c. Fat of the land: where resources survive but infrastructure is heavily damaged. For example, one might consider rural areas which are dependent on urban centers for external markets and institutional infrastructure. Self-regulating, informal market activities such as direct barter and self-help organizations are likely to arise.

d. Mad Max: where both infrastructure and resources suffer heavy damage. For example, such conditions may describe the remote suburbs of a largely-destroyed city following a nuclear exchange. Informal market activity, subject to some form of network protection, is a likely outcome. These conditions are also likely to be a haven for criminal market activities.

Guidelines should combine models and disciplines. Global disaster planning problems will necessarily involve questions addressed by different literatures and conceptual models. Thus, some guidelines must be followed in the research design to combine this diverse information. Three aspects of combining diverse literatures are particularly important for our research design problem.

First, it is important to recognize inconsistencies across and between levels of analysis. To study the rules governing exchange within each scenario, we referred to models of exchange from anthropology, economics, and sociology. Unfortunately, these models are rarely consistent with each other, especially with respect to the level of analysis. Thus, models of collective, individual, social-network, firm, and industry behavior were reviewed. One of the more difficult problems was determining which level of abstraction in these models was appropriate to provide useful information for the analysis without losing a realistic perspective. For example, we had to resist the reductionism of formal economic models that rely on the extreme version of utility theory. Excluding the social network factors, these models tend to represent market decisions using only the individual perspective, where all factors are reduced to individual costs and benefits.

A second consistency problem emerged in the combination of different areas of applied research. We based the characteristics of markets, with varying conditions and levels of participation, on existing evidence from four fields of study. These fields included formal market behavior in economics, past efforts at emergency rationing and market regulation, informal economies in peace and wartime, and primitive markets and trading. Many of these areas have been seen as incompatible by other researchers. However, we found that by transforming the information into the rules governing market behavior and exchange, we provided a common framework for the study areas. The rules approach allowed us to compare the very diverse applied research in a consistent and efficient manner.

Finally, pooling diverse research areas to address the global disaster planning problem reveals a number of unanswered questions as well as generating some new ones. The research design must acknowledge these questions to point out important areas of missing information underlying the recommended planning actions. For example, one surprising finding was that no rigorous and consistent definition of a free market could be found in any of the literature examined. Although we uncovered a number of very specific definitions, they ranged from the general idea of a sphere where supply and demand come together (Alchian and Allen, 1969) to only those transactions that are conducted among completely anonymous traders (Williamson, 1985). Thus, the first definition includes just about every exchange imaginable, while the second excludes any market where contracts or social networks are present. Finally, unanswered questions are inevitably raised by our rules framework since the rules derived from theoretical models are extremely difficult to test and the rules derived from applied work in other cultures may not be valid for the surviving U.S. population.

Beware of the ethnocentric influence of past research, existing theoretical or empirical information. As mentioned, an obvious strategy for studying the planning problem would have been to extrapolate from existing documented experience or models of the effects of natural disasters on market activity. However, such an approach suffers from a major shortcoming. The effects of natural disasters are usually so localized that the general system of law and institutions to enforce contracts and regulate transactions can be assumed to have survived, at least beyond the area of immediate devastation. Even where these mechanisms are impaired or destroyed locally, there will be, in the U.S. at least, a higher authority with external resources available to step in and restore order. Past experience of natural disasters suggests that new institutions governing distribution, exchange, the enforcement of contracts, and maintenance of property rights do not have the opportunity to establish themselves. In part, this is due to the limited duration of a natural disasters, an assumption that cannot be used to guide the planning for global disaster problems.

THE ETHICAL DIMENSION

In the course of this research we were confronted with a number of ethical problems. Therefore, we conclude with some thoughts on the researcher's moral dilemma when accepting or rejecting global disaster planning research. In general, these problems do not confront the natural disaster planner, whose role is regarded as necessary and, to some extent, a public good.

First, global disaster planning may appear to be admitting defeat in the sense that being prepared makes the disaster event more likely to occur. This concern is usually raised in the planning research for man-made global disasters such as nuclear war or carbon dioxide pollution induced weather changes. Second, because so many of the critical policy options are untestable in the pre-disaster world, there is an additional burden placed on the planner regarding the consequences of being wrong. Third, since research funds are scarce, funding of global disaster planning research displaces alternative research programs, possibly those that aim to study the prevention of global disasters.

In contrast, the positive aspects of accepting this research should be acknowledged. Thinking the unthinkable provides an opportunity to foster new ideas about old areas of research, even stimulating new criticisms of accepted conventions in research areas. There are spin-off contributions to other areas of study. For example, our work has important implications for research into economic development. Finally, there is the consideration that a moral obligation exists to use pre-attack research skills in order to leave global disaster survivors with the best possible information.

The positive and negative moral questions present an ethical dilemma for the analyst of global disaster planning. In addition, this dilemma parallels a general problem of all risk analysis: to what extent do we create new conflicts and stress in our society by examining low probability and high consequence events?

REFERENCES

Alchian, A., and Allen, W., 1969, *Exchange and Production: Theory in Use*, Wadsworth Publishing Co., Belmont, CA.

Dynes, R., Quarantelli, E. L., and Kreps, G., 1981, *A Perspective on Disaster Planning*, Third Edition, The Disaster Research Center, Ohio State University, Columbus, OH.

Williamson, Oliver E., 1985, *The Economic Institutions of Capitalism*, The Free Press, New York.

Risk Aversion in Agricultural Policy Analysis: A Closer Look at Meaning, Modeling, and Measurement

Michael D. Weiss
U.S. Department of Agriculture
Washington, DC

ABSTRACT

Estimating participation rates of farmers in voluntary farm programs requires an understanding of farmers' preferences among alternative probability distributions of income. Traditionally, agricultural economists have assumed that farmers' utility functions of income completely expressed their risk preferences. In particular, it has been widely believed that risk aversion is equivalent to the concavity of the utility function of income. However, this belief is not entirely accurate. While the equivalence between risk aversion and concavity holds when risk preferences are continuous, the equivalence can fail in one direction and *appear* to fail in the other direction when risk preferences are discontinuous. In this paper, we describe these results within a mathematically precise framework. In addition, we illustrate how our techniques could be used to measure farmers' risk responses in the presence of agricultural price supports.

KEYWORDS: Risk aversion, expected utility theory, behavior under risk, convex set of lotteries, agricultural policy, lottery space, risk preference ordering, induced utility function

INTRODUCTION

Risk is an ever-present feature of agricultural production. The unpredictability of weather, pest infestations, harvest-time prices, and other factors leads to great uncertainty in farmers' incomes, both individually and collectively. Thus, in their roles as entrepreneurs, farmers are continually faced with the need to assess risks. But this assessment of risks involves not merely the attempt to determine, by whatever scientific or unscientific means, the probability of various disagreeable (or agreeable) individual events; rather, it involves also the need to choose among entire alternative probability distributions of outcomes. For example, if a farmer in a region of normally adequate precipitation installs an irrigation system, his expenditures will increase by the cost incurred, but he will be protected against the small probability of a calamitous loss due to drought. In deciding whether to install the system, he must choose between a probability distribution of net income with a small but positive probability of a drought-induced calamity, and a distribution of net income with

New Risks
Edited by L. A. Cox, Jr., and P. F. Ricci
Plenum Press, New York, 1990

possibly a smaller expected value but essentially a zero probability of such a calamity. Similarly, in determining whether to participate in a farm program that may affect his probability distribution of income, a farmer must decide whether he prefers the program-induced income distribution to his free-market income distribution.

Officials responsible for agricultural policy seek to devise programs that will "improve" farmers' probability distributions of income. These programs rely on voluntary farmer participation, and their success depends crucially on policy makers' astuteness in predicting participation rates—in effect, farmers' preferences between the old and new risky environments. Thus, agricultural economists have long sought to understand farmers' preferences under risk. In this endeavor, they have relied largely on three mainstays of the economic theory of individual choice under risk: (1) the assumption that individuals have a utility function of income (a functional relationship, unique only up to an increasing linear transformation, showing the "utility" an individual receives from each level of income); (2) the theory of expected utility, which posits an individual faced with a choice among alternative probability distributions of income as preferring that distribution, G, whose expected utility $\int_{-\infty}^{\infty} u(t)dG(t)$ (u being the utility function of income) is largest; and (3) the concept of risk aversion, which (in its strong sense) refers to behavior in which an individual given a choice between receiving the random outcome of a probability distribution and, alternatively, receiving with certainty the expected value of the distribution, always prefers the latter. Agricultural economists have attempted to determine the presence and degree of risk aversion in farmers' behavior by estimating the Arrow-Pratt measure of risk aversion, an expression involving derivatives of the utility function u (see Varian, pp. 108-110).

These efforts of agricultural economists—and, indeed, similar efforts by general economists to model and measure behavior under risk—have been predicated on the widely held but unexamined belief that, within expected utility theory, risk aversion is equivalent to the concavity of the utility function of income. As we shall see, this belief is not entirely accurate. Rather, while the equivalence holds (in a weak form) when risk preferences are continuous, it may fail in one direction and *appear* to fail in the other direction when risk preferences are discontinuous. Importantly, the axioms of expected utility theory do not imply that risk preferences need be continuous (Weiss, 1985, 1987). Furthermore, an assumption of continuous preferences imposes the not-entirely-innocent requirement that behavior under certainty completely determine behavior under uncertainty. Such an assumption is not obvious on any empirical grounds.

Risk aversion is customarily illustrated or even defined through the use of a diagram displaying a utility curve (the graph of the utility function) lying in the income/utility plane. Unfortunately, this graphical approach is misleading, for it only shows the individual's preferences among certainties. In the general case, it conveys no information whatsoever concerning preferences among continuous probability distributions, such as normals. For example, it is perfectly possible for a farmer's risk preferences to be discontinuous—for his preferences among certainties not to determine his preferences among continuous distributions—even while his utility function of income is continuous. In such a case, concavity is not equivalent to risk aversion. Moreover, the use of Arrow-Pratt risk aversion coefficients to measure risk aversion becomes entirely spurious.

In what follows, we shall present a new, more fundamental way of looking at risk aversion. Our purpose is not only to clarify the relationship between risk aversion and concave utility, but, more generally, to contribute to the ongoing search for improved theories of farmers' behavior under risk by placing the key notion of risk aversion on a conceptually clearer and more rigorous foundation.

PRELIMINARY RISK CONCEPTS AND DEFINITIONS

The theory of individual choice under risk is concerned with the choices that individuals make when confronted with alternative risky prospects, called *lotteries* and represented mathematically as cumulative probability distribution functions.

For each r, the lottery F_r defined by $F_r(t) = 0$ if $t < r$, $F_r(t) = 1$ if $t \geq r$, is called *degenerate*. For purposes of empirical application, degenerate lotteries represent certainties. A lottery L is called *simple* if it can be expressed as a convex combination, $L = \sum_i p_i F_{r_i}$, of finitely many degenerate lotteries F_{r_i} (where each p_i is nonnegative and $\sum_i p_i = 1$). (Note that $\sum_i p_i F_{r_i}$ is an algebraic combination not of numbers, but of *functions*. It is a function whose value at any number t is a number $\sum_i p_i F_{r_i}(t)$.) A lottery is called *continuous* if it is continuous as an ordinary function.

An essential requirement in risk theory is that the individual's choice set of lotteries be *convex*. (A set is convex if, for any two points in the set, the line segment joining these two points itself lies in the set. This definition holds even in abstract settings, when the "points" are functions. See Schaefer, p. 37.) Note that any choice set of lotteries is contained in the set of all numerical-valued functions on the number line; moreover, the latter set is a *vector space* (Hoffman and Kunze, pp. 28-29) under the usual operations of addition/subtraction of functions and multiplication of functions by numbers. A choice set of lotteries is *convex* if it is convex as a subset of this vector space. Henceforth, we shall call a convex choice set of lotteries a *lottery space*. If **L** is a lottery space and V a numerical-valued function defined on **L**, we call V *linear* if $V(tL_1 + (1-t)L_2) = tV(L_1) + (1-t)V(L_2)$ whenever $0 \leq t \leq 1$ and L_1, L_2 are lotteries in **L**.

The notion of an individual's risk preferences is formalized as follows. A (*weak*) (*risk*) *preference ordering*, $\gtrsim$, on a lottery space **L** is a binary relation on **L** satisfying the following properties: (1) for any L_1, L_2 in **L**, either $L_1 \gtrsim L_2$ (read as "L_1 is weakly preferred to L_2") or $L_2 \gtrsim L_1$ (the completeness property); (2) if $L_1 \gtrsim L_2$ and $L_2 \gtrsim L_3$, then $L_1 \gtrsim L_3$ (the transitivity property). A preference ordering $\gtrsim$ is analogous to the "greater than or equal to" relation, $\geq$, used with numbers.

Given a preference ordering $\gtrsim$, we can define the corresponding *indifference relation*, $\sim$, on **L** by $L_1 \sim L_2$ (read as "the individual is indifferent between L_1 and L_2") if and only if $L_1 \gtrsim L_2$ and $L_2 \gtrsim L_1$. We define the corresponding *strong* preference ordering, $>$, on **L** by $L_1 > L_2$ (read as "L_1 is strictly preferred to L_2") if and only if $L_1 \gtrsim L_2$ but not $L_1 \sim L_2$. If **L** is a lottery space and $\gtrsim$ a preference ordering on **L**, we call the pair (**L**, $\gtrsim$) a *preference lottery space*. A *utility function*, U, for (**L**, $\gtrsim$) is a numerical-valued function defined on **L** such that, for any L_1, L_2 in **L**, $U(L_1) \geq U(L_2)$ if and only if $L_1 \gtrsim L_2$. (U is said to *represent* $\gtrsim$.) A utility function for (**L**, $\gtrsim$) is called *measurable* if it is linear on **L**. The concept of a measurable utility function is central to expected utility theory. Of course, the linearity property mimics the computation of an expected utility. The whole thrust of von Neumann and Morgenstern and Herstein and Milnor was to prove that, under certain plausible conditions, an individual's risk preference ordering can always be represented by a measurable utility function. We stress, however, that a measurable utility function need not take the form of an expected utility integral.

Next, we explain what is meant by continuous risk preferences. A sequence $\{G_n\}$ of probability distribution functions is said to *converge* to a distribution G if $\lim_{n \to \infty} G_n(t) = G(t)$ for each point t of continuity of G (Breiman, p. 159). Let **L** be a lottery space, U a numerical-valued function defined on **L**, and L a lottery in **L**. We say U is *continuous at* L if, whenever $\{L_n\}$ is a sequence of lotteries in **L** that converges to L, one has $U(L_n) \to U(L)$

as $n \rightarrow \infty$. If U is continuous at each lottery in **L**, we say U is *continuous*. We call a risk preference ordering $\gtrsim$ *continuous* and speak of continuous risk preferences when there exists a continuous utility function representing $\gtrsim$.

As we have seen, the utility functions of risk theory assign numerical values to *lotteries*. Where, then, do so-called utility functions of income, as displayed in the income/utility plane in most treatments of risk aversion and which assign numerical values not to lotteries but to *income levels* (numbers), fit in? To answer this question, let U be a numerical-valued function defined on a lottery space **L** that contains all degenerate lotteries. Then, we can define a numerical function u by putting $u(r) = U(F_r)$ for each r. We say u is *induced (on the number line) by U*. If, in particular, (**L**, $\gtrsim$) is a preference lottery space for which U is a utility function, we may call u the *utility function induced by U* and interpret it as (for example) a utility function of income. The function u is, in effect, the certainty part of U — u reflects the values of U under certainty, that is, at degenerate lotteries. It is important to note, however, that distinct utility functions (and, in particular, distinct *measurable* utility functions) representing incompatible risk preferences can have the same certainty part, i.e., can induce the same utility function on the number line (Weiss, 1987, pp. 42-43). Thus, unless additional restrictions are imposed, a utility function on the number line cannot be assumed to describe an individual's risk preferences unambiguously.

Finally, we define the various classically-studied types of response to risk. Denote the mean of a lottery L by $E(L)$. Then, the degenerate lottery $F_{E(L)}$ represents receiving, with certainty, the expected value of L. Let (**L**, $\gtrsim$) be a preference lottery space such that, for each L in **L**, $E(L)$ is finite and $F_{E(L)}$ is also in **L**. Then, (**L**, $\gtrsim$) (or $\gtrsim$) is called

(1) *weakly risk averse* if, for each L in **L**, $F_{E(L)} \gtrsim L$,
(2) *risk neutral* if, for each L in **L**, $F_{E(L)} \sim L$, and
(3) *weakly risk loving* if, for each L in **L**, $L \gtrsim F_{E(L)}$.

If $\gtrsim$ is replaced by $>$ in (1) or (3), then (**L**, $\gtrsim$) (or $\gtrsim$) is called *strongly risk averse* or *strongly risk loving*, respectively.

The preceding definition makes it clear that risk aversion is a purely ordinal concept. In particular, it does not depend on any notion of utility or expected utility.

RISK AVERSION AND CONCAVE UTILITY

We can now state a result clarifying the relationship between risk aversion and concavity of the utility function of income:

Let (**L**, $\gtrsim$) be a preference lottery space for which **L** contains every degenerate lottery and for which every lottery in **L** has a finite mean. Assume (**L**, $\gtrsim$) has a measurable utility function that induces on the number line a utility function (of income), u. Then:

1. If $\gtrsim$ is weakly risk averse (respectively, weakly risk loving; risk neutral), then u is concave (respectively, convex; linear). The analogous statement holds if "weakly" is replaced by "strongly," using strict-inequality notions of concavity and convexity.

2. If u is concave (respectively, convex; linear) and $\gtrsim$ is continuous, then $\gtrsim$ is weakly risk averse (respectively, weakly risk loving; risk neutral).

For a proof of this result, see Weiss (1987).

The preceding result provides conditions under which the traditional equivalence between risk aversion and concavity holds. We now ask: when can this equivalence *fail*?

To answer this question, we first examine part 2 of the result. It is natural to ask whether concavity of u implies risk aversion for $\gtrsim$ when $\gtrsim$ is *not* continuous. It turns out that it does not. In fact, given an arbitrary numerical function u, it is always possible to construct a measurable utility function U that induces u, yet represents a preference ordering that is *not* risk averse. Briefly, U is constructed by patching together two other utility functions. First, a measurable utility function U_1 is defined over a lottery space $\mathbf{L}_1$ consisting of certain continuous lotteries. U_1 is constructed in such a way that the corresponding preference ordering is not risk averse. Then, a measurable utility function U_2 is defined over the lottery space $\mathbf{L}_2$ of all simple lotteries in such a way that U_2 induces u (note that every degenerate lottery is simple). Finally, U is defined over the *convex hull* of the union of $\mathbf{L}_1$ and $\mathbf{L}_2$ (i.e., over the "smallest" convex set containing both $\mathbf{L}_1$ and $\mathbf{L}_2$; see Schaefer, p. 39). The definition of U is based on the fact that any lottery L in this convex hull has an essentially unique representation of the form $L = pL_1 + (1-p)L_2$, where L_1 is in $\mathbf{L}_1$, L_2 is in $\mathbf{L}_2$, and $0 \le p \le 1$; this representation justifies defining $U(L)$ as $pU_1(L_1) + (1-p)U_2(L_2)$. It can be shown that U induces u and that U represents a non-risk averse preference ordering. Since u was arbitrary, this example shows that concavity of u (or, indeed, *any* property of u alone) is insufficient to guarantee risk aversion.

Although part 1 of our main result correctly asserts that risk aversion always implies concavity, even this implication can *appear* to fail. For, using the same type of two-part construction as above, one can define a numerical function v, strictly convex over [0,1], and a measurable utility function U representing a weakly risk averse preference ordering, such that, for any continuous lottery L on [0,1] (i.e., for which $L(0) = 0$ and $L(1) = 1$), one has $U(L) = \int_0^1 v(t)dL(t)$. Here, U is expressible as a traditional expected utility integral with a strictly convex integrand. Since the integrand in such representations normally must be the induced utility function, this example *seems* to establish that risk aversion need not imply concavity.

However, the situation here is not normal, for U is discontinuous—it is defined differently over certainty (degenerate lotteries) and continuous uncertainty (continuous lotteries). As a consequence, the usual proof that v must be the utility function induced by U does not apply, and our example of an apparent failure of risk aversion to imply concavity of the induced utility function turns out to be merely an illusion. If there is a moral to this example, it is that, in the absence of continuous risk preferences, risk aversion can appear in unexpected forms.

IMPLICATIONS FOR IDENTIFYING AND MODELING RISK AVERSION

To understand how the previous examples relate to the separate problems of identifying and modeling behavior in the presence of risk, consider first an investigator who wishes to construct an economic model involving risk averse behavior. Within the *traditional* interpretation of expected utility theory, the investigator would, in effect, assume that measurable utility functions take the form of expected utility integrals, and he would adopt a concave numerical function (a von Neumann-Morgenstern utility function) as a generator of the risk behavior. This approach, however, carries the implicit assumption that the individuals under study have the same frames of reference toward certainty (degenerate lotteries) and continuous uncertainty (continuous lotteries), since the measurable utility of a lottery is determined, through the integral formula, by the utility values assigned to certainties.

If, on the contrary, the researcher does not wish to rule out *a priori* the possibility of different frames of reference—the possibility that an individual's risk preferences for certainties may differ from his risk preferences for continuous lotteries, so that his risk preferences are discontinuous—then the assumption of concavity for v would not guarantee risk averse behavior. For, as evidenced by our first example, even a strictly concave utility function of income can be induced by a measurable utility function representing a preference ordering that is *not* risk averse.

Our second example, however, suggests that some risk averse preference orderings might manifest the *illusion* of being determined by a nonconcave—or even strictly convex—utility function of income. For, as that example indicated, preferences among certain continuous lotteries can be determined by such an "apparent" utility function even when the full preference ordering is risk averse. In a work developed within the classical expected utility framework, Hildreth and Knowles cite several empirical studies (including their own) of individuals' risk preferences that produced apparently risk neutral or risk loving responses (in effect, nonconcavities in the induced utility functions) in cases where one might expect the decision makers' true preferences to be risk averse (Hildreth and Knowles, p. 33). They characterize these nonconcavities as inaccurate and suggest various possible explanations for their occurrence (see also Robison).

The theory that we have delineated here, however, provides an alternative hypothesis: the nonconcavities are legitimate; interpreted within the context of measurable utility theory, they do not contradict risk aversion. Rather, the respondents in the studies, while being risk averse, may have had discontinuous, two-part measurable utility functions (reflecting dichotomous behavior toward certainty and continuous uncertainty), and they may have responded to some queries in the studies as though using their continuous uncertainty rule associated with a nonconcave function v (as in our second example) rather than their certainty rule associated with the (concave) induced utility function, u. This question, along with the more general question of whether farmers' behavior under uncertainty may not be entirely determined by their behavior under certainty (resulting in discontinuous risk preferences and inapplicability of Arrow-Pratt risk aversion coefficients), would appear to merit further investigation.

MODELING RISK RESPONSE TO AGRICULTURAL PRICE SUPPORTS: AN ILLUSTRATION

Two-part measurable utility functions permit us to model an individual's risk behavior in such a way that the values his measurable utility function, U, assigns to certainties (degenerate lotteries) do not determine the values of U at continuous lotteries. This approach is more general than that of traditional expected utility theory, which assumes that the values of an individual's utility function at certainties determine its values at *all* lotteries.

To illustrate the way in which two-part measurable utility functions might be applied to a problem relating to agricultural policy, consider the introduction of a price support for an agricultural commodity. Assume that the market price is a bounded continuous random variable, P_0. Suppose that a support price, s, is introduced, so that, in effect, P_0 is truncated below at s, forming a new random price variable, P. We assume that the event $P_0 < s$ has positive probability. Also, if X is any random variable, we denote its cumulative probability distribution function by "F_X". (In particular, in the special case in which X is constant, F_X is merely a degenerate lottery as defined earlier.) Then

$$P = \begin{bmatrix} s & \text{if } P_0 < s \\ P_0 & \text{if } P_0 \geq s \end{bmatrix},$$

so that

$$F_P(t) = \begin{bmatrix} 0 & \text{if } t < s \\ F_{P_0}(t) & \text{if } t \geq s \end{bmatrix}$$

and (as is easy to show) $F_P = F_{P_0}(s)F_s + [1-F_{P_0}(s)]G$, where G is a continuous lottery defined by

$$G(t) = \begin{cases} 0 & \text{if } t < s \\ [F_{P_0}(t) - F_{P_0}(s)]/[1 - F_{P_0}(s)] & \text{if } t \geq s. \end{cases}$$

The equation for F_P characterizes the price risk faced by a farmer following the introduction of a support price. Note that $F_{P_0}(s)$ is merely the probability that the market price will be equalled or exceeded by the support price, F_s represents the hypothetical situation in which the support price occurs with certainty, and $G(t)$ is the conditional probability that the market price will lie at or below t if it exceeds the support price.

Now, to define a two-part measurable utility function for this situation, let $\mathbf{L}_1$ be the lottery space of all simple lotteries, $\mathbf{L}_2$ the lottery space of all continuous lotteries, and $\mathbf{L}$ the convex hull of the union of $\mathbf{L}_1$ and $\mathbf{L}_2$. Suppose U_1 and U_2 are measurable utility functions defined on $\mathbf{L}_1$ and $\mathbf{L}_2$, respectively. (U_1 could be thought of as initially being defined only at degenerate lotteries—to represent behavior under certainty—and then formally extended to all of $\mathbf{L}_1$ by linearity. The nondegenerate simple lotteries do not play any practical role in the present example.) Given any L in $\mathbf{L}$, there exist a number p satisfying $0 \leq p \leq 1$ and lotteries L_1 in $\mathbf{L}_1$ and L_2 in $\mathbf{L}_2$ (unique when $0 < p < 1$) such that $L = pL_1 + (1-p)L_2$. Thus, we can define a two-part measurable utility function, U, on $\mathbf{L}$ by $U(L) = pU_1(L_1) + (1-p)U_2(L_2)$. In particular, we have $U(F_P) = F_{P_0}(s)U_1(F_s) + [1-F_{P_0}(s)]U_2(G)$, so that we can use U_1 and U_2 to represent different behaviors over the certainty and uncertainty parts of F_P. It follows that, in attempting to estimate an individual's utility function from empirical observations of his risk preferences, we do not have to impose the possibly oversimplifying assumption that he assigns utility values to continuous uncertainties (continuous lotteries) by using the same rule he uses for certainties (degenerate lotteries). Rather, our two-part model allows the possibility that he uses two *different* rules. By contrast, the traditional expected utility approach, incorporating as it does a hidden assumption of continuous preferences, would require that the utility functions U_1 and U_2 operate according to a *single* rule.

In particular, in conjunction with the equation for the price risk F_P under price supports, our two-part model provides a theoretically more flexible approach to measuring the utility of price supports to farmers.

CONCLUSIONS

Sound agricultural policy requires an understanding of farmers' behavior under risk. Consequently, it requires an understanding of the presence and degree of risk aversion in farmers' behavior. However, the meaning of risk aversion has long been obfuscated by the entanglement of this concept with an unrecognized assumption of continuous risk preferences in expected utility theory. In this paper, we have shown how risk aversion can be defined independently of any notion of utility. We have also shown how it can be distinguished from the concavity of a utility function of income. Finally, to illustrate the application of our methods, we have presented a new approach to modeling the utility of agricultural price supports to farmers. This approach, unlike others based on traditional methods, does not require behavior under risk to be determined by behavior under certainty. It therefore can encompass a wider range of farmer decisions.

REFERENCES

1. Leo Breiman, *Probability*, Addison-Wesley, Reading, MA (1968).
2. I. N. Herstein and John Milnor, An Axiomatic Approach to Measurable Utility, *Econometrica* **21**:291-297 (1953).

3. Clifford Hildreth and Glenn J. Knowles, Some Estimates of Farmers' Utility Functions, *Technical Bulletin 335*, Agricultural Experiment Station, University of Minnesota (1982).
4. Kenneth Hoffman and Ray Kunze, *Linear Algebra*, Prentice-Hall, Englewood Cliffs, NJ (1961).
5. Lindon J. Robison, An Explanation of Apparently Risk Preferring Behavior by Risk Averse Decision Makers, paper presented at meeting of North Central Research Committee NC-113, St. Louis, MO, October 1982.
6. Helmut H. Schaefer, *Topological Vector Spaces*, Macmillan, New York (1966).
7. Hal R. Varian, *Microeconomic Analysis*, W. W. Norton, New York (1978).
8. John von Neumann and Oskar Morgenstern, *Theory of Games and Economic Behavior*, 2nd ed., Princeton University Press, Princeton, NJ (1947).
9. Michael D. Weiss, Discontinuous Utility in Agricultural Risk Modeling, paper presented at 1985 American Agricultural Economics Association Annual Meeting, Ames, IA, August 1985.
10. Michael D. Weiss, Conceptual Foundations of Risk Theory, *Technical Bulletin 1731*, U.S. Department of Agriculture (1987).

Evaluating Risks to Agricultural Production from Acid Deposition

P. D. Moskowitz, N. L. Oden, W. H. Medeiros and E. A. Coveney
Brookhaven National Laboratory
Upton, Long Island, NY

ABSTRACT

Although it has been established that agricultural yields can be affected adversely by ozone and other air pollutants, effects of existing levels of acid deposition on crops are less well understood. Evaluations of potential effects from growth chamber, greenhouse and field experiments have not identified any single crop as being consistently sensitive to acid deposition. Quantitative analysis of one crop (soybeans), which has demonstrated some sensitivity to acid deposition treatments in field settings, suggest that if current acid deposition levels were reduced by 50%, then U.S. soybean production would increase by approximately 1%. These estimates are based on the fundamental assumption that estimated dose-response functions are homogeneous across biologic, geographic and temporal space; an assumption not supported by recently developed experimental data. As a result, confidence in this conclusion is weak.

KEYWORDS: Agricultural production, acid deposition, risk, uncertainty

INTRODUCTION

Although it has been established that agricultural yields can be affected adversely by ozone and other air pollutants, effects of existing levels of acid deposition on crops are less well understood. Evaluations of potential effects must be based on growth chamber, greenhouse and field studies. These studies in aggregate have not identified any single crop as being consistently sensitive to acid deposition. However, one commercially important crop—soybeans—has demonstrated some sensitivity to acid deposition in experimental settings. Estimation of the benefits/costs to this crop from changing acid deposition levels presents fundamental problems common to all similar assessments because of the need to extrapolate dose-response data across biologic, geographic and temporal space. These problems are discussed below.

DATA

Dose-response data are needed to evaluate quantitatively, the sensitivity of soybeans to changing acid deposition levels. These data may be derived from growth chamber, greenhouse and field studies. The preferred data for assessing effects are those obtained

New Risks
Edited by L. A. Cox, Jr., and P. F. Ricci
Plenum Press, New York, 1990

from field experiments which exclude ambient rainfall, but otherwise mimic standard agronomic protocols. Table 1 provides an overview of field experiments on soybeans which used such protocols. These studies, in aggregate, form the foundation for any effort attempting to evaluate impacts of acid deposition on soybean yields. Table 2 presents estimated linear slope coefficients for each of these data sets.

FUNDAMENTAL PROBLEMS

There are at least three fundamental problems the risk analyst must face in attempting to use these data in an assessment. They all relate to the extrapolation of dose-response data over differently dimensioned space: biologic, temporal and geographic.

Biologic space is important because there is limited evidence which suggests that sensitivity to acid deposition treatments varies among varieties. In Table 3, within variety homogeneity significance tests are presented. Although all linear coefficients were homogeneous within varieties in the 1982-1984 data sets, this was no longer true when the 1985 data were introduced. This issue from an assessment perspective is very important because there are a large number of soybean varieties planted across the U.S. Experimental data are only available for seven varieties: Amsoy, Asgrow, Corsoy, Williams, Davis, and Forrest. These represent 24% of the total U.S. soybean acreage planted in 1983. They were used in 16 of the 28 soybean growing states; in these states, they constituted up to 62.7% (Missouri) of the total plantings. Because variety specific dose-response data are limited simplifying assumptions must be made about the sensitivity of the untested soybean varieties to conduct an assessment.

Geographic space is important because dose-response functions must be extrapolated from soybean field experiments conducted at only four sites to all areas in which soybeans are grown: 1822 counties and 28 states. As shown in Table 3, response of yield to pH differs significantly across sites. Because of this heterogeneity, extrapolation of the dose-response surface over large geographic space is not fully justified.

Temporal space is important because risk analysts often attempt to project how the affected receptor population will respond, given some future change in a predictor variable. Some experimental data suggest that temporally variant predictor variables may exist and confound such predictions. Asgrow, for example, was examined at only one site. Over the two years of study, the estimated linear response coefficient of yield to pH changed from 2.98 to −0.6, a difference which is statistically significant ($p<0.02$). Therefore extrapolation to future times may not be justified based on the available data for this variety.

There are several possible reasons for the observed heterogeneity. First, the differences may be due to outliers. Second, intentional differences in experimental procedures may exist. Third, it may be that yield is a function only of pH, but the function is not linear. Fourth, that the dose-response functions should include more predictor variables which interact with the pH effect. Of these possible explanations, the last seems most plausible since there are both theoretical and limited experimental data which suggest that such interactions may exist.

DISCUSSION AND CONCLUSIONS

In a previous study conducted by Moskowitz *et al.* (1986), impacts of current acid deposition levels on U.S. soybean yields were estimated by extrapolating dose-response data from experiments conducted between 1982-1984 over large biologic (i.e., varieties not studied), geographic (i.e., counties and states in which experiments were not conducted), and temporal space. Inherent in this analysis was the fundamental assumption that the

Table 1. Soybean Experiments Used in the Dose-Response Analysis

Variety	Investigator/ Location	Experiment Year*	Experiment Design
Amsoy	Banwart/Illinois	1983[1]-1984[2]-1985[8]	Split plot
	Evans/New York	1982[3]-1983[4]-1984[5]-1985[9]	Latin square
Asgrow	Evans/New York	1984[5]-1985[9]	Latin square
Corsoy	Evans/New York	1984[5]-1985[9]	Latin square
Davis	Shriner/Tennessee	1983[6]-1984[6]	Randomized complete block
Forrest	DuBay/North Carolina	1983[3,7]	Split plot
Hobbit	Evans/New York	1984[5]-1985[9]	Latin square
Williams	Banwart/Illinois	1983[1]-1982[2]-1985[8]	Split plot
	Evans/New York	1982[3]-1983[4]	Latin square

*Experiments are referred to in the text by author and year of experiment.

1. Banwart, W. L., 1984, Simulated Acid Rain Effects on Yield and Growth of Corn and Soybeans and on Soil Parameters, in Proc. of Peer Review Meeting, sponsored by the EPA Acid Precipitation/Agriculture Program, March 1984, U.S. Environmental Protection Agency, Corvallis, OR.

2. Banwart, W. L., 1985, Sensitivity of Soybean Varieties to Simulated Acid Rain, in Agriculture Effects Task Group F Peer Review Meeting, Brookhaven National Laboratory, February 4-7, 1985, North Carolina State University Acid Deposition Program, Raleigh, NC.

3. Evans, L. S., K. F. Lewin, and M. J. Patti, 1984, Effects of Simulated Acid Rain on Yields of Field-Grown Soybeans, *New Phytol.* **96**:207-213.

4. Evans, L. S., 1984, Effects of Simulated Acidic Rain on Yields of Field-Grown Soybeans, in Proc. of Peer Review Meeting, sponsored by the EPA Acid Precipitation/Agriculture Program, March 1984, University of Illinois, Urbana, IL.

5. Evans, L. S., G. R. Hendrey, and K. F. Lewin, 1985, Comparison of Yields of Several Varieties of Field-Grown Soybeans Exposed to Simulated Acidic Rainfalls, in Agriculture Effects Task Group F Peer Review Meeting, Brookhaven National Laboratory, February 4-7, 1985, North Carolina State University Acid Deposition Program, Raleigh, NC.

6. Shriner, D. S. *et al.*, 1985, Acid Deposition: Effects on Agricultural Crops, Final Report to the Electric Power Research Institute (Project PR 1908-2) Oak Ridge National Laboratory, May 1985.

7. DuBay, D. T., A. S. Heagle, W. W. Heck, and J. D. Elson, in preparation, The Effects of Simulated Acidic Rainfall and Ambient Rainfall Exclusion on the Growth and Yield of Field-Grown Soybean (Glycine Max, c.v. Forrest).

8. Banwart, W. L., 1986, Personal communication.

9. Evans, L. S., 1986, Personal communication.

Table 2. Estimated Coefficients and Standard Errors from the Linear Analyses

Variety	Study	Year	B_1	SE
Amsoy	Evans	1982	2.31*	1.00
		1983	2.88*	1.40
		1984**	3.11*	0.74
		1985	2.76*	1.28
	Banwart	1983	2.31*	0.81
		1984	0.83	0.73
		1985	−1.13	0.89
	Pooled***	--	1.34*	0.38
Asgrow	Evans	1984	2.98*	0.84
	Pooled	1985	−0.60	1.33
			1.96*	0.71
Corsoy	Evans	1984	1.29	0.77
	Pooled	1985	2.19*	1.10
			1.59*	0.64
Davis	Shriner	1983-84	−0.29	--
Forrest	DuBay	1983	−0.02	0.49
Hobbit	Evans	1984	3.15*	0.60
	Pooled	1985	0.36	1.40
			2.72*	0.55
Williams	Evans	1982	0.66	0.98
		1983	0.63	1.04
	Banwart	1983	0.38	0.81
		1984	0.34	0.73
		1985	−0.02	0.74
	Pooled	--	0.34	0.37

Note: Linear models are of the form $Y = B_0 + B_1 *$ pH, where Y = soybean yield in bushels/acre, B_0 = the "Y" intercept, B_1 = linear slope coefficient, and pH = the growing season average pH. SE = standard error of slope coefficient.

*Significant, $p \leq 0.05$.

**Not included in pooled linear model.

***Pooled coefficients have been included in this table even though the data suggest a heterogeneous response.

Table 3. Within-Variety Homogeneity Significances as of 1984 and as of 1985 for All Experiments in Which the Linear Model Sufficed

	X = 1984	X = 1985
Amsoy	0.39	0.02
Asgrow	----	0.02
Corsoy	----	0.50
Hobbit	----	0.07
Williams	0.99	0.98

estimated dose-response functions were homogeneous across these different dimensions. This assumption was weakly supported by the data available at that time.

The newly collected data invalidate this fundamental assumption. As a result, the uncertainty surrounding the original estimates is likely to be larger than originally estimated. Ideally, this uncertainty could be reduced by conducting controlled field experiments in every county where soybeans are currently grown. These experiments would investigate all soybean varieties of interest and all interactive effects. Of course, the existing state of affairs is far from this ideal. Cross-sectional approaches which have been used to develop dose-response functions from unplanned data sets provide an alternative approach for resolving some of these concerns. They also introduce a set of new problems (Moskowitz *et al.*, 1985).

Nevertheless, this source of uncertainty may be of important statistical significance, but in this particular instance is unimportant from a practical or policy perspective. In the original study, Moskowitz *et al.* (1986) estimated that 1982 acid deposition levels were reducing soybean yields by approximately 0.1 to 3%. They noted, that in comparison with other natural (e.g., drought/50%, insects/3%) and anthropogenic (e.g., ozone/10-19%) factors, the estimated impacts from acid deposition were small. If the new field data were incorporated into this assessment, the estimated impacts would be further reduced since the coefficients are smaller.

REFERENCES

Banwart, W. L., 1984, Simulated Acid Rain Effects on Yield and Growth of Corn and Soybeans and on Soil Parameters, in Proc. of Peer Review Meeting, sponsored by the EPA Acid Precipitation/Agriculture Program, March 1984, U.S. Environmental Protection Agency, Corvallis, OR.

Banwart, W. L., 1985, Sensitivity of Soybean Varieties to Simulated Acid Rain, in Agriculture Effects Task Group F Peer Review Meeting, Brookhaven National Laboratory, February 4-7, 1985, North Carolina State University Acid Deposition Program, Raleigh, NC.

Banwart, W. L., 1986, Personal communication.

DuBay, D. T., Heagle, A. S., Heck, W. W., and Elson, J. D., The Effects of Simulated Acidic Rainfall and Ambient Rainfall Exclusion on the Growth and Yield of Field-Grown Soybean (Glycine Max, c.v. Forrest), in preparation.

Evans, L. S., Lewin, K. F., and Patti, M. J., 1984. Effects of Simulated Acid Rain on Yields of Field-Grown Soybeans, *New Phytol.* **96**:207-213.

Evans, L. S., 1984, Effects of Simulated Acidic Rain on Yields of Field-Grown Soybeans, in Proc. of Peer Review Meeting, sponsored by the EPA Acid Precipitation/Agriculture Program, March 1984, University of Illinois, Urbana, IL.

Evans, L.S., Hendrey, G.R., and Lewin, K. F., 1985, Comparison of Yields of Several Varieties of Field-Grown Soybeans Exposed to Simulated Acidic Rainfalls, in Agriculture Effects Task Group F Peer Review Meeting, Brookhaven National Laboratory, February 4-7, 1985, North Carolina State University Acid Deposition Program, Raleigh, NC.

Evans, L. S., 1986, Personal communication.

Moskowitz, P. D., Medeiros, W. H., Oden, N. L., Thode, Jr., H. C., Coveney, E. A., and Rosenthal, R.H., An Assessment of Risks to U.S. Soybean Yields from 1982 Acid Deposition Levels, *Risk Analysis* (in press).

Moskowitz, P. D., Oden, N. L., Thode, Jr., H. C., Medeiros, W. H., and Coveney, E. A., 1985, Response to Comments on Crops Research Projects 1A and 1B, *The American Statistican* **39(4)**,Pt.1:261-263.

Oden, N., Moskowitz, P. D., and Medeiros, W. H., Soybeans and Acid Deposition: An Analysis of Dose-Response Data from Field Experiments, Brookhaven National Laboratory, Upton, NY (in preparation)

Shriner, D. S., Johnston, Jr., J. W., Taylor, Jr., G. E., Luxmoore, R.J., McConathy, R. K., McLaughlin, S. B., Heagle, A. S., Norby, R. J., Takemoto, B. K., DuBay, D. T., Abner, C. H., and Richter, D. D., 1985, Acid Deposition: Effects on Agricultural Crops, Final Report to the Electric Power Research Institute (Project PR 1908-2) Oak Ridge National Laboratory, May 1985.

Economic Principles of Risk Management in Production

Edna Loehman
Purdue University
West Lafayette, IN

ABSTRACT

Risk management in a production context refers to strategic choice of input and output mix ("self insurance") when production is risky. Production risk may be due to price risk (when market prices are not known at the time input decisions must be made) and/or due to the "state of nature" (e.g. weather risk or machine breakdown risk). Decisions made regarding production affect the level of its riskiness.

This paper develops necessary conditions for tests of optimality of input use in risky production; these conditions are more general than conditions which have previously been used (based on mean-variance preference assumptions or normality).

Principles are demonstrated using an agricultural application involving estimation of a stochastic production function. Results indicate that inputs in this application are not being chosen in terms of economic principles.

KEYWORDS: Risk management, production, economic principles, agriculture

INTRODUCTION

Risk management is a term often used to refer to choice of investment portfolios or use of insurance. However, in a production context, it can also refer to strategic choice of input and output mix ("self insurance").

Risk refers to the fact that outcomes are probabilistic rather than certain. (Here we do not distinguish between objective and subjective probabilities in defining risk.) Production risk may be due to price risk (when market prices are not known at the time input decisions must be made) and/or due to the "state of nature" (e.g., weather risk or machine breakdown risk). Decisions made regarding production affect the level of risk.

For example, a production process may be subject to machine breakdown. A producer may then decide how much labor input to allocate to running the machine to produce an output and how much to allocate to inspection and maintenance of the machine. Agricultural production subject to weather risk is another example; use of fertilizer will increase output per unit of land provided that there is sufficient rain but will decrease output with insufficient rain. The farmer must decide how much

New Risks
Edited by L. A. Cox, Jr., and P. F. Ricci
Plenum Press, New York, 1990

fertilizer to use. As another example, when it is not known what types of products (e.g. clothes) consumers will buy, a manufacturer may decide to diversify and produce goods of different types.

For production with certain outcomes and certain prices, economic principles or rules have been developed to define optimal input use and other production choices; e.g. marginal profit due to an input (e.g. labor) should equal zero at the optimum and, if less than zero, input use should be decreased under the condition of declining marginal product. Here, we seek to develop similar principles and rules which can be used to assess input use and output mix in the case of risky production, using observable production and price information.

Previous research in risk has tackled this type of problem with mean-variance analysis. However, it is well known that this analysis is valid only if the utility function used to describe preferences for income is quadratic or if random variables are normally distributed. The approach developed here does not require such heroic assumptions; that is, the possibilities of skewness of probability distributions and nonzero third derivatives of the utility function (e.g. decreasing absolute risk aversion) are explicitly allowed.

The concepts developed here are demonstrated using an example of agricultural production with weather, insect, and price risk.

DESCRIPTION OF PRODUCTION RISK

As mentioned above, risk in production can occur both due to price and the production process. In order to determine optimal decisions, it is required that the resulting possible profit levels due to these causes be defined; then preferences over these resulting possible profit outcomes can be applied to rank alternative risk management strategies (i.e. alternative input-output combinations). In the case of only price risk, profit (price times quantity produced minus input expenditure) is a random variable which is a linear transformation of the price random variable. The effect of production process risk on profit is generally more difficult to describe mathematically.

The "stochastic production function" is a description of output (y) as a transformation of stochastic variables (θ) representing the state of nature; this transformation is dependent on the level of economic inputs (x). As discussed by Just and Pope (1978,1979), the functional form chosen to describe this transformation defines the risk management effects of inputs. For example, if one assumes a separable relationship between inputs and the state of nature,

$$y = f(x) + h(\theta),$$

then no effect of inputs on variance of output or profit will be implied. Thus, in order to model the effect of an input on variance, interaction terms between the inputs and physical factors are required in the specification of the production function. However, if a multiplicative form

$$y = f(x)h(\theta)$$

is used, then all inputs with positive marginal products will increase the variance of output and profit. To allow for the possibility of all types of effects of inputs on variance, a function of a general form

$$y = f(x) + g(x)h(\theta)$$

is needed. This form is general enough to describe single or multi-stage production processes. However, it implies that the distribution of output has the same skewness as $h(\theta)$ and that input use does not affect skewness of output or profit.

Below, we demonstrate how to measure the production function and its distribution and how to test the optimality of observed production decisions.

OPTIMAL INPUT USE

The economic model for optimal production decisions gives the basis for the producers' choice of inputs and mix of outputs consistent with their own subjective preferences. In addition to choosing the inputs to produce an output, producers may produce a mix of several outputs which differ in terms of "riskiness" due to the state of nature. The effects on outputs due to the random state of nature and interactions among inputs with the state of nature are explicitly specified by the stochastic production function.

The most general representation of multiple output production would be represented by

$$F(y^1, ...y^n, x_1, ...x_m, \theta) = 0$$

where there are n outputs y^j and m inputs x_i. To simplify analysis, here we assume that processes for each output are independent. Stochastic production functions are then expressed as

$$y^j = f^j(\{x^j\}, \theta).$$

Such functions express the relation of output (y^j) to inputs (x_i^j) used and the realization of the state of nature. Also, it is assumed that there is an input (A) which is fixed in supply in the short run (e.g. land or some other space or capacity measure). We define the production function in terms of output per unit of this resource so that total output of j is the amount of the fixed resource allocated to this output times output per unit.

Price risk is also included since output prices are random variables when decisions about input use are made. Input prices (w) are not random; they are assumed to be known at the time that input decisions are made. The joint distribution of output prices and the physical state of nature is denoted by $g(p^1, p^2, ..., p^n, \theta)$. Income in each state of nature is then revenue for each output (price times quantity) minus input expenditure.

It is also assumed that a utility function $u(\Pi)$ over income in each state is used to define preferences. Assuming that inputs are chosen by expected utility maximization with the given distribution and production functions and fixed resource constraint $\overline{A}$, the following problem describes optimal production:

$$\underset{x_i^j, A^j}{\text{Max}} \int \cdots \int u\left(\sum_j A^j(p^j f^j(\{x^j\}, \theta - \sum_i w_i x_i^j))\right) g(p^1, p^2, ..., p^n, \theta)dp^1..dp^n d\theta$$

$$s.t. \sum A^j \leq \overline{A} .$$

(Each random variable is integrated over the domain of its distribution function.)

First order conditions for this problem identify the nature of optimal solutions. The definition of covariance (denoted by Cov) is used to write the conditions in terms of expected profit and expected marginal profit, for example:

$$\operatorname{Cov}\left(u', \frac{\partial \pi^j}{\partial x_i^j}\right) = E(u'\pi^j) - Eu'E\frac{\partial \pi^j}{\partial x_i^j}$$

where $\pi^j = p^j y^j - w \cdot x^j$ is the per unit profit. "Eu'" denotes the expectation of marginal utility taken over prices and the state of nature. Then, for variable inputs, optimality is defined by

$$\left\{Eu'\left[E\left(\frac{\partial \pi^j}{\partial x_i^j}\right)\right] + \operatorname{Cov}\left(u', \frac{\partial \pi^j}{\partial x_i^j}\right)\right\} A^j x_i^j = 0\,.$$

First order conditions which describe optimal allocation of the fixed resource over alternative outputs may similarly be written as:

$$\{Eu'[E(\pi^j)] + \operatorname{Cov}(u', \pi^j) - \lambda\}A^j = 0$$

where $\lambda > 0$ is the multiplier associated with the resource constraint.

RISK PREMIUMS AND OPTIMAL PRODUCTION

In economic theory, the utility function has only been used in a conceptual way to define an optimum since it is generally believed to be not observable (although some experimental psychologists may dispute this point). Another problem of the concept is that it is not in dollar units so it is not possible to use in a comparative way. Another concept, the certainty equivalent, is in monetary units and ranks preferences equivalently. The certainty equivalent is defined by

$$u(CE(x,A)) = Eu(\Pi(x,A,\theta))$$

where u denotes the utility function, CE denotes the certainty equivalent which is a function of input choices, Π denotes the profit function, and E denotes the expectation operator. The certainty equivalent may be further decomposed in terms of expected profit and a "risk premium" RP:

$$CE(x,A) = E\Pi(x,A) + RP(x,A)$$

It is well-known that with risk aversion, $u(CE) < u(E\Pi)$, so that RP is negative. The risk premium is also zero when no production occurs. Figure 1 illustrates $RP(x)$ in the one input case.

The production function and utility function define the possible combinations of RP and $E\Pi$:

$$RP(x,A) = -E\Pi(x,A) + u^{-1}(Eu(\Pi(x,A,\theta)))\,.$$

Figure 2 illustrates the relationship between RP and $E\Pi$ for persons of varying risk aversion. A more risk aversion person has a more negative RP value for a given level of $E\Pi$.

Taking derivatives with respect to an input x_i^j,

$$MRP_i^j = -E\Pi_i^j + \frac{E(u'\,\Pi_i^j)}{u'(CE)}$$

where the subscripts and superscripts denote partial derivatives with respect to input i used for output j. The marginal risk premium shows how the risk premium (or cost due to riskiness of production) is affected by increasing use of a given input. A negative value indicates that increased use of an input would produce a greater absolute risk premium while a positive value would result in a smaller absolute risk

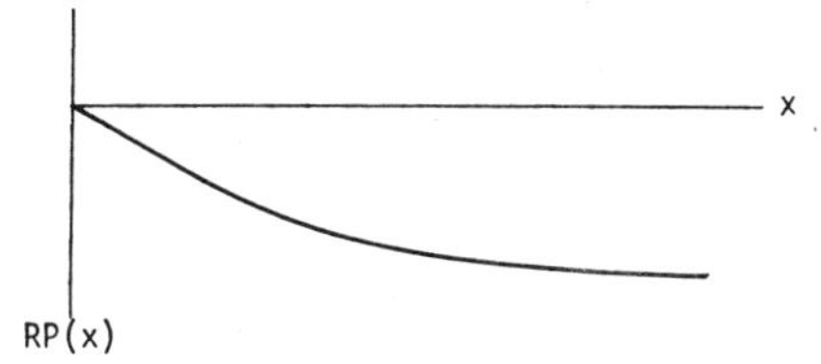

Fig. 1. Illustration of the risk premium as related to input use in the one input case.

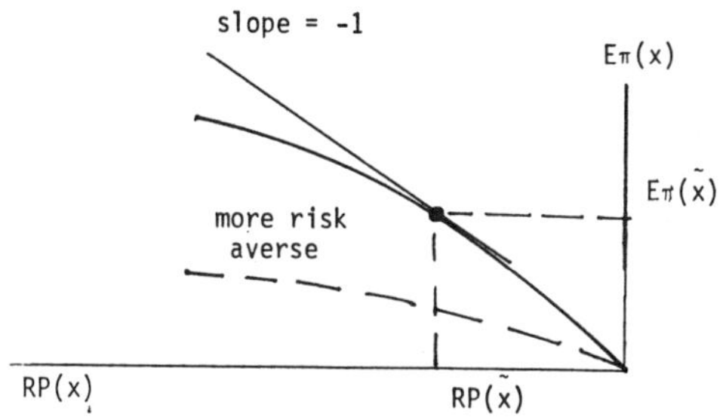

Fig. 2. The risk premium-expected profit frontier and optimal input use.

premium value since the risk premium is negative. Thus, the sign of the marginal risk premium leads to a classification system for inputs as risk premium increasing inputs or risk premium decreasing inputs.

Comparing the expression defining the marginal risk premium to the first order conditions above, we see that the optimum variable input use satisfies

$$MRP_i^j(x, A)) = -E\Pi_i^j(x, A) .$$

That is, at the optimum input use, MRP has the opposite sign to the expected marginal profit; in the one input case this means that this input must have a positive expected marginal product. Similarly, all inputs cannot have negative expected marginal products since then a positive risk premium would be implied.

The interpretation of the optimality condition is that at the optimum the risk premium cannot be simultaneously made less negative and the expected profit made greater by altering variable input use. That is, the optimum input use is on the expected profit-risk premium frontier where its tangent plane has slope −1 in all directions. In Fig. 2, persons with different risk aversion levels will make different input choices.

Comparing the MRP condition above to the corresponding first order condition, at the optimum,

$$MRP_i^j = \mathrm{Cov}\left(u', \frac{\partial \pi^j}{\partial x_i^j}\right) / Eu' .$$

(Note that a similar condition was proposed by MacMinn and Holtman but they used this as a definition of MRP whereas here this condition only holds at the optimum.)

With a fixed resource which is to be allocated over multiple outputs,

$$RP(x, A) = -\sum_j A^j \, E\pi^j + u^{-1}\left(Eu\left(\sum_j A^j \, \pi^j\right)\right)$$

so that

$$MRP_A^j = -E\pi^j + \frac{E(u' \, \pi^j)}{u'(CE)} \, .$$

Again, comparing this expression to the first order conditions, at the optimum,

$$MRP_A^j = -E\pi^j + \frac{\lambda}{u'(CE)}$$

and solving for $E\pi^j$ implies that at the optimum,

$$MRP_A = \text{Cov}(u', \pi^j)/Eu' \, .$$

Thus, an equivalent formulation of the first order conditions above in terms of the marginal risk premia at the optimum is then

$$\frac{\partial E\pi^j}{\partial x_i^j} = MRP_i^j$$

and

$$E\pi^j - E\pi^k = MRP_A^k - MRP_A^j$$

for two outputs j and k. Tests of optimality would be provided if the MRP signs were known, since then the signs of the expected marginal profits and difference in expected profits per unit of the fixed resource could be compared to the sign conditions implied above. Then, if a sign condition did not hold, it would be known that optimality would not be satisfied. If the sign conditions were satisfied, further information about preferences would be needed to confirm optimality.

DETERMINING THE SIGN OF MRP IN ONE CASE

We now consider under how the sign of the marginal risk premium might be determined based on observable profit effects while making only minimal assumptions about signs of the derivatives of the utility function. A one output case with no effect of inputs on skewness is presented here.

We use the first three terms of the Taylor series to approximate $u'(.)$ (or assume that the rest of the utility derivatives are zero); also assume $u' > 0$, $u'' < 0$ (risk aversion), and $u''' > 0$. Then

$$u'(\Pi) = u'(E\Pi) + u''(E\Pi)(\Pi - E\Pi) + u'''(E\Pi)(\Pi - E\Pi)^2/2$$

and

$$Eu' = u'(E\Pi) + u'''(E\Pi) \text{ var } \Pi/2$$

and

$$\text{Cov}\left(u', \frac{\partial\Pi}{\partial x_i}\right) = u''(E\Pi)\text{Cov}\left(\Pi, \frac{\partial\Pi}{\partial x_i}\right) + u'''(E\Pi)\text{Cov}\left((\Pi - E\Pi)^2, \frac{\partial\Pi}{\partial x_i}\right)/2 \, .$$

Now, we obtain equivalent expressions for the covariance terms:

$$\text{Cov}\left(\Pi, \frac{\partial \Pi}{\partial x_i}\right) = 1/2 \frac{\partial \text{ var } \Pi}{\partial x_i}$$

and in the case that skewness of profit (S) is not affected by input use x_i

$$\text{Cov}\left((\Pi - E\Pi)^2, \frac{\partial \Pi}{\partial x}\right) = \frac{S}{2}(\text{var } \Pi)^{1/2} \frac{\partial \text{ var } \Pi}{\partial x_i} .$$

Substituting for MRP_i, we obtain

$$MRP_i = \frac{\frac{1}{2}\left[u''(E\Pi) \frac{\partial \text{ var } \Pi}{\partial x_i} + u'''(E\Pi) (\text{var } \Pi)^{1/2} S \frac{\partial \text{ var } \Pi}{\partial x_i}\right]}{\left[u'(E\Pi) + u'''(E\Pi) \frac{\text{var } \Pi}{2}\right]} .$$

Sufficient conditions for when MRP is positive or negative can now be given. Under our assumptions about utility with negative or zero skewness, MRP takes on the sign opposite to the variance derivative term; that is, the absolute risk premium will become larger if an input increases the variance of profit and smaller if an input reduces the variance of profit. Thus, with negatively skewed profit, a profit variance increasing input must have positive expected marginal profit at the optimum while a profit variance reducing input under the same conditions must have a negative expected marginal profit.

Further insight about optimality requirements is obtained by considering profit variance assuming independence of price and the state of nature. Then, profit variance effects can be decomposed into

$$\text{var } \Pi = \left[\text{var } p \, (Ey)^2 + E(p^2) \text{var } y\right]$$

and marginal effects are given by

$$\frac{\partial \text{ var } \Pi}{\partial x_i} = \left[2 \text{ var } p \; Ey \; E\left(\frac{\partial y}{\partial x_i}\right) + E\left(p^2\right) \frac{\partial \text{ var } y}{\partial x_i}\right] .$$

Note that because of price risk, a negative marginal output variance effect is not sufficient to give a negative marginal profit variance effect. However, a positive output variance effect will give a positive profit variance effect; hence expected marginal profit should be positive for such an input.

MODELING AND ESTIMATION OF THE STOCHASTIC PRODUCTION FUNCTION

Here we consider use of inputs for agricultural production subject to weather, insect, and price risk. We use biological principles as a basis for the functional form of the stochastic production function. A biological growth model describes growth of a biological entity during a given growth period (t) by a differential equation:

$$\frac{dY_t}{dt} = \alpha_t Y_t$$

Discretizing and solving this model, we obtain a description of the final size (Y_T) of the entity as:

$$Y_T = Y_o \underset{t}{\pi} (1 + \alpha_t)$$

where Y_o is the initial size of the entity. This model may be applied to describe yield (aggregating from an individual entity to an acre basis) by interpreting Y_o to be the maximum yield potential (subject to management practices) and Y_T to be the final observed yield. Negative values of α_t would be associated with decreases from maximum yield.

Consistent with this growth model, Reetz *et al.* developed a multiplicative model of maximum yield potential; the form of this model was $Y = M \times W \times m$ where M denotes the maximal yield potential for an area, W denotes an index of weather and m denotes an index of economic management practices; values of W and m less than one would give a prediction of less than maximum yield, and factors set at one are not limiting to yield.

The form of the yield model here is based on the growth model with a maximum growth potential; multiplicative factors reflect reductions from the maximum. However, the form of the model also reflects assumptions about how inputs interact with risk, and thus the model is not strictly multiplicative. Inputs studied here are fertilizer (FCA), insecticide treatment ($\%\ ATI$), and herbicide treatment ($\%\ ATH$). The assumptions underlying model specification are that fertilizer is potentially risk premium increasing and insecticide is potentially risk premium reducing. For estimation of this model (using SPSS NONLINEAR regression), restrictions on the signs (but not sizes) of the coefficients are made to be consistent with these assumptions.

The general form of the model for corn yield (CY) is:

$$\begin{aligned} CY = {} & (b_1 + b_2\ FCA + b_3\ FCA^2 + b_4\ DU) \times (1 + b_5\ PSA + b_6\ PSI \\ & + b_7\ PCL + b_8\ PMU) \times (1 + Z \times (b_9 + b_{10}\ ln\ (1 + \%\ ATI))) \\ & \times W + b_{11}\ ln\ (1 + \%\ ATH); \end{aligned}$$

Z and W are stochastic variables, respectively, indices of insect damage and weather. Because of its form, in addition to separate effects of fertilizer, insects, weather, and soil, the model includes interaction terms between fertilizer and insects, fertilizer and weather, fertilizer and soil, and insects and weather, etc. More detailed variable definitions are given in Table 1.

The weather index is given by:

$$\begin{aligned} W = {} & (1 + (DTM - 19.)^2 \times (b_{12}\ DTMA + b_{13}\ DTMB)) \\ & \times (1 + (PJN - 4.29)^2 \times (b_{14}\ DJNPA + b_{15}\ DJNPB)) \\ & \times (1 + (TJN - 72.4)^2 \times (b_{16}\ DJNTA + b_{17}\ DJNTB)) \\ & \times (1 + (PJL - 4.55)^2 \times (b_{18}\ DJLPA + b_{19}\ DJLPB)) \\ & \times (1 + (TJL - 75.9)^2 \times (b_{20}\ DJLTA + b_{21}\ DJLTB)) \\ & \times (1 + (PAU - 3.98)^2 \times (b_{22}\ DAUPA + b_{23}\ )) \end{aligned}$$

where all coefficients are assumed to be negative. In defining the effects of weather on yield, we have used the Thompson agronomic model to guide the choice of weather variables; as in Thompson's work, weather variables are expressed as deviations from optimal values. Both temperature and rainfall for the critical growing period months are included. However, since weather conditions above and below optimum may not produce symmetric responses, the model includes separate coefficients for these cases. Coefficient values and t-statistics are given in Table 1.

Table 1. Model Coefficients Dependent Variable — Corn Yield (Bu)

Variable	Coefficient	*t*-value
Constant	59.51	2.13
FCA	0.1406	0.92
FCA^2	–0.0001396	–1.00
DU	–11.82	–1.46
PSA	–0.2507	–1.15
PSI	–0.0144	–1.24
PCL	–0.0775	–0.90
PMU	–0.9170	–0.50
ln (1 + % ATH)	58.85	3.05
Z	–0.00000291	–1.47
Z × ln (1 + % ATI)	0.00000684	1.45
$(DTM–19)^2$, above normal	–0.000405	–0.76
$(DTM–19)^2$, below normal	–0.0000292	–0.004
$(PJN–4.29)^2$, above normal	0	--
$(PJN–4.29)^2$, below normal	0	--
$(TJN–72.4)^2$, above normal	–0.00376	–0.044
$(TJN–72.4)^2$, below normal	–0.00189	–1.19
$(PJL–4.55)^2$, above normal	0	--
$(PJL–4.55)^2$, below normal	–0.0314	–5.87
$(TJL–75.9)^2$, above normal	–0.00392	–0.99
$(TJL–75.9)^2$, below normal	–0.00556	–0.54
$(PAU–3.98)^2$, above normal	0	--
$(PAU–3.98)^2$, below normal	0	--
$(TAU–74.15)^2$, above normal	0	--
$(TAU–74.15)^2$, below normal	0	--
$R^2 = 0.73$		

Variable definitions

FCA	county average of lbs. of fertilizer per acre
DU	dummy variable to test for error in 1978 in fertilizer measurement
PSA	fraction of sandy soil in county
PSI	fraction of silty soil in county
PCL	fraction of clay soil in county
PMU	fraction of muck soil in county
% ATH	fraction of acres in county treated with herbicides
INS	average number of corn borers per 100 stalks; an index of insects
DIN	dummy; a value of one indicates INS is above 40
Z	$(INS-40)^2 \times DIN$
% ATI	fraction of corn acres in county treated with insecticides
DTM	date in May when 50% of corn is planted
PJN, TJN	precipitation, temperature in June
PJN, TJL	precipitation, temperature in July
PAU, TAU	precipitation, temperature in August
DJNTA, DJLTA, DAUTA	dummies indicating temperature above normal; DJNTB, DJLTB, DAUTB indicate below normal
DJNPA, DJLPA, DAUPA	dummies indicating precipitation above normal; DJNPB, DJLPB, DAUPB denote below normal
DTMA	dummy indicating planting data later than normal; DTMB denotes below normal

Effects of the inputs on marginal products are obtained by taking partial derivatives of the yield equation and evaluating these expressions with the estimated coefficients. The marginal effects on yield due to insecticide treatment and fertilizer are given by:

$$\frac{\partial CY}{\partial \%ATI} = k_1\ W\ Z\ b_{10}/(1+\%ATI)$$

$$\frac{\partial CY}{\partial FCA} = k_2\ W\ (b_2 + 2b_3\ FCA)$$

where: $k_1 = (b_1 + b_2\ FCA + b_3 FCA^2) \times$ (soil index).

$k_2 = (1 + Z(b_9 + b_{10}\ ln\ (1+\%ATI))) \times$ (soil index).

Variance effects of inputs on yield are given by

$$\frac{\partial \text{ Var } CY}{\partial \%ATI} = 2 \text{ Cov}\left(CY, \frac{\partial CY}{\partial \%ATI}\right),$$

$$\frac{\partial \text{ Var } CY}{\partial FCA} = 2 \text{ Cov}\left(CY, \frac{\partial CY}{\partial FCA}\right)$$

Evaluating these with the estimated coefficients gave the results that fertilizer is mean and variance increasing (below the maximum yield level) and insecticide is mean increasing and variance reducing.

Since herbicide use is not assumed to be associated with risk, herbicide increases the expected marginal product by

$$\frac{\partial E(CY)}{\partial \%ATH} = \frac{b_{11}}{1+\%\text{ATH}}$$

but does not affect the variance of output.

A simulation package (SLAM) was used to evaluate the above mean and variance effects; 18 years of weather and insect data (1961–1978) were used with the specified functional form of the stochastic production function above. The resulting distributions for the weather index and yield were negatively skewed. [Similar results were found by Day (1965)]. Details of simulation are given in Loehman and Polymenopolous.

To demonstrate use of this production model for testing input use, data for two counties in Indiana were used. These counties are in different regions of Indiana so different weather distributions and insect distributions apply; Hamilton county has better weather index and insect damage distributions. Soil types are also different; Hamilton has better soil. In 1974, the average fertilizer use was 446 lbs. per acre in Starke and 411 pounds per acre in Hamilton; in both counties about 40% of the acres were treated with insecticides and 60% of the acres were treated with pesticides. The distribution of yield for 1974 input use for Hamilton County had a mean of 105 bu./acre with a standard deviation of 16 and the distribution for Starke had a mean of 83 bu./acre with a standard deviation of 11.

Table 2 shows the marginal mean and variance effects of inputs. Table 3 shows the values of expected marginal products, marginal costs, and expected marginal profits for Starke and Hamilton counties for 1974. Values of expected marginal

Table 2. Marginal Effects* of Inputs on Yield Mean and Variance, 1974 (bu/acre)

	Starke	Hamilton
Fertilizer, per lb.		
Change in mean	0.0102	0.0209
Change in variance	0.0258	0.1092
Insecticide, per fraction acre		
Change in mean	1.71	0.4588
Change in variance	–2.26	–0.3518
Herbicide, per fraction acre		
Change in mean	36.55	36.10
Change in variance	0	0

*Evaluated at 1974 input use levels.

Table 3. Marginal Values[a] ($ per acre), 1974

	Starke	Hamilton
Fertilizer (pounds/acre)		
Expected VMP	0.012	0.024
MC	–0.088	–0.094
Expected marginal profit	–0.076	–0.0331
Insecticide (fraction of acre treated)[b]		
Expected VMP	2.02	0.54
MC	–18.16	–16.32
Expected marginal profit	–12.98	–14.97
Herbicide (fraction of acre treated)[b]		
VMP	43.07	42.60
MC	–10.99	–11.56
Expected marginal profit	32.08	31.04

[a]Evaluated at 1974 input use levels, average corn price of 1.18.

[b]Assumes that each acre in a county has the same percent treated.

products were obtained from an estimate of anticipated price ($1.18, the average of ten previous years) times expected values of the marginal yield effects from Table 2. Marginal costs per acre were derived from expenditures given in the Census of Agriculture, assuming a linear cost per acre.

Results in Table 3 show that fertilizer use in both counties has a negative expected marginal profit per acre and so does not satisfy the requirement for optimal input use. Declining marginal product implies that use of fertilizer should be decreased. In both counties, herbicide use does not satisfy equality of marginal cost and value of expected marginal product required for a risk neutral input. Because of declining marginal product, herbicide use would need to be increased to equalize values of expected marginal product and cost. It is not possible to determine whether insecticide use is optimal without further information about risk preferences.

CONCLUSIONS

This paper has developed necessary conditions for tests of optimality of input use in risky production which are not dependent on mean-variance preference assumptions or normality. The type of test demonstrated here seems useful since it does not require specifying an exact functional form for the utility function.

We have also demonstrated an agricultural application involving estimation of a stochastic production function, simulation of the yield distribution and marginal input effects, and the use of this information for optimality tests. Results indicated that, probably because of information problems and complexity of economic decision making under risk, inputs are not being correctly chosen in this application. One might speculate that this same type of result could be obtained for many other production applications which have similar problems of information and complexity.

REFERENCES

Day, R. H., 1965, Probability Distributions of Field Crops, *J. Farm Econ.* **47**:713-741.

Just, R. E., and Pope, R. D., 1978, Stochastic Specification of Production Functions and Economic Implications, *Journal of Econometrics* **7**:67-86.

Just, R. E., and Pope, R. D., 1979, Production Function Estimation and Related Risk Considerations, *Am. Jour. of Ag. Econ.* **61**:276-284.

Loehman, Edna, and Polymenopolous, Alex, Stochastic Production and Evaluation of Input Choices, Staff Paper 85—11, Department of Ag. Econ., Purdue University, West Layfayette, In., September 1985.

MacMinn, R., and Holtman, A., 1983, Technological Uncertainty and the Theory of the Firm, *Southern Economic Journal* **50**:120-136.

Reetz, H. F., Holt, D. A., Dale, R. F., Hollinger, F. E., Daughtry, C. S. T., and Nelson, W. L., 1978, AGSERV — 78: Corn, Soybean, and Soil Moisture Simulation, Final Report to CDC Corporation, Purdue University, West Lafayette, IN.

Thompson, L. M., 1969, Weather and Technology in the Production of Corn in the U.S. Corn Belt, *Agronomy Journal* **63**:453-456.

Assessing Risk from Dermal Exposure at Hazardous Waste Sites

Elizabeth T. Hawkins, Elizabeth A. Ryan, and Susan L. Santos
E. C. Jordan Company
Wakefield, MA

ABSTRACT

The decision of which method, if any, should be used to assess risk from dermal contact should be made on a site-specific basis. At some hazardous waste sites, a purely qualitative approach or a simple screening process assuming 100% absorption is sufficient. However, at sites where dermal contact is considered an important exposure route, it is important to develop more refined estimates to provide the information needed in the risk management decision process. Uncertainties exist in the determination of all exposure parameters in any risk assessment. Risk assessors must therefore attempt to provide a means of reducing or at least accounting for some of the effects of uncertainties, and thus put developed risk estimates into context. This paper outlines one approach to refining risk estimates. As more information on dermal absorption from soils becomes available, the absorption and risk ranges discussed here can be further refined. Eventually, it is hoped that more research will be performed to determine actual soil adsorption/desorption and dermal absorption rates of individual chemicals, to further reduce the uncertainty associated with risk estimates.

KEYWORDS: Risk assessment, dermal absorption, hazardous waste sites

INTRODUCTION

When performing a public health risk assessment at a hazardous waste site, there are several exposure routes in various media that should be evaluated. Traditionally, the major emphasis has been placed on the ingestion of contaminated groundwater, as this route is the easiest to evaluate and tends to generate the greatest public concern. However, there are many sites for which the potential public health risk from contact with contaminated soils should be evaluated. Surface soil contamination is a common problem at hazardous waste sites where indiscriminate disposal practices may have resulted in spillage, with subsequent contaminant migration via overland transport or other contaminant fate and transformation processes. Contaminants found in the soils are often very different from those found in the ground or surface water, and pose different types of public health and remedial problems. Therefore this route should be evaluated separately.

It is essential to evaluate soil contamination at those sites with unrestricted or limited access, where people may actually come into contact with the soil. This is particularly

New Risks
Edited by L. A. Cox, Jr., and P. F. Ricci
Plenum Press, New York, 1990

important when the site may be accessed by sensitive populations such as children, who tend to come into contact with large amounts of soil through play activities. Direct dermal contact is also important when adults perform activities such as gardening or dirt biking on the site or in off-site contaminated areas where they may be exposed to large quantities of soil.

To assess risk from these sites, three potential soil exposure routes should be addressed: soil ingestion due either to inadvertent ingestion or pica, inhalation of fugitive dusts, and contact through direct dermal exposure. The focus of this paper is soil contact through dermal exposure; this route poses some unique and interesting problems. Although it is difficult to assess this route quantitatively, as risk assessors we feel that it is important to refine our risk estimates as much as possible to provide decision makers with the information necessary to make sound risk management decisions.

The subsequent sections of this paper contain a summary of the information that is needed to perform an assessment of risk from dermal contact with soils, the information that is available for such an assessment and the data gaps that exist. The methodology that we have used to date for refining dermal absorption estimates is then outlined. Finally, a case study is presented in which this methodology is utilized.

NECESSARY AND AVAILABLE INFORMATION

The risks associated with dermal contact are a function of the contaminant concentration in soil, the amount of soil contacted, the frequency of contact, and the dermal absorption rate of each contaminant. To provide a quantitative estimate of the risk from this route of exposure, numerical values must be assigned to each of these parameters. These parameters can then be substituted into the following formula:

$$\underset{(\mu g/g)}{\text{Concentration}} \times \underset{(g)}{\text{Amount contacted}} \times \underset{(exp./day)}{\text{Frequency of contact}} \times \underset{(\%/exposure)}{\text{Dermal absorption}} \times \frac{1}{\text{body weight (Kg)}}$$

This equation produces a body dose level in µg/kg/day. The body dose level can then be averaged over a lifetime and multiplied by the unit cancer risk to estimate incremental cancer risks for carcinogens, or divided by a relevant standard, criterion or guideline to estimate a risk ratio for noncarcinogens.

The values for these parameters vary in reliability and ease of attainability. While values for contaminant concentrations can be easily obtained from site analytical data, estimates for the other variables are more difficult to quantify. These inputs are often based on site specific factors such as the assumed activity and behavior patterns of exposed populations or the physical characteristics of the site. However, efforts by such researchers as Renate Kimbrough of the Centers for Disease Control (Kimbrough *et al.*, 1984) and John Hawley (Hawley, 1985) have helped refine quantitative estimates of the amount of soil contacted per exposure and the frequency of exposure contact. Because some uncertainty will always surround these numerical exposure parameters, these values are often varied to simulate a range of plausible exposure scenarios, including both the most probable and realistic worst-case exposure conditions. However, because there is little quantitative information related specifically to the dermal absorption rates of contaminants in a soil or sediment matrix, it is difficult to make an accurate estimate of the incurred body dose levels of contaminants resulting from dermal exposure.

Dermal absorption information is available for only a few of the many contaminants typically present at a hazardous waste site and much of the available information is based on a contaminant's absorptive behavior in a liquid or vapor phase. While these values are useful to provide qualitative information regarding absorption, they will tend to overestimate the potential absorption rates of contaminants in soils. This is because they do not incorporate the adsorption/desorption potential of a contaminant onto or off of soil particulates. As a result of inherent uncertainties and limited information, the risks associated with dermal contact with soils are often only qualitatively addressed. This provides no quantitative information relative to the potential magnitude of the risk, or the importance of this route of exposure compared with other potential routes of exposure at a hazardous waste site.

METHODOLOGY

In attempting to quantify the risk from dermal contact with contaminated soils, one possible approach for determining body dose levels in the absence of empirically derived absorption factors is to assume complete (100%) absorption of a contaminant across the skin barrier. The rationale for this approach is that it will generate a conservative worst-case estimate of the potential risks. However, the toxicological literature indicates that assuming unity leads to a gross overestimation of the actual dermal absorption of most compounds and thus does not provide a realistic estimate of risk even under worst-case exposure situations.

Another approach to provide a more reasonable estimate of the potential body dose level incurred from dermal exposure to contaminated soils is to refine the absorption factor of unity; this refinement can be based on the contaminant's physical, chemical and toxicological properties and on extrapolations from applicable scientific information pertaining to structurally similar compounds. By examining the log octanol-water partition coefficient, polarity, valence state and physical structure of a compound and by using applicable chemical-specific dermal absorption rates, it is possible to assign more realistic rates for dermal absorption that should decrease the uncertainty of the estimated body dose levels used to assess risk.

The methodology proposed does not assign chemical specific absorption rates for each contaminant detected, but rather provides a range of absorption percentages that encompasses the range of actual absorption rates for compounds in the three major chemical classes of volatile organic compounds, semi-volatile organic compounds (including PCBs and pesticides and inorganics). To apply this approach, it is assumed that chemicals within each class behave similarly both in their absorptive capacity to soil particulates, which in turn regulates their bioavailability, and in their absorption potential across the skin barrier, which in turn determines the incurred body dose level. Based on existing scientific literature, chemical-specific absorption rates can not be derived except in those very few instances where sufficient chemical-specific information exists, such as for TCDD. The uncertainties and limitations inherent in such absorption rates exceed their usefulness for risk characterization purposes. Using a range of values for a particular chemical class offsets the uncertainty in using a single absorption rate to calculate incurred body dose levels. A range of absorption rates provides a range of body dose levels within which the actual body dose level is expected to fail. Using a range also provides a limited sensitivity analysis by which uncertainties can later be evaluated.

The assigned absorption rates for each chemical class are based on applicable dermal absorption values present in the scientific literature. For example, much information has been generated for dermal exposure to TCDD-contaminanted soils. Poiger and Schlatter (1980) observed that between 0.07 and 3% of soil-bound TCDD was absorbed through an animal's intact skin. Based on these results, as well as other chemical and physical information on related compounds, the range of absorption values for semi-volatile compounds, pesticides and PCBs can be set at 1% and 10%. Similarly, based on scientific data generated for metals, the dermal absorption rates for inorganics can be set at 1% and 10%. These relatively low absorption percentages reflect the strong soil adsorptive potential of semi-volatile compounds, pesticides, PCBs and metals. The strong binding between compound and soil particulate reduces the amount of contaminant which is available for dermal uptake. Volatile organic compounds do not demonstrate a strong absorptivity to soil particulates. This fact combined with their relatively high lipid solubility suggests that these compounds are both bioavailable in a soil matrix and readily absorbed through the skin. For this chemical class, absorption percentages of 10 and 50 percent can be used to estimate incurred body dose levels.

To summarize, rather than calculate body dose levels for all compounds assuming 100% absorption, body dose levels are derived using a range of absorption values based on major chemical classes. For semi-volatile organics, PCBs, pesticides and metals, absorption percentages of 1 and 10 percent are used, and for volatile organic compounds, absorption percentages of 10 and 50 percent are used.

CASE STUDY

To illustrate the utility of the approach described above, a simplified, hypothetical case study is presented based on a hazardous waste site we previously evaluated. This site covered approximately 200 acres and had two highly contaminated groundwater aquifers beneath it. There were several drinking water wells downgradient from the contaminant plume and, although these wells were not yet contaminated, it became apparent that the plume would reach the wells in the near future. The first priority at the site was to perform a risk assessment on the water in the contaminated aquifers, to determine which (if any) of the contaminants would pose a risk to human health and to direct the study of remedial alternatives toward the cleanup of the most important chemicals. Under some circumstances, this assessment would have been sufficient; if, for example, the groundwater were the only contaminated media at the site, and/or access to the site was totally restricted, and/or the area was not highly populated. However, this site was in a highly populated area. Although a guard was on duty at the site, there was no fence and there was evidence of frequent trespassing by dirt bikers, hunters, and children. Both the surface and subsurface soils were highly contaminated. Because of the potential for human exposure it was decided that an analysis of the risk posed by the contaminated soil should be performed.

There are three ways by which we could have proceeded with the risk assessment. These methods are described below. Table 1 presents the results of these three methods. It should be noted that many more contaminants were found at this site than are shown in Table 1, and several distinct dump sites existed on the site with unique characteristics which had to be evaluated separately. This simplified table is meant to show the difference between the three methods and to illustrate why the third method is considered preferable under many circumstances.

METHOD I

The first method consists of an evaluation of the contaminants in a purely qualitative manner. In this method we would describe the major known toxic effects of each compound and the levels at which these effects have been seen (if known), to give decision makers an idea of the types of problems that these contaminants may pose. The main advantage of using this method is that the uncertainties in the dermal absorption factors become irrelevant. The disadvantage is that this method does not provide decision makers

Table 1. Risk Estimates Derived from Three Risk Assessment Methods

	Risk Estimates (Adult, Realistic Worst-Case)		
Contaminant	I (Qualitative)	II (100%)	III (Class-Specific Absorption Percentage)
Carcinogens:			
Methylene Chloride	--	2×10^{-7}	2×10^{-8} to 1×10^{-7}
Benzo (b) fluoranthene	--	8×10^{-5}	8×10^{-7} to 8×10^{-6}
Arsenic	--	5×10^{-4}	5×10^{-6} to 5×10^{-5}
Cadmium	--	6×10^{-5}	6×10^{-7} to 6×10^{-6}
Non-carcinogens:			
2-Butanone	--	0.2	0.02 to 0.1
Toluene	--	6.2	0.62 to 3.1
Fluoranthene	--	3.1	0.031 to 0.31
2-Methylphenol	--	1.9	0.019 to 0.19
Lead	--	0.4	0.004 to 0.04

with quantitative indices of the relative magnitude of the risks posed by different chemicals on the site or different source areas on the site. While at some sites this would be acceptable, at this site a quantitative analysis was indicated; the risk assessment was to be an integral tool for deciding the type and extent of remediation to be taken.

METHOD II

The second possible method includes the qualitative analysis described in Method I but also a quantitative risk assessment which uses site-specific factors to determine most of the exposure parameters and an absorption factor of 100% to provide a worst-case estimate of risk. The main advantage of this method is that it provides risk managers with the absolute worst-case risks associated with the contaminated soils. If the results of this very conservative approach showed that all the risks are extremely low and fall below the level at which decision makers consider taking action, this approach could be considered a valuable screening tool. However, as previously stated, a 100% absorption factor is not realistic, and overestimates the risks posed by these chemicals even under worst-case conditions. At this site, the risks posed by the chemicals, assuming 100% absorption (Column II of Table 1) are considered borderline; that is, all carcinogenic risks fall within the Superfund cleanup "target range" of 10^{-4} to 10^{-7} (ICF, 1985). The risk ratios for toluene, fluoranthene and 2 methylephenol are greater than one, but within an order of magnitude, and therefore can be considered borderline. The predicted body dose levels for these compounds could potentially result in adverse noncarcinogenic health effects. However, as stated above, these numbers are probably overly conservative. Therefore, it was decided that the risk assessment should be further refined.

METHOD III

In the method that was ultimately used for this site, we refined our risk estimates by using the class-specific range of absorption factors (1-10% absorption for semivolatile organics and inorganics and 10-50% absorption for volatile organics) to provide a range or risk, as shown in Column III of Table 1. While refined, these estimates are not indications of absolute risk. Rather, they represent ranges that should encompass the actual risks for each chemical. This approach gives decision makers more realistic information on which to make sound risk management decisions. It also helps to limit the effects that uncertainties will have on developed risk estimates. As shown, most of the carcinogens remain within the "target range." However, all of the noncarcinogenic risk ratios (except toluene) have now dropped below one. This is probably a more realistic picture of the actual risks posed by these contaminants through direct dermal contact with soil. The information developed with this method was used at the site to direct the study of remedial alternatives toward the cleanup of the most relevant chemicals in each of the unique source areas.

REFERENCES

Hawley, J. K., 1985, Assessment of Health Risk from Exposure to Contaminated Soil, *Risk Analysis* **5(4)**:289-302.

ICF, Inc., 1985, Draft Superfund Health Assessment Manual, U.S. EPA Office of Emergency and Remedial Response, December.

Kimbrough, R. D., Falk, H., Stehr, P., and Fries, G., 1984, Health Implications of 2, 3, 7, 8 Tetrachlorodibenzodioxin (TCDD) Contamination of Residential Soil, *J. Joxicel Environmental Health* **14**:47-93.

Poiger, H., and Schlatter, D., 1980, Influence of Solvents and Adsorbents on Dermal and Intestinal Absorption of TCDD, *Food and Cosmetic Toxicol.* **18**:477-481.

Schaum, J., 1984, "Risk Analysis of TCDD Contaminated Soil," U.S. EPA Office of Health and Environmental Assessment, November.

Bioaccumulation of TCDD in Lake Ontario Fish: Laboratory and Field Studies in Support of Hazardous Waste Landfill Risk Assessments

Neil S. Shifrin, Lisa W. Kaul
and Livia Benavides
Gradient Corporation
Cambridge, MA

ABSTRACT

As part of a risk assessment for remedial action at the Hyde Park landfill (Niagara Falls, N.Y.), the bioaccumulation factor (BAF) of 2,3,7,8-TCDD will be measured in the laboratory and in Lake Ontario fish. This three-year study is underway because it was decided that BAFs for TCDD used by agencies and found in the literature are substantially too low. The challenge of this study is to work with conditions representative of Lake Ontario waters which have TCDD concentrations below analytical detection limits. Despite such unmeasurable concentrations in water, TCDD has been found in Lake Ontario fish at concentrations as high as 160 ppt and in sediments at 15 ppt. Therefore, this study will focus on deriving a fish-to-sediment ratio to derive a modified version of the BAF. The results of this study will be used to finalize allowable loadings to the Niagara River and required remedies for the Hyde Park landfill, a site which contains approximately 1 ton of 2,3,7,8-TCDD. This study emphasizes the need for reliable estimates of environmental and food partitioning of chemicals in the risk assessment process.

KEYWORDS: Dioxin, TCDD, bioaccumulation

INTRODUCTION

Quantitative risk assessment for chemicals in the environment is sometimes criticized for being an uncertain science. Despite such uncertainties, quantitative risk assessment at least offers a systematic approach to decision making. Interim decisions can be made with existing information on fate, exposure and health effects models and at a future time can be modified, systematically, as our understanding is improved. The alternative to this process is to make decisions based on arbitrary action levels for environmental contaminant concentrations, decisions based on technology constraints, or to require "zero discharge" of contaminants—an aesthetically attractive concept but often of questionable technical feasibility.

New Risks
Edited by L. A. Cox, Jr., and P. F. Ricci
Plenum Press, New York, 1990

In August 1986, the Environmental Protection Agency (EPA) published several new policy documents on risk assessment. The need to address the uncertainties of risk assessment, head-on, is a clear theme in these documents. This paper describes how decision making at a dioxin-contaminated site, the Hyde Park landfill (Niagara, NY), relied on a risk assessment process that emphasized the resolution of at least one uncertainty—the bioaccumulation of dioxin (2,3,7,8-TCDD) in fish consumed by people in the Lake Ontario region.

BACKGROUND

The Hyde Park landfill was operated by a local chemical company in Niagara Falls, NY, from 1953 to 1975. In addition to depositing various chlorinated benzenes, toluenes, phenols, and aliphatics, the company dumped 0.6 to 1.6 tons of 2,3,7,8-tetrachlorodibenzoparadioxin in this landfill. The site is located approximately 2000 ft. east of the Niagara River, four miles below Niagara Falls. Two contamination plumes, one aqueous and the other a very dense organic phase (referred to as nonaqueous phase liquids, or NAPL), have entered the underlying bedrock. The aqueous phase plume currently extends all the way to the Niagara River and recent measurements of ground water seeps emanating from the Niagara River Gorge face indicate 2,3,7,8-TCDD concentrations of 0.18 ppt (NYS DOH, 1985). The NAPL plume extends towards the Niagara gorge but has not yet reached it.

The very large flow of the Niagara River (220,000 cfs) serves to dilute substantially this and the several other sources of TCDD to the river, resulting in water concentrations well below analytical detection limits. Calculations indicate that this dilution eliminates health concerns for drinking water and most other exposures (Rodricks, 1985) but the hydrophobic nature of TCDD results in bioaccumulation in fish to a level posing significant public health risks. The average 2,3,7,8-TCDD concentration in Lake Ontario fish is on the order of 30 ppt with a range of up to 160 ppt. Hence, extremely low, nondetectable concentrations of TCDD exist in Niagara River and Lake Ontario water while significantly high levels are found in fish.

During 1985 the chemical company, the U.S. EPA, and the State of New York negotiated a group of remedies for the Hyde Park landfill. This negotiation was based on data collected as a result of ground rules established in a court-approved consent decree signed by the parties in 1981. In brief, the major remedies include source containment and ground water pumping/controls to eliminate further NAPL migration and to minimize aqueous plume contamination to the Niagara River. Although requiring zero discharge was considered, it was concluded to be neither feasible nor necessary. Instead, an action level amounting to a maximum allowable mass loading of TCDD (and several other compounds) to the river was chosen after considering the results of a quantitative risk assessment for all chemicals found at the site. For TCDD the primary issue was bioaccumulation in fish.

The legal agreement binding the parties to these remedies is extremely flexible in terms of future modifications based on improved knowledge of contaminants, their concentrations, their fate, and their health effects. One uncertainty addressed specifically in the legal agreement was the effect of the value of the bioaccumulation factor (BAF) of TCDD in fish on the estimate of health risks due to fish consumption in the Lake Ontario region. As described below, the government negotiators felt that currently accepted values of the BAF were far too small, resulting in estimates of risk from Hyde Park contaminant discharges via the fish consumption pathway likely to be orders of magnitude too low. Extensive debates on this topic by the remedy negotiators led to an agreement that bioaccumulation research was needed to reduce the uncertainty of the TCDD BAF. When these results become available, the health risk assessment for this site will be re-examined and the TCDD maximum allowable loading will be modified, if appropriate.

BIOACCUMULATION OF TCDD IN FISH

The bioaccumulation factor (BAF) is defined as the ratio, at equilibrium, of the concentration of a contaminant in fish tissue to the corresponding concentration in water with sediments and food also present. This term is not to be confused with bioconcentration[a] (i.e., BCF), which is the same ratio but with contaminated food and sediments not present, and is generally lower than bioaccumulation (Schnoor, 1982). Bioaccumulation factors are more representative of actual environmental conditions.

Experimental evidence for the correct value of the BAF is improving but is not yet convincing. The Water Quality Criteria document for TCDD (EPA, 1984), clearly states the uncertainties but is worded in a way that has led people to use a BCF (not BAF) of 5,000 for risk assessments. As of 1985, laboratory experiments for BCFs or BAFs have yielded values of from 2,200 to 30,000 (Table 1). TCDD bioassay experiments are very difficult to design and each of the studies in Table 1 may have flaws that resulted in underestimates of bioaccumulation. The most prominent problem has been the inability of the experiments to achieve equilibrium, although more recent studies avoid this with kinetic uptake and depuration designs (ASTM, 1984).

There are three reasons to believe that currently available data on TCDD BAFs are too low. Although none of the reasons are conclusive in themselves, they offer a good argument that the story on TCDD bioaccumulation should *not* end with the conclusion that the true value for the BCF is 5,000.

Empirical Relationships Between Chemical Properties and BAFs

Chemicals accumulate in fish primarily in lipids. As a result, several researchers have attempted to correlate bioaccumulation (most often as represented by BCFs) with the water-octanol partition coefficient, K_{ow} (Neely *et al.*, 1974; Veith *et al.*, 1980; Oliver and Niimi, 1983; Kenaga and Goring, 1980; Mackay, 1982). The modeling of viable fish lipid by octanol is somewhat reasonable and more convenient because K_{ow} experiments are easier to perform than bioassays. However even K_{ow} experiments are difficult for highly hydrophobic compounds and, as is often the case with data for TCDD, uncertainty exists for the K_{ow} of this compound. Very recent data indicate a value (log scale) of 7.02 (Burkhard and Kuehl, 1986).

Table 2 presents seven regression equations and predicted BCFs for values of K_{ow} bracketing this recent measurement, an older measurement of 6.15 (Kenaga, 1980) and an estimate of 7.28 (U.S. EPA, 1984) based on structure-activity relationships (Hansch and Leo, 1979). The range of predicted BCFs in Table 2 is from 2,870 to 2,450,000. As an example using the recently reported value of 7.02 for log K_{ow}, Mackay's model presented in Table 2 would predict a BCF of 500,000, two orders of magnitude greater than the EPA "criteria value."

BAFs of Other Hydrophobic Compounds

If the correlations of BCF with K_{ow} are sound, it would stand to reason that experimental bioassays of other very hydrophobic compounds might yield BCFs of magnitude similar to the true value for TCDD. Table 3 presents experimentally measured BCFs for such compounds, including hexachlorobenzene, DDT, and PCBs with a range of 12,000 to 270,000. It might be noted in Table 3 that tetrachlorobiphenyl, a compound with

---

a. Unfortunately bioconcentration and bioaccumulation are sometimes used interchangeably which is not correct and is confusing. In this paper the terms "bioconcentration" or "BCF" are preserved when used this way in the source literature, although we feel that BAFs are more appropriate for future use in risk assessment.

Table 1. List of All Existing Experimentally Measured Bioconcentration Factors for TCDD in Fish

BCF	Species	Comments
4850	Mosquito Fish (Gambusia Affinis)	-Isensee, 1978 -Short exposure time (7D) -Static ecosystem -Soil treated with TCDD and put into aquatic eco system: no evidence of equilibrium between soil and water.
6970	Mosquito Fish (Gambusia Affinis)	-Isensee, 1978 -Average of several concentrations (0.05-239 PPT) -Short exposure time (3D) -Static ecosystem -Soil treated with TCDD
4875	Mosquito Fish (Gambusia Affinis)	-Yockim, Isensee & Jones, 1978 -Short exposure time (7D) -Static ecosystem -Soil treated with TCDD
2200	Channel Catfish (Ictarulus Punctatus)	-Isensee, 1978 -Average of several concentrations (0.05-239 PPT) -Static ecosystem -Short exposure time (6D) -Soil treated with TCDD
9270+/-1050	Rainbow Trout	-Branson *et al.*, 1984 -Short uptake period (6HR) in static system -Long clearance (139 D, T 1/2 = 58D)
>30,000	Fathead Minnows	-Adams *et al.*, (1985, pers comm) -Did not reach equilibrium during exposure time (28D) -Static system

*Other literature reviewed presented BCF that appeared to arise from the same experiments.

some similar structural components to TCDD, has a measured BCF of 72,950, an order of magnitude greater than the EPA "criteria value."

Niagara River/Lake Ontario Field Data

From 1978 to 1985, over 100 fish samples from western Lake Ontario have been analyzed for 2,3,7,8-TCDD resulting in an average concentration of 30 ppt (O'Keefe *et al.*, 1983; Ryan *et al.*, 1984; Fehringer, *et al.*, 1985; NRTC, 1984; NDS, 1986). This average includes many different species, compositing, and data-averaging approaches and should be

Table 2. Predicted Bioconcentration Factors for 2,3,7,8-TCDD Based on Estimated and Measured Values of the Octanol/Water Partition Coefficient

Equation	Reference	log K_{ow} Measured 6.15	log K_{ow} Estimated 7.28
$\log BCF = 0.542 \log K_{ow} + 0.124$	a	2,870	11,700
$\log BCF = 0.76 \log K_{ow} - 0.23$	b	27,800	201,000
$\log BCF = 0.79 \log K_{ow} - 0.40$	b	28,700	201,000
$\log BCF = 0.635 \log K_{ow} + 0.7285$	c	43,000	225,000
$\log BCF = 0.85 \log K_{ow} - 0.70$	b	33,700	308,000
$BCF = 0.048 K_{ow}$	d	67,800	915,000
$\log BCF = -0.869 + (0.997 \pm 0.056) \log K_{ow}$	e	183,000	2,450,000

*Modified from Ambient Water Quality Criteria for 2,3,7,8-Tetrachlorodibenzo-p-dioxin, 2/84, EPA 440/5-84-007.

[a] Neely, Branson, and Blau, 1974; trout muscle, flowing water.

[b] Veith, De Foe, and Bergstedlt, 1979; whole body, flowing water.

[c] Kenega and Goring, 1980; static water ecosystem.

[d] Mackay, 1982; whole body, flowing water.

[e] Oliver and Niimi, 1983; whole fish, flowing water.

viewed only as a general representation of Lake Ontario fish. The data range from not detected at 0.65 ppt (white sucker fillet; NDS, 1986) to 162 ppt (brown trout skinned fillet; O'Keefe *et al.*, 1983). Levels in eastern Lake Ontario ranged from not detected at 2 ppt to 107 ppt, and in the Niagara River from not detected at 1 ppt to 160 ppt, during the same time period.

Water measurements of TCDD have been unsuccessful at detecting this compound at detection limits in the range of 0.001-0.005 ppt (Hallet *et al.*, 1984). This absence of data precludes calculation of a BAF directly, but it is possible to estimate the concentration in water from sediment contamination data. Only a few sediment data exist but Onuska, Mudrock and Terry (1983) studied several depths down to 14 cm at 5 stations in Lake Ontario near the mouth of the Niagara River. A rough average sediment concentration of 10 ppt can be derived from their data. Using a sediment-water partition coefficient, K_p, of (log scale) 4.76 (Marple *et al.*, 1986 for the K_{oc} and an f_{oc} of 0.03) a Niagara River water concentration of 0.0002 ppt is suggested from liberal use of the scant data with the equation:

$$C_{water} = C_{sed}/K_p$$

Table 3. Experimentally Measured Bioconcentration Factors of Hydrophobic Compounds

Compound	BCF	Reference
Hexachlorobenzene	12,000-20,000	Oliver and Niimi, 1983
Pentachlorobenzene	13,000-20,000	Oliver and Niimi, 1983
Hexachlorobutadiene	17,000	Oliver and Niimi, 1983
Mirex	18,000	Mackay, 1982
Heptachlor	20,000	Veith *et al.*, 1979
Octachlorostyrene	33,000	Veith *et al.*, 1979
Chlordane	38,000	Veith *et al.*, 1979
PCB (1016)	43,000	Veith *et al.*, 1979
Aldrin	44,600	Metcalf *et al.*, 1973
Endrin	49,000	Metcalf *et al.*, 1973
DDE	51,000	Mackay, 1982
DDT	61,000	McCall *et al.*, 1983
Dieldrin	62,000	Metcalf *et al.*, 1973
Tetrachlorbiphenyl	72,950	McCall *et al.*, 1983
PCB (1248)	120,000	Nebekev, 1975
PCB (1954)	110,000-240,000	Nebekev, 1975
PCB (1260)	270,000	Nebekev, 1975

Although based on very limited data and assumptions of equilibrium, this estimated water concentration would result in a BAF of 150,000 using the average western Lake Ontario fish levels. Like the other approaches, this estimate results in a BAF significantly larger than the water quality "criteria value."

PLANNED STUDY TO REDUCE THE UNCERTAINTIES

Hopefully the survey of existing information and the deductive reasoning displayed above have at least demonstrated that the bioaccumulation factor for TCDD is not 5,000 but, instead, is possibly orders of magnitudes larger. Unfortunately that is all it proves and additional research must be performed to provide a better value for the BAF of TCDD.

This factor will have a significant impact on any estimation of health risk from Lake Ontario TCDD contamination because of its proportional effect on the calculation of lifetime dose via the ingestion pathway.

The study intended to reduce the uncertainties which is being sponsored by the government and corporate parties to the Hyde Park landfill Settlement has several components. These include uptake and depuration studies using lake trout in the laboratory, fish and sediment sampling throughout Lake Ontario, and a study of the sediment water partitioning of TCDD. This program is currently underway and will be completed in 1989.

The design of both the laboratory and the field portions of this study has been constrained by the fact that currently available analytical detection limits for TCDD preclude the direct measurement of this compound in waters representative of Lake Ontario. To overcome this, we have hypothesized that a fish-to-sediment ratio

$$BAF^* = C_{fish}/C_{sed}$$

could be used in place of the more conventional ratio:

$$BAF = C_{fish}/C_{water} \cdot$$

For this hypothesis to be valid, it must be shown that similar values of the BAF* can be derived from more than one sediment contaminant concentration. In such a case the BAF* could be used to predict fish contaminant levels, and therefore health risks, under future environmental contaminant loading scenarios. To do this, however, consideration must also be given to TCDD partitioning in the lake environment (e.g., sediment-water-biota-vapor), kinetics, and steady state. The study which is underway will provide the basic data on partitioning between fish and sediment and sediment and water. This will be of value for the development of an approach to model future TCDD loading scenarios in Lake Ontario. Below is a brief description of each study component.

Laboratory Uptake and Depuration Experiments

Flow-through tanks will be used to examine the uptake and depuration rate in lake trout exposed to various conditions as shown in the experimental matrix in Table 4. This will include the effect of two levels of contaminated sediments ("x" and "10x" in Table 4), clean vs. contaminated feedwater, and clean vs. contaminated food. The uptake period will last 90 days and depuration will extend at least to the point where one-half the maximum contaminant levels in fish are observed. Lake Ontario sediments will be used for these experiments—one batch from a highly contaminated area of the lake, one from a site less contaminated. Expected sediment TCDD concentrations are on the order of 100 and 10 ppt, respectively.

For the experiments involving contaminated feedwater, the equilibrium water concentration corresponding to each sediment contaminant level will be used, based on the best available estimate of K_{oc}. It is anticipated that these water concentrations will be below detectable limits, so precision metering and an accounting for system losses is planned. Current information suggests that up to half of the TCDD dosed to a system can be lost due to volatilization, adsorption to apparatus, and other as yet undefined mechanisms. For the experiments involving contaminated food, smelt taken from Lake Ontario will be used. One batch of these contaminated fish corresponding to each sediment contaminant level has been processed and frozen to be available for the entire uptake experiment.

TCDD analytical detection limit objectives for the study are 1 ppt for both fish and sediment samples. This will require extraordinary efforts and rigorous quality control, but

Table 4. TCDD Laboratory Experiment Exposure Matrix*

Test	Sediment	Food	Water
Sed-1a	x		
Sed-1b	x	x_{eq}	
Sed-2a	10x		
Sed-2b	10x	$10x_{eq}$	
Control sed-a			
Control sed-b		x_{eq}	
	Water Uptake Route		
Sed-2c	10x		$10x_{eq}$
Sed-2d	10x	$10x_{eq}$	$10x_{eq}$
No sediment			$10x_{eq}$

*x and 10x refer to two sediment concentrations, an order of magnitude apart; and x_{eq} and $10x_{eq}$ refer to water or food concentrations estimated to be in equilibrium with x and 10x sediments.

is necessary due to the realistically low concentrations anticipated for fish during the early portion of uptake. Whole and eviscerated fish will be analyzed at periodic intervals through the uptake and depuration period.

Sediment-Water Partitioning Experiment

Because the water concentrations in both the field and laboratory portions of this study are anticipated to be below detectable limits, it was felt to be necessary to reduce any existing uncertainties about the value of K_{oc} for TCDD. In this way water concentrations at least can be predicted, assuming equilibrium conditions.

These experiments go beyond the routine K_{oc} investigations in that an attempt will be made to account for partitioning effects of dissolved (also described as nonsettling, suspended, nonfilterable, or colloidal) material in the system. This is important to account for because of the extremely hydrophobic nature of TCDD.

A multiphased experimental design is planned by EPA and the State (NY) to investigate the following: (1) the relationship between the concentration of dissolved matter in an aqueous system and the solids concentration; (2) the partitioning of TCDD between the aqueous phase and the solid phase; and (3) the partitioning of TCDD between the aqueous phase and the dissolved matter phase. In all of these experiments, the solids used will be TCDD-contaminated sediments from Lake Ontario. In addition, these experiments will be performed for various equilibration times to examine kinetics and at various temperatures to examine temperature effects.

These experiments will require precise measurements of TCDD in aqueous and solid phases along with precise measurements of organic carbon and the amount of dissolved material. However, the results should reveal the most thorough understanding of TCDD partitioning in a natural aquatic system to date.

Fish and Sediment Sampling in Lake Ontario

Fish and sediments will be collected from Lake Ontario during the summer of 1987. Four fish species will be collected to provide data over a range of lipid content, feeding habits, and migratory habits. The species will be lake trout, brown trout, yellow perch and smallmouth bass. Lake trout were chosen because they will also be examined in the laboratory studies and because their migratory range covers the entire lake. Brown trout also offer a wide migratory range, have a relatively high lipid content, and are a commonly caught species. Yellow perch are probably the most commonly caught species in the Niagara River area. Small-mouth bass were chosen because of their sedentary habits. Approximately 80 fish samples will be analyzed for TCDD. These samples will each consist of composites of about 10 individual fish. In addition, the experimental design calls for the examination of whole fish vs fillets, viscera, and individual fish variability.

Sediments will be sampled around the entire perimeter of the lake and within the three major sedimentation basins of the lake. Approximately 80 sediment samples will be taken, most from the upper layer (0 – 3 cm) of the sediment bed with a few deeper cores (up to 30 cm) taken to examine historical deposition patterns. The depth of 3 cm for the majority of samples was chosen because this was assumed to be the biologically available portion of the bottom and because this depth represents approximately 30 years of deposition, which corresponds to the dumping history at the landfill.

These field data could be subject to several interpretations from a BAF standpoint. The basic assumption needed for any derivation is that the lake is in a state of quasiequilibrium and is approaching steady state. Of course, neither of these assumptions is strictly true for a receiving water such as Lake Ontario. Consideration also must be given to the fact that at any particular location, there is likely to be little correlation between sediment and fish contaminant levels, due to the migratory habits of the fish. The case of smallmouth bass might be an exception to this. The basic approach to using the data will be to derive a BAF (or a BAF*) from appropriately averaged data. For example, a ratio of fish to sediment data averaged over individual local areas might be appropriate for the smallmouth bass but whole-lake averages might be more appropriate for lake trout.

It is hoped that the BAF derived from these field data and supported by the laboratory studies can then be used to predict fish levels at any other sediment concentration resulting from other TCDD loadings to the receiving water. Such results might also be applicable to other aquatic systems receiving TCDD loadings. This better understanding of TCDD bioaccumulation will then result in reduced uncertainty for risk assessments to which it is applied.

CONCLUSION

The bioconcentration factor, and more appropriately, the bioaccumulation factor of TCDD are critical in risk estimations where fish ingestion is a concern. This paper and the research study it describes is intended to discourage the continued use of a BCF of 5,000 for TCDD. It is likely that the true value of the BAF, which should be used to determine the fate of TCDD in fish, is orders of magnitude larger than 5,000. Until the Lake Ontario studies are completed, BAF values of from 500,000 to 1,000,000 could be justified for use in risk assessments.

The Hyde Park landfill remedy negotiations served to stimulate research aimed at reducing one uncertainty in risk assessments involving TCDD. Other uncertainties remain for this and other compounds. Hopefully, the approach taken at this site will serve as a model to result in better, well founded remedies at the thousands of other landfills in need of attention.

REFERENCES

ASTM, Committee on Biological Effects and Environmental Fate, 1984 (May), Proposed New Standard Practice for Conducting Bioconcentration Tests with Fishes and Saltwater Bivalve Molluscs, American Society for Testing and Materials, 60 p.

Branson, D. R., Takahashi, I. T., Parker, W. M., and Blau, G. E., 1984, Bioconcentration Kinetics of 2,3,7,8-Tetrachlorodibenzo-p-dioxin in Rainbow Trout, Abstract presented at the Fifth Annual Meeting of the Society of Environmental Toxicology and Chemistry, November.

Burkhard, L. P., and Kuehl, D. W., 1986, N-Octanol/Water Partition Coefficients by Reverse Phase Liquid Chromatography/Mass Spectrometry for Eight Tetrachlorinated Planar Molecules, *Chemosphere* **15 2**):163-167.

Fehringer, N. V., Walters, S. M., Kozara, R. J., and Schneider, L. F., 1985, Survey of 2,3,7,8-Tetrachlorodibenzo-p-dioxin in Fish from the Great Lakes and Selected Michigan Rivers, *J. Agric. Food Che.* **33(4)**:626-630.

Hallett, D. J., Warry, D., Kuntz, K. *et al.*, 1984, An Investigation of Localized Toxic Chemical Contamination of the Niagara River Opposite the "S" Area Landfill, Environment Canada, Ontario Region, 47 p. April.

Hansch, C., and A. J. Leo, 1979, *Substituent Constants for Correlation Analysis in Chemistry and Biology*, John Wiley, New York.

Isensee, A. R., 1978, Bioaccumulation of 2,3,7,8-Tetrachlorodibenzo-p-dioxin, in: *Chlorinated Phenoxy Acids and Their Dioxins*, C. Ramel, ed.

Kenaga, E. E., and Goring, C.A.I., 1980, Relationships Between Water Solubility, Soil Sorption, Octanol-Water Partitioning, and Concentration of Chemicals in Biota, in: *Aquatic Toxicology: Proceedings of the Third Annual Symposium on Aquatic Toxicology*, J. G. Eaton, P. R. Parrish, and A. C. Hendricks, eds., American Society for Testing and Materials, Philadelphia, PA.

Kenaga, E., 1980, Correlation of Bioconcentration Factors of Chemicals in Aquatic and Terrestrial Organisms with Their Physical and Chemical Properties, *Environ. Sci. Technol.* **14**:553-556.

McCall, P. J. *et al.*, 1983, Estimation of Environmental Partitioning of Organic Chemicals in Model Ecosystems, *Residue Reviews* **85**:231.

Mackay, D., 1982, Correlation of Bioconcentration Factors, *Environ. Sci. Techno.* **16(5)**:274-278.

Marple, L., Brunck, R., Berridge, B., and Throop, L., 1986, Comparison of Experimental and Calculated Physical Constants for 2,3,7,8-Tetrachlorodibenzo-p-dioxin, in: Environmental Chemistry Division, Preprinted Extended Abstracts, New York, April 13-18, 1986, Vol. 26, No. 1, American Chemical Society 191st National Meeting.

Metcalf, R. R. *et al.*, 1973, Model Ecosystem Studies of the Environmental Fate of Six Organochlorine Pesticides, *Environmental Health Perspectives* pp. 35-44 (June).

Nebekev, A. V., 1975, Summary of Recent Information Regarding Effects of PCBs on Freshwater Organisms, Presented at EPA National Conference on PCBs, Chicago, November 19-21.

Neely, W. B., D. R. Branson, and G. E. Blau, 1974, Partition Coefficient to Measure Bioconcentration Potential of Organic Chemicals in Fish, *Environ. Sci. Technol.* **8(13)**:1113-1115.

New York State Dept. of Health, 1985, "NEWS" Release for Tuesday, July 30, 1985, Albany — RE: Results of Tests of Samples Collected from Groundwater in the Bedrock and Groundwater Seeps at the Niagara River Gorge Face.

O'Keefe, P., Meyer, C., Hilker, D., Aldous, K., Jelus-Tyror, B., Dillon K., Donnelly, R., Horn, E., and Sloan, R., 1983., Determination of 2,3,7,8-Tetrachlorodibenzo-p-dioxin in Great Lakes Fish, *Chemosphere* **12(3)**:3325-3332.

Oliver, B. G., and A. J. Niimi, 1983, Bioconcentration of Chlorobenzenes from Water by Rainbow Trout: Correlations with Partition Coefficients and Environmental Residues, *Environ. Sci. Technol.* **17(5)**:287-201.

Onuska, F. I., Mudrock, A., and Terry, K. A., 1983, Identification and Determination of Trace Organic Substances in Sediment Cores from the Western Basin of Lake Ontario, *J. Great Lakes Res.* **9(2)**:169-182.

NRTC, 1984, Report of the Niagara River Toxics Committee (NRTC), E.T. Wagner, Environment Canada, Co-Chairman, October.

Rodricks, Joseph V. [Environ Corp.], 1985, Affidavit of Joseph V. Rodricks in Support of Stipulation and Judgement Approving Settlement Agreement, Submitted to U.S. District Court (NY) Civil Action No. 79-989, December 11.

Ryan, J. J., Lau, P-Y., Pilon, C., Lewis, D., McLeod, H., and Gervals, A., 1984, Incidence and Levels of 2,3,7,8-Tetrachlorodibenzo-p-dioxin in Lake Ontario, *Environ. Sci. Technol.* **18(9)**:719-721.

Schnoor, J. L., 1982, Field Validation of Water Quality Criteria of Hydrophobic Pollutants, in: *Aquatic Toxicology and Hazard Assessment*, ASTM STP 766, J. G. Pearson, R. B. Foster, and W. E. Bishop, eds., p. 302(14), American Society for Testing and Materials.

The National Dioxin Study (NDS), 1986, U.S. Environmental Protection Agency, Tiers 3, 5, 6, and 7, Draft Report.

U.S. EPA, 1984, Ambient Water Quality Criteria for 2,3,7,8-Tetrachlorodibenzo-p-dioxin, PA-440/5-84-007, Office of Water Regulations and Standards, Washington, D.C.

Veith, G. D., DeFoe, D. L., and Bergstedlt, B. V., 1979, Measuring and Estimating the Bioconcentration Factor of Chemicals in Fish, *J. Fish Res. Board Can.* **36**:1040-1048.

Veith, G. D., K. J. Macek, S. R. Petrocelli, and John Carroll, 1980, An Evaluation of Using Partition Coefficients and Water Solubility to Estimate Bioconcentration Factors for Organic Chemicals in Fish, in: *Aquatic Toxicology: Proceedings of the Third Annual Symposium on Aquatic Toxicology*, J. G. Eaton, P. R. Parrish, and A. C. Hendricks, eds., American Society for Testing and Materials, pp. 116-119, Philadelphia, PA.

Yockim, R. S., Isensee, A. R., and Jones, G. E., 1978, Distribution and Toxicity of TCDD and 2,4,5-T in an Aquatic Model Ecosystem, *Chemosphere* **7(3)**:215-220.

Risk Assessment of Indoor Air Pollution by Termiticides

Bruce Molholt[a]
U.S. Environmental Protection Agency
Philadelphia, PA 19107

ABSTRACT

Chlordane and heptachlor, the major insecticides employed in the U.S. in 1986 for termite control, are EPA class B2 (probable human) carcinogens. Following injection in soil near homes, these termiticides have been found to enter indoor air at an average concentration in excess of 1 $\mu g/m^3$ and an individual lifetime cancer risk in excess of 10^{-4}. These results from EPA Region III are consistent with those of a previous national survey. In that millions of homes require termiticide treatment, it is suggested that widespread use of chlordane and heptachlor constitutes a significant public health risk. These termiticides should be replaced by noncarcinogenic alternatives.

KEYWORDS: Chlordane, heptachlor, termiticide, indoor air, carcinogen

INTRODUCTION

Termites are ubiquitous insects with an unusual capacity for property destruction. In many states property transfer must be accompanied by proof of termiticide application. By far the most frequently applied termiticides belong to the cyclodiene class with 6-12 chlorinations per cyclodiene molecule (Table 1). In this paper I will discuss four of the cyclodiene termiticides: chlordane, heptachlor, aldrin and dieldrin.

Chemically the four cyclodiene termiticides are similar to other chlorinated hydrocarbons in that they are stable, hydrophobic and volatilize poorly. Hence, soil application may last for 10-15 years (and is often guaranteed for such periods) and termiticide molecules rarely migrate with any alacrity either vertically through soil and into groundwater or horizontally and into homes. Yet a termite-impenetrable barrier demands soil migration from discrete injection sites.

Unfortunately, all four cyclodiene termiticides have been found to induce cancers in test animals and are considered by EPA to be class B2 "probable human carcinogens." From animal experimentation, each termiticide has been assigned a carcinogenic potency (q^* in $mg/kg/d^{-1}$) such that quantitative risk assessments from human exposure can be performed. The q^* values which will be used in this paper are chlordane (1.6), heptachlor (3.4), aldrin (11) and dieldrin (30) (see Table 1).

---

a. Present address: Environmental Resources Management, Inc., 855 Springdale Dr., Exton, PA 19341.

New Risks
Edited by L. A. Cox, Jr., and P. F. Ricci
Plenum Press, New York, 1990

Table 1. Risk Assessments for Cyclopentdiene Termiticides (August 1986)

Cyclopentadiene	$q^*(mg/kg/d^{-1})$	Source	Weight	#C1
Aldrin	11	CAG	B2	6
Dieldrin	30	CAG	B2	6
Endrin	--		E	6
Chlordane	1.6	HEA	B2	8
Heptachlor	3.4	CAG	B2	7
Mirex				12
Kepone	9	HEA	B2	10
Toxaphene	11	CAG	B2	7
Strobane				7
Endosulfan	--		E	6

In this paper I will assess carcinogenic risk for termiticide exposures both for normal and accidental scenarios. My calculations will assume that carcinogenic risk is a linear feature of total carcinogen exposure, that is, the product of both the concentration to which an individual is exposed and the total time of exposure.

CHLORDANE, HEPTACHLOR, ALDRIN AND DIELDRIN

The chemical structures of chlordane, heptachlor, aldrin and dieldrin are shown in Fig. 1. Chlordane, also known as octachlor, is a cyclopentadiene with eight chlorines, whereas heptachlor has one chlorine less. Similarly aldrin and dieldrin are both dicyclodienes, both containing six chlorine atoms. Dieldrin is aldrin epoxide and the major metabolic product of aldrin. The major metabolic product of both chlordane and heptachlor is heptachlor epoxide in which the right-most C=C bond is saturated with oxygen.

All four of these polychlorinated cyclodienes are broad spectrum insecticides with uncertain modes of action. Chlordane has been the most widely used with production peaking in 1974 at 10,000 metric tons. Restriction for all food crops and home and garden use has lessened chlordane production, but it is still (in 1986) the most prevalent termiticide employed. Technical grade chlordane contains much heptachlor and is often used in conjunction with aldrin.

GENOTOXICITY OF CHLORDANE AND HEPTACHLOR

Five studies in mice and two in rats show chlordane to be carcinogenic, the major response being hepatocellular adenoma. In one unpublished Japanese study, mammary fibro-adenomas were also seen to increase when female rats were fed 1 ppm technical chlordane. A volatile component of technical chlordane is mutagenic for *Salmonella* bacteria in the Ames test. In addition chlordane is twice as effective as Mitomycin C in inducing sister-chromatid exchanges in fish.

Heptachlor has been tested in mice (four studies) and rats (four studies) and also induces hepatocellular adenoma. Since the active genotoxic product of both chlordane and heptachlor metabolism appears to be heptachlor epoxide, it is not surprising that the two agents have similar genotoxic spectra. Heptachlor epoxide has been found mutagenic for human fibroblasts in cell culture.

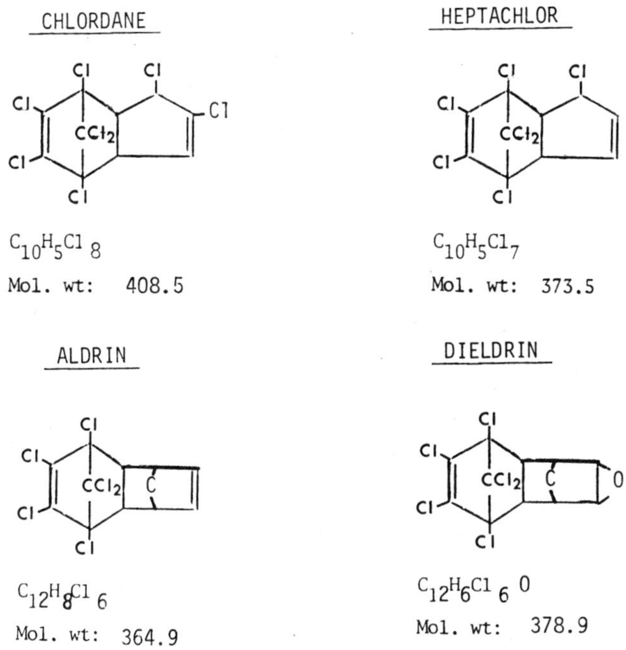

Fig. 1. Cyclodiene termiticide structures.

Both positive and negative dominant lethal tests have been reported for these termiticides. In one positive dominant lethal test with heptachlor, cytogenetic analyses of affected animals confirmed the presence of chromosomal abnormalities. In addition, it has been reported that chlordane crosses the placenta and interferes with T-cell development in transplacentally exposed fetuses.

RISK ASSESSMENTS IN REGION III

In the past three years two toxicologists in EPA Region III (Philadelphia) have been asked by nine homeowners to perform risk assessments in homes which had become inadvertently contaminated by cyclodiene termiticides. We have performed these assessments as a public service upon request and have made no attempt to collect all evidence of termiticide misapplication in the Region.

The results of these nine risk assessments are seen in Table 2. Lifetime unit cancer risks have been calculated assuming inhalation of 8 m^3 over a 14 hour day at the average air contamination levels detected. These upperbound unit risk values are remarkably consistent given the wide range of situations encountered and vary from 1.7×10^{-4} to 5.0×10^{-3}.

INCIDENTS REPORTED TO THE OFFICE OF PESTICIDE PROGRAMS

In 1984 Lois Rossi of EPA's Office of Pesticide Programs asked the 10 EPA Regions to report cases of termiticide misuse or misapplication for each state within their Region.

Table 2. Region III: Domestic Unit Cancer Risks from Termiticide Misapplication

#	Date	Toxicologist*	Termiticide**	Air Conc. (μg/m^3)	Unit Cancer Risk
1	12/84	SR	A	1.46	1.1×10^{-3}
2	11/85	SR	A	2.8	2.1×10^{-3}
3	12/85	BM	C H	5.7 2.8	1.2×10^{-3}
4	01/86	SR	C	22	2.4×10^{-3}
5	01/86	BM	C	2.1	2.3×10^{-4}
6	03/86	BM	A C H	4.23 1.43 7.30	5.0×10^{-3}
7	05/86	BM	C H	1.13 0.66	2.7×10^{-4}
8	06/86	SR	C H	0.57 0.49	1.7×10^{-4}
9	08/86	SR	C	1.9	2.0×10^{-4}

*SR = Sam Rotenbert
BM = Bruce Molholt

**A = Aldrin
C = Chlordane
H = Heptachlor

Note: All unit cancer risks are upperbound lifetime risks assuming 14 hours per day inhalation of 8 m^3 air for 70 years.

The response was mixed, state data being available for various periods of time or not at all. Table 3 summarizes the results of Lois Rossi's survey.

All told, accounts of 653 incidents were collected in the Rossi survey. Almost half of these occurred in Region II, largely in New York state. It should be noted that the 224 cases from New York were reported in just a two year period, showing a high degree of public awareness (and the subjective nature of the survey). Two of these 224 New York cases resulted in destruction of property by owners (one by bulldozer, one by fire). Subsequently both New York and Massachusetts legislatures banned all use of chlordane, heptachlor and aldrin in those two states.

Of the 653 total incidents, 326 were studied in further detail by Lois Rossi. Over half were caused by professional applicators, most of whom were uncertified. Most incidents were caused by improper surface application. These termiticides are registered strictly for soil application. Surface applications had often occurred within the home for control of other insects, such as cockroaches or fleas. These cases of improper surface application of cyclodiene termiticides inside domiciles resulted in the highest air contamination levels and greatest toxicity to residents.

Table 3. EPA Summary Termiticide Misuse/Misapplication by State (1984)

Region	State	Period	Incidents By State	By Region
I	MA	1981-83	15	
	CT	1976-83	30	45
II	NY	1983-84	224	
	NJ	1979-83	88	312
III	DE	1979-84	21	
	PA	1982-85	23	
	VA	1982-83	13	
	DC	1982-83	9	66
IV	FL	N/A	7	
	GA	N/A	15	
	NC	N/A	6	
	SC	N/A	11	
	MS	1974-84	8	47
V	IN	1979-84	46	
	OH	1983	48	94
VI	OK	1983	31	
	NM	1982-84	4	35
VII	NE	N/A	2	
	KS	N/A	7	9
VIII				0
IX	NV	1980-84	11	
	HI	1979-84	23	34
X	OR	1978-83	11	11
			TOTAL	653

REMEDIATION

The prognosis for remediation once homes have become contaminated by cyclodiene termiticides is dismal. These polychlorinated hydrocarbons are as stable in the home as they are in soil and cling tenaciously to all surfaces. Most cleaning strategies are useless; airborne contamination levels continue unabated for years.

The experience of the military is especially didactic. Barrack construction at an air force base placed concrete slabs over chlordane-treated soil. Chlordane migrated through cracks and crevices in concrete slabbing to such an extent that carpeting became wet and inhabitants became nauseous. The Air Force closed the barracks after measuring very high indoor air levels of chlordane. Then it conducted experiments in six barracks to discover which method of cleansing would maximally remediate chlordane contamination. Organic solvents, scouring, cleaning with strong detergents and other strategies all were of no avail. In each of the remediated barracks airborne chlordane levels were *higher* after cleansing than before.

CONCLUSION

Termite control is an important economic consideration for virtually every homeowner. However, given the potent human toxicity of those most normally employed termiticides, the potential health costs may well outweigh economic gains afforded by use of chlordane, heptachlor, aldrin or dieldrin. Our experience is that significant contamination by one or a combination of these termiticides is an all too frequent occurrence.

Alternative methodologies for termite control are either less effective or more costly than application of polychlorinated cyclodiene termiticides. These include application of alternative, noncarcinogenic termiticides (e.g., chlorpyrifos), biologic control (e.g., with nematodes) and biotechnologic control (e.g., with chitin synthetase inhibitors specific for termites). Biotechnologic control is in its infancy at present but promises to afford the most specific alternative for termite control in the future. Not only can specific enzyme inhibitors be tailor-made for insects, but for one specific species of insect, i.e., the termite.

In addition to causing human health risk from misuse, misapplication or accident, normal application of cyclodiene termiticides is associated with measurable carcinogenic risk. Homes to which chlordane, heptachlor or aldrin have been applied strictly according to label instructions, even by certified applicators, frequently have airborne levels of these compounds which entail a 10^{-5} to 10^{-4} lifetime risk for cancer from this source of exposure. In that an estimated 100 million persons are so exposed, the total lifetime burden of these carcinogens may be 1,000 to 10,000 cases of cancer in the U.S.

It is suggested that alternative termiticides not only afford public health protection in the case of accidental scenarios, but are in the best interests of a healthy society even under optimal conditions of application.

NOTE ADDED IN 1989: EPA banned further production of chlordane and heptachlor in January 1988 and all sales in April 1988.

Fuzzy Risk Analysis: Theory and Application

C. Preyssl
International Atomic Energy Agency
Vienna, Austria

ABSTRACT

So far only probability theory has been used in quantitative risk analysis for the fault tree method. In fuzzy fault tree theory, probability theory is postulated for modeling objective information based on randomness. When modeling subjective information based on fuzziness possibility theory is applied. The degree of subjectivity of each event is expressed by an index, which refers to a consideration of model uncertainty, whereas data uncertainty is displayed using distribution functions. After computing the probability and possibility of the TOP-event the index of their event yields the final result.

KEYWORDS: Fault tree analysis, possibility theory, uncertainties

INTRODUCTION

Risk analysis provides the only methods for the investigation of abnormal system behavior under rare or hypothetical events.

The gap between model and reality gives rise to a range of uncertainties in the analysis. Therefore quantitative fault tree analysis has to be based on a mathematical theory suitable for a correct modeling of these uncertainties.

So far in risk analysis probability theory was applied and uncertainties were represented as data uncertainties.[3,9,10] A more careful evaluation of estimation as opposed to statistically relevant data requires the development of a suitable mathematical model where the subjective character of input information remains transparent even in the results. Within the concept of fuzzy fault tree theory, the above requirements are fulfilled and uncertainties are modeled correctly.[1,2]

UNCERTAINTIES AND TYPES OF INFORMATION

Different sources of information for the quantification of system events can be identified.

Some typical events together with their basic characteristics would be:[1,4,6,8]

New Risks
Edited by L. A. Cox, Jr., and P. F. Ricci
Plenum Press, New York, 1990

1.	To toss a six with an ideal cube	mathematical *a priori* probability;
2.	Failure of a valve under normal conditions	many direct observations, good statistics;
3.	Failure of pump exposed to high level of radioactivity	no observations, only derived data, poor statistics;
4.	Catastrophic event	extremely rare, no statistics;
5.	Human error and behavior	diversity in observation, only estimations.

Clearly one can distinguish between objective information and subjective information concerning the events. As the events 1 and 2 contain accessible information based on randomness, they will be of the objective type. The events 3 to 5 contain information not directly accessible, and are characterized by fuzziness; human judgment will be involved in quantification. The degree of subjectivity in events 3 to 5 is increasing.

Two types of uncertainty can be defined in an analysis: model uncertainty and data uncertainty.

A separate treatment of objective and subjective information will be equivalent to the consideration of model uncertainty, whereas the degree of confidence in the data will be displayed when considering data uncertainty. These requirements can be satisfied by a mathematical model postulating the use of probability theory for objectives information and possibility theory for subjective information and their integration into one theory where model and data uncertainties are displayed separately.[1,6]

FUZZY MATHEMATICS

Fuzzy Set Theory [5]

The idea of a fuzzy set $\underset{\sim}{A}$ is based on the "degree of memberships" $\zeta\underset{\sim}{A}$ of an element x_i of the universe U to $\underset{\sim}{A}$.

$$\zeta\underset{\sim}{A}(x_i) \longrightarrow [0,1] \quad i \epsilon N.$$

So it is possible to define the degree of subjectivity of an information.

In this case $\underset{\sim}{A}$ is the set of subjective information. On all x_i one then has information which belongs with degree $\zeta(x_i)$ to $\underset{\sim}{A}$.

Let there be several $\zeta(x_\ell)$ which are combined to one $\zeta(y)$ using an operator $\odot$:

$$\zeta(y) = \underset{\ell=1}{\overset{k}{\odot}} \zeta(x_\ell).$$

Three different types of operators can be defined

$$\underset{\ell=1}{\overset{k}{\odot}} = \max_{1 \leq \ell \leq k}$$

$$\underset{\ell=1}{\overset{k}{\odot}} = \frac{\alpha}{k} \sum_{\ell=1}^{k} \quad 0 \leq \alpha \leq 1$$

$$\underset{\ell=1}{\overset{k}{\odot}} = \underset{1 \leq \ell \leq k.}{\min}$$

The first operator yields a higher weighting of large ζ, whereas the second and third varies between an equivalent treatment of objective and objective expressed by the arithmetic sum with $\alpha = 1$ and a higher weighting of objective expressed by $\alpha \neq 1$ or by the min-operation.

Fuzzy Measure Theory[5]

A fuzzy measure $g(B)$, expressing the grade of belonging of an unknown element $\underset{\sim}{x}$ to the classic set B.

$$g(B) \longrightarrow [0,1]$$

is defined as

$$\begin{aligned} g(\phi) &= 0 \\ g(\cup) &= 1 \\ g(B_1) \leq g(B_2) &\Longleftrightarrow B_1 \subseteq B_2 \\ g(B_1 \cup B_2) &= \zeta[g(B_1), g(B_2)] \geq \max\,[g(B_1), g(B_2)] \\ g(B_1 \cap B_2) &= \Omega[g(B_1), g(B_2) \leq \min\,[g(B_1), g(B_2)] \end{aligned}$$

There are two special forms of fuzzy measures, i.e., the probability measure $P(B)$ with the special properties

$$\begin{aligned} P(B_1 \cup B_2) &= P(B_1) + PC(B_2) - P(B_1 \cap B_2) \\ P(B_1 \cap B_2) &= P(B_1) \cdot P(B_2) \end{aligned}$$

and the possibility measure $\pi(B)$ with the special properties[7]

$$\begin{aligned} \pi(B_1 \cup B_2) &= \max\,\{\pi(B_1), \pi(B_2)\} \\ \pi(B_1 \cap B_2) &= \min\,\{\pi(B_1), \pi(B_2)\}. \end{aligned}$$

This yields for the complement # of B:

$$\begin{aligned} P(B) + P(\#B) &= 1 \\ \max\,\{\pi(B), \pi(\#B)\} &= 1. \end{aligned}$$

In general

$$\pi(B) \geq P(B)$$

will hold.

In order to represent uncertainties in the P or π values, distributions can be defined so that

$$\widehat{p}(P(B)) \longrightarrow \int_0^1 \widehat{p}(P(B))\, d\, P(B) = 1$$

$$\widehat{\pi}(\pi(B)) \longrightarrow \max_{0 \leq \pi(B) \leq 1} \widehat{\pi}(\pi(B)) = 1.$$

Fuzzy Algebra[1]

Let $M(x)$ and $N(x)$ be distribution functions in the $x-y$ plane, $y\epsilon[0,1]$. Using the definition of the $\tilde{b}$-cut,

$$M_{\tilde{b}} = \{x : M(x) = \tilde{b}\}$$

and when splitting the function $M(x)$ into two parts,

$$M(x) = M^u(x) \cup M^d(x)$$

$$\frac{d\,M^m(x)}{d\,x} \geq 0$$

$$\frac{d\,M^d(x)}{d\,x} \leq 0$$

one can define

$$M(x)\tilde{0}N(x) = \bigcup_{\tilde{b}} \bigcup_{a=u,d} [M^a_{\tilde{b}}\ 0\ N^a_{\tilde{b}}]$$

where the operators $\tilde{0}$ are correlated with the operators 0 of normal algebra as listed in Table 1.

Table 1. Operators Table

$\tilde{0}$	0
$\widetilde{\max}$	max
$\widetilde{\min}$	min
$\tilde{\oplus}$	+
$\tilde{\ominus}$	−
$\tilde{\odot}$	$\cdot$

A scalar $\underline{\lambda}$ will be treated as a needle function,

$$\underline{\lambda}0\ M(x) = \bigcup_{\tilde{b}} [\underline{\lambda}_{\tilde{b}}\ 0\ M_{\tilde{b}}] = \bigcup_{\tilde{b}} \left\{\lambda o x : M(x) = \tilde{b}\right\}$$

with the implicit definition of $\underline{\lambda}$

$$\underline{\lambda} = \bigcup_{\tilde{b}} \lambda_{\tilde{b}} \qquad \underline{\lambda}_{\tilde{b}} = \{x : x = \lambda\}$$

This yields automatically, for example,

$$M(x)\tilde{\ominus}M(x) = \underline{0}$$

$$1\tilde{\odot}M(x) = M(x).$$

FUZZY FAULT TREE THEORY

A fault tree of a system is a logic diagram showing the combinations of events leading to the TOP-event. For the computation of the TOP-event T, the basic events E_i have to be quantified, and a suitable mathematical formulation has to be defined for the logical operators. Fuzzy fault tree analysis will be based on fuzzy mathematics using both probability theory and possibility theory.

The concept of a "fuzzy measure" yields

$$\begin{aligned} E_i &\longrightarrow g(E_i) \\ E_i \text{ and } E_j &\longrightarrow g(E_i \cap E_j) \\ E_i \text{ or } E_j &\longrightarrow g(E_i \cup E_j) \\ E_i \text{ if } E_j &\longrightarrow g(E_i / E_j) \\ \text{not } E_i &\longrightarrow g(\# E_i). \end{aligned}$$

The basis for expressing information on the events E_i both as $P(E_i)$ and $\pi(E_i)$ can be either a relative frequency event rate $\lambda(E_i)$ or qualitative expression $Q(E_i)$ which then can be transferred into P and $\pi(E_i)$ using, for instance, Table 2.[4,7]

Table 2. Probability-Possibility Table

$P(E)$	$\lambda(E)$	$Q(E)$		$\pi(E)$
1.0	$2.0 - \infty$	Certain	Q1	1.0
$10^{-1} - 1.0$	$10^{-1} - 2.0$	Most probable	Q2	$0.9 - 1.0$
$10^{-2} - 10^{-1}$	$10^{-2} - 10^{-1}$	Quite probable	Q3	$0.7 - 0.9$
$10^{-3} - 10^{-2}$	$10^{-3} - 10^{-2}$	Probable	Q4	$0.5 - 0.7$
$10^{-4} - 10^{-3}$	$10^{-4} - 10^{-3}$	Rather improbable	Q5	$0.3 - 0.5$
$10^{-5} - 10^{-4}$	$10^{-5} - 10^{-4}$	Improbable	Q6	$0.2 - 0.3$
$10^{-7} - 10^{-5}$	$10^{-7} - 10^{-5}$	Highly improbable	Q7	$0.1 - 0.2$
$10^{-8} - 10^{-7}$	$10^{-8} - 10^{-7}$	Extremely improbable	Q8	$10^{-2} - 0.1$
$.0 - 10^{-8}$	$0.0 - 10^{-8}$	Next to impossible	Q9	$0.0 - 10^{-2}$
0.0	0.0	Impossible	Q10	0.0

Of course the mappings in Table 2 are based to a great extent on subjective evaluations.

The TOP-event T is computed using an algorithm yielding a matrix with elements $E_{k\ell}$. Events connected by OR are put into separate lines k, events connected by AND are put into separate rows ℓ. Starting with T the matrix finally contains only the basic events:[3]

$$\{E_{k\ell}\} = \{E_j\}.$$

It follows that

$$T = \bigcup_k \bigcap_\ell E_{k\ell} = \bigcup_k C_k,$$

where C_k stands for the minimal cuts, which implies $P(T)$ and $\pi(T)$ when using the corresponding operators in both theories.

Now one can define the TOP-interval I_T in π space[1]

$$I_T = [\pi_o(T), \pi_s(T)]$$

where

$$\begin{aligned} \pi(T) &= \pi_s(T) \\ P(T) &\longrightarrow \pi_o(T) \end{aligned}$$

using Table 2 for mapping $P(T)$ into π space.

In order to consider model uncertainties the concept of fuzzy set is applied to express the degree of subjectivity in the information on the basic events E_i in form of an SO index $\zeta(E_i)$:

$$E_i \longrightarrow \zeta(E_i)$$

with

$$\zeta(E_i) \in [0,1] \ldots \begin{cases} 0\ldots & \text{OBJECTIVE} \\ 1\ldots & \text{SUBJECTIVE} \end{cases}$$

The index is determined by judging how much randomness or fuzziness characterizes the event.

Starting from the bottom of the tree the "SO Index" of the TOP-event is obtained:

$$\zeta(T) = \underset{n}{\odot} \left\{ \zeta(E_\ell), \underset{j}{\odot} \left\{ \zeta(E_k), \ldots \underset{i}{\odot} \{\zeta(E_i)\} \right\} \right\},$$

where n is the number of logic gates combining the events. $\odot$ has to be chosen according to the weighting of "objectivity."

Integrated Fault Tree Theory

It is now possible to combine probability theory and possibility theory in one mathematical model using I_T and $\zeta(T)$. The final result in π space is the possibility π of the TOP-event T with index ζ^T, denoted $T^{\zeta T}$:[1]

$$\zeta^T \in [0,1] \longrightarrow \pi(T^{\zeta T}) \in I_T$$

where

$$\zeta = 1 \longrightarrow \pi_s(T)$$

$$\zeta = 0 \longrightarrow \pi_o(T).$$

This yields automatically

$$\pi(T^{\zeta T}) = \pi_s(T) + \Delta \cdot \zeta^T$$

where

$$\Delta = \pi_o(T) - \pi_s(T).$$

It should be mentioned that

$$s = \sqrt{s^2}$$

can be regarded as the degree of sensitivity to model uncertainty, which is an important characteristic of a fault tree. When s is zero it is not relevant whether objective or subjective information is available. Where safety systems with several redundant components are used, s may become large. Hence only objective information can yield a guarantee for more safety.

The concept of the fuzzy fault tree theory is summarized in Fig. 1.

So far model uncertainty was treated. Data uncertainty can be considered using distributions.

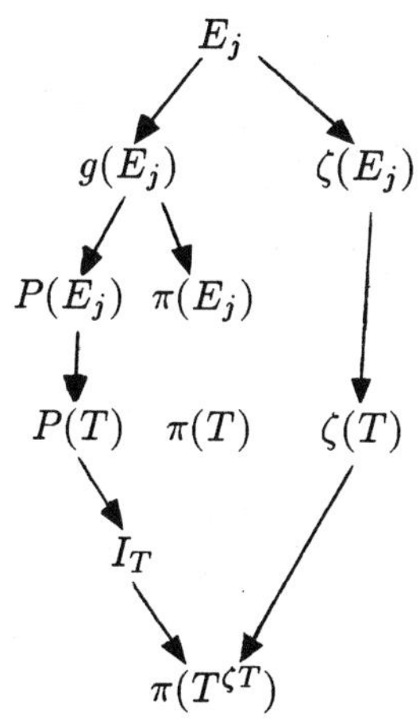

Fig. 1. Fuzzy fault tree logic.

For possibilities this means

$$\pi(E_i) \longrightarrow \widehat{\pi}(\pi(E_i))$$
$$0 \longrightarrow \widetilde{0}$$

where the necessary operators $\widetilde{0}$ of fuzzy algebra have to be taken.

For probabilities this means

$$P(E_i) \longrightarrow \widehat{p}(P(E_i))$$

and the probabilistic sum and product of distributions will be carried out as proposed in probability theory.

In this context it is useful to study the transformations

$$\widehat{\pi}(x) = k \cdot \widehat{p}(x)$$
$$\widehat{p}(x) = \ell \cdot \widehat{\pi}(x)$$

where

$$k = \frac{1}{\max\limits_{0 \leq x \leq 1} \widehat{p}(x)}$$

$$\ell = \frac{1}{\int_0^1 \pi(x)dx}$$

Considering model and data uncertainties, the TOP-event can be computed in the form of a distribution function:

$$T \longrightarrow \widehat{\pi}(\pi(T^{\zeta T})).$$

TOP-Event Classification

A spectrum of information is contained in the possibility distribution of the TOP-event $\widehat{\pi}(\pi(T^{\varsigma T}))$.[1]

The degree of sensitivity s of the fault tree structure and $\varsigma(T)$ reflect the need for objective information.

The maximality defined as the 1 cut of $\widehat{\pi}(\pi(T^{\varsigma T}))$

$$m = \widehat{\pi}_1 = \{\pi : \widehat{\pi}(\pi) = 1\} \quad m\epsilon[0,1]$$

should be as low as possible.

The potentiality, defined as

$$p = \int_m^1 \widehat{\pi}(x)\, dx \quad p\epsilon[0,1],$$

contains information on the degree of exceeding the peak of the distribution and should be small too.

The uncertainty, defined as

$$u = \int_0^1 \widehat{\pi}(x)\, dx \quad u\epsilon[0,1],$$

measures the uncertainty in the data.

The dominance of several basic events E_i can be defined for $\pi_s(T)$, giving the wanted information for $\widehat{\pi}(\pi(T^{\varsigma T}))$ using

$$\ell(T) = \bigcup_k \ell_k(E_j)$$
$$d\,(E_j) = \ell_k(E_j)$$

where $\ell(T)$ is the length of the $\widehat{\pi}(\pi_s(T))$-curve and $\ell_k(E_j)$ the fraction contributed by $\widehat{\pi}(\pi(E_j))$. The TOP-event most dominating basic event E_j^d can be determined from

$$\max_k \ell_k(E_j) \longrightarrow E_j^d.$$

The sensitivity to data changes can be studied as usual.

CASE STUDY

For reasons of simplicity a very simple fault tree — given in Fig. 2 — will be studied. In addition only model uncertainties will be considered, i.e., only mean values or peaks of distributions are used.

The algorithm determines the TOP-event as follows:

$$T = (A_1 \cap A_2 \cap A_3) \cup B_1 \cup B_2$$

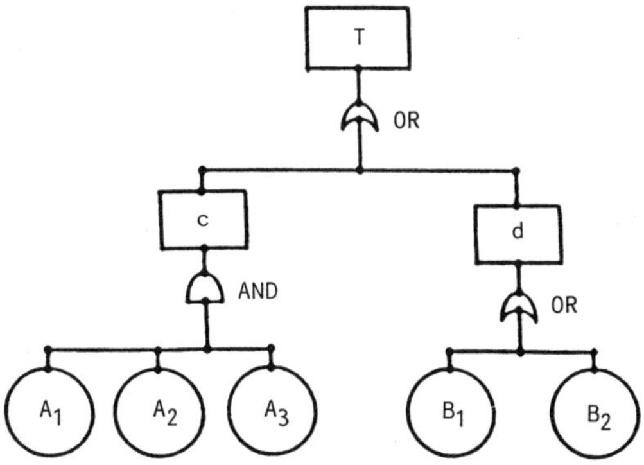

Fig. 2. Fault tree: Case study.

Two different cases will be studied. In the first case there are good statistics on all basic events A_i available, i.e. $\zeta'(A_i) = 1$:

$$\lambda(A_i) = 10^{-2} \quad i = 1, 2, 3.$$

In the second case only qualitative estimations exist, i.e. $\zeta''(A_i) = 0$:

$$Q(A_i) = Q3 \quad i = 1, 2, 3.$$

Subjectivity is not weighted, which means

$$\underset{i}{\odot} = \frac{1}{n}\sum_{i=1}^{n} \qquad \alpha = 1.$$

All data and intermediate results are given in Table 3.

Table 3. Results of Case Study

Event	π	P	ζ'	ζ''
A_i	0.8	10^{-2}	1	0
B_1	0.2	10^{-5}	1	1
B_2	0.3	10^{-4}	1	1
T	0.8	10^{-4}	1	0.5

The two different final results are

$$\pi(T^{\zeta'}) = 0.3$$
$$\pi(T^{\zeta''}) = 0.65$$

which demonstrates the need for objective information on a redundancy system to guarantee safety.

The high sensitivity $s = 0.3$ can be decreased to $s = 0$ when there is only one A. But when, for example, the number of events B_i is increased to $i \approx 10^3$, the sensitivity will increase as soon as the values of A are smaller than of B_i.

INTERPRETATION AND CONCLUSION

The fuzzy fault tree theory is characterized by a number of typical features which can be summarized as follows:

- For vague input information a more simple and less strict mathematical model is used, i.e. possibility theory as opposed to probability theory for real statistical data.

- The max-min operations are suitable to model estimations; they are more simple and less sensitive to data changes than probabilistic operators.

- Using max-min operators the ordering among numbers, i.e. subsystems, and not their absolute value is emphasized.

- Possibility values cannot be misused directly in terms of frequency.

- Even non-numerical expressions can be treated using a fuzzy logic.

- The min-operator is conservative and hence emphasizes the importance of redundancy, conditionality with respect to common mode and common cause and human error.

- When expressing data uncertainty, no analytic form of the possibility distribution is necessary. For the simple operation min and max, only the graphical representations have to be available. An analytic expression of the probability distributions will be very helpful.

- The information contained in the possibility distribution of the result can be used systematically for the classification of an event.

- The introduction of the SO index guarantees that input information is conserved during all transformations and is displayed in the result as output information.

- Normal probabilistic fault tree theory is included as a special case of fuzzy fault tree theory, i.e. when only objective information is available.

- The concept of combining subjective with objective information can be regarded as modeling man-machine interface.

- Each fault tree is characterized by the grade of sensitivity to the type of input information in addition to the sensitivity to data changes.

An important drawback of the fuzzy fault tree method is the need for more computational work as compared to the classical approach.

Finally it has to be pointed out that qualitative formulations have to be transformed into quantitative expressions, i.e., definition of π and of 0, and choice of ζ^i and $\odot$.

Although it can be shown that the theory is not sensitive to changes of the transformations, it has to remain evident that the construction of models is based on human judgment and therefore yields a certain individual arbitrariness.

REFERENCES

1. C. Preyssl, Unschärfe-Risiko, Analyse — Theorie und Anwendung, Thesis, Technical Unviersity Vienna, Austria, 1986.
2. C. Preyssl, H. Ruch, H. Böck, The Impact of Uncertainty on PRA in the Nuclear Fuel Cycle, Atominstitut der Oesterr, Universit\"aten, AIAU 85305 Vienna, Austria, 1986.
3. N. U. McCormick, *Reliability and Risk Analysis*, Academic Press, 1981.
4. Y. Nishiwaki *et al.*, Accidents and Human Factors, presented at the International Symposium on the Risk and Benefits of Energy Systems, Julich, FRG, IAEA-SM-273/7, 1984.
5. D. Dubois and H. Prade, *Fuzzy Sets and Systems*, Academic Press, 1980.
6. Y. Nishiwaki *et al.*, A New Approach Based on Fuzzy Sets Concept to Fault Tree Analysis and Diagnosis of Failure at Nuclear Power Plants, presented at IAEA Seminar on Diagnosis of and Response to Abnormal Occurrences at Nuclear Power Plants, Dresden, GDR, 1984.
7. H. Prade, A Simple Inference Technique for Dealing with Uncertain Facts in Terms of Possibility, presented at Information Processing in Artificial Intelligence and Operations Research, Cambridge, U.K., 1984.
8. A. D. Swain and H. E. Guttmann, Handbook of Human Reliability with Emphasis on Nuclear Power Plant Application, NUREG/CR-1278, 1983.
9. N. H. Roberts *et al.*, Fault Tree Handbook, NUREG-0492, 1981.
10. C. Preyssl and M. C. Cullingford, Uncertainties and Demonstration of Compliance with Numerical Risk Standards, presented at IAEA Seminar on Implications of Probabilistic Risk Assessment, Blackpool, U.K., IAEA-SR-111/23, 1985.

Assessment of the Acceptability of the Proposals for a Fixed Link Across the English Channel

J. C. Chicken
J. C. Consultancy Ltd.
United Kingdom

M. R. Hayns
United Kingdom Atomic Energy Authority
United Kingdom

ABSTRACT

This paper presents the results of a study in which the Risk Ranking Technique was applied to the evaluation of the acceptability of the four proposals for a fixed link across the English Channel. The reason for using the Risk Ranking Technique was that it gives a comprehensive assessment of the acceptability of the technical, economic, and socio-political aspects of the risks inherent in the proposals.

KEYWORDS: Channel Tunnel, risk, ranking, comprehensive risk assessment

INTRODUCTION

In January 1986, after considering four proposals for a fixed link across the English Channel — the largest civil engineering project ever undertaken by the private sector in Europe — the British and French Governments announced that they had found one proposal to be acceptable,[1] and the necessary treaty was signed in February 1986.[2]

This paper presents the results of a study of the acceptability of the four proposals using the Risk Ranking Technique. The paper is arranged in five parts: the nature of the Risk Ranking Technique, a comparison of the Risk Ranking Technique with other decision analysis techniques, the ranking of the four proposals, an assessment of the efficacy of the Ranking Technique, and, finally, conclusions about the use of the Ranking Technique and the problems of assessing the acceptability of novel projects.

THE NATURE OF THE RISK RANKING TECHNIQUE

Earlier development of the Risk Ranking Technique was reported at the SRA meetings in Knoxville in 1984[3] and Washington in 1985.[4] The incentive for developing the Risk Ranking Technique was to provide decision makers with a comprehensive assessment of the acceptability of a project.

New Risks
Edited by L. A. Cox, Jr., and P. F. Ricci
Plenum Press, New York, 1990

Essentially the technique consists of making an assessment of all the relevant factors associated with a project. For analytical purposes, the factors are divided into three groups: technical, economic and socio-political. The acceptability of each group of factors is scored on a scale 0 to 4, the higher the score the lower the acceptability. The overall ranking of the acceptability of a project is calculated by integrating the scores of the individual factors. Table 1 shows how the definition of risk rank is constructed.

The success of the ranking techniques depends, to a large extent, on having available a reliable analysis of the conditions of interest. Table 2 suggests a logical procedure for checking the reliability of the data available. Possible sources of errors include errors and bias in the data, lack of data points, and application of the data to conditions quite unrelated to the conditions from which they were generated.

Table 1. Construction of Risk Rank

Risk Rank	Acceptability	Control action	Total Score	Factor score range
1	Unlikely.	Unlikely any possible.	>6 -- 12	No factor >4.
2	Only if risk can be reduced.	Administrative and engineering.	>4 -- 6	No factor >3.
3	Subject to certain action.	Engineering or technical.	>2 -- 4	No factor >2.
4	Without restriction.	None required.	0 -- 2	No factor >1.

Table 2. Logical Analysis Requirement Check

Step	Nature of Check	Comment
1	Identify data.	The data required must be specified.
2	Assess reliability of data to be used.	Data reliability analysis procedure should be adopted.
3	Assess analytical procedures.	Determine if procedures and data are appropriate.
4	Assess inherent errors.	No confidence in the results unless errors are known.
5	Show how analysis will be refined as the project develops.	Analysis likely to require iterative refinement if novel features incorporated.

COMPARISON OF THE RISK RANKING TECHNIQUE WITH OTHER DECISION ANALYSIS TECHNIQUES

All that is attempted here is a comparison of the Risk Ranking Technique with the general pattern of other techniques. In all decision making there is concern about the accuracy of the predictions. It has been argued that subjective probabilities may be the most appropriate basis for decisions because, unlike objective probabilities, they are not based purely on past experience, which in any case may not be exactly relevant to new and novel circumstances.[5] Few if any circumstances are repeated exactly. Several techniques express the decision options in terms of payoff and probability and others in more arbitrary terms. In practice, in the evaluation of novel situations the results of such assessments generally have to be expressed in qualitative terms. The principle of insufficient reason is an attempt to determine the optimum decision, when there is no evidence. The principle assumes all options as equally probable, which makes the decision process into a risk problem. The method does not provide a comprehensive evaluation of all the factors involved.

The multivariate model technique, another approach to the problem, attempts to logically structure and coordinate the analysis of the views of several individuals or organizations on a particular issue.[6] The technique allows for various factors, their defining characteristics and error functions. If accurate data were available, the solution of the equations would be helpful, but in practice the size of the error function is likely to reduce the solution to general guidance. This problem of dealing with the fuzziness of the data has led to the development of the fuzzy logic approach.

Use has been made of the fuzzy logic approach to control cement kilns automatically[7] and consideration has been given to using fuzzy logic processing for assessing company profit margins, driving trains, guiding robot arc-welders and assessing compliance with safety goals.[8]

Unlike the Risk Ranking Technique, none of the other decision analysis techniques consider as comprehensively technical, economic and socio-political factors, nor do they identify in a comprehensive way what is acceptable. Moreover, the way the techniques allow for uncertainty is not always satisfactory and they do not present their findings in a way that is generally understood.

RANKING THE FOUR FIXED LINK PROPOSALS

The four channel crossing proposals considered were the Eurobridge, the Euroroute, the Channel Tunnel and the Channel Expressway. The main technical factors of the proposals are summarized in Table 3, the economic factors are summarized in Table 4, and the socio-political factors are summarized in Table 5.

In none of the proposals was there an attempt to quantify the magnitude or the probability of the technical risks involved. In some cases, it appeared that the problems were glossed over. For example, in one proposal it was not mentioned that "live" experience with the material proposed for the suspension bridge cables was only about one-seventh of the life proposed for the material.

On the basis of the available information, the ranking of the four proposals was built up, together with the option of not constructing a fixed crossing. The justifications for the scoring of the technical, economic, and socio-political factors are shown in Tables 6, 7, and 8, respectively. Evaluation of the economic implications of the proposals showed that the total cost of the proposals was likely to be a significant factor in the economies of the two countries. Possible variation in total cost is a factor that has to be considered, as experience

Table 3. Summary of Main Technical Factors of Interest

Proposal*	Construction Features	Hazard Concerns
Eurobridge	Suspension bridge of seven 5-km spans plus 6-m-diameter rail tunnel; traffic lanes enclosed in a Superferrolo tube; suspension cables 1.4-m-dia Parafil.	Suspension towers to withstand impact of a 250,000-ton ship; oscillation of spans; ventilation of traffic tube; explosions on the bridge; driver fatigue; life of components.
Euroroute	For road transport, two bridges from each coast to artificial islands; islands linked by tunnel on the sea bed; rail tunnel on sea bed coast to coast.	Ventilation of the tunnels; earth movement; resistance of the 34 protective caissons to damage by shipping; explosions in the tunnel; life of components.
Channel Tunnel	A three-tunnel system; two railway and one service tunnels; terminals for loading and unloading road vehicles on and off trains.	Ventilation; earth movements; life of components; explosions in the tunnels.
Channel Expressway	A twin tunnel system, each tunnel taking both road and rail traffic; would be the largest drive-through tunnel in the world.	Ventilation; earth movement; life of components; driver fatigue.

*See Ref. 9.

Table 4. Summary of Main Economic Factors of Interest

Proposal	Proposing Group	Estimated Cost	Completion Time	Suggested Toll	Return
Eurobridge	Laing, Brown & Root and ICI.	£5.9 bn.	5 years.	Comparable to sea ferries.	21% to 22%; payback 6-11 years.
Euroroute	Trafalgar House, British Steel and banks.	Motorway link £7.2 bn; rail link £ 3.5 bn.	5 years for bridge; 8 years for tunnel.	Comparable to sea ferries.	17% gross; complete payback in 15 years.
Channel-Tunnel	Five UK & two French construction companies, and three French banks.	Max. debt £4.75 bn allowing £1 bn for contingencies.	7 years from Government announcing decision.	10% below existing ferry tariffs.	19% rate of return; complete payback 15 in years.
Channel	British Ferries Ltd.	£2.5 bn, including construction period interest.	Expressway 5 years.	Cars 50% cheaper & lorries the same as ferries.	Return on equity 27%.

has shown that the cost of major novel projects often far exceeds initial estimates. For example, quite unrelated projects like the Sydney Opera House and Concorde escalated to about ten times the original estimate.[10] The Thames barrier was originally estimated to cost 23 million and actually cost 461 million.[11]

Table 5. Summary of Socio-Political Factors

Proposal	Public Consultation	Advantages for Public	Disadvantages for Public	Argument For	Argument Against
Eurobridge	No public inquiry; public may raise questions in parliament; simple poll showed 63% favored fixed link.	Reduces journey time; greater efficiency should improve employment.	Economic adjustment in Dover and Calais areas; loss of agricultural land for terminals.	Improve European employment.	High cost technology uncertain.
Euroroute	Same as for Eurobridge.	Same as for Eurobridge.	Same as for Eurobridge.	As Eurobridge above.	High cost driver safety & ventilation.
Channel Tunnel	Same as for Eurobridge.	Same as for Eurobridge.	Same as for Eurobridge.	Less risk of accident.	Road to rail change.
Channel Expressway	Same as for Eurobridge.	Has benefit of both road and rail.	Same as for Eurobridge.	As Eurobridge above.	Driver safety & ventilation.

Table 6. Technical Factor Rank Score of Proposals

Proposed	Technical Factor Concern	Score and Justification
Eurobridge	Limited data about life of Superferrolo and Parafil; no information about oscillation of bridge, earth movements, adequacy of ventilation, driver fatigue, water tightness and resistance to explosions.	Proposed score = 2. Considerable technical justification of new materials required.
Euroroute	Resistance to earth movements; ventilation; driver fatigue; bridge oscillation; tunnel water tightness; resistance to explosions.	Proposed score = 2. Considerable technical justification of design required.
Channel Tunnel	Earth movements; ventilation; water tightness; resistance to explosions; driver fatigue.	Proposed score = 2. Some justification of design required.
Channel Expressway	Earth movements; adequacy of ventilation; water tightness; resistance to explosions; driver fatigue.	Proposed score = 2. Some justification of design required.
No fixed link crossing	Revision of ferry regulatory requirements.	Proposed score = 3. Revision of requirements unlikely in the short term.

In assessing the socio-political factors, some objection by people living in the area of the terminals was found in the discussion in Parliament and the press. However, no carefully designed survey of public opinion was made and the tests made by local papers are considered to be of questionable value.[12]

The overall ranking of acceptability is shown in Table 9. The conclusion that appears to be justified from the overall ranking is that the Channel Tunnel proposal was the most acceptable and the no fixed link option the least acceptable. Also the Channel Tunnel proposal was shown to be subject to the least number of technical reservations.

Table 7. Economic Factor Rank Score of Proposals

Proposal	Economic Factor Concerns	Score and Justification
Eurobridge	£5.9 bn; no indication of possible variation.	Proposed score = 2. No consideration of variation given.
Euroroute	£5.2 bn.; proposer suggested total cost may reach £10.7 bn (£7.2 bn for the motorway link and £3.5 bn for the rail link).	Proposed score = 3. If variation in the range mentioned was realized, the project would be unacceptable.
Channel Tunnel	Maximum debt the project would incur would be £4.75, allowing £1 bn for unforeseen contingencies.	Proposed score = 1. This seems to include a reasonable allowance for contingencies.
Channel Expressway	£2.1 bn, excluding interest during construction; £2.5 bn, including construction period interest and fees; no discussion of variation.	Proposed score = 2. Although this is the simplest scheme, no allowance is made for cost variation.
No fixed link crossing	No costs are given, but for a comparable service, new ferries and terminals would be required.	Proposed score = 2. No figures given.

Table 8. Socio-Political Factor Rank Score of Proposals

Proposal	Socio-Political Factor Concerns	Score and Justification
Eurobridge	Changes in local environment and improvement in employment prospects; risks to drivers on bridge.	Proposed score = 1. No serious objection.
Euroroute	Changes in local environment and improvement in employment prospects; adequacy of ventilation.	Proposed score = 1. No serious objection.
Channel Tunnel	Changes in local environment and improvement in employment prospects; adequacy of ventilation.	Proposed score = 1. No serious objection.
Channel Expressway	Changes in local environment and improvement in employment prospects; adequacy of ventilation.	Proposed score = 1. No serious objection.
No fixed link crossing	Inadequate transport service.	Proposed score = 2. Some objection until service improves.

A CRITICAL ASSESSMENT OF THE EFFICIACY OF THE RANKING TECHNIQUE

Ultimately the governments reached their decision by evaluating the qualitative advice of a multidisciplinary assessment team. The published data gave the impression that there had been no consistent attempt to quantify the significance of the risks and uncertainties inherent in the proposals. Assessing which proposal is the most acceptable on relatively thin data is not an unusual position for a decision maker in the precontract stage. It would be easy to say the adequacy of the ranking was limited by the adequacy of the data

Table 9. Overall Ranking of the Acceptability of Proposals

Proposal	Total Score	Rank
Eurobridge	5	2
Euroroute	6	2
Channel tunnel	4	3
Channel expressway	5	2
No fixed link option	7	1

available in Ref. 9, but such a comment would be misleading. The assessment was made in a consistent way and the ranking scores allocated reflect an unbiased assessment of the acceptability of the salient features of the proposals. A benefit of the analysis is that even though comprehensive quantified data were not available, the structured form of the ranking technique allowed a logically defensible comparative assessment of acceptability to be made. This also implies that the quality of the ranking would be improved with better data. The study showed both strengths and weaknesses of the ranking technique.

The strengths shown are: (1) The ranking can still be made even though there is only qualitative data available. (2) The ranking process produces a logically defensible comparative assessment of the options. (3) The ranking process identifies points that should be clarified before a final decision is made. (4) Ranking presents the decision maker with an easily understood and non-emotive evaluation of acceptability. (5) The ranking system provides a way of codifying the views of experts.

The weaknesses shown are: (1) The rank divisions may be too coarse. (2) The relationship between ranking score and evidence is too arbitrary. (3) As used, the ranking process does not define evidence that has to be presented.

CONCLUSIONS

In our opinion, the study appeared to justify the following conclusions about the Ranking Technique: (1) The technique can make a useful, logical assessment of acceptability even when only qualitative evidence is available; in this context it can be useful in structuring analysis of expert opinions. (2) The Ranking Technique gives a more easily understood and a more comprehensive assessment than other decision analysis techniques. (3) The Ranking Technique can be applied to complex decisions. (4) No technique can give a reliable estimate of acceptability if it has to be based on poor qualitative information. (5) The Ranking Technique is worth development and testing.

REFERENCES

1. The Channel Fixed Link, Her Majesty's Stationery Office, Cmnd 9735, London, February 1986.
2. The Channel Fixed Link, Concession Agreement, Cmnd 9769, Her Majesty's Stationery Office, London, April 1986.

3. J. C. Chicken and M. R. Hayns, A Multi-Factor Criterion for Risk Acceptability, paper presented at the Annual Meeting of the Society for Risk Analysis, Knoxville, October 1984.
4. J. C. Chicken and M. R. Hayns, Development of the Non-Dimensional Method of Ranking Risks, in *Risk Assessment and Management*, pp. 481-490, Lester B. Lave, ed., Plenum Press, New York (1987).
5. A. Koutsoyiannis, *Non-Price Decisions: The Firm in a Modern Context*, pp. 564-569, Macmillan Press Ltd., London (1982).
6. G. Hilton, *Intermediate Polimetries*, Columbia University Press, New York (1982).
7. Definitely No Can Mean Maybe, *The Economist*, pp. 91-92, May 11, 1985.
8. A-R. F. Abdul-Fattah and W. H. Abdulfarj, Siting of Nuclear Power Plants in Saudi Arabia Using Fuzzy Decision Analysis, *Nuclear Technology* **58**:404-413 (September 1982).
9. Channel Fixed Link Environmental Appraisal of Alternative Proposal, Her Majesty's Stationery Office, London, December 1985.
10. P. Hall, *Great Planning Disasters*, Penguin Books, p. 7 (1980).
11. Hansard, House of Commons Official Report, **88(24)**, December 9, 1985, Col 707, Her Majesty's Stationery Office, London.

Risk Assessment and Management: A Regional Approach

R. Coulon, J. Aigueperse, and F. Anguenot
Commissariat à l'Energie Atomique
Fontenay Aux Roses Cedex
France

G. Bresson
Centre de Développement des Etudes et Applications
en Hygiène et Sécurité
Paris, France

ABSTRACT

As major human activities develop, those responsible for orienting these activities and making decisions must take more and more account, notably due to pressure from the public, of risks likely to arise, as well as technical, economic, social and political considerations.

It is therefore essential to make an objective assessment of these risks to which individuals are exposed naturally or because of their life styles.

Because most decisions are made on a regional level and because it is also at this level that the environmental characteristics, the way of life and the economic and social structure which determine the risks to which the regional population is exposed can be realistically defined, it is on this level that a comparative assessment is of maximum value.

This assessment should be as exhaustive as possible and should take into account all the risks whether natural or artificial, from whatever source (industrial, domestic, medical, etc.). It should take into account the exposure to various hazards, and the resulting consequences to health through the use of appropriate indicators and available knowledge of exposure-effects relationships. Subjective aspects of risk perception should also be considered.

A study program has been begun which covers the region called the "Grand Delta," which includes the Rhone valley and its banks to the Mediterranean coast in the South-East part of France. The paper presents the motivation and aims of this program and describes the method which has been used.

KEYWORDS: Risk assessment, oil-fired plant discharges, atmospheric pollution, radiological exposure, chemical exposure

New Risks
Edited by L. A. Cox, Jr., and P. F. Ricci
Plenum Press, New York, 1990

INTRODUCTION

During the last decades, the concept of security has undergone important changes. These appeared first in the professional field, where both the workers and managers came to realize that any improvement in the level of security was for the benefit of all. The nuclear industry has been the object of consistent but not always justified attacks from the ecologists; but it was from the very same industry that came the first efforts to demonstrate the necessity and the possibility of weighing the advantages and disadvantages of an undertaking in order to choose the desired level of safety.

Nevertheless, a great deal remains yet to be done, especially

- to have a more inclusive view of risks; an individual remains an individual whatever be the situation he may be in: at home, on travel, on holiday or at work, his security under all circumstances must be ensured with the same principles in mind.

- to have a more rational overview of relative risks; even when subjective considerations might occur, the coherent management of risk may be possible only from mature, objective analyses; this would require knowledge, methods, structure and above all, will.

The task may appear unbounded: it does not follow that it may be abandoned. It is in that sense that the Grand Delta program has been created. Its ambition is to contribute to progress in this field.

PRESENTATION OF THE "GRAND DELTA" PROGRAM

Motivation and Aims

The basic motivation for the creation of the Grand Delta program stems from the points made in the introduction.

At the outset, the aim of the study was only to increase knowledge of the relative importance of risks associated with the major human activities and to make this knowledge available to all those who could be interested, i.e., the general public, the socio-professional organizations, the political, administrative and technical bodies.

In view of the experience gained in the radiological field, it appeared that it was possible and useful to transfer this experience to other areas.

In a second step, it became apparent that a more effective use of the results of this work could be made in the context of risk management. Such use required a new research area in order to find common risk indicators and to adapt methods able to take these indicators into account and to integrate them in the management system with all the associated social and economic criteria.

The Approach Used

It soon appeared that the interest and the efficiency of the study would be increased if the risks taken into consideration were of concern to an existing and clearly defined population having interests in common.

Mainly because of lack of this point in view, numerous programs developed in the past were unable to provide clear and practical responses to the actual preoccupations of the population and of the responsible authorities.

For these reasons, we thought that the best way of achieving this aim was to carry out the study at a regional level. By "region" is meant an area owing its unity to factors which are of physical nature (climate, topography...) or of human nature (population, economy, administrative structure...).

In addition, the study was limited to the present situation or the near future, avoiding any consideration of the distant future for which a special approach must be adopted. Assessment of exposure over long periods raises a number of problems: the reliability of the assessments, the solidity/flimsiness of the assumptions made about the environment, the uses made by man, the characteristics of populations (size, distribution, life style...). Moreover, many risks correspond to the situation at a given moment and cannot be projected into the future. For these reasons, it seems more appropriate to look for a common representation covering the risks likely to occur during a given year.

Finally, it seemed more useful to test the possibilities, the difficulties and the impossibilities of the adopted approach, by applying it to an actual situation: so the Grand Delta program corresponds to a case study.

Choice of the Region

The region chosen as the basis for the study is located in the Southeast of France (Fig. 1): the main reason for this choice lies in the diversity of human activities to be found there, with regard to industry as well as agriculture, transports, and social development. For instance, different types of industrial activities are represented, such as chemical, metallurgical, food and more specially those connected with energy production: almost all the stages of the oil, coal and uranium cycles occur, from the extraction of raw materials to the management of waste.

Principles and Stages of the Study

In principle, every potential risk must be considered whatever its nature or origin, with no a priori exclusions. Thus, the study covered not only risks of natural origin but also those resulting from human activities, whether industrial, medical, agricultural, social, or domestic. Also included were radiological, physical and chemical hazards, which can occur either in normal or accidental situations and which concern people, whether as workers or members of the public, and the environment (Table 1).

However, in a first step, only the risks associated with industrial activities (and mainly energy-producing activities), whether of radiological or chemical nature, have been covered and only in the case of normal situations. Nevertheless, preliminary information about risks from sources other than industrial has been provided, for instance those due to fertilizers and pesticides in agriculture, radiation in medicine, and natural radiation.

All the assessments are made on an annual basis, the reference year being 1982.

Risk estimations can be performed in two different ways: from observations of statistical nature and, in the case of exposure to radiation or chemical pollutants, from assessments based on modeling.

In short, the work carried out in the context of the Grand Delta program has involved:

- gathering the information available about the characteristics of the region studied in relation to its climate, hydrology and geology, its agricultural produce, the distribution of its population, its way of life, its transport network and so on;

Fig. 1. The southeastern area of France and its three administrative regions.

- gathering the information available about the existing industries in the region: the type of activity, the location, personnel, materials processed, the effluents discharged and other factors;
- carrying out programs for the measurement of pollutants discharged by certain industrial plants such as dusts, gases, organic compounds, trace elements, radioactive elements and so on;
- collecting the results of the measurements monitoring existing environmental pollution, whatever its origin;
- establishing the bases of calculations that enable the levels of contamination and exposure to be evaluated from source-terms, model and transfer parameters, representation of distributions, methods of computation and other methodological issues;
- gathering the statistics available on accidents at work, occupational diseases, occupational exposure and other data;
- carrying out a first synthesis of all the observations and assessments and of the knowledge about exposure-effect relationships, and drawing preliminary conclusions.

Table 1. Classification of Hazards to Human Beings and the Environment in Normal or Accidental Situations

Origin	Nature Radiological	Chemical	Others
Natural	Radiation: - cosmic - terrestrial - internal	Volcanoes Sand storms	Earthquakes Floods
		Contributed by Man	
Domestic	- Construction materials: external radiation - Radon and its products: internal radiation - Consumer goods	- Culinary practices heating - Construction materials and equipment - Nicotinism	Fire Electrocution Asphyxie Noise
Leisure activities		Use of solvents and glue...	Drowning accidents
Industrial activities	Public: discharge of effluents and storage of wastes Workers: Occupational exposure	Public: discharge of effluents and storage of wastes Workers: Occupational exposure	Noise, vibration... Thermal modification of the environment
Transport	Air: cosmic radiation	Emission from motor vehicles	Noise, vibration... Traffic accidents
Medical	Public: diagnostic and therapeutic radiation Hosptial waste Workers: Occupational exposure	Public: treatment and medication Workers: Occupational exposure	Mistakes in diagnosis and treatment
Agriculture	Public and Workers Fertilizers	Public and Workers Pesticides, fertilizers Emission from motor vehicles	Noise, vibration...

Other tasks were carried out at the same time. These were concerned particularly with the perception of risk, with the economic aspects and, as has already been indicated, with risks from other sources (natural, medical, agricultural).

Finally, it should be pointed out that, given the size of the task, not all these aspects have been treated in the same depth. In general, more effort has been put into gathering the information, inquiring into its relevance, and preparing data banks. The use made of these data banks, particularly from the viewpoint of health and the economy, has not been carried as far as is needed and requires further work.

PRINCIPAL RESULTS OBTAINED

Most of these results concern the assessment of risks due to the atmospheric and liquid discharges from installations of the three energy-producing cycles existing in the region.

The system utilized is based on a grid with a 100 km^2 mesh covering the region. All the data are inserted in the grid — data concerning source terms, population distribution, agricultural productions, results of environmental monitoring as well as the data resulting from assessments, such as atmospheric concentrations of pollutants and exposures, and other relevant factors pertaining to the analysis. The principal results obtained can be briefly illustrated as follows.

Source Terms

The data on atmospheric and liquid discharges from installations of the nuclear fuel cycle have been obtained from the people in charge of these installations. In some cases, mainly for oil-fired and coal-fired power plants, the discharges have been evaluated by direct measurements in the stack or by measurements, after sampling, at the discharge point. The radiological and chemical composition of these releases is presented in Table 2.

Environmental Concentration

From the source-terms and using conventional models, the concentrations of the different pollutants in the main environmental components have been calculated in each mesh of the grid. Figures 2 and 3 present the atmospheric concentrations of SO_2 and U_{238} respectively.

Exposures

In the case of radioactive discharges, it is possible to assess exposure due to external radiation, inhalation and ingestion of water and foodstuffs. This is done on an individual basis, in order to determine the mesh where the "critical group" (that is, the population group which is the most exposed due to the different sources of releases) lives, and on a collective basis by calculating the product, in each mesh, of the average individual dose multiplied by the number of people, in order to assess the corresponding detriment for the regional population.

In the case of chemical pollutants, the exposure can also be expressed on an individual or collective basis; but, since a concept of "dose," as used for exposure to radiation, does not exist, the exposure is only founded on quantities such as the concentration of the pollutant in air or in a given organ or tissue. The distribution of the collective exposure due to the discharges of SO_2 is presented in Table 3.

Table 2. Discharges During 1982 from Oil-Fired Power Stations (kg for Chemical Pollutants and kBq for Radionuclides)

Substance Discharged		Fuel Oil: Aramon 700 MWe	Fuel Oil: Martigues-Ponteau 250 MWe	Fuel Oil: Loire-sur-Rhone 500 MWe
Gases	SO_2	6×10^6	24×10^6	7×10^6
	HF	17	--	--
Organic Vapors	Formaldehyde	21	863	250
	Nitrobenzene	76	283	82
	Benzene	10	0.7	0.2
	Toluene	45	0.3	0.1
	Xylene	87	--	--
	Styrene	48	0.2	0.04
	Acrolein	NM	164	47
	Benzaldehyde	--	298	86
Trace Elements	As	0.4	6	1
	Be	1	0.5	0.1
	Cd	0.4	4	1
	Co	10	--	--
	Cr	9	2	1
	Hg	2	--	--
	Ni	761	2976	862
	Pb	14	--	--
	Se	3	--	--
	V	2076	16×10^3	4741
Dust		67×10^3	570×10^3	165×10^3
Hydrocarbons	Anthracene	8×10^{-4}	--	--
	Benzo-a-pyrene	8×10^{-4}	--	--
	Pyrene	2×10^{-3}	--	--
Natural Radionuclides	^{40}K	5×10^3	338×10^3	10×10^3
	^{238}U	6×10^3	15×10^3	4×10^3
	^{230}Th	--	25×10^3	7×10^3
	^{226}Ra	3×10^3	74×10^3	22×10^3
	^{214}Bi	1×10^3	--	--
	^{214}Pb	2×10^3	--	--
	^{210}Po	--	202×10^3	59×10^3
	^{223}Ra	--	13×10^3	4×10^3
	^{232}Th	4×10^3	9×10^3	3×10^3
	^{228}Ac	1×10^3	42×10^3	12×10^3
	^{212}Pb	2×10^3	--	--
	^{208}Tl	1×10^3	--	--
	^{228}Th	--	89×10^3	26×10^3
	Total alpha	21×10^3	844×10^3	244×10^3

NM: Not measurable.

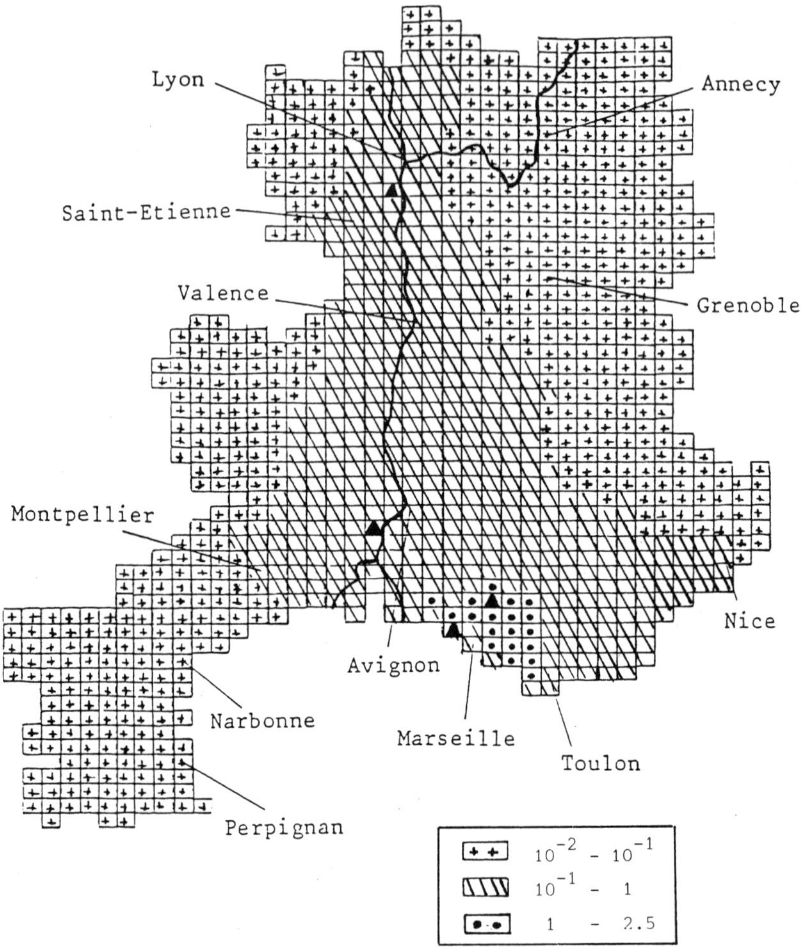

Fig. 2. Annual mean concentrations of SO_2 ($\mu g/m^3$) from fossil fuel fired power plants (▲).

Effects of Exposure

To express the consequences of exposure, relationships must be available that enable it to be translated into a probability that harmful effects will occur. These relationships can take very diverse forms (e.g., linear, quadratic, linear-quadratic). There may be no threshold, so that the probability of an effect appearing becomes zero only when the exposure is zero; or a threshold may exist, when the probability becomes zero below a certain finite level of exposure. In the case of exposure to radiation, there are generally accepted assumptions available for the dose-effect relationship: that of a linear relationship without threshold for so-called "stochastic" effects (solid cancers, leukemia) and of a relationship with a threshold value, above which the probability of appearance is unity, for "non-stochastic" effects (burns and acute syndromes).

For exposure to chemical pollutants, our knowledge is very limited and exposure-effect relationships are only available for a few pollutants — and even here there are great uncertainties. It is important to be extremely cautious in the use of exposure-effect relationships and to consider the values obtained as orders of magnitude rather than precise assessments: the use made of any results should take this consideration into account.

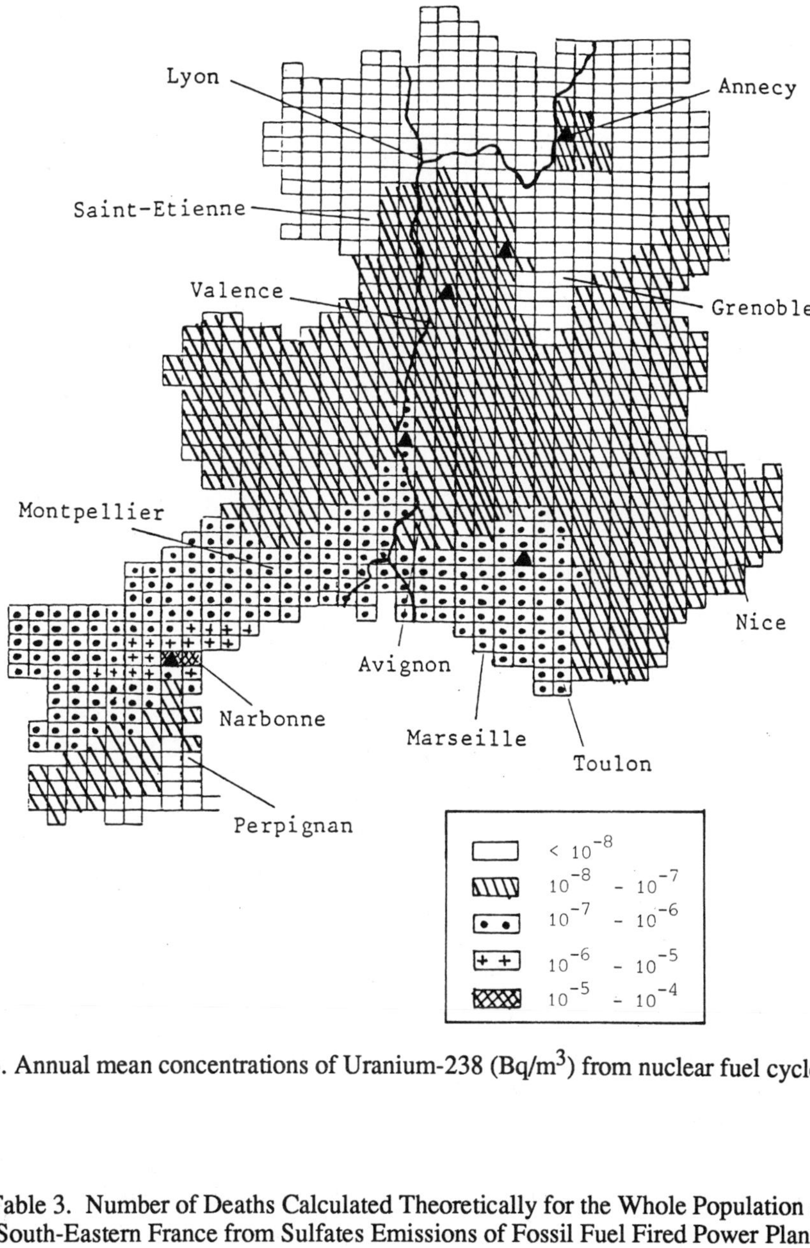

Fig. 3. Annual mean concentrations of Uranium-238 (Bq/m^3) from nuclear fuel cycle (▲).

Table 3. Number of Deaths Calculated Theoretically for the Whole Population of South-Eastern France from Sulfates Emissions of Fossil Fuel Fired Power Plants

Range of Concentration (μg/m^3)	Number of People Involved	Collective Exposure (man-μg/m^3)	Number of Deaths Calculated Theoretically*
$< 10^{-2}$	2,451,697	0.20×10^5	0.2 à 4
$10^{-2} - 10^{-1}$	7,979,323	0.17×10^6	2 à 34
$10^{-1} - 1$	0	0	0

*Mean annual risk of between 1 and 20 deaths for a population of 10^5 exposed to 1 μg/m^3 of atmospheric sulfates (Hamilton).

DIFFICULTIES, LIMITATIONS AND POSSIBILITIES OF THE STUDY

Difficulties and Limitations of the Assessments

The first difficulty concerns the acquisition of the data and, more specifically, those data relating to emission from the industrial establishments involved. Several alternatives are possible:

- with the agreement of managers, data specific to the installations concerned are used, where they exist, or else programs of measurement are set up in these installations. Both of these methods were used during the investigation.

- lacking those possibilities, it is necessary to rely on data from international publications, and this means that the investigation is no longer involved with a concrete case. This was the situation encountered mainly with certain operations in the oil fuel cycle.

The second difficulty concerns the exploitation of the data and in particular the limited validity of the models used. It is impossible in a regional study, with the prohibitive costs in time and equipment and the sheer physical problems involved, to measure the quantities of the pollutants in the environment directly. The use of models is thus inevitable in such a study, even though the ability of modeling to represent the situations accurately will always be difficult to test.

The third difficulty, certainly the most important, concerns the assessment of risks and associated detriment. This assessment requires the existence of exposure-effect relationships, for which there are very few for chemical pollutants.

Concepts of Risk and Limitations of the Study

The object of the assessments to be made is described by the word "risk."

This word has not, as a rule, been used in its everyday sense of "danger" or "harm" but with its more scientific meaning, namely "the probability that a harmful consequence of a particular event will occur during a given time." For example, it could be the probability of accidental death during one year for a person driving a car or engaged in some occupation, or the probability of contracting a fatal illness following a year's exposure to radiation or chemical pollutants.

If the size of the population taking part in the event concerned is known, then it is possible to determine the expected number of deaths from this cause and, if several events are involved, to make a comparison between them based on the criterion of the mortality rates.

However, mortality is not the only criterion involved, since the consequences of an event can also be non-lethal damage. This can be assessed in the same way as lethal damage, but it is not possible to compare events directly on the basis of the two different criteria (mortality and morbidity): this requires the use of other indicators such as "detriment," a term employed in radiological protection.

Still more complicated is the situation where the psychological state is affected or where it is the material goods that suffer, or the natural environment, or production processes. New criteria then have to be found to portray these situations accurately.

The objective of "comparative risk assessments" appears in the strict sense that was given to it, but only as a theoretical objective. The real aim must be to adopt a different meaning which can be expressed as "the attempt to establish results justifying a comparison between certain activities on the basis of the hazards they are capable of creating." This is then the ordinary meaning of "detriment" which, in what follows, will have to include the sense of the word risk.

Overall Assessment

The question to which the present investigation was intended to be an answer is this: Is it possible to establish a clear comparative assessment of the risks due to the industrial activities in a given region to which the population of that region are subjected and to be able to relate them to risks due to other activities? Moreover, is it possible to draw conclusions about the action to be undertaken to improve the protection of people and to plan socio-economic developments?

The investigation carried out, which has concerned mainly the production of energy and the exploitation of the results from it, does not permit a simple reply, that is, a mere "yes" or "no." A much more qualified reply might be:

a. It is not possible in the current state of knowledge to produce the desired overall assessment, that is, a complete, irrevocable answer accessible to everybody, including the uninitiated.

b. On the other hand, it is possible to ascertain the relative importance of risks, even without referring to a single indicator, particularly those which are bound to be of major concern.

c. It is also possible to establish a complete assessment of the risks resulting from exposure to ionizing radiation.

d. Points b and c therefore show that it is possible to provide the responsible authority and organizations with information that can be used to form an objective guide for some of their decisions.

e. Such information cannot be presented and used in a simple way: it must be accompanied by qualifications and reservations to prevent any misuse.

f. The gaps and uncertainties which make the reply not as clear as we would wish can be rectified, at least partially, so that in the future a less qualified reply can be envisaged.

Another important foreseeable aspect is that it will be possible to make appropriate assessments of radiological risks, subject to uncertainties concerning the models used to evaluate transfer factors and with some reservations about the validity of the dose-response relationship postulated. Another feature that has been demonstrated is the ability to transfer knowledge gained in the radiological field to the non-radiological one: that is particularly true with regard to the approach adopted for evaluating the effects on individuals.

CONCLUSIONS

The difficulties encountered and their influence on the definition of aims must not be allowed to discourage the execution of a study of this type; deficiencies, inaccuracies and uncertainties do not in the least prevent assessments being carried out which enable the hazards caused by various activities to be put into perspective. An objective consideration of such difficulties merely causes more modest limits to be put on the work undertaken.

In the light of the work carried out in the context of this investigation and taking into account the limitations deliberately imposed and the thoughts leading up to them, it would appear that the direction initially adopted, namely a regional approach integrating as completely as possible the various sources of hazard and risk, is a good one, even if it has not been possible, in the current state of knowledge, to formulate the results in a very satisfactory way. It also clearly emerges that the comparison should not constitute an objective in itself, either for its own sake or because of the possibility of carrying it out, but that the true aim is *the management of risk*, for which it is essential to define a framework into which all the necessary (or simply useful) elements can be fitted, particularly putting into perspective the different risks, the associated social costs and their perception by the various protagonists.

REFERENCES

1. J. Aigueperse, F. Anguenot, A. Bouville, and R. Coulon, Evaluation des conséquences sanitaires des rejets radioactifs et non radioactifs de divers types de centrales (nucléaire, charbon, fuel), VIème Congrès Mondial pour la Qualité de l'Air, Paris, France, 16-20 mai 1983, Ed., *SEPIC* **4**:5-12.
2. J. Aigueperse and F. Anguenot, Approche globale des risques au niveau d'une région: le projet Grand Delta, cas des filières énergétiques, IAEA-SM-273, Colloque International sur les risque et les avantages des systèmes énergétiques, Jülich, République Fédérale d'Allemagne, 9-13 avril 1984.
3. Commissariat à l'Energie Atomique-National Radiological Protection Board, Méthodologie pour l'évaluation des conséquences radiologiques des rejets d'efffluents radioactifs en fonctionnement normal, CCE (1979).
4. P. Pages and F. Rancillac, Modèle conversationnel pour l'évaluation des transferts atmosphériques de rejets continus; Rapport CEPN n° 61 (1983).
5. International Commission on Radiological Protection, *Recommendations of the ICRP*, ICRP Publication 26, Ann. ICRP 13, Pergamon Press, Oxford (1977).
6. United Nations Scientific Committee on the Effects of Atomic Radiations, Ionizing Radiations: Sources and Biological Effects, United Nations, New York (1982).
7. L. H. Hamilton, Health Effects of Electricity Generation, presented at the Conference on Health Effects of Energy Poduction sponsored by Chalk River Nuclear Laboratories, Ontario, Canada, September 12-14, p. 3 (1979).

Scientific and Technological Inputs to Nonroutine Decision Making by Three Groups: Scientists Government Officials, Corporate Managers

C. McLaughlin, N. Kathuria, R. Zmud and F. Lynn
University of North Carolina
Chapel Hill, NC

ABSTRACT

This paper discusses a number of factors influencing the way that science and technology information is requested, framed, transmitted and interpreted when the chips are down and the situation risky. Our research on the use of science and technology (S&T) information in policy analysis draws on research studies of three groups: (1) scientists, whose information-gathering has been well studied; (2) government officials; and (3) corporate managers. While S&T information search by technical professionals centers on efficiency, using informal gatekeepers or information stars, the other two groups do not focus primarily on S&T issues and would be expected to exhibit more idiosyncratic patterns of search. Government officials have policy analysis groups, expert panels, technical staffs, and personal contacts to draw on. Corporate executives have comparable staff personnel and groups and personal contacts to utilize. Yet both may be unable to use the best sources because of adversarial roles, procedural constraints, and cultural barriers. Detailed current S&T knowledge tends to reside at a much lower organizational level than decision making, and a number of mediating steps pass it through legal, political and technological filters before it reaches the decision makers. These filters may create problems which are accentuated when the problem is a nonroutine one, such as toxic shock, ice minus, or even Chernobyl.

KEYWORDS: Nonroutine decisions, information search, scientific information, policy analysis

INTRODUCTION

Many important and risk-related decisions are made by senior managers in business and government. How do senior decision makers who are not scientists acquire the scientific and technological (S&T) inputs for their decisions? This paper studies literature from a number of fields on how S&T information is gathered and processed by scientists, industrial executives, public officials, and policy analysts when they are faced with making a decision for a situation that is not routine. Its objective is a synthesis of a common way of looking at how key people in our society to gather and process S&T information under pressure.

NONROUTINE SEARCH

Many important issues are handled nonroutinely, because they carry unusual political risk, they are too urgent to allow normal research procedures to be followed, the public or the courts demand action, the organization lacks precedents and procedures to handle them, or some combination of these factors. The more extreme case, where there is a combination of (1) surprise, (2) considerable threat to organizational values, and (3) a short decision time, is referred to as a crisis.[1,2] We could have focused only on crises, but these have been studied intensively.[3]

The psychological literature associates crises with a loss of cognitive decision-making performance under stress, including selective filtering of information, reliance on past search experience, cognitive rigidity, reduced tolerance for ambiguity, premature closure, and reduced analysis of alternatives.[4,5] We have not excluded crises, but we have broadened the topic to a range of "nonroutine" situations in which there is insufficient time to conduct new scientific research for information because a decision is required.

S & T SEARCH

The search for S&T information by scientists and engineers is an everyday task. Therefore, they focus on efficiency, using individuals who have proven to be effective information sources in the past, variously called integrators,[6] information stars,[7] or gatekeepers.[8] These individuals actively keep up with science inside and outside the organization, and most fellow employees turn comfortably to them for information or leads to information. They have personal and telephone contact with key experts and, if they don't have the information, they know whom to call to get it.

Senior government officials and industrial executives, however, consult S&T sources less frequently. Government officials can turn to policy analysis groups, expert panels, technical staffs, and personal contacts for S&T information. Industrial executives have comparable staff personnel and groups and their own personal contacts to utilize. Yet both may be unable to use the best internal and external information sources because of adversarial roles,[9] procedural constraints, and cultural barriers. Detailed and current S&T information tends to reside at lower levels of the organization and must pass through a number of mediating steps which can act as legal, political and technological information filters. These filters may function effectively to reduce the information overload at the top.

SCIENTISTS AND INFORMATION

The originators and owners of S&T information — scientists and engineers — are expert at handling such information and their behavior has been studied by those interested in improving research and development (R&D) efforts.

Basic researchers rely more on external sources, on the literature, and on key individuals who are proven sources of information. Research laboratories are scanners of the scientific environment much as the senior executives are environmental scanners. The more applied technical units rely more on internal documentation sources and on informal communications within the firm. They are sharing experience, much of it proprietary, with their coworkers.[10,11]

Individuals exhibit multiple and simultaneous modes of information gathering, aimed both at problems and at competency development. People photocopy articles that might be useful, develop information networks with coworkers, and participate in the "invisible colleges" of their professions. When asked how they acquired useful specific pieces of S&T information, professionals reported that a third of the time the information

> was acquired because someone pointed it out. Finally, in about one-fifth of the cases, the respondent's intention in seeking information had been to develop his general competence rather than to acquire some particular knowledge.... In the transfer of information in industrial laboratories, information looking for the man seems to be nearly as frequent an occurrence as the man seeking information (Ref. 10, p. 17).

Organizations abound with solutions in search of problems and problems in search of solutions. People have problems and they have pet projects and favorite ways of handling problems.[12,13]

In research laboratories, a relatively small number of individuals serve as the hubs of these informal information networks.

> It is the supportive nature of informal channels, however, rather than their leading directly to a solution, which accounts for their importance in the innovation process. Because they are a prompt source of information, which in a sense is pre-digested, and because they allow instantaneous feedback, informal channels are a necessary complement to more formal channels of scientific and technical communication (Ref. 14, p. 73).

Bodensteiner found that (1) the more uncertain the individual is about needed scientific information, the greater the pressure to seek additional information, and that (2) when seeking that additional information, the individual will choose channels/sources with increased information carrying capacities and/or with increased opportunities for personal support.[15] A survey of 384 engineers in a large government R&D organization showed a strong correlation between technical uncertainty and a decision maker's selection of face-to-face sources of information.[16] They saw such sources as richer, having the ability to communicate inflections, gestures, pauses, and expressions. Fischer's research went further, emphasizing the importance of informal sources which allow personal confrontation, interaction, and reaction.[17]

This and other research also indicates that external sources become more important in uncertain situations.[17,18] As scientific professionals' uncertainty increases, uncertainty in terms of their professional expertise or the urgency and risk of the situation, they also seek their S&T information in less formal channels, channels with richer and more supportive information, with greater opportunity for interaction and even debate.

ORGANIZATIONS AND INFORMATION

Feldman and March have suggested that organizations not only process information for problem solving, they overconsume it to (1) provide employment, (2) scan the environment, (3) amass power, and (4) appear rational.[19] S&T information certainly is not exempt from these four other uses.

The concept of uncertainty also becomes considerably more complex in large bureaucratic organizations. There is technical uncertainty, insufficient data to answer a specific question like "What is the probability of the failure of the containment vessel with a 2 kiloton blast at ground level one-half mile away?" But not all uncertainty relates to a specific scientific value. There also is equivocality, defined as ambiguity, the existence of multiple and conflicting interpretations about an organizational situation. High equivocality means confusion and lack of understanding.... asking a yes-no question is not feasible. Participants are not certain about what questions to ask, and if questions are posed, the situation is ill-defined to the point where a clear answer will not be forthcoming (Ref. 6, pp. 556-7).

Daft and Lengel[6] also outline a hierarchy of media richness, including:

Rich, personal media	Group meetings
↓	Integrator
	Direct contact
	Planning
	Special reports
	Formal information systems
Less rich, impersonal media	Rules and regulations

The term *integrator* refers to a coordinator, a person who uses telephone and face-to-face means to interpret issues, resolve conflicts, and reduce equivocality about goals. The construct, *direct contact*, includes telephone, face-to-face, and even written letters or memos, provided they contain subjective as well as objective data.

These concepts help in understanding how organizations process information. They can help analyze uses of S&T information. Most searches that we observe involve a combination of technical uncertainty and equivocality.

THE EXECUTIVE AND INFORMATION

Much attention has been paid to how senior business executives acquire and process policy-related information. They spend much of their time scanning the environment for knowledge of events, trends, threats and opportunities. Yet little has been said about such individuals' use of scientific and technical information under high risk conditions.[20-22]

Kotter[23] has observed significant behavioral differences between functional managers and top managers. Senior executives spend much of their time first figuring out what to do despite uncertainty, ambiguity and information overload; they then get things done through a large and diverse set of people over whom they may have little direct control most of the time. Successful senior executives maintain a relevant and manageable agenda, build effective networks of people, and then use those networks to implement the agenda. Networks of cooperative relationships are carefully crafted to create the informal and formal contexts for pushing the agenda forward. This leads senior executives to present a large number of questions on a continuing basis to staff and line managers, including R&D managers.

Senior executives do not use data bases for reference.[21] Even in high-technology firms data bases and decision support systems seldom are used for policy-oriented decisions and most CEOs still operate much as they always have — working face-to-face with staff and lower-level managers. Those who use the computer appear to use it for personal productivity enhancement and not for decision analysis.[24-26] Senior executives do not appear to follow the normative model of decision making. It can be difficult to pinpoint exactly when a major decision gets made.[27] Many successful executives seem to act in an almost anti-decision mode, by asking probing questions and raising issues to key organizational participants, otherwise there is a brief, almost inconsequential problem-solving phase, followed by much emphasis on decision implementation.

Executives prefer verbal rather written information.[28,29] Information from personal sources (other managers, subordinates, friends, or customers) is highly valued, as is information obtained through meetings (scheduled or unscheduled) and committees. Telephone calls seemed to be valued extremely low, despite their high quantity. This may be because the incoming telephone call often jumps the priority queue and the medium is less rich. The most valued written source is the memo. Often it is hard to judge the importance of a piece of information because information use follows "chains of information" where receipt of one piece triggers the need to access others.[28]

Jones and McLeod suggest four executive roles, of which the "disturbance handler" comes closest to the subject of this paper, being defined as dealing with "situations and changes that are partially beyond his or her control when the pressures on the organizations (as the result of unforeseen events or a problem too long ignored) are too great to ignore" (Ref. 28, p. 226). This made up a surprisingly significant proportion of the executive's transactions (42%). Of the information used to handle disturbances:

> 37 percent originated from first- and second-level subordinates, the people most likely responsible for knowing the facts and most able to tap the informational networks quickly. Senior executives worked closely with their upper level subordinates in this decision role. Internal support units also provided a significant volume (16 percent) of the information but were not valued as highly as subordinate supplied information. At times executives must dig deeper in the organization for information. Although this does not occur frequently, such transactions appear critical (5.2) to resolving disturbances. At such times task forces and committees may play a key role in securing internal information for handling disturbances (Ref. 28, p. 240).

Organizational decision processes are not uniform or linear. Empirical studies show that decisions, even "unstructured" ones, go through identifiable stages or phases, although these stages are skipped if experience brings to mind usable existing alternatives or analyses. It is also clear that decision processes can recycle and experience delays.[13,30] "The most complex and novel strategic decisions seem to involve the greatest incidence of comprehension cycles. We found specific evidence of cycling and recycling in all 25 decision processes, with a total of 95 occurrences" (Ref. 30, p. 265).

Interrupts also occur because of unexpected constraints, political impasses, and the appearance of new options. Interrupts can stretch out the decision making process significantly.[30] Decision making can be stimulated by opportunities, problems or crises. Most decisions are solution-centered. Concrete ideas are sought as soon as possible and search and scanning are limited. Interrupts are more frequent in crisis situations and in government or quasi-governmental organizations.[13,30]

Where equivocality is high, there is a need for rapid information cycling among managers. Assumptions, scenarios, and outcomes are presented, critiqued, and reframed until consensus is approached and implementation commitments are made. As a result, increased emphasis is placed on the role of informal, typically face-to-face, "rich" media.[31]

SUPPORTING THE SENIOR EXECUTIVE

The senior executive does not scan the environment alone. Technical units and trade associations exist to serve as windows on the world for technical events, trends and opportunities. The choice of information sources professional employees use to scan the environment is related to perceived task complexity and the information source's perceived accessibility. Individuals with complex scanning tasks tend to use personal subscriptions, the company library and peers. Staff professionals were more involved in scanning the environment than line professionals.[32]

The organization may also filter the information by its selective attention to problems or by consciously or unconsciously biasing upward flows of information.[33,34] If senior executives do not collect the perceptions of the environment directly, they must factor in their own corrections for the amount of bias in the information.

In assessing the impact of S&T information, it is important to recognize that in the majority of cases technical expertise exists lower down in the organization in functional groups. It must almost always go up through one or more levels of the organization, each with successively less scientific background. Each such level is perforce a filter and potentially a biaser of the estimates of risks, probable events, and possible outcomes.

GOVERNMENT EXECUTIVES AND INFORMATION

The process of using S&T information in government carries with it the need for participation, consultation and public accountability that accompany all governmental processes. The corporation can make its technical decisions in relative privacy, but the government does not have that luxury. Doing the right thing, using the right process, giving the right appearances becomes an end in itself, right alongside making the right decision.[35,36] This means that in addition to the internal scientific staff and line units the government is likely to add the scientific advisory committee. Advantages of the committee or "Blue Ribbon Panel" are that it can attract highly qualified persons for a short period, it commands the attention of a wide audience, and the decision makers can accept or ignore its recommendations.[37]

Much scientific information that is funneled to the decision makers is from noted members of "invisible colleges" that scientists and engineers use to keep up-to-date well ahead of the literature. But this scientific establishment can be biased toward orthodox scientific views or toward one set of values.[37,38] They are subject to the same concerns and limitations as technical panels in the grants review process.[39]

Individual departments and agencies have their own research units and also make extensive use of contractors and advisers for S&T information. The survey by Caplan *et al.* of social science information usage found that 94% of social science information used in decision making was funded, conducted, or both, by government, some 86% by the using agency.[40]

Sabatier notes that technical information is collected as part of a process that is also collecting intelligence about:

1. the legal rules and regulations constraining its policy choices;
2. the past activities of the agency and its internal resources;
3. the preferences of important actors within the agency, its constituency groups, other agencies, and its executive, legislative, and judicial sovereigns; and
4. the probable reactions of important political actors to the substantive consequences of each policy alternative (Ref. 3, p. 397).

This parallels the reported behavior of senior business managers who are very concerned with constructing and manipulating their agendas and networks and spend much thought and effort on implementation concerns.

Sabatier suggests that the degree to which technical information is collected by agency staff depends on the resources available for that task, and on

Magnitude and size of risks,
Susceptibility to empirical analysis, and
Incidence of impacts (Ref. 3, p. 398).

Increased risk or increased visibility of consequences leads to increased information acquisition.

POLICY ANALYSTS

Meltsner describes the policy analyst as a "middleman, a broker between scientist and policymaker" (Ref. 41, p. 271). He suggests that the policy analyst is one of three types of advisors that the decision maker has: staff cronies, program managers, and policy analysts. He seems to see analytical skills and content as what differentiates the policy analyst from the other two. He suggests that the policy analyst will seldom meet academic criteria for quality of work, because that individual is a bureaucrat first and then a policy scientist. He attributes the problems of policy analysts in the bureaucracy to the facts that

1. they are members of an emerging profession without enforceable standards and sanctions;
2. they lack an adequate base of knowledge and associated theoretical paradigms;
3. they have tenuous communication networks; and
4. they are low resource, low status political actors (Ref. 41, pp. 11-12).

Jenkins-Smith suggests that two critical questions for the analyst are the existing degree of commitment the client has to a particular policy position, and the degree of "analytical uncertainty" involved. He defines the latter as situations with parameters or relationships about which little is known. Therefore, analysts can make plausible assumptions over a wide range of values and bias is quite possible.[42]

As an advisor, the policy analyst has a number of ways of transmitting information to the senior decision maker. These include:

Formal briefings
Formal reports
Memos
Phone calls
Preparing Congressional testimony
Speechwriting
Contacts in the elevator or hall
Drafts circulated to elites
Publishing in outside journals

Policy analysts have a wide range of duties, including answering Congressional inquiries; writing reports, speeches, procedures, papers; trouble-shooting; and being evaluators.

Meltsner suggests that the formal briefing is used not only to convey information but also to build support and consensus. He notes that the problem with the formal report is that it must communicate technical material in nontechnical terms and it must be persuasive.

Meltsner's observations look applicable a decade after they were published, except that he places most of the action in the office of the secretary. Things may have changed. In the 1980s the operating agencies are more important actors in policy analysis and so is the budget process of the Office of Management and Budget.

Because policy analysts are often several layers away from key decision makers, there has to be considerable concern about how S&T messages are delayed and modified as they move up through the organization. Such situations would appear to be prime candidates for message delays and message modification.[43]

THE CORPORATE POLICY ANALYSTS

Who are the individuals in the corporation with positions analogous to the policy analyst in government? There are corporate staff analysts as well as line ones. At the headquarters level there are corporate planners, R&D directors, occupational health and safety people, offices responsible for regulation, and the CEO's office. Yet the few policy analysts assigned to senior decision makers are usually just those who are there to assist with strategic planning. Strategic planning defines what businesses the company should be in. The strategies of how to succeed in specific business units is up to their executives at the group or division level.

The analogous corporate tasks also take place in the line and staff functions of the strategic business units. Compared with the government agencies, the analytical skills are more likely to exist in the line managers as well as the staff specialists. There are relatively few political appointees in the corporation. Most executives have moved up through the corporation in a series of line and staff positions. In times of stress the firm is very likely to turn to an ad hoc committee of individuals from both staff and line.

People are used to supporting rather specific performance objectives and are far less concerned with process behind the veil of corporate anonymity. If the firm is a user of agricultural products, it may have its market analysts and/or consultants like the Department of Agriculture. But they are buried in their respective business units and are buffered by intermediate decision makers in marketing, sales, advertising and general management, all of whom consider themselves knowledgeable too. At the corporate and divisional levels there may be planners, R&D, regulatory staffs, market researchers, and operations and engineering staffs. When trying to find where S&T-linked policy analysis takes place in the corporation, there are a lot of alternatives.

While policy analysis occurs both at the corporate headquarters and at the divisional level, S&T information is more likely to pertain to the division level with its specific technology and markets.

WHAT WE DO AND DON'T KNOW

Managers turn to personal, media rich, reliable, accessible S&T information sources at all levels and in all jobs. Search is limited by time and energy. Formal presentations and group meetings are media rich; formal reports are not. The effect of risk and time pressure is to narrow individual search, but when an issue is given prominence, staff search may widen. Much S&T search is scanning or anticipatory, but its specific impact is difficult to access. In government the information is compartmentalized by staff specialty in industry by strategic business unit. The key research issues are not about search, but about the way S&T information moves among bureaucratic layers and the way technically competent staff members acquire attributes of visibility (leading to richness), credibility and accessibility. While filtering is a legitimate approach to reduce information overload, how can we accomplish that end without reducing the quality of S&T inputs? How can managers "dig deeper" into the organization or S&T staff emerge at critical times? These are important questions for future research.

REFERENCES

1. C. F. Herrman, *Crisis in Foreign Policy: A Simulation Analysis*, Bobbs-Merrill, Indianapolis (1969).
2. R. S. Billings, T. W. Milburn, and M. L. Schaalman, A Model of Crisis Perception: A Theoretical and Empirical Analysis, *Administrative Science Quarterly* **25(3)**:300-316 (June 1980).
3. P. Sabatier, The Acquisition and Utilization of Technical Information by Administrative Agencies, *Administrative Science Quarterly* **23(3)**:396-416 (1978).
4. O. R. Holsti, Limitations of Cognitive Abilities in the Face of Crisis, in *Studies on Crisis Management*, pp. 39-55, C. F. Smart and W. T. Stanbury, eds., Butterworth, Toronto (1978).
5. A. W. Kruglanski, Freeze-Think and the Challenger, *Psychology Today*, pp. 48-49 (August, 1986).
6. R. L. Daft and R. H. Lengel, Organizational Information Requirements, Media Richness and Structural Design, *Management Science* **32(5)**:554-571 (May 1986).
7. W. A. Fischer and B. Rosen, The Search for the Latent Information Star, *R&D Management* **12(2)**:61-66 (1982).
8. T. J. Allen, *Managing the Flow of Technology: Technology Transfer and the Dissemination of Technological Information within the R&D Organization*, The MIT Press, Cambridge, MA (1977).
9. J. L. Badaracco, Jr., *Loading the Dice*, Harvard Business School Press, Boston (1985).
10. R. S. Rosenbloom and F. W. Wolek, *Technology and Information Transfer: A Survey of Practice in Industrial Organizations*, Division of Research, Harvard Business School, Boston (1970).
11. C. P. McLaughlin, R. S. Rosenbloom and F. W. Wolek, Technology Transfer and the Flow of Technical Information in a Large Industrial Corporation, 2 Vols., Report to the American Academy of Arts and Sciences, Committee on Space, Harvard Business School, Boston (March 1965) (mimeo).
12. M. D. Cohen, J. G. March and J. P. Olsen, The Garbage Can Theory of Administration, *Administrative Science Quarterly* **17**:1-27 (1972).
13. P. C. Nutt, Types of Organizational Decision Processes, *Administrative Science Quarterly* **29**:414-450 (1984).
14. W. A. Fischer, Scientific and Technical Information and the Performance of R&D Groups, *TIMS Studies in the Management Sciences* **15**:67-89 (1980).
15. W. D. Bodensteiner, Information Channel Utilization Under Varying Research and Development Project Conditions: An Aspect of Inter-Organizational Communications Channel Usage, doctoral dissertation, University of Texas, Graduate School of Business Administration, Austin, TX (1970).
16. W. E. Holland, B. A. Stead and R. C. Leibrock, Information Channel/Source Selection as a Correlate of Technical Uncertainty in a Research and Development Organization, *IEEE Transactions on Engineering Management* **EM-23**:163-167 (1976).
17. W. A. Fischer, The Acquisition of Technical Information by R&D Managers for Problem Solving in Non-routine Contingency Situations, *IEEE Transactions on Engineering Management* **EM-23**:8-14 (1979).
18. J. S. Blandin, W. B. Brown, and J. O. Koch, Uncertainty — Information Gathering Behavior: An Empirical Investigation, paper presented at 84th Annual Convention of the American Psychological Association, Washington, DC, September 1976.

19. M. S. Feldman and J. G. March, Information in Organizations as Signal and Symbol, *Adminstrative Science Quarterly* **26(2)**:171-186 (1981).
20. F. Aguilar, *Scanning the Business Environment*, Macmillan, NY (1967).
21. H. Mintzberg, *The Nature of Managerial Work*, Harper & Row, NY (1973).
22. G. Kefalas and P. P. Schoderbek, *Management Systems: Conceptual Considerations*, Business Publications, Dallas (1975).
23. J. P. Kotter, What Effective General Managers Really Do, *Harvard Business Review* **60(6)**:156-167 (November-December, 1982).
24. B. S. Mittman and J. H. Moore, Senior Management Computer Use: Implications for DSS Design and Use, Proceedings DSS-84, pp. 42-49, Dallas (1984).
25. W. R. King, Editor's Comment: CEOs and their PCs, *MIS Quarterly* **9(3)**:11-12 (1985).
26. R. W. Zmud, Supporting Senior Executives through Decision Support Technologies: A Review and Directions for Future Research," working paper, University of North Carolina, School of Business Administration, Chapel Hill, NC (1986).
27. D. J. Isenberg, Research on How Senior Managers Think: Implications for Designing Executive Support Systems, Proceedings, DSS-85, pp. 64-66, San Francisco (1985).
28. J. W. Jones and R. McLeod, The Structure of Executive Information Systems: An Exploratory Analysis, *Decision Sciences* **17(2)**:220-248 (1986).
29. O. A. El Sawy, Personal Information Systems for Strategic Scanning in Turbulent Environments: Can the CEO Go On-Line? *MIS Quarterly* **9(1)**:53-60 (March 1985).
30. H. Mintzberg, D. Raisinghani and A. Theoret, The Structure of 'Unstructured' Decision Processes, *Administrative Science Quarterly* **21**:246-275 (June 1976).
31. C. H. P. Brookes, A Corporate Intelligence System for Soft Information Exchange, working paper, University of New South Wales (1984).
32. M. J. Culnan, Environmental Scanning: The Effects of Task Complexity and Source Accessibility on Information Gathering Behavior, *Decision Sciences* **14(2)**:194-206 (April 1982).
33. R. M. Cyert and J. G. March, *The Behavioral Theory of the Firm*, Prentice-Hall, NY (1958).
34. A. Downs, *Inside Bureaucracy*, Little, Brown & Co., Boston, MA (1966).
35. J. L. Bower, Effective Public Management, *Harvard Business Review* **55(2)**:131-140 (March-April, 1977).
36. P. S. Ring and J. L. Perry, Strategic Management in Public and Private Organizations: Implications of Distinctive Contexts and Constraints, *Academy of Management Review* **10(2)**:276-286 (1985).
37. H. A. Averch, *A Strategic Analysis of Science and Technology Policy*, The Johns Hopkins University Press, Baltimore, MD (1985).
38. M. E. Rushevsky, The Misuse of Science in Governmental Decision Making, *Science, Technology and Human Values* **9(3)**:47-59 (1984).
39. B. Bozeman and L. V. Blankenship, Science Information and Government Decision Making — The Case of the National Science Foundation, *Public Administration Review* **39**:53-57 (Jan.-Feb., 1979).
40. N. Caplan, A. Morrison, and R. J. Stambaugh, *The Use of Social Science Knowledge in Policy Decisions at the National Level*, Institute for Social Research, University of Michigan, Ann Arbor, MI (1975).
41. A. J. Meltsner, *Policy Analysts in the Bureaucracy*, University of California Press, Berkeley (1976).
42. H. C. Jenkins-Smith, Professional Roles for Policy Analysts: A Critical Assessment, *Journal of Policy Analysis and Management* **2(1)**.
43. G. Huber, Organizational Information Systems: Determinants of Their Performance and Behavior, *Management Science* **28(2)**:138-155 (1982).

Confidence in Technologies: Interaction Between Publics and Industries

Amy K. Wolfe
Oak Ridge National Laboratory
Oak Ridge, TN

ABSTRACT

Confidence in the safety of industrial or technological activities is examined in three communities hosting industrial facilities, as well as in a series of communities affected by low-flying jet aircraft. The case studies produced a number of findings which have significant implications for risk communication, including (1) the existence of several audiences for risk communicators, even within a single community; (2) the appropriateness of different risk messages; and (3) the effect of historical and continuing interactions between industries and communities on the types of situations that escalate to crises as well as on the types of risk messages that are appropriate and trusted within industrial communities.

KEYWORDS: Risk communication, risk perceptions, case studies, industry, aircraft, environmental health (perception of)

INTRODUCTION

Risk communicators who inform the public, politicians, or news media about the potential hazards associated with chemicals, industries, or technologies often are faced with frustration or failure when their messages are not convincing to their intended audiences. These frustrations and failures are particularly distressing in short-term acute risk communication situations. Risk communicators, for example, tend to be extremely active during the planning or siting stages of a project or at times of crisis, when there is a malfunction or accident. Another less acute and less confrontational context in which risk communicators function involves planned attempts to alter long-term behavior patterns, such as current efforts to convince people to wear seat belts. In each type of situation, information usually flows in a single direction from risk communicators to their audiences; there is little or no exchange of information between these groups. Risk communicators apparently see themselves as educators who teach members of the public about the quantification of risks associated with products or technologies. Thus, the information disseminated by risk communicators is apt to involve descriptions of chemicals or technologies, quantitative or probabilistic risk assessments, comparisons of those probabilities to other risks, and assurances of safety (perhaps resulting from suggested changes in behavior). Why, then, are risk communicators' messages often not accepted by, or convincing to, members of the public?

New Risks
Edited by L. A. Cox, Jr., and P. F. Ricci
Plenum Press, New York, 1990

RISK COMMUNICATION: A COMPLEX PROCESS

Risk communication is a complex process that encompasses many components, including the following:

- the information to be communicated;
- the medium or media through which information is transmitted;
- the item or process that may pose hazards to people and the environment;
- uncertainties and arguments surrounding risk assessments;
- the corporate entity responsible for a potentially risky item or process (although attribution of responsibility may pose a problem);
- characteristics, especially credibility, of individual communicators;
- risk perceptions, and the methods risk communicators may use to alter those perceptions;
- characteristics of, differences between, and degree of involvement of audiences or publics; and
- differences in types of risks.

Each risk communication event is affected by the nature of these components, as well as by the social and political contexts in which communication occurs. Moreover, risk communicators and risk communication are not merely additions to a particular social and political setting; they arise from, and interact with, that setting. For this reason, risk communication often occurs in non-formal, unintended ways, through continuing relationships and interactions between industries and/or their spokesmen, the community structure, and members of the public. An industry that generally has open channels of communication with its host community, for example, may be viewed as well-intentioned and trustworthy should a crisis occur.

Awareness of both the complexity of the process and the many components of risk communication can be helpful to risk communicators in two interrelated ways. First, this understanding may help to alter the risk communicators' basic role of damage control to one of continuing interaction. Second, through such a role redefinition, this awareness may help contribute to more effective and overtly interactive risk communication.

CASE STUDIES

The case studies described in this paper focus on the social and historical contexts of risk perceptions in communities surrounding industrial facilities and on informal risk communication. The reason for this focus is to demonstrate the historically based and interactive nature of risk communication rather than to examine the numerous components of discrete, more formal risk communication events. (For a more complete discussion of these case studies, see Wolfe, 1986.)

Palmerton, Pennsylvania: A Pilot Study

Until the mid-1960s, the growth and development of the borough of Palmerton, Pennsylvania, founded in 1898, was nurtured by the New Jersey Zinc Company (of Pennsylvania). There are several reasons why Palmerton currently might be considered to be a risky place to live. For instance, sparse vegetation, soil contamination, and bone and joint diseases in domestic and wild animals in the vicinity have been attributed to heavy metal oxide emissions from the smelters. Palmerton's residents have been cautioned that ingestion of home-grown vegetables may be hazardous due to their high cadmium content. Finally, a two-mile long slag heap near one of the zinc plants has been placed on the Environmental Protection Agency's (EPA) Superfund list of hazardous waste sites.

Based upon newspaper accounts and interviews with a few of Palmerton's residents, community members apparently do not think that they are at risk as a result of past New Jersey Zinc Company (NJZC) operations. First, people have disagreed with the idea that their health has been affected by NJZC's operations, and state that no specific hazards have been identified. As proof that no ill effects have arisen from exposure to emissions, they point to the percentage of elderly people in the town. Second, based largely upon economic considerations, residents have argued against the EPA's placement of the slag heap on its Superfund list. They fear that studies and mitigation efforts that might have been required would doom the company to financial failure. Also, there seemed to be some resentment towards the outsiders—the EPA, state, and other officials—who focus on the community as a site containing hazardous waste and who claim that Palmerton's residents may be at risk.

Many factors seemed to operate together in Palmerton to influence confidence in the safety of NJZC operations. Among these factors was the strong link to the past; the deeply felt trust placed in the old NJZC by long-time residents was partly because of the benevolent and paternal way in which the company interacted with the town.

Rockwood and Oak Ridge, Tennessee

Rockwood and Oak Ridge, located in eastern Tennessee, were selected for comparative study because they approximate Weberian ideal types of traditional industry and newer, high-technology communities. Rockwood was founded by the Roane Iron Company in 1868 and was an iron smelting community. Oak Ridge was founded in the 1940s by the U.S. government as part of the Manhattan Project to help build an atomic bomb. Because of the many differences between the two communities, comparisons were not made point-by-point. Rather, comparisons were made both to discern how ecological and historical factors influence the generation of confidence or fear in industrial settings and to develop testable hypotheses for future research.

Like the residents of Palmerton, residents in Rockwood and Oak Ridge generally were confident in the safety of the industrial activities in their communities, despite information to the contrary. Rockwood, for instance, fell into the non-attainment category for ambient air quality standards. Oak Ridgers were told about potential contamination from harmful substances such as mercury and uranium, and problematic practices, like waste disposal techniques, in their community. Although confidence in the safety of the industries was expressed by both communities, some doubt was present in portions of the Oak Ridge community. Different factors, however, accounted for the risk perceptions in each community.

In Rockwood, the community's economic structure best explained the lack of concern about potential hazards from local industry. The founding company, the Roane Iron Company, went bankrupt during the Great Depression, precipitating a decade of economic catastrophe for community residents. Partly as a result of that difficult time, residents were quick to state that smoke from industrial activities means that people are working, and that smoke means progress. Secondary factors contributing to perceptions of risk in Rockwood were the location of most newer facilities at some distance from residential areas and a prevailing wind pattern which usually took the smoke away from the community.

Length of residence played an important role in differing perceptions of risk within the Oak Ridge community. Older, long-time residents tended to be more confident in the safety of the facilities than newer residents for two primary reasons. First, unlike more recent residents, long-term residents have had a sense of security in the community arising from wartime when access to Oak Ridge was severely limited; it was a fenced city, patrolled by armed guards. There was an implicit notion that the government, now in the form of the Department of Energy, would not knowingly allow citizens to be harmed.

Second, because pioneers in Oak Ridge were pioneers in the nuclear industry, they have had a strong personal, professional, and economic stake in both the community and the industry.

Other factors that influenced variations in perceptions of risk in Oak Ridge were occupation and place of work. Although scientists, technicians, and housewives, for example, were confident in the safety of local industrial facilities, their reasons for this confidence varied. In part, these groups of people have had different sources of trust and distrust within the community. Scientists tended to trust the technical competence of other scientists and may have had professional networks of communication through which they could check about various items. Older housewives often trusted particularly the scientists they either knew, or knew of, in the community who have remained in Oak Ridge. Technicians at the three Department of Energy facilities in the community may not have trusted scientists completely; rather, they have trusted themselves and their knowledge of working conditions.

In both communities, but especially in Oak Ridge, it is clear that risk communication is a continuing process, built on past interactions within the community. The historical relations between industries, the corporate entities responsible for those industries, and community residents help to set the social and historical context within which formal risk communication events occur. Variations in risk perceptions in communities like Oak Ridge illustrate that sectors of communities, or publics, may respond to different sorts of information from different sources and different risk communicators. Thus, there may be a number of audiences for professional risk communicators even in a single community. Compounding this situation, issues affecting a particular community may draw opposing responses from within and outside the community.

The Air Force Case: Low-level Training Flights

The Air Force Strategic Air Command (SAC) operates low-level training routes along which aircrews practice avoiding radar detection and simulate dropping bombs. These training corridors range in length from several hundred to over 1,000 miles and typically are eight miles wide. SAC tries to fly these aircraft over areas with relatively low population density. As a result, no single community is affected by overflights; planes may fly over scattered residences rather than recognizable communities. The people living under the routes usually receive no economic benefit from the overflights, and often they do not even know who is responsible for the flights despite the fact that the routes studied have been in operation for fifteen to twenty years. However, since SAC's goal is to operate these training routes, officials essentially have been forced to listen and respond to the concerns and issues expressed by members of the public. In many senses, therefore, SAC has been placed in the role of risk communicator. Officials must relay information about flight operations and respond to complaints and such concerns as those about adverse effects of excessive noise, disrupting intrusiveness of the aircraft, or fears about crashes.

DIFFERENT RISKS, DIFFERENT APPROACHES TO RISK PERCEPTIONS

Of the four cases studies discussed in this paper, the three that investigate industrial communities are similar in a number of respects. First, each of these studies involves perceptions of industrial risks in communities that were created for particular industrial enterprises. Second, in each case residents tend to be confident that the health and safety of community residents are not harmed by local industrial activities, by-products, and emissions. Third, each community also has a history of interaction and familiarity with particular industries as both operating and corporate entities. These three cases, however, are different from the fourth case, which investigates perceptions of, and responses to, low-flying jet aircraft.

Although historically such aircraft have flown along the specific routes studied, the scattered residences underneath flight corridors have existed independently of the flights. Residents may be familiar with the flights, but they often do not know either their purpose or that the Air Force is responsible for them. Finally, there has been little communication of any sort between the Air Force and the people living under the routes. These sorts of distinctions between the industrial case studies and the Air Force case help to form an explicit basis for comparisons between the studies.

MODEL OF COMMUNITY PERCEPTIONS OF ENVIRONMENTAL HEALTH

In the examination of the three industrial communities, the phrase "perceptions of risk" was taken broadly to encompass fear of industries or technologies as well as confidence and doubt in the safety of such enterprises. For the purpose of explaining why residents of these three case study communities generally were confident in the safety of their industries, existing social science studies of risk perceptions are inadequate on two counts (Wolfe, 1986). First, with few exceptions (notably Couch and Kroll-Smith, 1985; Homrighaus *et al.*, 1985; and Kroll-Smith and Garula, 1985), existing studies do not use history in an integrative way to help understand present community conditions and perceptions. However, perceptions of risk in each of the communities studied were grounded in the historical development of each community and in the historical interactions between industries, community structure, and community residents. Second, no consolidated, theoretical framework proposed in published social science risk perception research explains the perceptions found in each of the case study communities.

Because of these inadequacies, the model of community perceptions of environmental health was developed as a way of stepping back from existing studies and taking a more holistic and historically grounded approach to the study of risk. Development and refinement of the model occurred throughout the course of case study investigation. Until sufficient research has been conducted on perceptions of risk in industrial communities that will allow generalizations about sources of confidence or fear, adopting a flexible but inconclusive model such as the one proposed here may be a better alternative for practitioners than adopting a more rigid, predictive model.

As can be seen in Fig. 1, this model of community perceptions of environmental health combines both synchronic (i.e., at one point in time) and diachronic (i.e., through time) perspectives. The synchronic part of the model, labelled "ecological components," depicts the internal and external factors in industrial communities that may interact to affect perceptions of risks at any point in time. The diachronic facet of the model is historical. This perspective concentrates on the sequence and content of stages, during which the model's ecological factors operate, that lead to current perceptions of community environmental health problems. Although the model has both ecological and historical components, it should be viewed as a single, processual model that combines historical, cultural, technological, institutional, biological/medical, and ecological factors. This model is intended to structure the investigation of community-based perceptions of risk and to aid the formulation of testable hypotheses. Furthermore, the model places the notion of industrially related risk in a social and historical context.

APPLICATION OF CASE STUDY FINDINGS AND MODEL TO AIR FORCE CASE

Information and the model derived from the case studies of the three industrial communities are applicable, but not completely transferable, to the Air Force case. For example, the historical stages specified in the model are derived from historical research on the three industrial case study communities and represent periods that were crucial in the development of the communities (see Fig. 1). Because residential areas beneath training corridors developed independently from training activities and are not industrial communities, the particular historical stages of community development depicted in the model largely were irrelevant for the Air Force case. Nevertheless, both history and the type of community influenced responses to low-level overflights. History played a part in the response because many of the people who knew about the flights only learned about them after planes crashed along the routes. Crashes also were catalysts for concerns about future crashes and safety issues. The type of community was influential in that resort/recreation communities were adamantly opposed to the operation of low-level training routes in close proximity because of their intrusiveness and potential adverse impacts from noise. More rural agricultural areas, on the other hand, exhibited minimal formal opposition to the routes.

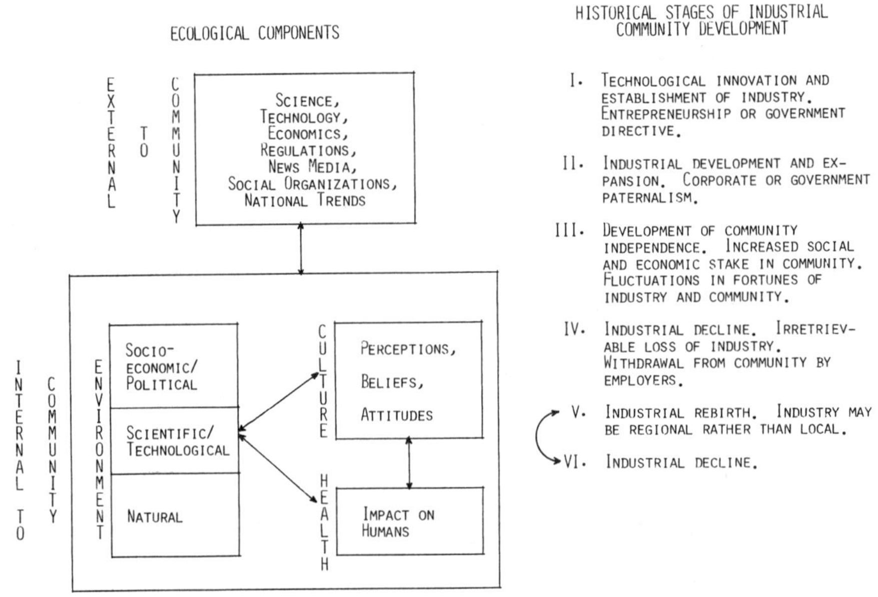

Fig. 1. Model of community perceptions of environmental health.

The ecological components depicted in the community perceptions of environmental health model are more applicable to the Air Force case than are the historical stages. Each of the ecological factors represented in the model operates in the low-level route setting. However, in the Air Force case, because the scientific/technological environment component (the industrial or technological operations of concern) was not housed in affected communities, it cannot be considered strictly an internal community component. Though housed outside the communities, aircraft intermittently were located directly over the areas in question. Therefore, the scientific/technological environment component (the particular aircraft flying over the communities) is a factor that simultaneously affects communities both externally and internally. In the industrial cases studied, the specific industrial facilities in question all were located permanently within the communities and thus were analyzed as an internal factor. The flexibility of the model is illustrated by its ability to account for such variations in communities, industries or technologies, and risks.

IMPLICATIONS FOR RISK COMMUNICATION

Examination of perceptions of risk in three industrial communities and under low-level Air Force training operations produced three findings that have significant implications for risk communication. First, several audiences for risk communicators exist, even within a single community. Second, different risk messages may be appropriate. Third, historical and continuing interactions between industries and communities affect the types of situations that escalate to crises as well as the types of risk messages that are appropriate and trusted within industrial communities.

REFERENCES

Couch, S. R., and Kroll-Smith, J. S., 1985, The Chronic Technical Disaster: Toward a Social Scientific Perspective, *Social Science Quarterly* **66(3)**:564-575.

Homrighaus, B. K., Couch, S. R., and Kroll-Smith, J. S., 1985, Building a Town but Preventing a Community: A Social History of Centralia, Pennsylvania, in *Proceedings of the Canal History and Technology Symposium*, Lance E. Metz, ed., Vol. 4, pp. 69-92, Center for Canal History, Easton, PA.

Kroll-Smith, J. S., and Garula, S., 1985, The Real Disaster Is Above Ground: Community Conflict and Grassroots Organization in Centralia, Pennsylvania, *Small Town*, pp. 4-11 (Jan.-Feb.).

Wolfe, A. K., 1986, Risk and Confidence in Industrial Communities: A Comparative History of Perceptions of Industrial Pollution in Oak Ridge and Rockwood, Tennessee, PhD dissertation, University of Pennsylvania. University Microfilms International, Ann Arbor, MI.

Ritualism in Communicating Risk Information[a]

Carolyn Needleman
Bryn Mawr College
Bryn Mawr, PA

ABSTRACT

Much of the concern over methods of risk communication has been essentially defensive: how to transfer risk information to the public in ways that minimize overreaction, in the sense of irrational fear, unwarranted litigation, and unnecessary institutional embarrassment. The converse possibility—that risks can be communicated ritualistically, in ways that fail to deliver positive social outcomes—has received less attention. However, since rationales for risk notification are usually derived from arguments for public health and social justice, the possibility of underreaction to communicated risk represents a problem no less serious than the problem of overreaction. Risk notification programs need to do more than avoid hysteria. They need to be capable of affirmatively motivating and empowering citizens to make appropriate, constructive responses to risk information, in order to protect their health and pursue their justified legal remedies. This paper reviews a number of sociological considerations relevant for designing such programs.

KEYWORDS: Risk communication, underreaction, hazard communication, risk notification, right to know

INTRODUCTION

In the late 1960's, a time of reflection on the basic premises of social policy research, there was a saying going around among social scientists: "Researchers usually study people for the purposes of the institutional system; it's now time to start studying the system for the purposes of people." The present paper examines how that principle might be applied in the field of risk analysis, particularly with regard to risk communication.

A major (and controversial) point of concern in risk communication interventions has been the possibility that people may overreact to the information they receive, responding in ways that cause trouble for institutions. That is, people might react to risk information with inappropriate anxiety, unjustified by the objective risk involved. Driven by irrational fears, they might descend in droves on clinics, seeking unnecessary medical services; they might demand unrealistically stringent regulatory standards; they might swamp the courts with unwarranted litigation; they might mobilize politically to block facilities and policies that pose low actual risks.

---

a. This article has been published in revised form in *Science, Technology & Human Values* **12(3&4)**:20-25 (1987).

Much less attention has gone to the opposite possibility, that the targets of risk communication may underreact. Having received some facts about the risks they face, they may then simply do nothing, individually or collectively, to use the information in ways that could protect their lives, health, legal rights and political interests. This kind of response is actually quite likely to occur, judging from what we know about people's ability to resist information which has unpleasant personal implications or which requires unfamiliar behavior.

As a relatively invisible and undramatic "nonevent" that causes no direct institutional problems, underreaction may go unanalyzed and even unnoticed. But to the extent that it occurs, it raises some serious questions about what we are trying to accomplish with risk communication. The point of this form of intervention is the transfer of information to those at risk, certainly, but to what ultimate end? If receiving the information is not clearly linked to behavioral responses with potential to reduce human suffering, what social good is risk communication doing? Does it become a ritualistic activity which merely fulfills legal obligations, providing the forms of public health protection and social justice without the substance?

SOURCES OF CONFUSION IN RISK COMMUNICATION

Answers to such questions are not well worked out, perhaps because the concept of risk communication is still fairly new in risk analysis. Only in the past few years has risk communication joined risk assessment ("the characterization of the potential adverse health effects of human exposure to environmental hazards") and risk management ("the process of evaluating alternative regulatory actions and selecting among them") as a third level of public policy effort and research interest.[1] The Environmental Protection Agency and the National Science Foundation sponsored the world's first conference on risk communication less than three years ago, in December of 1984.[2] The literature on the subject is still extremely sparse.[3]

In a way, the successive emergence of these three risk concepts—assessment, management and communication—represents an increasingly explicit acknowledgement of the inherently political nature of the field of risk analysis.[4] Politics are least visible at the first level of effort, risk assessment, which was established with and still maintains an aura of scientific objectivity. However, numerous writers have pointed out that even at this level, value assumptions and policy implications lie buried in the selection of analytical models and measures for evaluating risk.[5] The second level, risk management, was developed somewhat later to handle issues of social context. Here social values and political choices enter in more obviously around determination of "acceptable" risk levels and politically viable cost/benefit tradeoffs, but there is still a sense that rational planning in the public interest is possible.[6] At the third level, risk communication, those at risk are made aware of the hazards they face, and fashion individual and collective responses based on their perceptions of risk and their options for action. Here almost all of the critical issues are frankly political, as the nature of the audience moves along the continuum from presumably disinterested technical experts toward individual citizens with personal interests at stake.[7]

As an intervention aimed at influencing and empowering individual citizens, risk communication calls for a different perspective and some additional kinds of information not generally included in risk analysis done for other purposes. Unfortunately, the reality is that risk communication efforts often draw their content from already available risk assessment and risk management reports, without much modification. When this is done, no matter how technically sophisticated the information and analysis may be, the following four types of confusion are almost certain to result:

1. *Confusion over the basic premises of the risk information.* Analysis at the level of risk assessment and risk management revolves around questions of distribution, useful for allocating the costs, benefits and liability of various hazards in a way that achieves maximum equity within a national system. Within this framework, it may make sense to assume that some level of risk is inevitable, to study risk perceptions as a way of determining some social consensus about risk acceptability, and to decide how best to manage the distribution of risk for the general public good. But individuals at risk are likely not to be very interested in these distributional issues. Their concerns revolve around questions of protection—how unwanted risks (which will vary from person to person) can be reduced to zero probability. If people wishing information premised on protection from risk are instead given information premised on distribution of risk, they will frustrate the risk communicators by reacting "irrationally." They will demand "unrealistic" public policy to eliminate totally the risks they don't want in their own lives or don't think are justifiable in society, even while living comfortably with other, higher risks of their own choosing.[8]

2. *Confusion over the unit of analysis.* Risk assessment and risk management deal with large population aggregates. By contrast, people at risk tend to think in terms of impact on themselves and their friends and families as individuals, and may draw incorrect personal implications from aggregated statistical data. Their mistaken inferences are often interpreted as a "problem" in risk perception, showing the need for educating the public to interpret statistical probabilities and economic tradeoffs more rationally. But they are also a signal that some legitimate information needs of those at risk are going unmet, concerning the personal significance of the risk information.

3. *Confusion over time perspective.* Risk assessment and risk management analyses tend to be future oriented, because they are planning tools designed for system regulation and primary prevention of death, disease, and economic loss. They tend to focus on issues of consent and choice, concerning the acceptance or avoidance of health and safety threats that are not yet actual events—for instance, people's willingness to live in an earthquake prone area, accept a nuclear power plant in their neighborhood, wear seat belts, undergo elective surgery, or work in a hazardous occupation. Individuals who are the target of risk information, on the other hand, are likely to be at least equally concerned about their past exposures to hazards, where the choice of risk avoidance is no longer possible—for instance, having a history of smoking, a mother who took DES, long term exposure to radon in the home, contaminated drinking water over a number of years, past occupational exposure to toxic substances like asbestos. These concerns about past exposures, for which the only remedies are secondary prevention of disease and compensation for health damage, could be ignored as irrelevant to future regulatory and economic planning efforts. But to do so would be to neglect urgent risk information concerns of literally millions of Americans.[9]

4. *Confusion over behavioral assumptions.* The models used in risk assessment and risk management tend to make simplifying assumptions about risk-related behavior for the purposes of analysis—assuming, for instance, rational action by autonomous, competent individuals resulting in clear, permanent choices.[10] Individuals at risk live in a world where these assumptions do not hold true. They are subject to influences and limits from their social environment, they may periodically rethink their risk decisions, they may react to risk information with denial, cynicism or fatalism.[11] They may be unable to read written forms of risk information; recent estimates of adult illiteracy in the United States run as high as 10 to 15% of the population, or 27 million individuals.[12] They may be poorly educated, economically constrained, and vulnerable to intimidation.[13] Because of these

realities, the individuals who receive risk information may be blocked from understanding or making constructive use of it, unless support services, counselling and advocacy are provided along with the factual data about the risk itself.

The fact that the risk information offered may differ drastically from the risk information needed is more than merely unfortunate, or a matter of academic curiosity. It has economic costs, as well as costs in equity and quality of life. When the targets of a risk communication effort are left unable or unmotivated to take practical and timely steps to protect their health, opportunities for disease prevention are lost. In addition, if those who receive risk information are left unclear on how to seek due compensation for health damage already done, financial liability cannot be lodged with the party responsible for the harm; the economic burden falls instead on the victim or the general public.

WHAT COMES THROUGH IN RISK COMMUNICATION MESSAGES?

The difficulty of making risk communication meaningful at the individual level is dramatically illustrated in a study by Freimuth and Van Nevel.[14] These researchers studied the impact of the Asbestos Awareness Campaign, a 1978 federally sponsored effort to inform the general public—particularly those who had been shipyard workers during World War II—about the health risks they might be facing due to asbestos exposure. Communication was accomplished through public service announcements on radio and television, notices in print media, information to relevant special interest groups, and direct mail.

As envisioned by the National Cancer Institute, which supervised the program, the campaign had two quite different emphases. The primary objective was "to increase awareness of and information about the nature, extent, and seriousness of asbestos exposure for the risk target audiences and the general public." Accordingly, the materials prepared consisted largely of risk assessment and risk management information: explanations of how and where asbestos is used, how it harms health, and statistical patterns of asbestos related disease. However, in order to make the message "supportive and nonthreatening," announcements also had a secondary objective. They were to include instructions about three measures that individuals could take to reduce their personal risks of disease: (1) to quit smoking because of the strong synergistic effect of asbestos and cigarettes in causing lung cancer; (2) to get regular health checkups because early disease detection might catch some treatable asbestos-related illnesses in time to save lives; and (3) to seek further information.

The first part of the message conveys the health threat, in terms useful for public policy and planning purposes; the second part gives the individual some guidance in how to interpret the personal significance of the risk and how to take action on the information. The first part raises consciousness of the problem, in more or less alarming ways; the second restores to the individual some sense of control over the situation. Both parts of the message are important, but it is the second part that has the best potential for moderating anxiety and producing constructive individual responses.

Freimuth and Van Nevel found that, unfortunately, the second part of the message fared quite badly in the process of communication. Many radio and television stations declined to carry the public service announcements prepared by National Cancer Institute, or aired them only in off hours, so that this means of communication reached relatively few people. In the print media, which reached larger numbers, the second part of the message tended to get edited out. Most of the news stories analyzed dealt with the information in ways very hard to personalize. For instance, they emphasized the magnitude of the problem nationally, without estimating the numbers actually affected in their readership areas. Even in the high priority areas, very few news stories mentioned the idea that individuals could take any positive action to protect their health interests, such as getting a chest X-ray (8% of the stories), quitting smoking (7%), visiting a doctor (6%), or using protective equipment when working with asbestos (7%). (Information on legal recourse available to individual victims, not an objective of the campaign, was apparently not analyzed.)

In addition, a variety of other communications means were used, including 1.2 million pamphlets placed in supermarkets and discount stores, and 40 million fliers mailed directly as enclosures with social security checks, federal retirement checks, and federal paychecks. The pamphlets, entitled "Asbestos Exposure—What It Means, What To Do," carried both general information and guidance for personal health protective action.

To measure the effectiveness of the campaign in getting across the intended risk information, the researchers used a national probability survey conducted by the Gallup Organization, based on three waves, each consisting of 1500 personal interviews. The survey results showed impressive increases in public awareness that asbestos can be hazardous in shipbuilding. However, awareness of asbestos hazards in occupations other than shipbuilding did not increase much, and knowledge about personal behaviors that could lessen asbestos exposure and disease started low and remained low. For example, at the end of the campaign, only 5% of the sample identified quitting smoking as a risk-reducing step that an exposed individual might take. Although the surveys showed substantial continuing uncertainty about diseases, symptoms and workplaces associated with asbestos exposure, individuals in the target population were apparently not moved to seek further information. A minimum of 40 million people were given a phone number and address to contact for further information, but during the entire course of the campaign, fewer than 20,000 information requests were received.

The researchers conclude that, in general, the campaign was successful because it increased public awareness about asbestos exposure as a social problem. To the extent that citizens in a democracy need information in order to play an informed role in public policy, this is of course a significant outcome. However, the campaign conveyed to very few of those in the target population that they could or should do anything themselves, as individuals, to protect their own health. Since those who heard and understood the personal implications would not all go on to take the recommended actions, we can assume that the actual yield in terms of disease-reducing behavior was quite small indeed.

HAZARD COMMUNICATION AND HIGH RISK NOTIFICATION

I have described this particular program at length because it shows the kind of problem that we might anticipate in a number of other large scale risk communication efforts currently being established or proposed. Consider, for instance, OSHA's Hazard Communication Standard, in effect since November of 1985.[15] This rule requires detailed hazard labels on dangerous substances in use in certain industries. The labels, along with their more complete Material Safety Data Sheets (MSDS), are required to identify the hazardous substance's chemical content and to summarize available research on its health effects—essentially the kind of data appropriate for risk assessment and risk management planning. Outside of some information on first aid measures and appropriate personal protective equipment, there is little that workers can use directly to guide their personal responses to past or current exposures.

In principle, the link to individual action might be covered by the rule's worker education provision, which requires employers to train new workers about hazardous materials and to provide additional training whenever new toxic substances are introduced into the workplace. However, the wording of this provision is vague. Employers may choose simply to reiterate the MSDS's. They may intentionally exclude information and advice that management feels would go beyond the company's responsibilities or would not be in the company's interests.

Thus hazard communication, which then Assistant Secretary of Labor Thorne Auchter called "the most significant regulatory action ever taken by OSHA,"[16] is a necessary but not sufficient condition for effective risk communication. By itself, it may not improve worker health even with full compliance. The test of its success is not whether it makes information available as an end in itself, but whether it actually stimulates health-protecting individual behavior among the newly informed workers and their employers.

Similar problems may arise with the Right To Know legislation now in effect in many states, which gives hazard information to workers and in many cases also to community residents[17]; with the high-risk notification bills currently pending in the U.S. House and Senate, which propose giving risk information to 250,000 or more individual workers with known past exposures to toxic substances[18]; with the medical screenings being sponsored by some labor unions for their membership, such as the Sheet Metal Workers' current project to give medical examinations and risk information to 33,000 union members in the U.S. and Canada.[19] A number of cities—Philadelphia, for instance—have designed risk communication plans for thousands of municipal workers, raising the same issues.[20] In all these efforts, there is a chance that facts will be transferred, but not the facts that people can readily use.

MAKING RISK COMMUNICATION MORE USEFUL

What would a risk communication effort look like, if designed to avoid ritualism by maximizing individual health protecting responses? Below are listed some features that have been suggested as important in accomplishing this goal.[21] Not all of them could be spelled out clearly in a brief, one-shot, one-way notification message. Therefore, to be effective, risk communication should be seen as a more complex process involving follow-up intervention with support programs,[22] some of which are likely to be politically sensitive.

1. *The message would be personalized enough to provide a framework for individual action.* That is, it would be translated into implications and advice pertinent to individuals—what health symptoms to watch for; what kinds of individual behavior can minimize the risk of exposure and/or disease; what the individual's rights and legal recourse are; what benefit and compensation programs might apply in case of health damage, and any timing considerations that might be important, such as workers' compensation statutes of limitation; what resources are available in the local community for appropriate medical, legal and social service help; what the risk estimates are for specific worksites or geographic areas.

2. *The message would be repeated, using a variety of media, organizational vehicles, and authorities.* An appropriately staffed telephone hotline would be publicized for use by concerned individuals. Local sources of further information—such as relevant unions, COSH groups, health departments, specialized clinics, university-based programs, and government agencies—would be identified.

3. *The message would be reinforced by social support programs.* In recognition that the social context of information influences its significance for behavior, attempts would be made to create a climate of support for health protecting responses within the social networks of those at risk—their families, coworkers, communities, unions, civic and social associations, churches, etc. Because of possible negative psychological processes such as disease denial and displacement of anxiety onto

family members, counselling professionals (social workers, for instance) would be a key part of this effort.

4. *The message would recognize powerlessness as a significant barrier to both individual and collective action.* It would address the legitimate apprehensions of vulnerable high risk individuals by including clear advice on power-related problems such as retaliation and intimidation by employers or neighbors, social stigma, impairment of employability or insurability, loss of home value, and uncompensated medical expense. It would also offer guidance on resources for peer support, advocacy and organizing.

CONCLUSION

The points made here seem obvious. However, as a recent National Science Foundation review of risk communication literature reminds us, in this field many obvious principles are routinely disregarded in practice.[23] If the problem of ritualism in risk communication is allowed to drift into the background as something we all acknowledge but do nothing about, we will continue to lack funding for demonstration programs showing how to do a better job. We will continue to lack careful sociological research and evaluation of which methods are most effective in stimulating constructive behavioral responses to risk information. And we will continue to neglect the critically important roles that counselling, organizing and social services must play in making risk communication understandable and useful to those at risk.

REFERENCES

1. *Risk Assessment in the Federal Government: Managing the Process*, p. 18, National Academy Press, Washington, DC (1983); B. Ward, Communicating on Environmental Risk, *Environmental Forum* **4(9)**:7-11 (January 1986)>
2. M. Peltu, "The Role of Communication Media," in *Regulating Industrial Risks: Science, Hazards and Public Protection*, p. 142, H. Otway and M. Peltu, eds., Butterworths, London (1985).
3. V. Covello, Detlof von Winterfeldt, and P. Slovic, Risk Communication: An Assessment of the Literature on Communicating Information about Health, Safety and Environmental Risks, preliminary report to the Environmental Protection Agency, presented at the National Conference on Risk Communication sponsored by the Conservation Foundation, Washington, January 29-31, 1986.
4. H. Otway and K. Thomas, Reflections on Risk Perception and Policy, *Risk Analysis* **2(2)**:69-82 (1982).
5. N. A. Ashford and K. A. Gregory, Ethical Problems in Using Science in the Regulatory Process, manuscript; F. M. Lynn, The Interplay of Science and Values in Assessing and Regulating Environmental Risks, *Science, Technology and Human Values* **11(2)**:40-50 (Spring, 1986); B. Johnson, The Social Construction of Science and Hazard Assessment, presentation at the annual meeting of the Society for Social Studies of Science, Pittsburgh, October 23, 1986.
6. S. W. Samuels, The Fallacies of Risk/Benefit Analysis, *Annals of the New York Academy of Sciences* **329**:267-273 (October 26, 1979); R. E. Kasperson, Acceptability of Human Risk, *Environmental Health Perspectives* **52**:15-20 (October, 1983); R. C. Schwing and W. A. Albers, eds., *Societal Risk Assessment: How Safe is Safe Enough?* Plenum Press, New York (1980).
7. R. E. Kasperson, Six Propositions for Public Participation and Their Relevance for Risk Communication, *Risk Analysis* **6(3)**:275-281 (October 1986); A critical review of the literature on risk perception can be found in Cole, Gerald and Stephen B. Withey, Perspectives on Risk Perceptions, *Risk Analysis* **1(2)**:143-163 (1981).

Various aspects of risk perception and public policy are discussed in a special issue, Risk and the Social Sciences, *Risk Analysis*, **2(2)** (1982); see in particular the papers by Otway and Thomas, O'Riordan, Slovic *et al.*, Spangler, McLean *et al.*, and Covello *et al.*

8. N. Freudenberg, *Not in Our Backyards: Community Action for Health and the Environment*, Monthly Review Press, New York (1984); R. E. Kasperson and J. X. Kasperson, Determining the Acceptability of Risk: Ethical and Policy Issues, in *Proceedings, Risk: A Symposium on the Assessment and Perception of Risk to Human Health in Canada*, October 18 & 19, 1982, pp. 135-155, J.T. Rogers and D.V. Bates, eds., The Royal Society of Canada, Ottawa (1983).
9. D. Nelkin and M. S. Brown, *Workers at Risk: Voices from the Workplace*, University of Chicago Press, Chicago (1984); S. Epstein, L. Brown, and C. Pope, *Hazardous Waste in America*, Sierra Club Books, San Francisco (1982).
10. For example see W. K. Viscusi, *Risk by Choice: Regulating Health and Safety in the Workplace*, Harvard University Press, Cambridge (1983).
11. J. Fassenden-Raden and J. M. Fitchen, Community Perspectives on Communication of Risk, presented at the annual meeting of the Society for Social Studies of Science, Pittsburgh, October 23, 1986; J. F. Short, Jr., The Social Fabric at Risk: Toward the Social Transformation of Risk Analysis, *American Sociological Review* **49**:711-725 (December 1984); M. Douglas, *Risk Acceptability According to the Social Sciences*, Routledge & Keegan Paul, London (1986); M. Douglas and A. Wildavsky, *Risk and Culture*, University of California Press, Berkeley (1982); Nelkin and Brown, 1984 (see note 9).
12. J. Kozol, *Illiterate America*, Anchor Press/Doubleday, Garden City, NY (1985).
13. R. Kazis and R. L. Grossman, *Fear at Work: Job Blackmail, Labor and the Environment*, Pilgrim Press, NY (1982); D. M. Berman, *Death on the Job: Occupational Health and Safety Struggles in the United States*, Monthly Review Press, NY (1978); P. Brodeur, *Expendable Americans*, Viking, NY (1974); M. Gibson, *Workers' Rights*, Rowman & Allanheld, Totowa, NJ (1983).
14. V. S. Freimuth and J. P. Van Nevel, Reaching the Public: The Asbestos Awareness Campaign, *Journal of Communication* **31**:155-167 (1981).
15. Department of Labor, OSHA, Hazard Communication: Final Rule, *Federal Register*, 53280-53348 (Friday, November 25, 1983).
16. Quoted in OSHA news release, OSHA Issues Final Hazard Communication Standard, Office of Information, U.S. Department of Labor (Tuesday, November 22, 1983).
17. M. S. Baram, The Right to Know and the Duty to Disclose Hazard Information, *American Journal of Public Health* **74**:385-90 (1984).
18. H. R. 1309, The High-Risk Occupational Disease Notification and Prevention Act of 1985, Congressional Record, House of Representatives, H1037-1040 (Tuesday, March 5, 1985); S. 2050, The High-Risk Occupational Disease Notification and Prevention Act of 1986, Congressional Record, Senate, S1061-1064 (February 5, 1986).
19. Personal communication from Laura Welch, Project Coordinator and Director, Occupational Medicine Department, The George Washington University Medical Center, Washington, DC.
20. E. Bresnitz, A. Freedman, A. J. Henley, L. Levin, M. McDiarmid, C. Needleman, K. Rest, S. Shapiro, R. Sarrar, and R. Sokas, Worker Education and Medical Screening: Recommendations of the Task Force on Asbestos Exposure of Municipal Workers, report to the Commissioner of Health, Department of Public Health, Philadelphia, March 1985, available on request from the author at Bryn Mawr College, Bryn Mawr, PA 19010; summarized in Notes from the Field, *American Journal of Public Health* (June 1986).
21. S. W. Samuels, Workers At High Risk, in *Maxcy-Roseneau Public Health and Preventive Medicine*, J. M. Last, ed., Appleton-Century Crofts, NY (1980); P. Schulte and K. Ringen, Notification of Workers at High Risk: An Emerging Public

Health Problem, *American Journal of Public Health* **74(5)**:485-491 (May, 1984); U.S. Department of Health, Education and Welfare, NIOSH, The Right to Know: Practical Problems and Policy Issues Arising from Exposures to Hazardous Chemical and Physical Agents in the Workplace, U.S. Government Printing Office, Washington, DC (1977); C. Needleman, testimony on The High Risk Occupational Disease Notification and Prevention Act of 1985 (H.R. 1309), Joint Hearings before the Subcommittee on Health and Safety and the Subcommittee on Labor Standards, Committee on Education and Labor, pp. 240-258 November 13, 1985, U.S. Government Printing Office, Washington, DC (1986); C. Needleman, testimony on The High Risk Occupational Disease Notification and Prevention Act of 1986 (S. 2050), for the Subcommittee on Labor, Senate Committee on Labor and Human Resources, May 15, 1986.

22. Schulte and Ringen, 1984 (see note 21), p. 487.
23. Covello, von Winterfeldt and Slovic, 1986 (see note 3), pp. 35-36. See also M. Coon and P. Polakoff, Legal and Ethical Dilemmas of Worker Notification, in *Legal and Ethical Dilemmas in Occupational Health*, J. S. Lee and W. N. Rom, eds., Ann Arbor Science, Ann Arbor, MI (1982).

Summary of Panel Discussion on "The Role of Private and Public Sector Insurance and Compensation for Environmental Pollution"

Howard Kunreuther
The Wharton School of the University of Pennsylvania
Philadelphia, PA

Steve Wassersug
Environmental Protection Agency, Region III
Philadelphia, PA

ABSTRACT

This panel discussion brought together representatives from the insurance, government, waste management, industrial and academic sectors concerned with the problems affecting the market for Environmental Impairment Liability Insurance (EIL). Government laws mandate financial responsibility requirements for parties dealing with hazardous wastes, while legal developments and other problems have resulted in the collapse of one of the key methods of satisfying these requirements — the EIL insurance market. The causes of the crisis were detailed, with special emphasis on the impact of evolving tort doctrines. Remedies, including data needs were proposed and research priorities for better understanding and resolution of the problem were highlighted.

KEYWORDS: Insurance, liability, pollution, cleanup

INTRODUCTION

The market for Environmental Impairment Liability (EIL) insurance has dried up. Most insurers are not offering coverage today; reinsurers have withdrawn from the market. Hence manufacturers who must process or dispose of toxic substances cannot protect themselves against liability suits for pollution damages.

The problem is compounded because federal and state environmental regulations require treatment, storage and disposal facilities for hazardous wastes to meet financial responsibility requirements. In addition, to operate, all hazardous waste facilities are required to have third party bodily-injury and property damage coverage, either through private insurance, self insurance or some combination of both.

New Risks
Edited by L. A. Cox, Jr., and P. F. Ricci
Plenum Press, New York, 1990

With EIL insurance unavailable, facilities are in danger of being closed because they cannot meet these financial responsibility requirements.1[a] Similarly, the absence of insurance has significantly slowed the cleanup of existing sites. The situation will hopefully be ameliorated by the indemnification provisions of the CERCLA reauthorization (SARA) for EPA approved contractors. Appropriate risk assessment and risk management of cleanup operations, with insurance industry participation, will remain a critical issue in assuring that urgently needed cleanup work is accomplished with due speed.

This paper summarizes a panel discussion on this topic in which experts taking various points of view addressed these issues. The panelists were asked to discuss key problems and potential solutions in the environmental pollution area from their perspective. Attention was directed to the appropriate roles of the public and private sectors in dealing with these issues. Our summary of the panel discussion will focus on three aspects:

1. Nature of the problem

2. Causes of the current crisis

3. Proposed solutions and research needs

NATURE OF THE PROBLEM

Charles Powers, director of Clean Sites, Inc., provided a graphic example of the difficulty his organization has had recently in obtaining EIL coverage. Clean Sites was established as a consortium of industry and environmental organizations designed to support and police efforts to clean up existing sites. Although Clean Sites has never been sued, Powers said it is impossible to purchase pollution insurance today. He noted that Clean Sites, Inc. was involved with a PCB cleanup operation in North Carolina, at approximately one-fifth of the normal cost. The North Carolina site involved 180 potentially responsible parties, and in the normal process of cleanup, little would have been accomplished. In the future, it will be difficult for Clean Sites to undertake these cleanup activities without some type of insurance protection.

Jack Schramm, Director of Government Affairs, Waste Management, Inc., provided a perspective on the nature of the problem from the commercial waste services industry. He indicated that legislation and regulatory activity in the past ten years for the United States established a system of financial responsibility for owners and operators of hazardous waste facilities. To date (and more is coming) the EPA has established three major financial responsibility requirements:

- To ensure the availability of funds to close a facility (closure care).
- To ensure the availability of funds to monitor and maintain a facility for thirty years after closure (post-closure care).
- To provide liability coverage for sudden and non-sudden occurrences, until closure is completed.

The mechanisms used to comply with government financial assurance regulations, in addition to insurance, include surety bonds, trust funds, letters of credit, and the financial

---

a. This description of problems arising from the U.S. insurance crisis draws on remarks of Lee Thomas, Administrator of the U.S. Environmental Protection Agency, as presented at the Wharton School Workshop "The Role of Insurance and Compensation in Environmental Pollution Problems," May 15, 1986.

test/corporate guarantee (which is a means to use the company's own assets to meet these requirements.)

Schramm indicated that traditional transfer-of-risk insurance is not available to most waste service companies. Large firms have had to rely on self-insurance mechanisms—smaller firms who do not have the resources to self-insure are in danger of closing. The Resource Conservation and Recovery Act (RCRA) also authorizes surety bonds, but only a few have been made available for treatment-storage-disposal facilities. There have been few surety bonds available for closure care; post-closures have to be 100% collateralized.

Trust funds, given the opportunity costs of cash, are an unfavored mechanism. In addition, there are considerable administrative costs associated with the use of these funds. Letters of credit have been popular, but their use is decreasing. Banking practices stimulated by proposed treasury regulations are imposing increasingly complex limitations on them; furthermore, owners' indemnifications are being required. Thus, American borrowers are forced to seek lines of credit from offshore banks in Japan and elsewhere.

The financial test, the most favored mechanism, is effectively constricted because of the limiting "six times multiplier" requirement in its formula, hobbling its otherwise further potential use. As a result of such constraints, Schramm noted, it has been increasingly difficult for firms, particularly those with multiple facilities, to meet the financial responsibility requirements imposed by Congressional legislation. The system is in danger of overload and shorting out since more regulations are on the way (e.g., for corrective action) for Subtitle D facilities and possibly for TSCA facilities, all threatening to use the same already constrained mechanisms to provide financial assurance. The marketplace regulations and an inhospitable legal climate in the states, in combination, could lead to gridlock. Noncompliance with the regulations can mean the shutdown of firms in the commercial wastes services industry. The insurance industry has not been willing to offer the type of coverage which would be needed to satisfy the intent of the legislation due to the uncertain state of its potential liability. But, insurance is only one mechanism. The problem of financial assurance is much broader for the commercial waste service industry. It happens that all of the mechanisms provided by law to assure financial responsibility are effectively constrained. Fixing one mechanism is not enough. The entire system needs to be overhauled.

CAUSES OF THE CURRENT CRISIS

In his presentation Martin Katzman indicated that the breakdown of the pollution liability insurance market can be traced to ambiguities in the tort liability system. Until the 1970s, victims of nonsudden or gradual pollution had virtually no hope of collecting for their damage. Conventional statutes of limitation would have blocked claims for injuries that were manifested decades after alleged exposure. Proving negligence for events in the distant past was difficult. However, rapid evolution in the tort law in the past two decades has radically changed the position of the plaintiff. Katzman indicated three rulings that have made it extremely difficult for insurers to offer coverage against pollution damage:

- The discovery rule. In the conventional statutes of limitation, the period in which insurers were liable for potential future damages was only three to five years. Under the new tort law the discovery rule may make insurers liable for events occurring decades ago.

- Joint and several liability. This provision was designed for situations in which the contribution of each party to a waste site could not be reasonably established. Insurers fear that, when a provision is applied, liability for damages may be totally unrelated to the actions of the policy holder, especially when many of those who are potentially liable either cannot be identified or have gone out of business. Insurers fear the cost of damages will be levied primarily against the most affluent party—the deep pocket—or against the one with best of intentions, such as the firm which kept good records.

- The impact of ambiguity on markets. Insurers may have less general technical knowledge as well as less specific knowledge about a risk than the firm requesting insurance. Ambiguity on the part of sellers of insurance is reflected in a price substantially greater than its actuarial value (which may limit the demand for coverage).

Jack Morrison had an additional set of concerns facing the underwriter of insurance. A key issue for an underwriter is a certain degree of predictability. Given the current legal climate, CIGNA and other companies do not write pollution liability for several reasons:

- The interpretation of the term "sudden and accidental" (a clause in the insurance policy) by the courts. A recent court case in Jackson Township, New Jersey, interpreted the words "sudden and accidental" as synonymous with "neither expected nor intended". Hence the insurer was held responsible for paying third party pollution liability claims by citizens against the township even though the policy excluded all but "sudden and accidental" pollution and the pollution involved happened over a multi-year period.

- The replacement of the negligence theory with the concept of strict liability. Under the theory of negligence the plaintiff must identify who is responsible for an injury and whether this was due to carelessness or fault on his part. Under the concept of strict liability the court focuses on the risk itself rather than on the behavior of the defendant, such as a chemical company.

- The question of the interpretation of retroactive liability. Recent tort law has involved the principle of retroactive liability whereby a company may be held liable for damages from an operation which does not satisfy today's standards, even though the facility was constructed according to the best available technology at the time. In effect, the insurer must provide coverage against liabilities yet to be discovered.

Morrison also pointed out that joint and several liability has totally destroyed the underwriters' attempt to predict and evaluate risks associated with pollution.

Kirby Montgomery also indicated that existing legislation, current judicial decisions and the absence of past experience are at odds with fundamental principles of insurance. Specifically, the current state of tort law and interpretation of liability make it impossible to accurately assess potential bodily injury and property damage exposure and costs. Due to past decisions, insurers have a low degree of confidence that courts will uphold the coverage provisions and limitations as stated in insurance policy contracts.

PROPOSED SOLUTIONS AND RESEARCH NEEDS

Leslie Cheek proposed two solutions to the problems associated with liability for hazardous waste disposal.

1. *Possessory bodily/property damage.* Insurance would be provided against well-defined sudden accidents while potentially hazardous substances are in the possession of an insured business. Thus, if a generator's on-site holding tank

ruptures, he and his insurer would be liable for any bodily-injury or property damage caused by the release. Similarly, if a transporter's truck overturned en route from a generator's premises to a disposal facility, the transporter and its insurer would be liable for any bodily-injury or property damage caused by the spill. Cheek maintained that the policy would have to exclude damages for mental or emotional distress in the absence of clinically diagnosable injury. Costs such as medical surveillance should be explicitly excluded from coverage.

2. *Channel Cleanup Liability.* Cheek proposes that facility owners be required to accumulate reserves for onsite cleanup to cover the costs associated with accidents such as leakage of toxic substances. By instituting such a system insurers would only be liable for bodily'injury and property damage from a facility rather than for cleanup costs. Funds for these reserves could be raised through additional assessments to facility owners and to those making use of the disposal facilities.

A requirement to channel cleanup liability reserves would parallel current regulations that require sufficient funds be on hand to close the facility and to monitor and maintain it for a thirty-year post-closure period. Cheek feels that the future insurability of those involved in handling hazardous substances will depend heavily on making sure that insurers will never be called upon to pay the costs of onsite cleanup and/or damages to natural resources resulting from releases of hazardous substances. The above two mechanisms are steps in this direction.

Lee Nute stressed the importance of risk assessments of the health effects of toxic substances to clarify liability. He cited recent court cases in Tennessee and Missouri where testimony relied on health assessments of the potential effects of certain substances to stress the need for health monitoring and health care. Nute indicated that the costs of health assessments might be insurable under the reauthorized Superfund amendments passed in November 1986. Specifically, if there is evidence that the population may be at risk from a given substance, the EPA Administrator shall conduct assessments. Insurance may cover the assessments but not be related to the health care. If a causal link were established between a toxic substance and certain health effects, then the responsible party would either have to settle with the potential victim prior to a court case or have punitive damages imposed.

Montgomery added that the potentially responsible parties would still look to insurers for settlement money. He felt that both the shifting of costs to a broad spectrum of society and reducing costly litigation were prerequisites to solving the environmental pollution liability problem. His suggestions based on these two general principles include:

- Removing or greatly relieving retroactive liability of insurers for pre-1980 releases of hazardous substances.

- A voluntary insurance and/or reinsurance pool modeled after the Price-Anderson nuclear pools. This would encourage traditional insurers and reinsurers to reenter the market.

- Government excess support of a liability due to catastrophic losses such as found in the nuclear liability system. This would also increase predictability and would encourage greater participation by insurers and reinsurers.

- Shifting the cost of cleanup of past releases from insurers and their insureds to other elements of society. A tax, rather than litigation, is the most efficient way to spread costs and will allow for a speedier cleanup of hazardous waste sites. Private studies indicate that defense costs consume almost one-third of the Superfund dollar.

- Steve Wassersug felt that the Superfund Amendments and Reauthorization Act (SARA) sets an appropriate tone for accelerating cleanup activities by appropriating $8.5 billion to the task and imposing stricter regulations. He noted that one of the key issues in the Act is how to deal with uncertainty. The cost of cleanup has escalated, driven by concern over waste after it leaves a site, and this is leading to a much greater interest in the ultimate destruction of waste. New technologies such as biological stabilization and incineration are costly and it is estimated that new approaches to waste disposal translate into disposal costs which are 4 to 5 times higher than what they were five years ago. Wassersug estimated that one-third of the money for site cleanup would come from EPA while one-third of the sites would be cleaned up by potentially responsible parties.

- Katzman stressed the importance of data needs, indicating that the pollution liability insurance crisis was remarkable for its factual vacuum. Policymakers have almost no handle on the magnitude or frequency/severity distribution of third-party losses. Most of the figures offered by insurers refer to their potential cleanup liabilities, which from a public policy perspective are not losses but loss prevention expenditures. He noted that recent work by scientists such as Richard Peto indicates that environmental agents are only a minor cause of cancer, although juries may decide otherwise.

Katzman also highlighted the problems associated with standard occurrence-based policies whereby insurers are liable for damages when the individual was exposed to a potentially toxic substance. There are large uncertainties as to the magnitude of future claims from such an insurance policy. He stressed the need for widespread adoption of a claims-made policy, allowing the insurance industry to collect premiums to pay only for losses incurred during the current year. Under such policies, the insurer protects the company against claims filed during the policy year for accidents that occurred during an earlier time period. An aggregate limit is placed on coverage for the policy year to establish an upper limit on claims. These terms reduce uncertainty of future claims from pollution.

On the governmental side Katzman pointed out that the most pressing action today is making tort rules more predictable. One needed action is to extend the discovery rule to all toxic torts, which in some states are limited to a few substances like asbestos and Agent Orange. A more pressing one is the need to narrow the scope of joint and several liability by statute. One proposal would subject a company to liability approximating its market share for a particular activity (e.g., storing waste in a landfill).

Katzman indicated that joint and several liability has had some positive effects. Without it, mutual aid activities such as Clean Sites, Inc. never would have come into existence.

Katzman and others felt that some type of mutual insurance program would be an effective means of protecting firms subject to pollution liability risk. Companies in the same industry would join together in a pooling arrangement to protect themselves against potential liability claims to any of them. An example of such an organization is the American Casualty Excess Company (ACE), an organization formed in 1985. Marsh and McLellan serves as a broker for large companies in the chemical industry, each of whom provides a contribution to the pool. The insurance industry offers arbitration between the insurer and insured if a dispute arises.

Howard Kunreuther provided a summary of future research needs pertaining to the environmental pollution insurability problem. He indicated that long-term solutions to the problems of environmental liability must be preceded by better knowledge, both qualitative and quantitative. The following questions need to be directly addressed:

1. What data are available and what future data should be collected to understand the risks from environmental pollution?

2. What is known about the frequency and severity of past claims from insurance policies?

3. What are the implications of the absence of private insurers on financial responsibility requirements and long-term strategies of industrial firms?

4. What is the experience of alternative systems (e.g., mutual insurance) in filling the insurability gap?

There is a need for creative solutions for dealing with environmental liability questions so that groups like Clean Sites, Inc. and Waste Management, Inc. will feel comfortable in dealing with hazardous waste and that the insurance and reinsurance industries will want to offer protection against potential losses. The next few years are likely to bring some significant changes in dealing with environmental insurance. One key need is for different stakeholders, each of whom has his own agenda, to interact with each other and to continue an interactive dialogue.

The panel discussion at the Society for Risk Analysis Meeting in Boston provided a forum for such an interchange to take place. We hope that others will follow in the future.

Risk Management Through Market Incentives: Liability and Insurance

Martin T. Katzman
Oak Ridge National Laboratory
Oak Ridge, TN

ABSTRACT

Economists generally prefer markets to constraints in the management of external diseconomies. Theory suggests that insurance is the appropriate market mechanism for dealing with stochastic externalities, like pollution. Despite Federal encouragement in RCRA and Superfund, the market for pollution liability insurance collapsed after a brief emergence. Interesting questions are (a) Why has this market collapsed? and (b) Under what conditions can it revive? The adoption of the legal doctrine of joint and several liability results in de facto mutual liability. Consequently, most benefits of risk reduction are external to the firm. Emerging mutual institutional remedies to the diluted incentives for risk reduction are discussed.

KEYWORDS: Pollution, insurance, liability

INTRODUCTION

Economists are almost unanimous in their advocacy of market incentives over regulatory constraints in the management of external diseconomies. The economist's efficiency arguments apply no less to stochastic and accidental externalities, like oil spills, than to routine, predictable, and continuous externalities, like smoke emissions. While activities engendering continuous externalities can be charged a pollution tax, activities generating risks can be levied a risk-charge, like an insurance premium.

The argument in favor of the use of an insurance premium to achieve efficient levels of risk management can be stated rather simply (or, in the view of skeptics, "simplistically" or "simple-mindedly"):[1]

1. The actuarial value of third-party damages from a risky activity is computable. While risk analysis of rare events is a crude art, creating a competitive market in the setting of risk charges will create a demand for better risk analysis.

2. The party engendering the risk (the potential injurer) is charged this actuarial value, plus a "loading" for the transactions costs of risk analysis and loss adjustment.

3. The risk charge provides the potential injurer with incentives for efficient risk management. Not only does the potential injurer have an incentive to choose the

New Risks
Edited by L. A. Cox, Jr., and P. F. Ricci
Plenum Press, New York, 1990

least-cost method of achieving a given level of risk, the injurer has an incentive to equate the marginal cost of accidents (reflected in the risk charge) to the marginal cost of accident-reduction.[2] Moreover, if an activity is priced out of the market by the levying of a risk charge, then it is clearly an "unacceptable risk."

Although the argument in favor of societal risk management through risk charges is theoretically appealing to economists, market mechanisms for controlling either routine or stochastic externalities have earned little role in the American political system. The public generally rejects mechanisms like pollution charges as "licenses to pollute." While they have historically played a major role in the reduction of risks from fires and boiler explosions,[3] insurers do not relish their role as vicars of the regulators.

In this context, the espousal of market incentives in two acts governing third-party risks from hazardous chemicals is remarkable. The Resource Conservation and Recovery Act (RCRA) and Superfund Act require handlers of hazardous chemicals to establish financial responsibility for third-party damages. While larger firms can establish financial responsibility under RCRA by reference to a strong balance sheet, smaller firms can only satisfy this requirement by the purchasing of pollution liability insurance. Superfund explicitly mobilizes the assistance of the insurance industry in establishing the limits of responsibility for generators. The passage of RCRA and Superfund in 1976 and 1980, respectively, hopefully opened the way for an extraordinary new era in risk management, in which market incentives would facilitate the achievement of social objectives.

After a promising start in the late 1970s, the market in pollution liability insurance has virtually vanished. By simply decreeing "Fiat emptor, fiat vendor," public policy does not automatically create a market. Because the RCRA financial responsibility requirements were to be phased in slowly during the 1981-1985 period, the market floundered for lack of demand. Anticipating a growth of demand, at least a dozen underwriters planned to offer pollution liability insurance by the end of 1983. A domestic reinsurance pool encompassing around 40 insurers emerged. Since 1984, the market has floundered for lack of supply. Now only one company advertises pollution liability insurance.

The failure of the pollution liability insurance market reinforces Kenneth Arrow's classic observation that markets in contingent claims are rather sparse.[4] The interesting questions are (a) Why has this market failed? and (b) What conditions can be created to rejuvenate this market? In answering these questions, we assume that the problem of insurability goes deeper than the transient losses suffered by underwriters caused by short-sighted pricing practices during the recent period of high interest rates.

THE CAUSES OF MARKET FAILURE

Public policy has failed in the creation of insurance markets before, notably in the creation of flood insurance and nuclear liability insurance. The failure of the flood insurance market can be traced to both demand and supply. On the demand side, Howard Kunreuther's pathbreaking field work[5] and Paul Slovic's convincing experiments[6] show that individuals are generally unwilling to insure against rare events even at actuarially fair prices. Professional corporate risk managers, however, may behave more consistently with expected utility theory.[7] Supply has not failed to emerge because of controls on premiums aimed at making insurance "affordable". Controls were a factor in the withering away of supply in the flood insurance market.[8] Finally, the market has not failed because of the difficulties in *technical* analysis of complex environmental risks, from release through environmental pathways to human dose-response curves. There is a lively, competitive market in environmental risk analysis where the state of the art is being advanced rapidly.

Rather, the breakdown of the pollution liability market can be traced to ambiguities engendered by the tort liability system. The first assumption of the market-incentive model of risk management is that technically calculated losses translate one-for-one into losses to the potential injurer, and that the insurer can thus price this risk. This translation, however, is not automatic but is mediated through the tort system, which can make errors of omission or errors of commission.

Until the 1970s, victims of nonsudden or gradual pollution had virtually no hope of collecting for their damages. Conventional statutes of limitation would have blocked claims for injuries that were manifested decades after alleged exposure. Proving negligence for events in the distant past was difficult. The inability to trace the source of a generic molecule made identification of the defendant difficult. Blocking attempts to collect for demonstrable damages, the tort system provided potential injurers with poor market signals.

A rapid evolution in the tort law in the past two decades has radically changed the position of the plaintiff. The substitution of conventional statutes of limitation by the discovery rule, the displacement of the negligence standard by strict liability, and the diffusion of new concepts of joint and several liability increase the probability of plaintiff recovery.[9] The substitution of new toxic tort rules for the old has confounding efficiency implications.

Under conventional statutes of limitations, the discounting period for potential future damages is only 3 to 5 years, which meant that chemical polluters would have no signals about latent damages. The discovery rule requires potential injurers to consider the implications of their actions into the relevant future, thereby improving the price signals. For insurers, however, the discovery rule continues to cause severe problems for conventional occurrence-based insurance policies. Since such a policy is activated by the injuring event, insurers have found themselves liable for events that occurred decades ago.

The insurability problem *for future exposures* can be obviated by a claims-made policy which is activated at the time the injury is discovered. Under a claims-made policy, the underwriter must look at conditions in the past that may cause claims to be made in the current year. Premiums established by retroactive risk analysis, however, are poor incentives. They do not induce good risk-management practices today. Thus there is a contradiction between insurability and incentives.

The displacement of negligence by strict liability fully internalizes third-party damages. A rule of strict liability may enhance insurability by making the financial consequences of damages more predictable.[10] The adoption of strict liability clearly increases the magnitude of losses. Since these can be factored into premiums, strict liability per se causes no problem of insurability.

The crux of the insurability problem lies with joint and several liability. Through several landmark cases,[11] the right of the plaintiff to recover from chemical handlers engaged in the production of similar damaging substances was established. Under the doctrine of several liability, a plaintiff can recover 100 percent of the damages from a "deep pocket" defendant, even if the later was responsible for only a *de minimis* portion of the harm.

The joinder of defendants clearly increases the likelihood of a plaintiff recovering from *someone*, but it dilutes incentives for care. A sloppy, small, fly-by-night chemical handler has an incentive to imitate the processes and products or to share waste facilities with responsible "deep pocket" chemical handlers. Any firm knows that it will reap only a small portion of its loss-control expenditures in terms of reduced liabilities. By further "externalizing" the cost of chemical damages, joint and several liability creates a nightmare for insurers. While an underwriter might assess the direct risks of its own insured, it cannot easily assess the likelihood of its insured being joined in a suit for damages caused by another firm.

Why cannot the insurance industry simply raise its premiums to cover the eventuality that insureds may be held jointly liable for the damages caused by others? The answer may lie in the impacts of *ambiguity* on markets. Insurers are in an asymmetric position with respect to the insured in the analysis of risks of complex technologies. The insurer may have less technical knowledge generally as well as less knowledge about a specific risk, a scaled-up version of the "lemon problem."[12] The possibility of joinder increases the ambiguity of the loss.

How does ambiguity affect insurability? Ambiguity on the part of sellers of insurance is accommodated by charging a price substantially greater than its actuarial value.[13] Not only are buyers of insurance less ambiguous, ambiguity either depresses or has no effect on willingness to pay. Consequently, there may be virtually no demand at the equilibrium price.

WHAT ARE THE SOLUTIONS?

In order to narrow the range of potential solutions, the objectives of societal risk management should be restated: providing incentives for efficient levels of risk and loss control, victim compensation, and risk spreading among potential injurers. The recognition of these three goals eliminates several solutions. For example, a victim compensation fund, financed by taxes, would have a poorly honed deterrent effect. A complete reversion to the older tort rules would decrease victim compensation and deterrence.

A discussion of solutions should also separate financial problems of insurers resulting from a mismatch between old insurance contracts and recently discovered injuries. The allegation that regulators (EPA) are using the insurance industry to finance a social program (cleanup of hazardous waste sites) has different implications from retroactive and prospective viewpoints. While the financial impact of such a ($10 billion?, $100 billion?, $2000 billion?) cleanup can be financially devastating to insurers, the solution should focus upon the creation of institutions to manage risks that may arise from now on. The cost of cleanups, in the future, can simply be factored in the premiums.

Data Needs

The discussion of the pollution liability insurance crisis is remarkable for its factual vacuum. Policymakers have almost no handle on the magnitude of frequency/severity distribution of third-party losses. Most of the figures offered by insurers refer to their potential cleanup liabilities, which from a public policy perspective are not losses but loss-prevention expenditures. On potential third-party injuries themselves, policy analysis depends upon anecdote and headline. The important work of Peto suggests that environmental contaminants are only a minor cause of cancer,[14] although juries may decide otherwise. While other debilitating diseases may be related to chemicals in the environment,[15] pollution-engendered latent diseases may be a trivial source of damages and of liability, compared to other sources of disease and traditional, sudden environmental problems, like oil spills.

Further research on the probability-severity distribution of chemical risks is clearly needed in order to harness the insurance market to a regulatory purpose. Such research should include traces of typical environmental pathways of dispersion, the magnitude of human exposure and traditional dose-response analysis. Collaborative research funding by insurers, the chemical industry, government, and environmental groups may have considerable payoff in defusing or at least determining the scope of the problem. While cognitive psychologists warn against over-optimism about risk communication, research may at least reduce the ambiguity about the risk among professionals, like underwriters and corporate risk managers, who price and respond to prices, respectively.

Appropriate Roles of Government

It is important to emphasize what government should not do. First, the federal and state governments should not dilute the financial responsibility requirements. Firms must know that they cannot ignore third-party risks that exceed the net worth of the firm. Second, the widely publicized tort and insurance crises should not detract from the fact that the new toxic tort system has had positive deterrent effects. Despite their "finite reservoir of concern," top corporate managers have had their attention directed by the tort crisis to extremely low-probability/high-consequence events. Surveys of the Chemical Manufacturers Association indicate that there have been major reductions in waste generation though design changes in the production process.[16] The ambiguity of tort outcomes, fueled by unverifiable anecdotes about outrageous tort settlements, serves the function of accelerating the adoption of risk-reduction technologies. Waste-reduction in the production and use of toxic chemicals, of course, is an important part of the ultimate long-term solution. In considering proposals for toxic tort reform, the public should consider that incentives to invent and adopt waste-reduction technologies might be attenuated as a consequence.

The most pressing governmental action is to make the tort rules more predictable. One needed action is to *extend* the discovery rule to all toxic torts, which in some states are limited to a few substances like asbestos and Agent Orange. More pressing is the need to narrow the scope of joint and several liability by statute to something approximating a market-share rule. While the disincentive effects of joinder would remain, albeit in a weaker form, maximum liabilities would become more predictable.

Appropriate Roles for Private Action

There is considerable scope for private action in the solution of the insurance crisis. Underwriters have taken a major step in clarifying the rules by totally excluding pollution from the new Comprehensive General Liability policy and by spinning off all pollution claims to a new Environmental Impairment Liability policy. In the future, contractual (semantic) ambiguities about whether or not an occurrence was sudden or accidental can be avoided.

A more promising sign is the creation of mutual insurance by companies handling similar chemicals. Since joint and several liability is a form of involuntary mutual insurance, chemical handlers might be amenable to more formal arrangements. A mutual would have greater access to knowledge of risks and loss-control technologies than a general insurer, and hence a better basis for rating risks. Furthermore, adherence to industry standards of care could be a precondition for insurability. The recent establishment of mutuals by an asbestos firm, by a consortium of waste handlers, and by a group of chemical companies substantiates this prediction.[17]

REFERENCES

1. For a fuller presentation of the argument and further citations, see Martin T. Katzman, *Chemical Catastrophes: Regulating Environmental Risk through Pollution Liability Insurance*, Huebner Foundation Studies, Richard D. Irwin, Homewood, IL (1985). See also J.D. Hammond, Risk Spreading through Underwriting and the Insurance Institution, in *Societal Risk Assessment: How Safe*

is Safe Enough?, pp. 147-176, Richard C. Schwing and Walter A. Albers, Jr., eds., Plenum Press, NY (1980).

2. The equilibration of the marginal cost of accidents and the marginal cost of accident reduction is espoused as a criterion of negligence by Judge Learned Hand in *United States v. Carroll Towing Co.*, 159 F.2d 169 (2d cir. 1947).
3. For an interesting discussion of the public perspective, see Steven J. Kelman, Economic Incentives and Environmental Policy: Politics, Ideology, and Philosophy, *Incentives for Environmental Protection*, Chapter 14, Thomas Schelling, ed., MIT Press, Cambridge (1983). For an excellent statement of the industry position, see Leslie Cheek III, Risk-Spreaders or Risk-Eliminators? An Insurer's Perspective on the Liability and Financial Responsibility Provisions of RCRA and CERCLA, *Virginia Journal of Natural Resources Law* **2**:149-178 (Fall, 1982). In political economy, however, the distinction between the conscious purpose of actors and their social function is as venerable as the concept of the invisible hand.
4. Kenneth Arrow, *Essays in the Theory of Risk-Bearing*, North Holland Publishing, Amsterdam (1974).
5. Howard Kunreuther *et al.*, *Disaster Insurance Protection: Public Policy Lessons*, John Wiley and Sons, NY (1978).
6. Paul Slovic *et al.*, Preferences for Insuring against Probable Small Losses: Insurance Implications, *Journal of Risk and Insurance* **44**:237-258 (June 1977).
7. Martin T. Katzman, Creating Markets for Catastrophic Insurance: Experimental Evidence in the Pollution Liability Line, paper presented to the annual meeting of the American Risk and Insurance Association, Chicago, Aug. 18, 1986.
8. Fred B. Power and E. Warren Shows, A Status Report on the National Flood Insurance Program, *Journal of Risk and Insurance* **46**:61-76 (June 1979).
9. A lay presentation of the problems of conventional torts and the nature of the new toxic torts is in Martin T. Katzman, Chemical Catastrophes and the Courts, *The Public Interest*, No. 82, pp. 91-105 (Winter 1986). For more detailed references, see Katzman, *Chemical Catastrophes*, Chap. 3.
10. An argument like this is made by Richard A. Epstein, *A Theory of Strict Liability: Toward a Reformulation of Tort Law*, Cato paper No. 8, Cato Institute, San Francisco (1980).
11. E.g., *Sindell v. Abbott Laboratories*, 26 Cal. 3.d 588; Cal. 607 P.2d 924; 163 Ca;.Reptr. 132 (1980).
12. George Akerlof, The Market for Lemons: Qualitative Uncertainty and the Market Mechanism, *Quarterly Journal of Economics* **84**:488-500 (1970).
13. Robin M. Hogarth and Howard Kunreuther, Ambiguity and Insurance Decisions, *American Economic Review* **75**:386-390 (May 1985); Martin T. Katzman, Creating Markets.
14. Richard Peto, Why Cancer?, in *The Resourceful Earth*, Chap. 19, Julian L. Simon and Herman Kahn, eds., Basil Blackwell, NY (1984).
15. David E. Root and David W. Schnare, First Do No Harm: Diagnosis and Treatment of the Chemically Exposed, in *Chemical Contamination and Its Victims*, David W. Schnare and Martin T. Katzman, eds., Quorum Books, Westport, CT (1989).
16. Chemical Industry Watches Its Waste, *Chemecology*, p. 2 (June 1987); Waste Minimization, *Chemecology*, pp. 2-4 (Feb. 1988).
17. 17. Stephen Tarnoff, Asbestos Removal Firm to Form Insurer, *Business Insurance*, April 29, 1985; House Panel, Association Announce Efforts to Resolve Environmental Insurance "Circus," *Environmental Reporter* **16**:1766-1767 (Jan. 17, 1986), discusses the formation of Waste Insurance Liability, Ltd. The formation of a captive by sixteen chemical companies is sketched in *Insurance Review* **48**:19 (Sept. 1986).

A Comparison of the Human Exposure Model and the Air Emissions Risk Assessment Model

Lawrence B. Gratt
IWG Corp.
San Diego, CA

ABSTRACT

The Human Exposure Model (HEM) and the Air Emissions Risk Assessment Model (AERAM) are FORTRAN risk assessment computer models developed by the Environmental Protection Agency (EPA) and the Electric Power Research Institute (EPRI), respectively. HEM estimates public exposures to ambient air concentrations of pollutants emitted from stationary sources. AERAM calculates the risks to human health from toxics emitted from coal fired power plants. The AERAM model includes the capabilities of HEM, with the exception of the automated selection and use of population and climatological data (presently being implemented). AERAM surpasses HEM in the calculation of the source term, air dispersion modeling options, and risk.

KEYWORDS: Air emission, risk assessment, health risk

INTRODUCTION

The Human Exposure Model (HEM)[1] and the Air Emissions Risk Assessment Model (AERAM)[2] are computer models written in FORTRAN for risk assessment. HEM was developed using Systems Applications, Inc. models by EPA. AERAM was originally developed by A.D. Little[2] and modified by IWG Corp.,[3] both under EPRI sponsorship. Although the applications and the complexity of the two models are different, a comparison is made to contrast the similarities and differences for performing air emission risk assessments.

HEM is used to perform quantitative estimates of public exposure to ambient air concentrations of pollutants emitted from stationary sources and simplified risk estimates for screening purposes. Within HEM itself, there are two "sub-models," the Systems Applications Human Exposure and Dosage (SHED) model and the Systems Applications Human Exposure And Risk (SHEAR) model. The SHED model was developed to model major point sources on a large scale source category basis. SHEAR, on the other hand, was developed from SHED to handle multiple source exposure determinations for regional analysis. Additionally, the SHEAR model handles area sources. HEM is presently available only on a mainframe and is of limited modular design. Developmental versions of HEM that include additional capabilities such as SUPERSHEAR are not considered in this comparison.

New Risks
Edited by L. A. Cox, Jr., and P. F. Ricci
Plenum Press, New York, 1990

AERAM is used to estimate the risk to human health from toxics emitted from coal-fired power plants (CFPP's). The model calculates the risks using a four module computer code consisting of an emissions characterization module, atmospheric transport module, population exposure estimation module, and quantitative risk assessment module. AERAM uses EPA's Industrial Source Complex—Long Term (ISCLT) for the dispersion analysis and can model point, multiple point, line, area, and volume sources. AERAM's risk module allows for the modeling of different dose-response models fitted to animal toxicological or human epidemiological results. The AERAM code is available on both a personal computer and a mainframe computer.

COMPARISON OF METHODOLOGIES

The two models can be compared with respect to four risk analysis categories: source emissions, air transport, population exposure, and risk assessment.

Source Emissions Characterization

Both models use the geographic locations of the specific point sources. In SHED and AERAM these are input as geodetic (latitude/longitude) coordinates. While SHEAR also uses geodetic coordinates for its point source locations, SHEAR uses Universal Transverse Mercator (UTM) coordinates when placing area sources and locating the region to be analyzed. A separate computer utility is available for converting geodetic to UTM coordinates for input to the SHEAR model.

The HEM model does not perform emissions characterization, but starts with the input of annual emissions for each of the sources. The AERAM model performs emissions characterization for CFPP's based on specific power plant parameters in its Emissions Characterization Module. This Emissions Characterization Module may be modified to characterize the emissions for other sources (e.g., oil fired power plants). As in the HEM model, AERAM is capable of handling other source types. This is accomplished by bypassing the Emissions Characterization Module and entering the appropriate source data.

AERAM and HEM both require approximately the same source input parameters, which include emissions rate, stack height, stack diameter, gas exit velocity, and gas exit temperature. However, AERAM has the additional capability of handling pollutant distributions as a function of particle size.

The inclusion of pollutant distributions as a function of particle size allows the AERAM module to handle larger particles (above about 30 micrometers) without assuming that they act similarly to gasses or small particulates. The dispersion of particulates or droplets with significant gravitational settling velocities differs from that of gaseous pollutants and small particulates in that the larger particulates are brought to the surface by the combined processes of atmospheric turbulence and gravitational settling.[1] Additionally, gaseous pollutants and small particulates tend to be reflected from the surface, while larger particulates that come in contact with the surface may be completely or partially retained at the surface. As in the HEM model, the assumption that all particle sizes act as a single class may thus overestimate down wind concentrations due to higher reflection of larger particulates than may actually occur.

Air Transport

The AERAM Air Transport Module is the Industrial Source Complex—Long Term (ISCLT-5) model from the EPA UNAMAP series.[4] The HEM model incorporates a climatological dispersion algorithm similar to the EPA Climatological Dispersion Model (CDM) and, for area sources, the Atmospheric Box Model. The actual algorithms used by the CDM and the ISCLT model for plume rise, vertical variation of wind speed, and lateral and vertical plume spread are derived from the same basic equations.

Both models require the use of seasonal or annual meteorological STability ARray (STAR) data summaries. The number and type of stability categories in the STAR data used by the models are different. The HEM model uses seven stability categories: A, B, C, D-DAY, D-NIGHT, E, and F. HEM has the ability to consider chemical reaction and day-vs.-night dispersion. The AERAM model uses these categories, but combines D-DAY and D-NIGHT to form only six categories: A through F. The ISCLT model in AERAM has options for STAR data summaries with less than six categories. The HEM model requires data with 16 directions and six wind speeds, while AERAM has options to use additional data.

When using SHED, the STAR summary data can be automatically selected by HEM. This is accomplished by determining which STAR station is located nearest to the actual source. In most cases this is the best representation of the climatological conditions at the actual source location. If this is not the case, a STAR summary may be manually selected. This capability is not presently available in the AERAM model, but has been separately implemented.

One major advantage of the ISCLT air module is its ability to handle complex terrain. The HEM air transport algorithm treats all terrain as flat. The use of complex terrain adds additional realism to the modeling of plume behavior by the air transport module of AERAM. Another advantageous capability of ISCLT is its ability to handle a grid in either cartesian or polar coordinates. HEM must always use polar coordinates. Finally, the ISCLT model allows for the placement of up to 400 specific receptor locations in the grid system where they would be most helpful in the exposure analysis process (e.g., placing the receptors at the population centroids for each of the population units). The same polar grid locations (up to 480) are always calculated by HEM regardless of conditions.

With the exception of the automatic selection of STAR data summaries by SHED, the ISCLT model possesses capabilities similar to those of HEM. AERAM disperses up to 20 particle size classes.

Alternatives to the discussed models for generation of the pollutant concentrations are available. SHED and SHEAR can use concentration inputs from other models for the point source portion of the model. The capability to use separately generated ISCLT results is currently available.[1] AERAM can also use results of other air transport models.

Population Exposure

The exposure portions of both HEM and AERAM assign the concentrations determined by their air dispersion models to specific related populations. The two models use different approaches to perform this calculation. Because HEM determines its concentrations based on a polar coordinate grid pattern, spacing is non-uniform. Thus, at small radii the grid cells are much smaller than the smallest population group, and at large radii the grid cells are much larger than the population divisions. To pair the concentrations and populations HEM uses one interpolation scheme for small radii and another scheme for large radii. Populations are apportioned among the grid points at small radii, and the logarithm of the concentration-logarithm of the distance linear interpolation for population centroids at large radii. HEM automatically selects Bureau of Census data for the region near the source and assigns populations to the geographical centroids of these population units. In most cases, this automated process is advantageous and timesaving, but the way in which the population is assigned may mislocate people when the source is located near a large body of water. For small radii, if the highest concentration is located over a body of water and is used in estimating the exposure of the nearest population group, an overestimate of exposure may result.

The AERAM model allows for a cartesian coordinate system in which all grid spacing is uniform. This reduces the complexity of the interpolation process and requires only one interpolation scheme of population and concentration (if required). A polar grid system is also available in the AERAM model. The interpolation schemes for small and large radii required for HEM are not required for AERAM (even when using polar coordinates) because AERAM allows specific placement of receptors and the assignment of the associated specific populations. In practice, the receptors are placed at the centroids for each of the population units. The automated selection of appropriate population data is not presently integrated into AERAM but has been separately implemented. AERAM's automated placing of these receptors at the population centroids will reduce the potential for interpolation errors arising from the scheme used in the HEM model.

AERAM allows for the defining of subpopulations and their corresponding inhalation rates. This may be used to adjust for the activity level of these subpopulations or the difference in inhalation rates for age subgroups.

Risk

The risk determination performed by HEM uses a unit risk factor (constant) for the toxic substance analyzed. This corresponds to a linear, no-threshold dose-response relationship. In many cases, this relationship may result in an overestimate of the risk, especially at low doses. The use of generalized models for carcinogens may result in significant non-linear terms for response estimation. In such cases the estimate using non-linear models may yield very low resultant risks, whereas the use of a unit factor model may result in the need for expensive control devices to reduce the calculated risk.

When considering either safe doses or values of additional risk corresponding to certain low doses, the shape of the dose-response curve at low doses is of paramount importance.[5] It can be demonstrated that the selecting of different dose response functions can fit experimental data equally well and yet predict different results at low doses. The choice of models for low dose risk assessment should be based on more than statistical considerations alone. The choice of models containing mechanisms based on biological hypothesis may help discriminate among the various models available. The use of a unit risk factor places restrictions on how the dose response function behaves at low doses. Thus, the lack of statistical and biological agreement of the unit risk factor estimate may limit the use of the risk result and its regulatory significance.

In AERAM, several low-dose extrapolation techniques are available for extrapolation of animal toxicity data to humans. AERAM's Risk Module uses several of these statistical-biological models to relate the exposure presented to a population to the carcinogenic or toxic response of the exposed population. These models include the one-hit, multi-hit, log-probit, multi-stage dose-response functions, and the unit-risk approach. The estimation of human cancer risk from animal test data also involves a species to species conversion from animals to humans. AERAM also performs this conversion. If actual epidemiological data is available, the species conversion is omitted. The output of AERAM's Risk Module consists of human excess lifetime cancer risk estimates for the population and the maximum individual risk. There are separate risk estimates for each of the dose-response models. Confidence levels (based on the fit to the animal data) are calculated for the multi-stage model.

Functional Comparison

HEM was designed for large scale assessment of the risks involved on a nationwide basis. The HEM model is easy to use, requiring very few input parameters; it is also relatively fast to run. HEM's ability to select and use population and climatological data automatically reduces the time required to set up and perform risk assessment computations. On the other hand, a reduced number of input parameters, ease of use, and fast computation require some sacrifice in accuracy and the complexity of the methodology used. HEM's use of a simplistic final risk calculation is an example. Since both the human health and economic consequences of regulatory decisions can be enormous, it is important to apply the most scientifically valid risk assessment procedures available. HEM is really an exposure assessment model that calculates risk using the product of a unit risk factor and the exposure.

The AERAM model is capable of performing the same computations as HEM, with the exception of the built-in automated selection and use of population and climatological data. These options are presently being implemented. AERAM contains a significant number of capabilities which make it appropriate for accurate risk assessment. The modular design of AERAM allows for the replacement or enhancement of any module; its availability on a PC allows for economical use and portability.

NUMERICAL COMPARISON

A numerical comparison of the results from the AERAM and HEM models was performed. The HEM model does not contain individual modules with intermediate outputs. The comparison was made for only the final results of both models using arsenic emissions from the Mt. Tom Power Plant[6] as a sample case.

The Mt. Tom analysis was performed with actual data from the Mt. Tom coal-fired power plant in Holyoke, Massachusetts. The Mt. Tom plant burns bituminous coal from Pennsylvania containing 8.8 ppm arsenic. Emission controls for the plant consist of wet sluicing for the removal of bottom ash and cold-side electrostatic precipitation for flyash removal. AERAM computations using plant data resulted in an estimated annual arsenic output of 19.3 kg. The emissions source terms for arsenic, along with the parameters listed in Table 1, were used as input to the HEM model. AERAM requires these same parameters with other important inputs. Kowalczyk *et al*. give a complete description of the Mt. Tom Test Case and the AERAM input parameters.[6]

Table 1. Inputs to the HEM model for the Mt. Tom Power Plant

Latitude	42 20′ 00″ W
Longitude	72 37′ 00″ W
Ambient Temperature	294.26 K
Stack Height	113 m
Stack Diameter	3.05 m
Gas Exit Velocity	27.43 m/s
Temperature Gradient	
D Stability	0.000 K/m
E Stability	0.020 K/m
F Stability	0.035 K/m
Source Type	Urban
Unit Risk Factor[7]	
Arsenic	0.002912 cancers/lifetime/$\mu g/m^3$

Different inputs to HEM and AERAM were used for the meteorological data used in the dispersion analysis. Meterological data for AERAM input were obtained from an automated meterological tower at the Mt. Tom site. STAR summary data for Hartford, Connecticut, were selected and used by the HEM model. HEM did not have the option of using meterological data other than the STAR summaries presently contained in its data base.

The unit risk factor used by AERAM and HEM for arsenic is the combined result of several human studies. This unit risk factor, originally derived by the EPA,[7] was converted to appropriate units for input and is shown in Table 1. In the AERAM module calculations, several additional data sets were used. The arsenic carcinogenic risk assessment used three sets of data: two epidemiological studies and one toxicological study. The first set of data was from a study of a Taiwan village exposure to arsenic via drinking water.[8,9] The second set of data was from a toxicological study of tumorigenicity of arsenic trioxide in hamsters.[10] The third set uses epidemiological data from a study of lung cancer among pesticide workers exposed to inorganic chemicals.[11] Arsenic dose-response data for these three sets are shown in Table 2.

Table 2. Lung Cancer Dose-Response Data for Arsenic

Test Species Dose (µg/day)	Test Species Exposure Duration (weeks)	Actual Lifetime Of Test Subjects (weeks)	Dose Conversion Factor	Equivalent Lifetime Average Human Dose (µg/day)	Number of Test Subjects	Number of Test Subjects Responding
Arsenic[8,9]						
700	3500	3500	1	700	1000	290
450	3500	3500	1	450	1000	160
150	3500	3500	1	150	1000	45
0	3500	3500	1	0	1000	0
Arsenic[10]						
750	15	78	13	10069	10	3
536	15	78	13	7192	22	2
0	15	78	13	0	38	0
Arsenic[11]						
5333	1564	3650	0.429	2286	3	2
5332	1043	3650	0.286	1535	8	3
5331	521	3650	0.143	762	20	2
5330	130	3650	0.036	190	33	1
0	32	3650	0.009	0	58	1

The results of the comparison are shown in Table 3. A comparison of the arsenic runs for both models shows a consistency between the unit risk results, with the HEM estimate slightly lower than AERAM's. However, the estimated response using AERAM's various statistical-biological models and data sets shows a wide range of risk estimates. Probably the most significant advantage indicated by comparing the results is AERAM's ability to provide a broader view of the variability in the risk estimates. A single unit risk factor estimate cannot provide the analyst with much indication of the degree of confidence with which the risk is estimated. The use of several models with several data sets allows

Table 3. Arsenic Cancer Risk Results of the Mt. Tom Power Plant for the HEM-AERAM Comparison

Pollutant Data Set	AERAM Results One-Hit	Log-probit	Multi-Stage	Multi-Hit	Unit Risk	HEM Results Unit Risk
	Average Excess Individual Lifetime Cancer Risk*					
Arsenic						
Tseng[8,9]	1.7×10^{-8}	9.6×10^{-15}	1.1×10^{-8}	1.9×10^{-11}		
Ishinishi[10]	1.3×10^{-9}	1.6×10^{-18}	1.3×10^{-18}	$<1\times10^{-49}$		
Mabuchi[11]	1.1×10^{-8}	2.9×10^{-15}	1.8×10^{-9}	8.4×10^{-9}		
EPA[7]					6.8×10^{-9}	4.0×10^{-9}
	Maximum Excess Individual Lifetime Cancer Risk*					
Arsenic						
Tseng[8,9]	1.1×10^{-7}	4.6×10^{-13}	6.5×10^{-8}	2.3×10^{-10}		
Ishinishi[10]	8.0×10^{-9}	1.7×10^{-16}	1.1×10^{-16}	$<1\times10^{-49}$		
Mabuchi[11]	6.9×10^{-8}	1.4×10^{-13}	1.1×10^{-8}	5.1×10^{-8}		
EPA[7]					4.1×10^{-8}	1.4×10^{-8}

*Assumes zero background levels of arsenic inhalation for nonlinear models.

the analyst to select results based on closeness of biological mechanisms to the statistical methods used. For example, estimating dose-response of a substance for which toxicological data strongly indicate non-linear dose-response characteristics would not be best done using a linear model (such as a unit risk factor). Having several risk models available allows a range of results for decisions, including the regulatory significance of the results.

SUMMARY

The summary of the model comparison indicates that the AERAM model has more capabilities than HEM for air emission risk assessments. HEM is confined to a simplified relationship between exposure and adverse effects at an aggregate level using a unit risk factor to compute risk. The AERAM methodology includes a set of dose-response relationships along with the intermediate analysis of emissions, transport, and exposure as prerequisites to the final risk estimate. The flexibility of the AERAM code, the capability of varying the input data between each module and the ability to alter the way in which this data is processed makes the AERAM code more versatile than HEM for risk analysis. The AERAM code has been designed to allow for easy addition of new capabilities, including additional dose-response models, new transport techniques, and emissions from other types of sources.

REFERENCES

1. U.S. Environmental Protection Agency, User's Manual for the Human Exposure Model (HEM), EPA-450/5-86-001, Office Air Quality Planning and Standards, Research Triangle Park, NC, June 1986.
2. A. Q. Eschenroeder, G. C. Magil, and C. R. Woodruff, Assessing the Health Risks of Airborne Carcinogens, Research Project 1946-1, EA-4021, Arthur D. Little, Inc., Cambridge, MA, May 1985.
3. Mindware Corporation, Air Emissions Risk Assessment Model (AERAM) — User's Guide, Columbus, OH, July 1986.
4. J. F. Bowers, J. R. Bjorklund, and C. S. Cheny, Industrial Source Complex (ISC) Dispersion Model User's Guide, H. E. Cramer Co., Inc., Salt Lake City, UT, Dec. 1979.
5. P. F. Ricci, Editor, *Principles of Health Risk Assessment*, Electric Power Research Institute, Palo Alto, CA, 1985.
6. G. S. Kowalczyk, L. B. Gratt, and P. F. Ricci, Air emissions Risk Assessment Model: Benzo(a)pyrene and Arsenic Emissions from the Mt. Tom Power Plant, 85-34.4, 78th Annual Meeting of the Air Pollution Control Association, Detroit, MI, June 1985.
7. Environmental Protection Agency, Carcinogen Assessment Group's Final Risk Assessment of Arsenic, Carcinogen Assessment Group, EPA, Washington, D.C., 1980.
8. W. P. Tseng, H. M. Chu, S. W. How, J. M. Fong, C. S. Lin, and S. Yeh, Prevalence of Skin Cancer in an Endemic Area of Chronic Arsenicalism in Taiwan, *J. Natl. Cancer Inst.* **40**:453-463, 1968.
9. W. P. Tseng, Effects and Dose-Response Relationships of Skin Cancer and Blackfoot Disease with Arsenic, *Environ. Health Persp.* **19**:109-119, 1977.
10. N. Ishinishi, A. Yamamoto, A. Hisanago, and T. Inamasu, Tumorigenicity of Arsenic Trioxide to the Lung in Syrian Golden Hamsters by Intermittent Instillations, *Cancer Letters* **21**:141-147, 1983.
11. K. Mabuchi, A. Lilienfeld, and L. M. Snell, Lung Cancer Among Pesticide Workers Exposed to Inorganic Arsenicals, *Arch. Environ. Health* **34(5)**:312-320, 1979.

Comparisons Between Desktop Dispersion Calculations and Computerized Air Dispersion Model Results When Results Are To Be Used in a Risk Assessment

Carol Gevecker Graves
EA Engineering, Science, and Technology, Inc.
Sparks, MD

ABSTRACT

Air-dispersion modeling is used in risk assessment to estimate potential exposure concentrations. This paper compares two approaches to Gaussian air-dispersion calculations. The first follows Turner's Workbook approach using desktop calculations; the second incorporates a more complex computerized air model. A project to estimate the risk of developing a site containing PAH in the soil provided data for the comparison of the calculation approaches.

For the straightforward situation modeled here, the results were similar, both in terms of the distribution of the concentrations and the magnitude of the maximum concentration. Even though the cost was higher to obtain the computerized results, this approach may be justified in more complicated situations.

KEYWORDS: Air modeling, dispersion modeling

INTRODUCTION

One aspect of risk assessment is to determine the concentration of hazardous compounds in various media to which people are or will be exposed. Toward this end, samples are taken and chemically analyzed. For example, soil samples are collected and laboratory findings of a suspect compound expressed in parts per million. In an assessment of risk, these laboratory results would be used in an air dispersion model to estimate exposure concentrations under various conditions, such as during construction.

This paper briefly describes a project to carry out a risk assessment for an area under development in which the soil was suspected of containing hazardous compounds. Sampling confirmed contamination with PAH, and standard Gaussian air dispersion models were employed to estimate concentrations in the air under various construction scenarios. The purpose of this paper is to compare and contrast results obtained using desktop dispersion calculations with those obtained using computerized air-dispersion models. The paper is prepared from the viewpoint of a user of air models and not that of a developer of the models.

New Risks
Edited by L. A. Cox, Jr., and P. F. Ricci
Plenum Press, New York, 1990

THE PROJECT

The project that provided the data for this paper consists of a large urban tract in the process of being developed or redeveloped for multiple commercial uses—including shopping, hotel/apartments, parking, and open space/landscaping. For a long time, a portion of the site was used as a coal-gasification facility. A class of contaminates often found at former coal-gas sites, and in fact found here, are the polycyclic aromatic hydrocarbons (PAHs). Some of these PAHs are positive animal carcinogens and suspected human carcinogens. The organs primarily affected are the skin and the lungs.

One PAH, benzo(a)pyrene (BaP), is one of the more potent of the PAH carcinogens and is the most thoroughly studied. The unit risk factor for BaP developed according to U.S. Environmental Protection Agency (U.S. EPA) methods and based on an epidemiological study by Carnow and Meier (1973) is 4.8×10^{-5} for females and 7.1×10^{-6} for males. This risk is for a lifetime (70 year) exposure to a BaP concentration of 1 nanogram per cubic meter. Expressing this risk in another way which is more useful in understanding the exposure at this site, an exposure to a BaP concentration of 0.02-0.14 ng/m^3 over a lifetime increases the risk of developing cancer by 1×10^{-6}, the *de minimis* level of risk. (Remember that U.S. EPA's unit risk factor is an upper-bound estimate of the risk.)

PAH are naturally occurring and are produced by the inefficient combustion of hydrocarbons. Therefore, it is not surprising that they are found at former coal-gas sites. The population in general is environmentally exposed to PAH via auto and truck exhaust; oil, gas, or coal power generation; wood stoves; and barbecuing.

GAUSSIAN DISPERSION MODEL

One way to estimate possible exposures via the air, given concentrations of a contaminant in the soil from which an emission factor can be derived, is to use an air-dispersion model. These models are based on a Gaussian model of dispersion. The basic Gaussian dispersion model for the dispersion of continuous sources as suggested by Pasquill (1961) and modified by Gifford (1961) is as follows:

$$C = \frac{Q}{2\pi\sigma_y\sigma_z U}\left[e^{\frac{-y^2}{2\sigma_y^2}}\right]\left[e^{\frac{-(z-h)^2}{2\sigma_z^2}} + e^{\frac{-(z+h)^2}{2\sigma_z^2}}\right]$$

where

C = Concentration at point (x,y,z); X is the distance from the source in the downwind direction, y in the crosswind, and z in the vertical direction.

Q = Continuous uniform emission rate.

σ_y, σ_z = Standard deviations of the distribution of the concentration, C, in the crosswind and vertical directions.

U = Constant mean wind speed.

h = Height of plume centerline.
= Stack height + plume rise.

The σ's and the 2π term in the denominator together with the exponential terms in the brackets of this equation illustrate the derivation of the model from the Gaussian or normal distribution.

For this project, an emission rate for BaP of 1.1 milligrams/m^3 was used. This emission rate was derived considering the contaminated area under construction, an estimate of the dust generated during surface grading (the major dust-generating construction activity for this project), the projection of respirable particles (defined here as those particles less than 10 microns in diameter), and the concentration of carcinogenic PAH in these respirable particles as determined by laboratory analysis. The derivation of this emission rate is outside the scope of this paper. However, the same emission rate was used in both the desktop calculations and the computer model.

DESKTOP CALCULATIONS

Methods and applications for using the air dispersion model described in the previous section are presented in the *Workbook of Atmospheric Dispersion Estimates* developed by D. B. Turner of U.S. EPA (1970). The workbook presents computational techniques that use the Gaussian model to estimate atmospheric concentrations of contaminants from several types of sources. It suggests a method for simulating an area source such as that encountered in construction activities as a virtual point source. Estimates for σ_y, σ_z, and U in the equation can be derived from tables and figures in the workbook. For this exercise, the following values were derived according to the workbook and used in the dispersion equation:

Q = emission rate of 1.1 mg/m^3
σ_y = crosswind standard deviation of 21.8 m
σ_z = vertical standard deviation of 4.5 m
U = average wind speed of 5 m/s
z = o, i.e., concentration will be estimated at ground level
y = o, i.e., concentration will be estimated directly downwind
h = effective stack height of 10 m

This last assumption of an effective stack height of 10 meters assumes that the finer, respirable particles that are modeled here will be churned up by the surface grading activities and suspended at approximately 10 meters at the virtual point source. These particles will then disperse downwind.

The use of downwind (y=o) and groundlevel (z=o) receptors greatly simplifies the model equation so that it can easily be evaluated on a hand-held calculator. In fact, because the calculations are repetitious, a programmable calculator or a simple computer program makes them even easier to use.

Figure 1 presents the results of applying the values listed above in the dispersion equation, i.e., the calculated groundlevel, downwind concentrations. Concentration on the vertical axis is expressed in nanograms per cubic meter. The horizontal axis is the distance in meters downwind from the virtual point source. The maximum ground level concentration of 150 ng/m^3 occurs approximately 200 meters downwind from the source.

Recall the 1×10^{-6} probability of developing cancer from a lifetime (70 year) exposure to 0.02-0.14 ng/m^3 of BaP. For risk assessment purposes, this can be translated to a 1×10^{-6} increased probability from a 1-year exposure to 1.4-9.8 ng/m^3. It is possible that a 6-month or 1-year exposure from the construction scenario elevates the cancer risk to an area of concern. There should be a consideration of who might be exposed, for how long, and at what concentration.

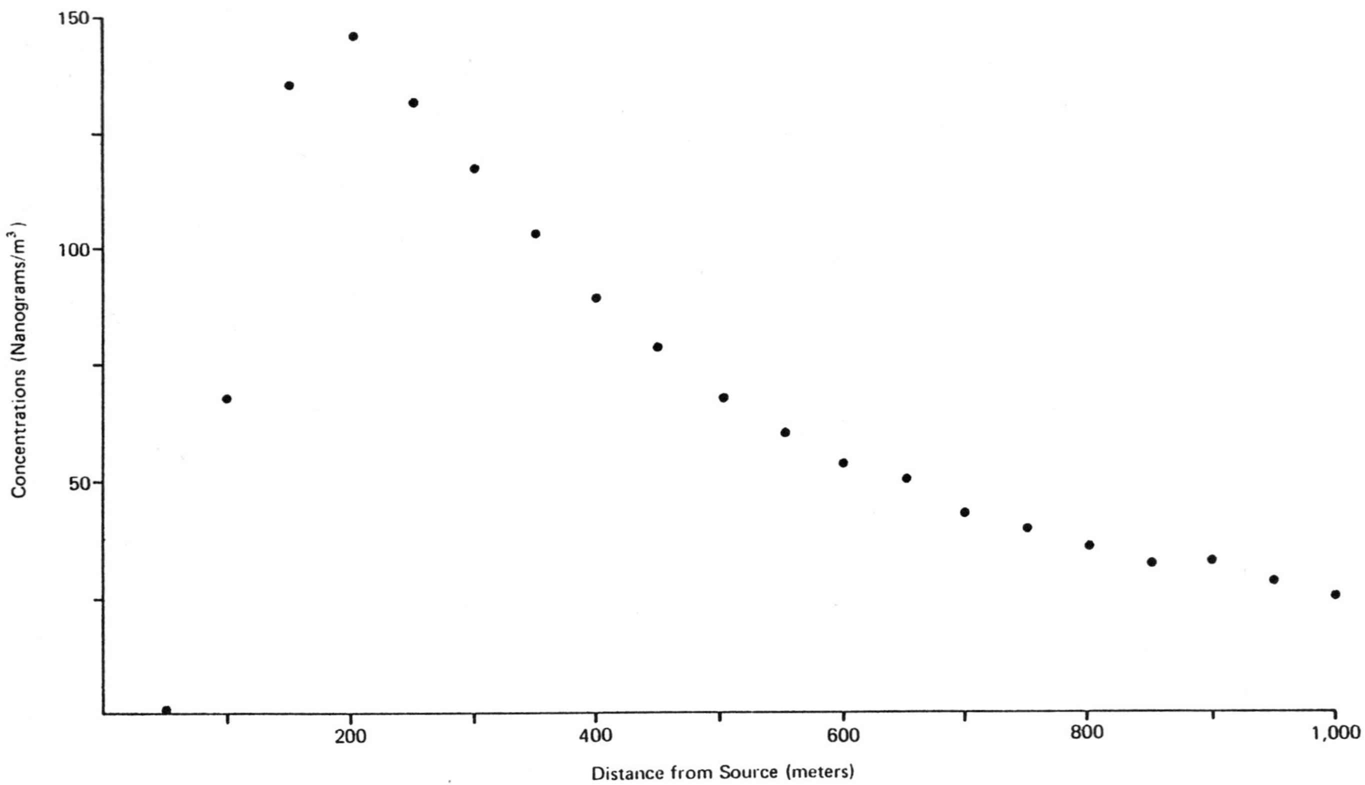

Fig. 1. Ground level, downwind concentrations resulting from desktop calculation of air dispersion model.

The concentrations shown in Fig. 1 are the downwind concentrations, and the wind blows in all directions as illustrated in Fig. 2. This wind rose illustrates the proportion of time the wind blows from each of the 16 compass points to the opposite compass point. For example, the wind at this site (based on 8-24 observations per day during 1975-1977) blows 0.025 of the time (2.5%) from east to west. The wind blows 0.107 of the time (10.7%) from west to east. The wind blows 0.081 (or 8.1%) of the time from north to south and approximately the same proportion (0.085 or 8.5%) from south to north. At this site, the wind blows relatively little from the east and much more regularly from the other compass directions. These wind rose data were obtained from the National Climatic Center (NCC) in Asheville, North Carolina, at a cost of $6.

If the downwind concentrations shown in Fig. 1 are distributed to the 16 compass directions according to the proportions shown in Fig. 2, then the pattern in Fig. 3 emerges. This figure was generated by using the Gaussian equation to calculate concentrations every 50 meters. Each concentration was multiplied by each directional proportion to obtain concentrations every 50 meters in each of the 16 compass directions. These concentrations were plotted using the SAS procedure PLOT to produce a contour plot. Boundaries were then drawn around areas of like symbols, i.e., areas of like concentrations.

As expected from the wind rose in Fig. 2, the PAH are distributed to the north, east, and south. Combining hand-calculator calculations from Turner's workbook with a simple distribution of these concentrations among the 16 wind directions, it was found that the areas of maximum concentration are within 150-250 meters of the source in the expected direction. This simple approach also shows that the maximum concentration is in the range of 13-16 ng/m^3.

COMPUTER MODEL

Can this estimate of concentration and picture of the distribution be improved upon using a large computerized model? To address this question, we used the Climatological Dispersion Model (CDMQC) developed for U.S. EPA (1973). This model is a predecessor of the Industrial Source Complex Long-Term Model (ISCLT) which is the current large computerized air dispersion model-of-choice for complex industrial sources (U.S. EPA, 1979).

In order to compare results, the same emission rate of 1.1 mg/m^3 and effective stack height of 10 meters were used with the computer model. The CDMQC also incorporates the joint frequency distribution of 16 wind directions, 6 wind speeds, and 6 stability classes for a total of 576 combinations of the occurrence of wind direction, wind speed, and stability class. These joint frequency data are analogous to the wind rose data used previously and can be obtained, as was the wind rose data, from the National Climatic Center. A computer printout of these data costs $24; the data on computer tape, along with mixing height and other data, are available for approximately $100. Constant estimates for mixing heights (average afternoon mixing height of 1,000 meters, average nocturnal mixing height of 100 meters) were obtained from Holzworth (1972), and mean annual temperature (12°C) was obtained from NCC (1981). The CDMQC could also have incorporated multiple point and area sources for multiple compounds, decay half-life for the modeled compounds, and background concentrations of the compounds.

The results of the computer modeling are shown in Fig. 4. Contours were not drawn in this figure as in Fig. 3, because the pattern was not as distinct. Areas of the lowest concentration were interspersed with areas of the next-to-lowest concentration. However, the general pattern shown in this figure is the same as that derived from the calculator results.

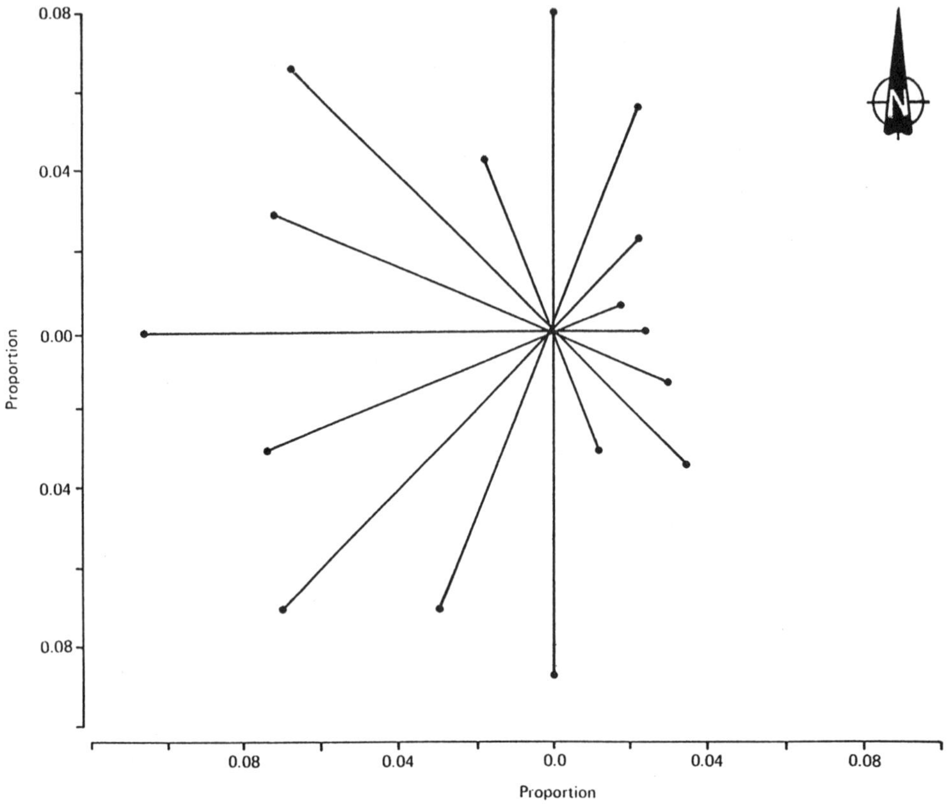

Fig. 2. Wind rose, 1975-1977. Radials show proportion of time wind blows from a direction to the opposite compass point.

COMPARISONS OF METHODS

In both Figs. 3 and 4, the areas of maximum concentration are to the north, east, and south and are within 200 m of the source. According to the computer model, the maximum concentration is 25.5 ng/m^3. The maximum concentration derived by combining the hand calculator results with the wind rose data was 15.5 ng/m^3. The two methods of calculation present the same picture and give numeric results which are essentially equivalent for the purpose of assessing the risk from a construction scenario.

A comparison of costs points out some differences between the two approaches. The wind rose data for the hand calculations cost \$6 versus \$24 for the wind speed/wind direction/stability class joint frequency data required for the computer model. The cost of the meteorologic data for the computer model would be even greater if more extensive data were requested on computer tape. The expenditure of personnel time was approximately equal for the two approaches. There was a difference in cost of computer resources. The hand-calculation approach did entail some computer costs for programming repetitive calculations and for plotting the results. The total cost was under \$100. The cost for running the computerized model, however, was just over \$600 on a large mainframe time-sharing system. This was the cost of running an already compiled version of the model.

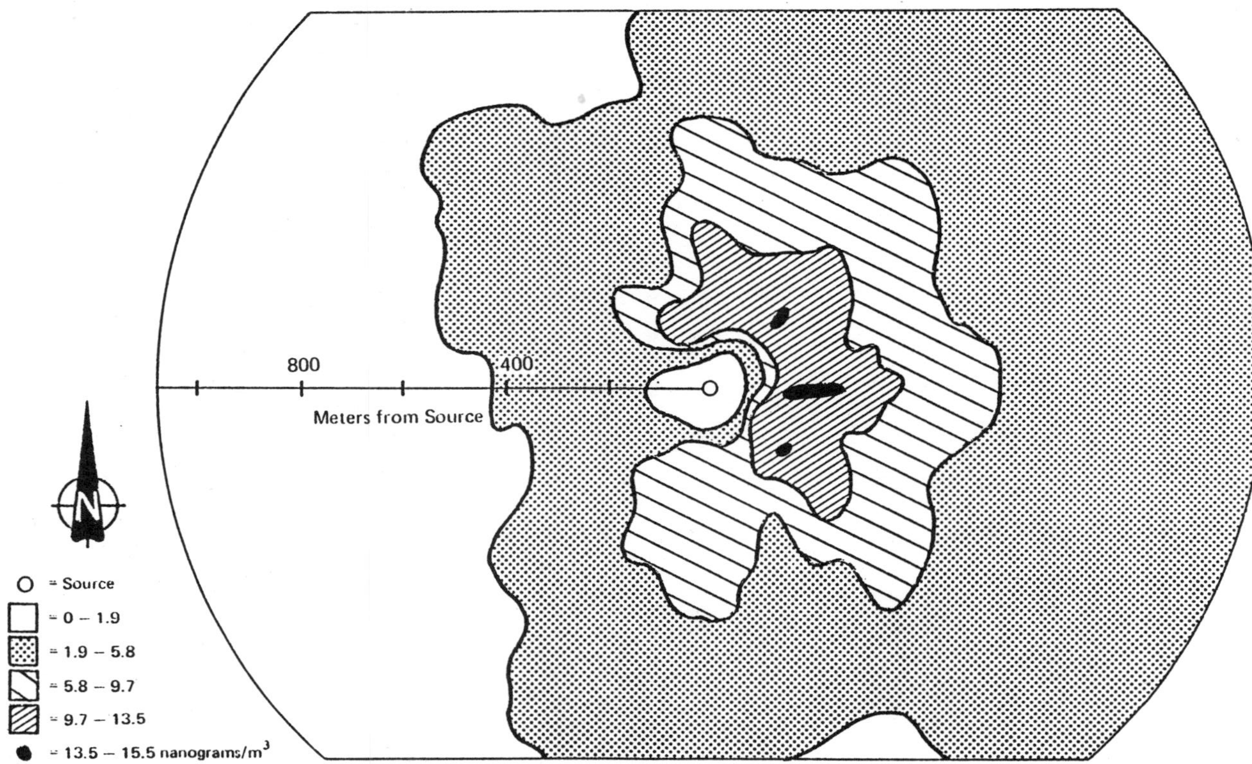

Fig. 3. Contour plot of desktop calculation results.

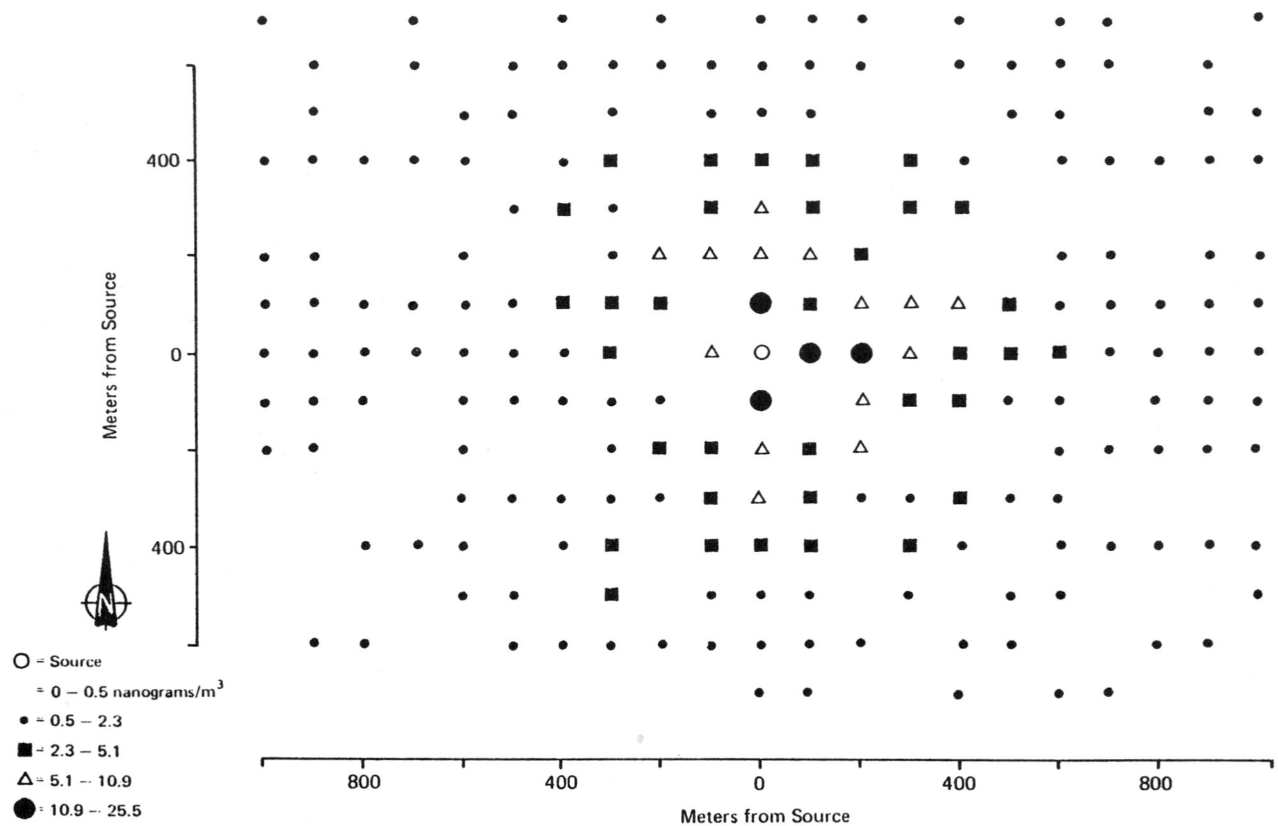

Fig. 4. Dot plot of computer model results.

CONCLUSIONS

For the purposes of this project, the hand calculation approach is sufficient. For the risk assessment being done here, the differences in the two methods are inconsequential. Both methods point to the same areas of concern and provide estimates of concentration which are similar. The hand calculations estimate the location and magnitude of areas of maximum concentration to a precision which is adequate for making judgments to manage the risk.

Would the more complex computerized model ever be useful in this type of assessment of risk? If the computer model were less costly to run, then the confidence provided by the computer model results might be worth additional cost. In fact, the cost of using the model is now reduced, because the model is available on an in-house computer where modeling options can be explored without incurring excessive project costs.

Aside from cost considerations, a computerized air-dispersion model is a necessity to estimate exposure concentration in a complicated situation. A computerized model would be necessary if there were multiple sources, especially if there were both area and point sources. Another situation calling for computer air modeling would be the presence of multiple hazardous compounds. The computer model can handle such aspects as decay factors, if these are important considerations. If the scenario dictated high emission rates, potentially greater health effects could result. In this case, one might want to buy the confidence provided by a computerized model, which incorporates a more complete meteorologic picture than that provided by the wind rose. However, if meteorologic data are missing, or if there is an emergency situation and results are needed quickly, then these results show that one can have confidence in the workbook approach.

REFERENCES

Carnow, B. W., and P. Meier, 1973, Air Pollution and Pulmonary Cancer, *Arch. Environ. Health* **27**:207-218.

Gifford, F. A., 1961, Uses of Routine Meteorological Observations for Estimating Atmospheric Dispersion, *Nuclear Safety* **2(4)**:47-51.

Holzworth, G. C., 1972, Mixing Heights, Wind Speed, and Potential for Urban Air Pollution Throughout the Contiguous United States, AP-101, Office of Air Programs, U.S. Environmental Protection Agency, Research Triangle Park, North Carolina.

Local Climatological Data Annual Summary for Newark, 1981, National Climatic Center, National Oceanic and Atmospheric Administration, Asheville, North Carolina.

Pasquill, F., 1961, The Estimation of the Dispersion of Windborne Material, *Meteorol. Mag.* **90(1063)**:33-49.

Turner, D. B., 1970, *Workbook of Atmospheric Dispersion Estimates*, Office of Air Programs, U.S. Environmental Protection Agency, AP-26, Research Triangle Park, North Carolina.

U.S. Environmental Protection Agency, 1973, User's Guide for the Climatological Dispersion Model, EPA-R4-73-024, Office of Research and Development, U.S. EPA, Research Triangle Park, North Carolina.

U.S. Environmental Protection Agency, 1979, Industrial Source Complex (ISC) Dispersion Model, User's Guide, Vol. I, EPA 450/4-79-030, Office of Air Quality and Planning Standards, U.S. EPA, Research Triangle Park, North Carolina.

Livermore Risk Analysis Methodology: A Structured Decision Analytic Tool for Information Systems Risk Management

Sergio B. Guarro
Lawrence Livermore National Laboratory
Livermore, CA

ABSTRACT

Computer and information systems have only recently emerged as subjects of attention for the performance of in-depth risk analyses. Grave episodes in the recent years have proven to the technical community and to the general public that very large losses can be sustained in both the financial and national security arenas as a result of the threats that can affect such systems.

The complexity of these systems and the importance of the information processed through them requires thorough analysis and systematic evaluation of the measures that are implemented, or that can be undertaken, to protect their value to the owners and users. The early approaches taken for the assessment of information systems risk were, however, either qualitative in nature or based on largely unstructured models.

The Livermore Risk Analysis Methodology (LRAM) is an attempt to supply the analyst with the tools needed to model logically and evaluate quantitatively the types of threats and consequences that are characteristic of complex information systems. It also seeks to provide managers and decision makers with explicit elements of judgment, such as risk reduction effectiveness, cost efficiency, and others, by which individual controls and protective measures can be a) related to risk scenarios, b) evaluated, c) chosen or discarded, and d) prioritized within a given budget framework.

The important characteristics of the LRAM methodology, and its setup for the evaluation of both monetary and non-monetary information systems risk components, are presented and discussed. Examples and discussion of critical implementation issues are also given.

KEYWORDS: Information systems, risk analysis, risk management, computer security

INTRODUCTION

The use of information systems has expanded in the last two decades at an unprecedented rate in all sectors of economic and social activity. As our dependence and reliance on the computers, networks and auxiliary devices that constitute such systems

New Risks
Edited by L. A. Cox, Jr., and P. F. Ricci
Plenum Press, New York, 1990

increases, so does our vulnerability to service disruptions and compromises that can affect them. It is thus not surprising that information systems risk assessment and risk analysis methods have found increased acceptance and application as tools for the security evaluation of these systems. The principal goal in the use of these tools is to develop an understanding of the nature and the severity of the risks to which owners and users of information systems are subject.

In the government sector the initial incentive for application of risk analysis techniques to computer systems was originated by a specific requirement[1] set in 1978 by the Office of Management and Budget (OMB) for all federal government agencies. This requirement has been recently reaffirmed and updated by OMB,[2] with useful qualifications reinforcing the concept that the management aspects of information system security remain of great importance. The private sector has also been implementing security assessment techniques similar or related to risk assessment at various levels of depth and sophistication.

Risk assessment methods vary in nature and depth. Their application to the evaluation of information security issues should be decided on the basis of their capability to provide justifiable and documentable answers to practical questions concerning the design and implementation of security controls in specific information systems. When prudently applied, risk analysis provides an objectively based approach to the problem of assessing and managing risk. As a decision making and risk assessment tool, it is not only capable of identifying potential losses that would be unacceptable for a given system, but it can be used to determine which specific security controls and countermeasures can be effective and cost justifiable. The Livermore Risk Analysis Methodology (LRAM) was developed to cover these objectives in a balanced and comprehensive way. After an initial discussion of some of the more important general issues concerning the application of risk analysis techniques to the information systems security field, this paper examines the LRAM models and procedures, from both the technical and risk management perspectives.

RISK MANAGEMENT, RISK ASSESSMENT AND RISK ANALYSIS

It is in the very nature of large and complex systems, whether they are based in hardware like engineering plants or in combinations of hardware and software like information systems, to be subject to multiple varieties of risk.

Risk results from the combination of the intrinsic value of an object with the effect that the materialization of any of the threats possibly applicable to such an object may have. This translates into a usually undesirable and sometimes extremely dangerous value loss potential for the owner of the object at risk. Because of the very complex nature of the systems to which consideration of risk applies, it is in most cases not realistic to respond to the existence and discovery of risk by seeking its total elimination. The originating sources of risk often reside beyond our reach, so that they cannot be simply eradicated altogether. Also, a complex system is usually subject to such a wide and varied spectrum of potential threats that a total elimination of the resulting risk would require allocation of an infinite amount of resources. The decision maker and system manager has thus to think in terms of risk management rather than in terms of risk elimination. In practice this means that he or she will have to:

a. understand the nature and severity of the various types of risk to which his system may be subject;

b. determine what level of risk can be considered acceptable in the environment in which the system operates, in view of the amount of resources that are expendable, either in the form of risk-induced losses or in the form of allocations for security controls and safeguards;

c. reduce, by selection, application and implementation of appropriate and cost effective controls, the existing risk to a level that can be identified as being acceptable or at least justifiable.

Risk management can be considered to be the broadest objective of a decision maker who needs to protect the valuable information assets in a system assigned to his administrative responsibility. To manage risk, the decision maker must understand, at least in a general sense, the security features of his system, and receive for this purpose appropriate input from analysts employing applicable methods of technical investigation. Among these is risk assessment. This term is today mostly used to indicate the analytical activities by which the nature of the threats potentially affecting a system and the severity of the consequences that may result from them are investigated and evaluated. The closely related term risk analysis is employed in a more technical context, usually to indicate the more detailed procedures employed by specialists to dissect risk into its more specific constituents, e.g., the relations between threats, assets, damage or loss consequences and countermeasures.

QUALITATIVE AND QUANTITATIVE ANALYSIS APPROACHES

In information systems risk analysis frameworks, it is generally recognized that risk is originated by the impact of threats, i.e. potentially harmful agents and events, on assets, i.e. the valuable constituents of the information systems themselves. It is also recognized that, in order to protect assets from threats, it may be necessary to introduce in a system sets of security measures and devices specifically designed for such a purpose. These measures and devices are referred to as controls, or, sometimes, safeguards.

If there is general consensus on the aforementioned concepts and definitions, dissent still exists on how to perform information systems risk assessments and risk analyses. Most of the controversy centers on whether risk assessment should be based on qualitative or quantitative techniques, and on what level of analytical sophistication can be considered adequate for the job.

To decide on this issue we should recall that the main purpose of risk analysis is to provide input to the decision maker so that he or she can effectively manage the risk affecting a given system. Thus, it is within the typical scope of an information system risk analysis to identify and assess the existing potential exposures to system damage and/or service disruptions. However, in accordance with the concepts of risk management that we have discussed in the previous section, the principal objective lies a few steps beyond this. Indeed, after the discovery of any serious potential danger, the administrator of an information system will normally seek means of reducing the extent of the damage that the system can possibly suffer. Faced with cyclical budget realities and administrative constraints, he or she will also be interested in determining how such a reduction of damage can be achieved in a justifiable and cost effective way.

Many information systems are large and complex because of the number of users and contained applications, or because of the presence of networking utilities which link them to other systems. Because of their complexity, meaningful security analyses of these systems cannot be carried out by means of unstructured approaches. Whenever the stakes in terms of potentially sufferable losses are high, more rigorous and formal analytical approaches are recommended. A recognition of this rationale can be found in the U.S. Office of Management and Budget (OMB) Circular Letter A-130,[2] which establishes information systems risk analysis guidelines for the Government sector:

"... Agencies shall establish and maintain a program for the conduct of periodic risk analyses at each installation to ensure that appropriate, cost effective safeguards are incorporated into existing and new installations . The objective of a risk analysis is to provide a measure of the relative vulnerabilities and threats to an installation so that security resources can be effectively distributed to minimize potential loss. Risk analyses may vary from an informal review of a microcomputer installation to formal, fully quantified risk analysis of a large scale computer system. ..."

The long lasting debate on whether information systems risk analysis should be performed by qualitative or quantitative means often obscures more important, although related, issues. For example, above and beyond the application of different possible evaluation philosophies and procedures, a stronger commitment on the part of risk assessment developers and users should be made towards basing analyses and reasoning on models and measures that have been objectively defined. This should be regarded by risk analysts as a primary goal, since only by realizing such a commitment will they be able to fulfill the fundamental risk analysis goal of communicating important findings to decision makers in a documentable, understandable and unambiguous way. In this view it is hardly useful to communicate to a manager that the "vulnerability" of a system is "high", if one has not provided an operative definition of "vulnerability", nor an indication of how the qualifier "high" should be objectively interpreted. Loosely defined risk parameters and qualifiers generate inconsistent interpretations and fail to provide a generally accepted yardstick for ranking, in absolute or relative terms, the risk variables of interest (such as possible losses, or effectiveness of implementable security controls).

For a concrete illustration of the above assertion, the reader is invited to consider a situation in which a risk analyst would be trying to elicit, from different system administrators in a financial organization, information regarding potential loss damages that could be induced by a certain type of threat. Administrator "A", who oversees a $150,000/year budget, assesses the possible losses as "high", whereas Administrator "B", who manages a $1,000,000/year budget, assesses the loss potential in his system as "medium". Could we expect these qualitative evaluations of "high" and "medium" to be consistent? We suspect not, since each administrator would be probably making his or her assessment with a bias derived from the amount of resources that he or she is accustomed to deal with under normal daily circumstances. Some consistency would be likely to be produced, however, if the two assessors had been advised in advance to use a rating of "high" for losses approximately falling, for example, above $100.000, a rating of "medium" for losses approximately between $10,000 and $100,000, and a rating of "low" for losses below $10,000.

A second order of considerations in favor of more formal analysis takes into account a type of induced benefit that should be no less important in the eye of a responsible information system manager than those linked directly to the choice and implementation of cost effective security measures. This benefit stems from the substantially increased level of understanding of the functions and inner workings of an information system that the performance of a thorough and structured analysis procedure, as required in a formal assessment, can bring to the analyst(s). This benefit is all the more significant for more complex systems and for organizations that make provisions for their system operation personnel to get involved in the security analysis activities, either as actual performers or more simply as reviewers. Companies in the private sector, traditionally sensitive to considerations more on the financial and monetary side, should not underestimate this "system knowledge benefit" and the increased level of organizational awareness that it may produce. Analyses based on unstructured methods often cannot be effectively reviewed and generate little know-how within the sponsoring organization.

GENERAL CONCEPTS OF QUANTITATIVE RISK ANALYSIS

Quantitative risk analysis can provide a systematic and documentable approach for the management and evaluation of information systems security. For this reason it should be considered the prime assessment tool for situations in which very valuable assets and important decisions are involved and a thorough support basis for these decisions is needed.

As a tool for both evaluation of information systems and effective communication of evaluation results to peers or managers, a quantitative risk analysis should be based on models and parameters whose definitions are objective and not dependent on context.

In general terms, quantitative risk analysis can provide the framework within which analysts can determine the magnitude of system security risk, evaluate the effectiveness of control options, and justify and budget system security expenditures. This requires the estimation of risk by means of quantifiable measures. A commonly adopted one is the rate of occurrence of undesired consequences (losses) per unit time, but other measures can and should be used, for example, to reflect aversion to single-occurrence large potential losses, as will be discussed in one of the following sections. Undesired consequences are defined by the scope of the analysis, and for information systems generally involve direct monetary losses, loss of the ability of the system or facility to fulfill its mission, disclosure, deletion, or alteration of important information, denial of service, and any other undesired effects on the organization operating the facility. Consequences are examined and quantified with respect to system assets, including hardware, software, information and personnel.

To determine the occurrence of consequences, initiators that have a direct adverse effect on system assets must be identified and the effectiveness of controls that prevent and/or mitigate these consequences must be analyzed. Initiators and their paths to each asset define the nature of the various system threats and are usually quantified using historical data and/or subjective judgment on their frequency of occurrence. The analysis of controls entails the determination of the probability of their failure and the cost effectiveness of additional controls or modifications.

One of the arguments that are frequently voiced against the use of quantitative risk analysis in the information systems security field is that reliable "actuarial" data on frequencies (i.e., probabilities of occurrence in any given unit of time) of specific scenarios and on the magnitudes of the sustained losses are not available for input into the developed risk models. Risk analysts must often use figures and values derived by application of their subjective estimates and judgment, and/or the opinion of experts in the field.

While it is desirable to use data from actual observations whenever possible, the use of subjective estimation in the quantification of probabilistic models is a natural means to express an uncertain state of knowledge about the associated events or phenomena, resulting from the scarcity or total unavailability of such data. Admittedly subjective estimation of parameter values to be fed into an objectively based model supporting a decision making process is preferable to basing decisions on arbitrary factors and subjective decision schemes, be they qualitative or quantitative. The use of an objectively based model will eventually lead to the collection of more precise and repeatable data for validation of the initial results obtained from it. However, if a reasonable model is not developed, the initial uncertainty on the risk scenarios of interest will never be reduced. Contrary to a common belief, the estimation of probabilities and other parameters reflecting uncertain knowledge does not have to be based on hard statistical data at all cost. The assessment of a probability reflects a degree of belief that a certain event will eventually occur. This degree of belief may or may not be initially supported by actual statistical evidence. The presence of such supporting evidence and its quality and quantity obviously affect the level of confidence that one can place in a given estimate, but its absence or

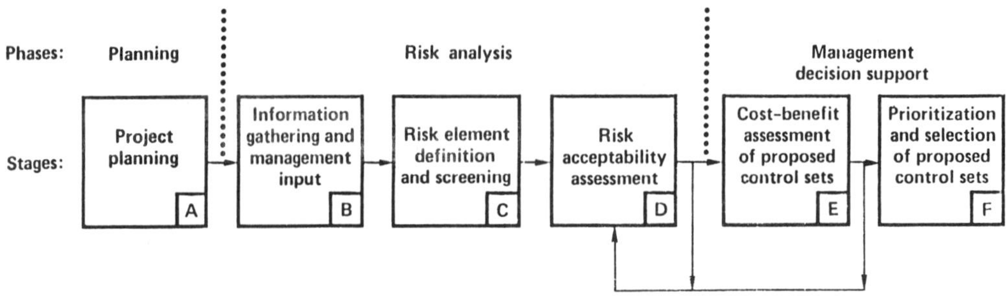

Fig. 1. Simplified LRAM process diagram.

scarcity should not preclude the estimate from being given and used with the appropriate cautions.

THE LIVERMORE RISK ANALYSIS METHODOLOGY

The Livermore Risk Analysis Methodology (LRAM)[3] was commissioned for development by the U.S. Air Force Logistics Command (AFLC) to the Lawrence Livermore National Laboratory (LLNL) in early 1985 and essentially completed in its fundamental lines by the end of that same year. Although principally meant to serve as a comprehensive decision making and risk management tool for the security design and administration of large integrated information systems, LRAM was also designed to be intrinsically flexible. Its application can therefore be tailored to the size and complexity of the system being analyzed and to the level of detail and resolution at which the output results are desired. LRAM has also been developed in the form of a PC-based and menu-driven software package, called ALRAM (for Automated LRAM). A general overview of the methodology is given in this section; a somewhat more detailed discussion of its features and procedures is contained in the next section.

Figure 1 presents in flowchart form the sequence of procedures in the LRAM approach. The principal phases of LRAM are indicated at the top of Fig. 1 as "Planning," "Risk Analysis," and "Management Decision Support." These three phases are in turn composed of six stages: (1) project planning (PP), (2) information gathering (IG), (3) risk element definition and screening (REDS), (4) risk acceptability assessment (RAA), (5) cost-benefit assessment of proposed control sets (CBA), and (6) prioritization and selection of proposed control sets (PS). The discussion of phases and stages given in the following is only meant to provide a high level understanding to a first time reader. The complete explanation of steps and procedures for execution of LRAM is documented in full detail in Guarro *et al.*, 1985. A comprehensive application example is also provided in that same reference document.

The project planning stage of LRAM is not significantly different from any other planning activity for relatively complex analysis projects entailing the performance of serial and parallel multiple tasks. An explicit discussion of this is therefore not provided here.

The actual LRAM analytical activities start out at the "information gathering" stage. The main objective of the analyst in this stage is to take inventory of the assets (hardware, software and data) composing the information system of interest and to make preliminary determinations regarding both their monetary value (e.g., replacement cost) and their non-monetary value attributes, such as criticality, classification and sensitivity levels. Other important information gathering activities include the listing and categorization of asset loss consequences and potential threats to assets. Some of these activities are for practical reasons procedurally executed in those later LRAM stages where the gathered information is actually processed and utilized (for example, in REDS or RAA).

In the "risk element definition and screening" (REDS) stage, individual risk scenarios, called in LRAM "risk elements" are fully identified and evaluated in terms of their potential severity. The definition of risk elements is performed in steps, with an initial focal point provided by the list of important system assets identified in the IG stage. "Generic risk elements" (GREs) must be first identified in REDS by determining which generic types of consequences may result from the loss or compromise of an asset, and which generic threat types and threat-to-asset paths can be considered applicable to each individual asset of interest. Generic risk elements can be identified and characterized by defining their constitutive components and attributes. The components of a risk element are the asset (or assets, if more than one can be affected by the same threat agent), the threat and threat-to-asset path, and the consequences (both direct and indirect) of the threat event. An example of GRE would thus be human intentional threat [generic threat] by direct physical attack [generic threat-to-asset path] to mainframe computer [asset] causing direct monetary loss and service interruption [generic consequences].

GREs constitute categories which will normally "contain" many "specific risk elements" (SREs) fitting the same generic description. They serve the purpose of structuring the analysis in an orderly way and hierarchically organizing the construction of the individual SRE scenarios that are actually used in evaluating system risk. As an example of SRE fitting the GRE category given above, saboteur [specific threat] pours water [specific threat-to-asset path] onto mainframe [asset] electric circuits causing electrical shorts [specific direct consequence]; unit cannot be repaired before two weeks and important applications must be transferred to other processing unit causing considerable service delays [specific indirect consequences].

Clearly, this SRE example is not the only one that can be identified within the given GRE type. The analyst has to develop other scenarios befitting the same generic risk element description, until he or she is satisfied that the set of SREs thus identified is in fact a reasonably comprehensive one, leaving, to the best of his or her knowledge, no significant scenarios unexplored within the given GRE category.

The assessment and evaluation part of the risk analysis is initiated by assigning a maximum possible loss (MPL) estimate to each SRE after the GRE and SRE definition activities have been completed. At various points in the process, those risk elements involving assets that are not material in value or whose loss consequences are not significant are screened out from further analysis. This keeps the analytical effort at a manageable level and focuses the attention of the analyst on those contributors that can produce the largest losses.

In the "risk acceptability assessment" (RAA) stage, security controls are identified and associated with the applicable risk elements. A security control is most effective when its probability of failing while being challenged by a threat is very small: control failure probabilities are thus taken into account to estimate a "loss potential indicator" (LPI) value, which models the reducing effect of controls on the previously identified and estimated SRE maximum potential losses. The effectiveness of controls which already exist in the system is evaluated by comparing the loss potential indicator value that can be obtained for a given SRE with a pre-defined "risk acceptability threshold." The selected threshold value is intended to reflect the opinion of management (in the organization performing the risk analysis) on the loss level that can be considered acceptable for each single manifestation of one specific kind of threat. If a risk element does not satisfy this risk acceptability test, it is retained in the analysis to identify additional security controls which could make it acceptable. Proposed additional controls are also evaluated in the RAA stage by applying to them and the associated SREs the same criteria used for evaluating existing controls.

In essence, the RAA stage concludes the "risk analysis" phase of LRAM with the identification of initially unacceptable SREs and of those "proposed controls" which, if applied, can make them acceptable. Acceptable SREs are risk scenarios whose loss potential in the face of the materialization of a specific threat does not exceed a management-defined threshold of acceptability. In the ensuing "management decision support" phase, the focus of the analysis moves from "risk acceptability" to "cost-benefit assessment" and "prioritization and selection of proposed controls."

The "cost-benefit assessment" (CBA) stage is centered around the evaluation of a "cost-benefit ratio" for each proposed control via a comparison of annualized benefits and costs. The "prioritization and selection" (PS) stage addresses the problem of prioritizing the choice and implementation of proposed controls in view of yearly budget allocation cycles and constraints. This is achieved by deriving for each control being considered a set of prioritization indices, which may include the cost-benefit ratio itself and/or other factors characterizing the attractiveness and desirability of a given control, for instance, its expected life and degree of flexibility (i.e., ability to be effective against multiple types of threats). A global prioritization scheme is finally developed and applied by assigning weights to the individual indices and combining them into one index whose values determine the final relative order of preference for all the proposed controls.

DISCUSSION OF THE LRAM MODEL AND BASIC PROCEDURES

As was stated in the previous section, the combination of a threat initiator, its propagation path, the affected assets, the possible resulting consequence and the applicable controls is defined in LRAM as a "risk element" (RE). The complete list of all credible and possible risk elements would describe the total "spectrum" of risk to the facility, system or subsystem being analyzed. Because a full description and characterization of total risk is not necessary nor practical, the LRAM approach does not seek to derive a total risk measure, but rather focuses on the risk produced by individual risk elements involving the occurrence of single event losses.

The construction and use of a comprehensive and structured risk model like the one employed in LRAM requires the performance of two complementary activities, namely, model definition and model evaluation. The order in which these activities were just now mentioned could also be the order of execution in an actual application procedure. In reality this is not practical, as the model developer usually needs to perform intermediate evaluations of certain portions of the model simply to decide how much further definition and development is warranted. In LRAM, sorting and screening of constituent components of the risk elements and certain combinations of such components is the mechanism used to optimize the definition/evaluation process and keep the analysis at a sufficient but manageable level of depth throughout the development process. By implementation of this mechanism the analysis can be kept focused on those risk elements of highest importance, i.e., having the highest perceived risk. The organization operating the facility being analyzed is the primary source of data for determining the sorting and screening levels.

Figure 2 shows the essential components of the LRAM risk model in both descriptive and quantitative representation terms. The descriptive part shows the logical relations between the threat initiator (e.g., a human attacker, such as a saboteur, or a natural agent, such as an earthquake), the potential target assets (e.g., computer and communication hardware, software or data), and the consequences that could result if the threat agent

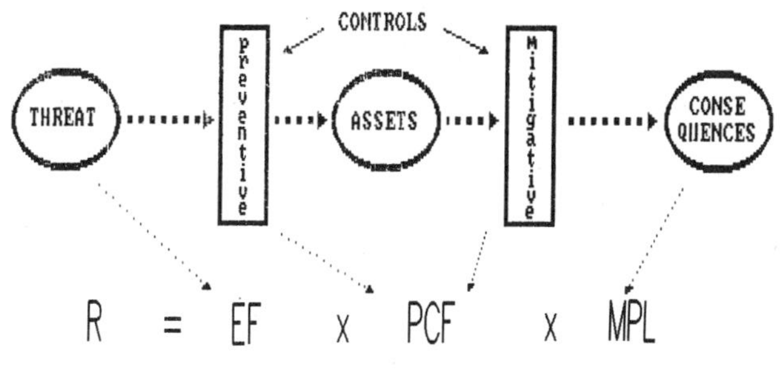

Fig. 2. Basic LRAM risk model.

reaches the assets (e.g., destruction of hardware, disclosure of information, loss of functionality, etc.). These logical links are in LRAM referred to as the "propagation path" from threat to asset to consequences. The existing security controls (sometimes also referred to as "safeguards" by security analysts) are supposed to check and contrast the progression of the threat initiator on this path: "preventive controls" are placed on the path from threat to assets and have the intended function of preventing the threat agent from affecting the assets in any significant way; "mitigative controls" are positioned along the path from asset to consequences and have the intended purpose of limiting and minimizing losses in those cases in which the preventive controls fail to keep the threat from affecting the assets.

The qualitative model just described can be quantified using the formula shown in the lower portion of Fig. 2, and which is repeated below:

$$R\,[REi] = MPL\,[Ci] \times PCF\,[PMCOi] \times EF\,[Ti]\,.$$

The formula indicates that an annualized measure of the risk (R) resulting from the *i*th risk element (REi) can be derived as the product of the "maximum potential loss" (MPL) that can be estimated to result from the unmitigated consequences (Ci) of the threat on the assets, times the probability of control failure (PCF) of the combined set of preventive and mitigative controls (PMCOi), times the "expected frequency" (EF) of the threat (Ti) — this last quantity often being also referred to as a "probability per year."

If the value indication contained in MPL is measured on an "equivalent dollar" scale, R in the formula above is expressed in the units of an annualized, equivalent-dollar loss expectancy. Special provisions are made in the LRAM procedures to combine in the equivalent-dollar measuring scale loss-value indications that are both monetary and non-monetary in nature. Examples of non-monetary loss values are the value of national security information disclosed to an unfriendly agent or the embarrassment caused by a hacker advertising a breakthrough into a supposedly secure system.

The definition of a risk element starts in LRAM with the identification of system assets functionally definable as a recognizable entity and with the definition of the worst credible consequences that could result if these assets were destroyed or otherwise compromised. Assets are then sorted into those having only monetary value and those that are critical, sensitive and/or classified. Those assets with direct monetary value only are screened at a materiality level. Monetary assets whose value exceeds the materiality level are retained in the analysis. In parallel to the asset screening process, consequences types are screened for their importance. Consequences that are not deemed important to the organization operating the facility or system of concern are eliminated from further consideration. This mechanism allows the flexibility to accommodate for different management and organization philosophies: for example, the significance of public embarrassment changes widely from organization to organization and from the government to the private sector. For the assets and consequences remaining in the analysis, the possible threats and their paths are identified to establish a risk element. A first form of risk element evaluation takes place at this stage, assuming that the threat has occurred and that no controls are present, in order to estimate a "maximum potential loss" (MPL) for the risk element. This estimate takes the form of a severity class value ranging from 0 to 6. Risk elements are then screened out if their MPL value is below a threshold established by management of the organization operating the information system being evaluated.

For each of the remaining risk elements, the applicable current controls are identified and their failure probability estimated. The MPL is then combined with the probability of control(s) failure (PCF) to establish a loss potential indicator (LPI) for each risk element. This parameter is also initially derived in severity class form, although its mathematical expression is actually given by the formula:

$$\text{LPI [REi]} = \text{MPL [Ci]} \times \text{PCF [PMCOi]}\ ,$$

corresponding to the first two factors in the complete risk model evaluation expression presented in this same section and in Fig. 2.

Both the concept of "loss potential indicator" and the procedures by which risk elements are screened for "acceptability" have been already discussed from a qualitative point of view in the methodology overview section. Those risk elements with an LPI less than a specific threshold defined with direct input from the system or organization management are eliminated from further consideration, as they do not need to be addressed for improvement beyond the current level of security.

The unacceptable risk elements (those with LPI equal to or greater than the threshold) require that new controls or control upgrades be proposed and new LPI values determined. For each upgraded RE the new LPI is subjected to the same screen as the original LPI. If it is found acceptable, the risk element with the new proposed controls is passed into the cost-benefit analysis (CBA) stage; otherwise, an iterative procedure is necessary until new controls or control upgrades are identified that will produce an acceptable LPI value for that risk element, or it is determined that no controls can be presently identified for that purpose.

New controls that have passed the "risk acceptability" test, in the sense just described above, are next evaluated in the CBA stage in terms of their cost-effectiveness. This analysis requires an estimate of the incremental cost of implementing the new or upgraded control(s) beyond the cost of the current controls, and an estimate of the reduction in risk over the initially unacceptable risk elements due to the application of the new or upgraded controls. A cost-benefit ratio (CBR) is developed by dividing the reduction in risk by the incremental cost for the new or upgraded controls. This ratio is then compared to a threshold value to determine if the new or upgraded controls are acceptable. A value equal to or somewhat greater than unity should normally be chosen for this CBR threshold, unity being the "break even" point indicating that a given control upgrade is expected to save one equivalent dollar worth of risk for every dollar spent on its installation and operation. If the controls being considered do not meet the chosen cost effectiveness threshold criterion, different new controls are proposed and the process of LPI acceptability and cost-benefit analysis is repeated to identify a better set or configuration of upgraded controls.

It should be noted that the calculation of a CBR value for a given set of upgraded controls requires the calculation of a risk reduction value. This implies the evaluation of the formula given above for the risk element risk value R [REi], first in the "baseline" configuration reflecting the current controls (CC), and then in all other configurations reflecting possible different sets of proposed controls (PC). The risk reduction for each proposed set of controls, referred to in the following as DR, is the difference between the baseline R value and the applicable R values for the upgraded configurations. In mathematical formulation:

$$DR\,[REi;PCj] = R\,[REi;CC] - R\,[REi;PCj],$$

where the notation REi;PCj is used to refer to the *i*th risk element with the *j*th set of proposed controls, and similarly "REi;CC" means the *i*th risk element with the set of current controls. The CBR value for the *j*th proposed control set is then calculated, according to the definition given above, as:

$$CBR\,[PCj] = \frac{\text{Sum over all i's}\,(\,DR\,[REi;PCj]\,)}{\dfrac{\text{Initial Cost}\,[PCj]}{\text{Expected Life}\,[PCj]} + \text{Annual Cost}\,[PCj]}\,.$$

It may be worthwhile to note again the iterative nature of the process for selection and evaluation of new proposed controls. This process is greatly facilitated by the use in the LRAM quantitative model of a specific parameter, i.e., the probability of control failure PCF, which measures, by a relation of inverse proportionality, the functionality and effectiveness of the controls associated with a particular risk element. More detailed LRAM procedures, not discussed here, are set to model the actual logical layout — e.g., in series and/or parallel logical blocks — of controls which concurrently apply to one risk element. A special section of the LRAM model is also dedicated to allowing the handling of mitigative controls (which unlike preventive controls operate on loss consequences rather than on the threat itself) in a standardized fashion fully coherent with the general model discussed here.

The new or upgraded controls that have an acceptable CBR are prioritized, selected and budgeted for installation. The prioritization process considers several indices chosen by the organization operating the facility in order to select and budget those new or upgraded controls that will best fulfill their security concerns. A brief example of possible indices to be considered in the prioritization scheme was given in the previous section. Such indices are combined via a weighted product to yield a global prioritization index (GPI) which is ultimately used to establish a preferred implementation order of the proposed controls. The first N proposed controls in the prioritized list that can be actually installed within the funding level of the presently available budget are selected and chosen for immediate implementation. Controls with lower application priority are delayed until more funding becomes available or can be negotiated on the basis of the risk analysis results.

RISK ELEMENT SCREENING CRITERIA

The reader may have noted that throughout the presentation given to this point the screening procedures intermixed with the LRAM model definition and evaluation processes are essentially based on three distinct kinds of criteria. These correspond to the three types of evaluation measures actually used in the screening, namely (1) loss value and consequence importance, (2) loss potential indicator value, and (3) cost-benefit ratio.

The first type of screening occurs in the REDS stage and is done first on monetary values and consequence "importance," then again at the end of the stage, once the asset and consequence components of the risk elements have been more formally defined and an MPL (maximum potential loss) value has been correspondingly evaluated. The purpose of these screens is to eliminate from the analysis those risk elements that cannot cause a single loss occurrence of significant severity, and to concentrate from the beginning the attention of the risk analyst on those that can. High loss per single occurrence events pose the most worrisome kind of risk to any organization, because in some cases they can produce damage from which it may be very difficult to recover.

The loss potential indicator (LPI) is used as the parameter for evaluation of the "risk acceptability" of a risk element, or, in other words, to determine whether the current set of controls applicable to a specific risk element is adequate or must be upgraded. A first evaluation of the upgrades is also based on the same criterion. Use of the LPI rather than the R parameter (i.e., the LRAM risk parameter expressing expected losses per year) for this judgment means that a decision on the adequacy of current and proposed controls is not going to be influenced by the estimated values of the expected frequency (EF) parameter which appears in the R formulation (given before in this section). This choice was explicitly made because, of the parameters employed in the LRAM model, EF is probably the one subject to the largest estimation uncertainty.

A second specific reason for basing the risk acceptability screening on LPI is that this choice guides the analyst into a conservative (i.e., risk averse) handling of risk scenarios with potentially very severe, though possibly infrequent, loss consequences. For example, assume that a risk element presents a maximum potential loss value (MPL) of $1,000,000, and that the current controls have an aggregated probability of failure (PCF) of 0.5. This yields an LPI value of $500,000, meaning that every time the threat initiator associated with that risk element manifests itself, that amount is the equivalent dollar loss that in the average can be expected. The high value of this loss calls the analyst's attention to the severity of the given scenario; if a reasonable risk acceptability threshold value had been chosen within the analysis framework, this risk element would probably be a good candidate for security control upgrades. Assuming now that the expected frequency of the associated threat (EF) were 0.001/yr (i.e., once in a thousand years), the resulting annualized risk measure, R, for the given risk scenario would then be an expected loss of $500 per year. If an acceptability decision were to be based on this figure alone, the estimated risk would be likely to be judged insignificant and not requiring application of additional security controls. A judgment of this kind may, however, give insufficient weight to the fact that the losses expected to result from the outlined scenario are not going to occur smoothly at the rate of $500 per year, but will be sustained in large $500,000 amounts, at average expected intervals of one thousand years. Due to the random nature of the event, there would actually be no guarantee that the loss could not occur the very next year (or month, or day). Thus, a prudent decision maker may want to see if the loss sustainable by one occurrence of the initiating threat could be reduced to a level more "survivable" for his or her organization. An actual $500 dollar loss event with an expected frequency of one occurrence per year would clearly not pose the same sort of "survivability" problem.

SEMI-QUANTITATIVE AND NON-MONETARY ESTIMATIONS

Until the end of the Risk Analysis phase of LRAM, one only needs to perform order of magnitude estimations. Up to that point the risk evaluation procedures are thus performed in a qualitative or, using perhaps a more accurate definition, "semiquantitative" fashion. Consequence "severity classes," corresponding to discrete ranges of values, rather than to exact values, are thus used. For those system assets whose loss value cannot be directly expressed in monetary (dollar) terms special verbally described, non-monetary consequence severity classes are used. These "non-monetary severity classes" have ranges that parallel the discrete dollar ranges used for consequences amenable to dollar estimation, and allow the analyst to convert all types of consequences to the same measuring scale (e.g., "severity class" discrete rankings, from Class 0 to Class 6). This is done by implementing equivalence criteria between monetary and non-monetary consequences that, like other important inputs to the LRAM analytical process, must be defined by responsible managers in the performing organization in order to assure consistency and documentability of the decision process.

Since a greater level of resolution is needed in the ensuing "Management Decision Support" phase, the semi-quantitative "severity class" approach for risk estimates is there replaced by a quantitative, point estimate based procedure which produces evaluation of consequences in "equivalent dollar" terms, as well as explicit estimations of threat frequencies and more precise control failure probability values. A complete quantification is a much smaller burden at this point of the analysis, since much of the previously generated data can be reutilized and the number of risk elements that are found unacceptable (and therefore carried through the final stages of LRAM) will usually be only a fraction of the initial comprehensive risk element list.

FINAL COMMENTS AND CONCLUSIONS

We argued at the beginning of this paper that quantitative risk analysis can be used very satisfactorily as a management tool in the process of making decisions related to the security of medium to large information systems. The analytical means and procedures for the achievement of this objective are available to the knowledgeable information security analyst, not only for the identification of problem areas where great risk potential exists, but also for the identification of the remedies and of the implementation strategies for these remedies in terms of budgeting and prioritization.

The Livermore Risk Analysis Methodology was developed to provide such a complete management tool. Although its full procedures may appear complex to people unfamiliar with risk assessment, the LRAM utilizes a very intuitive and simple risk model, based on parameters that directly and objectively relate to and illustrate the risk contributions of individual risk scenarios. In the risk model definition and evaluation, the portion of the model itself that relates to the effectiveness of security controls is explicitly and distinctly identified. This allows the designer or analyst of an information system to have direct analytical access to the one factor that he or she can most easily affect among those determining the ultimate magnitude of the risk.

The LRAM practical applicability and user productivity is greatly enhanced by the "ALRAM" automated version, which runs on personal computers and faithfully applies the methodology framework. To further help the user, the LRAM documentation includes sets of "generic" data—i.e., data not specialized for particular systems. It should be understood, however, that, as in almost all other computer security compilations, the nature of these data is "soft," as they largely represent expert opinion and approximate estimates. Data collection is probably the area where a concerted effort should be made in the future by all proponents and users of information systems risk analyses. Validation and comparison of the data at an aggregated level are possible now that complete and rigorous risk models are starting to be developed. It seems appropriate for the members of the information systems security and risk analysis communities to share and evaluate applicable databases in a more open and organized fashion than in the past.

REFERENCES

1. Office of Management and Budget Circular A-71, Transmittal Memorandum No. 1, July 27, 1978.
2. Office of Management and Budget Circular A-130, Dec. 12, 1985.

3. S. Guarro, A. Garcia, C. C. Wood, and P. G. Prassinos, LRAM-Livermore Risk Analysis Methodology for Information Systems Security, LLNL Report UCAR-10150, Dec. 1985.

LAVA: A Conceptual Framework for Automated Risk Analysis[a]

S. T. Smith
Los Alamos National Laboratory
Los Alamos, NM

J. J. Lim
Lim and Orzechowski Associates
Walnut Creek, CA

J. R. Phillips and R. M. Tisinger
Los Alamos National Laboratory
Los Alamos, NM

D. C. Brown and P. D. FitzGerald
U.S. Government
Washington, DC

ABSTRACT

At Los Alamos National Laboratory, we have developed an original methodology for performing risk analyses on subject systems characterized by a general set of asset categories, a general spectrum of threats, a definable system-specific set of safeguards protecting the assets from the threats, and a general set of outcomes resulting from threats exploiting weaknesses in the safeguards system. The *L*os *A*lamos *V*ulnerability and Risk *A*ssessment Methodology (LAVA) models complex systems having large amounts of "soft" information about both the system itself and occurrences related to the system. Its structure lends itself well to automation on a portable computer, making it possible to analyze numerous similar but geographically separated installations consistently and in as much depth as the subject system warrants. LAVA is based on hierarchical systems theory, event trees, fuzzy sets, natural-language processing, decision theory, and utility theory. LAVA's framework is a hierarchical set of fuzzy event trees that relate the results of several embedded (or sub-) analyses: a vulnerability assessment providing information about the presence and efficacy of system safeguards, a threat analysis providing information about static (background) and dynamic (changing) threat components coupled with an analysis of asset "attractiveness" to the dynamic threat, and a consequence analysis providing information about the outcome spectrum's severity measures and impact values. Each sub-analysis can be simplified or made complex, depending on the sensitivity and relative worth of the subject system. Personnel at the subject site see only an interactive questionnaire eliciting from them data about the presence and quality of the safeguards, the potential consequences of a successful threat, and the target organization's preference structure—the technical expertise is built into the model (and the computer code) itself. LAVA yields quantitative *and* qualitative insights: a pair of values (monetary and linguistic) express loss

a. Work performed under the auspices of the U.S. Department of Energy.

New Risks
Edited by L. A. Cox, Jr., and P. F. Ricci
Plenum Press, New York, 1990

exposure for each threat/asset/safeguards-function/outcome quadruple. Using LAVA, we have modeled our widely used computer security application as well as LAVA/CS systems for physical protection, transborder data flow, contract awards, and property management. It is presently being applied for modeling risk management in embedded systems, survivability systems, and weapons systems security. LAVA is especially effective in modeling subject systems that include a large human component.

KEYWORDS: Vulnerability assessment, automated risk analysis, threat analysis, hierarchical multilevel systems, LAVA

INTRODUCTION

LAVA (Los Alamos Vulnerability and Risk Assessment Methodology) is an original approach to risk management developed at the Los Alamos National Laboratory. It is a systematic methodology for assessing vulnerabilities and risks in complex safeguards and security systems. The associated LAVA software[1] is part of a research effort to provide tools to identify vulnerabilities and risks in large, complex systems whose modeling is generally intractable by other methods. LAVA is being implemented as a set of computer programs that run on a widely used class of personal computers; these programs operate on a variety of application system models and are executed in a team environment. Each LAVA application's implementation is designated as LAVA/XX, where XX is an identifier associated with the specific application.

Users are not required to be expert risk analysts to use LAVA/XX—that mathematical and analytical expertise already exists as a part of the methodology's software system. Expert knowledge about the structure and characteristics of safeguards and security systems is a part of the specific application model. The only knowledge required of users is information about that which they know best: their own facility, organization, assets, equipment, policies, procedures, and security practices. The LAVA software system elicits this information by means of automated questionnaires[2] administered to evaluation teams whose members have diverse backgrounds and responsibilities. LAVA/XX generates both general reports for management and detailed reports for operations staff from information obtained in the questionnaires.

The subject systems to which the LAVA methodology can be applied are massive, complicated systems characterized by a large human component, by large bodies of imprecise data (very little "hard" information and enormous quantities of "soft" information), and by often indeterminate events (events that may or may not have happened, or, if they happened, may not have been detected). The outcomes resulting from threats exploiting system vulnerabilities are often of a catastrophic nature, defying quantification in ordinary terms.

The methodology makes use of hierarchical multilevel systems theory, event-tree-like analysis, fuzzy set theory, decision theory, utility theory, knowledge-based expert-system theory, and natural language processing. The methodology gives both qualitative and quantitative insights into the vulnerabilities in the system of safeguards, yields an accurate picture of the state of the subject organization's safeguards system, and produces both qualitative and quantitative expressions for the system's loss exposure (risk).

The LAVA methodology has been applied successfully to modeling vulnerabilities and risks in computer-security systems,[3-6] plant control-room operation,[7] computer security for nuclear safeguards,[8-9] contract control systems, transborder data-flow systems,[10] and Government-property control systems. It is presently being applied for modeling risk management in embedded systems, in survivability systems, and in weapons systems security.

DEFINITIONS OF TERMS

Within the context of LAVA, some terms have a specialized meaning.

ASSET	An item or category of items having some intrinsic value to the subject organization; assets are acted upon by threats, leading to outcomes.
HARD DATA	Data or information that can be represented easily in quantitative terms, like a valve failure rate or the yearly average number of times that a known event occurs.
IMPACT	The consequence, cost, or effect upon the subject organization of an outcome of severity "X" resulting from a threat successfully exploiting a safeguards function vulnerability for a particular asset; the impact is given as a pair of values, one monetary (economic) and one non-monetary (linguistic).
LOSS EXPOSURE	The risk to the subject organization of a threat successfully exploiting a safeguards function vulnerability to produce an outcome of some calculated degree of severity; given as a pair of values, one monetary and one non-monetary.
OUTCOME	The (usually undesirable) event that occurs when a threat successfully exploits a vulnerability in a safeguards function for a particular asset. (Often wrongly confused with threat, for example, "the threat of total destruction;" total destruction is actually an OUTCOME.
OUTCOME SEVERITY	A measure of how successful the threat is for this specific outcome or how much damage occurs. Outcome severity is a function of the relative weakness of a safeguards function and the relative strength of the threat.
SAFEGUARDS	Policies, procedures, physical or logical devices, and so forth designed to prevent the undesirable outcomes by protecting the assets from the threats.
SAFEGUARDS FUNCTION	The functional representation of the protective mechanism that a safeguard or set of safeguards is intended to achieve.
SOFT DATA	Data or information that is difficult or impossible to quantify, like the value of a human life or of a highly classified document, or whether a subtle event has occurred or has been detected.
SUBJECT SYSTEM	The system or universe upon which the assessment is being performed (the SUBJECT ORGANIZATION is responsible for the subject system).
THREAT	A person, force of nature, thing, or idea (or category of same) posing some danger or menace to the assets; a threat is an active force (or actor) and is not to be confused with OUTCOME.
VULNERABILITY	A weakness or flaw in a safeguards function, for example, in a security system or procedural system, that can be exploited by a threat to cause harm to an asset or set of assets.

LAVA'S TECHNICAL BASIS

The LAVA methodology is a structured, modular systems approach to risk management. For any specific application, this approach consists of four phases: modeling the system, gathering the information necessary for LAVA's analysis, determining the subject organization's loss exposures from potentially three separate analyses, and "solving" the problem of risk management in terms of action.

The Modeling Phase

For a subject system, LAVA evaluates system vulnerabilities, analyzes the consequences of vulnerability exploitation, and calculates the set of loss exposures of the subject organization resulting from the consequence set. There are well-defined steps that we must take when modeling an application for LAVA. First we identify the subject system, delineate the characteristics of that system, and specify the scope of the analysis. Next we define the system assets, the set of undesirable outcomes, and the threats to the assets. Then we consider all the ways that the threats can interact with the assets so we can understand which of the outcomes might result from these interactions and what safeguards functions must be in place to protect the assets from the threats. We then determine what factors might affect the outcome's severity and what subfunctions will determine the performance level of the safeguards functions. We then design interactive questionnaires to model the specifics of the subject system.

Assets are items of value to the organization that must be safeguarded against harm or compromise. Assets include (but are not limited to) real property, equipment of all kinds, documents, personnel, information, reputation, the ability to do business, and anything else of value, including the organization itself. Assets possess properties affecting their value, such as sensitivity, criticality, compromisability, theftworthiness, timeliness, and so forth. In modeling assets for a LAVA application, we create asset categories—categories that treat similar assets as one, such as human-readable information (information in the form of reports, letters, computer-terminal displays, plots, or other human-readable form) or machine-readable information (information stored in coded form in a computer or word processor or on storage media for such machines, requiring a machine to translate the coded information into a form that can be understood by a human).

Threats are active forces posing some danger or menace to the assets. Threats can be people (insiders or outsiders), forces of nature, things, or ideas that can cause the assets harm. Again, in modeling a LAVA application's threat, we use broad threat categories, such as natural hazards, on-site humans, and off-site humans. Threats can be treated as always lurking in the background and not changing very much (the static or background threat) or, in the case of very sensitive applications in which the subject organization has access to certain intelligence information, threats can be treated as having two parts—a static component and a dynamic component changing with time. We will discuss the concept of dynamic threat further in the analysis section.[11]

We define an asset space, $\underline{A}$ and a threat space, $\underline{T}$ to describe the organization's assets and the threats to them, having in mind as we do this a specific set of undesirable outcomes. The asset and threat spaces comprise the specific categories modeling a specific application system:

$$\underline{A} = \{a_1, a_2, ..., a_i\}$$

$$\underline{T} = \{t_1, t_2, ..., t_j\} .$$

We consider the threats and assets in [threat, asset] pairs,

$$\underline{T} \times \underline{A} = \{t_1a_1, t_1a_2, ..., t_1a_i, t_2a_1, ..., t_ja_i\} ,$$

so that we can break down in a systematic way the kinds of threat-asset interactions that are possible for the application system. The kinds of possible interactions when coupled with the set of undesirable outcomes, represented as

$$\underline{O} = \{o_1, o_2, ..., o_q\} ,$$

determine what the safeguards function set should be.

The relationship of threats, assets, and safeguards functions is illustrated in the hierarchical multilevel disaggregation structures[12] (Figs. 1 and 2) taken from LAVA's application to computer security. In all LAVA models, there is a separate hierarchy for each threat; the top level is the threat, the second level contains the assets, and the bottom level lists the safeguards functions for each threat-asset pair. In the computer-security application, there are three threat categories: natural or random hazards, on-site humans, and off-site humans. There are four asset categories: the facility or physical plant, computer-related hardware, machine-readable information, and human-readable information. In this example application, there are six outcomes making up the set of undesirable outcomes that the total set of safeguards functions are designed to prevent: unauthorized access or use, modification or tampering, damage or destruction, disclosure, theft, and denial of use. Figure 1 is the hierarchy for the natural hazard threat; the assets are taken as a single target because the natural hazards cannot differentiate among the assets in terms of attractiveness. Figure 2 is the hierarchy for the on-site human threat; here the threat agent is able to differentiate among the assets and, hence, is able to mount a selective attack.

A safeguards system comprises a set of mechanisms protecting the assets from the threats. Such a system includes physical measures such as guards, fences, and dogs; mechanical things such as locks and security cradles; procedural controls such as rules, guidelines, and standard operating procedures (SOPs); electronic devices such as monitors, sensors, alarms, and closed-circuit television (CCTV); and technical measures such as passwords, intelligent cards, shielding, and so forth. Safeguards can be thought of as specific controls and countermeasures relative to each [threat, asset, safeguards function, outcome] quadruple.

Safeguards functions perform the objective of protecting the assets from the threats. We define the safeguards function space relative to the [threat, asset, outcome] triples wherein each [threat, asset] pair has a set of safeguards functions whose purpose is to thwart the threat-asset interactions and prevent the undesirable outcomes from occurring. This relationship can be represented as

$$\underline{F}(\underline{T} \times \underline{A}) = \{F_{ij1}, F_{ij2}, ..., F_{ijk}\} ,$$

where each of the k safeguards functions for the ith asset and jth threat is performed by a set of safeguards subfunctions that can be represented as

$$F_{ijk} <=> U_m (f_{ijkm}) ,$$

in which the symbol <=> means "is defined to be" and the symbol U_m is the union over m.

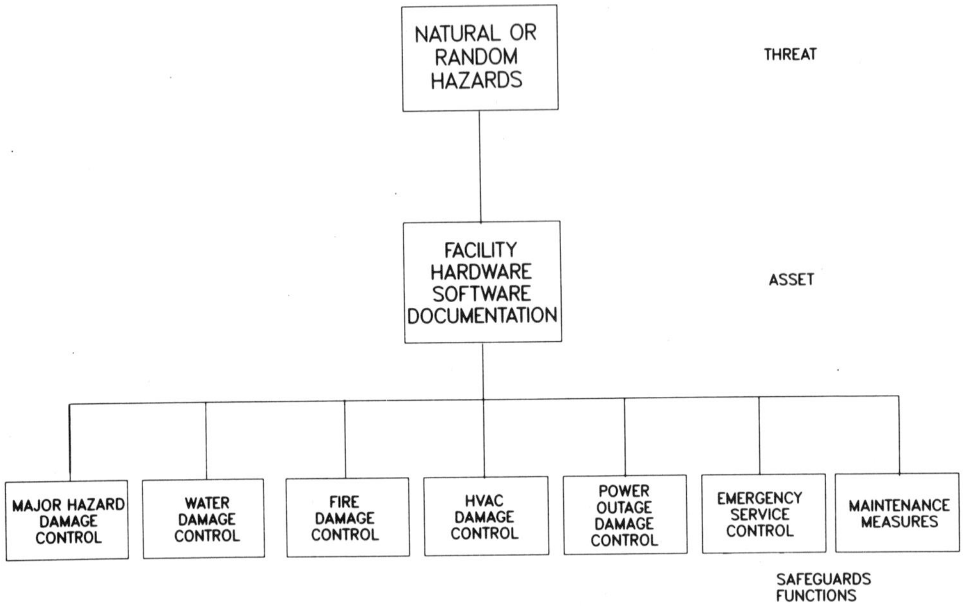

Fig. 1. Risk assessment structure for natural hazards threat.

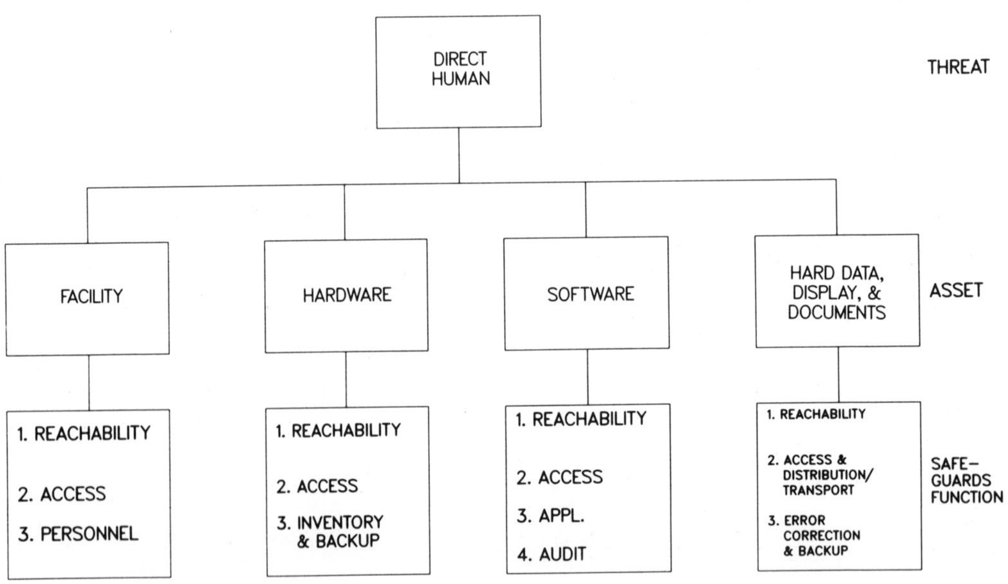

Fig. 2. Risk assessment structure for on-site human adversary.

Each safeguards subfunction is made up of elements that determine performance adequacy and can be represented as

$$f_{ijkm} = \Sigma_n \, (e_{ijkmn}) \, ,$$

and each element is further composed of attributes, α_{ijkmnp}, determining element completeness and additional information, i_{jkmnp}, about the elements and the attributes, represented as

$$e_{ijkmn} = \Sigma_p \, (\alpha_{ijkmnp}, i_{jkmnp}) \, .$$

The elements and attributes are, in fact, specific safeguards or counter-measures.

	Unauthorized Access or Use	Modification or Tampering	Damage or Destruction	Disclosure	Theft	Denial of use
Natural Hazards – Facility	0	1	1	0	0	1
Natural Hazards – Hardware	0	1	1	0	0	1
Natural Hazards – Software	0	1	1	0	0	1
Natural Hazards – Documents/ Displays	0	1	1	0	0	1
Direct Human – Facility	1	1	1	1	1	1
Direct Human – Hardware	1	1	1	1	1	1
Direct Human – Software	1	1	1	1	1	1
Direct Human – Documents/ Displays	1	1	1	1	1	1

Fig. 3. Outcome possibility matrix.

An outcome possibility matrix[13] relates the [threat, asset] pairs and the outcome set. (One can think of the outcome possibility matrix as a fuzzy matrix.) The values in the matrix represent the degree of possibility that the outcome in question could occur as a result of an interaction of the specific [threat, asset] pair. One can obtain the values from a fairly complex analysis, or one can simply assume that the outcomes are all either possible (a possibility degree of unity) or not possible for the potential threat-asset interactions (a possibility degree of zero). For most applications, the simpler case is more than adequate. Figure 3 shows the outcome possibility matrix for the computer-security application example we have been using throughout this paper.

Each outcome for each [threat, asset, safeguards function] triple can occur in varying degrees of severity. Outcome severity is a function of the relative weakness of the safeguards function (a performance measure of the safeguards function) and the relative strength of the threat. In the case where the dynamic threat is not analyzed separately, the threat strengths are assumed to be equal and unity. Outcome severity is expressed as a degree of membership in the fuzzy set of outcome severity, but it can also be translated into a linguistic descriptor to be used with the non-monetary measures for impact and loss exposure.

We also define a set of impacts to measure the effect that an outcome of some specific degree of severity would have upon the subject organization, where the outcome resulted from some threat-asset-vulnerability interaction. Impacts are given as pairs of values, one expressed in monetary terms for those components of impact that can be measured in financial terms, and one expressed in non-monetary or linguistic terms for those components of impact that are better measured another way. The impact set is represented as

$$\underline{I}_{ijkq} = \{I(M)_{ijkq}, I(L)_{ijkq}\},$$

which defines the consequences for the [threat, asset, safeguards function, outcome] quadruples.

Risk, or potential loss exposure, results from the interaction of three components: the threat component, measuring the relative potential strength of threat agents in producing a specific outcome by exploiting a safeguards function vulnerability; the vulnerability component, measuring the relative potential weaknesses in the safeguards functions; and the consequence component, measuring the relative potential costs of specific outcomes. A pair of loss exposures is given for each [threat, asset, safeguards function, outcome] quadruple, and is expressed as

$$\underline{R}_{ijkq} = \{R(M)_{ijkq}, R(L)_{ijkq}\},$$

where

$$R(M)_{ijkq} = f(V_{ijk}, S_{ijk}, O_{ijkq}, I(M)_{ijkq})$$

$$R(L)_{ijkq} = f(V_{ijk}, S_{ijk}, O_{ijkq}, I(L)_{ijkq})$$

in which V_{ijk} is a measure of the relative weakness of the *k*th safeguards function for the interaction of the *i*th asset and the *j*th threat; S_{ijk} is a measure of the relative strength of the *j*th threat against the *k*th safeguards function for the *i*th asset; O_{ijkq} is (usually) 1 or 0, indicating that the *q*th outcome can occur for the interaction of the *i*th asset, *j*th threat, and *ik*th safeguards function; and $I(M)_{ijkq}$ and $I(L)_{ijkq}$ are the monetary and non-monetary impact measures for the [threat, asset, safeguards function, outcome] quadruple.

The risk pairs can be aggregated to whatever level is desired. More aggregation provides a "bottom line" for upper management, along with a coarse indication of where the bottom line came from. However, less aggregation provides more detailed and specific information to those whose job it is to improve security and the overall risk posture. In general, we think that more aggregation has a tendency to lose information and to smear out important results; hence, we think that less aggregation is better for most purposes.

The Information-Gathering Phase

The information-gathering phase acquires the data for LAVA to operate upon by means of automated, interactive questionnaires. Information is collected about the organization's mission, its assets and the potential threats to its assets, its environment, the safeguards (or control) system it uses to protect its assets from the threats, and its value and preference structures. Interaction with the questionnaires is accomplished in a team environment: the safeguards vulnerability questionnaire is executed by management, operations, and security personnel; the dynamic threat questionnaire is executed by a group of people having access to the appropriate information; and the impact questionnaire is executed by a team made up of high-level management and operations personnel.

The subject organization's mission determines the necessary security level the safeguards system must achieve for adequate protection of the assets. This is essentially an assessment of the mission's sensitivity, its criticality, and its integrity requirements.

Many factors contribute to the organization's environment. Those having the most effect upon the analysis are geographical location, community environment, physical environment, and procedural environment. Geographical location determines, to a large extent, the potential for catastrophic natural events such as earthquakes, volcanic eruptions, hurricanes, tornadoes, floods from swollen rivers or burst dams, and so forth; it also indicates nearness to population centers, major transportation hubs, a ready source of spare parts, and so forth. Community environment describes the social, political, and intellectual climate in which the organization finds itself, including such factors as the presence of organized crime, political dissent, universities, and social and moral arbitrators. Physical environment describes the campus of the organization—the land, fences, buildings, and so forth. Procedural environment details the philosophy, policy, and procedures set forth by the organization's management.

The organization's value structure determines what effect a successful attack might have upon the organization. Consider, for example, a widely used computer, Brand X, whose cost is roughly half a million dollars. If the organization's sole computing power is this single computer, and if the organization depends heavily upon the computer for carrying out its daily business, the destruction of the machine would have a catastrophic effect upon the organization. If, on the other hand, the organization has a great many of these computers as well as several more powerful supercomputers and depends only marginally upon any one of the Brand X computers, the destruction of one of them, while unpleasant, would be far from catastrophic and indeed might be only inconsequential or a nuisance.

The Analysis Phase

The analysis phase consists of three separate analyses—two required analyses and an optional analysis. The first analysis uses information gathered about the subject organization's safeguards system to assess system vulnerabilities. The optional second analysis, if it is needed, uses information gathered about the dynamic component of the threat to assess current threat strength. The third analysis uses information gathered about the subject organization's values, preferences, and philosophies to evaluate the potential consequences of successful attacks. From these analyses come a pair (monetary and non-monetary) of loss exposures for each [threat, asset, safeguards function, outcome] quadruple.

1. From the vulnerability analysis, a value is calculated for each safeguards subfunction that represents the subfunction's degree of membership in the fuzzy set of safeguards-function effectiveness; its fuzzy complement[14] represents the vulnerability (or relative weakness) of the subfunction. (Recall that the degree of membership in a fuzzy set

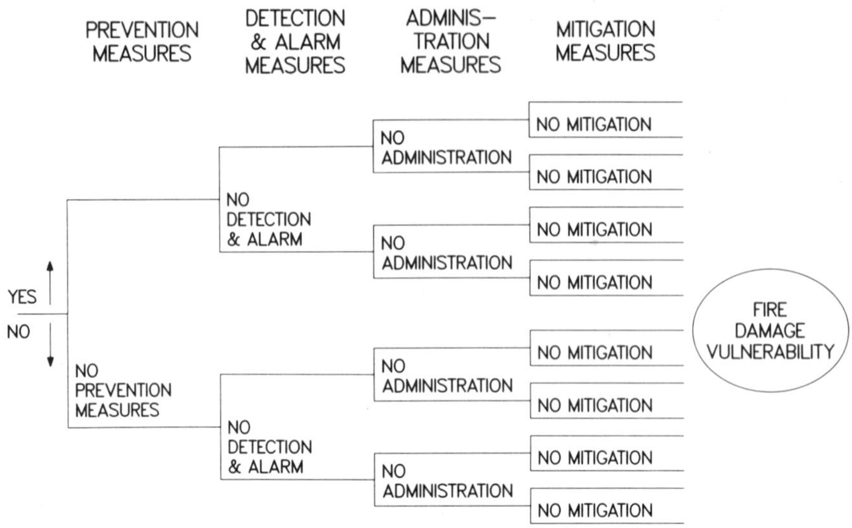

Fig. 4. Fuzzy tree for fire damage control safeguards function.

lies in the interval [0,1], where a zero degree of membership implies no membership and a value of unity for degree of membership implies full or complete membership; a set consisting only of these extremes represents a special case of fuzzy set, which is an ordinary set.)

Each safeguards function is represented as a fuzzy tree, whose branches represent the subfunctions for this particular function. For example, if the safeguards function is "Fire Damage Control," then it follows that appropriate subfunctions would be "Fire Prevention Controls," "Fire Detection Controls," "Fire Emergency Administration Measures," and "Fire Damage Mitigation Controls." This example is depicted in Fig. 4. Note that these trees are not probabilistic trees; the values that eventually appear on the branches are *not* probabilities but instead are degrees of membership in the fuzzy set of subfunction control performance. The subfunction values can be combined to produce a value for the safeguards function by taking their fuzzy union,[14] essentially equivalent to the maximum of the subfunction values.

A desirable set of safeguards that will accomplish the safeguards functions' objectives is modeled as an automated interactive questionnaire that elicits specific information about the presence and quality of the safeguards existing at the subject site. Each element e_{ijkmn}, attribute α_{ijkmnp}, or added information contribution is represented as a separate question. Just as the subfunctions are more or less equal within a specific safeguards function, element question within a specific subfunction are more or less equal. Further, each of the attributes within the element they modify is more or less equal. Any element safeguard can help to accomplish the objectives of more than one safeguards subfunction (a complex relational database keeps track of the numerous interrelationships), but each attribute is linked to a specific element.

In scoring the safeguards system questionnaire, each safeguards element has a maximum potential degree of vulnerability equal to 1. If the element is one whose mere presence implies adequate performance, the vulnerability degree is 0 if the element is present and 1 if it is not; such an element does not have attributes. If the element has attributes, this means that not only should the safeguard implied by the element question be present but also the safeguard has associated criteria to determine element quality and performance adequacy (and hence degree of vulnerability). Each of the performance criteria (attributes) are of about equal importance and have a maximum vulnerability value of $1/p$, where p is the number of attributes for the particular element under consideration. In this case, the element vulnerability value is given as

$$v(e_{ijkmn}) = \Sigma_p \, v(\alpha_{ijkmnp}) \,,$$

where the $v(\alpha_{ijkmnp})$ is either zero if the criterion expressed in the question is satisfied or $1/p$ if it is not.

Each safeguards element (and hence its associated attributes) can contribute to more than one safeguards subfunction. We use a relational database to keep track of which subfunctions are affected by each element, and after the safeguards questionnaire has been answered completely, the database is used to assist in the subfunction tabulations. The degree of vulnerability for each subfunction is the sum of the vulnerability values for all the elements that contribute to the subfunction, normalized to unity, or

$$V_{ijkm} = 1/n \, [\Sigma_n \, v(e_{ijkmn})]$$

where n is the number of safeguards elements contributing to the mth subfunction.

The vulnerability values for each of the subfunctions are interesting in themselves, but they also can be aggregated into vulnerability measures for the safeguards functions by taking their fuzzy union, obtained as the maximum of the values for the subfunctions that make up a safeguards function. This is the measure carried through into the loss exposure calculations.

The vulnerability component measures the relative weaknesses in the safeguards functions with respect to the spectrum of threat-asset pairs. In the vulnerability analysis, we assume that the threat is static (the dynamic component is zero), that all attacks are equally likely to occur, and that the consequences are extreme. The resulting vulnerability measure can later be made more realistic—it can be reduced or increased by including the dynamic threat and the "real" consequence measures when the risk measures are calculated. LAVA defines the vulnerability measure, V_i, as the complement of the membership function in the fuzzy set of "complete safeguards-function effectiveness," S_i (or $V_i = 1 - S_i$, where both V_i and S_i lie between zero and unity).

The LAVA software system includes an automated report generator that produces a report of the vulnerability analysis. The report includes rankings of the vulnerabilities by safeguards function and subfunction, a breakdown of vulnerabilities by threat category, and bar charts and scatter diagrams illustrating these. It includes a detailed report listing the missing or inadequate safeguards for each of the safeguards subfunctions, providing information to those whose mission is to reduce vulnerabilities.

2. A dynamic threat[11] analysis can be performed if the subject system is extremely sensitive to a changing threat and if the subject organization has access to the kinds of information the analysis requires.

The dynamic threat takes into account possible threat agents and their potential attack goals. The magnitude of the threat is determined from the motivation, capabilities, and opportunities of the various threat agents with respect to the target(s) of the attack.

There are several broad categories of threat agents having a variety of goals. Possible categories of threat agents might be, for example:

(a) information gatherers (e.g., spies or hostile intelligence services),
(b) terrorists,
(c) anti-"X" radicals or extremists (where "X" could be almost anything!),
(d) representatives of organized crime,
(e) other criminals (non-malicious criminals and pranksters),
(f) insiders (employees, contractors, etc.),
(g) outsiders with access, and
(h) Mother Nature.

The threat agents in each of these categories all act for different reasons, and so they may differ widely in motivation, capability, and opportunity. Similarly, the goals of the attacks may vary, but all categories of goals may be used by all categories of threat agents. Some possible goal categories are

(1) information and/or material collection (e.g., espionage),
(2) sabotage,
(3) theft, embezzlement, fraud,
(4) damage or destruction,
(5) extortion,
(6) disrupting business or mission,
(7) surmounting an intellectual challenge.

Clearly, more than one of the categories may be the goal of a single attack, and a single attack may be perpetrated by more than one category of threat agent.

The threat component measures the relative strength of identifiable threat agents in terms of motivation, opportunity, and capability with respect to the spectrum of assets, the corresponding safeguards functions, and the set of possible outcomes. *Motivation* is a measure of how much effort or what part of his resources a threat agent is willing to expend on an attack and how dedicated he is to carrying out the attack. *Capability* is a measure of the resources—knowledge (training), information (intelligence), funds, skills, equipment, armament, personnel—the threat agent has at his disposal. *Opportunity* is a measure of how easy it is for the threat agent to achieve an enabling proximity for an attack, how easy it is for him physically to reach the object of attack, how easy it is for him to attack or to access the object, how easy it is for him to travel undetected (both in the neighborhood of the object of attack and from afar to get near the object), and so forth. *Opportunity is separate and different from potential site vulnerabilities.* Figure 5 illustrates the tree structure for the dynamic threat analysis.

The approach to assessing the dynamic part of the threat component by considering categories of threat agents and possible categories of attack goals is parallel to the approaches used for both the vulnerability analysis and the consequence analysis. Potential scenarios are modeled implicitly as the relationship between the threat-asset pairs and the safeguards functions in the vulnerability analysis, and as the relationship between the assets and the threat elements (motivation, capability, and opportunity) in the threat assessment. Similarly, the attack goals are modeled implicitly in the capability component of the dynamic threat measure and are approximately equivalent to the outcomes used in the consequence analysis.

An interactive questionnaire models the contributors to the dynamic threat in terms of specific threat groups. A degree of strength is calculated for each group based on motivation, capability, and opportunity relative to a specific [threat, asset, safeguards function, outcome] quadruple. A relational database keeps track of which threat groups can

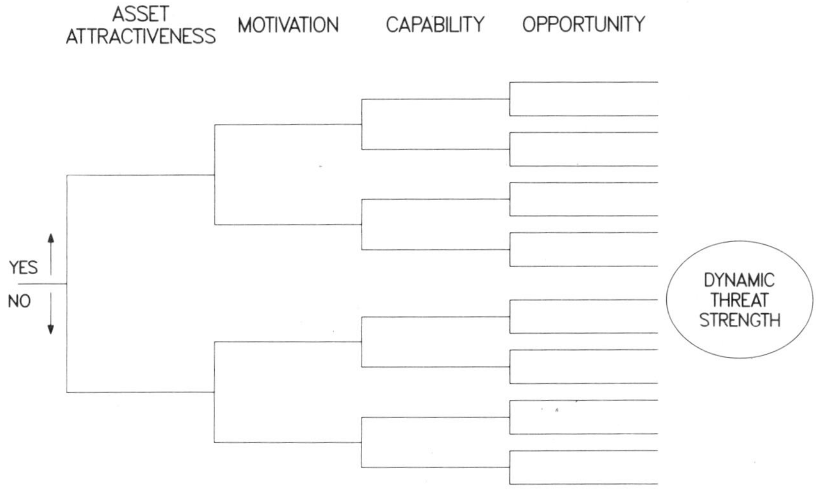

Fig. 5. Assessment structure for dynamic threat.

affect each quadruple so that an overall or total value for the dynamic threat strength can be calculated for each quadruple, to be used in the loss exposure calculations.

3. Evaluating the consequences, or impacts, is the third analysis in LAVA. The object of this analysis is to determine the effect that an outcome of a successful attack would have upon the subject organization.

The consequence (or impact) component measures the potential costs (both monetary and non-monetary) of a threat successfully exploiting a safeguards function's vulnerability with respect to the severity of the exploitation's outcome. The outcome severity metric results from combining the results of the previous two analyses, the relative weakness of the safeguards function and the relative strength of the dynamic threat.

LAVA's consequence measures are given as pairs of monetary and non-monetary descriptors: the monetary descriptor $I(M)_{ijkq}$ is used when the consequences can be given in terms of monetary cost (dollars, pesos, francs, pounds, kroner, etc.), and the non-monetary descriptor $I(L)_{ijkq}$ is used when the consequences can be given only in terms of intangible cost (such as "catastrophic" as a linguistic or non-monetary measure of irreparable reputation loss). The consequence values are obtained from another interactive questionnaire. Figures 6 and 7 illustrate the monetary and non-monetary consequence tree structure.

The monetary consequence values come about as the sum of the costs of investigation and followup, business interruption, future business loss, fraud or embezzlement, and business termination. Each of these contributors is calculated from questions about specific costs, such as costs for reprimand or legal action; costs associated with lost time, interim operation, and full replacement, repair, and recovery; costs for employee replacement and training; direct losses from damage and destruction, waste, fraud, or embezzlement; and so forth.

The non-monetary (linguistic) consequence values are expressed as the maximum of the set of values derived for the cost of adverse public reaction, the cost of organizational disruption and/or loss of morale, the cost of embarrassment and/or potential reputation loss, and the intangible cost associated with unsafe or restrictive working conditions. The

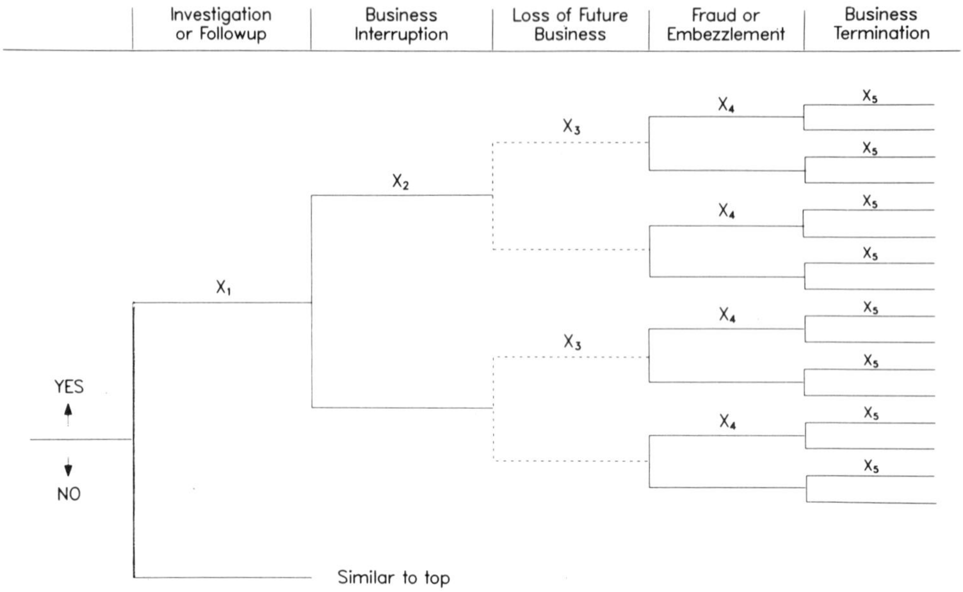

Let I_M represent the Monetary Consequence Factor.
Then I_M is quantified by:

$$I_M = X_1 + X_2 + X_3 + X_4 + X_5$$

Fig. 6. Monetary consequence tree.

linguistic values may include, for example, such measures as "catastrophic," "nuisance," and so forth.

Solving the Risk Management Problem

The best analysis does no good if appropriate action is not taken to improve the organization's risk posture. A thorough cost/benefit analysis provides a basis for optimally selecting a safeguards set to be improved or added to the existing safeguards system. The analysis not only should include how much the proposed action path will cost and how much it will reduce loss exposure, but also should include well-thought-out implementation plans and the effect that the implementation will have on the organization in terms of morale, training (or retraining), throughput reduction, and so forth.

Our plans for LAVA's future include automating the cost/benefit analysis and the optimal safeguards selection process. However, at this time, these features are left to the subject organization's analysts and management.

CONCLUSIONS

We have discussed LAVA, a methodology for evaluating vulnerabilities, threats, and risks in large, complex systems that often are too intractable to be handled adequately and reasonably by other methods. LAVA's measures for loss exposure (risk) for each [threat, asset, safeguards function, outcome] quadruple are functions of the relative weaknesses of the safeguards function (vulnerability), the relative strength of the threat, and the consequences of the outcome of a successful attack.

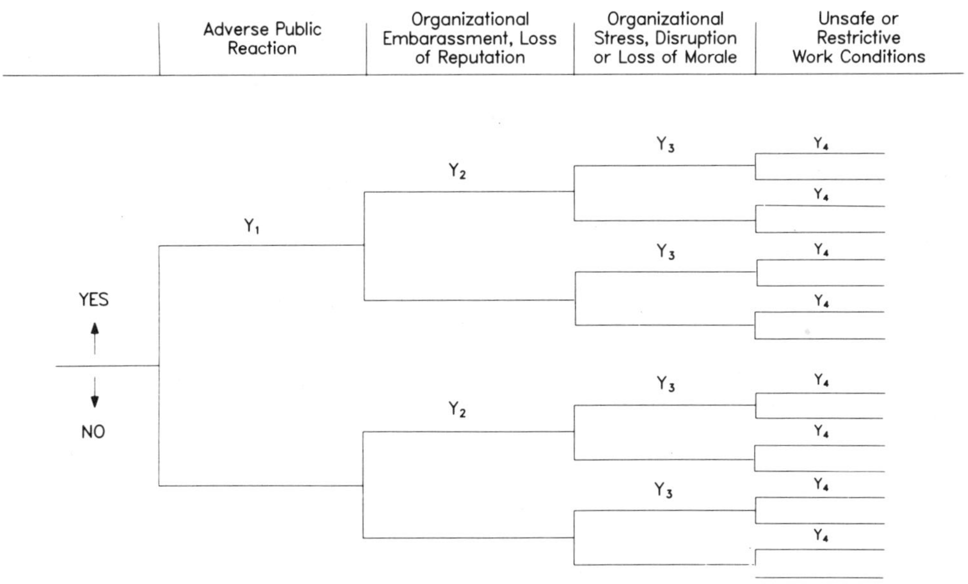

Let I_{NM} represent the Nonmonetary Consequence Factor.
Then I_{NM} is quantified by:

$$I_{NM} = \max \ (Y_I) \text{ , where } I = 1, 2, 3, 4.$$

Fig. 7. Non-monetary consequence tree.

The baseline vulnerability measures are derived with the assumption that the background (static) threats are the only operable threats (that is, their relative strength is 1). In evaluating the severity of the outcome of a successful attack when a vulnerability exists in a safeguards function, the effect of the vulnerability can be *increased* by the "real" threat measure if the information necessary to the threat analysis is accessible and available.

Our automated procedure can be used to elicit and analyze information to determine vulnerabilities and risks inherent in many application systems, such as computer systems, material control and accounting systems, physical protection systems, security systems, plant process-control systems, and a host of others. The subject system is modeled as interactive questionnaires that include inherent *a priori* decisions about the desirability of or aversion to applications-specific features of the system, about the subject system's threat environment, and about the value structure of the subject organization. Event-tree-like structures are used for organizing the analyses. Performance measures and associated decisions are evaluated in both qualitative and quantitative terms, giving greater insights into system performance than a strictly quantitative analysis can.

LAVA's approach is technically sound, accurate, interactive, secure, and portable. The accuracy is derived from exhaustive and comprehensive questions provided by the developer of a specific application. The interactive conversational questionnaire in natural language is straightforward and easy to use. The expertise and rigor built into the model, as well as the model's logical structure, makes decision-making simple. The functional structure of the model clearly indicates what safeguards are missing and provides a rationale for selecting the safeguards to add. The LAVA software system is compatible with standard IBM-PC software, making the system portable. Because it is interactive and self-contained, the organization using this methodology and the associated software does not require the services of outside consultants, hence adding another layer of security to the entire safeguards system.

REFERENCES

1. S. T. Smith and J. J. Lim, LAVA: An Automated Computer Security Vulnerability Assessment Software System (Version 0.9), Los Alamos National Laboratory document LA-UR-85-4014 (December 1985).
2. S. Sudman and N. M. Bradburn, *Asking Questions: A Practical Guide to Questionnaire Design*, Jossey-Bass, Inc., San Francisco (1982).
3. S. T. Smith and J. J. Lim, An Automated Method for Analyzing Computer Security Risk, *Proc. 7th DOE Computer Security Group Conference*, New Orleans, LA, April 17-19, 1984, Los Alamos National Laboratory document LA-UR-84-438.
4. S. T. Smith and J. J. Lim, An Automated Method for Assessing the Effectiveness of Computer Security Safeguards, *Proc. IFIPS Second International Congress on Computer Security*, Toronto, Canada, September 10-12, 1984, Los Alamos National Laboratory document LA-UR-84-2744.
5. S. T. Smith and J. J. Lim, An Automated Interactive Expert System for Evaluating the Effectiveness of Computer Security Measures, *Proc. 7th Department of Defense/National Bureau of Standards Computer Security Conference*, Gaithersburg, MD, September 24-26, 1984, Los Alamos National Laboratory document LA-UR-84-1580.
6. S. T. Smith and J. J. Lim, Framework for Generating Expert Systems to Perform Computer Security Risk Analysis, *Proc. First Annual Armed Forces Communications and Electronics Association Symposium and Exposition on Physical and Electronics Security*, Philadelphia, PA, August 19-21, 1985, Los Alamos National Laboratory document LA-UR-85-1933.
7. S. T. Smith and J. J. Lim, Assessment of Computer Security Effectiveness for Safe Plant Operation, *Trans. Am. Nucl. Soc.* **46**:525-526 (1984), Los Alamos National Laboratory document LA-UR-84-99.
8. S. T. Smith and J. J. Lim, An Automated Procedure for Performing Computer Security Risk Analysis, *Proc. Sixth Annual Symposium on Safeguards and Nuclear Material Management*, European Safeguards Research and Development Association, Joint Research Centre, Ispra, Italy, ESARDA 17, pp. 527-530 (1984).
9. S. T. Smith, D. C. Brown, T. H. Erkkila, P. D. FitzGerald, J. J. Lim, L. Masasgli, J. R. Phillips, and R. M. Tisinger, LAVA — A Conceptual Framework for Automated Risk Assessment, *Nucl. Mater. Manage.* **XV**:256-259 (Proceedings Issue) (1986).
10. S. T. Smith, J. J. Lim, and J. Lobel, Application of Risk Assessment Methodology to Transborder Data Flow, Handbook on the International Information Economy, Transnational Data Report, Springfield, VA LA-UR-85-2416 (November 1985).
11. S. T. Smith, J. R. Phillips, D. C. Brown, and P. D. FitzGerald, Assessing the Threat Component for the LAVA Risk Management Methodology, *Proc. 9th DOE Computer Security Group Conference*, Las Vegas, NV, May 6-8, 1986, Los Alamos National Laboratory document LA-UR-86-741.
12. M. D. Mesarovic, D. Macks, and Y. Takahara, *Theory of Hierarchical Multilevel Systems*, Academic Press, New York and London (1970).
13. K. J. Schmucker, *Fuzzy Sets, Natural Language Computations, and Risk Analysis*, Computer Science Press, Rockville, MD (1984).
14. L. A. Zadeh, K.-S. Fu, K. Tanaka, and M. Shimura (Eds.), *Fuzzy Sets and Their Applications to Cognitive and Decision Processes*, pp. 1-39, Academic Press, New York (1975).

A Probabilistic Risk Assessment Program for Analyzing Security Risks

Louis Anthony Cox, Jr.
U.S. WEST Advanced Technologies
Englewood, CO

ABSTRACT

An open question in security threat and risk assessment has been how to integrate detailed security information, e.g., from checklists, into useful summaries and overall assessments of a system's or facility's vulnerabilities. Threat assessment methodologies in this area have typically focused more on whether systems and facilities meet applicable security standards than on quantitative assessments of how secure they are, and have provided little insight on the effectiveness of specific countermeasures in reducing critical vulnerabilities.

This paper describes an automated threat assessment methodology (ATAM) and personal computer modeling system developed for the United States Air Force to meet these needs. ATAM is a tool for performing probabilistic risk assessment (PRA's) that include human intrusion threats as well as more traditional accident sequences. Three major software components are discussed: an interactive graphics package that allows users to easily model specific systems and facilities at multiple levels of abstraction; a set of goal resolution algorithms and a simulation routine for identifying high-likelihood intrusion strategies or "threat paths" and calculating the vulnerabilities of specified targets; and a report generation system that displays and ranks threats in various damage categories and that provides graphical answers to user queries.

The ATAM software has been fully implemented and tested, and is designed to run on any IBM-compatible PC. This paper concludes with a discussion of practical implementation issues and the potential uses of such automated threat assessment tools in the risk analysis process.

KEYWORDS: Probabilistic risk assessment, computer security, automated threat assessment

INTRODUCTION

In the January 1985 meeting of the Federal Information Systems Risk Analysis Workshop,[1] a number of important challenges for developers of security threat risk analysis methodologies were identified. Among these were integration of qualitative and quantitative approaches to threat assessment, with emphasis on new qualitative methods; use of computer-based methods to reduce the time and costs of the automated data

New Risks
Edited by L. A. Cox, Jr., and P. F. Ricci
Plenum Press, New York, 1990

processing (ADP) risk management process; development of tools for addressing multiple (ideally, all) phases of the ADP risk management process, from identification and initial priority ranking of threats and vulnerabilities to cost-benefit evaluation and selection of appropriate countermeasures and safeguards; and development of sophisticated algorithms and methodologies for dealing with nonquantifiable (e.g., "intangible") losses, interdependencies among assets and countermeasures, and uncertainties in data.

Since August of 1985, the Air Force Computer Security Program Office (AFCSPO) has been funding development of a PC tool to respond to some of these needs. This tool, called ATAM (for "Automated Threat Assessment Methodology") has been built at Arthur D. Little, Inc., based on pioneering development work at MITRE corporation, and has been beta tested by the Air Force since April 1986.

The ATAM system focuses on threat identification and ranking, quantitative estimation and display of vulnerabilities, and identification of critical threat paths or intruder attack strategies leading to high vulnerabilities of assets in different threat categories. It provides a sophisticated facility for organizing and integrating qualitative and quantitative information about physical and logical (electronic) security threats. ATAM also provides a "what-if" capability for predicting the effects of different countermeasures or safeguards on asset vulnerabilities. However, it does not evaluate residual risks for "acceptability," or actively identify and recommend appropriate countermeasures. Thus, it is a tool for automated threat assessment and risk analysis, rather than a complete risk management system.

Within this limited scope, however, the ATAM system introduces several technical and methodological innovations that make it a particularly powerful and flexible analytic tool. It combines three principal technical methods of analysis:

- a "fault tree analysis" approach of the type traditionally used in probabilistic risk assessment to calculate the reliabilities of complex systems (such as nuclear reactors);
- a "goal resolution" algorithm to model the behavior of an intelligent intruder; and
- a simulation routine for doing sensitivity analyses.

These tools are accessed through an interactive graphics package and menu system to allow both rapid on-line identification, screening, and analysis of security threats, and automatic preparation of detailed reports on all threats in any desired threat category. Examples of threat categories include physical property damage, information disclosure, information modification, loss of data integrity, and interruption or denial of service.

Perhaps the most important distinguishing feature of the ATAM system is that it explicitly addresses the traditionally hard-to-analyze risks to physical and information assets from deliberate human intrusion or attack. By simulating the behavior of an intelligent intruder using a simple "goal resolution" heuristic, ATAM reveals threat paths and attack strategies that make a facility or system vulnerable to physical or logical attack.

METHODOLOGY

Representing Threats

ATAM is based on three key ideas:

- Threats can be described as sequences of activities and events.

- Abstract activities (such as "intruder enters room x without being detected") can be broken down into logical combinations of more specific activities (such as "intruder enters room x by the door OR intruder enters room x by the window, AND intruder bypasses alarms and other physical access controls and safeguards for room x").
- The probability that an intruder can successfully complete a goal activity is the probability that he can successfully complete at least one path (sequence of activities and events) leading through that goal activity.

Figure 1 gives a simple toy example of an ATAM model.

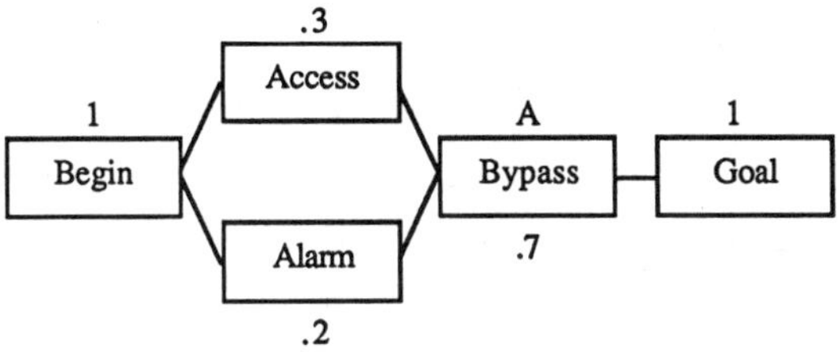

Fig. 1. A simplified ATAM model.

Activities are represented by boxes called "nodes" and are connected by arcs that represent precedence constraints. (Realistic models may contain hundreds of nodes rather than only five.) Each node is labelled with a short (six characters or less) mnemonic such as "access" or "alarm" that indicates the nature of the activity. In the software, positioning a cursor over a node also causes a short phrase or sentence (up to sixty characters) describing the activity (e.g., "Intruder gains access to target asset," or "Intruder successfully bypasses alarm system without being detected") to appear at the bottom of the screen. Each node is also attached to other nodes by means of logical precedence constraints. Thus, an ATAM model is represented graphically by a network, oriented from left to right.

For example, the activity called "Bypass" in Fig. 1 (short for "Bypass target defenses") can be attempted only after the two activities "Access" and "Alarm" have successfully been completed. The fact that both must be completed first is indicated by the "A" symbol over the "Bypass" node, which indicates that this node is "and-related" to its predecessors. If "Bypass" could be reached by successfully completing either "Access" OR "Alarm," then an "O" would appear over "Bypass," to indicate that it is "or-related" to its predecessors.

The overall interpretation of Fig. 1 (ignoring the numbers for a moment) is that an intruder can get from "Begin" to "Goal" only by successfully completing the three intervening activities, "Access," "Alarm," and "Bypass." Moreover, "Bypass" can be attempted only after "Access" and "Alarm" have been completed. The order in which he should try "Access" and "Alarm" is something that the intruder must decide.

The mechanics of building an ATAM model are straightforward. Starting from a single blank node, the user uses commands such as "Insert," "Delete," "Link," and "Unlink" and an "And/Or" assignment command to create new nodes and arrange them in the desired logical configuration.

Node Data

In the ATAM system, the need to integrate quantitative and qualitative information is addressed by associating with each activity (network node) three numerical attributes and up to four descriptive labels. These "node data" are entered by positioning the cursor over a node and invoking an "Edit" window with slots for the attribute values and labels (and also for the six-character mnemonic name and the sixty-character comment line) that will be attached to that mode.

The three numerical attributes for an activity are as follows:

- an estimated *success probability*, specifying the conditional probability that an intruder can successfully complete the activity, given that he starts it;

- an *expected duration* for the activity; and

- for goal activities, a numerical index of the *"value"* of completing that activity.

Both values and probabilities are assessed subjectively. As with any quantitative risk assessment methodology, useful estimation of these quantities presents a number of potential methodological problems. Fortunately, several formal methods for obtaining useful subjective estimates have been developed and validated in various applications over the past years.[2,3] In these approaches, a cardinal value scale is derived analytically from ordinal comparisons or classifications combined with consistency checks. The same approaches can be used to help obtain subjective probability estimates by replacing importance or value rankings with likelihood rankings. While formal methods such as these can make easier the task of obtaining useful subjective estimates, they do not eliminate the need for human judgment.

Each goal activity contains not only numerical attributes, but also up to four descriptive labels. Each label consists of an eight-character "threat description code" (TDC) which specifies the nature of a threat in terms of four elements: its *source* (e.g., a disgruntled employee), *target* (an asset), *effect* (damage to the asset of some type, e.g., theft or destruction), and *mechanism* (how the damage is done). Mechanisms are specified only in limited detail (e.g., "protected file overwritten by a borrowed program with read-access rights"). However, they suggest predecessor activities that may be attached to the labelled node.

ATAM allows a node to carry multiple threat description code (TDC) labels, since a single activity, such as setting fire to a magnetic tape library, sometimes involves multiple effects in terms of the threat taxonomy, e.g., "destruction of property" and "destruction of information."

Modularity and Calculations

As previously mentioned, real ATAM models tend to be fairly large, involving hundreds of nodes. To cope with this complexity, ATAM makes heavy use of "modules," or subnetworks of nodes that can be treated as single (aggregate) nodes with respect to the rest of the network representing a system or facility.

For example, Fig. 2 shows an abstract version of a network with seventeen nodes.

(For simplicity, this figure represents each node by a single letter, instead of by a labelled box, and indicates that nodes A and B are "and" nodes by cross-tie marks.) Rather than showing all seventeen nodes at once, however, the ATAM package divides this network into seven "modules" (also called "complex nodes") at three different "levels of

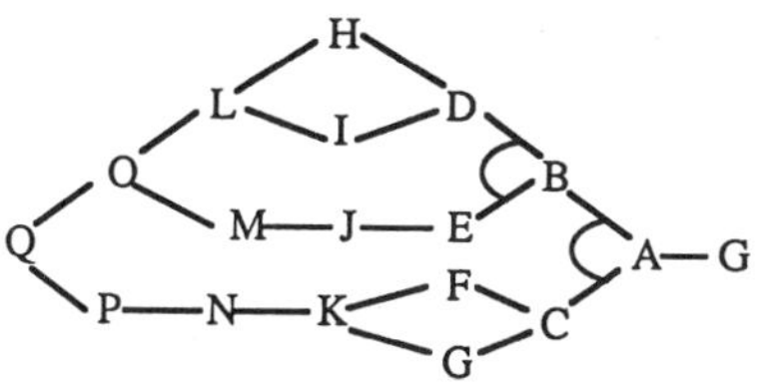

Fig. 2. Abstract logical structure of an ATAM model.

of abstraction." Only one module at a time can normally be viewed, although a special "Levels" command will draw a tree diagram of all the modules in an ATAM model and the inclusion relations among them. Figure 3 shows the different modules in Fig. 2 as partially nested boxes of "simple" nodes.

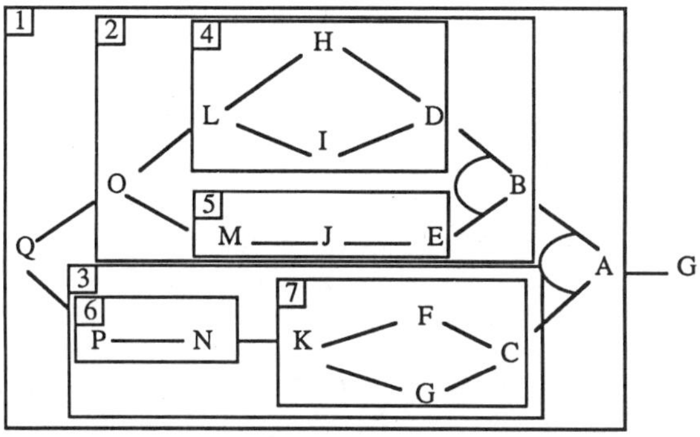

Fig. 3. Modules of the model in Fig. 2.

In Fig. 3, each node's "depth" or "level" in the abstraction hierarchy is found by counting the number of boxes that enclose it. At abstraction level 1 there is a single module, labelled "1," in addition to the simple node G: thus, there is a total of two nodes (one simple and one complex) at this level of abstraction. At level 0, there would be only one (complex) node, representing the entire model considered as a single node.

Movement among levels is controlled by two opposite operations called "pan" and "zoom." For example, zooming on module 1 in Fig. 3 would bring the sub-network (or "layout") consisting of nodes Q and A and modules 2 and 3 (each shown as a single node) onto the screen. Zooming on module 2 would then bring up the layout made up of nodes O and B and modules 4 and 5. Zooming on module 4 would bring up the four-node layout consisting of the simple nodes L, H, I, and D. Each zoom operation increases the depth of detail by one level.

The opposite of zooming is panning. For example, panning on node J, which would appear on the screen as part of the three-node serial layout consisting of M, J, and E, would again produce the layout consisting of 0, 4, 5, and B. However, panning does more than abstracting away from a layout by one level. When the user pans away from the layout showing a module's contents, two quantities are calculated:

- the probability that an intruder could successfully make his way through the module whose contents are shown in the layout (i.e., that he could successfully complete some path from one end of the layout to the other); and

- the expected time that it would take him to work his way through the module, assuming that he eventually succeeds and that he uses an optimal attack strategy.

The calculated success probability and expected duration are automatically entered into the corresponding data slots for the module as a whole. As soon as a module has been created and panned away from, it can be treated as a single node. In fact, it can be named (using a "Name" command) and then used as a building block elsewhere in the network, or even in other ATAM models. Modules representing generic activities or activity networks, such as logging onto a standardized computer system, can be stored in an ATAM library. Using a "Retrieve" command, copies of modules in the library can then be inserted wherever appropriate when a new model or portion of a model is being built. Thus, the effort required to develop an activity module representing some standardized activity or activity sequence need not be duplicated.

The success probability for a module as a whole is found as follows. For an "and" module, success probability is the product of the success probabilities of its components, while for an "or" module, it is the complement of the product of the complements of the success probabilities of its components. (The "complement" of a probability p is defined as $1-p$.) In applying these rules, a chronological sequence such as M-J-E in Fig. 3 is considered to be an "and" module.

Unlike success probability, the expected duration and value for a module depend on which path through it — e.g., L-H-D or L-I-D in module 4 — the intruder tries first. To simulate his decisions at such junctures, we need a model for the intruder's behavior, as captured in a planning strategy. ATAM assumes that the intruder always acts to minimize the expected remaining time until he either reaches his goal or finds that it is unreachable. Thus, if his goal in Fig. 3 is assumed to be G, then his strategy in the rest of the network will be always to take the path that minimizes the expected remaining time to reach G (or to discover that G cannot be reached, e.g., because neither B nor C can successfully be completed). In module 4, for example, he will try H first in preference to I only if the ratio of success probability to expected duration is greater for H than for I. This rule — that paths through a module should be chosen in order of decreasing ratio of success probability to expected duration — applies to any "or" node. Essentially the same rule applies to "and" nodes, but with failure probability (=1 – success probability) replacing success probability. ATAM uses these two criteria to optimally order the nodes in any module, and then calculates the expected duration for the module as a whole from the resulting optimized sequence of nodes. Explicitly, the expected duration for a module is calculated by weighting the cumulative duration up to each node in the sequence by the probability that the intruder will succeed (for an "or" module) or fail (for an "and" node) for the first time at that node in the sequence, and then summing these weighted durations. Expected values are calculated in the same way. It is assumed that the intruder attempts all paths within a module before switching to another module, reflecting an assumption that switches across modules are relatively time-consuming. The total probability of successfully reaching G, however, is independent of such sequencing assumptions.

The calculation of expected durations and success probabilities for modules is performed automatically by the "pan" option whenever a layout is collapsed into a single complex node. The same computational procedures are also used to calculate a facility's vulnerabilities to different types of threats. We define the *vulnerability* of a facility or system with respect to a particular goal activity (namely, the damaging of some asset in a specified way) as the probability that an intruder could successfully complete the goal activity. ATAM produces a variety of reports that reveal system or facility vulnerabilities in different damage categories. These are described below.

Using the ATAM System in Risk Analysis: Diagnostic Reports

Once an ATAM model has been constructed and its goal activities labelled, it can be used to answer a number of questions diagnosing a system's or facility's vulnerabilities. The risk analyst uses the threat description code (TDC) to specify the category of threats that he wants analyzed, and uses a report generation menu to select the type of report that he wants on threats in the selected category. Given an eight-character TDC code (or a piece of one) as input, the system searches through the entire network representing a system or facility and identifies the goal nodes carrying labels that match the specified piece of code. It then prepares reports on each of the corresponding activities.

Three basic types of reports may be selected: a *scatter plot* report, a *best path* report, and a *simulation* report.

For an overview of system vulnerabilities, the analyst would select the "scatter plot" option for each of the major damage categories, e.g., property damage, information asset damage, etc. Figure 4 shows what the scatter plot report looks like. For each goal node in the specified threat category, the system plots the "vulnerability" of that goal activity, i.e., the probability that it could successfully be carried out, against its "value" or relative importance. As previously discussed, the value axis may be different for different, not necessarily commensurable, damage categories.

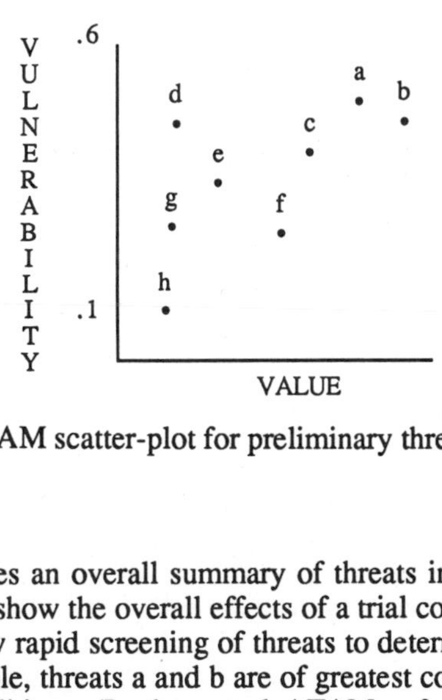

Fig. 4. An ATAM scatter-plot for preliminary threat screening.

The scatter plot provides an overall summary of threats in a given category. It has two primary applications: to show the overall effects of a trial countermeasure on threats in a given category; and to allow rapid screening of threats to determine which are of greatest concern. In Fig. 4, for example, threats a and b are of greatest concern, since they have the highest values and vulnerabilities. (In the actual ATAM software, two-character "node handles" or identifiers are plotted.) Nodes g and h are of least concern, other things being equal. To see the effects of any countermeasure or set of countermeasures, e.g., the installation of a proposed new access control system, the risk analyst may use a special "trial" feature that allows the scatter plot to be calculated with and without the proposed changes in place. This feature supports selection of cost-effective control measures by showing their predicted effectiveness in reducing vulnerabilities. It does not, however, select or recommend a set of countermeasures or eliminate the need for human judgment in making the trade-off between cost and security.

To gain a deeper understanding of a system's or facility's vulnerabilities, the risk analyst has three options. He can formulate more specific classes of threats using the TDC, e.g., to study the threats posed by different types of personnel or different mechanisms of attack, and prepare scatter plots (or node reports) for these classes. He can also obtain a "best path" report on each goal node in any specified class of threats. Finally, he can use a simulation routine to obtain a plot of the vulnerability of any specified goal node as a function of the amount of time allowed for an intrusion.

The "best path" report uses the planning heuristic discussed in "Modularity and Calculations" to find the "best path" (or "optimal" attack strategy) leading to a specified goal node. Returning to Fig. 1, suppose that the success probabilities for each node are as shown, i.e., 0.3 for "Access," 0.2 for "Alarm," and 0.7 for "Bypass." Starting from "Begin," the intruder must decide which of "Access" and "Alarm" to try first. Of course, he must eventually complete both if he is to get to "Bypass." If the expected durations for "Access" and "Alarm" are the same, however, then he should try "Alarm" first (since it is the one more likely to fail). If he is ultimately going to fail, he wants to find this out as soon as possible. Thus, the "best path" through this module would be Begin, Alarm, Access, Bypass. The "best path" report in ATAM lists the sequence of nodes (and their descriptions) along the "best path" to each goal in the class specified by the user.

The purpose of the best path report is to identify likely attack strategies that an intruder might use in pursuing different goals. However, the best path to a goal node is an incomplete strategy: it does not specify what the intruder should do if he finds that he is unable to complete some activity along the best path. To gain a final level of insight into the vulnerability of a node, the risk analyst may invoke a routine that simulates sending the intruder along the best path a large number of times and that dynamically replans for him (by calculating the "best path" with respect to the remaining feasible network) whenever he finds that he is unable to successfully complete some activity. The simulation routine produces a graphical report showing the growth in vulnerability of a target node as a function of the duration of an intrusion.

From this plot, it is possible to answer questions such as "What is the least amount of time needed to reach goal G?" and "What is the probability distribution of the amount of time that it would take an intruder to reach goal G?" As the amount of time allowed to reach the goal node increases in the simulation, the probability that the goal will be reached (its vulnerability) approaches the level shown in the scatter plot report.

DISCUSSION: USE AND LIMITATIONS OF THE ATAM SYSTEM

The ATAM modeling system described in this paper is a risk analyst's tool, rather than a comprehensive, stand-alone analysis system. It can integrate large amounts of qualitative and quantitative information to provide graphical and tabular summaries of a system's or facility's vulnerabilities to different types of threats. And it enables the risk analyst to obtain answers to questions about the dominant sources, targets, effects, and mechanisms of threats and plausible attack paths or strategies. It also allows "what-if" analysis of the effects of proposed countermeasures. However, it is not a substitute for human intelligence and judgment. ATAM can answer questions, but it requires an analyst to decide which questions to ask. It will compute the predicted effects of different proposed countermeasures but will not take the initiative in recommending appropriate ones. A profitable direction for further development would be the addition of an expert system that would use ATAM's computational facilities as a tool, but that would generate follow-up questions from the answers to initial exploratory questions (e.g., from scatter plots for the major damage categories), follow promising lines of inquiry, and recommend appropriate countermeasures based on an analysis of predicted effects in multiple damage categories (and on user-specified value trade-offs and risk attitudes). Similarly, the user interface could be substantially enhanced by automating the process of constructing ATAM models based on the answers to security-related and system configuration questions, and by allowing for natural-language queries using a restricted vocabulary, rather than requiring the user to use a threat description code. These features are implemented in a separate front end program delivered to AFCSPO in August 1987.

Despite its current limitations, ATAM incorporates a number of features that may potentially be useful in the design of future automated threat assessment methodologies. Among the most important are:

- Use of transferable modules to reduce duplication of effort. Since ATAM runs on the Z-100, the Z-150, and any IBM-compatible PC, models and module libraries are highly portable. To the extent that standardized and generic systems and configuration are used across facilities, a single ATAM library may provide users with many of the building blocks they need to rapidly model their own facilities.

- Integration of quantitative, qualitative, and descriptive information. The ATAM system uses TDC labels to describe threats, networks to represent configuration structure, and numerical attributes to store subjectively assessed value, duration, and probability data. All three types of information are used in preparing an output report such as a scatter plot.

- Explicit articulation and documentation of assumptions and estimates. An ATAM model contains all the numbers and data on which calculated vulnerabilities and other reports are based. Anyone who disagrees with a calculated vulnerability or expected duration estimate can examine the model to find out exactly how it was calculated.

- Quantification of vulnerabilities. The ATAM system is geared not toward deciding only whether a facility is in compliance with applicable standards, but to estimate quantitatively the residual risks, as reflected in the vulnerabilities of different targets, so that they can usefully be compared and evaluated for acceptability.

- Modeling the human element. ATAM explicitly models the decision behavior of an intelligent intruder, although in a very simple and heuristic way. While the behavioral model of a single-minded intruder pursuing one goal at a time can be refined, the idea of examining intrusion possibilities and attack strategies from the viewpoint of a planning attacker may lead to insights that other approaches cannot provide.

In addition to these features, ATAM allows the analyst to formulate and obtain answers to specific questions (using the TDC) and to look behind the summary statistics, using the best path and simulation options, to obtain a better understanding of asset vulnerabilities. This ability to back up summary numbers with plausible attack strategies and simulations of intrusions directed at specific goals allows the analyst to examine problem areas in depth before selecting or recommending countermeasures.

The final ATAM system, including an initial library of modules for physical and electronic intrusion, was delivered to the USAF at the end of August of 1986. How effectively it can be used in practice will depend in large measure on the USAF's success in identifying standardized generic systems and configurations and distributing relevant modules modeling them to potential users. At present, it is intended that the ATAM tool be made widely available to personnel who are currently performing risk analyses manually, using check lists and similar tools. Ideally, ATAM will not only reduce the number of man-hours needed to complete such analyses, but will contribute to better integrated, more thorough, and hence more useful assessments of threats and vulnerabilities.

REFERENCES

1. *Minutes of the Federal Information Systems Risk Analysis Workshop*, January 1985, Defense Technical Information Center, Alexandria, VA, 1985.
2. J. S. Dyer and R. K. Sarin, Relative Risk Aversion, *Management Science* **28**:875-886 (1982).
3. T. Gear *et al.*, A Unified Approach to the Acquisition of Subjective Data in R&D, *IEEE Transactions on Engineering Management* **29(1)**:11-19 (1982).

A General Purpose Computer Shell for Risk Analysis

Lance J. Hoffman
The George Washington University
Washington, DC

ABSTRACT

A microcomputer-based general purpose risk analysis shell is described which significantly decreases the effort involved in producing quantitative reports and in carrying out sensitivity analyses. It supports many quantitative methodologies and allows users to customize the package to suit their specific needs.

Example scenario files are supplied with the system, though it is expected that most users will develop files based upon their own specific models of risk. There are two kinds of users: system managers and end-users. System managers, using a menu-driven interface, set up the model files and boilerplate reports. End-users answer (one-line or longer) questions supplied by the system manager, and can easily perform sensitivity analyses or print reports with elicited and computed values imbedded. All formulas are completely visible and changeable by system managers, yet end-users cannot change the models. Thus, specific formulas or values can be locked in before the model is distributed for use at multiple field sites.

A user can try out values, perform sensitivity analyses, and print reports all from the same program. Moreover, the system can invoke any other program or operating system command as a subroutine, thus allowing easy interfacing with common word processing and spreadsheet packages.

The system is available as a commercial product which runs on the IBM Personal Computer and equivalent computers.

KEYWORDS: Computer security, computer shell, computer aids, quantitative methodology, sensitivity analyses, menu, report generator, boilerplate prose, conditional printing, spreadsheet, DEMOS, training

INTRODUCTION

RISKCALC® is a general purpose shell used to perform computer security risk analyses. It is a program which runs on any IBM Personal Computer (or compatible system) with 512K bytes of memory. It actually has much broader applicability than computer security risk analysis, but since almost all of its use to date has been in this area, we limit ourselves here to this topic.

RISKCALC, by carrying out routine computations and producing reports efficiently and easily, significantly reduces the effort in producing quantitative reports and carrying out sensitivity analyses. It does not, however, replace the analyst. The most difficult task of developing a model of the system under study remains a challenge.

An important aspect of the package is that it comes in two versions — one for the "system manager" and one for the "end user." System managers, using a menu-driven interface, set up files encapsulating assumptions of various models; they also set up files with boilerplate reports into which RISKCALC will later insert elicited and computed values. End users answer questions (Fig. 1) supplied by the system manager (in the models) and can easily perform sensitivity analyses (Fig. 2) or print reports with elicited and computed values imbedded (Fig. 3). All formulas are completely visible and changeable by system managers, yet end users cannot change the models (unlike spreadsheet packages). Thus, specific formulas or values can be locked in before the model is distributed for use at multiple field sites (Fig. 4).

```
                      EXAMPLE RISK ANALYSIS MODEL
                      File Name:encrypt.rc
                         RiskCalc Version 5.0

    VARIABLE                            VALUE            PREVIOUS   %CHG
  1 HARDWARE VALUE                 500,000.00          500,000.00      0
  3 DATA VALUE                   1,000,000.00        1,000,000.00      0
  4 DATA PBLM LIKELIHOOD                 1.00                1.50    -33
  5 PROGRAMS VALUE                 200,000.00          200,000.00      0
  7 TECHNICAL EXPOSURE              12,250.00           17,250.00    -29
  8 TECH COST FACTOR                     1.00                1.00      0
  9 PHYS PLANT VALUE             1,000,000.00        1,000,000.00      0
 11 OTHER ASSETS VALUE           4,000,000.00        4,000,000.00      0
 13 GRAND TOTAL A.L.E.            $102,250.00         $107,250.00     -5
 14 organization name          XYZ CORPORATION

 VARIABLE NUMBER (OR RETURN): ________________________
```

Fig. 1. Question answering for sensitivity analysis.

FEATURES

The user can try various values in answering questions and performing sensitivity analyses, and can print reports, all with the same program. In addition, spreadsheet, graphics, word processing, and other packages (indeed, any IBM PC program) can be invoked from RISKCALC.

All operations are presented in a menu; the user can select the operation he or she desires by striking one key. There is a constant on-screen display which illustrates the current and most relevant values of the fourteen key variables (as defined by the system manager) of the current model. Thus, the user can always "keep his eye on the ball"; it is harder to fall into the trap of spending a lot of time making changes in subsystems which have little effect on the big picture.

DEMONSTRATION REPORT 14 -- RISKCALC INCREMENTAL MODEL BUILDING

REP14 6-20-1986

The following data was used for this report:

Hardware value	$30,000,000
Media value	$100,000
Cost to relocate processing in case of major disaster	$3,000,000

Minor Threats

Threat	Number per year
Network failure	10
Lost listing	24
CPU failure	2

SYSTEM	LOSS EXPECTANCY FROM MINOR THREATS			TOTAL LOSS EXPECTANCY
	DATA INTEGRITY	DATA CONFIDENTIALITY	OPERATIONAL RELIABILITY	
Process A	5,000	10,000	12,000	27,000
Process B	7,000	20,000	9,000	36,000
TOTAL	12,000	30,000	21,000	63,000

Major Threats

Threat	Number per year	Damage factor
Flood	0	80%
Fire	0.02	60%
Earthquake	0	25%
Water leak	0.10	10%

SYSTEM	LOSS EXPECTANCY FROM MAJOR THREATS			TOTAL LOSS EXPECTANCY
	DATA INTEGRITY	DATA CONFIDENTIALITY	OPERATIONAL RELIABILITY	
Process A	200,000	3,000,000	350,000	3,550,000
Process B	670,000	5,600,000	390,000	6,660,000
TOTAL	870,000	8,600,000	740,000	10,210,000

REP14 6-20-1986

Based on the above data, the following annualized loss expectancies are computed:

Major Threat	A.L.E.
Flood	64,496
Fire	483,720
Earthquake	20,155
Water leak	$403,100
TOTAL MAJOR THREAT ALE........	$971,471

Minor Threat	A.L.E.
Network failure	210,000
Lost listing	90,000
CPU failure	66,000
TOTAL MINOR THREAT ALE........	$366,000
GRAND TOTAL ALE..............	$1,337,471

Fig. 2. Possible RISKCALC report with elicited and computed values imbedded.

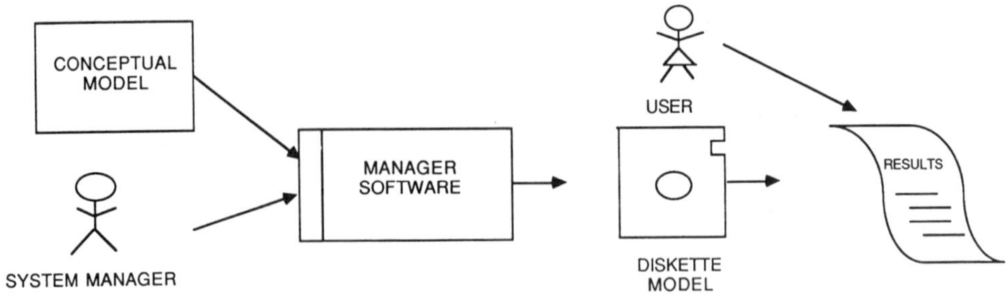

Fig. 3. Use of RISKCALC software by system manager and by user.

```
                    EXAMPLE RISK ANALYSIS MODEL
                    File Name: TEMP$.RC
                       RiskCalc Version 5.0

NAME:          TECHNICAL EXPOSURE
QUESTION:
FORMULA:       (#1*#2+#3*#4+#5*#6)/100
VALUE:         12250.00
DISPLAY:       Y

VARIABLE NUMBER (OR RETURN): 7
Currently a Formula. Formula, Question, Str, Calc-Str, or M-V? F
```

Fig. 4. System manager definition of a variable for later use.

Several model files are provided to describe typical installations, from which the system builder can build a more exact model. A report generator is available which produces risk reports in a form suitable for use by management; it allows the easy creation of a file which intersperses boilerplate prose with values of variables, names of variables, questions asked, formulas, previous values, and percent changes. In addition, text can print conditionally depending on the values of variables.

A common argument against doing any risk analyses is that "it costs too much" and "we can't control the level of effort." With RISKCALC, the user may, if he or she wishes, merely insert the disk into the computer system, turn the machine on, and have the default risk analysis file (as specified by the system manager) loaded in. He or she can insert a few values and produce a report; change one or two values to see what will happen; and produce other reports again with a few more keystrokes. On the other hand, analyses can be many pages long and can take hours of sensitivity analysis.

No structure is superimposed on the model used, so while one can use event trees or other common risk analysis structures, one need not. The tool described here is not limited to a particular methodology; indeed, it supports most quantitative methodologies. In some cases it is more appropriate to use a spreadsheet package, invoking it from RISKCALC with one keystroke, and then, once it is done computing its values, to save them in a file and then invoke a program to move the appropriate values from that file into a RISKCALC file for display, manipulation, sensitivity analysis, or printing.

RISKCALC is a shell and a tool. It does not address the issues of setting up the model and getting decent data. But once a model is defined theoretically, putting it into the computer is easy. The "system manager" need only specify for each variable a name, an associated formula (or question and default value if a value is to be elicited), and whether that variable's values (current and previous) should always be displayed in the window in the middle of the screen (see Fig. 1). No actual programming need be done.

AVAILABILITY

The system, which runs on the IBM Personal Computer and compatibles with 512K of memory and one floppy disk drive (though a hard disk is preferred), is available as a commercial product. It is in use at several large organizations doing risk analysis. One California based vendor of computer products has used it successfully for another application—it has given an "end user" diskette to its customers to allow them to determine how much they will save by buying a given product.

PREVIOUS WORK

RISKCALC is not the only computer aid for risk analysis. In addition to spreadsheets and other commercially available packages,[2] acknowledgement should be made of the previous work of Henrion and Morgan[6] on their DEMOS System. Some differences between it and RISKCALC are the following:

1. RISKCALC stands alone on an IBM PC, whereas DEMOS requires a larger Berkeley UNIX 4.2 environment.

2. RISKCALC has fewer built-in functions, though it does allow creation of libraries of functions. On the other hand, RISKCALC is much easier to interface with other systems to provide a more general package (and access to potentially more functions). In particular, DEMOS does not have the capability of invoking other programs such as spreadsheets and word processors and obtaining values from them for further use with the original model.

3. RISKCALC also allows Boolean and string as well as numeric values and allows system managers to easily specify long (multi-line) questions. It also integrates word processing reports with computed and elicited values which DEMOS apparently does not.

Future plans include incorporation of fuzzy variables and functions[3] and the addition of an expert system at the back end[8] which would suggest various alternative solutions, similar to work being carried out by Dr. Joseph Fiksel of Teknowledge, Inc. of Menlo Park, California. However, these expert systems should be subject to very careful review[1]; computer security may be an essentially new methodological area of risk analysis where distinctive analytical techniques are indicated.[7]

LIMITATIONS OF ALL COMPUTER AIDS

Henrion and Morgan made some very important points in an earlier paper:[6] "No computer aid can force an inexperienced analyst to do a thorough and faultless job of model development and analysis, nor can it reduce the level of creativity and judgment necessary for a good analysis. But it can greatly reduce the effort required for the more mechanical aspects of modeling, and so free the analyst to devote more time to the fundamental issues of structuring the problem, developing an adequate set of models, and clearly communicating them and their implications to others."

Indeed, one thing we have found while training RISKCALC users is that training in the design of the model (much more than in the use of package) is necessary. In computer security there is still very little understanding or knowledge of previous work on risk analysis as a general (not security-specific) topic, and most of the training end users get with automated packages is very problem-specific.

Henrion and Morgan also pointed out that a danger in using these computer aids is that they "enlarge the scope for incompetent analysts to do shoddy work that looks superficially impressive." They also discuss briefly, but well, the danger caused by institutional and political pressures for quick answers.

Automated risk analysis brings with it the promise of easy data input, prepackaged models, faster and more accurate sensitivity analysis, and standardized output reports. However, if misused, it can result in reams of paper containing misleading or wrong results which either focus attention on the wrong problems or create insurmountable credibility problems for the analyst presenting the analysis, or both. There is a need now and for the foreseeable future for real expertise to accompany the model building process. Indeed, the non-computer security risk analysis community evidenced an apprehensive view of expert systems in this area.[1] Perhaps these systems should carry their own risk analysis with them.

SUMMARY

We have discussed a general purpose shell used to perform risk analysis which runs on the IBM Personal Computer or compatibles. After a presentation of its features, we compared it with previous work and noted limitations of all computer aids.

REFERENCES

1. Lance J. Hoffman, Risk Analysis and Computer Security: Bridging the Cultural Gaps, *Proc. 9th National Computer Security Conference*, pp. 156-161, National Bureau of Standards, Gaithersburg, MD (1986).
2. Robert E. Johnston, Playing the Odds, *Infosystems*, pp. 32-37 (April 1, 1987).
3. Kurt J. Schmucker, *Fuzzy Sets, Natural Language Computations, and Risk Analysis*, Computer Science Press, Rockville, MD (1984).
4. S. T. Smith, J. R. Phillips, B. C. Brown, and P. B. FitzGerald, Assessing the Threat Component for the LAVA Risk Management Methodology, *Proc. 9th DOE Computer Security Conference*, Las Vegas, Nevada (May 1986).
5. S. B. Guarro, A. A. Garcia, C. C. Wood, and P. G. Praffinos, LRAM-Livermore Risk Analysis Methodology for Information Systems Security — A Quantitative Methodology for the Systematic Identification and Reduction of Risk, Lawrence Livermore National Laboratory, Livermore, CA, UCAR-10150 Draft–Rev. 1, Dec. 1985.
6. Max Henrion and M. Granger Morgan, A Computer Aid for Risk and Other Policy Analysis, *Risk Analysis* **5(3)**:195-207 (1985).

7. Rex V. Brown, Managing Diffuse Risks for Adversarial Sources (DR/AS) with Special Reference to Computer Security, *Proc. 9th National Computer Security Conference*, pp. 162-167, National Bureau of Standards, Gaithersburg, MD (1986).
8. L. J. Hoffman, Smoking Out the Bad Actors: Risk Analysis in the Age of the Microcomputer, Report GWU-IIST-87-16, Department of Electrical Engineering and Computer Science, The George Washington University, Washington, DC, October 1987.

A Preliminary Model of Radon Exposure

George O. Rogers
Oak Ridge National Laboratory
Oak Ridge, TN

Norman P. Hummon and Daniel J. Strom
University of Pittsburgh
Pittsburgh, PA

ABSTRACT

This paper develops a preliminary model of exposure to radon gases. It considers indoor and outdoor exposure to radon in residential and nonresidential settings. Differential exposure associated with level of activity, age, duration and seasonal variation are also considered. Dose equivalents to bronchial epithelium are estimated as the result of type of exposure, activity differentials, age differences, estimated length of exposure, and seasonal variation. The preliminary model of radon exposure draws on a variety of existing data relating to the social, technical and physical health characteristics of radon exposure.

KEYWORDS: Radon progeny, age dependency, time use, exposure, seasonal variation

INTRODUCTION

Exposure to naturally-occurring radon gas and its short-lived radioactive progeny has been used as a baseline radiation level for comparison with exposures from nuclear technologies and human-made sources of radiation. More importantly the problem of radon in the home is particularly significant because of the amount of time spent in the home, its significant health effects, and the social definition of the home as a "safe haven." Radon gas in homes is currently believed to be the cause of more deaths than all other types of radiation exposure combined, whether from natural or human-made sources (NCRP, 1984). The concentrations of radon in the home are an important consideration in determining exposure, as home "activities" account for the majority of people's time (Gesell, 1983; Nero, 1983; Cohen 1985). As appreciable time is also spent in non-residential buildings, radon concentrations in other indoor locations has also been considered (Cohen *et al.*, 1984). This paper considers indoor exposure to radon and its progeny, both in residential and non-residential settings, as well as outdoor exposures. In addition, this paper examines differential exposure associated with level of activity, age, duration and seasonal variation of exposure to radon and its progeny.

New Risks
Edited by L. A. Cox, Jr., and P. F. Ricci
Plenum Press, New York, 1990

BACKGROUND

Radon gas is present everywhere. It results from the radioactive decay of radium, which is part of the decay series beginning with uranium. "Significant amounts of radon—a natural radioactive gas—accumulate in our houses simply because we tend to build them on the largest source of radioactivity around: the ground" (Nero, 1986:28). A noble gas, radon has a half life of under four days, does not react chemically with surrounding materials, but can migrate into sources of air and water. Radon decays to a sequence of isotopes with short half lives: polonium-218, lead-214, bismuth-214, and polonium-214 (Nero, 1986:30). The health effects of radon are principally associated with these short-lived progeny (NCRP, 1984) because they are chemically active: depositing on airborne particulates, walls, furnishings and even the bronchial epithelium if inhaled. Once in the lungs, these isotopes are likely to complete the decay process to the next long-lived isotope, lead-210. This decay process irradiates the tissue surrounding the point of deposition. High concentrations of radon and its progeny have been associated causally with elevated lung-cancer rates among miners (Archer, 1978; Lundin *et al.*, 1971; Sevc *et al.*, 1976). The extent to which these high dose findings can be extrapolated to people exposed to lower doses is not known. This paper assumes a linear, no-threshold dose response model (NCRP, 1984).

TIME USE AND EXPOSURE LIKELIHOODS

Because exposure to radon varies with the amount of time spent in locations with various radon concentrations, exposure probabilities can be estimated from data on how and where people spend their time. Time budget analysis allows the analyst to determine the proportion of time spent in various locations that are subjected to different exposure rates. This paper focuses on the estimation of direct exposure to an ongoing hazard—radon and its progeny. The approach uses reliable data concerning the use of time by Americans to develop probability estimates of where people are and what they are doing for important exposure differentials.

In 1975, the Survey Research Center at the University of Michigan administered a time budget survey to a national probability sample of U.S. households (Robinson, 1977). The same households participated in a second panel of the same survey in 1981 (Juster *et al.*, 1983). In the 1975 survey, 1519 households were surveyed, which included 1519 respondents and 887 spouses. In 1981, attrition in the panel reduced the sample sizes to 620 households, with 620 respondents, and 376 spouses. The 1981 survey added the time budgets of children in the households. A comparison of 1975 with 1981 results indicates that the attrition in sample sizes caused little, if any, bias in the results. When demographic variables are controlled, these data indicate that the time budgets of U.S. households were amazingly stable over this period of time. The results in this study are from an analysis of the 1981 panel data. For both the 1975 and 1981 surveys, four waves were administered, one during each season of the year. For each wave, respondents and spouses were asked to construct a one day (24 hour) log of activities describing all activities the person engaged in during the previous day. Over the four waves of each survey, respondents reported their activities for two weekdays, and two weekend days. The 1975 survey contains 7207 person-days of data and the 1981 contains 3350 adult-days and 881 children-days. Most of the published reports of these University of Michigan data are based on an aggregated "synthetic week" (Stafford and Duncan, 1978 and 1980; Stafford, 1980). The two weekdays and two weekend days are combined and weighted to estimate how Americans spend time in an average week during the year. For some risk analyses the synthetic week approach does not provide enough detail about the daily schedules of people, which are critical in determining probability distributions of people at specific locations at fixed points in time. The preliminary model of radon exposure presented here examines the probability distributions for fixed locations, summing over seasonal and annual periods, to describe the probability distribution for those periods. The raw time log records contain, in part, the

Table 1. Time Budgets: U.S. Population —
Percent of Time by Place, Season and Age Groups

Season	Age	Home Asleep	Home Awake	Indoors Other	Out-doors	Transit	NEC	Total
Winter	3-5	42.19	19.84	12.26	1.19	2.19	22.34	100
Winter	6-15	38.40	22.06	25.27	2.93	2.91	8.42	100
Winter	16+	34.35	32.70	23.82	2.43	5.30	1.40	100
Spring	16+	35.72	31.40	21.05	5.38	5.88	0.57	100
Summer	3-5	44.91	17.45	11.85	1.28	3.31	21.21	100
Summer	6-15	41.17	23.51	19.39	4.55	4.10	7.28	100
Summer	16+	36.18	33.05	19.54	4.37	5.82	1.04	100
Fall	16+	33.60	31.02	27.74	1.76	5.50	0.37	100

following items: respondent identification number, day of week, month, date of interview, activity code (a typology of 233 detailed activities), time activity began, time activity ended, secondary activity code, and elapsed time for activity. Typically, about 30 records describe the activities for a respondent for each day. These raw data are processed in two ways to form a period-activity data structure.

The Michigan time budget activity codes cover 233 detailed activity types. The detail is best illustrated by a few examples. Activities in the home, such as meal preparation, are broken down into several subcategories including cooking, cleanup, and other. Transit activities include subcategories of travel to and from work, travel in search of employment, to and from shopping, and to and from day care facility. Both primary activities and secondary activities are recorded. Primary activities dominate a particular period, while secondary activities are engaged in simultaneously with primary activities, e.g., listening to the radio while preparing a meal. To use these data in the estimation of radon gas exposure, they are collapsed from 233 activity codes to 6 broader categories reflecting an activity-location structure. The collapsed activity codes are: 1) at home asleep, 2) at home active, 3) indoors other, 4) outdoors, 5) in transit and 6) not elsewhere classified (NEC).[a] Table 1 summarizes the percent of time spent in these activities by season of the year and age groups. Thus:

> The most dominant feature of the average annual time budget is being at home asleep. Comprising 34.8 percent of an individual's total daily time on average.... The second most dominant feature of the annual time use budget is the time being active in the home. This primary activity comprises 23.9 percent of the typical day. Together, in-home active and asleep categories account for 58.8 percent of the total daily activity of adult Americans (Hummon *et al.*, p. 368, 1987).

MODELING RADON EXPOSURE

The predominant contributions to absorbed dose to the bronchial epithelium are from the short-lived radioactive progeny of radon 222, namely, polonium 218 (halflife, 3.05

a. The not elsewhere classified category accounts for time that is not attributable to one of the other major categories, or time gaps in the respondent's time log.

Table 2. Absorbed Dose to Bronchial Epithelium from Radon Progeny Depends on Ten Factors in Addition to the Radon Concentration (Adapted from NCRP, 1984)

	Unattached fraction of polonium 210
	Degree of radioactive equilibrium among progeny
	Particle deposition models
	Particle size distribution
	Physical dose calculation
*	Breathing pattern as a function of activity level
*	Bronchial morphometry as a function of age
	Mucociliary clearance rate
	Mucus thickness
	Location of target cells
*	Denotes the factors emphasized here.

Table 3. Pittsburgh Radon Concentrations

Location:	Home Basemt	Home 1st Fl	Home 2nd Fl	Indoors Other	Out-doors
Radon-222, pCi/L	6.33	2.37	2.01	0.49	0.34
Average:	Annual	Annual	Annual	Annual	Summer
GSD	2.10	2.35	2.60	2.11	1.64
Winter/Summer = 1.6 (Cohen, 1986)					

minutes), lead 214 (26.8 minutes), bismuth 214 (19.7 minutes), and polonium 214 (0.00016 seconds). The absorbed dose can be inferred from concentration measurements of radon 222, but such inference requires the knowledge or assumption of many factors, as shown in Table 2 (adapted from NCRP, 1984). We examine the dose as a function of location, age, and level of physical exertion for a hypothetical population exposed to radon 222 levels such as those measured in Pittsburgh, Pennsylvania.

While some authors have reported no significant variations of indoor radon with seasons (George and Breslin, 1980), most authors report three-fold variations from summer to winter (UNSCEAR, 1982, Annex D). These variations are due to the decreased indoor-outdoor air exchange rate in the winter, and open windows in the summer. Cohen reports that the average ratio of indoor radon concentration from winter to summer is 1.6. We assume that spring and fall values are geometric means of these values, and equal to the annual average value. For scaling average annual indoor radon concentrations, we use factors of $1.6^{1/2}$, 1.0, $1.6^{-1/2}$, and 1.0 for winter, spring, summer, and fall, respectively (Cohen and Gromicko, 1986). Outdoor radon levels are taken as constant (UNSCEAR, 1982). Radon concentrations at home, in public buildings and outdoors are presented in Table 3. These concentrations, which account for location and season of the year, are used throughout this work (Cohen *et al.*, 1984; Cohen, 1985).

Table 4. Scaling Factors for Various Levels of Physical Activity

	Volume	Frequency	Correction Factor*
Resting	1	1.0	.64
Light Activity	2	1.5	1.00
Heavy Work	3	2.0	1.44

*Correction factor is derived from Hofmann *et al.* (1979), and is discussed in detail in the text.

The amount of radioactivity inhaled depends on the number of breaths per minute and the volume per breath; both are functions of the level of physical activity (Table 4). The physical activity correction factors in Table 4 are derived from the work of Hofmann *et al.* (1979) by computing the product of respiratory frequency, tidal volume, and fractional retention of radon progeny in the tracheo-bronchial (TB) region and the pulmonary (P) region of the lung, as specified in the ICRP lung model (ICRP, 1975).

We have used the values of Hofmann et al. for the TB region (as opposed to the pulmonary region) in our calculations, since most radiogenic lung cancers have been seen in the upper airways, primarily in Weibel generations 2-10 (NCRP, 1984). The more recent calculations of Harley and Pasternak (1982) were not used because they do not encompass a comprehensive range of breathing rates, that is, there are no values available for heavy work. At least some time is spent by average people in heavy work such as shoveling snow, mowing grass, raking leaves, bicycling, jogging, exercising, etc., and the higher intake rate should not be ignored. The Harley and Pasternak numbers show only an average of 1.34 times as much dose for "active" adult males as for "resting" adult males, while the Hofmann numbers show 1.56 times as much dose for "light activity" and 2.25 times as much dose for "heavy work."

The two fundamental components of the model are (a) the concentration of radon gas and its progeny in various locations, and (b) the disposition of chemically active progeny attached to airborne particulates in the lung. This analysis uniquely uses the time budget data to estimate the extent to which people spend time in areas likely to have concentrations of radon progeny on the one hand, and level of activity in which they are engaged on the other. Table 5 summarizes the mapping of radon gas concentrations and physical activity into the time budget activities in particular locations. These assumptions link the fundamental elements of the model via the time budget activity and reflected location. They are interpreted as percent of time spent in the activity exposed to locational concentrations (A) or physical activity distributions (B). For example, being at home asleep is 100% resting, and 60% takes place on the second floor, 35% on the first floor and 5% in the basement. The conditions set forth in Table 5 represent a baseline model.

Age-dependent Dose Modification Factors (ADMFs) have been adapted from Hofmann et al. and are shown in Table 6 (Hofmann *et al.*, 1979). These factors are consistent with those of the NCRP (NCRP, 1984) and take account of age-dependent differences in bronchial morphometry, mucociliary clearance rate, breathing rate and deposition. They are normalized to a model 30-year old male used in radiation dosimetry and known as "Reference Man" (ICRP, 1975), so that a dose calculated for Reference Man can be multiplied by the ADMF to estimate the dose to an individual of a different age.

Some additional miscellaneous assumptions complete the model. We assumed a 50% equilibrium factor, resulting in a conversion factor of 0.005 WL/[pCi/L of radon-222]. The

Table 5. Mapping Radon Concentration and Physical Activity Data into Time Budget Data (Baseline Model)

	Percent of Time Budget Activity							
	A. Location					B. Physical Activity		
	Home							
Activity	Basement	1st Floor	2nd Floor	Other Indoor	Outdoors	Resting	Light Activity	Heavy Work
Home Asleep	5	35	60			100		
Home Awake	15	70	15				90	10
Indoors Other				100			80	20
Outdoors					100		50	50
Transit					100		100	
NEC				100			100	

Table 6. Age-dependent Dose Modification Factors (Adapted from Hofmann *et al.*, 1979)

Age	ADMF
0-2	1.7
3-5	2.2
6-16	1.9
16+	1

WL is the "Working Level," a measure of potential alpha energy concentration in air (NCRP, 1984). For Reference Man in an environmental atmosphere, we used the NCRP factor of 0.7 rad/WLM, where the WLM is the "Working Level Month," exposure to one WL for a 170-hour "occupational month" (NCRP, 1984). To correct for environmental exposures, we used a factor of 8766 hours per year divided by 170 hours per occupational month = 51.6 occupational months per environmental year, followed by the self-evident conversion of 0.25 year/season. We have adopted the dose-equivalent to absorbed dose Quality Factor, Q = 20 rems/rad (NCRP, 1984).

COMPUTATION AND ESTIMATION

A dose equivalent for each age group and each time budget activity was computed by mapping the radon concentrations versus location in Table 3 into the time budget activity categories in Table 5, part A. Each location-weighted time budget activity was further modified by the physical activity percentages in Table 5, part B, using physical activity correction factors in Table 4. The result was multiplied by the ADMFs in Table 6.

Table 7. Seasonal Annual Dose Equivalents (rems) to Bronchial Epithelium in Pittsburgh

Years of Age	Season				Total
	Winter	Spring	Summer	Fall	
3-5	3.58		2.18		
6-16	3.14		2.04		
16+	1.91	1.49	1.21	1.47	6.07

A seasonal dose equivalent was computed by summing overall time budget activities, and an annual dose equivalent was computed by summing overall seasons.

The resulting exposure to radon and its progeny is 6.07 rems annual adult dose equivalent to the bronchial epithelium (Table 7). The present estimates are limited to Pittsburgh (due to the lack of detailed radon concentration data for the nation). The estimated rate is 2 times greater than existing national estimates of 3 rems annual adult dose equivalent (NCPR, 1984). The first floor radon concentrations in Pittsburgh (Cohen, 1985) are very similar to reported rates in other parts of the country, such as Central Maine, and the Pacific Northwest (Nero *et al.*, 1986), but the distribution by location, required to link the time budget data, is not available.

While estimates of exposure for children under 3 to 5 years of age and 6 to 16 years of age in the spring and fall were not possible (due to the lack of time budget data for children in the spring and fall) the winter and summer estimates alone provide 5.76 and 5.18 rems dose equivalence. These winter and summer estimates combined represent half the year, even though they do not represent half the annual dose equivalent. Even so it is apparent that the exposure for children is very high, if they are exposed at all during the spring and fall quarters of the year. For example, an additional .75 rem in each of these quarters, a dose equivalent of more than a third lower than any in Table 7, would result in estimates of 7.26 and 6.68 rems per year, respectively.

Analysis indicates that minor modifications in the model assumptions, say less than 10% of the values in Table 5, create very little variation in the resulting estimates. For example, shifting physical activity while home awake from 90% light activity and 10% heavy work, to 80% light activity and 20% heavy work, or to 100% light activity changes the resulting estimates about +2.5%.

The baseline model results in an annual adult dose equivalent of 6.07 rems to the bronchial epithelium (Table 7), while these two models estimate the annual dose at 6.22 and 5.92 respectively. Even fairly significant alterations in the assumptions have relatively minor impact on the resulting dose estimates. By assuming that no time is spent in the basement, where concentrations are highest, and equally splitting time spent at home awake between 1st and 2nd floors, even though 2nd floors are most often bedrooms where more limited waking time is spent, and ignoring whether half the houses in the United States even have a second floor, the annual adult dose equivalent declines to 5.19 rems, or a decline of about 14.6 percent. Similarly, equally proportioning the time spent home awake among the basement, 1st floor and 2nd floor, a condition that comes closer to reflecting the trend of basement game and family rooms, including exercise facilities, raises the dose equivalent rate about 13% to 6.87 rems per year for the average adult.

Table 8. Percent of Dose Received by Time Budget Activity, Age Group and, Season

Age	Season	Home Asleep	Home Awake	Indoors Other	Out-doors	Transit	NEC	Total
3-5	Winter	44.6	42.3	4.6	0.4	0.5	7.7	100
3-5	Summer	48.6	38.1	4.6	0.4	0.8	7.5	100
6-16	Winter	40.0	46.3	9.3	0.8	0.7	2.9	100
6-16	Summer	41.2	47.4	6.9	1.3	0.9	2.4	100
16+	Winter	30.9	59.4	7.6	0.6	1.1	0.4	100
16+	Spring	32.6	57.9	6.8	1.4	1.2	0.2	100
16+	Summer	32.1	59.2	6.1	1.1	1.2	0.3	100
16+	Fall	31.2	58.0	9.1	0.5	1.2	0.1	100

This model of radon exposure is unique because it links the examination of where people are and what they are doing with the greatest contribution to the dose. Table 8 presents the percent of dose received by activity/location, season, and age group. The dominant location of radon exposure is clearly in the home. Children receive more than 85% of their overall exposure at home. Adults receive around 90% of their overall exposure at home. Adults receive the largest amount of exposure while at home awake, while children receive the majority of their overall dose while sleeping. This stems from differential amount of time spent at home asleep by adults and children (Table 1). The adult dose is heavily concentrated while awake at home (59.4% in winter[b]). Using the winter season as a bench-mark, the home asleep category accounts for the second largest exposure rate (30.9%) for adults. Other indoor locations account for 7.6% of the exposure, while other activities including transit, outdoors and not elsewhere classified account for the remaining part of the dose (2.1%) in winter. The primary source of the seasonal exposure variation was found to be the differential concentrations in the homes. The variation in time budget activity by season of the year does have an impact, although it is quite small. Hence, the overall estimates of exposure are largely dependent on time budget activity, with significant variations by age of individual and season of the year.

CONCLUSIONS AND IMPLICATIONS

The most important finding of this preliminary model of exposure to radon and its progeny is that the majority of the dose is received at home. For adults, activities while awake at home dominate exposure at home, while for children time spent sleeping seems to be most important. The differences in exposure by age stem from two primary sources: age-dependent dosimetry and age dependent time use. These differences lead to a significant variation among the age groups available, even though the model was unable to characterize exposure for "infants," due to the lack of time use data. Seasonal variations in time use do not significantly affect proportions of doses received, although seasonal variation in radon concentration engenders significant seasonal variation in dose equivalent. Concern over radon and its progeny in the home is certainly justified by the exposure estimates presented herein.

---

b. The other seasons are similar although lower. All seasons are reported in Table 8.

Measurement of radon concentrations in homes is extremely important. This preliminary model indicates that sample measurements should be taken where dose is greatest. For example, if there are no children in the home, measurements should be taken in the living areas of the home. If children, particularly young children, are living there, measurements in sleeping areas are very appropriate. Finally, because radon concentrations are quite large in the basement, families spending significant time in basement living areas should be encouraged to place samplers in these areas. Because radon and its progeny are a naturally occurring phenomenon, they represent a national problem, affecting millions of households, perhaps being second only to smoking as a cause of lung cancer. EPA has recommended a limit of 0.02 WL for homes (USEPA, 1986). An examination of policy alternatives is appropriate, but beyond the scope of this paper. Such alternatives might include building code limitations on basement use, or requiring ventilation systems in new houses in affected areas, or providing tax incentives for families to retrofit existing homes, limiting or regulating underground housing in some areas. The estimates of exposure to radon are indeed significant; further study including the examination of policy alternatives is certainly indicated.

REFERENCES

Archer, V. E., 1978, Summary of data on Uranium Miners, Oak Ridge National Laboratory Report, ORNL-5348, pp. 23-25.

Cohen, B. L., Kulwicki, D. R., Warner, Jr., K. R., and Grassi, C. L., 1984, Radon Concentrations Inside Public and Commercial Buildings in the Pittsburgh Area, *Health Phys.* **47**:399-405.

Cohen, B. L., 1985, Survey of One-Year Average Rn Levels in Pittsburgh Area Homes, *Health Phys.* **49**:1053-1059.

Cohen, B. L., and Gromicko, N., 1986, Surveys of Radon Levels in U.S. Houses, *Health Phys.* (under review).

George, A. C., and Breslin, A. J., 1980, The Distribution of Ambient Radon and Radon Daughters in Residential Buildings in the New Jersey-New York Area, in *Natural Radiation Environment III*, CONF-780422, Vol. 2, pp. 1272-1292, National Technical Information Service, Springfield, VA.

Gesell, T. F., 1983, Background Atmospheric ^{222}Rn Concentration Outdoors and Indoors, *Health Phys.* **45**:289.

Harley, N. H., and Pasternak, B. S., 1982, Environmental Radon Daughter Alpha Dose Factors in a Five-Lobed Human Lung, *Health Phys.* **42**:789-799.

Hofmann, W., Steinhausler, F., and Pohl, E., 1979, Dose Calculations for the Respiratory Tract from Inhaled Natural Radioactive Nuclides as a Function of Age — 1, *Health Phys.* **37**:517-532.

Hummon, N. P. *et al.*, 1987, Time Budget Analysis and Risk Management: Estimating the Probabilities of the Event Schedules of American Adults, in *Risk Assessment and Management*, pp. 365-374, L. Lave, ed., Plenum Press, New York.

ICRP 1975 International Commission on Radiological Protection, 1975, Report of the Task Group on Reference Man, ICRP Publication No. 23, Pergamon, Oxford.

Juster, F. T., *et al.*, 1983, 1975–1981 Time Use Longitudinal Panel Study, The University of Michigan, Institute for Social Research, Survey Research Center, Ann Arbor.

Lundin, E. F. *et al.*, 1971, Radon Daughter Exposure and Respiratory Cancer: Quantitative and Temporal Aspects, NIOSH, HIEHS Joint Monograph No. 1.

National Council on Radiation Protection and Measurements (NCRP), 1984, Evaluation of Occupational and Environmental Exposures to Radon and Radon Daughters in the United States, NCRP Report No. 78, NCRP, Bethesda, MD.

Nero, A. V., 1983, Airborne Radionuclides and Radiation in Buildings, *Health Phys.* **45**:303.

Nero, A. V., 1986, The Indoor Radon Story, *Technological Review*, pp. 28-40.

Nero, A. V. *et al.*, 1986, Distribution of Airborne Radon-222 Concentrations in U.S. Homes, *Science* **234**:992-997.

Robinson, J. P., 1977, *How Americans Use Time: A Social-Psychological Analysis of Everyday Behavior*, Praeger Publishers, New York.

Sevc, J. *et al.*, 1976, Lung Cancer in Uranium Miners and Long Term Exposure to Radon Daughter Products, *Health Phys.* **30**:433-437.

Stafford, F., 1980, Women's Use of Time Converging With Men's, *Monthly Labor Review*, pp. 57-59 (December).

Stafford, F., and Duncan, G. J., 1978, Market Hours, Real Hours and Labor Productivity, *Economic Outlook USA*, pp. 74-76 (Autumn).

Stafford, F., and Duncan, G. J., 1980, The Use of Time and Technology by Households in the United States, *Research in Labor Economics* **3**:335-375.

United Nations Scientific Committee on the Effects of Ionizing Radiation (UNSCEAR), 1982, Ionizing Radiation: Sources and Biological Effects, United Nations, New York.

U.S. Environmental Protection Agency, 1986, Citizens Guide to Radon, OPA-86-004, Washington, D.C.

Power-Frequency Fields: Risk Without Measure

H. Keith Florig
Carnegie-Mellon University
Pittsburgh, PA

ABSTRACT

The issues of whether and how to regulate people's exposure to power-frequency fields are complicated by evidence suggesting that, if such fields can adversely affect human health, those effects may not increase monotonically with either the duration or the intensity of exposure. This paper clarifies this situation by presenting a taxonomy of the many physical quantities that describe different aspects of the relationship between people's encounters with fields and possible health effects. In particular, the concept of "dose" is defined in the power-frequency context. This paper also describes a strategy for regulating people's exposure to transmission line fields that does not depend on the availability of a "more is worse" measure of dose.

KEYWORDS: Sixty hertz, exposure, dose, regulation, transmission lines, electric and magnetic fields

INTRODUCTION

Power-frequency (50/60 Hz) electric and magnetic fields are produced by all components of the power system including high voltage transmission lines, neighborhood distribution circuits, household appliances, and wall wiring. These fields can induce electrical currents in the body by three different mechanisms. The changing electrical charges on power lines or household appliances can, by Coulomb's law, attract or repel free charges in the body. The changing magnetic fields of power lines and appliances establish eddy currents in the body by Faraday's law. Finally, the electric and magnetic fields of power lines and appliances can induce voltages on non-energized conductors such as automobiles, fences, and the chassis of ungrounded appliances. Contacts with any of these objects typically results in a power-frequency current flowing into the body through the contact point. Enough is known about these induction mechanisms and the intensities of fields produced by various sources that one can estimate the magnitude of body currents induced in different exposure situations.[10] Examples of such estimates are given in Fig. 1.

Only a couple of decades ago, no one worried about the possibility that power-frequency fields might affect human health. Today, this concern has a standing position on the public agenda and consumes regulatory, legal, and research resources at a rate of several millions of dollars per year. Concerns about the health effects of power-frequency fields have their most direct origins in the introduction of 765 kV transmission lines in the late 1960's. These lines not only transport several times more power than the lower voltage

New Risks
Edited by L. A. Cox, Jr., and P. F. Ricci
Plenum Press, New York, 1990

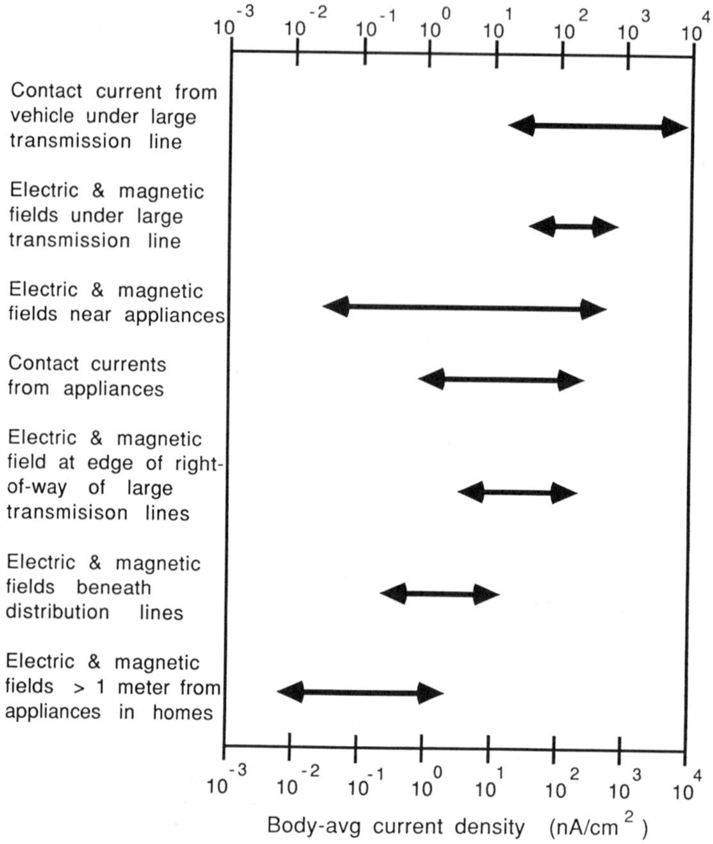

Fig. 1. Ranges of current density induced in the body under seven common exposure conditions. Body-averaged current density is expressed in nano-amperers per square centimeter.

lines that they superseded; they also have greater environmental impacts due to aesthetic and corona effects. In addition, the electric shock and spark discharge hazard from induced voltages on large ungrounded conductors located near 765 kV lines is greater than that from lower voltage lines. The attention drawn to transmission lines by these "side effects" resulted in a closer examination of all aspects of the health and environmental impacts of power transmission. From this examination emerged concerns that transmission line fields might have a direct influence on human health.

Since the mid-seventies, regulatory and legal proceedings involving this issue have arisen in many states and a number of public and private organizations have sponsored research aimed at determining whether or not power-frequency fields can affect human health. To date, the U.S. Department of Energy, the Electric Power Research Institute, New York State, and a several utility companies have spent over $70 million ('86 dollars) on this research.[3,12] The results of this work have been summarized in a number of recent reviews.[1,9,22] Many of the bioeffects studies performed over the last decade found no effect of field exposures. In some cases, experimental artifacts indicated an effect when, in fact, none was present. There have been a number of epidemiological studies suggesting a statistical link between exposure to electromagnetic fields and cancer. Studies by Wertheimer and Leeper[20,21] suggest that distribution line magnetic fields are leukemogenic, and work by Milham[15] and others suggests that workers in the electric and electronics industries suffer an increased incidence of certain cancers. These studies have been criticized on a number of grounds, including failure to control for possible confounding variables and uncertain techniques for classifying subjects by their field exposure.

Besides the many studies with either negative or controversial results, there have been a number of apparently artifact-free *in vitro* and laboratory animal experiments that show that power-frequency electric and magnetic fields *can* induce changes in some biological systems under some circumstances. One of the most interesting aspects of these results is that many reported effects are not related to the intensity or duration of exposure in any simple way. There are windowed effects that appear at some intensities and frequencies but not at others,[4] effects with clear intensity thresholds,[6] effects that appear only after several weeks of exposure but then do not change with subsequent exposure,[23] effects that do not change with increasing field strength,[13,23] effects that appear only for a short period, after which the system adapts and the effect disappears,[18] and effects that appear only for pulsed waveforms with very specific shapes.[2] None of these results demonstrate that exposure to weak power-frequency fields has an adverse impact on human health. These results do suggest, however, that if such adverse effects are ever found, they may not be monotonically related to either exposure duration or exposure intensity.

Power-frequency fields are thus an unusual agent from a risk assessment and regulatory perspective. Although research has not demonstrated that health effects exist, neither has it shown that health effects are negligible. If power-frequency fields do affect people's health, the existing bioeffects literature does not support any particular index of that risk. One cannot make the assumption so common in environmental toxicology that "if some is bad, then more is worse." Traditional regulatory strategies such as cost-benefit and cost-effectiveness, therefore, cannot be applied to the problem of regulating people's exposures to power-frequency fields.

This paper offers two contributions to the problem faced by those involved with making decisions about whether and how to regulate exposures to transmission line fields. First, the concepts and quantities that describe people's exposures to power-frequency fields are clarified using a generic taxonomy of agent encounter and effect. In particular, the concept of "dose" is defined in the power-frequency context and a clear distinction is made between fundamental and speculative descriptors of exposure. The second goal of this paper is to describe an approach to regulating exposures to transmission fields that makes use of available exposure data yet does not depend on estimates of relative health impact.

RELATIONSHIP BETWEEN ENCOUNTER, DOSE, AND EFFECT

Uncertainties about the relationship between exposures to power-frequency fields and possible health effects make it difficult to distinguish between fundamental and speculative descriptors of exposure events. This section draws on work by other investigators and on precedents in environmental toxicology and radiology to derive a generic taxonomy of the quantities that describe people's encounters with a given agent and the resulting health effect. This taxonomy allows one to order the various quantities that describe power-frequency exposures. Most importantly, the concept of power-frequency dose can be defined in a general sense.

Previous Efforts to Structure Encounter-Effect Relationship

At least two other investigations have dealt explicitly with the problem of defining the relationship between field encounter and effect. The International Radiation Protection Association (IRPA) recently identified three different types of quantities that are commonly used to describe non-ionizing radiation and its interactions with biological tissue and other matter.[11] These are (1) "radiometric" quantities that describe the physical characteristics of the field (e.g., power density), (2) quantities that describe the interaction of the field with matter (e.g., attenuation, reflection, absorption), and (3) dosimetric quantities adequate for the specification of exposure of biological objects to non-ionizing radiation (e.g., specific absorption rate (SAR) for microwave exposures that cause tissue heating). While these concepts have clear applications for high power exposures in the far field of radio-frequency sources where biological effects due to tissue heating are well established,[17] the applicability of these classifications to power-frequency exposures is not clear. The IRPA states that

> Due to the poor understanding of interaction mechanisms, the dosimetric concepts are not fully developed for extremely low frequency fields. At the moment, exposure conditions are often quantified by a statement of the unperturbed external electric and magnetic fields strengths and the duration of exposure.[11]

While field strength and duration of exposure may provide an unambiguous description of exposure conditions in laboratory experiments in which field intensity and other electrical characteristics do not change with time, these quantities are not sufficient for characterizing people's encounters with power-frequency fields, since the kinds of fields to which people are exposed are typically non-uniform in both space and time.

Silva, Zaffanella, and Hummon[19] have described an "activity systems model" for estimating the time-integral of the equivalent[a] unperturbed electric field to which people are exposed. This work identifies three key inputs for characterizing exposures to power-frequency electric fields. These are the locations and unperturbed field profiles of sources, the locations of people relative to sources, and various activity-related factors that account for changes in shielding, body posture, and the impedance of the body to ground.

The Generic Model

Toxicology and radiology have firmly established models of the relationship between a person's encounter with a particular agent and the health effect resulting from that encounter. These models have various elements in common. These common elements can be said to form a generic taxonomy of encounter and effect that can be applied to any agent, including power-frequency fields. This generic encounter-effect taxonomy is illustrated in Fig. 2. One of the fundamental determinants of exposure events is the *environmental concentration* of the agent as a function of space and time. The particular measure of environmental concentration depends on the agent. Measures used for chemical agents include mass per unit volume of air or water. Measures used for radioactive agents include energy deposition per unit mass of air and disintegrations per second per unit mass of material.

A second fundamental determinant of population exposures is *people's locations relative to sources* as a function of time. The confluence of these two elements—concentration and people's locations—permits an *external exposure history* to be described for any given person. A person's external exposure history is the time function of the environmental concentration (i.e., concentration external to the body) of an agent to which that person is exposed.

---

a. The "equivalent" electric field is defined in terms of a reference measure of exposure intensity which Silva *et al.* take to be the electric field averaged over the torso of a person standing in a uniform vertical field and grounded through the feet.

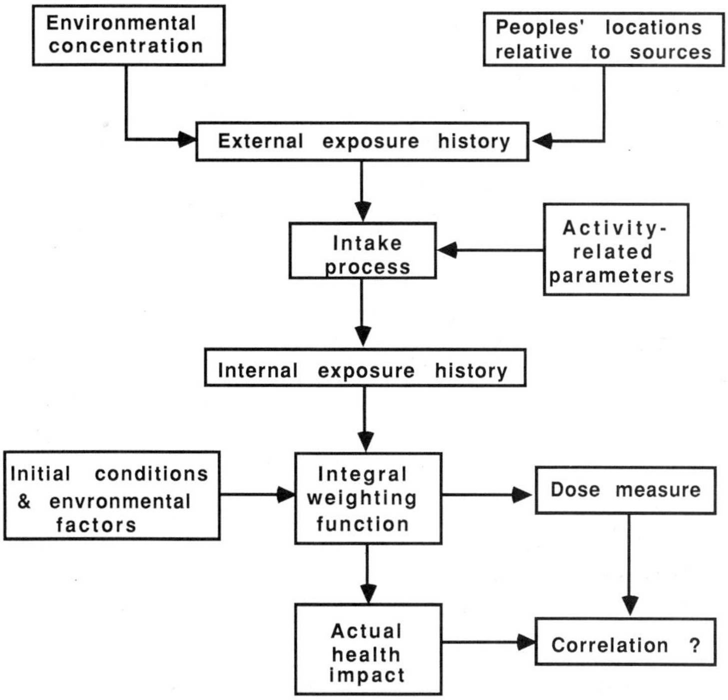

Fig. 2. Generic model of the relationship between encounters with a given hazardous agent and the resulting health impact.

The next element in the link between encounter and effect is the *intake process*. This relates environmental concentrations of an agent to concentrations within the body or to the influx of the agent across the environment-body boundary. The intake process accepts two inputs, the external exposure history and various *activity-related parameters*. Examples of activity-related parameters include breathing rate for exposures to air pollutants and body posture for exposure to external gamma radiation. The intake model produces a time function of the concentration of an agent within various organs or a time function of the rate of influx of an agent into the body. This time function can be called an *internal exposure history*. Examples of internal exposure histories include time functions of the ingestion rate of a chemical toxin, the concentration of a toxic chemical species in a particular organ, and the energy deposition rate associated with an exposure to ionizing radiation.

The effects induced by exposures to most harmful environmental agents can depend on a number of factors that are independent of the internal exposure history. These include the *initial conditions* of the exposed biological system (e.g., phase of circadian cycle, age, sex) and the behavior of or exposure to other *environmental factors* during the exposure period (e.g., weather, light/dark cycle, diet). In theory, there exists an *integral weighting function* that can exactly map initial conditions, the time function of exposure to other environmental factors, and the internal exposure history into a specific level of *health impact*. Since it is often difficult, if not impossible, to consider initial conditions or environmental influences in making estimates of the health impact of a particular agent, people generally settle for a procedure that predicts health impact based only on some convenient measure of the internal exposure history. This satisficing measure of the internal exposure history is referred to as *dose*. The utility of any particular dose measure is determined by how accurately it *predicts* effect.

Applying the Generic Model to Power-Frequency Fields

The various quantities that describe encounters with power-frequency fields can be classified according to the framework in Fig. 2. Measures that describe the environmental concentration and external exposure history of power-frequency electric fields include the unperturbed field and, in cases in which the body is close to the source, the source potential or charge.[b] Magnetic field intensity and flux density serve as measures of the environmental concentration and external exposure history of the magnetic field. The "environmental concentration" of contact currents is described by the short-circuit currents and locations of all sources of contact current. The external exposure history for contact currents is the time function of the contact currents to which a person is exposed, assuming that all contacts draw the full short-circuit current.

The intake component of a power-frequency exposure model accepts as inputs these external exposure histories and various activity-related factors such as the orientation of the body axis relative to the field, body posture, and the impedance of the body to ground. For encounters with power-frequency fields, the internal exposure history can be specified by the vector current density within the body as a function of position and time. Other possible measures of internal exposure history such as the internal electric field, internal magnetic field, and charge on the body's surface can be derived from the current density function. It is apparent that, if power-frequency exposures can affect human health, *the time function of internal vector current density is the only power-frequency information needed to predict the effect.* That is, the current density function completely captures all of the power-frequency-related factors that might be related to effect.

How does one choose the means for collapsing an internal exposure history, a function of both time and position within the body, into a single number or "dose" that is a strong predictor of effect? In toxicology and radiology there are often compelling theoretical considerations that suggest particular measures of dose. The proportionality between reaction rate and the concentration of a reactive chemical species suggests that the quantity ingested is likely to correlate strongly with effect for hazardous chemical agents. The intuitively appealing hypothesis that the biological effects of ionizing radiation should be related to the number of ionization events leads to a measure of radiation dose defined by energy deposition per unit mass of tissue. Innumerable experiments have demonstrated a high correlation between these measures of dose and the health impacts of exposures to these agents.

In the case of exposures to power-frequency fields, there are no strong theoretical bases for choosing a dose measure. As mentioned above, the available experimental data suggest that if power-frequency fields do adversely affect human health, such effects may not be highly correlated with either the intensity or the duration of exposure. That is, the measures of power-frequency dose that best correlate with effect may be non-linear functions of the internal exposure history. Other experimental evidence suggests that measures of dose from power-frequency fields may be poor predictors of effect unless they also consider the local geomagnetic field.[4] These considerations indicate the speculative nature of calculations that carry the encounter-effects model of Fig. 2 beyond the stage of the internal exposure history. This should serve to caution policy-makers about using measures of dose to draw conclusions about (1) the relative health impacts of exposures from transmission line fields and fields from other sources or (2) the relative biological effectiveness of typical exposure histories of people and exposure histories of laboratory animals.

---

b. The unperturbed field can be a poor indicator of exposure intensity when the body is close to the source because (1) the unperturbed field may be non-uniform over the distance of a body dimension, and (2) the charge on the source may be affected by the presence of the body.

Despite the fact that no measures of power-frequency dose have yet been shown to correlate strongly with some public health endpoint, there are several roles that dose estimates can play in both research and policy-making. First, laboratory experimenters or epidemiologists are justified in applying various dose measures to the exposure histories of their subjects to determine which measure best "explains" the differences observed between exposed and control groups.

Second, dose estimates can help characterize exposures when detailed exposure histories are not available. Measurements of exposure histories are complicated by limitations on data storage and by limitations on the locations in or on the body at which monitoring devices may be placed. Many investigators have adopted a compromise approach to measurement of exposure histories in which the instrumentation records only certain aspects of the entire exposure history. Deno,[7] for instance, developed an electric field "dosimeter" to measure the amount of time that the device spends in various ranges or bins of perturbed field intensity. The device is worn either on an arm band or on a hard hat. More recently, Deno and Silva[8] designed and tested a more sensitive device with an active sensing area that covers the entire torso. Lo and colleagues[14] have built an instrument that takes occasional samples of the local 60 Hz magnetic fields. The active sensing area of this device is about 5 cm^2.

The "doses" recorded by these devices provide a valuable characterization of certain aspects of human exposures to power-frequency fields that are otherwise difficult to estimate. Since these devices record only time-integrated and spatially selective aspects of the exposure history, however, their data should not be used as a basis to compare the level of possible health impacts from different sources unless these aspects of exposure history are shown to correlate with effect. One useful application for these devices in the absence of such correlations is to calibrate theoretical exposure models.[5]

A third use for estimates of dose in the absence of biological evidence in support of any particular measure of dose is to prioritize the gathering of data on the various factors that determine population dose (other than the definition of dose itself). These factors have been described in detail elsewhere.[10] Examples include the population density around transmission lines, the amount of time that people spend indoors, and the degree to which houses shield their occupants from the electric field of nearby overhead lines. Priority factors can be identified by ranking the inputs to exposure models by the uncertainty that each contributes to a hypothetical measure of population dose. One can expect the ranking of factors by their contribution to uncertainty in dose to be relatively insensitive to the choice of dose measure.[10]

A final use for estimates of dose in the absence of biological evidence in support of any particular dose measure is to evaluate the similarities between the exposure histories of people living along transmission line corridors and the exposure histories of those not exposed to transmission line fields. Such evaluations are an important part of the "similarity" approach to regulating people's exposures to transmission line fields. This approach is described below.

SIMILARITY-BASED CONTROL OF TRANSMISSION LINE EXPOSURES

If human exposures to power-frequency fields have a deleterious impact on health, the current evidence on biological effects does not allow one to assume that those health effects increase with either the duration or the intensity of exposure. Traditional regulatory approaches such as cost-effectiveness and cost-benefit, therefore, are not applicable to the problem of regulating exposures to transmission line fields. This section describes an alternative regulatory strategy that makes use of available exposure data but does not require a dose measure that is correlated with effect. This alternative, which is referred to below as "similarity" or "equity-based" control, is described by Morgan *et al.*,[16] who write:

> Almost everyone is exposed to electric appliances, wiring, and power-distribution lines, but only a few people live next to transmission lines. In the absence of clear scientific evidence, those few might argue that they are being asked to bear greater exposures than the rest of us. On the grounds of equity, arguments might be made for designing transmission facilities so that the electromagnetic fields to which they expose nearby people are "comparable" (similar) to the fields of other common sources.

The principle of this approach is that if the bioeffects science does not show either that the risk of exposure to power-frequency fields is significant or that the risk is insignificant, then society should not expect persons living along transmission corridors to accept exposures that are substantially different from exposures that people voluntarily accept from sources in the household environment. While this strategy does not guarantee that transmission line risks are zero, it does assure that the risks to people living near transmission line risks are comparable to risks assumed by others.

The equity-based strategy as described by Morgan *et al.* need not be limited to comparisons based on the intensity attribute of field exposure. The Florida Electric and Magnetic Fields Science Advisory Commission (FEMFSAC), on which Morgan served, observed that the definition of "comparable" exposure is scientifically arbitrary.[9] That is, there is no scientific basis for choosing what aspects of exposure (intensity, temporal characteristics, target organ, etc.) should be made similar.

While neither explicitly states so, both Morgan and colleagues and the FEMFSAC imply that, under equity-based exposure control it is sufficient to make exposures from transmission lines *alone* similar to exposures from household sources. A more equitable strategy is to demand that the *combined* exposure from transmission line and household sources of a person living along an transmission line be made similar to exposures from household sources of those not living along transmission lines.

In the sections below, a formal approach to similarity-based control is presented and the extent of its practical applicability is explored. For simplicity, the analysis is restricted to only electric field exposures. Extension to magnetic fields is straightforward.

Definition of Similarity

What does it mean for exposures to those living near and far from transmission lines to be similar? A formal definition of "similarity" is proposed below.

Let $J(r,t)$ be the vector current density electrically induced in a person as a function of time t, where r is position within the body. This function completely defines a person's internal exposure history. Let $J_T(r,t)$ be the internal exposure history for a person living along a transmission line. That is, $J_T(r,t)$ represents induction from both transmission line and household sources

$$J_T(r,t) = J_{\text{line}}(r,t) + J_{\text{house}}(r,t) .$$

Let $J_{H,i}(r,t)$ be the internal exposure history due to only household sources for the ith person living far from any transmission line. Finally, let $\boldsymbol{H}$ be the collection of all of the $J_{H,i}(r,t)$.

"Similarity" requires that $J_T(r,t)$ be a plausible member of $\boldsymbol{H}$. That is, if $J_T(r,t)$ does not stand out in an ensemble of exposure histories formed by $J_T(r,t)$ and a sample of the

$J_{H,i}(r,t)$ drawn from $\boldsymbol{H}$, then one can claim that the transmission line resident whose exposure is represented by $J_T(r,t)$ is not subjected to exposures that are noticeably different than exposures associated with only household situations.

But how can the differences between $J_T(r,t)$ and the household exposure histories in $\boldsymbol{H}$ be quantitatively measured? One approach is to define a measurement operator that can collapse each exposure history into a scalar. Let $\boldsymbol{O}_k$ be the $\boldsymbol{k}$th of N proposed measurement operators, such that

$$\boldsymbol{O}_k\,[J(r,t)] = a_k\,,$$

where a_k is a scalar. Applying $\boldsymbol{O}_k$ to each of the internal exposure histories in $\boldsymbol{H}$,

$$\boldsymbol{O}_k\,[J_{H,i}(r,t)] = a_{H,i,k}\,,$$

one can form a probability distribution $\boldsymbol{A}_{H,k}$ on the resulting collection of the $a_{H,i,k}$. This distribution can be said to characterize $\boldsymbol{H}$ under the particular measure, $\boldsymbol{O}_k$. If $\boldsymbol{O}_k[J_T(r,t)] = a_{T,k}$ lies within $\boldsymbol{A}_{H,k}$, then one can conclude that $J_T(r,t)$ is similar to household exposures under the measure $\boldsymbol{O}_k$.

The degree to which $a_{T,k}$ lies within $\boldsymbol{A}_{H,k}$ may be quantified using the concept of a membership function borrowed from the theory of fuzzy sets.[24] One can define a membership function, μ, that accepts as inputs the distribution, $\boldsymbol{A}_{H,k}$, of a given household exposure measure and the value, a_k, of the $\boldsymbol{k}$th measure of some exposure history and produces as output a number between zero and one that specifies the "houseness" of the exposure history for which a_k is a measure. A plausible membership function might be as follows:

$$\mu(a_k, \mathbf{A}_{H,k}) = \left[2\int_{-\infty}^{a_k} \mathbf{A}_{H,k}\, da\right]^{1/2} \text{ for } \int_{-\infty}^{a_k} \mathbf{A}_{H,k}\, da \leq 1/2$$

$$\mu(a_k, \mathbf{A}_{H,k}) = \left[2\int_{a_k}^{\infty} \mathbf{A}_{H,k}\, da\right]^{1/2} \text{ for } \int_{-\infty}^{a_k} \mathbf{A}_{H,k}\, da > 1/2\,.$$

In this example, the closer $\mu(a_k, A_{H,k})$ is to 1, the closer a_k lies to the median of $\boldsymbol{A}_{H,k}$, and the more the exposure history that generated a_k can be said to look like the household exposure histories in $\boldsymbol{H}$, at least under the particular measure $\boldsymbol{O}_k$ used to generate $\boldsymbol{A}_{H,k}$.

Since we require that the similarity approach is applied only when no robust dose measure for human health effects is available, there are no preferred choices for the form of any particular $\boldsymbol{O}_k$. Under what collection of measurement operators, then, should similarity be evaluated? One approach is to evaluate similarity under a number of measures chosen to capture different attributes of exposure histories. Alternatives might include the maximum current density at any point in the body at any time, body and time-averaged current density, maximum body-averaged current density at any time, maximum time-averaged current density anywhere in the body, average current density in a particular organ, measures with intensity and time thresholds, measures that depend on the time-derivative of current density, and measures that are frequency-dependent.

Since the value of the membership function, μ, will in general be different for each measure $\boldsymbol{O}_k$, one needs a procedure for combining values of μ for different k. If one places roughly equal weights on each integral measure, one could judge similarity based on the average value of the membership function for all measures used. That is, if $\boldsymbol{O}_k$ is the $\boldsymbol{k}$th of N proposed measures of $J(r,t)$, and if $\mu(a_T, A_{H,k})$ is the membership function that measures the extent to which $\boldsymbol{O}_k(J_T[r,t]) = a_{T,k}$ lies within the distribution $\boldsymbol{A}_{H,k}$ obtained by applying

O_k to the household exposure histories in $\boldsymbol{H}$, a quantitative measure of how similar $J_T(r,t)$ is to the household exposure histories in $\boldsymbol{H}$ could be represented by

$$\bar{\mu}_T = \sum_{k=1}^{N} \mu(a_{T,k}, \mathbf{A}_{H,k})/N \text{ where } 0 \leq \bar{\mu}_T \leq 1 .$$

Of course, one might want to assign weights to each $\mu(a_{T,k}, A_{H,k})$ in computing $\bar{\mu}_T$. One might feel, for instance, that exposures in the torso should be weighted more heavily than exposures in the toe, or that changes in the direction of rotation of the phasor that describes currents induced by three-phase lines are likely to be less important than changes in the rms magnitude of that phasor. If w_k is the weight assigned the $\boldsymbol{k}$th of N proposed measures of $J(r,t)$, the measure of similarity can be written as

$$\bar{\mu}_T = \sum_{k=1}^{N} W_k \; \mu(a_{T,k}, \mathbf{A}_{H,k}) / \sum_{k=1}^{N} W_k \text{ where } 0 \leq \bar{\mu}_T \leq 1 .$$

The range of values of $\bar{\mu}_T$ for which $J_T(r,t)$ is considered similar to the household exposure histories in $\boldsymbol{H}$ can be determined by considering the distribution obtained when the average value of the membership function is computed for many individual *household* exposure histories. The average value of the membership function for the $\boldsymbol{i}$th household exposure history is

$$\bar{\mu}_{H,i} = \sum_{k=1}^{N} W_k \; \mu(a_{H,i,k}, \mathbf{A}_{H,k}) / \sum_{k=1}^{N} W_k .$$

Given the distribution of the $\bar{\mu}_{H,i}$ one need only choose some percentile of this distribution as the threshold for similarity determination. If $\bar{\mu}_T$ is smaller that this threshold, one can declare that $J_T(r,t)$ is not similar to the collection of exposure histories in $\boldsymbol{H}$.

Discussion

The conditions under which it is proper to apply the similarity approach are quite specific and deserve reiteration. These are:

1. The agent is not a known hazard. (If it were, one could presumably identify a dose measure and apply a cost-benefit approach to control.)

2. It cannot be shown that the risk of the agent is *de minimus*. (If the risk was *de minimus*, regulation would not be required.)

3. There is no basis for assuming that, if the agent were a hazard, a "more is worse" dose measure would apply. (If a "more is worse" dose measure applied, a cost-effectiveness approach could be used.)

Regulatory programs based on the similarity approach should be abandoned once any of these three conditions no longer hold. If, for instance, a theory of power-frequency bioeffects gained enough experimental support that the risks of power-frequency fields could be shown to be negligible, then controls would no longer be needed. If follow-up studies to the epidemiological work of Wertheimer and Leeper[21] show that cancer incidence is indeed a function of magnetic field intensity, then exposure controls could presumably be based on one of the traditional regulatory approaches.

The advantage of the similarity approach over a purely political approach to regulating exposures to transmission line fields is that the similarity approach makes explicit use of available information about people's exposures to fields from transmission lines and household sources without making unwarranted assumptions about the validity of any single measure of dose.

One problem with the similarity approach is that the data required to analyze similarity under many randomly chosen measures are unobtainable from a practical standpoint. Physical and resource constraints will limit the kinds and numbers of measures that can be used. This adds a degree of arbitrariness to the definition of similar.

Another source of difficultly in applying the similarity approach lies in deciding what sources are admissible for defining people's exposure histories. Are perceptible or painful contact currents from appliances, though fairly common, to be considered part of one's voluntary exposure to household sources? The electromagnetic environment of the home contains contributions from televisions, dimmer switches, microwave ovens, and other sources at frequencies other than power-frequency. Should these other frequencies be included in the exposure history?

The last objection to the similarity approach that will be discussed here is one raised by the Florida Electric and Magnetic Fields Science Advisory Commission. It argued that "precise 'equity' or 'comparable exposure' of the public to all actual, potential, or perceived risks is not possible in practice nor compellingly desirable in theory."[9] If one restricts the application of the similarity approach to only those agents that meet the three criteria listed above, the number of candidates is likely to be very small and perhaps manageable. These criteria admit only potential risks for which the "more is worse" assumption does not apply.

REFERENCES

1. American Institute of Biological Sciences, Biological and Human Health Effects of Extremely Low Frequency (ELF) Electromagnetic Fields, prepared for the U.S. Navy, 1985. Available from the National Technical Information Service, NTIS ADA152731.
2. C. A. L. Bassett, S. N. Mitchell, and S. R. Gaston, Pulsing Electromagnetic Field Treatment in Ununited Fractures and Failed Arthrodeses, *Journal of the American Medical Association 247* **5**:623-628 (1982).
3. Bioelectromagnetics Society, Bioelectromagnetics Funding Survey, *Bioelectromagnetics Society Newsletter*, 68 (May/June 1986).
4. C. F. Blackman, S. G. Benane, J. R. Rabinowitz, D. E. House, and W. T. Joines, A Role for the Magnetic Field in the Radiation-Induced Efflux of Calcium Ions from Brain Tissue in Vitro, *Bioelectromagnetics 6* **4**:327-337 (1985).
5. V. L. Chartier, T. D. Bracken, and A. S. Capon, BPA Study of Occupational Exposure to 60-Hz Electric Fields, *IEEE Transactions on Power Apparatus and Systems PAS-104* **3**:733-744 (March 1985).
6. C. F. Dalziel, The Threshold of Perception Currents, *IEEE Transactions on Power Apparatus and Sytems 73* 990-996 (August 1954).
7. D. W. Deno, Monitoring of Personnel Exposed to a 60 Hz Electric Field, in *Biological Effects of Extremely Low Frequency Electromagnetic Fields*, Phillips, R., M. Gillis, W. Kaune and D. Mahlum, Eds., Technical Information Center, U.S. Department of Energy, 1979. Proceedings of the Eighteenth Annual Hanford Life Sciences Symposium, Richland, Washington, CONF-781016.
8. D. W. Deno and J. M. Silva, Method for Evaluating Human Exposure to 60 HZ Electric Fields, *IEEE Transactions on Power Apparatus and Systems PAS-103* **7**:1699-1706 (July 1984).
9. Florida Electric and Magnetic Fields Science Advisory Commission, Biological Effects of 60-Hz Power Transmission Lines, prepared for the Florida Department of Environmental Regulation, Tallahassee, FL.

10. H. K. Florig, Population Exposure to Power-Frequency Fields — Concepts, Components, and Control, Ph.D. Thesis, Department of Engineering and Public Policy, Carnegie-Mellon University, 1986.
11. International Radiation Protection Association, Review of Concepts, Quantities, Units, and Terminology for Non-Ionizing Radiation Protection, *Health Phys.* **49(6)**:1329-1362 (December 1985).
12. R. Kavet, Biological Effects of Electric Fields: EPRI's Role, *IEEE Transactions on Power Apparatus and Systems PAS-101* **7**:2115-2121 (July 1982).
13. A. R. Liboff, T. Williams, Jr., D.M. Strong, and R. Wistar, Jr., Time-Varying Magnetic Fields: Effect on DNA Synthesis, *Science* **223**:818-820 (February 24, 1984).
14. C. C. Lo, T. Y. Fujita, A. B. Geyer, and T. S. Tenforde, A Wide Dynamic Range Portable 60-Hz Magnetic Dosimeter with Data Acquisition Capabilities, *IEEE Transactions on Nuclear Science NS-33* **1**:643-646 (February 1986).
15. S. Milham, Mortality in Workers Exposed to Electromagnetic Fields, *Environmental Health Perspectives* **62**297-300 (1985).
16. M. G. Morgan, H. K. Florig, I. Nair, and D. Lincoln, Power-line Fields and Human Health, *IEEE Spectrum 22* **2**:62-68 (February 1985).
17. National Council on Radiation Protection and Measurements, Biological Effects and Exposure Criteria for Radiofrequency Electromagnetic Fields, NCRP Report No. 86. Available from NCRP Publications Office, Bethesda, Maryland.
18. R. S. Rosenberg, P. H. Duffy, and G. A. Sacher, Effects of Intermittent 60-Hz High Voltage Electric Fields on Metabolism, Activity, and Temperature in Mice, *Bioelectromagnetics 2* **4**:291-303 (1981).
19. J. M. Silva, L. Zaffanella, and N. Hummon, An Activity Systems Model for Estimating Human Exposure to 60 Hz Electric Fields, *IEEE Transactions on Power Apparatus and Systems PAS-104* **7**:1923-1929 (July 1985).
20. N. Wertheimer and E. Leeper, Electrical Wiring Configurations and Childhood Cancer, *American J. of Epidemiology* **109**:273-284 (1979).
21. N. Wertheimer and E. Leeper, Adult Cancer Related to Electrical Wires near the Home, *International J. of Epidemiology* **11**:345-355 (1982).
22. World Health Organization, United Nations Environment Programme, and the International Radiation Protection Association, ISBN 92-4-154095-8, Volume: Environmental Health Criteria 35—Extremely Low Frequency (ELF) Fields. World Health Organization, Geneva, Switzerland, 1984.
23. B. W. Wilson, L. E. Anderson, D. I. Hilton, and R. D. Phillips, Chronic Exposure to 60-Hz Electric Fields: Effects on Pineal Function in the Rat, *Bioelectromagnetics 2* **4**:371-380 (1981).
24. L. A. Zadeh, Fuzzy Sets, *Information and Control* **8**:338-353 (1965).

Computer Security Risk Analysis

Lance J. Hoffman
The George Washington University
Washington, DC

ABSTRACT

Twenty-four experts in risk analysis and computer security spent two and a half days at an invited workshop and concluded that there are nine areas where significant problems exist which currently limit the effectiveness of computer security risk analysis. These are discussed in this report, and suggestions are given for future research to advance the state of the art.

The specific problems and issues addressed are standard definitions, guidelines on when to do risk analysis, risk communication, need for test beds and baseline studies, case data collection, desirability of a general risk model as a conceptual framework, lack of metrics, difficulties in transferring knowledge between the fields of risk analysis and computer security, and the appropriateness of various efforts to automate the risk analysis process.

KEYWORDS: Computer security, metrics, automation, annual loss expectancy, diffuse risks, adversarial sources, risk communication, test beds, uncertainty, fuzzy sets, Dempster-Shafer calculus

INTRODUCTION

This paper presents the conclusions of an interdisciplinary workshop on risk analysis in computer security which brought together 24 experts to conduct an interdisciplinary assessment of risk analysis applied to computer security. It presents problems and issues that need further treatment or resolution before significant progress can be made in computer security risk analysis.

COMPUTER SECURITY

Computer security refers to the technological safeguards and managerial procedures which can be applied to computer hardware, programs, and data to assure that organizational assets and individual privacy are protected [NBS74]. Computer security was first recognized as a problem in the mid-1960's when time-shared and multiprogrammed computer systems emerged [Turn81]. It was shown then that the existing operating systems were unable to prevent determined users from making unauthorized accesses to or modifications of other users' data or programs, or from intercepting data on

New Risks
Edited by L. A. Cox, Jr., and P. F. Ricci
Plenum Press, New York, 1990

communications lines between terminals and computers or between computer systems. The problem was especially acute for computer applications in national security.

Technical computer system security can be regarded as consisting of the following subareas: software, hardware, communications, and data base security. Each of these may involve theoretical aspects, design and implementation problems, and testing or verification of correctness. In addition, there are questions of security policies, assessment of security risks, effectiveness, costs, and auditing [Mart73, Hoff77, Denn82]. With the increased realization that most computer security problems and solutions are not technical, and with the increasing number of real-world systems at risk, the risk analysis process has received heightened attention.

POTENTIAL CONTRIBUTIONS OF RISK ANALYSIS TO COMPUTER SECURITY

While risk analysis is an interdisciplinary area, computer security specialists have in general not used models and techniques from other fields to the extent possible; one hope of the workshop was to cross-fertilize computer security with risk analysis so that both fields might benefit.

It was realized at the inception that there were problems to be overcome: most previous risk analysis literature dealt with intentional as opposed to random events; many thought that the types of losses in computer security were harder to quantify than those in areas to which traditional risk analyses had already been applied. Unlike traditional risk analyses which deal with concrete consequences (such as money or lives lost), often computer security concerns are diffuse and intangible (e.g., military advantage, competitive advantage, privacy protection). Asset values were more difficult to arrive at, since data can be an ambiguous asset whose value varies significantly over time and by use. And perhaps more so than in other areas, operational priorities put real-world constraints on what computer security measures will be used. Still, the existing risk analysis literature may be able to shed light on the costs of breached security.

In the past, the computer security community has often used ill-fitting adaptations of risk analysis methods that were developed for problems which were significantly different. The majority of computer security risk analyses have used annual loss expectancies (ALEs), a method well-suited to and used by insurance companies. However, unlike insurance risks, computer security risks often are multiple and not readily specified (by money, injury, or death). Often the potential losses are intangible—related to national defense, corporate goodwill, or other nonmonetary assets. Unlike traditional risk analysis problems, computer security problems tend to lie in a relatively uncharted area, that of diffuse risks from adversarial sources where the objects at risk and the nature of the risk may be diffuse and where the source of the risk may be a malevolent adversary. These risks might be characterized as points on the right of Fig. 1.

At the workshop, the experts agreed that there was a large literature on decision processes, analytic processes, and uncertainty assessment. Valuation issues were also discussed, including eliciting value tradeoffs, noneconomic cost estimates and multi-attribute techniques, which may have some bearing on computer security. On the other hand, some were doubtful that risk analysis can provide "gems" for computer security. Several participants mentioned the idiosyncratic nature of typical real world problems.

There has been little of the basic research done to quantify commonly needed area-specific parameters that has characterized the more established applications of risk analysis. For example, we have very little prior data on which to base estimates of the likelihood of adverse events (comparable to the studies of component failure rates in safety risk analysis).

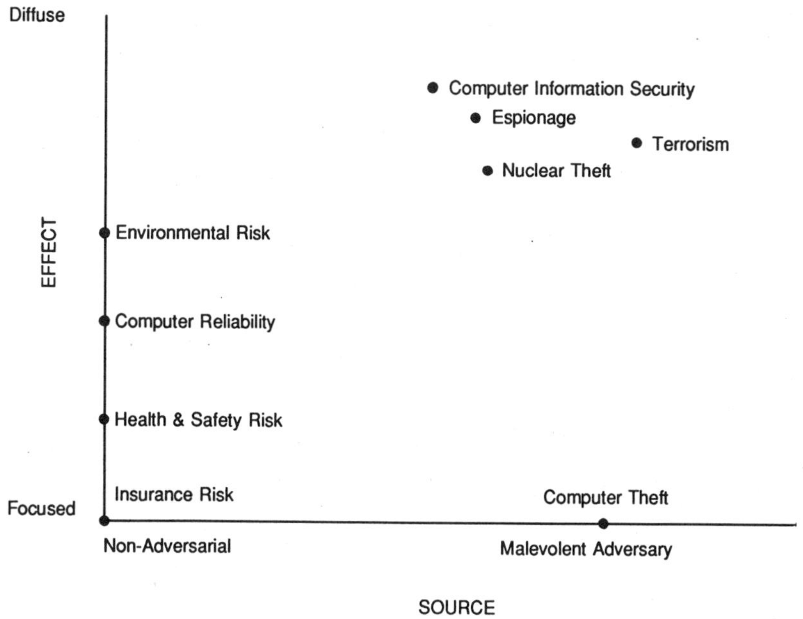

Fig. 1. Categorization of risk analysis problems by type and source of risk [Brow86].

In addition, there has been little research on value tradeoffs to guide computer security policy decisions (comparable to the roughly $1 million value put on a human life in airline safety risk analysis, and $1,000 cost equivalence of a man-rem of exposure used in nuclear regulation).

The paucity of this basic research to date is not surprising when one notices that risk analysis is a multidisciplinary field, often requiring a blend of skills, and that the development of the field must cross disciplinary boundaries and breach semantic barriers, as discussed below.

MAJOR PROBLEMS IN COMPUTER SECURITY RISK ANALYSIS

There are several areas where significant problems exist which currently limit the effectiveness of computer security risk analysis. The problems appear to be tractable; by bringing resources to bear on them, the quality of computer security risk analysis can be significantly improved. This section describes the specific areas and suggests appropriate actions to take.

Semantic Problems Due to a Lack of Standard Definitions

A critical area where research is needed is that of standard definitions in risk analysis for computer security. As with all workshops which bring together workers from different fields, this one saw a considerable amount of time spent resolving semantic problems and definitions of terms. While there are accepted terms in both fields, sometimes the same term means two different things, depending on the field; in some cases, there are differences among workers even in computer security; in a few cases, there are out and out conflicts between the commonly accepted definitions in risk analysis and usage in computer security. As an example, one discussion group spent about half their initial session wrestling with the distinction between "methodologies" and "implementations." By and large, however, these conflicts appear resolvable if addressed promptly; both fields are relatively new and the workshop participants, leaders in these fields, were quite willing to work together to agree on one common set of terms.

Several years ago the National Bureau of Standards issued a computer security glossary which contains several hundred definitions; a more recent one is [NCSC85] from the National Computer Security Center. Even so, in the workshop we had some trouble with communication even among computer security people. There does not appear to be any widely used formal glossary of terms for the risk analysis field in general. It is absolutely necessary that the two disciplines communicate well; therefore, harmonization of existing definitions is needed and a common glossary of terms would be helpful.

Absence of Guidelines on When to Use Risk Analysis

Another important issue is the question of when to use risk analysis. Too often in the past, computer security practitioners have either avoided initiating a risk analysis due in part to fear of its cost or, alternatively, initiated full-fledged analyses which slavishly used methodologies better suited for other problems and, as a result, cost more and produced less of value than desirable and possible. Apparently no guidelines exist regarding when a risk analysis or a specific methodology should be initiated or terminated.

It is inappropriate to use certain kinds of risk analysis when the potential benefit gained from such use is too small. A related issue is how far to go. One may often need a relatively simple or even cursory analysis and nothing more. Alternatively, one may begin an analysis, suspend it for a time, and then resume it as events warrant. Finally, in some cases a full-blown exhaustive analysis may be called for.

Communicating Risk Management Options to Decision Makers

The results of risk analyses are often not communicated effectively by computer security professionals to their management. Too often risk analyses, once done, sit unread upon a shelf. This is often the case even when the results are appropriate and accurate and the analysis was done efficiently. Typically this happens because top management was not convinced of the need to take any corrective action. Often management reacts to events rather than planning protective measures in advance. The first major national conference on communicating risk to the public took place in January 1986 (the National Conference on Risk Communication, Mayflower Hotel, Washington, D. C., January 29-31, 1986, sponsored by The Conservation Foundation, National Science Foundation, Environmental Protection Agency, American Industrial Health Council, and the University of Southern California). Efforts to improve this situation may go farther than anything else to mitigate, in the long run, the real problems risk analysts are asked to address.

Lack of Test Beds and Respected Available Studies

There is a critical lack of test beds and of well-known, respected impartial risk analyses of computer system security to use as examples. It is difficult to evaluate the performance of various risk analysis methodologies or tools without a suitably rich test bed. With one or two exceptions, such a research asset does not exist and the field's growth and maturity are hindered by incomplete testing and information.

A few test beds and pilot studies which incorporated traditional risk analysis techniques with real problems from computer security (and other application areas) and using real data would be very important in advancing research progress by helping us to develop a general model and conceptual framework for computer security risk analysis based on real world experience. Specific research related to concrete problems in computer security can and should be undertaken on specific test beds with the notions of generalization, methodological development, and demonstration in mind, while at the same time carefully focusing the efforts. The scope must be narrow enough to be manageable; one would hope that at the end the result could be a highly visible successful application of known techniques and models to a real computer security problem. Later efforts can be held up to that standard.

Problems with Available Data

Risk analysis and computer security experts have long bemoaned the fact that there is very little real-world case data available on which to base estimates or assessments. Real world case data collection would help matters, both in the general risk analysis case and in the specific computer security case. Examples of such data are the relative frequencies of different types of security breach and the measurable impact on security of specific incidents and of various risk management measures. Eliciting this information is not easy, and relatively highly trained individuals must be available to encode the data for later use.

Desirability of a General Risk Model as a Conceptual Framework

A general risk model which could be used by both the risk analysis and computer security communities would provide significant benefits to both communities in the areas of testing, methodology evaluation, and completeness of analysis. Such a conceptual framework should be very flexible, able to handle numerous types of risk analysis computer security problems and all external policies imposed on the problem. It should be easily refinable as new knowledge (for example, from test beds and pilot studies) becomes available or new constraints appear. It should be acceptable to both communities and rich enough to represent just about all computer security situations. Without such a model, "we will continue to have methods that are as different as apples, oranges and pears and which will produce results which cannot be compared" [Katz85].

While a formal model is necessary to define all of the functional relationships, it must also allow for approximations to those functions we do not know how to define. This will allow us to implement tools in the near term that represent simplifications of more complex interrelationships. As we obtain more insight about interrelationships, we should be able to replace the simplistic representations with more complex ones. Furthermore, a model must allow alternative methods for different purposes; it should allow for appropriate combination of qualitative and quantitative input data; and it should be consistent when applied to the same problem by two different teams of people using the same data.

The interrelationships between threats, threat frequencies, vulnerabilities, safeguards, risk, outcomes, etc., should all be described in a formal way so that a common understanding of the risk analysis process emerges [Katz85]. Such a model might not be highly mathematical, since it would ideally make effective use of case-based and quasi-statistical data.

Dearth of Metrics for Risk Assessment and Risk Management

One significant lack today is metrics for risk analysis and risk management. There is no currently accepted set of criteria against which all methods can be compared. It is difficult evaluate or to convey the advantages and disadvantages of a given methodology or tool when no accepted evaluation metric exists. Until such a set of criteria is developed, we can expect proliferation of various methodologies, most of which are adaptations of previous ones (even if the previous ones have serious deficiencies).

Appropriateness of Automation

Recently, there has been a proliferation of computer security risk analysis tools and products [Hoff85, Fiks85, Henr85] which are particularly useful in getting the risk analysis started, allowing quick sensitivity analyses, and producing reports. Despite these advantages, the risk analysis community was quick to caution against premature development or use of quantification, automation, or expert systems. They were concerned that "Issues of modeling, uncertainty assessment and judgment of value require the kinds of thinking that software tools can't provide right now." These tools should be used only for routine calculations, such as implementing the logic of fault trees.

All agreed that there should not be a rush to computerize and that automated tools should not claim or imply more than is there; in essence, automated tools are fine as a means, not as an end. Some dangers are involved here also [Henr85], including the lack of credibility when the systems don't deliver what they promise and the locking in of inappropriate methodologies by premature computerization and inflexible software.

RESEARCH AREAS

All participants agreed that rather than awarding a set of grants or contracts in specific areas, funding sources should pick projects which attack specific problems which are likely to generalize and work them well, possibly from several points of view. They emphasized that care should be taken not to award grants or contracts in currently "hot" areas, such as expert systems development, at the cost of starving research which could prove more important in the long run. Some major areas which could be profitably explored further for computer science include these:

1. *Value functions*. The computer security application area lacks clearly defined value functions which trade off risks against benefits. As an example, an initial attempt at a multi-attribute utility function related to congressional options on a number of issues in information security is given in [Brow86].

2. *Treatment of dependencies*. Another area of concern is the treatment of dependencies and common cause events; ignoring interdependencies may lead to unrealistic assessments of risk that cause the analysis to lack credibility.

3. *Non-binary events (partial failures, etc.)*. There is no generalization of binary logic to multiple states that handles reliability with degraded performance (e.g., a safeguard that works some of the time or which partially works). This is another potential area to explore.

4. *Representation of deliberate events*. Traditionally, risk analysis has treated random or "accidental" events as opposed to those triggered deliberately (e.g., sabotage). Some options which might suitably augment previous methods include the use of simulation with people, game theory and empirical evidence.

5. *Uncertainty*. Even though the risk analysis community has already made some important progress in the area of uncertainty by using probability distributions to quantify uncertainty about exposures and severity of effects [NRC75, Howa76, Morg84, Henr85, Cox81], a number of important questions with respect to uncertainty remain unanswered in the area of computer security. For example [Henr86], should risks be quantified? Should all uncertainties about numerically expressed risks be quantified? Are linguistic expressions of severity and uncertainty ("rarely," "likely") sufficient, or are they inevitably bedeviled by ambiguities? Are probabilities always the best approach? What of fuzzy sets, Dempster-Shafer

calculus, and various other approaches to representing uncertainty, both quantitative and qualitative, developed by researchers in artificial intelligence and expert systems? Studies are needed which compare these approaches (on both theoretical and practical criteria), assess their merits and drawbacks, and start to develop guidelines about which may be appropriate under a given set of conditions.

REFERENCES

(Brow86) Brown, R. V., 1986, Managing Diffuse Risks from Adversarial Sources (DR/AS) with Special Reference to Computer Security: Ideas for a New Risk Analysis Research Area, Working paper 86-1, Decision Science Consortium, Inc., Falls Church, VA (January).

(Cox81) Cox, D. C., and Baybutt, P., 1981, Methods for Uncertainty Analysis: a Comparative Survey, *Risk Analysis* **1**:251-258.

(Denn82) Denning, D. E. R., 1982, *Cryptography and Data Security*, Addison-Wesley, Reading, MA.

(Fiks85) Fiksel, Joseph, 1985, *Automated Threat Assessment for Computer Facilities*, A. D. Little, Inc., Cambridge, MA.

(Henr85) Henrion, Max, and Morgan, M. Granger, 1985, A Computer Aid for Risk and Other Policy Analyses, *Risk Analysis* **5(3)**:195-208.

(Henr86) Henrion, Max, Private communication, January 30, 1986.

(Hoff77) Hoffman, Lance J., 1977, *Modern Methods for Computer Security and Privacy*, Prentice-Hall, Inc., Englewood Cliffs, NJ.

(Hoff85) Hoffman, Lance J., 1985, PC Software for Risk Analysis Proves Effective, *Government Computer News* **4(18)**:58-59 (September 27).

(Howa76) Howard, R. A., Matheson, J. E., and Miller, K. L., eds., 1976, Readings in Decision Analysis, Decision Analysis Group, Stanford Research Institute, Menlo Park, CA.

(Katz85) Katzke, Stuart, Summary of Key Issues, in USAF85.

(Mart73) Martin, J., 1973, *Security, Accuracy, and Privacy in Computer Systems*, Prentice-Hall, Inc., Englewood Cliffs, NJ.

(Morg84) Morgan, M. G., Morris, S. C., Henrion, M., Anaral, D., and Rish, W. R., 1984, Technical Uncertainty in Quantitative Policy Analysis — A Sulfur Air Pollution Example, *Risk Analysis* **4(3)**:201-216.

(NBS74) National Bureau of Standards and Association for Computing Machinery, 1974, Executive Guide to Computer Security.

(NCSC85) National Computer Security Center, 1985, COMPUSECese Computer Security Glossary, NCSC-WA-001-85, Ft. Meade, MD (October).

(NRC75) NUREG-75/014, 1975, Reactor Safety Study, an Assessment of Accident Risks in United States Commercial Nuclear Power Plants, WASH-1400 Study, Nuclear Regulatory Commission, Washington, DC (October).

(Turn81) Turn, R., ed., 1981, *Advances in Computer Systems Security*, Artech House, Inc., Dedham, MA.

(USAF85) Minutes of the Federal Information Systems Risk Analysis Workshop, 22-24 January 1985, Air Force Computer Security Program Office, Gunter AFS, AL [available through Defense Technical Information Center, Alexandria, VA].

Diffuse Risks from Adversarial Sources: An Emerging Field of Risk Analysis

Rex Brown
Decision Science Consortium, Inc.
Reston, VA

ABSTRACT

An essentially new methodological area of risk analysis is emerging, in which the risks are multiple and diffuse and the source of risk is a human adversary. It is typified by computer information security, sabotage, nuclear theft, terrorism and military combat. This paper defines the distinctive methodological needs for assessing and managing this type of risk, relates them to the current state of the art in older branches of risk analysis, and decision analysis, and suggests promising avenues for new developments. Issues covered include formulating risk consequences, evaluating risk consequences, predicting adversarial behavior and complex risk aftermaths, determining institutional and social values, and choosing and implementing options. Innovative technical approaches discussed include game theory, macromodels, plural analysis, multiattribute utility, step-through simulation, and interactive decision theory.

KEYWORDS: Diffuse risks, adversarial risks, computer information security, decision theory, macromodels, plural analysis, step-through simulation, multi-attribute utility theory), decision analysis, consequence evaluation

INTRODUCTION

Risk analysis, as a distinct field of inquiry, has been steadily evolving over the past few decades in terms of the complexity of risks it addresses, and it thus requires increasingly ambitious analytic tools.

Risk situations might be characterized along two dimensions: the source of the risk and the effect of the risk. The source might be represented along a continuum between nonadversarial and a malevolent adversary. The effect of the risk might range between focused (or exact) and diffuse (or inexact). These dimensions of risk analysis are shown conceptually as the two axes of Fig. 1.

Simple Focused Risk, Nonadversarial Source

Historically, risk analysis research first addressed the simplest type of risk: focused risk whose source is nonadversarial, typified by the risk faced by insurance companies. Risk is focused in that it can be expressed along a single, easily measured dimension — money — and the bearer of the risk is a single entity, such as a corporation or an individual.

New Risks
Edited by L. A. Cox, Jr., and P. F. Ricci
Plenum Press, New York, 1990

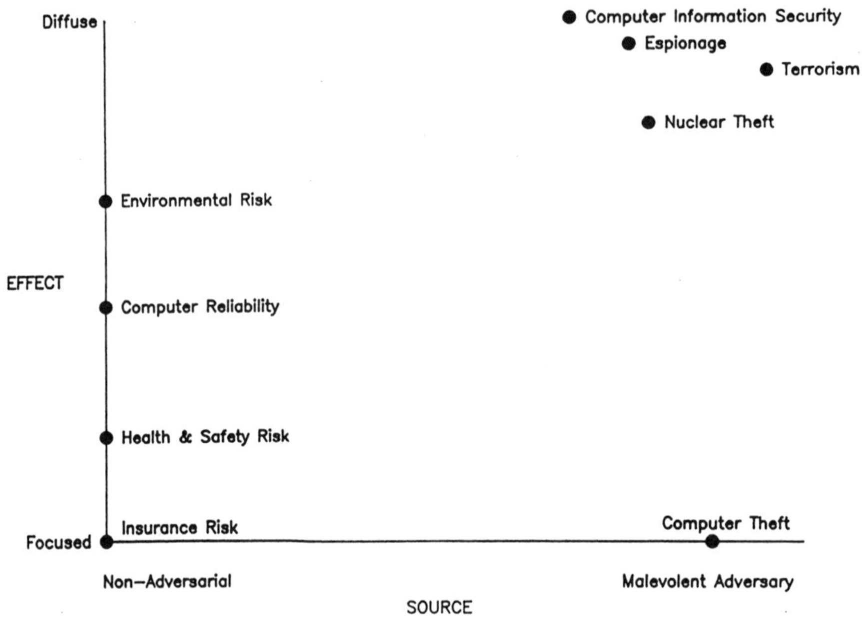

Fig. 1. Categorization of risk management problems by effect and source of risk.

In some of these cases, such as life and health insurance, risk assessment is simplified by the availability of substantial historical records which permit uncontroversial determination of probabilities and require no theory more ambitious than classical or frequentist statistics. In other cases, where essentially unique events are involved, human judgment must play a significant role in the assessment, which calls for new concepts such as personal probability. An example is the recent case where Lloyd's of London insured against the discovery of the Loch Ness Monster (for a manufacturer of Scotch who had offered a million pound reward). However, this still remains the simplest case of risk analysis.

Multiple Focused Risk, Nonadversarial Source

Within the last ten years, risk analysis research has expanded to consider more complex risks, ones that are somewhat more diffuse than the simple risks addressed by insurance companies. This risk is typified by health or safety risks of the type addressed by many governmental regulations.

The risk is somewhat more diffuse than the first case, because the objects at risk are multiple (people) and the risks themselves are multiple (various health effects). However, the risks are defined along easily measured dimensions (such as the number of deaths and days of disability) and the objects at risk are easily specified (such as definable human populations).

The source of this risk is still nonadversarial and is often tied to evolving technology (such as new drugs or nuclear power plants). Although, as with life insurance, quantification of the risks can often be anchored to observed frequencies, human judgment invariably has a significant role to play (for example, in predicting events that have never occurred, such as a reactor core meltdown).

This health and safety area has now achieved some considerable measure of technical maturity in predicting, evaluating and managing the risks involved. Progress has been helped along by significant research funding support, for example, from the Risk Analysis Program at NSF. This area of risk analysis uses, among other techniques, personalized (Bayesian) probability for quantifying risks, and multiattribute utility theory (MAUT) for trading off health consequences and economic cost. However, the version of MAUT commonly used is heavily reliant on attributes which are quantified in natural units (such as number of deaths).

The literature in this area is now quite extensive, and the state of the art is reasonably represented in Refs. 1-6.

Diffuse Risk, Nonadversarial Source

Alongside health and safety risk, interest has begun to focus on risks to the environment. It is a substantially more diffuse risk, but still nonadversarial (i.e. technological). Multiple ill-defined effects are experienced by often equally ill-defined objects at risk (e.g., environmental interest groups). Environmental risk analysis has been spurred during the past decade or so by the National Environmental Protection Act (NEPA), the establishment of the Environmental Protection Agency, and the resulting requirement for environmental assessments and environmental impact statements for a wide range of projects.[7] Like most health and safety risk analysis, work in this field is heavily motivated by government regulation.

DIFFUSE RISKS FROM ADVERSARIAL SOURCES (DR/AS)

More recently, attention has been focusing on risks in which the objects at risk and the nature of the risk are diffuse and the source of risk is a malevolent adversary. Such risks might be termed "diffuse risks from adversarial sources" (DR/AS). They are the main focus of this paper and are characterized by the top right corner of Fig. 1. Computer information security is a prime example of this type of risk. Other examples include theft and sabotage at nuclear and other energy facilities, espionage and terrorism in its many forms.

Multiple effects are not necessarily diffuse. One of the multiple effects of environmental risk might be the destruction of a wildlife sanctuary, which is quite focused. Compare this with the breached security effect of a computer system, which can only be evaluated by considering a typically complex pattern of "aftermaths" leading, for example, to possibly harmful uses of information by potential enemies of the United States.

As shown in Fig. 1, not all risk analysis problem areas fall cleanly in or out of the DR/AS category. For example, theft of proprietary data stored in a computer system ("Computer theft" in Fig. 1), as a subcategory of computer security, has the element of an adversarial source (e.g., the thief), but the risk itself may be monetary (and to this extent has much in common with the financial risk facing insurance companies). Conversely, there are areas where the risk is diffuse (such as certain kinds of environmental impact), but the source is inanimate and technological and in this respect similar to health and safety risk analysis.

A Primitive Risk Analysis Field

Available analytic methodology is not well adapted to the new levels of complexity introduced by this class of risk. There have been isolated instances of promising methodological innovation developed in the process of solving specific practical problems, for example, consequence evaluation for nuclear safeguards. However, little has been done to unify or generalize them. Systematic approaches have been developed on analogous problems (for example, modeling adversaries in negotiation and competitive situations, and modeling complex future scenarios in military planning). However, they have not been adapted for, and are rarely applied to, the problem of analyzing and managing risk.

Although nearly all of recent risk analysis literature has been on physiological risks from technological sources, plus a little on environmental risks and on natural hazard sources (as typified by the coverage of the journal *Risk Analysis*), DR/AS risk analysis has not been entirely lacking. It has, however, been piecemeal and typically case-specific. We at DSC and others have developed ad hoc risk analyses for nuclear safeguards against theft, malevolent acts against energy facilities, and international monitoring of nuclear proliferation. Methodological innovations have been developed and applied, but they have not been systematized or codified for general use.

Relevant Developments

Each of the two parts of DR/AS can draw on technical developments in analogous fields outside risk analysis. Analytic techniques for modeling diffuse future effects are being developed for defense planning, which needs to take into account the unfolding of complex military scenarios. Various forms of scenario specification and simulation have been devised including step-through simulation,[8,9] which economizes on mental burden. These methods are, on the whole, at an early stage of development and have not been adapted to problems of DR/AS.

A distinctive aspect of "adversarial source" of risk is the role of motivation and perception, which interact in complex ways with risk management efforts. For example, where there are several alternative ways for an adversary to realize his threat, a risk manager's success in blocking one path, if perceived by the adversary, may cause the latter to reassign his effort in other directions. Models which inversely relate the probability of adversarial actions to the probability of their being thwarted have been applied to the diversion of nuclear material by states intending to thwart non-proliferation agreements.[10] Analytic approaches to this class of problem have been attempted in non-risk fields, notably game theory.[11,12] Prescriptive decision analysis models have been used to predict adversarial and other human behavior; for example, NATO response to an impending Warsaw Pact attack.[13]

Interactive decision theory, which incorporates concepts from both decision analysis and game theory, has been developed for negotiation applications and also has suggestive analogies with the case of DR/AS risk analysis.[14,15]

Computer Information Security as a Special Case of DR/AS

Computer security risk analysis has recently attracted a good deal of attention, especially in military and commercial contexts. Although, as we have noted, some computer security risks (e.g., of computer theft) have focused effects (money), and others have nonadversarial sources (e.g., computer reliability), computer information security is a prime example of DR/AS. There may be a human adversary, such as a "hacker,"[16] and the effects of the threat are diffuse because once data is compromised, it may be impossible to precisely specify the effects. Moreover, multiple risks must be considered, and these risks are not always easily quantified (as contrasted with money, injuries, or deaths).

The tools currently available for risk analysis within the broader computer security community draw very little on previous risk analysis work, partly because the computer security community is generally unaware of this work and partly because computer security problems, especially those of DR/AS, have not always been readily amenable to many of the traditional risk analysis techniques. The threats will vary from one installation to another; the countermeasures available to handle the threats include not only technical measures, but also physical and administrative security techniques; there are very little case data available.

The computer security community has, in the main, been using ill-fitting adaptations of risk analysis methods that were developed for significantly different problems. One of these is the Annual Loss Expectancy (ALE) method based on practices in the insurance business, where the risk of concern is that of losing money in insurance claims. Usually these methods are not appropriate in situations where the cause of the risk is a human adversary and where the effects of the threat are diffuse.

Special attention needs to be paid to the distinctive methodological needs of computer security as a legitimate area of risk analysis research in its own right. A DOD-sponsored workshop on computer security risk analysis, chaired by Lance Hoffman of George Washington University,[17] was recently held. It was suggested there that a general conceptual model for computer security can be developed and used to model unauthorized disclosure, destruction, modification of data and denials of service from the point of view of risk analysis. The problem of setting a value on intangibles deserves to be examined, as do problems involved in characterizing and propagating uncertainties[18] which are to date almost unrecognized by the computer security community. Also of interest are problems in communicating computer security risks to various risk management actors (e.g., Congress and facility managers) and constituency groups (e.g., segments of the general public, system manufacturers and vendors, end-users, government administrators).

DISTINCTIVE RESEARCH NEEDS

We suggest that DR/AS is a class of risks that is distinguished from the more conventional areas of risk analysis in similar ways, such that they could be usefully studied together, leading to the development of a common methodology. The dominant distinguishing features shared by this class of problem are the nature of the risks and their sources.

An appropriate unifying methodological perspective is that of personalized decision analysis, which incorporates human judgment in quantifying uncertainty and value in the process of prescribing action.[19,20] A schematic outline of one such analysis is given in Fig. 2.

Specific pieces of new methodology need to be developed to address the most serious deficiencies in the current state of the art applied to DR/AS risk analysis problems, as discussed above. They can be illustrated in the context of computer security and other DR/AS examples, and address both of the standard divisions of risk analysis: risk assessment and risk management.

Risk Formulation

We are concerned with risky situations where the possible consequences are diffuse, i.e., they cannot readily be characterized by a few simply specified standard events or measures, such as monetary costs, a core-melt accident, or a number of deaths. The question: "What is risk?" cannot be simply formulated in operational terms and, indeed, the appropriate formulation may vary with the situation and defy standardized definition.

Alternative methods and principles for formulating risk deserve investigation. At this time, it appears that four possibilities show a high degree of promise. First, risk might be specified in a "macromodel" containing few high-level abstract attributes that could be expected to span the range of concerns in given risk assessment. For example, computer security risk might be specified in regard to such attributes as national security, economics, privacy, cost, civil liberties, and others. As another example, a macromodel of the risk of nuclear material theft (shown in Fig. 3) might specify risk in regard to the attributes of material stolen, deaths, damage, possession of material by adversary at any time, apprehension of adversary, and penetration of the safeguard system.

Second, the macromodel might be extended to represent the risk from the points of view of several different constituencies. For example, the risk at a U.S. government computer facility might be represented from the point of view of the facility manager, the U.S. legislature, and several segments of the public. As another example, the risk of nuclear material theft might be represented from the points of view of the facility manager, a government regulator, and society, as illustrated schematically in Fig. 2.

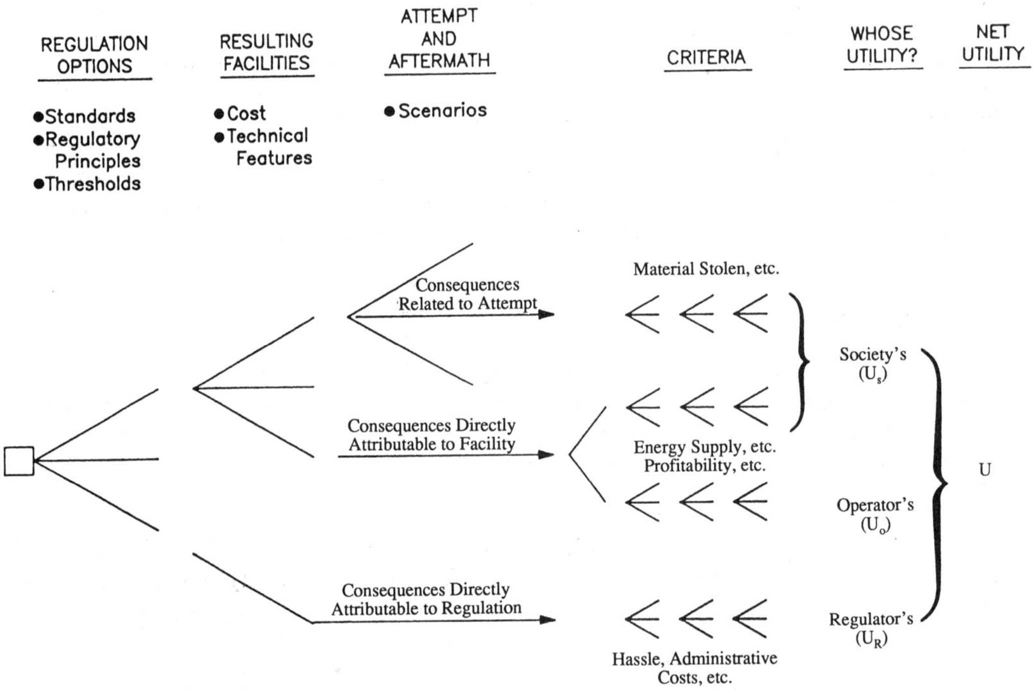

Fig. 2. Decision analysis in framework for regulatory decision with several constituencies.

Third, since macromodels may be too highly aggregated and broad to represent all important details of the risk, a series of "feeder" models may be developed and incorporated into the modeling process to provide detailed analyses of the most important aspects of risk. For example, a macromodel of nuclear power plant risk might use a feeder probabilistic risk assessment (PRA) to provide a detailed analysis of the magnitude of accidental exposure to radiation.

Fourth, since no single specification of a diffuse risk is always adequate for all purposes, techniques of "plural analysis"[21] should be investigated. Plural analysis involves pursuing two or more separate approaches to the same problem and then formally reconciling or pooling the different results.

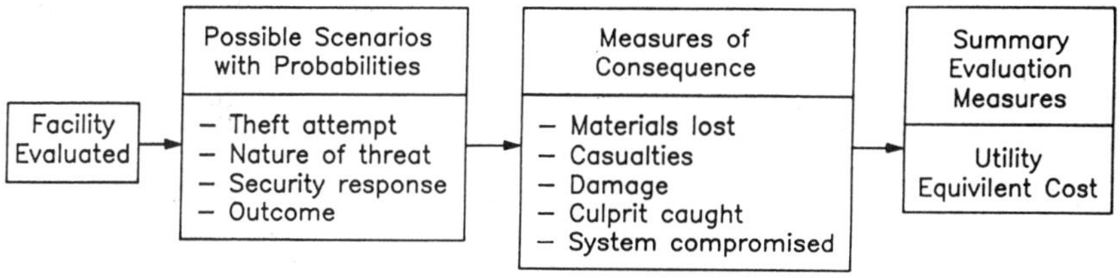

Fig. 3. Nuclear theft risk model.

Consequence Evaluation

In addition to problems of formulating risks, situations with diffuse risks pose problems in evaluating the consequences of the risk. The problem is due primarily to the need to characterize multiple attributes of risk. Thus, not only do the individual attributes need to be assessed, but comparisons across different categories of risk must be determined. For example, in computer security risk, total risk might be characterized in terms of national security, economics, privacy, cost, civil liberties, and others. An assessment of total risk requires a method to compare a level of risk on one attribute with the level of risk on another attribute. Multiattribute utility analysis[2] offers a promising method for the development of such comparisons.

It is worth exploring general ways both of defining appropriate scales for multiattribute utility analysis and of deriving value parameters that compare different attributes. The methods can be exercised in the context of computer security. A tentative example is presented in Ref. 22, which evaluates alternative national computer security policies. One might also build on other related work which involved developing an index of hazard for radioactive waste, as reported in Ref. 23. This involved field work to elicit value judgments from three sources: members of the general public, the responsible Government administrator, and technical experts. The analysis can be either weakly or strongly quantified (see Figs. 4 and 5, respectively).

Modeling the Aftermath of a Risk Event

An alternative method to macromodels for handling diffuse risk is Monte Carlo simulation, where complex possible consequences of aftermaths to a risk are represented as a sampling of possible complete paths. However, for diffuse risks, the conventional Monte Carlo simulation requires specifying probabilities for all possible contingencies and poses an unmanageably heavy burden. We have developed an alternative, called "step-through simulation," in the context of diffuse consequences of military actions, where an expert and a model interact in producing each trial.[9,10]

Modeling Adversarial Behavior

The methodological significance of the "adversarial source" of risks is that anticipation of deliberate, hostile human action requires special assessment techniques which are not appropriate for inanimate, or at least nonadversarial, sources of risks (c.f. human error in the operation of nuclear plant). A major avenue to be explored is modeling the decision processes of the adversary.

Game theory[11,12] is one implementation of this idea, though the need for restrictive assumptions on the rationality of behavior and extensive information available to the adversary severely limit the practicality of this approach.

ATTRIBUTES	Risks Addressed								Option Costs				Other Attributes			Net Evaluation
	Nat. Sec.		Econ. Loss			Priv–acy	Serv–ices		Direct			Ind. 'tax'	Bur. Upset	Democ. Value		
	Def.	Civ.	Gov.	Bus.	Pub.				Gov.	Bus.	Pub.					
WEIGHTS	***	**	**	*		*	*		**			**	*	**		
OPTIONS																
0. Do Nothing																
1. Cancel 145		–											–			–3
2. Impl. 145		+	–	+		–	+		–			–	–	–		–7
1. Modify 145																
1. Civil. Input		+		–			+		–				+	+		+3
2. Civil. Agency																
1. NBS+			+	+					–				+			+3
2. New			++	++		+	+		––				–			+3
3. Govt.–wide Ag.																
1. CSC+	–		+	+					–				––	+		–2
2. Broader	–		+	+		++			––				––			–3

Fig. 4. Qualitative evaluation of organizational options for information security.

ATTRIBUTES[b]	Risks Addressed								Option Costs				Other Attributes			Net[e] Evaluation
	Nat. Sec.		Econ. Loss				Gov.		Direct			Ind. 'tax'	Bur. Upset	Democ. Value		
	Def.	Civ.	Gov.	Bus.	Pub.	Priv.	Serv.		Gov.	Bus.	Pub.					
WEIGHTS[c]	10	2	1	.2	.5	2	2		1	.2	.5	1	.5	1		
OPTIONS[d]																
0. Do Nothing	0	–30	–20	–10	0	–50	–30		0	0	0	0	0	0		–2.41
1. Encryption	0	+20	+5	0	0	+20	0		–20	–10	0	–30	0	0		+.33
1. DES	0	+25	+10	0	0	+20	0		–50	–20	0	–35	–10[z]	0		–.30
2. Stronger																
2. R&D on Encryp	0	+20	+10	+20	0	+20	0		–30	–10	0	–15	0	0		+.47
3. Drop Satellite & Microwave	0	+20	+10	–10	0	+20	0		–40	–20	0	–10	–10	0		–.27
4. Dedicated Lines	0	+20	+10	–10	0	+20	0		–60	–20	0	–10	–10	0		+.09

Notes

a. See guidelines on completing form.
b. Dimensions of concern to government.
c. Relative importance of stakes under each attribute, or of $1B swing for costs & losses.
d. Scores are % of potential deterioration from present.
e. Weighted sum of scores.

Comments

z. NSA upset.

Fig. 5. Quantitative evaluation of civilian telecommunications security options.[a]

A more promising alternative is to anchor prediction of an adversary's behavior to that which a decision-analytic model of his choice would indicate. This has been used in a study concerned with probabilistically predicting a NATO response to an impending Warsaw Pact attack.[13] The literature for predicting deliberate, but nonadversarial, human action can also be reviewed for applicability.

A key element in this prescriptive approach to prediction which needs substantial development is handling the slippage between prescription and prediction, acknowledging the fact that the adversary may not behave as the decision analysis of his choice would indicate. The psychological work of Duncan Luce[24] and the interactive decision analysis work of Howard Raiffa[14] provide a starting point. There has also been some more specific work on this problem, in the context of predicting nuclear theft behavior of malevolent acts against energy facilities.[26]

CONCLUSION

The object of this paper has been simply to get out, for comment and suggestion, some preliminary ideas on what might constitute a fruitful new area of risk analysis. We believe it will call for distinctive — and major — research and methodology development, on a scale comparable to that which has been devoted in recent years to health and safety risk analysis.

REFERENCES

1. V. T. Covello and J. Menkes, Issues in Risk Analysis, in *Risk in the Technological Society*, AAAS Selected Symposium 65, C. Hohenemser and J. X. Kaserson, eds., pp. 287-301, Westview Press, Boulder, CO (1982).
2. R. L. Keeney and H. Raiffa, *Decisions with Multiple Objectives: Preferences and Value Tradeoffs*, Wiley, NY (1976).
3. L. B. Lave, ed., *Quantitative Risk Assessment in Regulation*, Brookings Institution, Washington, DC (1982).
4. P. Ricci, L. Sagan, and C. Whipple, eds., *Technological Risk Assessment*, Alphen an den Rijn, Sijthoff and Noordhoff, The Netherlands (1983).
5. W. D. Rowe, *An Anatomy of Risk*, Wiley, NY (1977).
6. W. E. Vesely, Guest ed., Special issue on nuclear probabilistic risk analysis, *Risk Analysis* **4(4)** (December 1984).
7. R. C. Schwing and W. A. Albers, eds., *Societal Risk Assessment*, pp. 181-216, Plenum Press, NY (1980).
8. J. W. Ulvila, R. V. Brown, and L. S. Randall, Step-Through Simulation: A Method for Implementing Decision Analysis, Technical Report 76-18, Decisions and Designs, Inc., McLean, VA, November 1976 (NTIS No. AD A036969).
9. J. W. Ulvila and R. V. Brown, Step-Through Simulation, *Omega: The International Journal of Management Science* **6(1)**:25-31 (1978).
10. R. V. Brown and J. W. Ulvila, A Weighted Average of Diversion Path Detection Probabilities as a Measure of Detection Capability, Draft paper for the International Atomic Foreign Agency, Decision Science Consortium, Inc., Falls Church, VA, July 1982.
11. R. D. Luce and H. Raiffa, *Games and Decisions*, Wiley, NY (1957).
12. M. Shubik, *Game Theory in the Social Sciences*, M.I.T. Press, Cambridge, MA (1982).
13. R. V. Brown, C. W. Kelly, III, R. R. Stewart, and J. W. Ulvila, A Decision-Theoretic Approach to Predicting the Timeliness of NATO Response to an Impending Attach (U), *Journal of Defense Research*, pp. 126-135, Special Issue 77-1 (Crisis Management) (May 1977).

14. H. Raiffa, *The Art and Science of Negotiation*, The Belknap Press of Harvard University Press, Cambridge, MA (1982).
15. J. W. Ulvila, Decisions with Multiple Objectives in Integrative Bargaining, Doctoral dissertation, Harvard University, Graduate School of Business Administration (1972); Dissertation Abstracts International, 1979, 4, 1594A, University Microfilms No. 79-20985.
16. S. Levy, *Hackers: Heroes of the Computer Revolution*, Anchor Press/Doubleday, Garden City, NY (1984).
17. L. J. Hoffman, A Research Agenda for Computer Security Risk Analysis (draft), Report on a DOD-Sponsored Workshop, The George Washington University, Department of Electrical Engineering and Computer Science, Washington, DC (January 1986).
18. R. V. Brown, Assessment Uncertainty and the Firmness of Information: A decision-Oriented Methodology, Decision Science Consortium, Inc., Falls Church, VA (May 1986).
19. H. Raiffa, *Decision Analysis*, Addison-Wesley, Reading, MA (1968).
20. R. V. Brown, A. S. Kahr, and C. R. Peterson, *Decision Analysis for the Manager*, Holt, Rinehart, and Winston, NY (1974).
21. R. V. Brown and D. V. Lindley, Plural Analysis: Multiple Approaches to Quantitative Research, *Theory and Decision* **20**;133-154 (1986).
22. R. V. Brown, Personalized Decision Analysis as an Expert Elicitation Tool: An Instructive Experience in Information Security Policy, Report to OTA - Task 2, Technical Report No. 85-9, Decision Science Consortium, Inc., Falls Church, VA (February 1985).
23. S. R. Watson, An Index of Hazard from Radioactive Waste, Technical Report, Decision and Designs, Inc., McLean, VA (1977).
24. R. D. Luce, *Individual Choice Behavior: A Theoretical Analysis*, Wiley, NY (1959).
25. G. A. Hill, Societal Consequences of Malevolent Situations: Implications for Safeguards Policy, Technical Report 81-8, Decision Science Consortium, Inc., Falls Church, VA (May 1982).

Uncertainty and Variability in Environmental Risk Assessment: A Framework for Analysis

Kenneth T. Bogen
Lawrence Livermore National Laboratory
University of California
Livermore, CA

Robert C. Spear
School of Public Health
University of California
Berkeley, CA

ABSTRACT

Communication of the impacts of uncertainty and inter-individual variability on risk prediction, typically by means of complex tables relating given sources with corresponding risk-impact factors, is often confusing and does not lend itself to quantitative use of this information in decision making. We describe an integrated, quantitative approach to incorporating, to the greatest practicable extent, both uncertainty and variability into risk prediction models. By this approach, predicted individual risk R, for $0 \le R \le 1$, is modeled as the function $P(\mathbf{V},\mathbf{U})$, in which $\mathbf{V}$ and $\mathbf{U}$ are vectors of variables whose distributions model uncertainty and inter-individual variability, respectively. Individual risk R is thus treated as a variable distributed in both an uncertainty dimension and a variability dimension, whereas population risk I (the number of additional cases caused by R), for $0 \le I \le n$, is purely uncertain. I is shown to follow a compound Poisson-binomial distribution, which in low-level risk contexts can often be well approximated by a corresponding compound Poisson distribution with parameter $\bar{R}$, equal to the conditional expectation of R with respect to the variables $\mathbf{V}$ only. For very large n, the limit $\lim_{n \to \infty} I = n\bar{R}$ may be adequate to characterize uncertainty in I. The proposed analytic framework is illustrated with an application to a risk assessment for a particular environmental carcinogen exposure scenario.

KEYWORDS: Uncertainty, inter-individual variability, environmental models

INTRODUCTION

There is increasing interest in quantifying the impact of uncertainty and variability in environmental health risk assessment.[1-4] In the words of former U.S. Environmental Protection Agency Administrator William D. Ruckleshaus:[5]

New Risks
Edited by L. A. Cox, Jr., and P. F. Ricci
Plenum Press, New York, 1990

> First, we must insist on risk calculations being expressed as distributions of estimates and not as magic numbers that can be manipulated without regard to what they really mean. We must try to display more realistic estimates of risk to show a range of probabilities. To help do this, we need new tools for quantifying and ordering sources of uncertainty and for putting them into perspective.

But current quantitative treatment of uncertainty and variability in an environmental health regulatory context, where such treatment exists, relies largely on tabulations of possible risk levels associated with alternative risk modeling assumptions. As an alternative to this approach, an explicit framework is proposed here to quantitative treatment of both uncertainty and variability in risk assessment. The proposed approach is first described in general terms and then presented formally. The approach is illustrated with a brief application to a particular environmental cancer risk problem involving 1,2-dibromochloropropane (DBCP) contamination of ground water in Fresno County, California.

AN EXPLICIT APPROACH TO UNCERTAINTY AND VARIABILITY

We use the term "uncertainty" to refer specifically to a lack of knowledge concerning some risk-related characteristic, either in a statistical sense (regarding some predictable form of inaccuracy or imprecision in parameter estimates) or in a scientific sense (regarding missing or ambiguous information or gaps in scientific theory required to make predictions on the basis of causal inferences). In contrast, the term "variability" is used to refer specifically to inter-individual heterogeneity with respect to some risk-related characteristic that is distributed in a population at risk.

With these two definitions in mind, we propose a more quantitative approach to uncertainty and variability in health-risk assessment. This approach differs from prevailing regulatory approaches by focusing on probability distributions of both individual and population risk as the final product of risk assessment, using the following procedures to the greatest feasible and reasonable degree allowed by the best available data and scientific/engineering judgment:

1. Use subjective probability distributions to characterize scientific uncertainties.

2. Quantitatively model all identified statistical uncertainties.

3. Quantitatively model all relevant source of inter-individual heterogeneity.

4. Use an appropriate, quantitative methodology to calculate, from all quantified sources of uncertainty and variability in risk components, the corresponding probability distributions characterizing (1) uncertainty and variability in predicted individual risks and (2) uncertainty in predicted population risk.

The actual extent to which these procedures are applied to the various components of any given risk assessment will, of course, depend on the particular hazard, toxicological data base, exposure data base, target population and risk-management priorities involved. For example, Procedure I is applied easily whenever available inference options are logically or reasonably limited to a finite set of identifiable, plausible alternatives. Instead of automatically selecting, say, the most conservative option, alternatives could be assigned subjective probability weights according to the best available data and scientific judgment. After applying Procedure I, there may be residual scientific uncertainties that cannot be quantified using subjective probability assignments. Risk assessment must then proceed conditional on a residual set of qualitative inference bridges. The point is that by applying

Procedure I, the size of this residual set may be substantially reduced, providing the basis for a less arbitrary, more objective approach to inference option selection in the risk assessment process. This, in turn, helps to delineate more clearly the process of risk management from that of risk assessment, which is an important step in achieving a greater degree of coordination and accountability in both of these processes.[2]

PROPOSED ANALYTIC FRAMEWORK

"Risk" is defined here as the probability of incurring (or dying from) a particular class of disease(s) caused by the environmental agent(s) under consideration. If there is background incidence for this disease class, then "risk" represents incremental risk imposed by a given exposure scenario. "Individual risks" refer to a set of independent risks faced on one or more individuals in an exposed population, whereas "population risk" refer to the probability distribution of the actual health impact (i.e., the total number of additional cases) that may be generated by the imposition of a given set of individual risks. The constant n always represents the number of individuals in an exposed population. The variable R represents risk to individuals, such that $0 \leq R \leq 1$. The integer random variable I represents the total number of (additional) individual cases arising from a particular exposure scenario, such that $0 \leq I \leq n$ and $0 \leq I/n \leq 1$. Using this notation, we may restate our general problem as that of characterizing the uncertainty represented by the cumulated distribution functions (cdf's), $F_R(r)$, and $F_I(i)$.

In our general model, a given risk imposed on a given individual in a population at risk is some function P of either, or both, of two types of variable (or model parameter). The first type we call "uncertain variables," referring to random variables used to model uncertainty (as defined above), i.e., to model unknown constants pertaining to all individuals at risk. In contrast, we define an "empirical variable" to be a mathematical reference to a given set of n specified constants reflecting inter-individual variability, i.e., a reference to the values of some quantity that is distributed among the members of a population at risk.

Our general model, then, of risk to an ith individual in a population at risk is

$$R_i = P(\mathbf{v}_i,\mathbf{U}),\ i = 1,2,\ldots,n \tag{1}$$

where the random variable R_i is represented as a function of a vector of uncertain variables $\mathbf{U} = \{U_1,U_2,\ldots\}$, and a vector of particular values $\mathbf{v}_i = \{v_1,v_2,\ldots\}$ of the corresponding vector of empirical variables $\mathbf{V} = \{v_1,v_2,\ldots\}$. Accordingly, the risk R to an individual selected at random from the population at risk is given by

$$R = P(\mathbf{V},\mathbf{U}) \tag{2}$$

With this analytic framework, uncertainty and variability in risk may be conveniently characterized from the individual and population perspectives.

Individual Risk

For the case in which no empirical variables are in the risk model, all individuals face an identical risk which is uncertain, i.e., $R_1 = \ldots = R_n = R = P(\mathbf{U})$. Hence, the cumulative probability distribution function (the pdf) $F_R(r)$ is the distribution of risk faced by all n individuals.

When no uncertain variables are in the risk model, risk to the ith individual is the constant $r_i = P(\mathbf{v}_i)$. In this case, R is an empirical variable whose distribution $F_R(r)$ is defined simply by $(i-1)/(n-1)$ for the ordered individual risks $r_{(i)}$.

When R is a function $P(\mathrm{V,U})$ of both empirical and uncertain variables, the distribution of uncertainty in risk to the ith person at risk is simply $F_{R_i}(r_i)$. In this case, the n individual risks $\{R_1,...,R_n\}$ are all random variables which may be unambiguously ranked with respect to one another to yield an (increasing) ordered set $R_{(1)},...,R_{(n)}$ if, and only if, the inequalities

$$F_{R(1)} \leq F_{R(2)} \leq ... \leq F_{R(n)} \tag{3}$$

hold for all r where $0 \leq r \leq 1$, in which case it makes sense to refer to uncertainty in the risk $R_{(i)}$ to the "ith highest risk" individual (e.g., the risk $R_{(n)}$ to the "maximum risk" individual) in an unconditional sense. If the inequalities in Eq. (3) do not hold, it will nevertheless always be possible to rank n individual risks which are all evaluated at a specified probability level, i.e., conditional on some particular level of uncertainty.

Population Risk

Uncertainty in population risk, characterized by $F_I(i)$, is a function of both the risks faced by all the individuals in that population and whatever uncertainty is associated with those individual risks. In the case that $R = P(\mathrm{U})$, R is a particular uncertain risk faced independently by each member of the entire population at risk. We may reasonably assume that health effects arising in different individuals due to the imposition of low-level environmental health risk are independent events. The total number I of people affected may be viewed as the outcome of n independent Bernoulli trials in which the probability of an occurrence at each trial is uncertain and has the a priori distribution $F_R(r)$, (i.e., $I = \sum_{i=1}^{n} X_i$ where X_i are independent random variables equal to 1 with a probability R and equal to 0 with a probability $1-R$). If R were equal to the constant r, then the probability mass function (the pmf) $p_I(i)$ would simply be the binomial pmf with parameters n and r. In the present case, the relevant binomial pmf is therefore the conditional pmf $p_I(i|r)$. By Bayes' rule, $p_I(i)$ is defined, therefore, as the compound binomial pmf:

$$\begin{aligned} p_I(i) &= \int_0^1 p_I(i|r)\, f_R(r)\, dr \\ &= \int_0^1 \binom{n}{i} r^i\,(1-r)^{n-i}\, f_R(r)\, dr \end{aligned} \tag{4}$$

in which f is the probability density function (pdf) corresponding to F. For example, if uncertainty in R's true value is modeled by the standard Beta distribution

$$f_R(r) = \frac{1}{B(\alpha,\beta)} = r^{\alpha-1}\,(1-r)^{\beta-1}\,,$$

then the conjugate relationship between the Beta and binomial distributions yields a simplified form for the probability that $I{=}i$:

$$p_I(i) = \binom{n}{i}\,\frac{B(\alpha+i,\beta+n-i)}{B(\alpha,\beta)} = \frac{\binom{\alpha+i-1}{i}\binom{\beta+n-i-1}{n-i}}{\binom{\alpha+\beta+n-1}{n}} \tag{5}$$

where $i{=}0,1,...,n$.[5,6] (For example, Eq. (5) has been used recently to characterize uncertainty in population risk to small groups of people exposed to risk from microbially contaminated drinking water.[7]) In the context of environmental health risk assessment, the size n of an exposed population is very often large, so that it is of interest to consider the limiting form of $F_I(i)$ as $n \rightarrow \infty$, which is easily shown to be simply $F_R(i/n)$ regardless of the form of $f_R(r)$, provided that $f_R(r)$ is a pdf and $0 \leq R \leq 1$.[6] That is, as $n \rightarrow \infty$, the asymptotic of distribution population risk uncertainty reduced in form to the *a priori* distribution of individual risk.

For the case in which $R = P(\mathbf{V})$, I can be again treated as the number of occurrences after n independent Bernoulli trials, but in this case the probability r_i of an occurrence on the ith trial is drawn exhaustively and without replacement from the set of known probabilities $\{r_1,...,r_n\}$. That is, I here represents the sum of n independent variables equal to 1 or 0 with the probabilities r_i and $(1-r_i)$, respectively, for i=1,2,...,n. In this case, $F_I(i)$ is what is referred to as a "Poisson binomial" distribution with parameters r_i, i=1,2,...n, which has expectation $E(I)=n\bar{r})$ and variance $\sigma_I^2 = n[\bar{r}(1-\bar{r}) - \sigma_R^2]$, where $\bar{r} = \sum_{i=1}^{n} r_i / n$ and $\sigma_R^2 = \sum_{i=1}^{n} (r_i - \bar{r})^2/n$.[7] (Note that here σ_I^2 is less than the variances $n\bar{r}$ and $n\,[\bar{r}(1-\bar{r})]$ for the corresponding Poisson and binomial distributions with parameters, $n\bar{r}$ and $\{n,\bar{r}\}$, respectively.) The exact pmf for this distribution is simply

$$p_I(i) = \sum_{\{C_i\}} \left[\left(\prod_{j=1}^{i} r_{c_j} \right) \left(\prod_{k \neq \{C_i\}} (1 - r_k) \right) \right] \tag{6}$$

where $\{C_i\}$ represents the set $\{c_1,...,c_i\}$ comprising one of the $\binom{n}{i}$ combinations of the integers $\{1,2,...,n\}$ taken i at a time, and c_j is the jth element of that particular combination. This distribution is difficult to evaluate exactly for intermediate values of i when n is large, unless the r_i are substantially non-unique.[8] It is known that the Poisson binomial distribution asymptotically approaches a corresponding Poisson distribution with parameter $n\bar{r}$ as $n \rightarrow \infty$, provided r is scaled such that $n\bar{r}$ remains finite and $\max(r_i) \rightarrow 0$ as $n \rightarrow \infty$.[9-11] However, the present application of the Poisson binomial distribution involves definite values of n and r_i which are not "tending" toward anything. Nevertheless, an approximation theorem of Le Cam shows that the Poisson binomial distribution is well approximated by the corresponding Poisson distribution under certain conditions.[13] Specifically, given a Poisson binomial cdf $F_I(i)$ associated with the given risks $\{r_1,...,r_n\}$, and given the corresponding Poisson cdf $\hat{F}_I(i)$ with parameter nr, and defining $r_{(n)}$ and the difference d by

$$r_{(n)} = \max(r_i)$$
$$d = \max(|F_I(i) - \hat{F}_I(i)\,|\,),$$

Le Cam proved, *inter alia*, the following inequalities

$$d < 2n(\bar{r}^2 + \sigma_R^2) \tag{7a}$$

$$d < 9r_{(n)} \tag{7b}$$

$$d < 16\bar{r}\left[1 + \sigma_R^2/\bar{r}^2\right] \quad \text{for } r_{(n)} \leq 0.25 \tag{7c}$$

which can be used to bound the error introduced by using $\hat{F}_I(i)$ to approximate $F_I(i)$. In many environmental health contexts, these inequalities are sufficient to guarantee that d is small.

In the case where $R = P(\mathbf{V},\mathbf{U})$, that is, the case of mixed empirical and uncertain variability, $F_I(i)$ represents the distribution of the sum of n independent random variables equal to 1 or 0 with the uncertain probabilities R_i and $(1-R_i)$, respectively, for i=1,2,..., n. Each of the uncertain individual risks $\{R_1,..,R_n\}$ is defined by Eq. (1). Using Bayes' rule, we see that $F_I(i)$ is in this case the compound Poisson binomial cdf defined by

$$F_I(i) = \int_0^1 ... \int_0^1 F_I\ (i|\mathbf{r})\ f_{\mathbf{R}}(\mathbf{r})\ dr_1 ...\ dr_n \tag{8}$$

in which $\mathbf{r}$ and $\mathbf{R}$ are used to denote the particular values $\{r_1,...,r_n\}$ and the random variables $\{R_1,...,R_n\}$ respectively, and in which $F_I(i \mid \mathbf{r}) = \sum_{k=0}^{i} p_I(k)$ where $p_I(k)$ is defined by Eq. (6). Since the multiple integral in Eq. (8) is even more difficult to evaluate than the exact Poisson binomial distribution, it is convenient to employ a distribution $\hat{F}_I(i)$ that may be used to approximate $F_I(i)$ in this case, analogous to the Poisson approximation $\hat{F}_I(i)$ that was used above. Here, our new approximation $\hat{F}_I(i)$ shall be defined as the following compound Poisson distribution

$$\begin{aligned}\hat{F}_I(i) &= \int_0^1 \hat{F}_I(i|\bar{r})\ f_{\bar{R}}(\bar{r})\ d\bar{r} \\ &= \int_0^1 \sum_{k=0}^{i} \frac{(n\bar{r})^k e^{-n\bar{r}}}{k!}\ f_{\bar{R}}(\bar{r})\ d\bar{r}\end{aligned} \tag{9}$$

where $\bar{R}$, which may be thought of as the uncertain risk to a hypothetical "mean-risk individual" in a population at risk, is simply the conditional expectation of risk R with respect to only the empirical variables, $\mathbf{V}$, i.e.,

$$\bar{R} = E_{\mathbf{V}|\mathbf{U}}(R) = \frac{1}{n} \sum_{i=1}^{n} P(\mathbf{v}_i, \mathbf{U})\ . \tag{10}$$

When it is very difficult or impossible to evaluate $f_{\bar{R}}(\bar{r})$ using Eq. (10) (as may be the case in realistic contexts involving large n, a complicated exposure-response function P, and disparate stochastic models for variability in $\mathbf{V}$ and uncertainty in $\mathbf{U}$), then $f_{\bar{R}}(\bar{r})$ may be approximated using a Taylor series expansion of the function $P(\mathbf{V},\mathbf{U})$ in order to derive an approximate expression for $\bar{R}$ as a function of $\mathbf{U}$. For example, where $\bar{R}_k$ denotes a kth-order approximation of $\bar{R}$, this approach yields

$$\bar{R}_1 = P(\bar{\mathbf{v}}, \mathbf{U}) \tag{11a}$$

$$\bar{R}_2 = \bar{R}_1 + \frac{1}{2} \sum_j \frac{\partial^2\ P(\mathbf{V},\mathbf{U})}{\partial \mathbf{V}_j^2}\ \sigma^2_{\mathbf{V}_j} \quad \text{for independent } \mathbf{V}_j \tag{11b}$$

in which $\bar{\mathbf{v}}$ is the vector of arithmetic means $\{\bar{v}_1,\bar{v}_2,...\}$, and the partial derivatives with respect to V_j are evaluated at the corresponding mean values $\bar{v}_j$. Since the right-hand sides of Eqs. (11a-b) involve only a single function of $\mathbf{U}$, these expressions are much simpler to evaluate than Eq. (10), which involves n functions of $\mathbf{U}$, although this simplification in characterizing uncertainty in $\bar{R}$ is obtained only at the expense of introducing error associated with the Taylor series approximation used (which, however, may be limited to an acceptable degree by demonstrating that an increase in k over some initial value does not significantly change $f_{\bar{R}_k}(\bar{r}_k)$ evaluated over a given percentile range of interest). Given the expression for an acceptable approximation $\bar{R}_k$, a corresponding density function $f_{\bar{R}_k}(\bar{r}_k)$ can be derived and used in place of $f_{\bar{R}}(\bar{r})$ in Eq. (9) to evaluate $\hat{F}_I(i)$, using a Monte Carlo approach if necessary.

In the case that $f_{\bar{R}}(\bar{r})$, and hence $\hat{F}_I(i)$, can be evaluated precisely, it can be shown that the error introduced by using $\hat{F}_I(i)$ to approximate $F_I(i)$ is again bounded in accordance with the Le Cam inequalities (7a-c) in which the particular values d, $\bar{r}$, σ_R^2, $r_{(n)}$ and r_i are replaced by corresponding uncertain variables (i.e., by D, $\bar{R}$, etc.), such that this error may now be specified only in probabilistic terms. In the present case of mixed empirical and uncertain variability, it can also be shown that $\lim_{n \to \infty} F_I(i) = F_{\bar{R}}(i/n)$, by recalling that the variance of the normalized Poisson binomial cdf $F_{I/n}(\frac{i}{n} \mid \mathbf{r})$ is less than that of the corresponding Poisson cdf $F_{I/n}(\frac{i}{n} \mid \bar{r})$ by an amount equal to σ_R^2/n, and that both approach 0 as $n \to \infty$.[8]

APPLICATION; CANCER RISK ASSOCIATED WITH INGESTION OF DBCP-CONTAMINATED GROUND WATER IN FRESNO COUNTY, CALIFORNIA

Here we apply the analytic framework developed above to an assessment of increased cancer risk associated with lifetime ingestion of trace amounts of the nematocide 1,2-dibromo-3-chloropropane (DBCP) in ground water by residents of Fresno County, California. To predict DBCP cancer risk, we have used a model of the form $R = P(\{C,A\}, \{U,H,Q\})$ in which C and A are empirical variables and U, H, and Q are uncertain variables, using the terminology introduced above. Here C represents lifetime, time-weighted average (TWA) concentration of DBCP in ppb specific to given census tracts within Fresno County; U represents the uncertainty associated with any given value of C such that actual DBCP concentration is modeled as the product CU; A represents absorption or intake of DBCP as a function of lifetime TWA water consumption in liters/kg-day; and H represents uncertainty regarding the extrapolation of a test animal dose to its tumorigenic equivalent in human dose. In the present model, then, lifetime TWA dose D_i of DBCP in mg/kg-day to an ith of n exposed individuals is given by

$$D_i = \frac{c_i \, a_i \, U H}{1000 \text{ ppb/mg} - \text{liter}} \tag{12}$$

in which c_i and a_i represent the ith individual values of the empirical variables C and A, respectively, for i=1,2,...,n. Finally, Q represents the low-dose carcinogenic potency that DBCP is presumed to have in humans based on an analysis of DBCP carcinogen bioassay data. Here Q is the linear parameter in dose within the following model for individual risk:

$$R_i = 1 - \exp\left(-\left(QD_i + \sum_{j=2}^{g} \hat{Q}_j D_i^j\right)\right) \tag{13}$$

in which D_i is defined in Eq. (9); g is the number of non-control dose groups in the bioassay used; $\hat{Q}_j$, j=2,3,...,g, are constants estimated from the bioassay data; Q is a variable whose uncertainty distribution is estimated from this data; and R_i is the increased "composite" risk that an ith exposed individual will contract any of a set of specified tumor types as a consequence of lifetime TWA dose D_i, given that no such occurrence would take place in the absence of this exposure. This risk model is based on a version of the "linearized multistage" model

$$R_{ik} = 1 - \exp\left(-\left(q_{1k}d_i + \sum_{\substack{j=0 \\ j\neq 1}}^{g} \hat{q}_{jk}d_i^j\right)\right) \tag{14}$$

widely used to describe and extrapolate the occurrence probability of tumor type k for an ith exposed individual from a given set of dichotomous experimental tumor incidence data.[13-15] Here, $\hat{q}_{jk}$, j=0,2,3,...,g, are maximum likelihood parameter estimates, and q_{1k} is an uncertain variable (the "potency" coefficient for tumor type k) whose distribution is estimated from the data. Under the assumption that different tumor types occur independently in test animals, the increased "composite" risk to individual i of incurring one or more of any t significant tumor types observed in a bioassay given by Eq. (13), in which $Q = \sum_{k=1}^{t} q_{1k}$ and $\hat{Q}_j = \sum_{k=1}^{t} \hat{q}_{jk}$.[8]

To parameterize our tumor-specific dose-response models, we used data from a study of male rats exposed by ingestion to DBCP summarized in Table 1.[16-18] To avoid reliance on asymptotic estimation procedures, "exact" distributions for q_{1k} were derived by repeated calculation of q_{1k} from Monte Carlo-simulated input data,[19] and that for Q (see Fig. 1) was derived by Monte Carlo addition (justified here in that the incidence data for tumor-type multiplicity happen to be consistent with the independence hypothesis stated above).[8]

Table 1. Tumors in male albino rats exposed to DBCP*

Dose (d) in TWA mg/kg/day	0	.24	.80	2.4
No. animals	49	47	49	46
Segregated incidence				
Tumor type (k)				
1. Stomach	0	0	3	20
2. Liver	0	1	2	6
3. Kidney	0	1	4	11

*Bioassay data derived from Refs. 17 and 18.

The empirical distribution of projected lifetime TWA concentrations C used (see Fig. 2) was derived from a detailed reanalysis[8] of data on DBCP contamination in water from wells in Fresno County, originally compiled by Jackson *et al.*[20] The distribution of the concentration uncertainty factor U was estimated from the concentration data to be reasonably well represented by

$$U = \begin{cases} 1 + 0.082\ Z & \text{if } C \geq 0.005 \text{ ppb} \\ T & \text{if } C < 0.005 \text{ ppb} \end{cases}$$

where .005 ppb was the concentration detection limit for DBCP in the data base, Z is a standard normal random variate, and T is a random variate uniformly distributed between 0 and .005 ppb.[8]

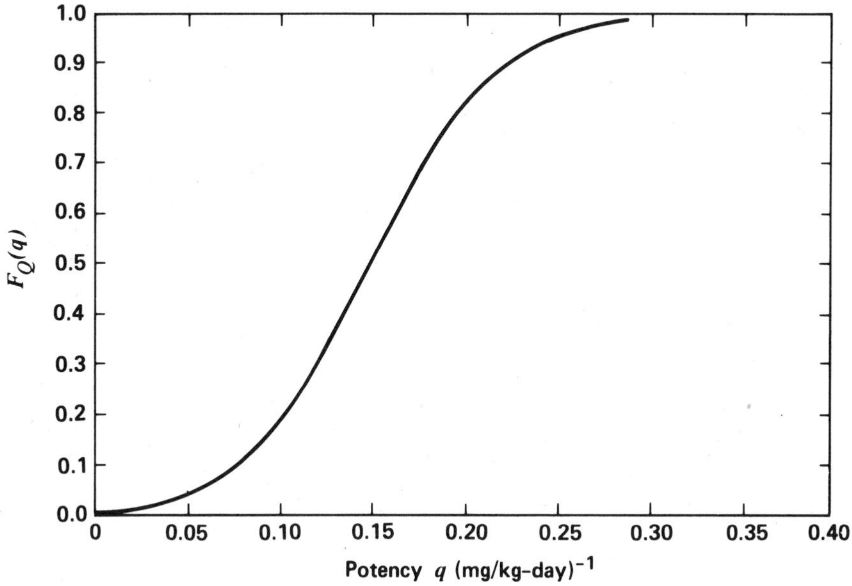

Fig. 1. Cumulative (uncertainty) distribution $F_Q(q)$ of composite carcinogenic potency, Q, associated with parameter estimation error using a "multistage" extrapolation model to fit bioassay data on male rats exposed to DBCP (see Table 1).

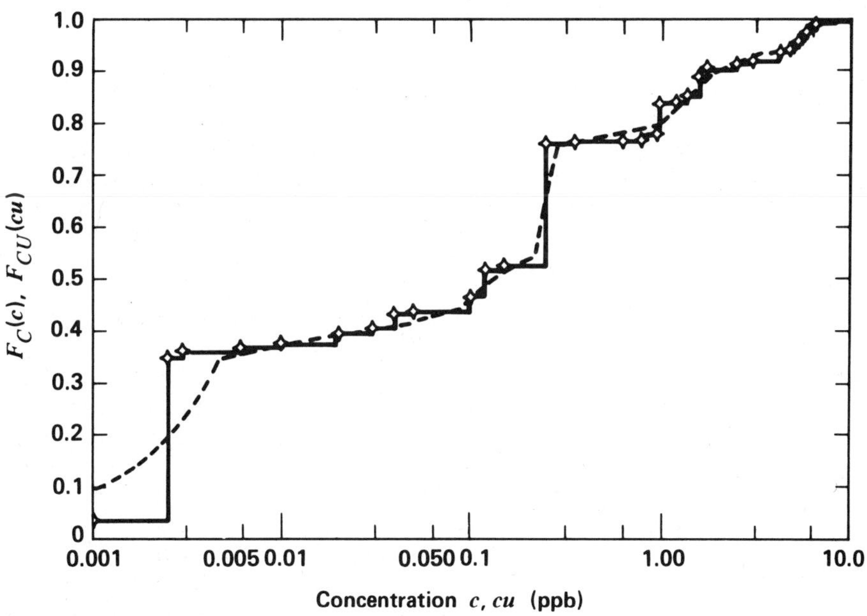

Fig. 2. Cumulative (empirical) distribution $F_C(c)$ of mean DBCP concentration c (solid curve), and cumulative (uncertainty) distribution $F_{CU}(cu)$ of randomly sampled DBCP concentration, CU (dashed curve), in ground water from wells in Fresno County, California.

The distributions of inter-individual heterogeneity in lifetime TWA water consumption A for males and females were modeled to account for a large body of water consumption data and relevant physiological constraints. On this basis, we approximated male and female consumption pmf's as symmetric triangular with means equivalent to 1.95 liter/70 kg-day and 1.4 liter/60 kg-day, respectively, with ranges equal to 40% of the corresponding means, and the average ratio of males to females was taken to be 51/49 (see Fig. 3).

To account for uncertainty in the calculation of response-equivalent doses when extrapolating from rat dose-response data to predicted human response, we used an interspecies dose equivalence uncertainty factor H equal either to 1 or to $(70\ \text{kg}/0.5\ \text{kg})^{2/3}$ with equal probability, based on two plausible extrapolation methods often used.[14,15,22]

Increased risk to selected individuals in our modeled Fresno County cohort of size n=508,954 was calculated using the methods outlined above. Figure 4 shows the calculated individual risks: R to a randomly selected individual, $R_{(n)}$ to the "maximum risk" individual, $R_{(.95\ n)}$ to the "95th percentile risk" individual, and $\overline{R}$ to a hypothetical "mean risk" individual. (Note that inequalities (3) happen to hold in our model for $R_{(i)}$, given sufficiently large i, at most uncertainty levels λ in $F_{R(1)}^{-1}(\lambda)$, so that the correlation between C and U does not, for practical purposes, render $R_{(.95n)}$ or $R_{(n)}$ ambiguous.) The corresponding approximate population risk distribution, $\hat{F}_I(i)$ and its limiting form as $n \to \infty$ are shown in Fig. 5. Using the Le Cam inequalities to bound error on $\hat{F}_I(i)$, as discussed above, it can in this case be shown that $\hat{F}_I(i)$ approximates the true population risk distribution $F_I(i)$ very well (Prob $\{D > 0.04\} < 0.01$).[8]

DISCUSSION

The additional information concerning uncertainty and variability in predicted risk provided throughout the proposed analytic framework is potentially useful to the risk manager, for it necessarily increases the extent to which the efficiency of alternative risk management policy options may be judged. For example, the proposed framework facilitates the identification of controllable sources of variability and uncertainty in risk and thus allows a quantitative analysis of the impact of their potential removal on predicted risk. One result of this is that previously downplayed aspects of health risk assessment, such as detailed exposure assessment, may be found to have such a profound impact on predicted risk that they might demand greater attention from risk managers than they currently receive.

The application of our approach in the context of environmental risk management would certainly highlight the distinction between individual and population risk and so confront the risk manager with the need to define risk acceptability criteria from a broader and more complex set of alternatives, involving aspects of the predicted distributions of both individual and population risk. Several questions arise regarding how such criteria might be framed. For example, how should individual risk considerations be weighted relative to population risk considerations? While it is only population risk that may be realized in the form of new cases of health impairment, the predicted empirical distribution of individual risks imposed that flows from quantitative uncertainty analysis raises—in a dramatically highlighted fashion—vexing issues concerning equity in health-risk burden. Interesting attempts have been made to address these issues in the context of point estimates of individual vs population risk.[23] But how is an acceptable distribution of predicted individual risks defined, and how should superimposed uncertainty influence this definition?

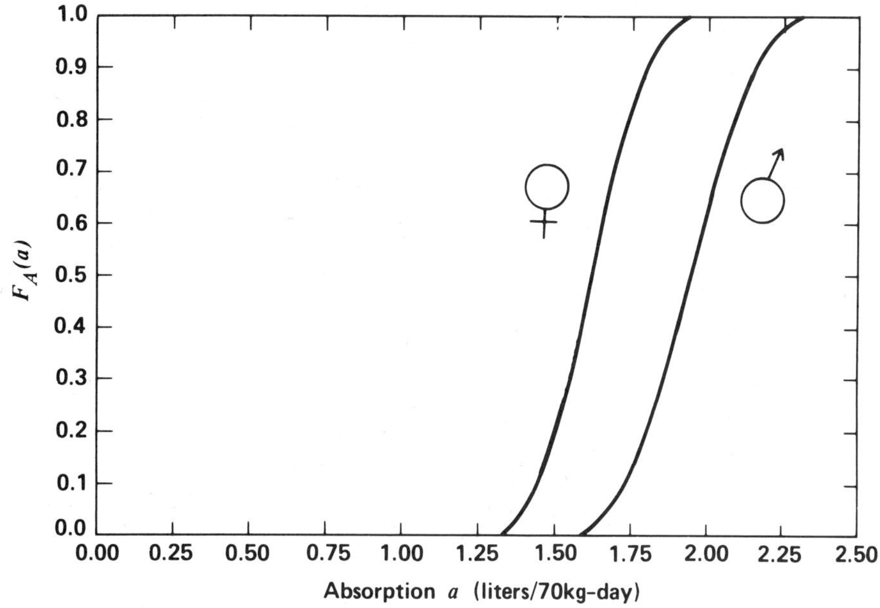

Fig. 3. Cumulative (empirical) distribution $F_A(a)$ of lifetime, time-weighted average total fluid intake, A.

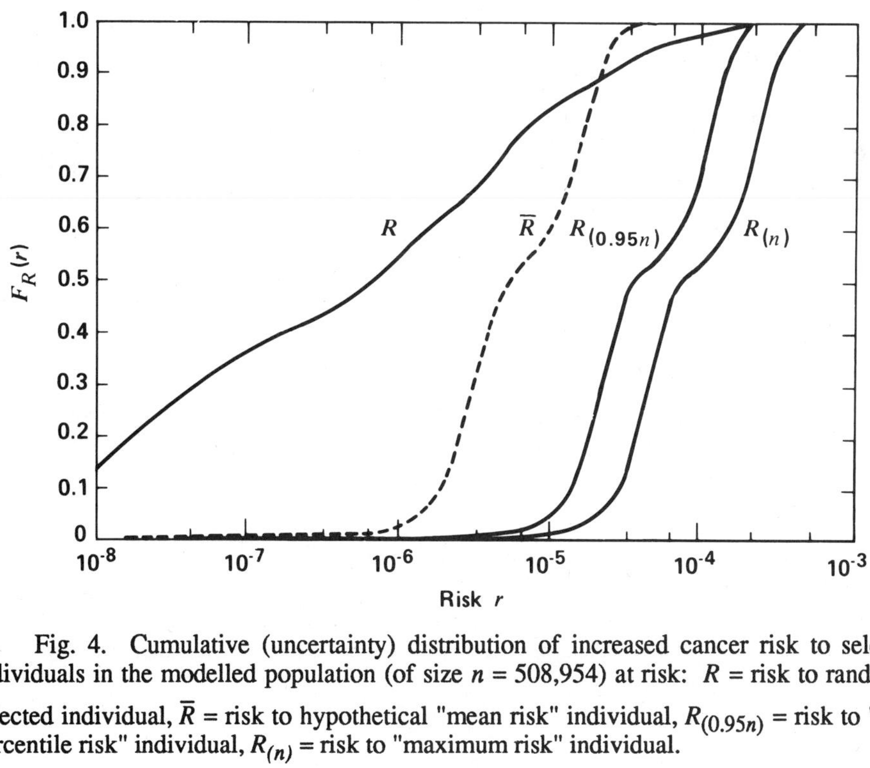

Fig. 4. Cumulative (uncertainty) distribution of increased cancer risk to selected individuals in the modelled population (of size n = 508,954) at risk: R = risk to randomly selected individual, $\overline{R}$ = risk to hypothetical "mean risk" individual, $R_{(0.95n)}$ = risk to "95th percentile risk" individual, $R_{(n)}$ = risk to "maximum risk" individual.

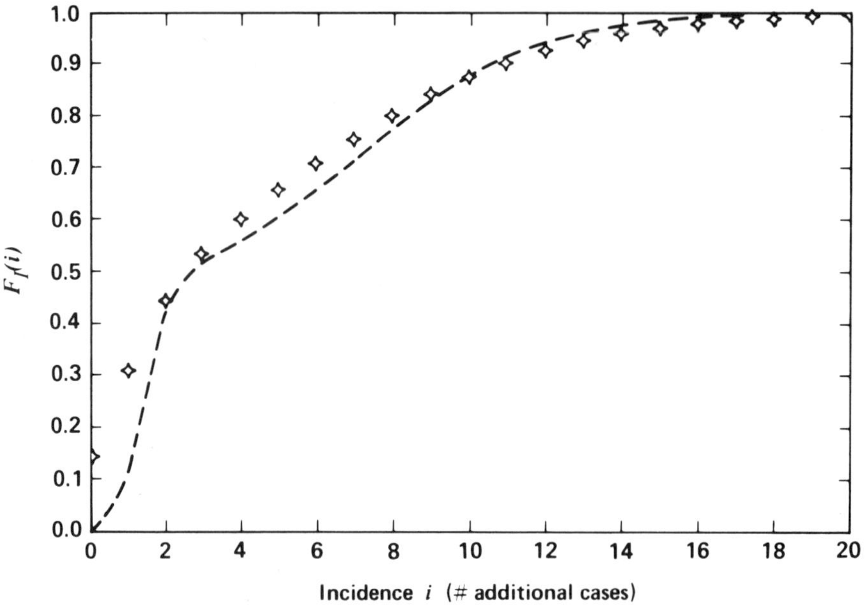

Fig. 5. Approximate cumulative (uncertainty) distribution $\hat{F}_I(i)$ of increased population risk, I, associated with the individual risk distributions shown in Fig. 4, equivalent to the compound Poisson distribution with parameter $\bar{R}$ (shown with the points <>); also shown (----) is the corresponding asymptotic distribution $\lim_{n\to\infty} F_I(i) = F_{n\bar{R}}(n\bar{r})$.

In forcing risk managers to deal with these and other issues, the use of a quantitative approach to uncertainty and variability, such as that proposed here, is bound to stimulate the political, economic, legal, and moral debate surrounding the problem of defining appropriate acceptability criteria for regulated environmental risks. But, by implementing a more explicit and systematic approach to uncertainty and variability, this approach may also do much to depoliticize the risk assessment phase of the regulatory process, and to focus the policy debate on how best to do risk management.

REFERENCES

1. National Academy of Sciences, Committee on Prototype Explicit Analyses for Pesticides, *Regulating Pesticides*, National Academy Press, Washington, DC (1980).
2. National Academy of Sciences, Committee on the Institutional Means for Assessment of Risks to Public Health, *Risk Assessment in the Federal Government: Managing the Process*, pp. 36-37, National Academy Press, Washington, DC (1983).
3. U.S. Environmental Protection Agency, Proposed Guidelines for Carcinogen Risk Assessment, *Fed. Regist.* **49(227)**:46294-46301 (1984).
4. U.S. Environmental Protection Agency, Proposed Guidelines for Exposure Assessment, *Fed. Regist.* **49(227)**:46304-46312 (1984).
5. W. D. Ruckleshaus, Managing Risk in a Free Society, *Princeton Alumni Weekly*, pp. 18-23 (March 7, 1984).
6. M. Dwass, *Probability and Statistics*, pp. 365-370, W. A. Benjamin, Inc., New York, (1970).
7. M. G. Kendall and A. Stuart, *The Advanced Theory of Statistics*, 4th ed., Vol. 1. pp. 134, 157, 251, 348, Hafner, New York (1977).

8. K. T. Bogen, Uncertainty in Environmental Health Risk Assessment: A Framework for Analysis and an Application to a Chronic Exposure Situation Involving a Chemical Carcinogen, Dissertation, University of California, Berkeley, CA (1986).
9. R. V. Mises, Uber die Warscheinlichkeit seltener Ereignisse, Z. *Agnew. Math. Mech* **1**:121-124 (1921).
10. B. O. Koopman, Necessary and Sufficient Conditions for Poisson's Distribution, *Proc. American Math. Soc.* **1(6)**:813-823 (1950).
11. J. E. Walsh, The Poisson Distribution as a Limit for Dependent Binomial Events with Unequal Probabilities, *J. Operations Research Soc. America* **3(2)**:198-209 (1954).
12. L. LeCam, An Approximation Theorem for the Poisson Binomial Distribution, *Pacific J. Math.* **10(4)**:1181-1197 (1960).
13. K. S. Crump, An Improved Procedure for Low-Dose Carcinogenic Risk Assessment from Animal Data, *J. Envir. Pathol. Toxicol.* **5(2)**:675-684 (1981).
14. U.S. Environmental Protection Agency, Water Quality Criteria Documents, *Fed. Regist.* **45(231)**:79350 f. (1980).
15. E. L. Anderson and the Carcinogen Assessment Group of the U.S. Environmental Protection Agency, Quantitative Approaches in Use to Assess Cancer Risk," *Risk Analysis* **3(4)**:277-295 (1983).
16. Hazelton Laboratories America, Inc., 104-Week Dietary Study in Rats: 1,2-Dibromo-3-chloropropane, Vols. 1 and 2, (basis for subsequent EPA cancer-risk assessment, 1977).
17. U.S. Environmental Protection Agency, Office of Research and Development, Quantitative Risk Assessment of Dibromochloropropane (DBCP), memorandum from R. E. McGaughty to J. Cotruvo dated May 6, 1981.
18. Shell Oil Company, Assessment for DBCP, memorandum from C. S. Smith to C. E. Stehr dated December 19, 1983.
19. C. J. Portier, D. G. Hoel, and J. VanRyzin, Low-Dose Rate Extrapolation Using the Multistage Model, *Biometrics* **39(4)**:867:906 (1983).
20. R. J. Jackson, C. J. Green, J. T. Thomas, E. L. Murphy, and J. Kaldor, Literature Review on the Toxicological Aspects of DBCP and an Epidemiological Comparison of Patterns of DBCP Drinking Water Contamination with Mortality Rates from Selected Cancers in Fresno County, California 1970-1979, report to the California Department of Food and Agriculture by Epidemiology Studies Section, California Dept. Health Services, Berkeley, CA (1982).
21. International Commission on Radiological Protection, *Report of the Task Group on Reference Man (No. 23)*, Pergamon Press, New York (1975).
22. M. D. Hogan and D. G. Hoel, Extrapolation to Man, in *Principles and Methods of Toxicology*, A. W. Hayes, ed., Raven Press, New York (1982).
23. P. Milvy, A General Guideline for Management of Risk from Carcinogens, *Risk Anal.* **6**:69-79 (1986).

An Experiment in Combining Estimates of Uncertainty

Bruce Tonn and Richard Goeltz
Oak Ridge National Laboratory
Oak Ridge, TN

ABSTRACT

This research explores how subjects combine estimates of uncertainty. Four major computer programs written in Common Lisp asked the subjects questions about the likelihoods of single events and conjunctions of independent events. The results suggest that the subjects did not widely use the product rule, appeared to use a minimum rule, and varied the combining heuristic as the magnitude of the likelihoods varied. The results suggest that inaccurate risk perceptions and judgments may, in part, result from deficient reasoning about uncertainty. Future research should encompass dependent events and time and explore other methods of representing uncertainty.

KEYWORDS: Uncertainty, decision making, risk analysis, expert systems

INTRODUCTION

Research in decision making is characterized by a diversity of approach (Hammand *et al.*, 1980). Economic research in decision making usually assumes that people are completely rational and typically focuses on how decisions ought to be made (Keeny and Raiffa, 1976). Sociologists and political scientists address how groups of individuals arrive at decisions. Psychologists have been interested in describing how people actually make decisions (Edwards, 1954). Recently, researchers in the field of artificial intelligence (AI) have also become interested in decision making as it pertains to the way experts render judgments and the way autonomous agents ought to make decisions in real-time environments. Risk analysis and management requires knowledge about how uncertainty is represented and utilized in all these viewpoints. This article focuses more narrowly on aspects of uncertainty that are common to the fields of psychology and risk analysis.

Psychologists have found that human beings have difficulty in making decisions under uncertainty. People do not have the information processing capabilities to consider comprehensively and weigh all decision options pertinent to a problem (Simon, 1979). We tend to ignore statistical data in favor of more intuitive heuristics (Fong *et al.*, 1986). The representativeness heuristic, for example, yields "an assessment of the degree of correspondence between a sample and a population" (Tversky and Kahneman, 1983). This heuristic is very useful for judging the identity of objects such as chairs, where base rate data may be largely irrelevant. However, as Tversky and Kahneman (1982) illustrate,

New Risks
Edited by L. A. Cox, Jr., and P. F. Ricci
Plenum Press, New York, 1990

people will often use the representativeness heuristic at the cost of excluding important base rate information.

In addition, psychologists have found that decision making could be biased for a number of other reasons. Decisions could be affected by the limited information available to the decision maker (Combs and Slovic, 1979). Sometimes decisions are influenced when the data in a problem are given in such a way that decisions appear to be incorrectly anchored around the data. At other times, decisions may be influenced by the way the problem is framed (Tversky and Kahneman, 1981) or by the degree of confidence the decision maker may feel (Fischhoff and MacGregor, 1982).

The topic of this paper is the way individuals mentally combine estimates of uncertainty. For example, given the likelihood of event A and the likelihood of event B, what is the likelihood of event $A\&B$? Psychological research has shown that individuals have difficulty following the axioms of probability theory, which at a minimum says that $P(A) + P(B) \geq P(A\&B)$. Tversky and Kahneman (1983) found that given certain problems, individuals will assert that $P(i) > P(i\&j) > P(j)$. This finding held in numerous contexts, from personality perception to medical judgment, from sporting judgment to forecasting. In a different set of studies, Kahneman and Tversky (1979) found that individuals will tend to underestimate high probabilities and overestimate low probabilities, and will irrationally favor choices that have certain outcomes.

Future research in this area must attempt to explore separately the processing of uncertainties and value estimates in decision making. Kahneman and Tversky (1984) state that decisions have four components: outcomes, probability of winning and/or losing, a context/frame, and a decision process. In most studies, subjects are given problems which contain all four elements (e.g., Kahneman and Tversky, 1979) and it becomes impossible to determine whether the subjects are biasing uncertainty cognitions, outcome cognitions, or both. Thus, we must learn how individuals process uncertainties independent of outcome information. To date, psychological research has shown that people do not make decisions using normative theory, but it also is valuable to know exactly how people actually combine uncertainties.

One reason why psychologists have yet to describe how people reason under uncertainty is a lack of competing theories. Bayesian theory so dominates normative theory that it appears difficult to conceptualize uncertainty in different ways. To facilitate this research, new theories of uncertainty being developed in the artificial intelligence subfield of expert systems were studied. Expert systems are computer programs which are designed to make decisions as human experts would.

They are typically composed of a knowledge base and an inference engine. The former usually consists of a set of if-then rules which carry an estimate of uncertainty that relates how confident the expert is that the 'if' part actually relates to the 'then' part. Using a medical example, the estimate of uncertainty might relate symptoms to a diagnosis. The inference engine processes the rules and the accompanying estimates of uncertainty. Thus, the inference engine needs explicit rules about how to combine estimates of uncertainty.

The AI literature contains numerous proposals concerning alternative combination methods. Many researchers still defend the probability/Bayesian approach (e.g., Cheeseman, 1985). Others have found probability wanting because it is: (1) impractical because data do not exist or experts do not have the time to estimate all combinations of possible conditional probabilities (Quinlan, 1986); and (2) theoretically deficient because it does not handle in a logical way uncommitted probability mass (Shafer, 1976). In our research, it is possible that the new proposals may explain observed reasoning under uncertainty behavior better than Bayesian theory. Alternatives to probability theory have names such as certainty factors (Shortliffe, 1976), endorsements (Cohen, 1985), Dempster-

Shafer theory (Shafer, 1976; Shortliffe and Buchanan, 1984), and fuzzy set theory (Negoita, 1985). In the analyses presented below, we explored the use of probabilities and plan on conducting experiments to indicate the use of the other methodologies.

The first section of this paper presents the methods used in this research. The remainder of the paper presents our results, which indicate that individuals do not use traditional probability theory when combining estimates of uncertainty but do use minimum-type rules.

METHOD

The survey instrument consisted of four main computer programs written in Common LISP on a VAX 8600 computer. We found that the computer approach offered the opportunity to design flexible experimental instruments and to incorporate features to check for suspicious data. Statistical analysis was conducted using SAS on an IBM 3033.

The general approach consisted of dividing the survey into two separate sessions on the computer. The first session focused on acquiring from subjects their estimations of the likelihood of single events (e.g., likelihood that their television will break); the second session on estimations of compound events. In this way, we could use the single event estimations, apply a combining rule found or suggested in the literature, and compare the predicted value with the combination value given by the subject. The second session occurred at least one day after the first session because we did not want the subjects to remember their exact answers from the first session so that they could blindly apply a combining rule that they might have been taught, such as the product rule.

In the first session, subjects were allowed to review their answers after all questions had been asked. The answers were listed in ascending order, from least to most likely, and if an answer required editing, the program allowed the subjects to identify the incorrect answer and change the answer. The new answer would then appear in a resorted table. There were no time limits at any point in the exercise.

In the second session, subjects were presented with questions that inquired about the likelihood of compound events. Previous answers were not supplied and answers were not allowed to be reviewed. However, the program did perform analysis on the data to determine whether answers were inconsistent. To understand how this was done, we must describe how the second session programs worked. Specifically, the programs had the goal of filling a 6×6 array, where the X and Y coordinates represented points on the likelihood continuum from 0.0 to 1.0. The purpose was to find single likelihood answers that fit into each of the six "compartments" and then put them together into combination questions so that we could determine whether or not combining rules were dependent on the magnitudes or order of the likelihoods being combined. Inconsistencies arise when a subject's array does not exhibit a monotonically increasing pattern from the 0.0,0.0 point to the 1.0,1.0 point. For example, let X and Y represent subject-estimated likelihoods for events A and B. Let Z represent the subject-estimated likelihood for the event $A\&B$. The Zs should monotonically increase as the Xs and Ys increase; otherwise, there are logical inconsistencies in the data.

If an inconsistency was found, then the program would ask the subject if he or she agreed with the answer to the combination question. If yes, then the program would search for additional inconsistencies. If no, then the program would ask for agreement on the two single likelihood estimates that comprised the question. If either of these answers was no, then the program would ask the single likelihood question(s) again. All cells in the array that were dependent on the now-changed single likelihood estimates were refilled by the program asking new combination questions.

SUBJECTS

Twenty-one subjects completed both sessions for the frequentistic contexts. Nineteen subjects are professional or technical employees of Oak Ridge National Laboratory. The remaining two are also professionals. The subjects were self-selected based on their desire to contribute to the research and on their time constraints. Each session lasted approximately 30-40 minutes.

RESULTS: COMBINING FREQUENCY-BASED ESTIMATES OF UNCERTAINTY

This study explores the way subjects combined uncertainty estimates in a frequentistic context. That is, the likelihood questions inquired about the estimated frequency of an event occurring. In this case, events were 31 activities associatable to subjects' everyday lives. We might expect that these subjects combine such estimates using rules similar to those of probability. If individuals do not use the product rule, our next goal is to explore what rules or heuristics they might have used.

For a single event, the subject is instructed to estimate the likelihood of the event happening on any given day. The subjects are allowed to answer in two paradigms, probability or chances. For example, the likelihood of going to the grocery store might be stated as probability 0.67 or chances 2 in 3. They are equivalent mathematically but conceptually one individual may find thinking in terms of chances easier than thinking in terms of probability.

Table 1 contains the 31 single event questions asked of each subject and the means for the answers over the 21 subjects. The set of events was designed to be as independent as possible. In the second session questioning, generally described above, the computer posed questions that inquired about the likelihood of the conjunction of two of the 31 single events. We want to compare observed x, y and z combinations with specific rules. For this part of the study, we created a set of six rules to compare to the observations (Table 2).

Rule 2 is the product rule, the normatively correct method to combine the likelihoods of two independent events. Rule 1 is an extreme version of rule 1, where the subject may cognitively overmultiply the x's and y's together. Rule 4 is a simple minimum rule. The literature suggests that individuals have difficulty in thinking about low likelihood events so we hypothesized that given a very low X and a very low Y, subjects may just choose the minimum of the two. Rule 3 represents an anchoring heuristic, by which the subject would anchor the combined estimate of the minimum value and effectively subtract some likelihood mass. Rule 5 is straightforward averaging. Rule 6 is the certainty factor rule mentioned above. We will find that this rule describes data produced in evidential experiments to a much better degree (Tonn and Goeltz, 1986).

Table 3 presents aggregate results for the observed combinations and for the six suggested rules. The rules are in an ascending order, with rule 1 predicting the lowest z values and rule 6 predicting the highest. Column 1 contains the means for all the x and y combinations for all individuals for the rules and the observations. Thus, each number is calculated over approximately 21*36 data points. The observed mean (0.134) falls between rules 3 and 4, the anchoring rule (0.127) and the minimum rule (0.145). The averaging rule (5) and the certainty factor rule (6) display means much greater than the observed mean and these relationships hold through the rest of the analysis. Rules 5 and 6 do not adequately describe the subjects' combining of frequentistic data.

Table 1. Means of the Likelihoods of Events Happening on Any Given Day (N=21)

No.	Question Description	Mean
1	That it will rain or snow	.33
2	That you will be in an automobile accident	.016
3	That you will receive in the mail an incorrect bank statement	.011
4	That your television will break	.046
5	That you will go to a grocery store	.31
6	That you will have a good day at school or work	.67
7	That you will receive a phone call from friends or relatives	.46
8	That you will be awakened at night by outside noises	.14
9	That you will have a sandwich for lunch	.31
10	That you will listen to the radio	.84
11	That you will throw your garbage out	.45
12	That you will eat dinner out	.17
13	That you will get some exercise	.58
14	That you will read a daily newspaper	.63
15	That you will write a letter to your parents	.021
16	That you will receive a phone call asking for money	.064
17	That you will take a shower or bath	.96
18	That you will be hit by an object falling from the sky	.00065
19	That you will suffer from a headache	.078
20	That you or someone you live with will wash some clothes	.44
21	That you will watch a sports show on television	.19
22	That you will floss your teeth	.56
23	That you or someone you live with will use the oven	.57
24	That you will watch a news program on TV	.55
25	That you will drink at least one glass of milk	.52
26	That you or someone you live with will vacuum your home	.28
27	That you will go to a movie theater	.094
28	That you will write a check	.39
29	That you will drink at least one beer	.16
30	That you will replace a light bulb in your home	.10
31	That you will take a nap	.12

Table 2. Definitions of Six Frequentistic Rules

Rule	Label	Formula
1	Product rule squared	$(X * Y)^2 = Z$
2	Product rule	$X * Y = Z$
3	Anchor off minimum	$[X*Y + \mathrm{MIN}(X,Y)]/2 = Z$
4	Minimum	$\mathrm{MIN}(X,Y) = Z$
5	Average	$(X + Y)/2 = Z$
6	Certainty factor	$X + (1 - X) * Y = Z$

Table 3. Means of Observed z Values and Predicted z Values for Six Rules Over 10 Cases for the Frequentistic Data*

Rule	1 All	2 High/High	3 High/Low	4 High/Med	5 Low/High	6 Low/Low	7 Low/Med	8 Med/High	9 Med/Low	10 Med/Med
z Observed	.134	.50	.025	.229	.023	.0044	.018	.254	.023	.126
1. Produce rule squared	.050 (.64)	.374 (.48)	.00016 (.16)	.041 (.33)	.00019 (.47)	1.39*10-7 (.36)	2.3*10-5 (.17)	.040 (.53)	1.5*10-5 (.14)	.0041 (.40)
2. Product rule	.110 (.75)	.572 (.46)	.00062 (.22)	.175 (.38)	.0064 (.53)	9.7*10-5 (.35)	.0022 (.32)	.174 (.54)	.0019 (.22)	.054 (.45)
3. Anchor off minimum	.127 (.76)	.606 (.46)	.0073 (.22)	.203 (.36)	.0071 (.55)	.0015 (.38)	.0056 (.36)	.202 (.52)	.0057 (.24)	.111 (.39)
4. Minimum	.145 (.76)	.64 (.44)	.0084 (.23)	.231 (.32)	.0077 (.56)	.0029 (.38)	.0089 (.37)	.231 (.47)	.0095 (.23)	.168 (.36)
5. Average	.334 (.66)	.756 (.42)	.375 (.09)	.491 (.37)	.380 (.15)	.0083 (.30)	.118 (.11)	.490 (.45)	.117 (.06)	.236 (.45)
6. Certainty factor	.557 (.49)	.941 (.23)	.744 (.08)	.806 (.31)	.753 (.12)	.017 (.30)	.235 (.10)	.810 (.32)	.232 (.05)	.419 (.44)

*Numbers in parentheses are correlation coefficients between the observed z values and the particular predicted z values.

In the aggregate (col. 1), the probability rule (2) also does not sufficiently describe decision making behavior. We were not surprised by this finding because we hypothesized that combining heuristics may be dependent on the magnitudes of the x's and y's being combined. Indeed, this is the case. In column 2 are the means for combining high x's and y's, (i.e., those that are greater than 0.45). In this case, the observed mean (0.50) falls between rules 1 (0.374) and 2 (0.572), suggesting that, conceptually speaking, some heuristic resembling the product rule was being applied. This is the only case where the product rule appears to apply, however.

In all cases where either x or y is a low likelihood (less than or equal to 0.05), the data indicate that the subjects used a rule very close to rule 4, the minimum rule. The presence of one low value appears to create a conceptual floor for the combined estimate. This is a fairly robust observation because the observed means are essentially equal in columns 3, 5, 7, and 9, four of the five cases where either x or y is a low likelihood. However, the observed mean for the low/low column 6 (0.0044) is much lower than for the high/low (0.025) and low/high (0.023) columns, 3 and 5 respectively. Thus, apparently the subjects did adjust their decision making in the low/low case but not enough to indicate that a rule other than the minimum rule was used.

Columns 4, 8 and 10 present results for combinations with medium likelihoods (between 0.05 and 0.45). The results suggest that the heuristics being used fall between rules 3 and 4, with a very strong correlation to rule 4. Thus, in most cases, the aggregate data indicate that subjects use something akin to a minimum rule to combine estimates of uncertainty, except when two high likelihoods are being combined.

One of the strengths of using a computer to gather data is that it can gather and organize large amounts of data from each individual. Thus, it facilitates data analysis aimed at determining what any one individual may be doing with respect to combining uncertainty heuristics. Given that our programs could not automatically learn individuals' combining heuristics, our approach in this research is to cluster individuals by the kinds of rules they may use. To facilitate this methodology, we aggregated answers given by each individual into just three categories, a high/high category, one for all the combinations with a medium likelihood but no low likelihood, and one for all combinations with a low likelihood.

Next, for each observation by each individual within each category, we developed a scale number that places the observation in a space defined by the six predicted combinations. Thus, an observed z that is exactly equal to the prediction of rule 2 would be valued at 2.0. If the observed z falls close to rule 2 but between rules 2 and 3, then the observation may be valued at 2.2. In this way, we can characterize the kinds of rules that the subjects may be using to answer any question in the three categories of combinations.

Next, we performed a cluster analysis, using Ward's method, over the means for the characterizations of the observations using variables defined over the three categories (Table 4). The results indicate that there is wide variation among individuals in heuristics used to combine uncertainty estimates in the frequentistic context. Over four clusters, we can see that cluster 2 is the closest to using the probability rule, where no mean characteristic exceeds 2.5. However, the high/high category mean is very low, almost to rule 1.

In contrast, cluster 4 subjects exhibit Rule 2 behavior in the high/high category but exhibit rule 4 behavior in the other two categories. These subjects appear to be using very different heuristics conditional on the magnitudes of the likelihoods being combined. Subjects in clusters 1 and 3 exhibit consistent behavior over the three categories; cluster 1 subjects lean toward rules between rules 3 and 4, whereas cluster 3 subjects lean toward rules between rules 2 and 3.

Table 4. Cluster Analysis of Frequentistic Data*

Cases of Likelihood Combinations	Means/Cluster			
	1 (N=9)	2 (N=5)	3 (N=3)	4 (N=3)
1. Low-high, low-low, low-med high-low, med-low	3.54	2.50	2.65	4.13
2. Med-med, high-med, med-high	3.21	2.38	2.61	4.00
3. High-high	3.83	1.1	2.69	1.22

*Ward's method was used. R^2 for four clusters = 0.78. The values in the table refer to the mean of the observed Z values for each cluster over each case when placed on a six point scale defined over the rules listed in Table 2.

DISCUSSION

Our results illustrate three major points. One, they suggest that in the frequentistic context the subjects did not widely use the theoretically correct heuristic, the product rule. Instead, the data indicate that, in the aggregate, the subjects' heuristics were very dependent upon the magnitudes of the likelihoods being combined. An extreme version of the product rule describes heuristics used when both estimates are relatively high. A minimum rule describes heuristics used when at least one of the likelihoods is low. On a disaggregated level, the data indicate that individuals apply heuristics very idiosyncratically. A cluster analysis identified groups of individuals who appear to use something akin to the product rule, another which uses the minimum rule, and yet another group which uses heuristics that vary widely given different likelihood magnitudes.

Two, with respect to risk analysis, these results suggest that people's methods of processing uncertainty may significantly impact their risk perceptions. Some people may overestimate combinations; others may underestimate combinations. Each kind of mistaken estimate leads to inaccurate beliefs about the likelihoods of complex events.

Three, this research also demonstrated the utility of using the computer to collect data associated with decision making under uncertainty. The programs allowed the subjects to review answers and also probed the consistency of the subjects' answers. Thus, in some sense the data may be of higher quality than that obtainable from pencil and paper studies. Moreover, the programs tailor-made the second sessions for each subject, conditional on the answers in the first session. This flexibility in the collection of data is not possible with paper and pencil tests.

Future research should explore more complex uncertainty problems. The present research neglected to incorporate explicit time periods and did not attempt to capture and represent dependency between events or pieces of evidence.

More work can be done to improve the computer programs. Most important would be the development of a true machine-learning algorithm that could determine within reason the exact heuristics used by each subject. The algorithm would have to create and test hypotheses, using as well as possible its ability to manipulate data quickly and ask unique questions. Also, we found that getting subjects to use the VAX system was not as easy as we had thought because most did not have accounts on the system and we had to strictly control the formation and access to the data files. Sophisticated user interfaces could be developed on AI workstations, but the costs would be increased inaccessibility of the programs to subjects. Additional thought is required to identify the best delivery system for computer-based psychological research.

Finally, work should begin to incorporate gains and losses so that data can be collected to support or reject the various models of economic decision making under uncertainty that have been proposed in the literature. Based on research continued along the path reported herein, we should have a better idea about the heuristics individuals use to combine the uncertainties and thus be better able to separate uncertainty biases from value biases. That is, we will know more about whether people underweight high gains and overweight loses, for example, versus over-combining high probabilities and under-combining low probabilities. Results of such research should help us to better train people to make better decisions under uncertainty, given that we know more precisely the kind of heuristics being used.

REFERENCES

Cheeseman, P., 1985, In Defense of Probability, *Proceedings of the Ninth International Joint Conference on Artificial Intelligence*, pp. 1002-1009.

Cohen, P., 1985, *Heuristic Reasoning About Uncertainty: An Artificial Intelligence Approach*, Pitman, Boston.

Combs, B., and Slovic, P., 1979, Causes of Death: Biased Newspaper Coverage and Biased Judgments, *Journalism Quarterly* **56**:837-843.

Edwards, W., 1954, The Theory of Decision Making, *Psychological Bulletin* **51**:380-417.

Fischhoff, B., and MacGregor, D., 1982, Subjective Confidence in Forecasts, *Journal of Forecasting* **1**:155-172.

Fong, G., Krantz, D., and Nisbett, R., 1986, The Effects of Statistical Training on Thinking about Everyday Problems, *Cognitive Psychology* **18**:253-292.

Hammand, K., McClelland, G., and Mumpower, J., 1980, *Human Judgment and Decision Making—Theories, Methods, and Procedures*, Praeger, New York.

Kahneman, D., and Tversky, A., 1984, Choices, Values, and Frames, *American Psychologist* **39**:341-250.

Kahneman, D., and Tversky, A., 1979, Prospect Theory: An Analysis of Decision Under Risk, *Econometrica* **47**:263-291.

Keeny, R., and Raiffa, H., 1976, *Decisions with Multiple Objectives: Preference and Value Tradeoffs*, John Wiley & Sons, New York.

Negoita, C., 1985, *Expert Systems and Fuzzy Systems*, Benjamin/Cummings, Menlo Park, Calif.

Quinlan, R., 1986, INFERNO: A Cautious Approach to Uncertain Inference, in *Expert Systems: Techniques, Tools, and Applications*, P. Klahr and D. Waterman, eds., Addison-Wesley, Reading, Mass.

Shafer, G., 1976, *A Mathematical Theory of Evidence*, Princeton University Press, Princeton.

Shortliffe, E., 1976, *Computer-Based Medical Consultations: MYCIN*, Elsevier, New York.

Shortliffe, E., and Buchanan, B., 1984, A Model of Inexact Reasoning in Medicine, in *Rule-Based Expert Systems*, B. Buchanan and E. Shortliffe, eds., Addison-Wesley, Reading, Mass.

Simon, H., 1979, Rational Decision-Making in Business Organization, *American Economic Review* **69**:493-513.

Tonn, B., and Goeltz, R., 1986, Combining Estimates of Uncertainty: A Psychological Analysis, Oak Ridge National Laboratory, Oak Ridge, Tennessee.

Tversky, A., and Kahneman, D., 1983, Extensional Versus Intuitive Reasoning: The Conjunction Fallacy in Probability Judgment, *Psychological Review* **90(4)**:293-315.

Tversky, A., and Kahneman, D., 1982, Judgments of and by Representativeness, in *Judgment Under Uncertainty: Heuristics and Biases*, pp. 84-100, D. Kahneman, P. Slovic, and A. Tversky, eds., Cambridge University Press, New York.

Tversky, A., and Kahneman, D., 1981, The Framing of Decisions and the Psychology of Choice, *Science* **211**:453-458.

Linear-Nonparametric Low-Dose Disease Risk Estimation

David W. Gaylor
National Center for Toxicological Research
Jefferson, AR

ABSTRACT

Various mathematical functions have been used for low-dose extrapolation of carcinogenic risk estimates. Generally, experimental data fit various mathematical models about equally well in the experimental dose region, but low-dose risk estimates may vary by orders of magnitude for the various models. In the absence of convincing biological arguments, there generally is no basis for selecting the appropriate mathematical model for low-dose extrapolation. If the low end of the dose response curve is convex (curving upward), it is possible to obtain an upper limit on the risk at low doses by using linear extrapolation in the low-dose region.

In this paper,no model is used to fit the experimental data and only dose response monotonicity is assumed. A low-dose linear extrapolation procedure is proposed which is based upon nonparametric upper confidence limits on the observed risk for the combination of experimental doses that gives the smallest low-dose slope compatible with the data.

The purpose of the technique is to provide a simple, widely applicable procedure for obtaining upper limits on low-dose risk estimates without use of any mathematical model for fitting a dose response curve. In general the predictions of "safe" dosages by the linear-nonparametric method were within a factor of 2 of the "safe" dosages predicted by the linear-multistage procedure for 14 sets of widely varying toxicological bioassays. This indicates that employment of the multistage model in the experimental data range had little influence on the "safe" doses predicted. If one does not wish to assume a mathematical model in the experimental dose range, only slightly lower estimates of "safe" doses result from the nonparametric procedure.

KEYWORDS: Risk estimation, low-dose extrapolation, nonparametric

INTRODUCTION

Animal bioassays often are conducted in order to estimate the disease risk from exposure to toxic substances. Usually, it is necessary to perform animal bioassays at dosage levels exceeding human exposure levels in order to detect the presence of disease with relatively small numbers of animals. This requires extrapolation of results at relatively high experimental dosage levels to lower levels of human exposure. A discussion of the

New Risks
Edited by L. A. Cox, Jr., and P. F. Ricci
Plenum Press, New York, 1990

problem of extrapolation from animals to humans will not be undertaken. The discussion here is limited to estimating low-dose risk for the experimental animals.

Various mathematical functions have been proposed for low-dose extrapolation of carcinogenic risk estimates.[1] Generally, experimental data fit various mathematical models about equally well in the experimental dose region, but risk estimates may vary by orders of magnitude for the various models.[2] Hence, no model can be selected on the basis of empirical fits of bioassay data. In the absence of convincing biological arguments, there is no basis for selecting an appropriate mathematical model for low-dose extrapolation. Also, it is not feasible to directly measure the dose response relationship precisely for low level risk. Hence, animal bioassay data cannot provide probable estimates of risk at low levels.

However, if the low end of the dose response curve is convex (curving upward), it is possible to obtain an upper limit on the risk at low doses by using linear extrapolation in the low-dose region.[3] For example, Gaylor and Kodell[3] used a multistage model,[4] in the experimental dose range to fit a dose response curve and then used linear extrapolation only below the lowest experimental data point. Because estimation is generally imprecise for doses that produce a disease risk of less than 1%, a slight modification was suggested. Farmer *et al.*[5] used linear extrapolation below the lowest experimental dose or the dose estimated to produce a 1% excess risk, whichever is larger. Similar results for upper confidence limits on low-dose risk are obtained by including the largest linear term in the multistage model consistent with the data at a specified confidence level as described by Howe and Crump.[6] Gross *et al.*[7] propose successively discarding observations beginning with the highest dose level until a nearly linear fit of the dose response data is obtained. Conceivably, all data points except the lowest dose could be discarded.

Several authors have presented arguments for linearity of responses at low doses.[8-10] In such cases, low-dose linear extrapolation may be no more conservative than conventional use of statistical confidence limits.

In this paper, no model is used to fit the experimental data and only monotonicity is assumed. The low-dose linear extrapolation procedure proposed is based upon nonparametric upper confidence limits on the observed risk at the experimental doses. The procedure for obtaining nonparametric upper confidence limits on the observed risk at the experimental doses is very similar to the method of "justifiable" combinations presented by Mantel and Bryan.[11]

The purpose of this paper is (1) to provide a simple, widely applicable procedure for obtaining upper limits on low-dose risk estimates without use of any mathematical model for fitting a dose response curve and (2) to compare the performance of this procedure with the linear-multistage procedure for a variety of dose response toxicological data, including but not limited to carcinogens.

LINEAR-NONPARAMETRIC LOW-DOSE EXTRAPOLATION

Method

Low-dose linear extrapolation does not mean fitting a straight line through the experimental data to obtain an estimate of risk. Rather, low-dose linear extrapolation only attempts to provide an upper limit on the potential risk at low dosages. In the low-dose range in which the dose response is convex (curving upward), a line from a point in this range connected to the spontaneous background disease risk at zero dose lies above the true dose response curve. Because of inherent experimental variation, an additional degree of conservatism is introduced by extrapolating from an upper confidence limit on the observed excess tumor rate at a low experimental dose. The following steps are employed in the linear-nonparametric low-dose extrapolation procedure.

Step 1. Consider the experimental data: x_i responders out of n_i animals at dose d_i (i=1,2,...,k). Calculate estimates of the proportions of diseased animals, $\tilde{p}_i = x_i/n_i$. Impose the condition of a nondecreasing proportion responding with increasing dose. Using the procedure described by Bartholomew,[12] pool animals from adjacent doses for which $\tilde{p}_{i+1} < \tilde{p}_i$ giving $\hat{p}_i = (x_i + x_{i+1})/(n_i + n_{i+1})$ at d_i in place of $\tilde{p}_i$ and $\tilde{p}_{i+1}$. Continue pooling adjacent decreasing proportions until an array of nondecreasing proportions is obtained:

$$\hat{p}_1 \leq \hat{p}_2 \leq \ldots \leq \hat{p}_j \leq \ldots \leq \hat{p}_\ell, \quad \text{where } \ell \leq k.$$

Step 2. Identify the portion of the dose response curve that is likely to be concave by the following procedure: Starting with the largest dose, find the slope between adjacent doses, $(\hat{p}_{j+1} - \hat{p}_j)/(d_{j+1} - d_j)$. Continue calculating slopes between lower pairs of doses as long slopes are increasing, that is, until a lower slope than the slope from the previous higher pair of doses occurs. This will tend to identify the concave portion of the dose response curve, if any, in the higher dose region. Designate the lowest dose from the above procedure as d_m. This is the largest dose observed considered to be in the convex portion of the dose response. Note that no extrapolation is performed if there is no experimental dose point greater than zero identified for d_m.

Step 3. Calculate the response at all doses and all combinations of doses for which the lowest dose is in the convex region.

Step 3a. Consider all the $\hat{p}_j$ for j=1,2,...,m.

Step 3b. Form all pairs (j, j') of doses for which the lowest dose is $\leq d_m$ (i.e., $j \leq m$) and combine the data giving

$$\hat{p}_{j,j'} = (x_j + x_{j'})/(n_j + n_{j'}) .$$

Note: j' may be greater than m.

Step 3c. Form all triplets (j, j', j'') of doses for which the lowest dose is $\leq d_m$ (i.e., $j \leq m$) and combine the data giving

$$\hat{p}_{j,j',j''} = (x_j + x_{j'} + x_{j''})/(n_j + n_{j'} + n_{j''}) .$$

Note: j' and j'' may be greater than m.

Step 3d, 3e,.... Continue combining data until all combinations are formed for which the lowest dose is $\leq d_m$. Thus, a set of justifiable combinations of data are obtained similar to the procedure suggested by Mantel and Bryan.[11]

[In general, let $\hat{p} = \Sigma x/\Sigma n$ denote the estimate for a combination.]

Step 4. Calculate an approximate 95% confidence limit for the excess response above the spontaneous background rate, $\hat{p}_1$, for each $\hat{p} = \Sigma x/\Sigma n$.

$$\left(\frac{\Sigma x + 1.5}{\Sigma n + 1.5} - \hat{p}_1\right) + 1.645 \sqrt{\frac{\left[\frac{\Sigma x+1.5}{\Sigma n+1.5}\right]\left[1 - \frac{\Sigma x+1.5}{\Sigma n+1.5}\right]}{\left(\Sigma n + 1.5\right)} + \frac{\hat{p}_1(1-\hat{p}_1)}{n_1}}$$

In order to maintain conservatism, no adjustment to the background, $\hat{p}_1$, is made. (See the Appendix for the development of this approximation.)

Step 5. Calculate the upper confidence limit on the slope for each combination

$$\mathrm{UCL}(\hat{s}) = \mathrm{UCL}(\hat{p}) \,/\, d_j$$

where d_j is the lowest dose in each combination and $d_j \le d_m$ (i.e., the lowest dose is in the observed convex region).

Step 6. Denote the minimum UCL($\hat{s}$) by s_m. For low-dose extrapolation (interpolation), the upper bound on the proportion of responders at dose d is $P = s_m d$. Conversely, for a given upper limit on the risk, P_0, the corresponding dose is $d = P_0/s_m$.

Hence, the procedure determines the combination of experimental doses which gives the smallest upper confidence limit on the low-dose slope which is compatible with the data. Conversely, for a given upper limit on risk, the procedure gives the largest dose which is consistent with the data without utilizing a parametric model.

EXAMPLE

Consider the following example to illustrate the linear-nonparametric procedure. The procedure finds the combination of data that gives the lowest upper confidence limit on the low-dose slope. Suppose the proportions of animals with tumor are

Dose (ppm)	0	100	150	200	400
Proportion	1/50	2/25	1/25	4/50	15/50

Step 1. Since the proportion is smaller at 150 ppm than at 100 ppm, combine the two doses, giving an increasing array of proportions:

Dose (ppm)	0	100	200	400
Proportion	1/50 = 0.02	3/50 = 0.06	4/50 = 0.08	15/50 = 0.30

From this step on, the combined data of the 100- and 150-ppm dose groups will be referred to as the 100-ppm dose group. This provides conservatism for that dose.

Step 2. The slopes between the adjacent doses, starting with the largest dose, are (0.30–0.08)/(400–200) = 0.0012, (0.08–0.06)/(200–100) = 0.0002. Because the second pair gives a lower slope, the largest dose likely to be in the convex region is d_m = 200 ppm. It is impossible to determine whether or not the largest dose, 400 ppm, is in the convex region.

Step 3. The responses for all combinations of doses for which the lowest dose is in the convex region are

Dose (ppm)	100	100 + 200	100 + 400	100 + 200 + 400	200	200 + 400
Proportion	3/50	7/100	18/100	22/150	4/50	19/100

Step 4. For the data from 100 ppm, the approximate upper 95% confidence limit for the proportion of excess tumors is

$$\left(\frac{3+1.5}{50+1.5}-\frac{1}{50}\right)+1.645\sqrt{\frac{\left[\frac{3+1.5}{50+1.5}\right]\left[1-\frac{3+1.5}{50+1.5}\right]}{50+1.5}+\frac{\left[\frac{1}{50}\left(1-\frac{1}{50}\right)\right]}{50}}=0.140.$$

For the data from 100 and 200 ppm combined,

$$\left(\frac{7+1.5}{100+1.5}-\frac{1}{50}\right)+1.645\sqrt{\frac{\left[\frac{7+1.5}{100+1.5}\right]\left[1-\frac{7+1.5}{100+1.5}\right]}{100+1.5}+\frac{\left[\frac{1}{50}\left(1-\frac{1}{50}\right)\right]}{50}}=0.120.$$

The approximate upper 95% confidence limits for the proportion of excess tumors are calculated in a similar fashion for the other combinations of doses.

Step 5. The approximate upper 95% confidence limits on the low-dose slope based on the data from the 100-ppm dose is 0.140/100 = 0.00140. Similarly, the upper confidence limits on the slopes from the dose combinations are

Dose (ppm)	100	100 + 200	100 + 400	100 + 200 + 400	200	200 + 400
Upper slope	0.00140	0.00120	0.00244	0.00193	0.00082	0.00128

Step 6. The approximate upper 95% confidence limit on the low-dose slope based upon the data from the 100-ppm group is 0.00140. Pooling the data from the 100- and 200-ppm doses raises the point estimate of the slope, but the increased sample size results in a lower upper confidence limit of 0.00120. Even though the response is higher at the 200-ppm dose than at the 100-ppm dose, the upper limit on the slope is lowest at this dose, 0.00082. Thus, for purposes of low-dose risk extrapolation, since the goal is to find the largest "safe" dose, the minimum upper limit on the estimate of excess tumors is used which is compatible with the data, that is, $P = 0.00082 * \text{dose}$. Hence, the risk is estimated to be less than $8.2*10^{-4}$ at 1 ppm. This low-dose extrapolation procedure is accomplished without assuming any parametric dose response model.

RESULTS

The method of linear-nonparametric extrapolation described in the previous section was applied to 14 sets of bioassay data compiled by the Food Safety Council.[13] First, data from adjacent response inversions were pooled to give monotonic proportions. The observed convex portion of the dose range was identified. All admissible combinations of data were formed and the minimum upper 95% confidence limit on the slope (risk) was calculated. The data and these results are tabulated in Table 1. For each set of data, the doses predicted to have less than a 1 in 1 million disease rate in the rodent test animal are given in Table 2. Results from the linear-multistage procedure of Gaylor and Kodell[3] are also given for comparison with the linear-nonparametric procedure. GLOBAL82, Howe and Crump,[6] was used to fit the multistage confidence limit on the risk at the lowest response, whichever was larger, with linear extrapolation below that dose. For risks of 1 in 100,000, all of the doses would be a factor of 10 higher, giving the same relative doses between the two procedures.

Table 1. Toxicological Data Sets, Monotonic Proportions, and Convex Regions

NTA: Kidney tumor						
Dose	0%	0.02	0.20	0.75	1.50[a,b]	2.00
Proportion	0/127	0/48	0/48	1/91	2/91	12/48
Monotonic *p*	0	0	0	0	0.022	0.250
Aflatoxin B1: Liver tumor						
Dose	0 ppm	1	5	15[a,b]	50	100
Proportion	0/18	2/22	1/22	4/21	20/25	28/28
Monotonic *p*	0	0.068	--	0.190	0.800	1.000
Ethylene Thiourea: Fetal anomaly						
Dose	0 mg/kg	5	10[b]	20[a]	40	80
Proportion	0/167	0/132	1/138	14/81	142/178	24/24
Monotonic *p*	0	0	0.007	0.173	0.798	1.000
2,3,7,8-Tetrachlorodibenzodioxin: Intestinal fetal anomaly						
Dose	0 mg/kg	0.125	0.25	0.50[a,b]	1.00	
Proportion	0/24	0/38	1/33	3/31	3/10	
Monotonic *p*	0	0	0.030	0.097	0.300	
Dimethylnitrosamine: Liver tumor						
Dose	0 ppm	2	5[a,b]	10	20	
Proportion	0/29	0/18	4/62	2/5	15/23	
Monotonic *p*	0	0	0.065	0.400	0.652	
Vinyl chloride: Liver angiosarcoma						
Dose	0 ppm	50	250[a,b]	500[b]	2500	6000
Proportion	0/58	1/59	4/59	7/59	13/59	13/60
Monotonic *p*	0	0.017	0.068	0.119	0.220	0.217
Hexachlorobenzene: Rib anomaly						
Dose	0 mg/kg	10	20	40[a,b]	60	
Proportion	0/80	4/79	8/91	15/87	25/96	
Monotonic *p*	0	0.051	0.088	0.172	0.260	
Botulinum toxim - type A: Death						
Dose	.010 ng	0.015	0.020[b]	0.024[b]	0.027[b]	0.030
Proportion	0/30	0/30	0/30	0/30	0/30	4/30
Monotonic *p*	0	0	0	0	0	0.133
Dose	0.034	0.037	0.040[a]	0.045	0.050	
Proportion	11/30	10/30	16/30	26/30	26/30	
Monotonic *p*	0.350	--	0.533	0.867	0.867	
Bischloromethyl ether: Respiratory tumor						
Dose	10[c]	20	40	60[a,b]	80	100
Proportion	1/41	3/46	4/18	4/18	15/34	12/20
Monotonic *p*	0.024	0.065	0.222	0.222	0.441	0.600
Sodium saccharin: Bladder tumor						
Dose	0.01%	0.10	1.0	5.0[a,b]	7.5	
Proportion	0/25	0/27	0/27	1/25	7/29	
Monotonic *p*	0	0	0	0.040	0.241	

Table 1. Continued

Ethylenethiourea: Thyroid cancer						
Dose	0 ppm	5	25	125[b]	250[a]	500
Proportion	2/72	2/75	1/73	2/73	16/69	62/70
Monotonic p	0.023	--	--	0.027	0.232	0.886
Dieldrin: Liver tumor						
Dose	0 ppm	1.25[a,b]	2.50	5.00		
Proportion	17/156	11/60	25/58	44/60		
Monotonic p	0.109	0.183	0.431	0.733		
DDT: Liver hepatoma						
Dose	0 ppm	2	10	50[a,b]	250	
Proportion	4/111	4/105	11/124	13/104	60/90	
Monotonic p	0.036	0.038	0.089	0.125	0.667	
Span Oil: Cardiac lesion						
Dose	0%	5	10	15[a,b]	20[b]	
Proportion	1/10	1/10	4/10	4/10	5/10	
Monotonic p	0.100	0.100	0.400	0.400	0.500	

[a]The maximum doses, d_m, in the convex region.

[b]Dose(s) giving the minimum upper 95% confidence limit on the slope, s_m.

[c]Number of exposures for 6 hours of 100 ppm and the background tumor incidence assumed to be zero.

DISCUSSION

In general, the predictions of "safe" dosages by the linear-nonparametric method were within a factor of 2 of the "safe" dosages predicted by the linear-multistage procedure. Utilizing the multistage model in the experimental data range generally resulted in only a slightly higher predicted "safe" dose. This indicates that employment of the multistage model in the experimental data range had little influence on the "safe" doses predicted. Hence, if one does not wish to employ a mathematical model in the experimental dose range, little is lost by using the nonparametric procedure suggested.

APPENDIX

Various procedures and approximations are available for calculating confidence limits on differences of proportions. Where x out of n animals respond, an approximate upper confidence limit of the following form based upon the normal approximation was considered:

$$\frac{x+c}{n+c} + 1.645 \sqrt{\left(\frac{x+c}{n+c}\right)\left(1-\frac{x+c}{n+c}\right) / (n+c)}$$

where $c = 1.5$ was selected to provide a good approximation to exact limits for small x. This is very close to an approximation given by Blyth.[14] This approximation provided slightly conservative confidence limits and higher than exact limits when there were one or

Table 2. Dosages Predicted to Have Less Than a Risk of One in a Million in the Rodent Test Animal

Substance	Response	Dose Units	Linear Nonparametric	Linear Multistage
NTA	Kidney Tumor	%	2.1×10^{-5}	3.7×10^{-5}
Aflatoxin	Liver tumor	ppb	3.8×10^{-5}	4.1×10^{-5}
ETU	Fetal anomalies	mg/kg	2.7×10^{-4}	4.7×10^{-4}
TCDD	Intestinal anomaly	mg/kg	2.1×10^{-6}	4.0×10^{-6}
DMNA	Liver tumor	ppm	3.5×10^{-5}	4.5×10^{-5}
VC	Liver angiosarcoma	ppm	0.2×10^{-2}	1.3×10^{-2}
HCB	Rib anomaly	mg/kg	1.6×10^{-4}	1.7×10^{-4}
Botulinum	Death	ng	0.5×10^{-6}	1.0×10^{-6}
BCME	Respiratory tumor	exposures*	1.3×10^{-4}	1.6×10^{-4}
Saccharin	Bladder tumor	%	2.7×10^{-5}	4.8×10^{-5}
ETU	Thyroid carcinoma	ppm	1.8×10^{-3}	2.6×10^{-3}
Dieldrin	Liver tumor	ppm	6.7×10^{-6}	6.7×10^{-6}
DDT	Liver hepatoma	ppm	3.1×10^{-4}	2.8×10^{-4}
Span Oil	Cardiac lesion	%	2.4×10^{-5}	2.4×10^{-5}

*Number of 6-hr exposures to 100 ppb.

more responders ($x \geq 1$) regardless of the sample size. Even when there were no responders, $x = 0$, the approximate and exact upper 95% confidence limits were certainly close enough for use in low-dose extrapolation:

n	exact	approximate
10	0.259	0.294
20	0.139	0.160
50	0.058	0.068
100	0.030	0.034

For the difference of two proportions, a value of $c = 1.5$ was chosen for the dosed animals. In order to maintain conservatism and to provide a good approximation, particularly for the large number of cases in which no disease was observed in p_1, the control animals, the proportion for the background was not inflated (i.e., $c = 0$ was used for the control animals).

REFERENCES

1. D. Krewski and C. Brown, Carcinogenic Risk Assessment: A Guide to the Literature, *Biometrics* **37**:353-366 (1981).

2. Food and Drug Administration, Advisory Committee on Protocols for Safety Evaluation: Panel on Carcinogenesis Report on Cancer Testing in the Safety Evaluation of Food Additives and Pesticides, *Toxicol. Appl. Pharmacology* **20**:419-438 (1971).
3. D. W. Gaylor and R. L. Kodell, Linear Interpolation Algorithm for Low-Dose Risk Assessment of Toxic Substances, *J. Environ. Pathol. Toxicol.* **4**:305-312 (1980).
4. P. Armitage and R. Doll, Stochastic Models for Carcinogenesis, *Proc. Fourth Berkeley Symposium on Mathematical Statistics and Probability*, pp. 19-38, L. M. LeCam and J. Neyman, eds., University of California Press, Berkeley, CA (1961).
5. J. H. Farmer, R. L. Kodell, and D. W. Gaylor, Estimation and Extrapolation of Tumor Probabilities from a Mouse Bioassay with Survival/Sacrifice Components, *Risk Analysis* **2**:27-34 (1982).
6. R. B. Howe and K. S. Crump, GLOBAL82: A Computer Program to Extrapolate Quantal Animal Toxicity Data to Low Doses, K. S. Crump and Co., Inc., Ruston, LA (1982).
7. M. A. Gross, O. G. Fitzhugh, and N. Mantel, Evaluation of Safety for Food Additives: An Illustration Involving the Influence of Methyl Salicylate on Rat Reproduction, *Biometrics* **26**:181-194 (1970).
8. K. S. Crump, D. G. Hoel, C. H. Langley, and R. Peto, Fundamental Carcinogenic Processes and Their Implications for Low-Dose Risk Assessment, *Cancer Res.* **36**:2973-2979 (1976).
9. H. A. Guess, K. S. Crump, and R. Peto, Uncertainty Estimates for Low-Dose Rate Extrapolations of Animal Carcinogenicity Data, *Cancer Res.* **37**:3475-3483 (1977).
10. R. Peto, Carcinogenic Effects of Chronic Exposure to Very Low Levels of Toxic Substances, *Environ. Health Perspectives* **22**:155-159 (1978).
11. N. Mantel and W. R. Bryan, Safety Testing of Carcinogenic Agents, *J. National Cancer Institute* **27**:455-470 (1961).
12. D. J. Bartholomew, A Test of Homogeneity of Means Under Restricted Alternatives, *J. Royal Statistical Soc.* (Series B) **23**:239-281 (1961).
13. Food Safety Council, Report of the Scientific Committee, *Food and Cosmetics Toxicol.* (Supplement 2) **16**:1-136 (1978).
14. C. R. Blyth, Approximate Binomial Confidence Limits, *J. Am. Statist. Assoc.* **81**:843-855 (1986).

Methods Used in Probabilistic Risk Assessment for Uncertainty and Sensitivity Analysis

Ronald L. Iman
Sandia National Laboratories
Albuquerque, NM

ABSTRACT

Probabilistic Risk Assessment (PRA) plays an important role in the nuclear reactor regulatory process, and the assessment of uncertainties associated with PRA results is widely recognized as an important part of the analysis process. However, uncertainty analysis and sensitivity analysis in the context of PRA are relatively immature fields. A review of available methods for uncertainty analysis and sensitivity analysis in the context of a PRA is presented. This review first treats methods for use with individual components of a PRA and then considers how these methods could be combined in the performance of a complete PRA. In the context of this paper, the goal of uncertainty analysis is to measure the imprecision in PRA outcomes of interest, and the goal of sensitivity analysis is to identify the major contributors to this imprecision.

KEYWORDS: Latin hypercube sampling, Bayesian techniques, differential analysis

INTRODUCTION

Probabilistic Risk Assessment (PRA) is playing an increasingly important role in the nuclear reactor regulatory process.[1] The assessment of uncertainties associated with PRA results is widely recognized as an important part of the analysis process.[2,3] One of the major criticisms of the Reactor Safety Study[4] was that its representation of uncertainty was inadequate.[5] The desire for the capability to treat uncertainties with the MELCOR risk code[6] being developed at Sandia National Laboratories is indicative of the current interest in this topic. However, uncertainty analysis and sensitivity analysis in the context of PRA are relatively immature fields.

In this paper, available methods for uncertainty analysis and sensitivity analysis in a PRA are reviewed. This review first treats methods for use with individual components of a PRA and then considers how these methods could be combined in the performance of a complete PRA. In the context of this paper, the goal of uncertainty analysis is to measure the imprecision in PRA outcomes of interest, and the goal of sensitivity analysis is to identify the major contributors to this imprecision.

There are a number of areas that must be considered in uncertainty analysis and sensitivity analysis for a PRA: (1) information, (2) systems analysis, (3) thermal-hydraulic phenomena/fission product behavior, (4) health and economic consequences, and (5)

New Risks
Edited by L. A. Cox, Jr., and P. F. Ricci
Plenum Press, New York, 1990

display of results. The synthesis of these different areas into a complete PRA is a complex task and involves a large variety of disciplines and techniques. Each of these areas will be reviewed briefly.

INFORMATION

Information designates all inputs used in a PRA such as data, modeling assumptions, and expert opinion. Information constitutes the diverse set of "raw material" that underlies the other areas in a PRA. Due to this diversity, a wide variety of techniques from classical and Bayesian statistics[7,8] is used both to calculate quantities of interest and to quantify the uncertainties inherent in such calculations. The outcomes of such techniques include estimates of expected values, variances, distribution functions, credible intervals and confidence intervals. A fundamental problem in the information area is that some questions do not lend themselves to quantification (e.g., completeness problems, branch points in some physical processes).

A variety of viewpoints exists on the use of Bayesian techniques.[9,10] However, many areas that must be considered in PRA lack the necessary data for the use of classical statistical techniques and so there is little recourse but to use subjective techniques. This necessity is indicated by the extensive use of these techniques in several PRAs[4,11,12] and continues in new PRAs as shown in a series of new reports.[13] As a result, considerable literature has developed on the use of Bayesian techniques in PRA.[7,14,15,16] Closely associated with the use of Bayesian techniques is the need to solicit and quantify expert opinion.[17,18] "Fuzzy set" theory has been proposed as an alternative to Bayesian techniques for use in PRA[19] but has not been widely tried.

SYSTEMS ANALYSIS

Systems analysis involves the development, quantification, and analysis of event and fault trees. The ultimate outcomes of such analyses are dominant accident sequences and their associated frequencies of occurrence. Uncertainty analysis techniques in this area have focused on estimating the uncertainty in the occurrence of events of interest. Several recent reviews of analysis techniques for this area are available.[2,20,21] Methods that have been used include (1) analytic calculation of distribution functions, (2) propagation of discrete probability distributions,[22,23] (3) methods using moments,[24,25] (4) Bayes/Monte Carlo methods,[7] (5) the Maximus method,[26] and (6) Monte Carlo techniques.[20,21] Outcomes of such analyses include estimated distribution functions and confidence intervals.

The analytic calculation of distribution functions is unlikely to find extensive use in PRA due to its numerical complexity. The propagation of discrete probability distributions has been used in several PRAs.[11,12] However, considerable care must be exercised in combining the distributions; otherwise, the number of estimated points in the propagated distributions can become unmanageable. The methods using moments are limited to quantities that can be expressed as relatively simple models (e.g., low order Taylor series); for complex propagations, such methods have the potential to become unmanageable. The Maximus method cannot be fully validated but present information indicates that it is an effective tool for the propagation of binomial and Poisson failure data. Monte Carlo techniques appear to offer effective ways to propagate distributions through event and fault trees, provided the simulation costs do not become excessive. A recent comparison study[21] recommends Monte Carlo techniques. It is possible that use of Latin hypercube sampling[27] rather than random sampling might improve the efficiency of such propagation.

A new development for performing uncertainty and sensitivity analysis on fault trees is based on a matrix representation of the top event.[28] This matrix approach provides a basis for simplifying computations and has been incorporated into a new computer program called the Top Event Matrix Analysis Code (TEMAC).[28] This program utilizes Monte Carlo input to make the following calculations: (1) an uncertainty interval for the frequency of the top event, (2) uncertainty intervals for sensitivity measures such as partial derivatives, risk reduction, and risk increase, (3) uncertainty importance[7] for base and initiating events, and (4) uncertainty intervals for each of the cut sets that jointly constitute the top event.

THERMAL-HYDRAULIC PHENOMENA/FISSION PRODUCT BEHAVIOR

Thermal-hydraulic phenomena/fission product behavior is represented by a collection of models for the degradation of the reactor core, failure of the pressure vessel, core-concrete interactions, fission product transport in the primary system and the containment, and the effectiveness of the reactor containment system. The outcome of the modeling effort in this area is a set of estimated radionuclide releases to the environment for the dominant accident sequences along with estimated probabilities for these releases. Further, such modeling is often used to characterize behavior at branch points for containment event trees. This area is different from the systems analysis area in that the models tend to be based on physical phenomena and hence are more complex than the logic-oriented models used in systems analysis. Several reviews of analysis techniques in this area are available.[30,31] Methods which have been used include (1) differential analysis,[32] (2) response surface methodology,[33] and (3) Monte Carlo simulation. Less used methods include the Fourier amplitude sensitivity test[34] and Kalman filtering.[35] Possible outcomes of such analysis techniques include estimates for both variability and causes of variability.

Differential analysis is based on expanding the model prediction of interest as a Taylor series. This series is then used as a surrogate for the model in uncertainty and sensitivity analysis. Two problems associated with differential analysis are the local nature of the expansion and the possibly significant expense in evaluating the necessary partial derivatives. However, specialized techniques based on adjoint procedures[36,37] and Green's function methods[38] can sometimes be used to reduce the amount of required computation. When a suitable implementation is possible, differential techniques can provide insight in a cost-effective manner with respect to the importance of many individual independent variables. Response surface techniques are based on building a replacement for the model under consideration with the use of experimental designs and regression analysis. The resultant regression model is then used as a replacement for the original model in subsequent uncertainty and sensitivity analyses. The advantage of this approach is that it utilizes an experimental design to determine the dominant independent variables. The chief disadvantage is the difficulty of developing a fitted response surface that can be used as a replacement for a complex model. A recent comparison illustrates problems in the use of response surface techniques.[30] Monte Carlo techniques are easy to implement but may be expensive to use when many runs of a slow-running model are required. Such expense can be reduced by the use of variance reduction techniques, such as Latin hypercube sampling.[27,39] It is felt that an approach to uncertainty and sensitivity analysis based on Latin hypercube sampling has a number of advantages, including sampling of the full range of each variable, direct estimates for distribution functions, expected values and variances, and various measures of sensitivity such as standardized regression coefficients and partial correlation coefficients.[40] There are numerous examples of the application of the techniques indicated here in contexts related to PRA.[41,42,43]

CONSEQUENCE ANALYSIS

Health and economic consequences are predicted by models for environmental radionuclide movement and associated human radiation exposure. Typically, the outcomes of this modeling effort are distribution functions for quantities of interest (e.g., early fatalities). These distribution functions result from the treatment of weather as a stochastic variable. The same analysis techniques available for use with thermal-hydraulic phenomena/fission product behavior can also be used with health and economic consequence modeling. In this area, it is important to recognize that distributions due to weather conditions are not representations of uncertainty in the modeling process but rather are representations of inherent variability which exists in nature. Several recent examples of uncertainty and sensitivity analysis related to consequence analysis are available.[44,45] Commonly, analysis efforts in this area have used parametric studies.

DISPLAY OF RESULTS

There are a number of ways in which uncertainty might be displayed at various stages in a PRA. One of the simplest is to present a range of possible values. Such a range might result from a propagation of ranges obtained from assessments of expert opinion for a number of individual independent variables. Similarly, if distributions are known for the independent variables, then distributions can be estimated for dependent variables. When the quantity of interest is itself a distribution, this results in a family of distributions.[44] In turn, such families of distributions can give rise to new variables (e.g., means, quantiles). However, it must be recognized that the quality of such distributions depends on the quality of the distributions which are propagated. Confidence intervals can also be estimated (e.g., through the use of the Maximus methodology). Credible intervals can be estimated using a Bayes/Monte Carlo approach.[7] Various propagation procedures can also be used to estimate expected values and variances; however, such parameters are not robust measures of model behavior. Even worse, expected values and variances may not adequately represent variability (e.g., discontinuities, tails of distributions). Ultimately, representation of uncertainty in a PRA involves determining a set of representative accident sequences along with the uncertainties in the frequencies and consequences associated with these sequences.[15]

There are also a number of ways in which sensitivity results might be displayed in a PRA. Parametric studies show the results of varying one or a few variables at a time; however, it is hard to investigate the relative effects of a large number of variables with this technique. Many sensitivity analysis methods involve generating partial derivatives or regression coefficients for dependent variables of interest. In turn, various normalizations (e.g., with respect to mean or variance) provide insight into relative variable importance. Further, for analyses based on stepwise regression, R^2 values and the order in which individual variables enter a regression model provide insight with respect to variable importance. Correlation coefficients and fractional contribution to variance can also be used to indicate variable importance. The uncertainty importance measure[7] shows how much the uncertainty in the dependent variable is reduced by having knowledge of one of the inputs. When Monte Carlo techniques are used, scatterplots can often provide insight into the influence of individual variables, especially when the dependent variable is discontinuous. A recent study contains comparisons of a number of these techniques.[30]

SYNTHESIS

The treatment of uncertainties in the context of an entire PRA is initially quite intimidating due to the scale of a PRA and the resulting difficulty of visualizing the manner in which the individual pieces fit together. A recently proposed matrix representation for a

PRA is useful in this regard because it facilitates a conceptual grasp of a PRA and the way its individual pieces are combined.[46] With this approach, the analysis in a PRA for a particular damage type x (e.g., early fatalities, latent cancer fatalities, economic cost) can be represented by the matrix product

$$\phi^x = \phi^I M C S,$$

where $\phi^x = [\phi_\ell^x]$ is a vector such that ϕ_ℓ^x is the frequency of occurrence of damage level x_ℓ for damage type x, $\phi^I = [\phi^I_i]$ is a vector such that ϕ^I_i is the frequency (occurrences per reactor year) of initiating event i, $M = [m_{ij}]$ is a matrix such that m_{ij} is the probability that initiating event i will lead to plant state j, $C = [c_{jk}]$ is a matrix such that c_{jk} is the probability that plant state j will lead to release category k, and $S = [s_{k\ell}]$ is a matrix such that $s_{k\ell}$ is the probability that release category k will lead to damage level x_ℓ. The vector ϕ^x defines the distribution function for damage type x and thus may be thought of as the final result of a PRA with respect to this particular type of damage. Further, various products involving ϕ^I, M, C and S are possible and have interpretations that are relevant to various parts of a PRA. The Indian Point PRA provides additional background on this approach as well as an example of its application [see Ref. 12, Appendices 8.5.1-8.5.9].

The individual terms in the product $\phi^I MCS$ arise from different parts of a PRA. Specifically, ϕ^I comes from the analysis of initiating events, M comes from the use of event and fault trees to determine possible plant states, C comes from the use of containment event trees and appropriate physical models to determine radionuclide releases to the environment, and S comes from the use of consequence analysis models to determine distributions for consequences of interest conditional on given radionuclide releases. The techniques and examples indicated previously have typically been used in analyses that were equivalent to the study of elements in the matrices ϕ^I, M, C and S. With respect to such elements, the goal of uncertainty analysis is to indicate how much variability exists, and the goal of sensitivity analysis is to determine what causes this variability. When uncertainty and sensitivity analysis are considered in the context of an entire PRA, the goals are the same. However, the scale of the problem is much larger and necessitates breaking the analysis into individual parts that are more tractable. The previous matrix representation provides a conceptual and computational bridge between the parts of a PRA and the entire PRA.

A few of the problems that might be considered in the context of uncertainty and sensitivity analysis for a PRA are briefly indicated. If "best estimate" values are determined for the elements of ϕ^I, M, C and S, then differential techniques can be used to determine the relative effects of these elements on the elements of ϕ^x. If distributions can be determined for the elements of ϕ^I, M, C and S, then Monte Carlo techniques can be used to estimate the resultant variability in ϕ^x. A given set of values for the elements of ϕ^I, M, C and S actually leads to an estimate of the distribution function for x; thus, interesting possibilities arise for displaying uncertainty in terms of families of curves. If the elements of ϕ^I, M, C and S can be expressed as functions of other variables, then sensitivity analysis techniques can be used to determine the effects of these variables on ϕ^x and its elements. The matrix representation of a PRA facilitates viewing it as a whole and assessing whether or not consistent representations of uncertainty are being used. If prior analyses indicate that correlations exist between elements in ϕ^I, M, C and S, it should be possible to preserve this in later uncertainty analyses. Overall, the matrix representation for a PRA facilitates the implementation of a complete and coherent analysis. In this regard, an uncertainty or sensitivity analysis in a PRA must be carefully planned. It must first be decided what information (i.e., insight) is desired and then an analysis must be planned that will lead to the desired result.

REFERENCES

1. M. Ernst, Probabilistic Risk Assessment (PRA) and Decision Making under Uncertainty, in *Proceedings of the Ninth Annual Statistics Symposium on National Energy Issues*, October 19-21, 1983, pp. 177-188, Report No. NUREG/CP-0053, LA-10127-C, Los Alamos National Laboratory, Los Alamos, NM (1984).
2. U. S. Nuclear Regulatory Commission, *PRA Procedures Guide*, Report No. NUREG/CR-2300, Vols. 1 and 2, Washington, DC (1983).
3. W. E. Vesely and D. M. Rasmuson, PRA Uncertainties and the Roles of Sensitivity and Uncertainty Analyses, in *Proceedings of the Ninth Annual Statistics Symposium on National Energy Issues*, October 19-21, 1983, pp. 75-102, 189, Report No. NUREG/CP-0053, LA-10127-C, Los Alamos National Laboratory, Los Alamos, NM (1984).
4. U. S. Nuclear Regulatory Commission, *Reactor Safety Study — An Assessment of Accident Risks in U. S. Commercial Nuclear Power Plants*, Report No. WASH-1400 (NUREG-75/014), Washington, DC (1975).
5. H. W. Lewis *et al.*, Risk Assessment Review Group Report to the U. S. Nuclear Regulatory Commission, Report No. NUREG/CR-0400, U. S. Nuclear Regulatory Commission, Washington, DC (1978).
6. J. L. Sprung *et al.*, Overview of the MELCOR Risk Code Development Program, in *Proceedings of the International Meeting on Light Water Reactor Severe Accident Evaluation*, pp. TS-10.1-1 to TS-10.1-8, American Nuclear Society, La Grange Park, IL (1983).
7. S. C. Hora and R. L. Iman, Bayesian Analysis of Learning in Risk Analyses, Report No. SAND84-1348, Sandia National Laboratories, Albuquerque, NM (1985).
8. V. Barnett, *Comparative Statistical Inference*, 2nd ed., Wiley, New York (1982).
9. G. W. Parry and P. W. Winter, Characterization and Evaluation of Uncertainty in Probabilistic Risk Analysis, *Nucl. Safety* **22**:28-42 (1981).
10. R. G. Easterling, Letter to the Editor: Bayesianism, *Nucl. Safety* **22**:464 (1981).
11. Commonwealth Edison Company, Zion Probabilistic Safety Study (1981).
12. Power Authority of the State of New York and Consolidated Edison Company of New York, Inc., Indian Point Probabilistic Safety Study (1982).
13. Sandia National Laboratories, Analysis of Core Damage Frequency from Internal Events, Vols. 1-7, Report No. NUREG/CR-4550, Albuquerque, NM (1986).
14. G. Apostolakis, Bayesian Methods in Risk Assessments, in *Advances in Nuclear Science and Technology 13*, pp. 415-465, Plenum Press, New York, NY (1981).
15. G. Apostolakis, Data Analysis in Risk Assessments, *Nucl. Eng. Des.* **71**:375-381 (1982).
16. S. Kaplan and B. J. Garrick, On the Quantitative Definition of Risk, *Risk Analysis* **1**:11-27 (1981).
17. D. A. Seaver and W. G. Stillwell, Procedures for Using Expert Judgment to Estimate Human Error Probabilities in Nuclear Power Plant Operations, Report No. NUREG/CR-2743, SAND82-7054, Sandia National Laboratories, Albuquerque, NM (1983).
18. H. F. Martz, M. C. Bryson, and R. A. Waller, Eliciting and Aggregating Subjective Judgments: Some Experimental Results, *Proceedings of the 1984 Statistical Symposium on National Energy Issues*, Seattle, October 1984, pp. 63-82 (1985).
19. S. D. Unwin, "A Fuzzy Set Theoretic Foundation for Vagueness in Uncertainty Analysis," *Risk Analysis* **6(1)**:27-34 (1986).
20. P. S. Jackson *et al.*, Uncertainty Analysis of System Reliability and Availability Assessment, *Nucl. Eng. Des.* **68**:5-29 (1981).
21. H. F. Martz *et al.*, A Comparison of Methods for Uncertainty Analysis of Nuclear Power Plant Safety System Fault Tree Models, Report No. NUREG/CR-3263, LA-9729-MS, Los Alamos National Laboratory, Los Alamos, NM (1983).

22. S. Kaplan, On the Method of Discrete Probability Distributions in Risk and Reliability Calculations — Applications to Seismic Risk Assessment, *Risk Analysis* **1**:189-196 (1981).
23. S. Ahmed *et al.*, A Method for Propagating Uncertainty in Probabilistic Risk Assessment, *Nucl. Tech.* **59**:238-245 (1982).
24. G. Apostolakis and Y. T. Lee, Methods for the Estimation of Confidence Bounds for the Top-Event Unavailability of Fault Trees, *Nucl. Eng. Des.* **41**:411-419 (1977).
25. J. D. Murchland and G. G. Weber, A Moment Method for the Calculation of a Confidence Interval for the Failure Probability of a System, *Proceedings of the 1972 Annual Reliability and Maintainability Symposium*, pp. 565-577, Institute for Electrical and Electronic Engineers, New York, NY (1972).
26. F. W. Spencer and R. G. Easterling, Lower Confidence Bounds on System Reliability Using Component Data: The Maximus Methodology, Report No. SAND84-1199C, Sandia National Laboratories, Albuquerque, NM (1984).
27. M. D. McKay *et al.*, A Comparison of Three Methods for Selecting Values of Input Variables in the Analysis of Output from a Computer Code, *Technometrics* **21**:239-245 (1979).
28. R. L. Iman, A Matrix-Based Approach to Uncertainty and Sensitivity Analysis for Fault Trees, *Risk Analysis* **7(1)**:21-33 (1987).
29. R. L. Iman and M. J. Shortencarier, A User's Guide for the Top Event Matrix Analysis Code (TEMAC), Report No. NUREG/CR-4598, SAND86-0960, Sandia National Laboratories, Albuquerque, NM (1986).
30. R. L. Iman and J. C. Helton, A Comparison of Uncertainty and Sensitivity Analysis Techniques for Computer Models, Report No. NUREG/CR-3904, SAND84-1461, Sandia National Laboratories, Albuquerque, NM (1985).
31. D. C. Cox and P. Baybutt, Methods for Uncertainty Analysis: A Comparative Survey, *Risk Analysis* **1**:251-258 (1981).
32. P. M. Frank, *Introduction to System Sensitivity Theory*, Academic Press, New York (1978).
33. R. H. Myers, *Response Surface Methodology*, Allyn and Bacon, Boston, MA (1985).
34. G. J. McRae *et al.*, Global Sensitivity Analysis — A Computational Implementation of the Fourier Amplitude Sensitivity Test (FAST), *Comput. Chem. Eng.* **5**:15-25 (1982).
35. R. J. Meinhold and N. D. Singpurwalla, Understanding the Kalman Filter, *Amer. Stat.* **37**:123-127 (1983).
36. D. G. Cacuci *et al.*, Adjoint Sensitivity Analysis of Extremum-Type Responses in Reactor Safety, *Nucl. Sci. Eng.* **83**:112-135 (1983).
37. E. M. Oblow, An Automated Procedure for Sensitivity Analysis Using Computer Calculus, Report No. ORNL/TM-8776, Oak Ridge National Laboratory, Oak Ridge, TN (1983).
38. J.-T. Hwang *et al.*, The Green's Function Method of Sensitivity Analysis in Chemical Kinetics, *J. Chem. Phys.* **69**:5180-5191 (1978).
39. R. L. Iman and W. J. Conover, Small Sample Sensitivity Analysis Techniques for Computer Models with an Application to Risk Assessment, *Communication in Statistics* **A9(17)**:1749-1842 (1980); and Rejoinder to Comments, pp. 1863-1874 (1980).
40. R. L. Iman *et al.*, An Approach to Sensitivity Analysis of Computer Models, Parts 1 and 2, *J. Qual. Tech.* **13**:174-183 and 232-240 (1981).
41. J. C. Helton, R. L. Iman, J. D. Johnson, and C. D. Leigh, Uncertainty and Sensitivity Analysis of a Model for Multicomponent Aerosol Dynamics, Report No. SAND84-1307, Sandia National Laboratories, Albuquerque, NM (1985)
42. N. D. Cox, Comparison of Two Uncertainty Analysis Methods, *Nucl. Sci. Eng.* **64**:258-265 (1977).
43. D. H. Nguyen, The Uncertainty in Accident Consequences Calculated by Large Codes Due to Uncertainties in Input, *Nucl. Tech.* **49**:80-91 (1980).

44. D. J. Alpert *et al.*, A Demonstration Uncertainty/Sensitivity Analysis Using the Health and Economic Consequence Model CRAC2, Report No. SAND84-1824, Sandia National Laboratories, Albuquerque, NM (to appear).
45. A. Bayer *et al.*, The German Risk Study: Accident Consequences Model and Results of the Study, *Nucl. Tech.* **59**:20-50 (1982).
46. S. Kaplan, Matrix Theory Formalism for Event Tree Analysis: Application to Nuclear-Risk Analysis, *Risk Analysis* **2**:9-18 (1982).

Can We Bring Quality of Management into PRAs?

D. Okrent and S. Arueti
University of California, Los Angeles
Los Angeles, CA

ABSTRACT

The basic assumption of this study is that utility and plant corporate management has a significant role in nuclear power plant safety. From this point of view we try to identify paths through which management effectiveness affects plant safety, partly in terms of measurable parameters. Some of the available data is analyzed in light of the proposed parameters, in order to examine possible correlations. As a last step, preliminary proposals are made of methods for including management performance as a variable in future probabilistic risk assessment (PRA) studies.

KEYWORDS: Nuclear power plants, quality of management, performance indicators, probabilistic risk assessment (PRA), reactor safety

SIGNIFICANCE OF MANAGEMENT AND ITS PERFORMANCE

Probabilistic risk assessment (PRA) does not currently factor the quality of nuclear power plant management into the evaluation of core damage frequency or risk to the public. To some extent, certain aspects of management quality may be reflected in the number of scrams, the unavailability of some components due to an abnormally high number of human errors, etc. Such data could enter into a PRA via the use of plant specific data to update generic data. However, many of the more important influences of management quality on safety cannot be captured in this fashion.

We expect plant management to play the role of the plant's brain. We do not expect management to play a direct significant role in performing "reflexive" operations such as the early steps in emergency shutdown.

Although the prevention and the proper handling of safety related events must receive primary emphasis, separation of safety from other aspects of plant operation, can be artificial in certain respects. A positive correlation is expected between safety and certain economic parameters. The same parameters that affect public safety, such as a low number of scrams, rare failures of complete systems and a well trained staff affect operational reliability and economic effectiveness. The same management which is, say, very effective in assuring good economic performance is usually not expected to be negligent in performing tasks related to safety, and vice versa.

Hence some of the important activities of plant management are the following:

New Risks
Edited by L. A. Cox, Jr., and P. F. Ricci
Plenum Press, New York, 1990

a. Provide an overall safety philosophy and approach and strategy for plant operation.

b. Provide means for accomplishment of goals: budget, appropriate personnel, allocation and scheduling of time for training and preventive maintenance.

c. Assure that direct responsibility exists and that there is an effective communication system.

d. Carefully monitor plant activities, including those of quality control (QC) and engineering groups, review procedures and technical specifications, and assess adequacy of lessons learned from prior operational experience in this and other plants.

e. Foresee and assure preparedness for non-standard situations: transients, external events, accidents-both within and beyond the design basis.

SOME QUESTIONS THAT RELATE SAFETY TO MANAGEMENT QUALITY

There are more than one set of questions that can be used to examine the quality of management with respect to reactor safety. A first set, which relates to the existence of a pervasive safety culture and a strong, in-depth knowledge of plant design and behavior, might be like the following in nature:

a. Is there a pervasive safety awareness throughout the licensee's organization?

b. Does the Station Manager have a deep technical knowledge of the safety design and operational characteristics of the plant?

c. Are there strong engineering organizations, both on site and at the home office of the operating organization, each of which knows the safety design characteristics in detail, understands the operational behavior of the plant over a very wide range of transients and abnormal conditions, knows the characteristics and phenomena associated with severe core accidents, and *knows* the alternatives available for accident management?

d. Does the on-site engineering organization analyze its own operating experiences, and are appropriate modifications implemented?

e. Does the on-site or home office engineering organization take prompt and full advantage of nationwide operating experience?

f. Are the shift technical advisors really properly qualified to act as detached diagnosticians during complex transients?

A second set could relate to the structure of management and assignment of responsibility, somewhat like the topics used in reviews made by the Institute for Nuclear Power Operation (INPO):[1]

a. Is there an organizational structure wherein the responsibilities are clearly defined, all aspects of nuclear support are covered, and assignments to off-station staff are minimized?

b. Is there adequate involvement and commitment, on the part of management, to safe and reliable station operations?

c. Does the senior nuclear manager receive proper support, and does he have the authority and control necessary to carry out his responsibilities?

A third set could relate to what are frequently called "performance indicators."[2] Some performance indicators that can be used, are: Systematic Assessment of Licensee Performance (SALP) ratings; Scrams; Engineered Safety Features (ESF) actuations; Safety system failures and challenges; Forced outages; Enforcement actions; Licensee event reports (LERs); Unit availability.

Performance indicators usually have the advantage that they can be quantified, whereas the first two sets of questions can only be evaluated with difficulty, and then usually subjectively. On the other hand, the quantitative performance indicators, of necessity, are far from an adequate indicator of the safety quality of management. They bear only indirectly, or not at all, on many major facets of safety. One limited data point to this effect is the Soviet information that Chernobyl 4 had had an excellent operational record from the point of view of plant availability.

OBJECTIVES OF THE PRESENT STUDY

The study is to be conducted at three levels, each of which is assigned its goals.

Level 1 deals mainly with parameters that may generally be classified as symptoms of management performance. Such a set of symptoms is largely reflected in the performance indicators discussed above. We analyze both safety and economy related parameters, since they are interrelated. These symptoms are discussed in the following section. Our goal is to identify such symptoms, to analyze relevant available data, and to look for statistical correlations between various symptoms, as well as between symptoms and subjective opinions on the performance of plant management.

Level 2 is intended to identify root causes for good/bad performance, and to find possible correlations between the causes and the symptoms. Such causes are chosen (in part) from tasks that we believe to be directly related to good management, such as: good preventive maintenance, a good communications system, up-to-date procedures and technical specifications, appropriate plant personnel, well prepared outages, etc. Example of parameters that may be related to such root causes are question sets 1 and 2.

In Level 3 of the study we plan to try to propose a methodology that could account for management quality in PRA studies.

We are aware that an important step in data analysis for a PRA is the assessment of plant specific data. One might argue that this step accounts for the effects of management quality. However, it may not allow for trends. More importantly, the recent operating experiences at Davis Besse[3] and Rancho Seco,[4] and the shutdown of all the Tennessee Valley Authority (TVA) nuclear power plants,[5] have all pointed up major management deficiencies which, in little or no way, would be reflected in a currently performed PRA. That the quality of safety management cannot only be important but vital has, of course, been vividly demonstrated by the accident at Chernobyl 4.[6] Hence the question really should not be "Can we bring the quality of management into PRA?" but "How are we going to do it and do it plausibly?"

We believe that issues like those posed in the first two lists of questions are vital to any adequate treatment. In the first phase of this study, unfortunately, we are unable to address either of these lists, although INPO should have information bearing directly on the second one.

LEVEL 1: METHODOLOGY AND RESULTS

Methodology

On 5/22/86 the U.S. Nuclear Regulatory Commission (NRC) listed the best 15 and the worst 17 nuclear plants in the USA.[7] The differences were attributed mainly to management. We used this report as an initial point to identify four pairs of nuclear units for our study. Those pairs satisfy the following conditions:

a. Each pair consists of one unit from the list of "best 15" and one from the list of "worst 17".

b. The units in each pair are as similar as possible to each other, i.e. same vendor, roughly same power and same age.

c. We avoided choosing new plants to prevent analyzing events that are mainly results of "childhood diseases".

As a result we chose the following units: Kewaunee, McGuire 1, Millstone 1, Oconee 3, Pilgrim 1, Rancho-Seco, Sequoyah 1, Turkey Point 4.

Each of the pairs is separately analyzed based on existing public data, namely: NRC precursor studies,[8-10] NRC gray books[11] computerized descriptions of LERs.[12]

Up to this point only part of the data for four of the eight units has been analyzed. At this stage we refer to these units as units 1A, 1B, 2A, 2B.

Data are classified on the following bases:

a. Parameters that are mentioned in the literature as measures of plant achievement. These are: forced outage rate and availability factor.

b. Based on fast scanning of the data, we identified several parameters which might point towards the effectiveness of plant management. Those are the number of LERs related to incomplete procedures, the number of human errors (which may point to quality of training), the number of actual events versus faulty equipment identified during test or surveillance (may point to the adequacy of surveillance and preventive maintenance programs), the number of events related directly to lack of maintenance (e.g. lubrication), the number of events related to communication or to administrative controls, and the number and severity of precursors as identified in references 8 through 10.

Another parameter which we have examined in preliminary fashion is the number of repetitive events. We consider two types of repetitions: Events occurring at the same system, and events of the same nature (e.g. same cause of failure).

Results

Partial results for four of the eight units are summarized in Tables 1 and 2, and Fig. 1. Table 1 classifies the LER's since 1983 for the four plants according to their significance as measurements for management performance. Thus we divide the types of events to reports on incomplete specifications and procedures, actual events that occurred, identification of non-calibrated instruments and design deficiencies. In each case we identify whether human error was involved. Another aspect that we look at is whether the fault described in the event was identified by the local personnel before causing damage, or it caused an actual event or was identified by external enforcement procedure such as NRC surveillance.

Table 1. Classified Licensee Event Reports

Event Type	Human Error (h) / Non Human Error (n)	Unit	NRC or QA	Actual Event	Plant Personnel Test Surveillance Review Inspection etc.	Other
Incomplete Specs Procedures Records etc.	h	1A	0	1	1	
	h	1B	2	2	3	
	h	2A	0	4	2	
	h	2B	0	4	0	
	n	1A	0	0	1	
	n	1B	3	1	4	
	n	2A	0	1	0	
	n	2B	0	0	1	
Safety Related Actual Events	h	1A	0	6	1	
	h	1B	2	22	2	
	h	2A	0	20	1	
	h	2B	0	24	0	
	n	1A	0	15	11	
	n	1B	1	43	22	
	n	2A	0	16	15	
	n	2B	0	37	3	
Other Actual Events	h	1A		0	0	
	h	1B		3	1	
	h	2A		7	1	
	h	2B		3	0	
	n	1A		1	3	
	n	1B		5	1	
	n	2A		3	1	
	n	2B		3	1	
Instrument Calibration	h	1A			0	
	h	1B			1	
	h	2A			0	
	h	2B			0	
	n	1A		1	21	
	n	1B		1	3	
	n	2A		2	1	
	n	2B		1	0	
Design Deficiency		1A		0	1	
		1B		4	3	
		2A		6	1	
		2B		0	2	
N.A.		1A		1		4
		1B		0		7
		2A		0		4
		2B		3		0

Table 2 includes a list of possible indicators and counts the number of events that could be related to each indicator. It seems from those tables that there is a considerable agreement among most of the indicators. Except for the number of forced shutdowns and forced outage rate, there is good evidence that although two of the units belong to the category of "best 15" and two to the "worst 17", in fact only unit 1A shows outstandingly good indicators. Among the other three, unit 1B looks worse than the other two for most parameters, while the image of 2A is only slightly better than that of 2B.

The precursor data[8-10] includes too few points, over a non-continuous period of time, to enable a reliable statistical analysis for each of the units. It gives, though, an indication of the performance of all the "good" plants compared to the "bad" plants as a group over the years for which data has been collected. Over those years 26 severe precursors have been identified in the four "bad" plants, versus 11 precursors in the "good" plants. If we attach to each event a severity index which is its relative location in the list of precursors and draw two cumulative distributions of index vs. number of events (as illustrated in Fig. 1), we find a consistent frequency distribution that predicts a lower risk in operating the "good" units then in operating the "bad" ones. Some sort of consistency can also be found in the economic parameters. For three of the four pairs, each of the "good" plants had a much

Table 2. Events Related to Various Indicators

	Actual Events	Human Errors	Maintenance Errors	Human Errors Related to Non Licensed Operators or Contractors	Events Related to Maintenance	Common Mode Failures	Multi Failures
1A	24	9	2	0	6	5	1
1B	76	38	12	5	19	10	5
2A	53	35	13	1	14	0	0
2B	72	31	7	6	11	2	3

better performance then its counter-part both in terms of availability and of forced outage rate. The fourth pair had among the best measures in both terms.

LEVELS 2 AND 3: DIFFICULTIES AND PROPOSED METHODOLOGY

Neither field nor controlled tests (as is done for human factors study) can be directly applied to management performance, for the following reasons:

a. The tasks of the management team are not specific and vary in time and conditions.

b. Each action may affect many parameters in different ways.

Table 2. Continued.

Unit	Events Identified by External Enforcement Procedures	Number of Reactor Trips	Events Attributed to Administration	Number of ESF Actuations	Incomplete Procedures	Forced Outage Rate	Events Attributed to Bad Communication	Availability Factor
1A	0	6	4	3	2	2.1	1	86
1B	8	8	21	8	15	6.7	2	89
2A	0	10	10	8	7	0.4	1	84
2B	0	17	5	12	5	14.6	3	65

c. Most of the actions have a long range (as opposed to an immediate) effect.

d. Usually it is difficult to separate the effect of management actions from other attributes such as hardware design, aging, other human errors, etc.

A major difficulty in performing Levels 2 and 3 is lack of appropriate data sources. Most of the parameters discussed below require data which are either not ordinarily collected, or if collected, not made public. Most of the data that we could identify for our purposes were either lacking and incomplete, or inconsistent from one utility to another, thus not enabling comparisons. Some examples:

- Personnel classification does not have the same basis.
- In most cases data are collected per utility or per multi unit plant (where a plant may have completely different units).

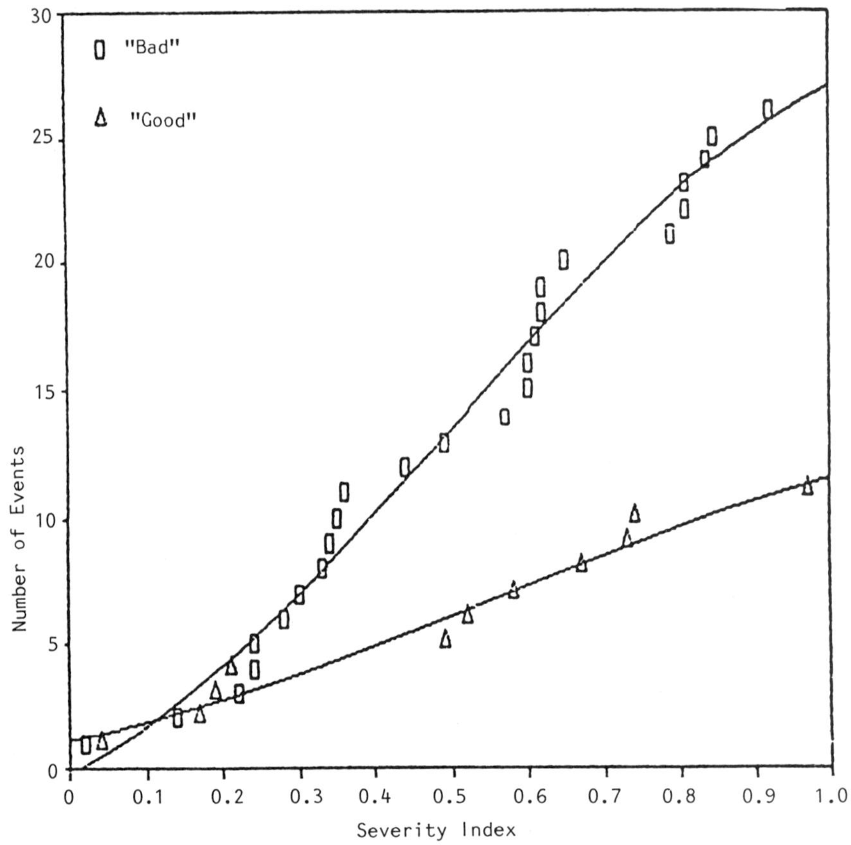

Fig. 1. Frequency of Precursors.

Due to those difficulties, Level 2 requires several steps of work. We will first have to collect and classify data according to their relevance and consistency. We will then continue in two directions:

- Use the appropriate data for identifying correlations between possible root causes and the symptoms that are studied in Level 1.

- Identify additional information that is necessary for more significant results to be derived, the form of this information and its relative importance.

The difficulties discussed above necessitate inserting a great deal of subjectivity into the incorporation of management quality into PRA. Several alternative approaches to that process are:

a. One might use the empirical information on severe core damage accidents in the US (TMI 2) or in the world (TMI 2 and Chernobyl 4), relate all this experience to bad management quality and use it to construct a generic distribution of management quality versus severe core damage frequency. This approach, though, has drawbacks.

b. A second approach would assign higher levels and frequencies of radioactive material releases to plants whose operators have poor knowledge and skill with regard to severe accident management. Study would be required to evaluate how far management can aggravate an accident, and what the chance of their doing so would be.

c. Poor management with regard to maintenance could be reflected in a PRA by an increased probability of otherwise rare common mode failures in the future.

d. Poor management resulting in a lack of diagnostic skill could be reflected in a PRA by an increased rate of errors of commission, and an increased time for repair during an event.

e. Poor management could make unlikely transients or accidents more likely: e.g., poor maintenance could affect scram reliability, and poor control of water chemistry could lead to pipe cracking in boiling water reactors (BWRs) or steam generator tube rupture in pressurized water reactors (PWRs). This can be reflected in the PRA by an increased frequency of initiating events.

f. Poor management during construction could lead to deficiencies in anchorage of vital components.

The above considerations suggest that a global evaluation of management may not be sufficiently discriminating. Management at a single nuclear plant may be good in some aspects and poor in others, each leading to an expectation different than the average (or generic) estimates in each category significant to risk.

In a preliminary effort to treat such effects, we plan first to apply a qualitative approach through which as many as possible relations between symptoms and parameters affecting them will be identified. Two alternative methods are being considered:

- The Analytical Hierarchical Process (AHP) method[13] which is based on hierarchical levels, each level consisting of parameters influencing the level above it.

- Digraph (directed graph) method which does not identify hierarchical levels, but rather relates all the parameters to each other by cause-effect relations.

For a second quantitative stage we plan to try the Multi Attribute Utility (MAU) method,[14] which is flexible enough to utilize both objective data and subjective expert opinions. By applying such methods, we intend to try to formalize subjective assessments of plant management which have already been performed in studies such as SALP reports and the Markey report,[15] and to take advantage, as much as possible, of objective information.

SUMMARY

This paper presents the current stage of a study on management performance and its role in plant safety. We partially studied four of eight units, found evidence of correlation between some performance measures, which was not consistently positive.

We plan to continue the study through Levels 2 and 3, in which root causes and methods for incorporating the results in PRA will be studied.

The paper addresses difficulties in achieving these goals, and suggests in preliminary form a general methodology for including management quality in PRA.

REFERENCES

1. P. Beard, Results of INPO Management Review of Nuclear Power Stations, Remarks to the ANS/ENS Topical Meeting on Thermal Reactor Safety, San Diego, CA, February 4, 1986.
2. V. Stello, Jr., Performance Indicators, NRC paper SECY-86-144, May 5, 1986.
3. NUREG-1154, Loss of Main and Auxiliary Feedwater Event at the Davis-Besse Plant on June 9, 1985, August 1985.
4. USNRC Report, Abnormal Occurrences for First Quarter, CY 1986, AO 86-2 Loss of Integrated Control System Power and Overcooling Transient, Rancho Seco, Dec. 26, 1985.
5. V. Stello, Jr. Status of Staff Actions Regarding TVA, NRC paper SECY-86-IC, May 14, 1986.
6. INSAG Summary Report, Post-Accident Review Meeting on the Chernobyl Accident, IAEA, Vienna, 8/30/1986-9/5/1986.
7. S. Diamond, Management cited as Key to Nuclear Safety in U.S., *The New York Times*, May 23, 1986.
8. J. W. Minarick and C. A. Kukielka, Precursors to Potential Severe Core Damage Accidents: 1969-1979, A Status Report, NUREG/CR-2497, June 1982.
9. W. B. Cottrell *et al.*, Precursors to Potential Severe Core Damage Accidents: 1980-1981, A Status Report, NUREG/CR-3591, July 1985.
10. J. W. Minarick *et al.*, Precursors to Potential Severe Core Damage Accidents: 1985, A Status Report, NUREG/CR-4674, to be published.
11. NRC, Licensed Operating Reactors, Status Summary Report, NUREG-0020, Feb. 1983 through May 1986.
12. Licensee Event Reports (LERs) of analyzed plants.
13. J. L. Saaty, A New Approach to Performance Measurement, the Analytic Hierarchy Process, *Design Methods and Theories* **13(2)**:64-72 (1979).
14. R. L. Keeney and H. Raiffa, *Decision with Multiple Objectives*, John Wiley & Sons (1976).
15. NRC, Attachment to Testimony of United States Nuclear Regulator Commission, July 16, 1986, U.S. House of Representatives Subcommittee on Energy Conservation and Power of the Committee on Energy and Commerce, Subcommittee chairman: Edward J. Markey.

Studies with a New Risk Reduction Model

Jussi K. Vaurio, Jaana Isaksson and Ulf Linden
Imatran Voima Oy
Loviisa Power Station
Loviisa, Finland

ABSTRACT

Risk reduction models have been developed based on the rationale that learning is due to failure events rather than passage of time. A practicable moment matching technique is presented for estimating the parameters of a particular continuous fractional learning model. Several applications on component, system and plant levels are presented, demonstrating a broad validity of the model.

KEYWORDS: Reliability growth, learning curves

INTRODUCTION

The fundamental concept of reducing risk by learning is that the reliability of an item will improve as failure modes which degrade its function are discovered and eliminated by redesign. Some systematic failure modes may be revealed early in the development of an item, whereas others may take a considerable time to show themselves.[1] The difficulty and time taken to rectify or reduce such failure modes by redesign will also vary. Therefore, the rate at which reliability will grow will depend upon the rate at which

- systematic failure modes are detected and failure causes investigated,
- failure information is made available to the responsible authority and to other similar plants,
- redesign or other changes are made to overcome the reported failures, and
- the modifications are introduced and checked to ensure that they are effective.

The process indicates that the learning that leads to risk reduction and reliability growth is due to failure events rather than passage of time. Perhaps the most universally applied time-dependent model is the Duane model[2] that provides a straight line on log/log graph paper for cumulative number of failures vs. cumulative exposure time. It has the conceptual drawback that it implies infinite failure rate (hazard rate) at time zero. Sheppard[1] has recently shown that Duane estimators can be very sensitive to early data, to the selection of the time zero, and to the way design modifications are incorporated, and can

New Risks
Edited by L. A. Cox, Jr., and P. F. Ricci
Plenum Press, New York, 1990

lead to erroneous conclusions about learning. Even the three-parameter IBM model has been found unsatisfactory[1] in several cases. A collection of other potential models can be found in Ref. 3. Balaban has compared several of these models, and he also found Duane model results disappointing.[4]

This paper introduces a new reliability growth model that is based on the concept of learning from events rather than time. A continuous version of the model is developed for predictive purposes.

Applications to fires and reactor "trips" at nuclear power plants and to failures of integrated circuits demonstrate the validity of the model.

LEARNING MODELS

Whenever there is a rate parameter (function) $r(t)$, such that $dn = r(t)dt$ is the probability of an event occurring in time dt, this is also the expected number of events in time dt. The expected cumulative number is

$$< n(t) >= \int_o^t r(t')dt' \,, \tag{1}$$

and in any time interval (t_1, t_2) the number of events has a Poisson distribution[3] with mean $<n(t_2)> - <n(t_1)>$.

The Fractional Learning Model (FLM)

The fundamental assumption in this FLM model is that all learning is based on events such as failures, accidents or human errors. Because of modifications made in design, manufacturing, material, etc., after each event, the expected rate of new events is smaller after an event than before it. Let the accumulated total exposure time prior to event i be T_i, and the time between events $i-1$ and i be $t_i = T_i - T_{i-1}$ (i=1,2,...n; T_0=0). Thus, the actual cumulative number of events, $n(t)$, is a step function in time with increments of 1 at times T_i and the final value n=$n(T)$ at the time of observation and analysis T ($T \geq T_n$). More specifically, in the FLM model we assume that the failure rate is

$$h(t) = Ac^i \text{ for } T_i \leq t < T_{i+1}, \; i = 0, 1, ..., n. \tag{2}$$

The common factor A is the event rate parameter before the first event, and c is the fraction to which the rate is reduced after the lessons from an event have been learned and implemented. If a history of events and times has been recorded, the values of A and c can be estimated and used in predictions for the future. Equations for the maximum likelihood estimation and for the accuracy of such estimators can be found in Ref. 5.

The FLM is a suitable model for severe accidents. For smaller routine failures a more continuous hazard rate model is appropriate.

A Continuous Fractional Learning Model (CFLM)

The failure rate of Eq. (2) can be written as $h(t) = A\, c^{n(t)}$ where $n(t)$ is the cumulative number of events at time t. Assuming a large number of realizations of the process, one can define the average failure rate

$$\bar{h}(t) =< Ac^{n(t)} >\approx Ac^{<n(t)>} \tag{3}$$

where the last form is the first order approximation when c^n is developed into a Taylor series. When Eq. (3) is substituted for $r(t)$ in Eq. (1), we obtain an integral equation

$$< n(t) >= \int_o^t A\, c^{<n(t')>} dt' \,. \tag{4}$$

With the condition $<n(o)> = 0$ this yields

$$< n(t) >= \frac{A}{a}\, ln\,(1 + at),\ a = A\, ln(1/c), \tag{5}$$

which has the derivative failure rate

$$h(t) =< \frac{dn}{dt} >= A/(1 + at), \tag{6}$$

where $a = A\, ln(1/c)$. The failure accumulation model defined by Eqs. (5) and (6) is called a Continuous Fractional Learning Model (CFLM). It is not only a convenient fitting function but has a rationale based on learning from events.

The remainder of this paper is focused on estimating the values of A and a from observed data.

Estimation of Parameters

The idea in the moment matching method is to determine the free fitting parameters A and a so that the observed number of events $n(t)$ equals the expected $<n(t)>$ of Eq. (5), and the observed integral $I(t)= \int_o^t n(t')dt'$ equals the expected value

$$\int_o^t < n(t') > dt' = \frac{A}{a^2}\, [(1 + at)\, ln\,(1 + at) - at] \,. \tag{7}$$

The ratio of this integral and $<n(t)>$ is independent of A. Based on these results, we have the following procedure for determining a and A:

1. Plot the observed $n(t')$ as a function of time t' up to the current time t, and determine $n(t)$ as well as the area below, the integral $I(t) = \int_o^t n(t')dt'$.

2. Calculate the value of the ratio $Z = I(t)/[t\, n(t)]$.

3. Determine the value s (the product $s = a{\cdot}t$) from the condition

$$Z = 1 + \frac{1}{s} - \frac{1}{ln(1 + s)} \tag{8}$$

   by iteration, or from Fig. 1.

4. Determine the value of $a = s/t$.

5. Determine the value of A,

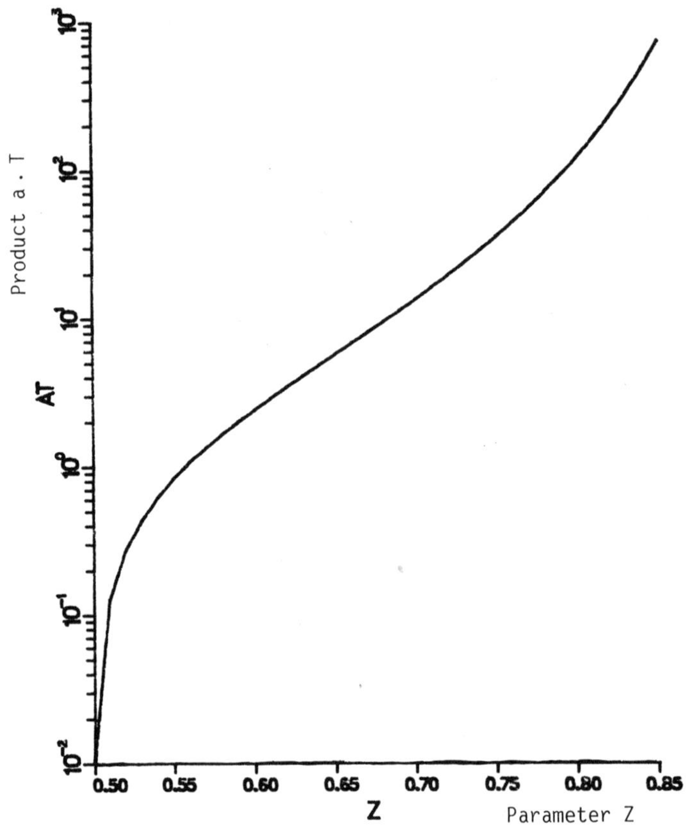

Fig. 1. Parameter $a \cdot T$ as a function of the observable parameter $Z = I(T)/[Tn(T)]$.

$$A = n(t)a/ln(1+s). \tag{9}$$

6. Plot the function $\langle n(t)\rangle = \frac{A}{a}\, ln(1+at)$ in the same picture with $n(t)$ and verify the validity of the model.

7. Calculate $c = \exp(-a/A)$ (if needed).

In spite of the simplicity of the method there are several alternatives or interpretations actually included in this formalism.

Learning From Own Failures Only (LOF). When a unit (a plant or a system) is learning from its own failures only, the learning process is age-dependent, i.e. the time variable t is age (τ) for a single unit. When multiple identical units are available, they can be plotted in the same figure, all $n(\tau)$ beginning at τ=0. Average $n(\tau)$ and $I(\tau)$ may be determined, and used in the procedure to obtain A and a for an average unit.

One should notice, in particular, that if older units are decommissioned before reaching age (τ), they contribute to the exposure time and to the failures only until their decommissioning age is reached.

Learning from All Failures (LAF). When a unit is learning from all failures that occurred before the current calendar time t_c, no matter which unit the failures took place in, the time variable t is the sum of the exposure times of all units before t_c, and $n(t)$ the corresponding cumulative number of failures in all those units before t_c. Old units cease to contribute more to both t and $n(t)$ after decommissioning, and new units begin to contribute when t_c passes their commissioning time.

On-Line Estimates (OLE) and Post Estimates (PE). In both cases, LOF and LAF, it is possible to perform two different kinds of estimations. First, it is possible to make just one estimation at the current time of the study, $t = T$, and plot the resulting $\langle n(t) \rangle$ for all $t<T$, as described in Step 6 of the procedure. This result is called a Post Estimate (PE) because it estimates only once, using all data up to T for estimating $\langle n(t) \rangle$ for all $t \leq T$.

Secondly, it is also possible to carry out the estimation procedure for A and a at many different points of time t ($t<T$), and let such $A(t)$ and $a(t)$ define $\langle n(t) \rangle$ of Eq. (5). This is called an On-Line Estimate (OLE), because it provides a new estimate at every moment t as time goes on.

The comparison of PE and OLE provides some clues as to the validity of the model. A good agreement indicates a valid model, while large discrepancies indicate that fitted parameters (A,a) are not truly independent of time, i.e., the model may not be valid.

APPLICATIONS

Fires in Nuclear Power Plants

As a first numerical example, consider 24 fires which occurred in 17 nuclear power plants which were in operational status during the period from March 1968 to June 1978. The times to fire and times of operation have been published in Table 1 (Set A) of Ref. 6. If one assumes that the plants learn lessons from fires in all other plants, too, and applies the LAF-procedure, the post estimate curves of Fig. 2 are obtained. The parameters obtained, $A = 0.026$/mo., $a \cdot T = 2.1$, $c = 0.99$ indicate very limited learning between the plants.

When the same data is analyzed assuming that each plant learns only from its own fires, i.e. using the LOF procedure, the results indicate strong learning as a function of age. The observed cumulative number of fires per plant is presented in Fig. 3, together with three models. The first model, CFLM-LOF fitted at the age t=50 months, is very accurate prior to age 50. A significant change in the fire event pattern occurs at $t = 50$. It could be due to the Browns Ferry fire or another significant event causing changes in safety requirements. The second model is the Weibull model obtained in Ref. 6, and the third model is a CFLM-LOF fitted at $t = 90$ months. None of the models seems to fit all data well on the whole age range. While the initial mean time between fires is $1/A = 14.5$ months, the average over the whole period is about 58 months.

Failures in Integrated Circuits (IC)

Peck has recently analyzed early failures in semiconductor devices and concluded that the failure rate is not constant and that it does not follow the well known Duane plot (straight line in a log-log scale).[7] He suggested a special interpolation technique to obtain instantaneous failure rates without introducing a new parametric model.

In this example the rate parameter of a population is changing in time because the failed units (high failure rate units) are removed from the population after their first failure. The population is initially very large (623,500 ICs) so that the reduction in the rate is not

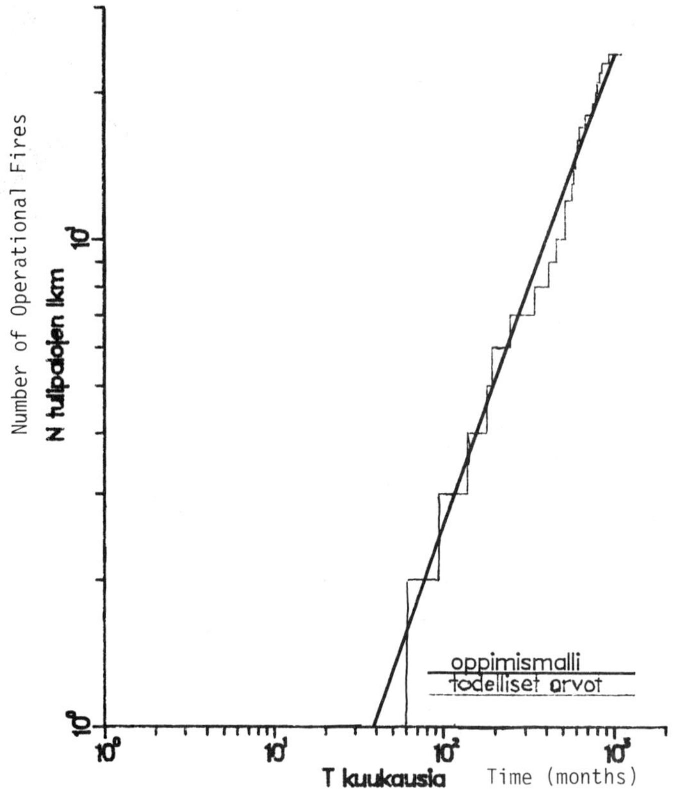

Fig. 2. Observed and model-estimated numbers of failures in nuclear power plants.

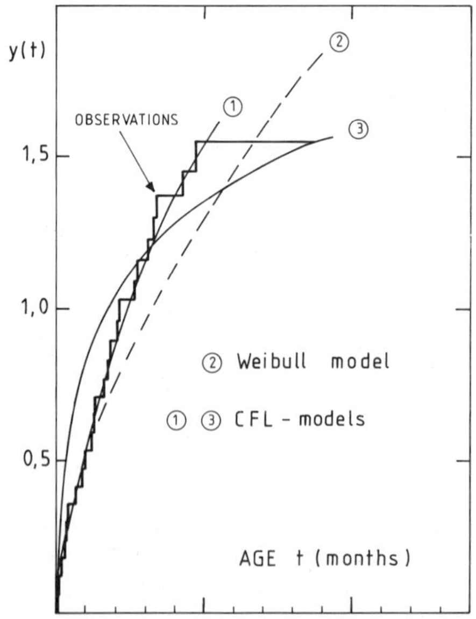

Fig. 3. Observed and expected number of operational plant fires per plant, vs. age t; (1) $y(t) = 1.083\ ln\ (1+0.0638\ t)$; (2) $y(t) = 0.1284\ t^{0.592}$; (3) $y(t) = 0.3468\ ln\ (1+0.961\ t)$.

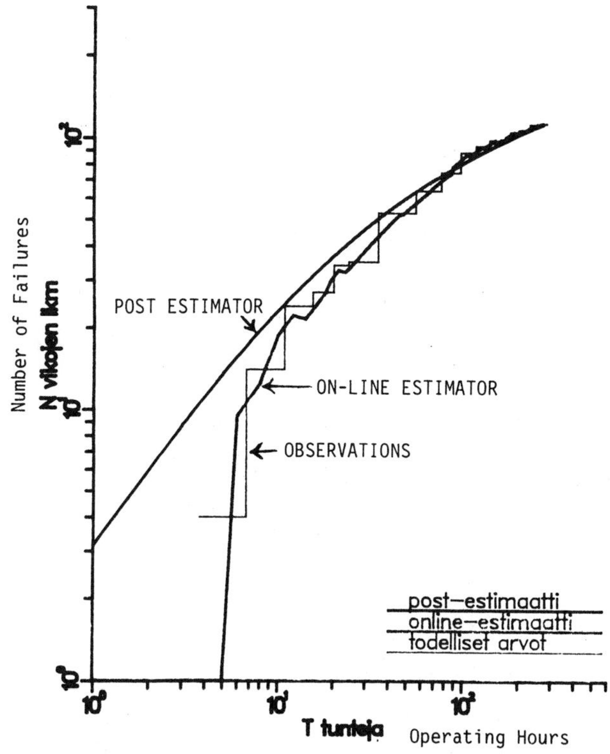

Fig. 4. Cumulative number of failures of integrated circuits.

due to depletion of the population but to the fact that high failure rate individuals are removed. Thus, the process is not exactly "learning" but rather selection or elimination.

Nevertheless, the same model Eq. (5) has been tried on the data presented in Table 1 of Ref. 7. Both the Post Estimate and the On-Line Estimate for the cumulative number of failures are presented in Fig. 4, together with observed data. PE and OLE deviate significantly for operating times less than ten hours but agree reasonably well for longer periods of operation.

Reactor Trips

Reactor scram experience on Swedish nuclear power plants has been studied by Laakso.[8] The cumulative numbers of reactor trips are presented in Fig. 5. They clearly reflect the decreasing trend of the derivatives, consistent with Eqs. (5) and (6). The corresponding model estimators are presented in Fig. 6. The curves have been obtained assuming that each plant learns from its own trips only, using the LOF-PE procedure.

Assuming that all plants learn from trips at other plants, too, the results of Fig. 7 are obtained using the LAF-PE procedure. There is a systematic discrepancy between the observed values and the model in this case, indicating that the assumption is not supported by evidence, at least within the framework of the continuous fractional learning model.

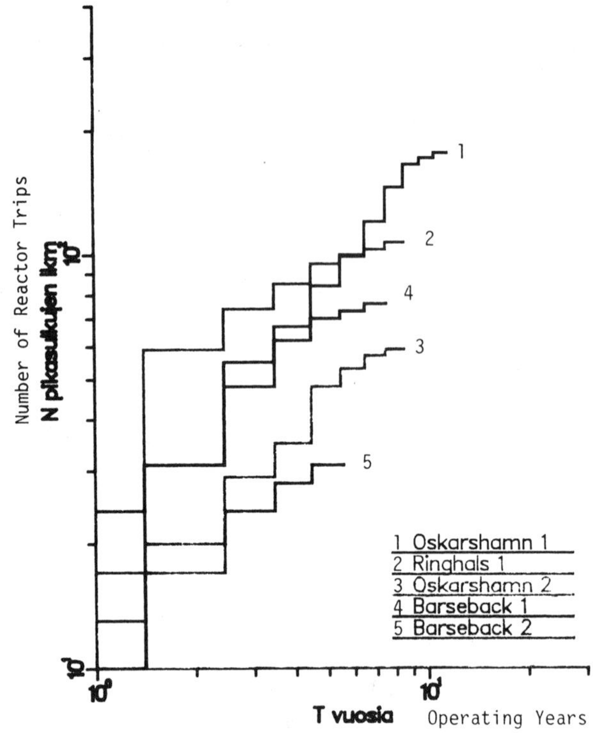

Fig. 5. Cumulative number of reactor trips on Swedish nuclear power plants.

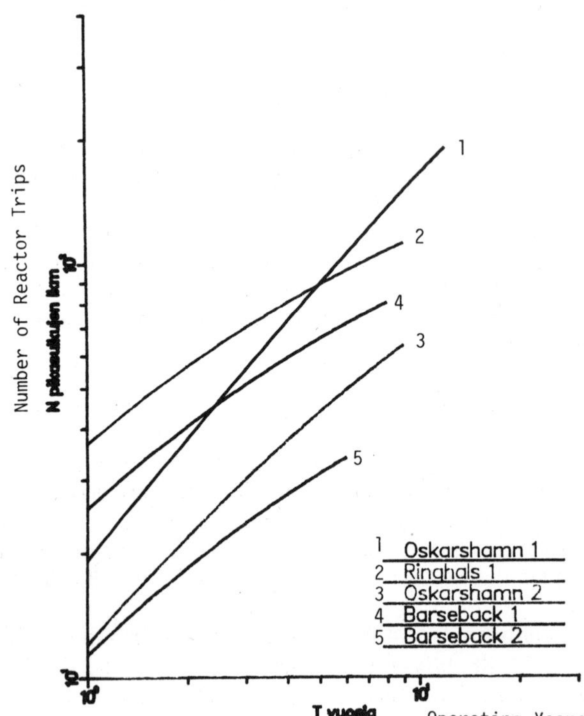

Fig. 6. Continuous fractional learning model estimators for the number of reactor trips, assuming the LOF-PE procedure.

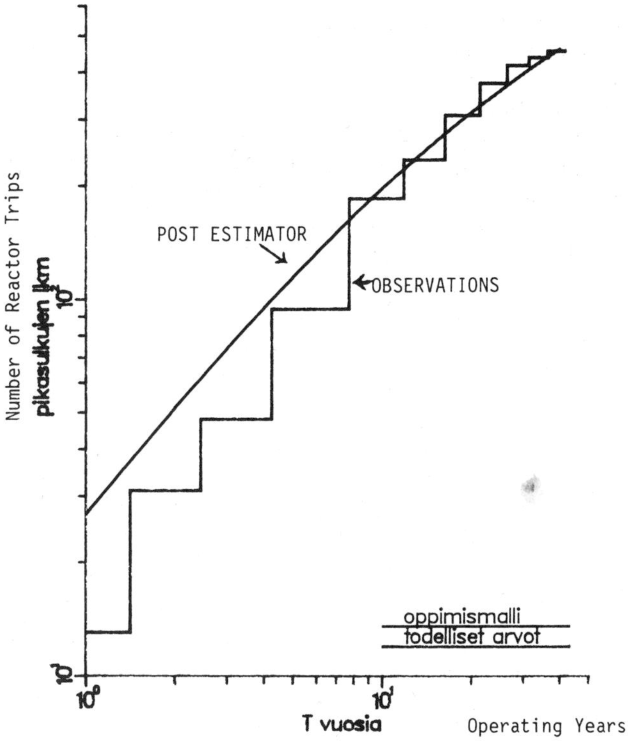

Fig. 7. Total number of reactor trips on Swedish nuclear power plants with a model assuming interplant learning (LAF).

CONCLUSIONS

A continuous fractional learning model has been introduced, and a moment matching technique has been developed to estimate parameters of the model. The fundamental assumption that learning is actually due to events, rather than time, is built into this time related model. The model allows different assumptions made and tested about the learning process, using a single observed data set. It has been applied to several failure data sets, and it seems to fit the experience with a variety of equipment on component, system and plant levels. The predictive capabilities of the model are obvious when the fit is good. Future development should include an additional constant background failure term in Eq. (6).

REFERENCES

1. J. M. Sheppard, Are Reliability Growth Models Planning and Monitoring Tools for Development? *Rel. Engng*. **11**:1-12 (1985).
2. J. T. Duane, Learning Curves Approach to Reliability Monitoring, *IEEE Trans. Aerospace* **2**:563-566 (1964).
3. Reliability Growth Management, MIL-HDBK-189, 109-120 (1981).
4. Harold S. Balaban, Reliability Growth Models, *J. Environ. Sciences*, pp. 11-18 (January/February, 1978).
5. J. K. Vaurio, Learning from Nuclear Accident Experience, *Risk Analysis* **4**:103-115 (1984).

6. R. W. Hockenbury *et al.*, Occurrence Rates of Fires in Nuclear Power Plants, *Nucl. Eng. Design* **66**:233-240 (1981).
7. D. S. Peck, Extension of the Duane Plotting Technique, *IEEE Trans. Rel.* **R-34(2)**:102-106 (1985).
8. K. Laakso, Systematisk erfarenhetsaterforing av driftstorningar pa blockniva i karnkraftverk, doctoral thesis, Helsinki University of Technology, 1984.
9. J. K. Vaurio, "Learning Curve Estimation Techniques for Nuclear Industry," *Proc. Int. CNS/ANS Conf. Numerical Methods in Nuclear Engineering*, Montreal, Canada, September 6-9, 1983.

Hits on MRS

Herbert Inhaber
NUS Corporation
Gaithersburg, MD

ABSTRACT

The proposed Monitored Retrievable Storage (MRS) facility for temporary handling of high-level nuclear wastes and spent fuel rods in Oak Ridge, Tennessee, has generated considerable controversy. Opponents state it would become a *de facto* final repository; proponents state it would help the overall waste program and pave the way for an eventual solution to the waste problem. Although risk perception clearly underlies many of the attitudes taken, calculations of risk have played little part in the debate. This is in spite of the fact that the Department of Energy (DOE) has devoted considerable effort to those calculations. Rather, the personal knowledge of the analysts seems to have been more persuasive than their paper studies. This suggests a different approach to altering perceptions of risk.

KEYWORDS: High-level waste, risk perception, monitored retrievable storage, citizens' participation, spent fuel, repositories

INTRODUCTION

Monitored Retrievable Storage (MRS) as a concept in dealing with high-level nuclear wastes first came to prominent public view in the National Waste Policy Act of 1983 (NWPA — Public Law 97-425). Under Subtitle C, Congress found that the MRS was an option for management of these wastes and/or spent fuel rods.

RISKS AND HIGH LEVEL WASTES

Many members of the public think that high-level wastes (HLW) are extremely risky, in some general sense. Conversely, many scientists who study the subject believe that, with the proper measures, risk can be made very small. There have been few studies of how these conflicting perceptions drive the United States high-level waste program and auxiliary programs. The object of this paper is to discuss one aspect of this program, the proposed MRS in Oak Ridge, Tennessee, and to determine how, if at all, risk analysis played a part in public and private attitudes. No effort will be made here to trace all aspects of the MRS controversy. Such a delineation would require far more space than is available.

New Risks
Edited by L. A. Cox, Jr., and P. F. Ricci
Plenum Press, New York, 1990

In addition, the viewpoint here is impressionistic, rather than solely analytical. The MRS odyssey has received few if any analytical studies, and the task of its detailed consideration has only begun.

UNANIMOUS APPROVAL

However, there is one salient point about the MRS which deserves mention in any evaluation of risk perception. The proposed installation is apparently the only one which has received unanimous approval from local elected officials in the United States. These officials comprised the Oak Ridge City Council and the Roane County Commission. The proposed site for the MRS will be in Roane County, within the city limits of Oak Ridge.

The suggested final repository site in Hanford, Washington, has also received general approval from local officials, but it was not unanimous. As far as is known, this unanimity is unique in the history of the U.S. HLW program, and may even be unique on a world scale. This suggests that risk perceptions of HLW differ radically from community to community.

While local officials endorsed the MRS concept without exception, they did attach a number of economic, environmental, managerial, and other conditions to it. Most of these conditions were suggested by the Clinch River MRS Task Force, a group of appointed technical and non-technical citizens of the area. That is, the original proposal by the Department of Energy was substantially modified by this task force and the elected officials to whom they reported. Their study was probably one of the most comprehensive done on HLW by any citizens' body anywhere.

It would take far too much space to evaluate the positions the Task Force took and how they arrived at them. However, in terms of risk the Task Force had an unusual approach. One of its leaders[1] has stated publicly that speaking with those responsible for risk analysis was more persuasive than the analyses themselves. The Department of Energy had sponsored tours of various DOE laboratories throughout the country so that Task Force members could investigate technical questions for themselves.

The statement was coupled with another to the effect that, while DOE had made many calculations of risk available to the group, the latter had spent little if any time analyzing these paper studies. Whether this was due to lack of interest or inability to follow the often complex calculations, or both, was not clear.

In any case, the conclusions are evident, at least in regard to the MRS. Discussing potential risk with analysts was seen as more persuasive than reading the same information presented in written (or computer) form. This lesson may be applicable to other risk perception studies, where risk data is presented as given, and respondents can accept or reject it. The results of these studies might be strongly different if the sources of the data could be presented to the respondents.

"HIGH TECH/HIGH TOUCH"

The situation has been described outside the orbit of risk analysis by John Naisbitt. In his best-selling book "Megatrends,"[2] he used the term "high tech/high touch." Naisbitt said that when a new technology is introduced, there must be a counter-balancing human response, that is, high touch, or the technology will be rejected.

In the context of the MRS, although the Department of Energy, its proposer, may not have explicitly followed the "high tech/high touch" formulation, this approach is the essence of the result. The proposal might not have received the approval of the Task Force and the local officials if a less personalized attitude had been taken.

CONGRESSIONAL ORIGINS

The origin of the MRS Congressional subtitle is obscure. The original NWPA bill, after being considered by Congress for some years, was reported out of the House Subcommittee on Energy Research and Development without any MRS proposal attached.[3] According to some sources, Senator Bennett Johnston of Louisiana was anxious to have his state removed from the list of potential final repositories. Because of the presence of salt domes in Louisiana, it had been mentioned in press speculation as a candidate for this repository. He therefore proposed Subtitle C, an MRS study. While the reasoning has never been set down on paper, apparently if the MRS were to be located near Oak Ridge, probably on the basis of DOE facilities, DOE would conversely be less likely to locate the repository in that region. Congress would never approve both an MRS and repository in the same region, the argument went.

In a sense, Senator Johnston was successful. Louisiana did not appear on the list of three candidate repository sites. Washington, Nevada, and Texas did. However, the MRS has yet to be approved. The Governor and Legislature of Tennessee have given notice of disapproval.

One may search Subtitle C in vain without finding any trace of the words "risk, hazards" or their synonyms. This is remarkable, especially since the MRS, as later detailed by DOE, was supposed in some sense to reduce the risk to the population of radiation exposure.

What is found instead are such phrases as "long-term," "for the foreseeable future," and "as long as may be necessary." These phrases produced some of the most bitter controversies between opponents and proponents. Proponents characterized the MRS as essentially short-term. That is, because of the large back-log of spent fuel in reactors around the country, the MRS would admittedly be busy in the first few years of its existence. However, as the back-log was disposed of, the MRS would then be used less. As reactors presently in operation were retired, the rate of incoming material to the MRS would diminish greatly.

Opponents took the opposite tack. They noted no use of language indicating short-term operation in the NWPA. On the contrary, "long-term," they contended, meant what it said. In fact, they said, any building of the MRS or its successful operation would militate against construction of one or more final repositories. If an MRS is built, why have a repository at the same time, especially if the MRS can operate indefinitely according to Congressionally mandated language?

The two sets of arguments are not capable of resolution here. In any case, even if a rational and defensible choice could be made here, the history of the HLW program in the U.S. suggests many more turnings in the road before the question is settled definitively. The fact that over four decades have elapsed since the first HLW was produced in this country, and none has been buried permanently since that time, indicates that solutions are as rare as snowstorms in July. Another interpretation is that, due to the complexity of the issues, it takes much time for a consensus to emerge from either Congress or the public. With the passage of definitive HLW legislation in 1982, it appeared that this consensus had emerged. It hadn't.

OAK RIDGE CHARACTERISTICS

As noted above, the unanimous approval given the MRS by local officials is apparently unique. It might be argued that because Oak Ridge and its environs are so nuclear technology-oriented and already so dependent on DOE funding, the results were a foregone conclusion. This contention would be only part of the truth.

There are some points in opposition to this argument. First, while it is true that Oak Ridge probably has one of the highest proportions of Ph.D.'s of any city in the nation, there are apparently few if any on the City Council. The question was put to the elected representatives, not solely to Ph.D.'s in the sciences.

Second, even if it could be contended that the Oak Ridge City Council is generally cognizant of matters nuclear, the same cannot be said of the Roane County Commission. This mostly rural county contains a small part of Oak Ridge, but its elected representatives are strongly non-technical and more typical of traditional Tennessee.

Third, the credibility of DOE in the Oak Ridge area has deteriorated in the last few years. The revelation of environmental problems produced decades ago about which little if anything was done for years contributed to this decline. Whereas most proposals of DOE's predecessor, the Atomic Energy Commission, were greeted with approval, most originating from DOE now meet with skepticism.

Fourth, the city and surrounding areas are making strenuous efforts to break away from their "all-nuclear" image. This has gone as far as proposals to eliminate the nuclear symbol from the city's logo, used on official stationery. This trend probably would have come about eventually, but has been accelerated by the lessening of national interest in nuclear power as an energy source.

Fifth, the proposed MRS, although a massive facility, would not bring enormous numbers of jobs to the area. One estimate is of 700 positions, compared to a total Oak Ridge population of 28,000. This probably would increase direct employment in the city by abut 5%.[4]

On the other hand, Rep. Edward Markey has estimated the total MRS cost at $1 billion, a substantial investment. If this Federal facility is taxed at the same rate as industrial property, as was proposed by the Clinch River MRS Task Force, it would greatly increase the city and county tax base.[5]

The five points sketched above suggest that while the MRS certainly has positive aspects for Oak Ridge, there are too many negative actors for blind approval.

The only poll taken in Oak Ridge did indicate approval of the MRS by 64% to 18%, with the rest undecided. While this was a substantial margin in favor of the facility, it was far from unanimous. As a historical footnote, this may be one of the few opinion polls, if not the only one, to favor a HLW storage site. The NIMBY — not in my back yard — syndrome evidently does not operate everywhere.

WHY OAK RIDGE?

The rest of this paper will be devoted to other aspects of the MRS study. One of the key questions in the debate was, why Oak Ridge?

Skeptics contended that Oak Ridge would provide an automatic "yes" vote, and that other large federal energy reservations, such as that at Los Alamos, New Mexico, were too close to the likely Western repository to make logistic sense. DOE contended otherwise.

Under the NWPA [Section 141(b)(4)], the Department was obligated to suggest at least three sites. When all three were located in Tennessee, the consternation among many state officials was intense.

DOE stated that no political considerations were involved and that mathematics determined the sites. When the proposal was first announced on April 9, 1985,[6] Ginger King, speaking for the DOE Office of Civilian Radioactive Waste Management (OCRWM) said, "Knoxville, Tennessee (was used) as a center point (of) a circle with a radius of approximately 100 miles ... Knoxville was chosen as the middle point because it was determined to be in the center of an area in which much of the nation's commercial spent nuclear fuel is generated." In addition, the proposed site had to have substantial areas available to the Federal government, which reduced the potential sites to a small number.

In principle, then, site selection is a matter for mathematical analysis. Geographers and those who choose industrial sites are familiar with the techniques.

OBJECTIONS FROM CONGRESS

Albert Gore, Jr., junior Senator from Tennessee, did not accept this explanation. He asked Ben Rusche, the director of OCRWM, in a series of Congressional hearings and statements, if it was merely a coincidence that Knoxville was at the center of the map. Early on, he concluded[7] that DOE was "misrepresenting the facts." He was helped somewhat by an editorial cartoon in *The Nashville Tennessean*, the most prestigious newspaper in the state. It showed a sad-looking government official pointing to a U.S. map, distorted so that Tennessee appeared in the middle, rather than the more logical Kansas. The official says, "Can we help it if your state happens to be in the exact nuclear center of the whole country?"[8]

The debate on the mathematical center of the present fuel rod distribution raged between Gore and Rusche for some months. Eventually Rusche stated that the true mathematical center to minimize transportation risk would have been in southern Ohio, not Tennessee. This admission strengthened the hand of Tennesseans who felt the Federal government was foisting an unwanted facility on them.

This history is given here because it is apparently the only specific example in the public debate over MRS where risk calculations were mentioned explicitly. The sources of DOE's calculations, or Gore's counter-calculations, were not, however, specifically cited in press reports. Thus disinterested risk analysts, or mathematically-trained members of the public, could not check claims and counterclaims.

The three largest concentrations of reactors in the U.S. are in the Midwest, the New England states, and the Southeast. Intuitively, one could hypothesize that the mathematical center of these reactors would be somewhat north of the Southeast. However, an exact determination would depend on the number and type of spent fuel rods at each reactor and the specific model used to weight each rod. Senator Gore, not a mathematician, did not need that training to see that it was unlikely, although not impossible, that Knoxville would be at a mathematical minimum in terms of transport risk.

MAPS AND THE MRS

As students of risk perception know, concepts of risk can be transmitted in many ways, most of which do not involve detailed calculations. Prime among these was another map circulated by Frank Cochrane, Public Service Commissioner.[9] Cochrane announced for the Democratic nomination for governor and made opposition to the MRS his major plank. He later dropped out of the race.

The map he circulated widely showed Oak Ridge connected by straight lines to all the 90 or so reactors with an operating license, and the approximately 40 with a construction permit. This gave the nation the look of a spider web, with Oak Ridge representing the hapless fly at the center. Most people who saw that map would have been at least somewhat perturbed.

Yet a similar map could be made for practically any centralized waste facility. Lines can be drawn from the waste origins to the site, and their spider web effect is likely to disturb many residents near the proposed site. Risk perception analysts have yet to devise a method of educating the public that this "spider web" is an inevitable consequence of moving any waste to a centralized treatment or disposal facility.

EDITORIAL CARTOONS

The cartoon mentioned above was only one of many that appeared throughout the state's newspapers. Although journalists may say that news stories and editorials have the greatest impact in forming public opinion, others would contend that editorial cartoons have at least equal ability to shape this opinion.

With one exception, until November 1985, the date of the Oak Ridge attitude survey, there was apparently no cartoon published in Tennessee newspapers that showed the MRS in a neutral light, let alone a favorable one. The exception occurred after a Ralph Nader visit to Jackson,[10] when he criticized the state for its "arrogance" in promoting nuclear power. He said that a nuclear waste facility should be built here as a type of punishment for this arrogance. In so doing, he became one of the few national figures outside the nuclear industry to advocate building an MRS for any reason.

Nader went on to say, however, that he doubted "whether you have geologically stable structures, so you are safe by the ancient accident of how this state was formed geologically in the Ice Age." There may have been some confusion in Nader's mind between a temporary site, such as the MRS, and a permanent repository. The former will be above ground, at least as envisioned by DOE, and thus not much dependent on geology. The latter will be thousands of feet below the surface, and its effectiveness in keeping the wastes from seeping into a water table will depend in part on geologists' estimates of the stability of underlying rock. While the surface of many areas in North America was disturbed by Ice Age glaciers, the subterranean rock was formed millions of years before.

Editors everywhere do not appreciate out-of-staters criticizing their state. Nader was denounced the next day[11] in an editorial saying he was a "much travelled figure clad in baggy ideas, railing at figments of his imagination with drivel running from his mouth." The accompanying cartoon was the only major one showing MRS in a non-negative light. Nader is portrayed in cowboy garb with a maniacal look on his face, riding a garbage can labelled "nuclear waste" over the state. The implication is that an outlander is using the issue either to have fun at Tennesseans' expense or for his own political purposes.

Almost all other cartoons on the subject were highly negative, indicating that efforts of the nuclear industry to convince at least journalists that nuclear wastes can be handled and stored safely have been less than fully effective. One cartoon by Tom Oliphant showed two skeletons in the outlines of human bodies, in front of a field of small cylinders labelled, "Tennessee Nuclear Waste Dump." They are telling a visitor, "Yes, this *is* Tennessee. How did you know?" A later one by Bissell in *The Tennessean*[12] shows two highways meeting. Prominently displayed at the junction is a cross-buck similar to that at railway crossings. The difference is that the cross-buck, titled "N-Waste Crossing," has a skull perched on top. The skies lower ominously over the lifeless scene, and the caption reads, "For Every Tennessee Road."

Not content with a downbeat image, the cartoonist for *The Jackson Sun*[13] showed what he thought would be the brighter side of the MRS. The Grand Ole Opry would have a new slogan, "The Oldest Continuous Radioactive Program in the U.S.A." Seemingly half the barns in the state now bear a slogan for a tourist attraction, "See Rock City," painted on their sides. The cartoon shows this amended to, "See Rod City — temporary home of spent fuel rods." Showing a puffy-nosed boozer hoisting a jar of corn liquor, the cartoon has a caption which reads, "A new meaning will be given to the term, "Tennessee 'Shine!'" The quaffer says, "Puts a real glow on your face." And to top it off, a new state flag is proposed, one with only the international symbol for radiation on its furls.

CONCLUSIONS

Where does the MRS go from here? At the time of writing, almost all state politicians from outside the Oak Ridge area have gone on record as opposing the MRS. At a debate in October, 1986,[14] both the Democratic and Republican gubernatorial candidates spoke against it. Presumably either, if elected, would fight the MRS.

However, the NWPA gives Congress a veto over a veto. If a state opposes a repository on its territory — it is not legally clear if this power extends to an MRS — Congress can override that veto. Whether or not it will depends on many factors, risk being only one of them.

So while Tennessee as a whole has gone on record against the MRS, Congress will undoubtedly look carefully at the pro-MRS votes and opinion surveys from Oak Ridge and Roane County. One incentive that Congress has to approve the MRS in Oak Ridge is that if it is located there, it will not be in the home districts of other members of Congress. While observers and analysts disagree on almost everything concerning HLW, they do agree on one matter: the maneuvering will go on into the indefinite future.

REFERENCES

1. Elizabeth Peelle and John H. Reed, MRS Task Force Process: Dealing with Local Aspects of Risk, Institute for Energy Analysis, Oak Ridge, Tennessee, May 29, 1986.
2. John Naisbitt, *Megatrends*, 1981.
3. Marilyn Lloyd, U.S. Representative from Third District of Tennessee, August 1986, private communication.
4. *Knoxville Journal*, Oak Ridge Likely Site of Waste Dump, April 24, 1985.
5. *Knoxville News-Sentinel*, Abandoned Oak Ridge Site Studied for N-Wastes, March 22, 1985.
6. *The Washington Post*, Spent Nuclear Fuel Facility Sought, April 9, 1985.
7. *Knoxville News-Sentinel*, DOE's Nuclear Map Wrong, Gore Says, April 28, 1985.
8. *The Nashville Tennessean*, May 22, 1985.
9. Paris (Tennessee) *Post-Intelligencer*, State Might Package Nuclear Waste, June 23, 1985.

10. Memphis *Commercial Appeal*, Nader Chides State on Nuclear Power, Cites 'Arrogance,' May 23, 1985.
11. Memphis *Commercial Appeal*, Nader's Meltdown, May 24, 1985.
12. *The Nashville Tennessean*, August 12, 1985.
13. *The Jackson Sun*, May 29, 1985.
14. Oak Ridge, *The Oak Ridger*, Few Sparks in Debate, October 8, 1986.

A Psychological Perspective on Accident Evolutions and How to Arrest Them in Nuclear Power Plants

Ola Svenson
Lund University and University of Stockholm
Stockholm, Sweden

ABSTRACT

An accident in a nuclear power plant is regarded as the result of an interaction between an operator and the system and is modelled in an accident evolution model. This model contains an initiating event and several links at which the accident evolution may be arrested. The accident evolution may be analyzed from a predominantly psychological perspective or a predominantly technical perspective, one possibly leading to different results and recommendations than the other. In this paper a psychological cognitive perspective is chosen in an illustration of how a qualitative human reliability analysis may be designed.

KEYWORDS: Human factors, nuclear power, cognition, human reliability analysis

INTRODUCTION

A nuclear power operator controls a system that produces great amounts of valuable electricity per hour in operation. Unfortunately, in the event of an accident, the system can cause great damage to the plant, people, and the environment. So, as in other industries, the goals of the operators are to manage the system as efficiently as possible and as safely as possible. This means that the facility should be delivering power as many hours as viable and that safety should be as high as reasonably achievable.

This paper focuses on safety and in particular on sequences of events that may, or may not, lead to an accident. Furthermore, the goal of the paper is also to draw attention to the possibility of improving the safety of nuclear power plants through analyses of the human as a problem solver and decision maker. Starting with characteristics of human problem-solving processes and the particular weaknesses of these processes (and ways to compensate for them), it is possible to lay the foundation for an analysis of the operator-plant interaction that points out potential operator weaknesses and suggests improvements to avoid or eliminate operator errors.

New Risks
Edited by L. A. Cox, Jr., and P. F. Ricci
Plenum Press, New York, 1990

ACCIDENT EVOLUTION AS AN INTERACTION OF HUMAN AND TECHNICAL SYSTEMS

Most of the major accidents that can be envisioned in a nuclear power plant are the result of an unfortunate and unlikely sequence of interactions between the operators and the technical system. This is a result of the technology being so well designed that a single failure or disturbance is very unlikely to lead to an accident in itself.

Figure 1 illustrates the interaction between the human and technical systems. Starting with a human (operator) action, the system responds, and then the operator responds to the system, and so on until the system, as illustrated, may get out of control and an accident follows. However, the accident evolution may be stopped at any point, as is illustrated by the parallel lines after H3. This way of conceptualizing the process leading to an accident is analogous to the hazard evolution model presented by Kates, Hohenhemser, Kasperson and Kasperson (1985). The accident evolution concept is orthogonal and complementary to the different-level perspective on hazard management presented by Svenson and Fischhoff (1985).

Exemplifying this from the nuclear power context and the Davis-Besse incident (NUREG, 1985), it is possible to shorten the story and start the description for illustrative purposes with both main steam isolation valves spuriously closing. (This happened after one of the main feed water pumps had tripped, followed by the reactor and turbine automatically being tripped on high reactor coolant pressure.) After the complete loss of main feed water (T1), an operator error followed (H1). The system reacted to this error by signaling that both steam generators had experienced a steam-line break or leak; the system responded as designed by isolating both steam generators (T2). It was then found that the operator earlier had pushed the wrong two buttons (H1), and this error was corrected (H2). The isolation valves in each auxiliary feedwater train closed but failed to reopen automatically (T3) (in response to H2). In response to this reaction from the technical system, a sequence of operator actions not described here (H3) stopped the development towards an accident.

In most full or partial accident evolutions, several interactions between the human and technical systems occur. Using the accident evolution model illustrates that an initiating factor may be identified early in the development. It also illustrates that it is not of any value to identify a single cause of an accident. However, it is of great value to identify points at which the accident evolution may be arrested, because this may teach us how to make the man-technology interaction safer.

CONTRASTING THE PSYCHOLOGICAL AND THE TECHNICAL PERSPECTIVES

In fact, the model in Fig. 1 can be used to illustrate two extreme perspectives. The first, from the psychological human system side, involves accepting the human system as it is, with all its natural merits and faults. This perspective argues that because the human system is imperfect and can never become perfect, all the errors in the control of the system should be attributed to the technical system. The technology is therefore seen as causing an accident if it does not correct or eliminate human malfunctioning.

The other extreme perspective, from the technical system side, involves accepting the technological system with its merits and faults. It assumes that the technology is as safe as it can reasonably be and that if an accident occurs, all the errors should be attributed to malfunctions of the human system.

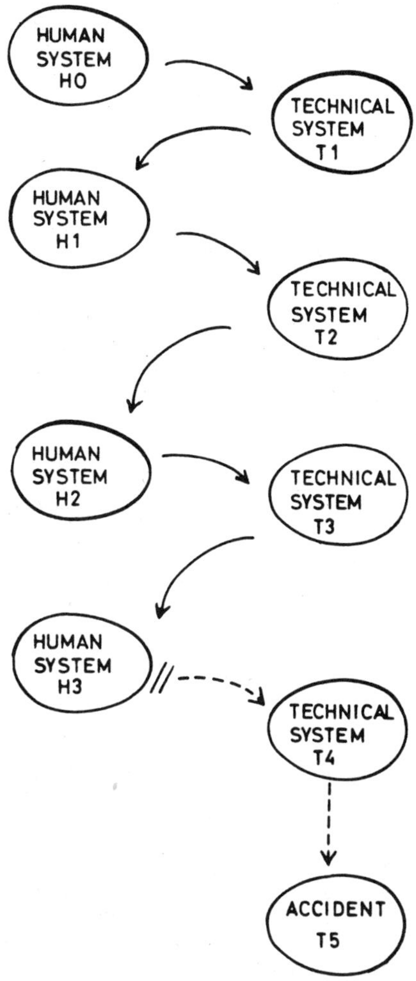

Fig. 1. A model for an accident evolution. The parallel bars indicate that the process towards an accident is arrested by the human system.

The above perspectives were made "one-eyed" for illustrative purposes. The viewpoints held by most people are actually compromises of one or the other. However, the model illustrates that developing a risk management procedure for a nuclear power plant wholly from a psychological approach may lead to different emphases on the factors contributing to an accident evolution process than would be the case if it were developed wholly from a technical approach.

The model also illustrates a potential dilemma: a psychologist may determine that most of the factors contributing to an accident belong to the technical system but he cannot himself serve as an expert in eliminating the technical shortcomings. Correspondingly, a technician may find that most of the factors contributing to an accident belong to the human system, but he cannot serve as an expert in eliminating the psychological shortcomings. As a result a conflict over the causes of an accident may develop and the recommended actions may be very general and non-professional (i.e., someone with no qualifications may attempt

Table 1. Safety-Related Questions Starting from the Psychological and Technical Perspectives

Psychological Perspective	Technical Perspective
1. Which imbalances and errors may occur in the human system?	1. Which imbalances and errors may occur in the technical system?
Weaknesses due to organization, performance shaping factors, cognitive representation, etc.	Weaknesses due to technical design and materials; risk analysis (e.g. PRA).
2. What are the effects of no technical correction of operator error?	2. What are the effects of no operator correction (error of omission) technical error?
Operator does not know that he made an error and will act as if he had not.	Risk analysis (e.g. PRA connected to HRA).
3. What are the effects (from the psychological viewpoint) of "erroneous" or improper reaction from technical system?	3. What are the effects of erroneous operator action (error of commission)?
Operator gets the wrong picture of the system and will act accordingly. (Human factors work on displays and controls is one way to improve interaction).	Technical risk analyses on assumptions of plausible reasons for human actions.

to improve a technical sub-system, or someone with no qualifications may attempt to improve the training program for operators).

ASKING QUESTIONS ABOUT THE PSYCHOLOGICAL AND TECHNICAL SYSTEMS

Table 1 illustrates how different perspectives lead to different kinds of questions related to the management of a nuclear power plant. Three sets of parallel questions are asked about both the human system and the technical system. The first set is concerned with possible weaknesses of the systems themselves. In order to illustrate a possible approach to increased safety, the remainder of this paper will be devoted to this first question from the psychological perspective. In particular, human problem solving will be examined, and phenomena illustrating the human as a problem solver will be presented, along with corresponding examples from the nuclear power context. First, however, the remaining questions for each perspective in Table 1 will be discussed briefly.

Probabilistic Risk Analysis (PRA) was designed to explore and assess the risks of errors in the technical system (Question 1, technical perspective) and is treated in other contributions of this volume. The effects of an operator not acting when he should (Question 2, technical perspective) have also been studied in great detail in many PRA studies in which the human factor has been integrated via outputs of Human Reliability Analysis (HRA). Question 3 under the same perspective includes operator errors of commission: operator does something actively that he should not do. This type of error is harder to handle than errors of omission as it is necessary to imagine what reasons there may be for the operator to make a mistake and what sequence of mistakes he may be able to make in relation to the system.

The second question from the psychological perspective is quite straightforward. The third question from that perspective concerns the adaptation of the technical information and control systems to the operator. Human factors work intends to answer this question and to improve the design so that the interface between man and technical system is optimized (Kinkade and Anderson, 1984).

SOME CHARACTERISTICS OF HUMAN PROBLEM SOLVING PROCESSES

This section explores problems under the psychological perspective question 1 in Table 1. In particular, cognitive aspects of the operator-system interaction will be presented and discussed.

A GENERAL PROCESS MODEL

Figure 2 illustrates human information processing in a generic model. From left to right, input information leads to reaction, in this case motor reaction as in a nuclear power plant. On top of this horizontal line there are three different types of memories. The first of these, iconic or echoic memory, has the capacity to store almost all the visual or auditory input during a very limited period of time (less than one up to a few seconds). The short-term memory (STM) is a working memory in which we solve problems and think actively. In other words, STM represents "focused thinking" that makes use of present input and facts that we know "by heart" and access from long term memory (LTM). For example, thinking about the causes of an alarm in a control room involves some thinking in STM and some retrieval of facts about the power plant that have been stored in LTM (or in an external memory (EM) which may be anything from a computerized manual for a specific situation to a piece of scrap paper). On top of the memories one finds motivation and performance shaping factors. Some of the latter are fatigue and stress; they have been treated in more detail by Swain and Guttman (1983). The effects of variation in these factors tend to affect the whole human system.

The bottleneck in human information processing is the STM capacity, which inhibits us from the really complex thinking needed in many situations. Because of this restriction, the way in which a problem situation is structured cognitively is of ultimate importance for the problem solving process. Thus, the way in which a problem solver interprets and understands a problem in a nuclear power plant greatly impacts the way the problem is solved (or not solved).

Human problem solving should be regarded as goal driven; that is, the problem solver wants to transform the present state to another (goal state). This applies whether the problem is one of chess or that of stabilizing a nuclear process.

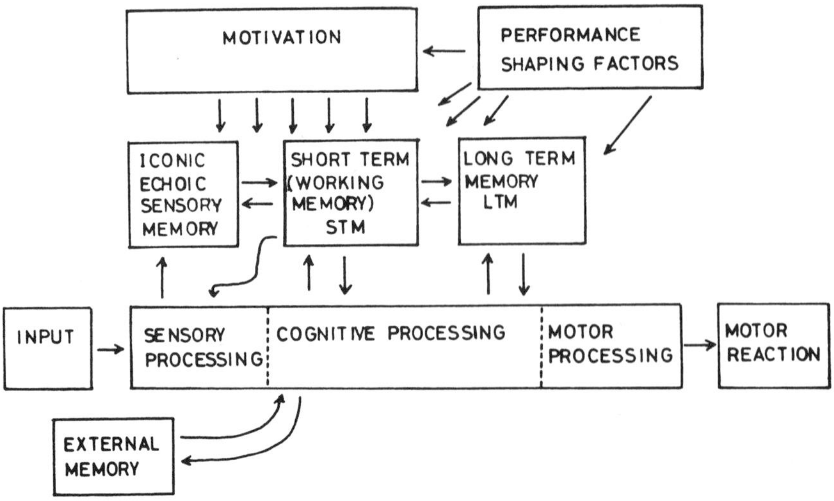

Fig. 2. A simple flow chart model of the human operator as a processor of input information in order to perform a motor reaction.

SOME HUMAN PROBLEM-SOLVING PHENOMENA

A complete exposition of interesting phenomena associated with human problem solving cannot be given here, and the interested reader is referred to a project under way (Svenson, 1987; Rasmussen, Duncan, and Leplat, 1987). Table 2 lists some characteristics of human problem-solving processes that may pose difficulties in the nuclear power setting. All problems considered here are related to the individual and not to the group.

ANALYZING OPERATOR-SYSTEM INTERACTIONS THAT MAY INVITE HUMAN ERRORS

Starting from a human systems perspective, the operator's task in controlling the system — in particular during transients — can be analyzed. Ways in which this can be done are, for example, to adjust relative error frequency estimates for the current performance-shaping factors conditions (cf. Swain and Guttman, 1983), or to analyze the operator's working situation through the use of a check list indicating the cognitive difficulty of the task on a set of attributes proposed by (Gertman, 1986) (e.g., conflicting data on instrument? no? yes? how serious?).

It is also possible to start with only one or a small set of psychological phenomena, like those in Table 2, and try to find situations in which they may occur, for example, during a transient. In this way, weak links in the system-operator interaction may be identified and their effects either eliminated or controlled. In the following, a few of the listed psychological phenomena related to human problem solving will be identified in documented accident evolutions. The section will only serve illustrative purposes and no attempt will be made to cover the field systematically.

During the Chernobyl accident, the actions of the operator trying to stabilize an unstable reactor show that this process required all his STM capacity (No. 1 in Table 2). In the absence of the regular alarms and emergency systems that were switched off, he did not process the information that he read from the computer printout that indicated that just one and a half minutes before the accident excess reactivity of the reactor required immediate

Table 2. Examples of Phenomena in Human Problem Solving That May Cause Problems in Operator-System Interactions in a Nuclear Power Plant

Phenomenon	Effects of relevance to nuclear power context. (All effects stronger under adverse performance shaping factor conditions.)
1. Limited STM capacity.	Operator overload if information is not easily coded, which may lead to neglect of important information.
2. Conflicting goals may lead to the domination of just one goal.	If the situation necessitates the trade-off between different goal parameter values, it becomes more difficult for the operator and there is a risk that he will concentrate on one goal only.
3. Slips and errors of action occur even in highly automatized processes.	If the operator is disturbed during well-known (even automated) sequence of behavior, he may fall back to another (even better learned) behavior pattern and commit an error, or forget parts of the original sequence, or stop watching the effects of an initial act.
4. Delayed feedback very difficult to handle.	If the situation involves delayed feedback information, there is a risk that the operator will have great difficulty in controlling the system.
5. Nonmonotonous approaches to the goal state are difficult if the whole solution process is not planned beforehand.	If the approach towards the goal involves a temporary detour (e.g., increasing pressure or temperature), this is not natural to the operator unless he has full understanding of the complete problem-solving process.
6. Inferring missing information from related parameter values.	If two parameter values correlate very strongly, there is a risk that only one of the the parameters is attended to and that information unavailable for one will be inferred from information for the other. There are also other ways in which nonavailable information is inferred.

[Note: All examples focus on individual behavior and no group effects are included here.]

shutdown (USSR State Committee on the Utilization of Atomic Energy, 1986). The operator was under severe stress created by a long delay of the planned test of a turbogenerator and an unstable reactor suffering from "xenon poisoning." For two hours all his attention had been given to a reactor threatening to shut down completely. In this situation the operator seems to have concentrated on just one goal: to keep the nuclear process going in order to perform the test. The other always present goal, to keep the process safe, was in conflict with and completely dominated by the first goal, according to the data available on the accident. Thus, both Nos. 1 and 2 in Table 2 were applicable in this accident evolution as well as other cognitive phenomena.

A subsequent report from a Swedish reactor related an operator error during a leak test of plugged tubes in the steam generator, which illustrates No. 3 in the table. Here, the reactor operator was active in the test (starting a pump to increase the pressure), but he was also active in trying to determine why a normal flow could not be established somewhere else in the plant. Just after he had started the pump to increase the pressure, he got a phone call concerning the latter problem. Only 4 minutes and 45 seconds later did he again watch the meter for the pressure, which was then 50% too high (Swedish State Power Board, 1986). A delayed reaction (No. 4 in Table 2) is in itself harder to manage, and this is illustrated through the unfortunate phone call which distracted the operator so that he lost track of the process first initiated.

For reasons of space, the following citation [from Lydell, Stampelos, and Stetkar, 1984, p. 4-20] concerning a bleed and feed cooling situation will be the last illustration (No. 5).

> The situation surrounding bleed and feed cooling is evaluated as being a moderately high stress situation. There is a sense of urgency to provide adequate core cooling and the operators must rely on a procedure that is a backup and, therefore, not frequently used. If the operators have not had adequate theoretical and practical training with periodic refreshers addressing this procedure, there exists the chance that when the PORVs are opened, the operators will misinterpret the sudden loss of subcooling (due to the sudden drop in pressure) as an indication of an improper action or condition and return the plant to its previous condition. If no further action is taken, the PORVs will lift due to charging flow and/or plant heatup.
>
> The opening of the PORVs will again cause plant pressure to drop to saturation pressure; subcooling will again be lost. All the time that this is going on, core temperature is rising due to the low energy removal rate. Therefore, as time passes, the saturation pressure rises and the pressure drops from succeeding PORV opening decrease. The final result is that an inadequate energy removal rate will exist leading to possible core damage.

Note that the phenomena listed in Table 2 are not necessarily even logically independent of each other. Furthermore, they tend to go together when the human factor fails. Performance-shaping factors, in particular stress of different types, tend to increase the risk of the human failing quite substantially.

A COGNITIVE HUMAN FACTORS ANALYSIS OF OPERATOR-SYSTEM INTERACTIONS

The present psychological perspective on human reliability analysis aims at spotting potentially weak links or dangerous situations in the operator-system interaction. Instead of trying to improve the error frequency estimates across a number of similar situations, it attempts to identify when and why the human may make an error. The logic behind this is that general knowledge about human behavior may be utilized in a qualitative sense so that situations which may pose particular problems in managing a nuclear power plant can be identified. A situation may involve a cognitively difficult task in a non-routine situation when stress is imposed on the operator. Even though all three of these conditions may be necessary for the initiation of an accident evolution, it may be wise to deliberately change such situations so that the effects of at least two of the three factors are decreased (e.g., through changing the problem for the operator, providing extensive training on this particular task, or changing the factors creating stress).

Focusing on the cognitive factors, situations in the operator interaction with the plant may be identified that seem to resemble some of the situations in which particular psychological phenomena (like those in Table 2) occur. If such situations have been identified, it is possible (a) to investigate whether a possible error may constitute the link in an accident evolution and, if this is the case, (b) to change the situation so that the likelihood of the possible human error is decreased.

A checklist according to which an analysis of an operator's work situation can be made could be of help in this analysis. However, there is a risk that a checklist blinds the analyst so that he stops his own creative search for unusual situations that may invite human errors. Also, we do not yet have enough knowledge about which cognitive phenomena are most relevant for the nuclear power setting. This, and the fact that we do not know about how the phenomena we have identified interact in creating difficulties for the operators, make it difficult to create a checklist with more than very crude scale levels (e.g., doesn't apply, may apply, applies). Furthermore an evaluation of the checklist that does not follow a simple (weighed) sum for each situation is preferred. Individual analyses of different situations may be needed in which questions such as the following (Waessman, 1986) are included: Why may the operator want to do this? What is his goal in this situation? What may turn this situation into a perilous one?

CONCLUDING COMMENTS

This paper has presented accident evolution as the result of a series of interactions of the human and technical systems. Depending on the prevailing perspective held by an analyst, the more important parts of an accident evolution may be attributed to the technical or human component. Whatever the perspective on an accident evolution, the most important thing is to halt the process before it results in an accident.

Starting from a cognitive perspective, a method for human reliability analysis was suggested that involves the use of psychological knowledge for spotting situations that are particularly difficult for the operator. Following this, operator-system interactions in which the difficult situation may occur have to be analyzed to see if these interactions may lead to a major incident or accident. If that is the case, proper measures have to be taken. Correspondingly, starting from a technical perspective, the framework presented in this paper encourages the search for technically crucial parts of a possible accident evolution and the attempt to figure out reasons why an operator would take action which may lead to a crucial and dangerous state of the technical system (Waessman, 1986).

The approach presented in this paper provides not only a tool for improvements in existing human reliability analyses but also a knowledge base which can be used when designing programs for nuclear power operator training.

This study was supported by a grant from the Swedish Nuclear Power Inspectorate.

REFERENCES

Gertman, D., 1986, Operator Information Needs During Accidents: Influence of Cognitive Elements and Cognitive Shaping Factors (CSF), OECD Halden Reactor Project, Norway, HWR-175.

Kates, R. W., Hohenhemser, C., Kasperson, R., and Kasperson, J. X., eds., 1985, *Perilous Progress: Managing the Hazards of Technology*, Westview Press, London.

Kinkade, R. G., and Anderson, J., eds., 1984, Human Factors Guide for Nuclear Power Plant Control Room Development, Electric Power Research Institute, Palo Alto, CA.
Lydell, B. O. Y., Stampelos, J. G., and Stetkar, J. W., Human Reliability Analysis in Contemporary Probabilistic Risk Assessment Studies, Pickard, Lowe and Garrick, Inc., PLG-0349, Irvine, CA.
Rasmussen, J., Duncan, K., and Leplat, J., Eds., 1987, *New Technology and Human Error*, Wiley, Chichester.
Swedish State Power Board, 1986, Overtryckning av G2 i samband med thetsprovning, Ringhalsverket PR-Driftrapport No. 93/86 (in Swedish).
Svenson, O., 1988, Cognitive Psychology, Operator Behavior and Safety in the Process Industry with Emphasis on Nuclear Power Plant Applications, Cognition and Decision Research Unit, Department of Psychology, University of Stockholm, Report No. 29.
Svenson, O., and Fischhoff, B., 1985, Levels of Environmental Decisions, *Journal of Environmental Psychology* 5:55-67.
Swain, A. D., and Guttman, H. E., 1983, Handbook on Human Reliability Analysis with Emphasis on Nuclear Power Plant Applications, U.S. Nuclear Regular Commission, NUREG/CR-1278, Washington, DC.
U.S. Nuclear Regular Commission, 1985, Loss of Main and Auxiliary Feedwater Event at Davis-Besse Plant on June 9, 1985, NUREG-1154, Washington, DC.
USSR State Committee on the Utilization of Atomic Energy, 1986, The Accident at the Chernobyl Nuclear Power Plant and Its Consequences, Information compiled for the IAEA Expert's meeting 25-29 August, 1986, draft of working document.
Waessman, P. O., 1986, Personal communication.

Chernobyl Accident — Emergency Monitoring and Protection Measures in Poland

Zbigniew Jaworowski
Fysisk Institutt, Universitetet I Oslo
Oslo, Norway

ABSTRACT

About 58 hours following the Chernobyl accident, the 140 stations of Poland's Service of Measurements of Radioactive Contaminations (SPSP) converted to an emergency mode for reporting ambient gamma-ray dose rates and radioactivity levels in the environment and foodstuffs to the Central Laboratory for Radiological Protection (CLOR) in Warsaw. In addition, vertical profiles of radioactivity in the troposphere and stratosphere over Poland were obtained daily, and measurements of the concentrations of particular radionuclides were carried out at several stations. Based on these reports and recommendations of the National Atomic Energy Agency, the Governmental Commission began distributing stable iodine to everyone below the age of 17 on April 29 and began measuring thyroid doses in approximately 1200 persons. The consumption of milk from cows fed with green fodder was also discouraged. The Chernobyl emergency was a serious test for the Polish radiation monitoring system and demonstrated that it could meet the challenge. It also showed that the International Atomic Energy Agency (IAEA) could assume the role of an international information center during periods of crises.

KEYWORDS: Chernobyl, radiation monitoring, stable iodine distribution, Polish radiation emergency system, radionuclides in troposphere and stratosphere

Early on the morning of Monday, April 28, 1986, observers at the Mikolajki station of the Service of Measurements of Radioactive Contaminations (SPSP) noted a 700-fold increase in non-natural atmospheric radioactivity and a 10-fold increase in the ambient gamma-ray dose rate. In accordance with the "Instruction of the SPSP," the station immediately notified the Central Laboratory for Radiological Protection (CLOR) in Warsaw (a national coordinating center for the SPSP system) of the increased dose rates and then changed to an emergency mode. That is, instead of replacing the aerosol filter only once a day and measuring the radioactivity after one hour and then again after 120 hours (to allow for the decay of short-lived natural activity), station personnel began replacing the filter every two hours and measuring the radioactivity immediately. The results, together with the ambient gamma-ray dose-rate readings, were reported every two hours.

The first information on findings in Mikolajki reached CLOR at about 9 AM on April 28, and very soon similar information arrived from several other SPSP monitoring stations in northern Poland and in Warsaw. At 10 AM the same day, CLOR initiated the emergency

New Risks
Edited by L. A. Cox, Jr., and P. F. Ricci
Plenum Press, New York, 1990

mode at 24 stations, and, beginning on the morning of April 29, at all 140 stations of the SPSP system.

The SPSP is composed of various types of stations, which are located throughout the country and are administered and sponsored by several ministries: 49 stations by the Ministry of Environmental Protection and Natural Resources, 39 stations by the Ministry of Health and Social Welfare, 30 stations by the Ministry of Agriculture Forestry and Food Management, 11 stations by the Ministry of Construction and Area and Communal Management, 2 stations by the Ministry of Foreign Trade, 3 stations by the National Atomic Energy Agency, and 9 stations by various scientific institutions. All these stations are equipped with basic radiation measuring instrumentation by CLOR, and they report their measurements directly to CLOR. Their personnel training and work programs are also controlled by CLOR.

Nine of the SPSP facilities are situated at synoptic meteorological stations and under normal conditions perform continuous air sampling. They measure both gross beta activity and gamma-ray dose rate. The other stations collect samples for total deposition measurements. Samples of milk, meat, agricultural products, grass, soil, surface and tap water and liquid effluents from industrial installations are monitored for gross beta activity at intervals ranging from one measurement per month to one per year, depending on the type of samples.

All 140 SPSP stations are equipped with ZAPKS-1, a basic radiation measurement set (Polish made) for continuous gamma-ray dose-rate counting. The detection range of the scintillation counter is 0.01 mR/h to 100 R/h, with graphic registering of results and emergency signaling at a preset level. The set also performs beta counting of environmental samples with a plastic scintillation counter in a lead casing. Four other types of GM and scintillation counters are also used by most of the stations. The 10 stations which monitor radioactivity in milk are prepared to measure concentrations of I-131 by a quick radiochemical method. During the Chernobyl emergency, they also performed simple gamma-ray measurements of liquid milk. Concentrations of Ce-144, Cs-137, Sr-90 and of other man-made and natural radionuclides in the environmental samples, foodstuffs, and human tissues are normally determined at CLOR and 10 other stations by radiochemical methods.

Gamma spectrometric measurements of environmental samples are performed routinely at CLOR. During the Chernobyl emergency, five other institutes with gamma spectrometry capabilities joined with the SPSP system. Analysis of tritium is routinely performed at two of the institutes and of C-14 at one. The activity levels in the populace are measured by two whole-body counters and by the radiochemical analysis of post-mortem material. Simple emergency measurements are also made with hand-held scintillation counters in cases of internal iodine-131 contamination. Stationary thyroid counters are in operation at CLOR and at two other institutes.

Vertical profiles of radioactivity in the troposphere and stratosphere over Poland are routinely measured on a monthly basis at altitudes of 0, 3, 6, 9, 12, and 15 km. For these measurements, 400- to 1000-m^3 (STP) air samples are collected by airplanes, and the radionuclides are determined by radiochemical analysis and gamma spectrometry. During the Chernobyl emergency, daily monitoring of the vertical distribution of activity was started over eastern Poland on April 29 and was continued until May 2; subsequently measurements were taken at intervals depending on the meteorological situation. In addition, after the evening of April 28 the airborne teams from CLOR and from the Armed Forces monitored dose rates, ground depositions and thyroid accumulations of radioiodine all over the country.

Table 1. Number of Sampling Sites and Number of Samples Measured Between April 28 and May 31, 1986, by SPSP System in Poland

Type of Sample	Number of Sampling Sites	Number of Measurements
Total deposition	67	1374
Soil	79	1172
Grass	93	1481
Surface water	126	1416
Tap water	70	839
Vegetables	186	2456
Milk	193	2393
Meat	36	515
Fish	19	144
Eggs	11	122
High-altitude aerosols	--	50
Ground-level aerosols	9	3670
TOTAL	889	15632

In an emergency, the SPSP stations report their coded monitoring results to CLOR by telex, telegram or telephone every two hours. The use of an 18 numbers-per-measurement code saves time in the transmission of information. This system is designed to ensure a quick warning in case of radioactive contamination of large areas. It offers the possibility of a prompt estimate of the population's exposure to radiation and provides for an immediate changing of the whole SPSP system from normal monitoring to an emergency mode. It appears that during the Chernobyl emergency the SPSP system fulfilled these expectations. Table 1 illustrates the monitoring capacity of the system.

In the morning hours of Monday, April 28, a large number of reports started to reach CLOR, where a six-person operational group was formed to analyze incoming information. That afternoon a similar group was formed at the National Atomic Energy Agency. From the gamma-spectrometric analysis of an air filter collected at 1 PM in Warsaw, it was clear that the composition of radionuclides in the air was typical of a reactor fission rather than of a nuclear explosion, with the radioiodines and Te-132 contributing about 80 percent of the activity. At 8 PM the group prepared the first preliminary report on the radiological situation in the country. At 9:30 PM a second report was prepared. Both reports were sent to the president of the National Atomic Energy Agency.

The gross beta activity concentrations in the air measured during the afternoon of Monday, April 28, in northern and central Poland were 550 Bq m^{-3} in Mikolajki (15,000 times higher than two days earlier), 170 Bq m^{-3} in Gdynia, 87 Bq m^{-3} in Warsaw, and 0.79 Bq m^{-3} in Poznań. The gamma-ray dose rates in these locations ranged between 0.1 and 2.5 mR h^{-1} (this last value was later corrected to 0.17 mR h^{-1}). At that time the stations in southern Poland did not report a large increase in gross beta activity in the air, except for the meteorological observatory at Snieżka Mtn. (altitude of 1602 m) in the southwestern part of the country, where an activity of 1.8 Bq m^{-3} was found. However, a southward trend in radioactivity migration was already identified by Monday evening.

As was found later by the analysis of the previously collected air filters and the automatically registered gamma-ray dose-rate measurements, the first wave of radioactivity from the Chernobyl reactor reached the Mikolajki region at about 9 PM on April 27 and the Warsaw region about 5 hours later. The maximum contamination of various components of the environment occurred at various times after the Chernobyl accident. The maximum gamma-radiation-dose rate of 0.45 mR h^{-1} and the maximum gross beta activity concentration in air of 571 Bq m^{-3} were observed on April 29 in Mikolajki in northeast Poland, when the national average was 104 Bq m^{-3} (Table 2). On the same day the maximum values of gross beta activity were found in milk (2000 Bq per liter) and in the total fallout (43,000 Bq m^{-2}). The maximums in the other components of the environment appeared one to five days later — on April 30 in grass (105,000 Bq kg^{-1}), on May 1 in green vegetables (253,800 Bq kg^{-1}), on May 2 in surface water (417 Bq per liter) and in tap water (111 Bq per liter), and on May 4 in the soil (16,400 Bq kg^{-1}).

Before Chernobyl the national average of gross beta activity in the air from man-made sources, i.e., from weapons tests, was 0.1 Bq m^{-3}, so the maximum increase of atmospheric radioactivity in Poland after the Chernobyl accident reached a factor of about 6000. The average pre-Chernobyl gamma-ray dose rate in Poland measured 1 m above the ground was 0.012 mR h^{-1}; the maximum value on April 29 was about 40 times higher. By far the highest increase in gross beta activity was observed in the total fallout (by a factor of about 14,000) and the next highest in green vegetables (by a factor of 1200); the lowest increases were observed in the soil (by a factor of 34) and in milk (by a factor of 48).

Gamma-ray dose rate and gross beta measurements were the principal tools used in Poland for the quick assessment of the changing amounts of radioactive contamination; however, measurements of the concentrations of particular radionuclides were carried out at some stations.

The concentration of I-131 in ground-level air in Warsaw on April 29 was 11.5 Bq m^{-3}; and that of Cs-134,136,137 was 4.8 Bq m^{-3} (Tables 3 and 4). These elevated activities in the air were confined to the ground level, with a thousand times lower activities at altitudes between 3 and 12 km. The next day greatly increased activities were detected in the air between the ground level and an altitude of 3 km, with comparatively low activities between 6 and 12 km and increased activities at 15 km. The increase at 15 km was probably not an effect of the initial explosion but was caused by a nonviolent transport through the tropopause, similar to the upward transport of Ra-226 and stable lead from the troposphere to the stratosphere observed earlier (Jaworowski and Kownacka, 1976).

A sudden fall in the concentration of radioactive iodine and cesium in the atmosphere occurred on May 1 and lasted several days. Then on May 7 and 8 the concentrations of these nuclides at ground level increased by an order of magnitude. A similar increase in ground-level air had been observed in Stockholm three days earlier (Jensen and Lindhé, 1986), apparently after new releases of activity from the damaged reactor May 2 to 5 (National Committee, 1986). The high concentrations of I-131 and radiocesium at ground level on May 7 and 8 were accompanied by high contamination in the whole tropospheric profile over Poland (up to 9 km) and also at stratospheric altitudes of 12 and 15 km.

The dispersion of radioactivity in the upper troposphere and the lower stratosphere probably resulted from two mechanisms. The first was an explosion of hydrogen (Semenov, 1986), 2.2 tons of which were probably released in the chemical reaction of the coolant water with zirconium and graphite present in the core (IPSN, 1986). The second was the thermal rise of the plume, the energy source for which was the reactor's residual power of approximately 20 MW, the oxidation of zirconium of 50 MW, and the burning of graphite of approximately 8 MW (IPSN, 1986). The explosion is believed to have projected about half of the released activity up to the altitude of about 7 km (Knox, 1986). The thermal rise may have transferred radioactive material to a somewhat lower elevation

Table 2. Gamma-Ray Dose Rates and Gross Beta Activities in Poland Between April 28 and May 30, 1986[a]

Date	Gamma-Ray Dose Rate (mR/h)		Beta Activity (Bq/m^3), Air		Beta Activity (Bq/liter)						Beta Activity (Bq/kg)						Beta Activity (Bq/m^2)	
					Milk		Surface Water		Tap Water		Grass		Soil		Green Vegetables		Total Fallout	
	Max.	Avg.	Max.	Avg.	Max.	Avg.	Max.	Avg.	Max.	Avg.	Max.	Avg.	Max.	Avg.	Max.	Avg.	Max.	Avg.
Pre-1986	--	0.012	--	0.1	--	42	--	0.7	--	0.4	--	223	--	481	--	132	--	3
28.04	0.08	0.05	87[b]	30.1	241	87	12.0	4.4	--	--	--	--	--	--	--	--	570	305
29.04	0.45	0.23	571	104.1	2000	235	40	14	48	21	50000	10315	2870	1230	--	--	43000	4570
30.04	0.35	0.19	514	91.2	1770	225	89	17	74	16.0	105000	25861	12958	2053	6847	3821	19514	5307
1.05	0.22	0.12	110	3.9	945	177	103	25	101	7.8	87000	28328	8701	2198	253800	19505	15000	5236
2.05	0.11	0.06	4.9	1.1	990	235	417	38	111	7.5	52000	37060	6300	3607	82000	13939	23000	2084
3.05	0.05	0.03	6.7	1.1	583	250	59	9.6	37	7.9	49000	19940	15955	2310	17000	10728	2968	817
4.05	0.05	0.03	34.5	3.5	1640	420	149	8.5	13	4.2	91000	18716	16400	2159	19000	2590	4236	586
5.05	0.07	0.04	8.2	1.1	844	138	156	18.3	11.1	3.9	46000	13519	15695	2604	27800	3370	5182	698
6.05	0.07	0.04	5.8	1.1	1328	210	57.0	7.3	27.8	8.4	35000	12363	8000	1616	45800	4262	2885	214
7.05	0.07	0.04	14.9	3.2	953	171	41.1	4.3	55.9	7.3	71300	14498	4100	1723	8500	5270	578	303
8.05	0.07	0.04	19.4	3.8	1021	185	29.8	4.4	13.0	2.7	17800	5824	8250	2261	11500	2010	9631	1719
9.05	0.07	0.04	1.4	0.6	574	189	98.3	3.7	5.7	2.2	9870	4556	7620	2883	9502	1862	11828	1733
10.05	0.04	0.02	3.4	0.3	433	312	19.6	6.4	19.9	3.0	20000	4137	7400	1605	5100	1424	7400	913
11.05	0.035	0.02	0.6	0.3	474	195	114.9	9.2	5.9	1.4	6900	3097	3400	2253	4350	1174	572	186
12.05	0.035	0.02	0.6	0.2	290	162	114.9	6.0	3.4	2.3	4890	2287	4650	1810	4778	1620	520	171
13.05	0.032	0.02	1.2	0.3	500	223	51.1	2.6	5.0	1.2	6140	2254	3660	1736	5600	930	787	87
14.05	0.028	0.02	1.2	0.3	220	121	7.1	2.8	5.0	1.7	18200	2834	4680	1598	5600	988	787	67
15.05	0.03	0.02	3.2	0.3	219	123	23.3	2.5	2.7	1.0	6086	1217	1756	1172	1628	860	66	63
16.05	0.042	0.02	0.7	0.1	219	112	8.0	1.9	4.2	1.1	5090	1207	1950	927	3210	570	333	102
17.05	0.042	0.03	3.9	0.3	158	113	4.8	1.4	2.5	0.8	8980	1444	1240	1077	2900	411	331	40
18.05	0.039	0.025	5.8	0.4	293	110	31.8	3.1	3.5	0.9	4180	1073	4160	999	2960	531	125	42
19.05	0.04	0.025	2.1	0.2	184	98	4.3	1.9	3.1	0.9	23680	1433	1441	1166	2960	393	235	29
20.05	0.037	0.024	1.5	0.2	200	94	12.3	1.3	14.0	1.0	21620	1134	7195	1249	1200	393	152	27
21.05	0.038	0.024	2.5	0.3	192	88	3.0	1.1	3.5	2.0	3320	679	5293	1019	7400	534	140	19
22.05	0.037	0.024	2.8	0.3	121	95	3.2	1.0	4.0	1.0	7400	816	4000	904	2230	434	215	26
23.05	0.036	0.024	2.5	0.3	108	93	4.1	1.9	1.9	0.8	10820	1194	3698	908	1480	319	601	97
24.05	0.036	0.02	0.6	0.3	147	74	6.6	1.5	2.7	0.9	4000	749	1878	1202	893	220	79	20
25.05	0.036	0.02	0.13	0.1	203	77	3.9	1.2	1.0	0.7	2100	600	2470	870	1807	251	852	64
26.05	0.035	0.02	0.5	0.2	149	73	1.9	0.9	1.2	0.5	5872	554	6349	769	2904	295	188	38
27.05	0.034	0.02	0.4	0.3	173	70	3.6	1.6	1.7	0.7	1645	617	3718	734	2013	289	181	24
28.05	0.034	0.02	0.4	0.2	136	69	2.3	1.0	0.9	0.5	2540	379	2924	899	2633	349	92	23
29.05	0.032	0.02	0.3	0.1	70	59	3.6	1.6	1.1	0.6	977	574	1208	830	1338	339	723	96
30.05	0.031	0.018	0.3	0.1	83	61	3.5	1.8	2.4	0.7	340	292	3170	931	1930	319	142	46

[a]After report of Polish Governmental Commission, 1986, based on preliminary data. The updated highest gamma-ray dose rate was 0.17 mR/h.

[b]In Warsaw.

Table 3. Concentrations of Particulate I-131 in Troposphere and Lower Stratosphere (Bq 100 m^{-3} STP) (Tropopause Levels in Parentheses)[5]

Altitude km	29 Apr. (11.1)	30 Apr. (11.5)	1 May (11.5)	2 May (11.7)	5 May (11.4)	7 May (11.7)	8 May (11.7)	11 May (10.4)	13 May (12.4)	20 May (-)
0 Warsaw	1149	519.5	25.6	28.5	17.8	200.1	138.6	0.86	0.28	0.15
0 Mikolajki	--	807.0	43.9	41.9	22.5	179.7	195.0	2.79	0.67	0.30
1	--	63.65	94.43	28.18	6.52	97.85	61.36	1.55	0.19	0.00
3	2.17	133.55	--	0.53	1.31	9.71	19.25	3.90	0.10	0.18
6	1.26	2.33	--	--	26.66	27.72	34.05	0.10	0.04	0.09
9	1.27	4.26	--	--	62.70	86.61	25.84	0.72	0.09	0.02
12	2.56	3.66	--	--	0.66	5.81	0.27	0.74	0.00	0.07
15	--	15.51	--	--	1.73	5.04	1.26	0.00	0.00	0.00

Table 4. Concentrations of Cs-134,136,137 in Troposphere and Lower Stratosphere (Bq 100 m^{-3} STP) (Note: Ratio of I-131 activity to Cs-134,136,137 activity in parentheses; calculated after correcting I-131 concentrations with 0.45 filter efficiency factor.)[5]

Altitude km	29 April	30 April	1 May	2 May	5 May	7 May	8 May	11 May
0 Warsaw	485.4 (5.3)	188.0 (6.1)	4.90 (11.6)	1.73 (36.6)	1.43 (27.7)	42.98 (10.3)	32.85 (9.4)	0.20 (9.6)
1	--	12.64 (11.2)	21.56 (9.7)	2.10 (29.8)	0.44 (32.9)	9.43 (23.1)	20.96 (6.5)	0.58 (5.9)
3	0.68 (7.1)	42.23 (7.0)	--	0.06 (19.6)	0.19 (15.3)	1.34 (16.1)	4.28 (10.0)	0.74 (11.7)
6	0.97 (2.9)	0.45 (11.5)	--	--	1.10 (53.9)	1.41 (43.7)	3.22 (23.5)	0.03 (7.4)
9	0.26 (10.9)	1.74 (5.4)	--	--	2.02 (69.0)	2.71 (71.0)	2.19 (26.2)	0.15 (10.7)
12	0.79 (7.2)	1.95 (4.2)	--	--	0.17 (8.6)	0.34 (33.9)	0.07 (8.6)	0.17 (9.7)
15	--	3.61 (9.5)	--	--	0.09 (42.7)	0.95 (11.8)	0.89 (3.1)	0.37

(STUK, 1986). The presence of highly contaminated air at an altitude of 9 km and also in the stratosphere between May 5 and 8 is clearly connected with the second thermal phase of radioactivity release from the reactor and needs further elucidation.

In the period studied we observed two types of radioactive clouds over Poland, characterized by the differing radioiodine to radiocesium ratios. On April 29 and 30 we observed that at most altitudes the I-131 to Cs-134,136,137 ratio was lower than 9.8, which would have been the ratio in the reactor core inventory 4 days after the chain reaction had stopped (IPSN, 1986). On the other hand, levels much higher than the core inventory values of this ratio were found after May 2. On May 5, i.e., 10 days after the chain reaction stopped, the core inventory ratio would have been 6.3, but in the tropospheric air between May 2 and 8, it ranged between 6.4 and 71.0. Differences in the temperature and in other conditions of release between these two periods may have caused this phenomenon.

Because the activities following initial deposition indicated that the radioiodines in milk might reach concentrations above the intervention level (1000 Bq per liter), at 1:40 AM on April 29, the National Atomic Energy Agency recommended that from that day on, all over the country, the consumption of milk from cows fed with green fodder should be banned and that this milk should be used only after industrial processing. The Agency also recommended that cows should stay in barns and not be fed with green fodder. Distribution of preaccident powdered milk was recommended for children up to 3 years of age. In addition, the Agency considered prohibiting the consumption of green vegetables, meat and fish with activity levels above 5000 Bq kg^{-1}. The Agency recommended that in 11 northern and central provinces stable iodine be distributed to all children up to the age of 16. These recommendations were accepted by the Governmental Commission for Assessment of Nuclear Radiation and Prophylactic Measures, which was formed early on the morning of Tuesday, April 29. The Governmental Commission, taking into account the changing meteorological situation, the contamination from the Chernobyl reactor, and also the socio-psychological factors, decided that stable iodine should be distributed throughout the country to everybody below the age of 17.

The Ministry of Health and Social Welfare took over the responsibility for stable iodine prophylaxis. It decided that a single dose of iodine should be administered in a solution (2 g KI + 1 g I per 100 g of distilled H_2O). Children up to 1 year of age received 15 mg of iodine, those from 2 to 6 years received 30 mg, and those from 7 to 16 years received 60 mg.

Administration of stable iodine was started on the evening of April 29 in the Bialystok region and on April 30 in the rest of the country. It was estimated that about 10 million children and adolescents, i.e., 98 percent of the population group up to 16 years, received prophylactic doses of stable iodine. In addition, several million adults received this treatment, although this was not recommended by the authorities. The administration of stable iodine prophylaxis on such a scale was a difficult task. Apparently its fast and efficient implementation was made possible by the decision to administer the iodine in a dissolved form which could be prepared and then distributed by each of 3348 pharmacies in the country. The iodine solution was distributed also by hospitals, schools, nurseries, and other agencies.

According to a preliminary estimation, the average external gamma-ray dose received by the adult inhabitants of about 50 percent of the territory of Poland was 0.15 mSv. In the highly contaminated regions (approximately 25 percent of the country), the dose was 0.44 mSv. Thyroid doses were measured in approximately 1200 persons. The average dose commitment is tentatively estimated as ranging from 1 to 10 mSv. In a limited number of cases in the highly contaminated regions, the measured thyroid doses were about 100 mSv in adults and about 800 mSv in children. The highest thyroid dose found in an adult was

230 mSv. Such high doses were found among persons who consumed the contaminated milk of cows grazing in pastures.

The Chernobyl emergency was a serious test for the Polish radiation monitoring system. It demonstrated that the system is sufficient to stand up to that type of challenge. From the first days of the emergency, our laboratory had the opportunity of exchanging information with similar institutions in other countries and with the International Atomic Energy Agency, which, at least in our eyes, assumed the role of an international information center. Because of this opportunity, we were able to compare the levels of radioactivity and projections of the radiological situation in Poland with those in other countries, which was useful and reassuring.

REFERENCES

1. IPSN, 1986, Institute de Protection et de Surete Nucleaire, Paris, Report No. IPSN 2/1986 (1986)
2. Jaworowski, Z., and Kownacka, L., 1976, *Nature* **263**:303-304.
3. Jensen, M., and Lindhé, J.-C., 1986, *IAEA Bulletin* **28(3)**:30-32.
4. Knox, J., and Beardsley, T., 1986, *Nature* **321**:187.
5. Kownacka, L., and Jaworowski, Z., 1987, *Acta Geophysica Polonica* **35**:101-109.
6. National Committee of the USSR of Nuclear Energy, 1986, Report presented at IAEA Experts Meeting, Vienna, August 25-29 (Polish translation).
7. Polish Governmental Commission for Assessment of Nuclear Radiation and Prophylactic Measures, 1986, Warsaw.
8. Semenov, B. A., 1986, Address by the Representative of the USSR to the Board of Governors of the IAEA, 1986.05.21 (unpublished).
9. STUK, 1986, Finnish Centre for Radiation and Nuclear Safety, Report STUK-B-VALO 44/1986.

The Accident at Chernobyl: A Report on Risk Management at a Local Hot Spot in West Germany

M. Deicher, A. Ernst, H. Hofsaess [a]
G. Lindner,[b] **and E. Recknagel**
University of Konstanz
Konstanz, F.R.G.

C. Hohenemser
Clark University
Worcester, MA

ABSTRACT

Local fallout in the days following the accident at Chernobyl reached levels 30 to 40 times natural background at locations as far as 1500 km from the accident. In the absence of coordinated responses at the state and national levels, spontaneously generated local risk management efforts were undertaken in many places. We describe here our own local risk management activities in a locally hot region of southern West Germany. Based on the achievement of high credibility with the local population and the institution of effective protective action in the food chain, we argue that in future radiological disasters risk management should retain a strong decentralized component.

KEYWORDS: Chernobyl, radionuclide dispersal, risk management, radiation protection, Konstanz region

INTRODUCTION

The reactor accident at Chernobyl in the Soviet Ukraine on April 25-26, 1986 released the largest quantity of radionuclides ever freed in one technological accident. According to the U.S.S.R. State Committee on the Utilization of Atomic Energy,[1] the accident released at least 3-4% of the radioactive inventory of the 3,200 MWth reactor—a unit that had been in operation for about two years without significant fuel change. Chernobyl marks, therefore, the first nuclear reactor accident that approaches what has been described as a "worst case" release, i.e., an event in which a large reactor with a mature fuel inventory breaches confinement or containment and releases a significant fraction of the core's radionuclide content.

---

a. Present address: University of North Carolina, Chapel Hill, NC.

b. Present address: College of Technology Ravensburg-Weinganten, Weinganten, F.R.G.

New Risks
Edited by L. A. Cox, Jr., and P. F. Ricci
Plenum Press, New York, 1990

Favorable weather, siting, effective evacuation, and the dispersion of early releases to high altitude all contributed to holding prompt casualties in the Soviet Union to a small number—31 at last report.[1] Air concentrations and ground deposits as far away as 1000-2000 km, however, frequently exceeded protective action guidelines applicable to the ingestive pathway. Since European risk management institutions were by and large unprepared for an accident of the magnitude and transnational character of Chernobyl, it was necessary to improvise appropriate responses to the fallout. Even the first step—monitoring and risk identification—proved difficult for most countries because a prepared network of measuring stations did not exist.

In some countries, such as Sweden[2] and the Netherlands[3] improvisation was prompt and relatively efficient because well-equipped radiation safety laboratories coupled quickly to relevant public officials. In other cases, such as West Germany and France, it took many days to map the fallout and to institute adequate public information campaigns and appropriate countermeasures. The spectrum of response ranges from Switzerland, which failed to respond at all, to the West German state of Hessen, which set standards at almost impracticably low levels (i.e., 20 Bq/l for iodine-131 in milk).

If anything is clear from the European experience, it is that protective action planning at large distances from a reactor accident is an issue whose time has come. The need for such planning is not only driven by public fear and confusion, but also by early collective dose estimates. Whereas most of these estimates agree that Chernobyl will impose an incremental cancer risk for individuals on the order of 0.01% or less, and hence epidemiologically undetectable, the same estimates show that the total social burden of cancer will increase by thousands or even tens of thousands of cases.[1,4] In light of this, there can be little question that the public will demand appropriate protection.

As a step in the direction of defining the requirements for protection at a distance, we review here what is now known about the physical characteristics of Chernobyl fallout and describe the experience of our group in developing a local response in the Konstanz area—a locally "hot" region in southern West Germany. We also compare the responses of local government to state and federal responses and discuss optimal parameters for risk management in the case of future radiological and other disasters

PHYSICAL CHARACTERISTICS OF CHERNOBYL FALLOUT

Radionuclide Release

The accident at Chernobyl has been described in some detail at a review conference sponsored by the International Atomic Energy Agency. According to Soviet scientists,[1] the accident occurred during a series of tests designed to establish the capability of the reactor's turbines to provide power to emergency cooling pumps following emergency shutdown. To simulate this situation, operators attempted to run the plant at about 700 MWth prior to a sudden planned shutdown. In the course of this, they inadvertently dropped to 200 MWth and encountered instability due to xenon poisoning.

Instead of shutting down altogether when they encountered the xenon poisoning instability as required by safety regulations, they continued the test by violating additional explicitly stated safety rules. Since the reactor has a positive reactivity coefficient at low power, this eventually created a "runaway" condition in which local temperatures rose to well over 2000°C in a few seconds. The result was a steam explosion that severed the top of most of 1661 vertical pressure tubes and dropped the fuel changing crane onto the core. The steam explosion was followed by a second (hydrogen) explosion and a vigorous fire that burned for several hours.

The initial explosions and fire began at 1:23 a.m. on April 26, and carried radioactive debris to an altitude of 2-5 km. Release of radionuclides continued for over nine days and reached peak values during the first and ninth day. Whereas the first day peak was caused by the initial explosion, the second originated from heat accumulation in the destroyed core, which had been covered with 5000 tons of sand, clay, boron, and lead in order to contain the fire. The daily release, decay corrected to May 6, is mapped in Fig. 1 and sums to about 50 MCi of activity.[1]

The data of Fig. 1 were obtained through aerial surveys inside and outside the 30-km-radius emergency zone but are restricted to ground-deposited activity within the geographic limits of the Soviet Union. Not included are about 40 MCi of gaseous xenon-133 and 0.6 MCi of gaseous krypton-85, as well as activity that was surface deposited or is still airborne activity outside the Soviet Union. Since the reactor core contained about 1400 MCi of radionuclides on May 6, the ground deposits in the U.S.S.R. amount to 3.5% of the core. Adding gaseous fission products brings the release to 6.4%; adding surface deposits outside the Soviet Union places the total release fraction as high as 8-10%.

The release fractions for various isotopes differed substantially from those expected in a worst case pressurized water reactor (PWR) accident. In particular, owing to the release through explosion, Chernobyl had a lower fraction of some volatile fission products and a higher fraction of less volatile fission products than a corresponding PWR accident.[1,5] To illustrate the difference, we show in Fig. 2 the release fractions as a function of volatility class for the two cases.

Dispersal of Radionuclides

Some of the available dispersion data are summarized in Fig. 3, in which we have plotted local exposure rates at a time 7-10 days after the accident as a function of distance from Chernobyl. The exposures are presented as dimensionless ratio, E, defined as the reported rate in millirem/hour, divided by the natural background in the same units. Numbers for Soviet locations are taken from the Soviet report to the IAEA,[1] whereas numbers for Western Europe are obtained from our own compilation.[6] As indicated by the drawn line, the recorded peak exposure falls off with distance, r, roughly as $E \sim r^{-1.4}$. Despite scatter in the data caused by timing of the cloud's passage and non-uniformity of release and dispersion, Fig. 3 permits a rough interpolation for non-measured stations.

It is seen that the data fall into two families—those describing locations where rain occurred during the cloud's passage and those which represent essentially dry deposition. At a given distance, rainfall locations appear to have exposure levels that are approximately 15-20 times higher than dry locations. Because dispersal was dominated by long-range transport of small particles, considerably further fractionation of radionuclides occurred during dispersal. This had the consequence that fallout received in Western Europe had quite a different composition than the material released by the reactor. In particular, fallout deposited in southern West Germany showed a marked relative deficiency in alkaline earths (barium, strontium) and transuranics (neptunium, plutonium), as shown in Fig. 2. This means that Western Europe was largely spared the most dangerous radionuclides present in the core—strontium-90 and the various isotopes of plutonium. The same can not, unfortunately, be said about regions of the Soviet Union close to the accident.

EVOLUTION OF LOCAL RISK MANAGEMENT IN THE KONSTANZ REGION

The concise technical description of the accident and the fallout we have given is the result of months of intensive study and owes its clarity to compilation and synthesis of data from many sources. In the days immediately following the accident such a description was beyond the reach of anyone, including the governments of all affected countries. To track

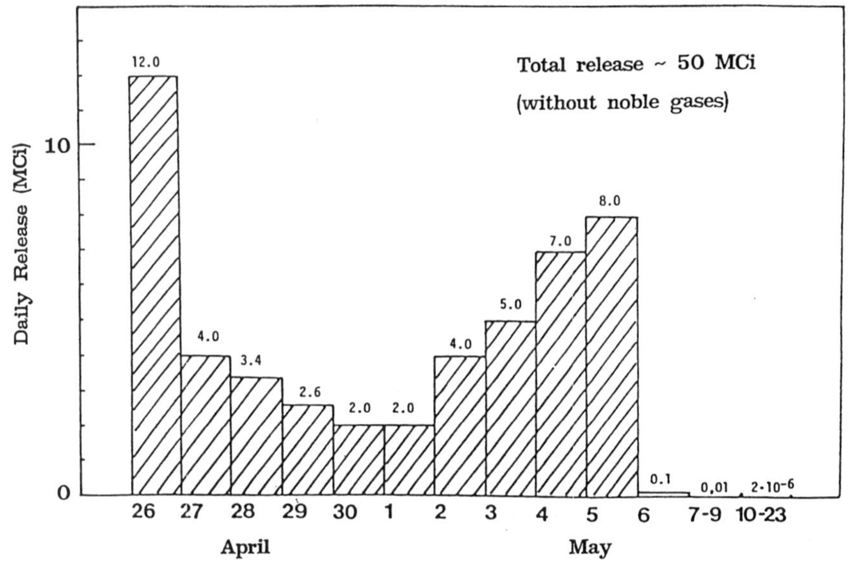

Fig. 1. Radioactive emission from Chernobyl as a function of time. Emission data are normalized to May 6 and are based on ground deposits within the Soviet Union (from Ref. 1).

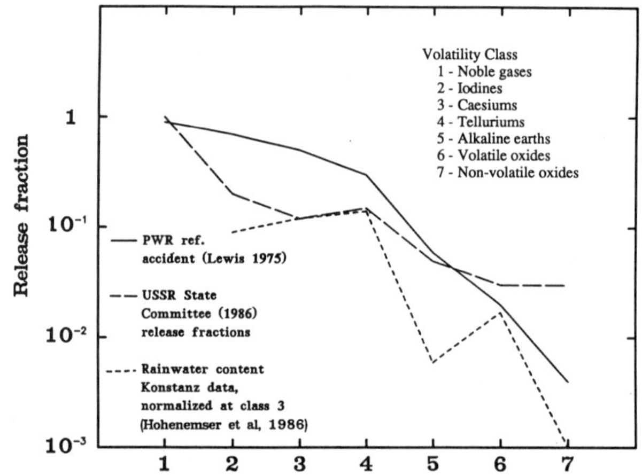

Fig. 2. Release fraction under three conditions as a function of volatility class. The solid curve describes a theoretical reference accident for a pressurized water reactor. The dashed curve describes the Chernobyl release. The dotted curve describes the radioactivity measured in Konstanz normalized to Chernobyl at Class 3.

the evolution of understanding and the local risk management response in the Konstanz region, we recall the situation in the days immediately following the accident.

Early Discoveries

When Sweden announced the arrival of a radioactive cloud on Monday, April 28, few of us were listening or deeply concerned. We began to pay attention on Tuesday, April 29,

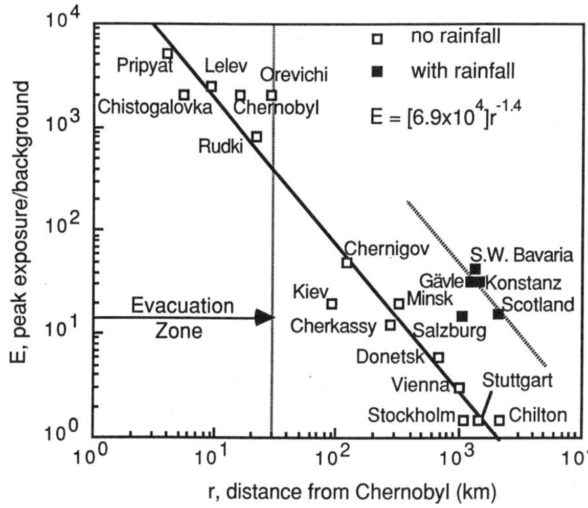

Fig. 3. Exposure levels, due to Chernobyl in Eastern and Western Europe, measured 7-10 days after the accident. The open symbols describe dry deposition whereas the closed symbols describe rain deposition.

when the Soviet news agency TASS announced a serious reactor accident in the Ukraine. Our concern was reinforced by a telephone call from the editor of the local newspaper, *der Suedkurier*, seeking our comments about the Soviet report. In view of weather conditions which promised east wind in the coming days, we began discussions about the advisability of making fallout measurements. At 1500 km from the accident we were, however, highly skeptical that significant quantities of radioactivity could be measured.

On the afternoon of Wednesday, April 30 heavy rains in the form of thunderstorms fell along the northern frontier of the Alps, including the Konstanz region. Despite our skepticism, we carried a portable geiger counter outside after the rain had stopped to see if anything could be found. We were shocked to find that the counter, which normally reads 8-10 counts/sec inside the building, jumped to 250-400 counts/sec outside. After some checks, we concluded that we had been literally showered with radionuclides in amounts equivalent to 30-40 times natural background.

By nightfall, members of the research group had begun measurements with a high resolution Ge counter, and by the next morning we had a "textbook" gamma ray spectrum from which we could identify more than 15 well known fission products. Figure 4 illustrates some of our early results. Figure 5 shows the spectrum for milk five days after the cloud passed. Though our usual pursuits involved basic research in physics and biology, our familiarity with nuclear spectroscopy quickly made us aware that the level of radionuclide contamination outside our laboratory was far from "acceptable." For example, we knew that the quantities we were measuring approached levels of contamination beyond which radioisotope laboratories would have to be legally closed. Particular findings in the period immediately following April 30 were concentrations of iodine-131 and tellurium-132 in rainwater of 30,000 Bq per liter and surface concentration of cesium-137 of 8,000-12,000 Bq per square meter. The latter level gave us pause, because by checking the literature, we found it exceeded by a factor of 10-15 the cesium-137 deposition in the peak year of weapons testing fallout in West Germany.[6]

In the first few days after April 30, almost all detailed information about the Konstanz fallout was derived from our own measurements. Data about fallout constituents

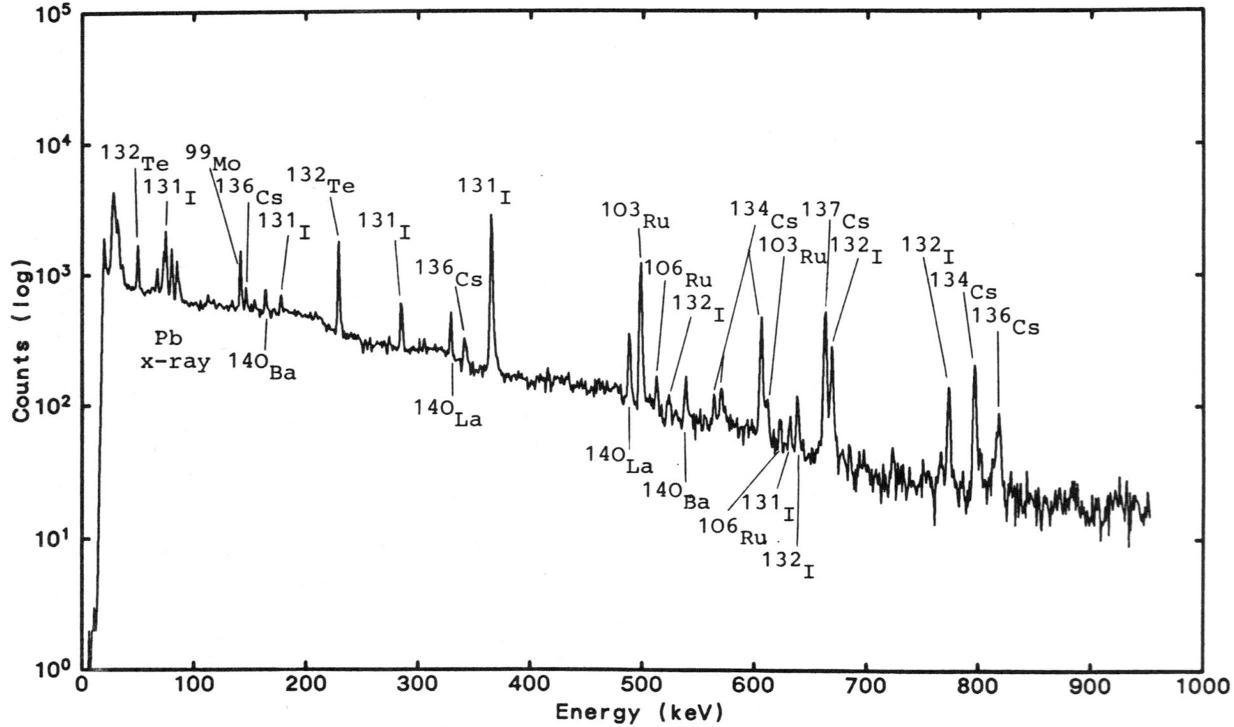

Fig. 4. Gamma spectrum obtained from a grass sample in Konstanz on May 9, 1986.

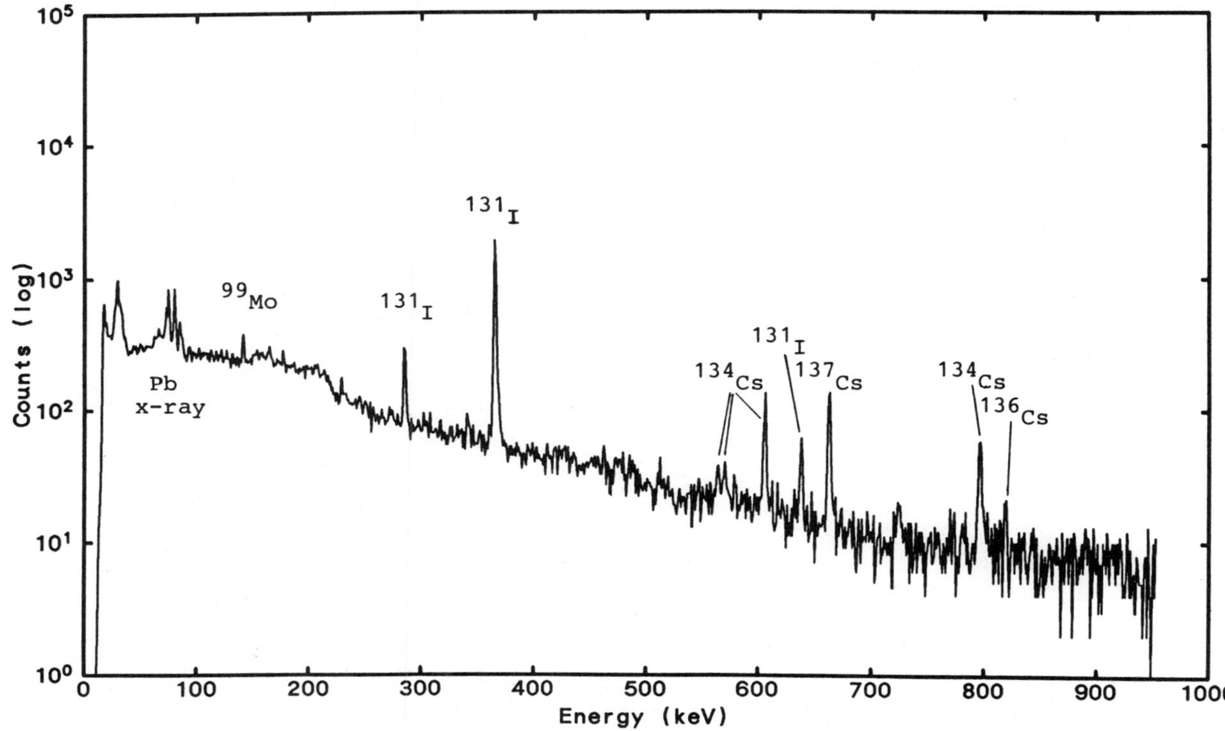

Fig. 5. Gamma spectrum obtained from cow's milk in Konstanz on May 10, 1986. In contrast to Fig. 4, only iodine and cesium isotopes are present.

were obtained with a Ge detector; food contamination was checked with a number of NaI detectors; ground-level contamination was measured with a portable geiger counter designed for detecting gross beta activity; contamination of persons was checked with a stationary personnel monitor. These measurements were supplemented by gross-beta measurements conducted by the Konstanz fire department.

On the whole we were puzzled by our own results because they indicated radionuclide burdens that were much larger than those reported in scattered press reports for other regions of West Germany. The situation was finally clarified on May 8, eleven days after the accident, when the state government of Baden-Württemberg provided a map which compiled gross-beta activity measurements by fire departments throughout the state. As seen in Fig. 6, the map indicates radiation levels in the southeast corner of the state near Konstanz, that were as much as a factor of 20 higher than those found in the northwest part of the state. Subsequently we were able to correlate high radiation levels in Fig. 6 with the presence of rainfall during the passage of the cloud. This correlation is also illustrated in Fig. 6. In retrospect, early recognition of the importance of rain would have enabled a much more effective handling of the crisis.

Despite the extent of our measurements, our information during the first week was far from complete. In particular, we had no way of determining total radioactivity in air, and we could make only crude estimates of the degree to which pure beta and alpha emitters (e.g., strontium-89, strontium-90, and plutonium-239), were present in the fallout. Our estimate that these activities were largely absent was confirmed only much later.[7] Furthermore, we had no reliable prognosis about the possible arrival of further activity from Chernobyl. These considerations led us to be particularly cautious in advising the local population—some would even say we were overly cautious.

The Structure of Local Risk Management

Local risk management in Konstanz was based on the "minimum principle" (Minimierungsprinzip) of West German radiation protection law,[8] which, freely translated, demands that efforts must be made to keep exposure "as low as reasonably achievable (ALARA)." Except for dependence on this principle, the structure of local risk management developed independently of state and national authorities during the period from April 29 to May 5. Figure 7 chronicles the structure of the first week's activities and indicates which local institutions were involved. During the first three days, local action was dominated by the interaction between the local daily newspaper, *der Suedkurier*, and our group. On May 2 the county commissioner (Landrat) formed an advisory committee, which in subsequent days became the center for information exchange and decision making. Throughout the early period a variety of means of reaching the public played a critical role. Included were numerous telephone calls, public lectures, and articles in the local newspaper.

The first contact with the state government of Baden-Württemberg came on May 5, eight days after the accident and a week after the beginning of the crisis in Konstanz. At that time our laboratory was asked officially to do what it had been doing unofficially for a week—namely to help with assessment of radiation levels in food. Subsequently, there was no significant change in the self-evolved local decision making structure, though contact with higher levels of government and regional media became more frequent. Financial support from the state government eventually made possible expansion of monitoring capacity and monitoring personnel.

Our public statements made in the course of local risk management efforts followed three guidelines: (1) rapid publication of all data, with suitable qualifications; (2) description of physical phenomena in heuristic terms understandable to the public—e.g., the description of particulate fallout as "powder sugar"; (3) repeated emphasis that the central

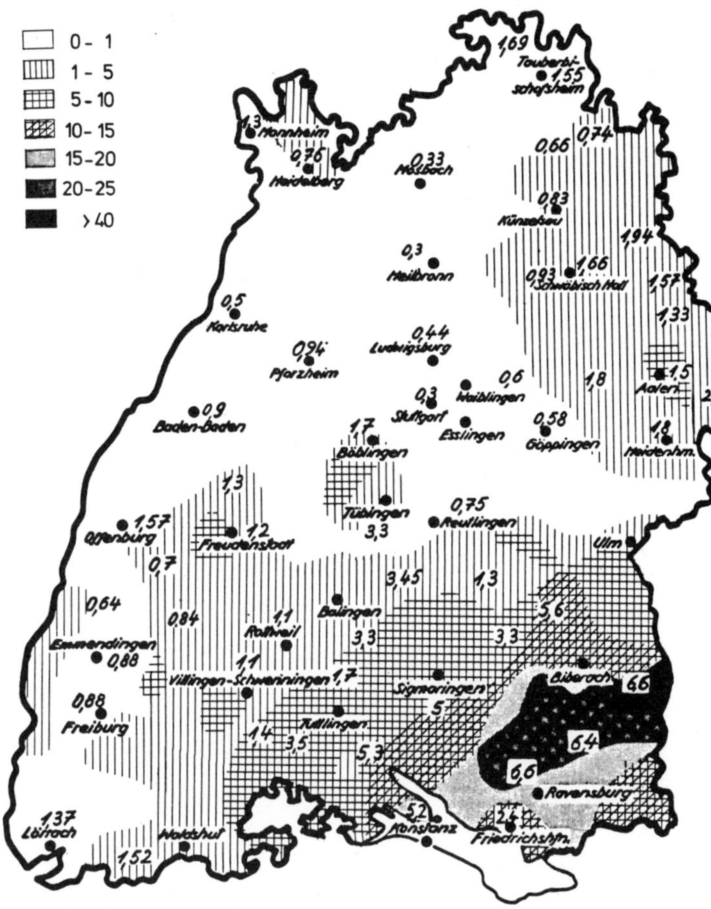

Fig. 6. Illustration of the correlation of activity level with rainfall. The numbers indicate ground level iodine-131 contamination in the state of Baden-Württemberg on May 1, 1986 in Bq/cm^2. The shaded area indicated rainfall in mm during the period of April 30 to May 3, 1986.

objective for everyone should be avoiding ingestion of breathing of radioactivity. With help of these guidelines, it became possible for many lay people to self-design their own protective actions.

We found that many, indeed, did achieve a degree of independent judgment of the situation. For example, citizens became knowledgeable about key protection action guidelines for foodstuffs, including the annual intake limit of 1800 Bq for iodine-131 and 20,000 Bq for the mixture of cesium-134 and cesium-137. With these guidelines in hand and information of the radioisotope content of local food, citizens could decide for themselves what to eat and what to avoid.

The Content of Local Risk Management

Specific action guidelines that were published during the first weeks are given in Table 1. Many of these were supported by specific experiments. Thus we found that washing did not remove activity from shoes, asphalt, and vegetables; hence, we advised that shoes be removed before entering houses and that locally grown vegetables not be

Date | **Communication** | **Action/Reaction**

Tue, April 29

Writing an article for the local newspaper

Wed, April 30

Announcing a lecture, radioactivity measurements, information to weather observ. station

Thu, May 1 (holiday)

Radioactivity measurements, information to local newspaper, telephone service

Fri, May 2 (no school)

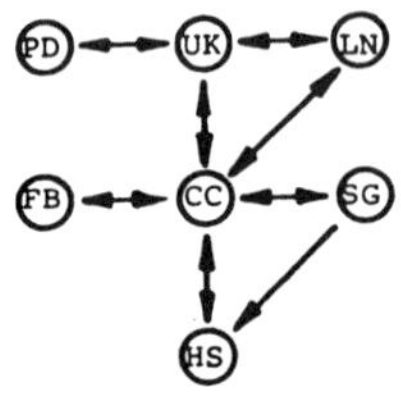

Radioactivity measurements, appointment of an advisory committee of the county commissioner, article to newspaper, public lecture, telephone service

Sat, May 3 (no school)

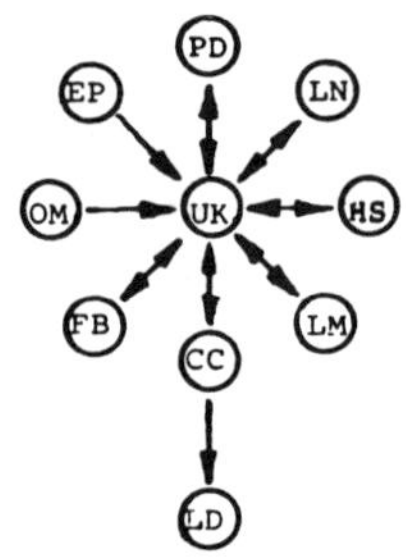

Radioactivity measurements, communications

Sun, May 4

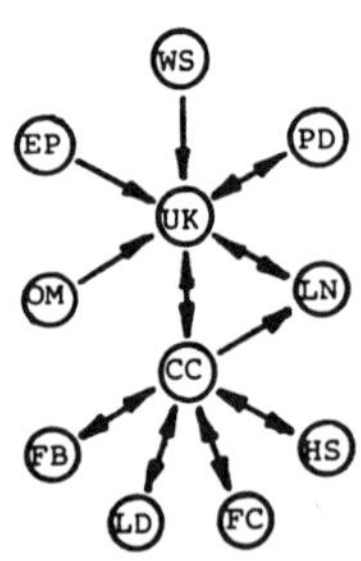

Begin of milk monitoring, advisory commitee, press-release

Mon, May 5

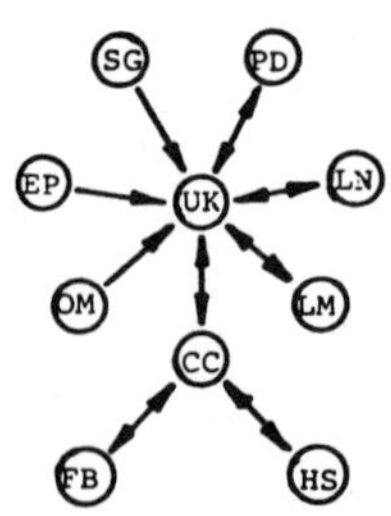

Milk monitoring; vegetable monitoring, advisory committee, press release, appointment by state government as monitoring station

Fig. 7. Interaction diagram for hazard management during the first week after the arrival of the cloud in Konstanz. Abbreviations are as follows: CC-County Commissioner; EP-environment protection groups; FB-fire brigade (local); FC-food control service; HS-health services (local); LD-local dairy; LM-local mayor; LN-local newspaper (*der Suedkurier*); OM-other media (radio, television, other newspapers); PP-political parties; SG-state government; UK-University of Konstanz, Recknagel *et al.*; WO-weather observation station; WS-water supply station; PD-direct communication with the public.

Table 1. Summary of Specific Recommendations Made by Local Hazard Managers

1. Dairy cows should not be permitted to graze on open pasture.
2. Vegetables should be washed; later prohibition of vegetable sales, and that all vegetables be plowed under rather than harvested.
3. No rainwater collected on April 30 should be consumed.
4. Children should be kept indoors as much as possible.
5. Children should wash thoroughly (especially the eyes) after being outdoors.
6. Kindergarteners should curtail sandbox play.
7. Sporting events on hardtop surfaces should be curtailed because of dust formation.
8. Schools should monitor children during recess in order to avoid injuries. Schools should consider indoor sport.
9. Sidewalks should not be swept.
10. Street cleaning with sprinkler trucks should be increased.
11. Lawn mowing should be postponed. Later recommendation to mow and follow instructions for disposal of grass cuttings.
12. Children should not engage in gardening chores in order to avoid injury and dust build-up.
13. People should not wear street shoes indoors.
14. Hunting season should be postponed.

consumed. We also found that air filters had a high radio-particulate content, implying that a substantial proportion of the fallout was in particulate form; for this reason, we recommended that children minimize outdoor play and that most dust-forming activities be curtailed. Through experiments at vegetable farms, we found that the bulk of activity lay on the ground and that only a small fraction (about 20%) adhered to the plants. For this reason, disposal of vegetables was rejected, and farmers were advised to plow under the entire crop. Because of the high activity found on grass, farmers were warned to keep milk cattle from open pasture and to continue to feed stored hay.

Perhaps the most significant intervention in Konstanz was the locally conceived monitoring system for milk. All trucks carrying raw milk collected from farmers were sampled for iodine-131, and, if any contained more than 100 Bq/1, the milk was processed into storable commodities, such as dry milk or cheese. As shown in Fig. 8, average values for milk in Konstanz remained below 100 Bq/1, indicating that milk farmers, to a large extent, followed the recommendation of feeding cattle stored hay. The low iodine-131 values in milk were further consistent with the annual intake limit of 1800 Bq specified by West German radiation protection law.[8] In contrast, the neighboring canton of Thurgau, Switzerland instituted no measures to keep cows off open pasture and recorded iodine-131

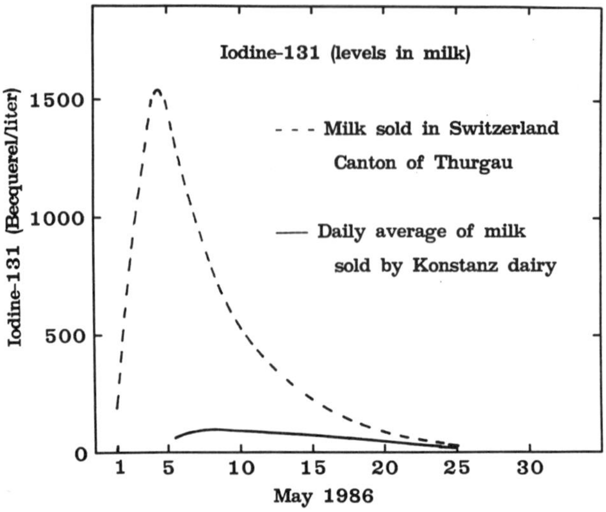

Fig. 8. Comparison of iodine-131 in milk in Konstanz and the neighboring regions of Switzerland. Both regions had similar fallout level, but only Konstanz established milk controls.

levels that peaked at 1500 Bq/1, as also shown in Fig. 8. An important and fortuitous factor in the Konstanz milk policy was, of course, that stored hay was available to most farmers. In addition, it is interesting that despite the low levels of iodine, milk demand in Konstanz dropped drastically during the crisis.

Another significant intervention in Konstanz concerns locally grown vegetables, which are a major source for the entire region. Due to information distributed via the *Suedkurier*, citizens knew early in the crisis that vegetables had ^{131}I contamination well above the eventually announced federal guideline of 250 Bq/kg (see Fig. 9). As a result, the local vegetable market had already collapsed by May 4, when an official ban on vegetables sales was instituted by the state. With no market available, vegetable farmers followed local recommendations to plow under their entire spring crop in the hope of an early replanting, as well as compensation from the federal government. A total of over two million heads of lettuce, 65,000 kg of spinach, and 1.3 million kohlrabi were destroyed during the middle of May. Starting on May 21, the vegetable market was reöpened, and, after some weeks, product demand had recovered to normal levels.

In addition to major control programs, the county commissioner and his advisory board solved a number of practical problems that resulted from the crisis. These included opening of special repositories for contaminated grass cuttings, recommendations for handling contaminated air filters in hospitals and institutions, and actions to prevent the use of contaminated sewage sludge in the community composting plant.

Not all control measures of which we conceived were workable. For example, although we knew that the first hay cutting for the next winter would be contaminated, we were unable to find a solution. In particular, our suggestion to cut the grass well before maturity, and dispose of it could not be carried out because of the resistance of farmers and the unavailability of a sufficiently voluminous disposal area. As a result, there is now a serious problem with contaminated hay in the Konstanz area. Our only useful initiative on this problem was an international warning issued through a letter published in the journal *Nature*.[9]

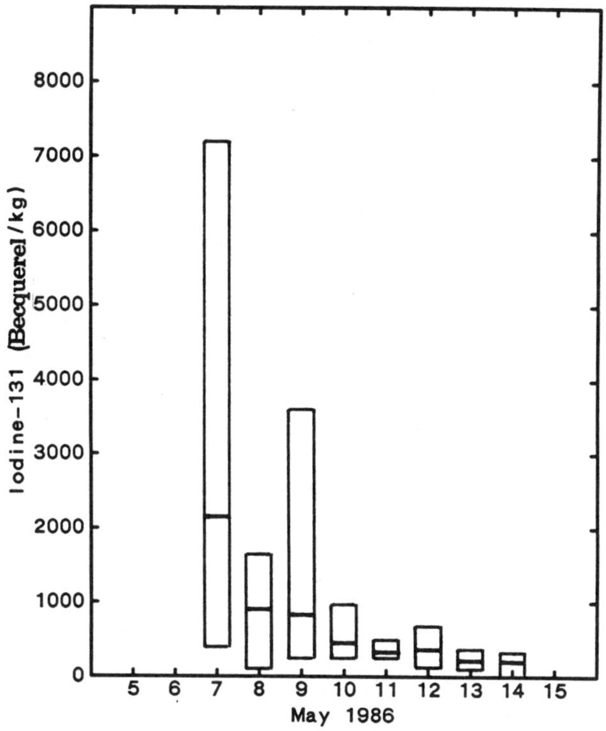

Fig. 9. Time dependence of iodine-131 in lettuce in Konstanz. The bars indicate scatter of daily measurements, the heavy line the daily average.

Risk Communication

Starting immediately after the first reports of the accident, the local population in Konstanz developed, what appeared to us, an almost insatiable appetite for detailed information. We can explain this in part because West Germany has an active anti-nuclear movement which has sensitized the public to the hazards of ionizing radiation. But even in this context, the depth of concern in Konstanz was surprising.

The most important means for communication was the local newspaper, *der Suedkurier*. This paper adapted to the crisis by opening its pages to extensive reports, questions, and answers about the accident and its local impact. Given the tendency of many newspapers to introduce errors and distortions because of deadline pressure, the accuracy of the *Suedkurier* was remarkable. One reason for this was the editor's personal interest in the material, much of which he checked himself. A second reason was that a large fraction of the published material was directly written by the primary sources instead of reporters. On one occasion we prepared a compilation of composite answers to 500 letters of inquiry received by the paper. On other occasions we wrote detailed articles on the Soviet reactor, radioisotope content in food, and radiation risk assessment. These writings were both technically demanding and cognizant of the ambiguity of some of the issues. The paper also printed reports of guidelines generated by the advisory committee to the county commissioner and other sources.

A second key source of information for the local citizenry was direct contact with individuals involved in measurements, risk assessment, and public policy discussions. We believe that these meetings, more than anything else, generated what we interpret as a high level of public confidence in local actors. Members of our group presented more than 50 public programs in May, with an additional 50 in the following two months. The first public presentation occurred on May 2, when over 2000 townspeople listened inside and outside the University's largest auditorium. Later, public meetings were sponsored by a wide range of interest groups, including political parties, farmers, and church groups, and had average audiences of 100 or more. We did nothing to organize these meetings ourselves, responding only to invitations received. To deal with the large number of meetings we trained our students in relevant subject matter and public speaking. In most presentations the public demanded a high level of technical detail. The question periods were often longer than the initial talks, and the worries presented, particularly by young mothers, were often poignant. Throughout the process we were repeatedly amazed by people's willingness to grapple with relatively complicated issues. In many cases we had to admit our own inability to answer sophisticated questions.

A third means of communication with the public was the telephone. As our number became known, it soon became steadily occupied. With the subsequent establishment of a special number for information services, we began to handle a large number of individual questions and concerns. In these telephone conversations members of the public expressed fear, confusion, and helplessness. In some, though not all cases, we were able to counter these feelings via specific suggestions for exposure reduction or explanations of the rather small individual risks involved in local radiation exposure. A number of people called to request monitoring personal samples. Although we tried to respond, we soon found ourselves overwhelmed and had to reject most such requests.

As our work proceeded, regional and national media (radio, television, and larger newspapers) showed an increasing interest in our activities and became a fourth outlet for our results. In the case of television, strict time limits made it difficult to describe our findings with sufficient detail and qualifications. In contrast to local communications, we do not know much about the effectiveness of these outlets.

Did We Overreact?

At this writing, six months after the accident, questions have arisen about the appropriateness and adequacy of local risks management in Konstanz. The most frequently heard critique is that local risk managers overreacted and that guidelines and interventions were too numerous and far-reaching. On the other hand, there are those who argue that we have been too lax and have not done justice to the ALARA principle of West German radiation law. Whereas it is obviously difficult for us, as chief actors in the story, to make a balanced judgment, we believe it is important that we respond to these issues.

An important source of our caution was that early information available to us was limited. In the first week of May it was not known whether further emissions from Chernobyl would be received. It was also not known to what extent strontium-90 was present in the local fallout. An additional cause of caution was our use of the West German radiation protection law,[8] which sets forth annual isotopic organ dose limits based on expected *cancer incidence*. Both of our major actions—the control of milk and the ban of local vegetables—had the specific objective of staying within annual dose limits set out by West German law. This approach to radiation protection turns out to be considerably more conservative than the concept of "effective equivalent dose" used by the International Commission for Radiation Protection.[10,11] The ICRP approach weights organ doses with factors that reflect expected *cancer fatalities* and, accordingly, gives thyroid doses a nearly negligible weight of 0.03.

A specific case of caution, which has been questioned, concerns the recommendation that pregnant women avoid eating local fish, which in July reached peak cesium-137 and cesium-134 values of 600 Bq/kg. Even though this is relatively low compared to the annual intake limit of 21,000 Bq, we justified our caution because fish in some cases forms a large part of the diet, and because the further increase in cesium-137 and cesium-134 levels could not be ruled out. As it turned out, fish contamination in Lake Konstanz stabilized later in the summer at cesium-137 and cesium-134 levels of 300-600 Bq/kg.

The discovery in late May that the fallout contained particulates with high specific beta activity[9,12] suggests that we may not have exercised sufficient caution in relation to inhalation dose. For example, it may have been advisable to warn more extensively against open air activities, and it may have been helpful to institute more intensive street washing efforts.

In the final analysis it is important to understand that our information campaigns provided the means for autonomous decision making by citizens. In particular, it permitted individuals to avoid particular foodstuffs that were found to have high specific activity levels. This, in fact, created a powerful market mechanism for ensuring radiation protection, albeit with some considerable losses for affected merchants and producers.

THE REACTION OF THE STATE AND FEDERAL GOVERNMENTS

The actions of the state and federal governments form an interesting contrast to the local risk management we have just described. As seen in Table 2, the Bonn government established import limitations, as well as protective action guidelines (grenzwerte) for milk and leafy vegetables. Early in the fallout episode it also announced that there is no reason to limit outdoor activities. Import limitations went into effect on the day after the cloud passed, and protective action guidelines for milk and leafy vegetables came on the third and fifth day after the cloud, respectively.

The state government in Stuttgart went considerably farther than standard setting and, in fact, converted the federal guidelines into several concrete actions. On the day the cloud passed it warned against iodine tablet use. On the third day it warned against open pasture feeding of milk cattle and recommended washing leafy vegetables to prevent fallout ingestion. On the fourth day it warned against direct sale of fresh milk, and on the fifth day, in recognition that washing was futile, it banned the sale of locally grown leafy vegetables altogether.

Compared to the federal government's actions, local efforts in Konstanz were clearly more cautious. While the Bonn announced that there was not reason to limit outdoor activities, local citizenry in Konstanz heard warnings to limit the outdoor play of children. These warnings were based on local radiation levels of which the federal government was apparently unaware when it issued its statements.

Compared to state actions, local risk management in Konstanz moved more rapidly and effectively. Local warnings against consumption of leafy vegetables were published on the third day after the cloud, so that when the state banned sales on the fifth day, the market had already collapsed. Local warnings about vegetables recognized the ineffectiveness of washing and avoiding the confusion created by the state's washing directive. Finally, local efforts established milk control at one fifth of the federal standard suggesting that the federal standard was inconsistent with the ALARA principle.

The problem of standard setting is a general and difficult one. In keeping with the ALARA principle of West German radiation law, a good approach might be to set local acceptance levels that are *as low as reasonably achievable* in the local context. The function of state and federal standards would then be to serve as some sort of upper limit. Apparently following this logic, several states adopted lower iodine-131 levels for milk than the Bonn government. The most stringent levels appeared to have been set in states in

Table 2. Chronology of the First Ten Days

Date	International	Federal	State	Local
4/26	Explosion at Chernobyl, begin emission			
4/27	Evacuation at reactor			
4/28	Detection of radiation in Sweden, USSR announces accident	Activation of monitoring		
4/29	Emission continues	Issuance of public statement	Activation of monitoring	Article in paper on accident
4/30	Radioactive cloud crosses West Germany	Formation of working units	Formation of working units	Activation of monitoring, rain-water measured
5/1	Continued emissions in Chernobyl	Import quotas & border control established	Press release, establish re-search program	Monitoring of ground contamination
5/2	No further fallout in South Germany	Radiation protec-tion guideline for iodine-131 set at 500 Bq/1 in milk	Recommend no open grazing, wash vegetables	Publication of monitoring results in press warn against leafy vegetables, open forum at University
5/3	continued emissions		Warn against direct milk sale, advise dairies hold to federal iodine standard	Press report on leafy vegetable contamination
5/4	continued emissions	Guideline for iodine-131 for vegetables set at 250 Bq/kg	Prohibit sale of leafy vegetables	Begin control of milk at local dairy
5/5	Emission at Chernobyl terminate		Universities asked to make measurements	Monitoring vege-tables, state measuring begins

which the opposition to nuclear power was most strongly established, as for example in Hessen.

Individuals also set their own standards. There is good evidence that many people considered the federal standard for milk too high with the consequence that milk sales dropped precipitously during the crisis. This raises the interesting question whether market losses, in fact, would have been smaller with a more stringent federal standard.

In the case of cesium-137 Bonn set no standard. As a result, a "black market" of unofficial guidelines developed which served to orient the public. For example, we find from discussions with participants in local risk management that when milk cesium-137 levels reached 30 Bq/1, people become reluctant to use it for small children. Although this level cannot be justified in terms of West German radiation protection principles (which set an annual intake limit of 27,000 Bq cesium-137), it nevertheless plays an important public policy role. For example, the state government of Baden-Württemberg is currently working on plans to keep cesium-137 levels in milk below 30 Bq/1 by controlling the use of cesium-137 contaminated hay in the winter season of 1986-87. It is also interesting that, in the absence of federal and state standards, public perception arrives at different standards for different food stuffs. Thus, the acceptance level for cesium-137 in berries appears to be as high as 200 Bq/kg, whereas fish and wildlife did not become unmarketable until levels reached as high as 600-1000 Bq/kg. This suggests that an "informed market place" is able to correct for low intake of some foods, like meat and berries, by adopting higher acceptance levels.

Perhaps the largest point of contrast between federal and state actions and the local work in Konstanz was the unavailability of direct measurements in early state and national pronouncements. Instead, central authorities concentrated on general statements which had the clear intent of minimizing the danger of the crisis. It is an open question whether this approach was, in fact, counterproductive, leading directly to the loss of credibility that occurred—particularly among individuals who have the ability and tools to form their own judgment.

In any case, we attribute the high credibility of local risk management in Konstanz to early and detailed publication of data. We also note with interest that, in contrast to the early days of the crisis, the state government in Stuttgart recently has been considerably more forthcoming with specific data, presumably in recognition of the fact that this enhances its credibility with the public.

LESSONS FOR THE FUTURE

Chernobyl was an accident with a broad transnational impact. How should society respond to similar accidents in the future, whether they arise in the nuclear or chemical industry, or at military installations?

An appropriate reaction demands, first of all, immediate and complete exchange of information, coupled with aid to the country in which the accident occurred. These responses must be assured through a worldwide international agreement and are currently under discussion at appropriate international organizations such as the International Atomic Energy Agency and the World Health Organization.

Beyond this, it seems clear that individual nations must improve their own response. Many countries have plans for responding to technological disasters that occur within their borders, and this includes the Federal Republic of Germany. But few, if any, countries are prepared to respond to disasters outside their own borders. Nor are there plans for dealing with non-acute exposures that result from accidents that have occurred at great distances.

The unquestionable lesson of Chernobyl is that large-scale disasters that convey quantities of harmful particulates through the atmosphere can produce significant and highly variable impacts at large distances from the source. The variability of these impacts is further a strong function of local weather and terrain. From the very structure of such impacts it is plausible, if not obvious, that local assessment and response should be a key ingredient for managing such a disaster.

The first institutional response recorded in Bonn was the establishment of a ministry for "Environment, Nature Conservancy, and Reactor Safety." The next step, taken in October 1986, was legislation to provide radiation protection for the public.[13] The new law foresees *centralized* collection, analysis and evaluation of data, and envisions that protective actions will be prescribed by Bonn. The tasks and duties of local governments are not explicitly described, but it appears that they are not expected to conduct their own analysis and evaluation of data. We believe that Bonn's centralized initiatives are in conflict both with the structure of large atmospherically coupled disasters. They also appear to ignore the positive experience of local risk management, such as we have described for Konstanz.

Under centralized control, the clear danger exists that effective decision making and action will be significantly delayed because of the complexity of the national picture, and the high variability of local conditions. This will increase fear and confusion among the general public and will assure that handling of local problems will not occur in the most timely and focused manner. Consider an analogy. If Konstanz were threatened by flooding because of excessively rapid runoff from the Alps, should the local emergency brigade stand by until the central office of flood control has finished calculating the magnitude of the expected flood? Or should it go ahead and pile up the sandbags?

Based on our experiences with Chernobyl, we emphatically advise against a centralized emergency management system, or even one that is dominated by centralized authority. Instead we recommend the formation of local commissions for handling radiological protection that is not now specifically included in emergency plans around existing West German reactor sites. Such local commissions should handle most of the issues confronted in the recent Chernobyl episode in Konstanz. In particular, it should be the responsibility of local commissions to report in detail to the public. It would also be their job to provide local data to the state and federal governments. The setting of general guidelines and the handling of purely state and national issues, such as import limitations, would remain, as they are now, the responsibility of centralized authority.

The advantages of local commissions would lie in the ability to respond rapidly and flexibly to exposure patterns with high spatial and temporal variability. An additional advantage is the potential for providing rapid and accurate communication to locally affected populations. This will forestall the loss of credibility that so often accompanies the waffling and delays of centralized organizations that attempt to deal with too much complexity in a general way.

In establishing local commissions, consideration should be given to the notion that universities and other organized centers of learning may provide one of the best places for conducting unbiased measurements and analyzing risk management options. Thought should also be given to how large local management regions should be. It is our guess that the best size is not much larger than a few counties (Landkreise). Whatever is done, it is clear that equipping local monitoring and measuring groups may require significant additional investment in equipment and personnel, especially in those cases where local educational institutions can not provide appropriate facilities.

An important condition for the success of local commissions is a framework for establishing protective action guidelines within each of several exposure regimes. The framework should rest on the federal radiation protection law and the annual intake limits it sets forth. In the case of unavoidably large exposures in which annual limits must, out of necessity, be surpassed, prior planning should identify a series of emergency exposure regimes, within each of which the ALARA principle of the radiation protection law could be applied. For each exposure regime it would then be useful to develop catalogs of possible countermeasures appropriate for the local units. These catalogs should be sensitive to the season, local weather, and other particulars of local geography.

SUMMARY AND CONCLUSIONS

We have described the release and dispersal of radioactivity from the largest radiological accident ever recorded. We have noted that it led to a highly variable exposure at large distances from the accident. Based on our experience with locally organized radiation protection, we believe future planning for managing the hazards of large radiological accidents should involve a strong decentralized component with sufficient autonomy to take local measurement, inform the local population, and design effective local countermeasures.

NOTE ADDED MAY 6, 1989

This article reflects the experiences with improvised local risk management at Konstanz within the very first months after the Chernobyl nuclear reactor accident. Three years later, the knowledge about radionuclide release and dispersal has considerably gained in profoundness, and detailed reports are published with respect to radiation protective action on the local, state, and federal level in West Germany as well as in other European countries (see e.g., C. Hohenemser, The Accident at Chernobyl: Health and Environmental Consequences and the Implications for Risk Management, *Annual Review of Energy* **13**:383 (1988); G. Lindner and E. Recknagel, Tschernobyl — Auswirkungen auf die Bodensee-Region, Thorbecke, Sigmaringen, 1988; Nuclear Energy Agency Group of Experts, Chernobyl and the Safety of Nuclear Reactors in OECD Countries, OECD, Paris, 1987; Nuclear Energy Agency, The Radiological Impact of the Chernobyl Accident in OECD Countries, OECD, Paris, 1987; Ministerium für Ernährung, Landwirtschaft, Umwelt und Forsten des Landes Baden-Württemberg, Dokumentation der Landesregierung Baden-Württemberg über die Auswirkungen und Maßnahmen zum Kernkraftunfall in Tschernobyl, Stuttgart, 1987; Bundesminister für Umwelt, Naturschutz und Reaktorsicherheit, Auswirkungen des Reaktorunfalls in Tschernobyl auf die Bundesrepublik Deutschland, Gustav Fischer Verlag, Stuttgart, New York, 1987).

Recently, a map became available showing the dispersal of long lived cesium-137 radioactivity within the Soviet Union due to the Chernobyl accident (J. Israel, Chernobyl: Past and Future, Prawda No. 79, March 20, 1989), which exhibits strong variations on a local scale in the European part of the Soviet Union and, in our opinion, corroborates the need of local components of risk management as put forward in our argumentation.

It should also be noted that the guidelines of the West German "Strahlenschutzverordnung" adopted in our early radiation protection efforts were derived from accumulated life span doses and are thus more appropriate to chronical exposures than to single events. On the other hand, at the time of the accident no other national guidelines based on well-documented scientific foundations were available. The federal contamination guidelines for milk and vegetables introduced by ad-hoc argumentations some days after passage of the radioactive cloud unfortunately were not set in relation to those existing guidelines. Recently, the new West German "Strahlenschutzvorsorgegesetz" dealing with response to accidental radionuclide contaminations was complemented by contamination guidelines established by the European Community.

In the state Baden-Württemberg problems with hay and other stored cattle feed contaminated by cesium radionuclides could be minimized by a prevention program of the state government including (1) the determination of representative transfer factors into milk by feeding experiments, (2) the release of detailed feeding recommendations and advices to farmers, which were accommodated to regional variations in agricultural practices, (3) the establishment of a milk monitoring program covering all dairies in the state throughout winter 1986/87, and (4) arrangements with dairies about the handling of milk with unacceptable contamination. Our group participated both in planning and realization of this program. In spite of peak cesium-137 contaminations of up to 4000 Bq/kg in hay, the cesium-137 level in milk available to the public from dairies never exceeded 30 Bq/l throughout the winter 1986/87.

In the following major long-term problems from Chernobyl contamination remained for deer from some forests of the southeastern part of the state with the highest fallout levels, fishes from some small eutrophic lakes from the same region (but not from the large Lake Constance, where cesium-137 fish contamination has dropped below 10 Bq/kg until 1988) and with some kinds of mushrooms.

REFERENCES

1. USSR State Committee on the Utilization of Atomic Energy, The Accident at the Chernobyl Nuclear Power Plant and its Consequences, information compiled for the IAEA experts' meeting, Vienna, August 25-29, 1986 (draft), and International Atomic Energy Agency, Summary Report of the Post Accident Review Meeting on the Chernobyl Accident, Report by the International Nuclear Safety Advisory Group (INSAG), Safety Series No. 75-INSAG-1, IAEA, Vienna, 1986.
2. M. Jensen and J. C. Linhe, Activities of the Swedish Authorities Following the Fallout of the Soviet Chernobyl Reactor Accident, Swedish National Institute for Radiation Protection, Box 60204, S-10401 Stockholm, Sweden, May 14, 1986.
3. J. Bonhof, RIVM Rapport on Geval Tsjernobyl, Staatstoezicht op de Volkszonheid, Zerikelaan 25, 9747 AA Groningen, Netherlands, May 16, 1986.
4. F. von Hippel and T. B. Cochran, Chernobyl—Early Thoughts about the Long-Term Health Consequences, *Bull. At. Scientists* (September 1986).
5. H. W. Lewis *et al.*, Report to the American Physical Society by the Study Group of Light Water Reactor Safety, *Rev. Mod. Phys.* **47**:S1-S123, Suppl. 1 (1975).
6. C. Hohenemser, M. Deicher, A. Ernst, H. Hofsaess, G. Lindner, and E. Recknagel, "Chernobyl: An Early Report," *Environment* **5**:6 (1986).
7. Gesellschaft fuer Strahlen-und Umweltforschung, Umweltradioaktivitaet und Strahlenexposition in Suedbayern durch den Tschernobyl-Unfall, GSF-Bericht 16/1986, Gesellschaft fuer Strahlen-und Umweltforschung GmbH Muenchen, Ingolstaedter Landstrasse 1, D-8042 Neuherberg, West Germany.
8. Strahlenschutzverordnung, Verordnung ueber den Schutz vor Schaeden durch ionisierende Strahlen, Bundesgesetzblatt no. 125, p. 2905, October 13, 1976.
9. C. Hohenemser, M. Deicher, H. Hofsaess, G. Lindner, E. Recknagel, and J. I. Budnick, Agricultural Impact of Chernobyl: A Warning, *Nature* **321**:817 (1986).
10. ICRP, International Commission on Radiological Protection, Report 38, Pergamon Press, Oxford (1979/80).
11. D. A. Baker, G. R. Hoenes, and K. J. Soldat, Food—An Interactive Code to Calculate Internal Radiation Doses From Contaminated Food Products, BNLWL-Sa-5523 (1976).
12. L. Devell, H. Tovedal, U. Bergstroem, A. Appelgren, J. Chryssler, and L. Andersson, Initial Observations of Fallout from the Reactor Accident at Chernobyl, *Nature* **321**:192-193 (1986).
13. Government of W. Germany, Entwurf eines Gesetzes zum vorsorgenden Schutz der Bevoelkerung gegen Strahlenbelastung, Drucksache 428/86, 5300 Bonn 1, F.R.G., 1986, and Strahlenschutzvorsorgegesetz vom, December 19, 1986, Bundesgesehblatt, p. 2610 (1986).

The Comparative Ranking of Risks

Michael C. Cheok, Geoffrey D. Kaiser,[a] and Gareth W. Parry
NUS Corporation
Gaithersburg, MD

ABSTRACT

This paper summarizes work done for the Federal Emergency Management Agency to investigate the feasibility of a quantitative risk ranking scheme to be applied at the county level for risks from a range of natural and technological hazards. In particular it addresses some of the issues that arise in attempting to construct such a scheme.

KEYWORDS: Natural and technological hazards, risk ranking, complementary cumulative distribution functions (CCDFs)

INTRODUCTION

If it were possible to derive a quantitative ranking scheme whereby the risks from a range of natural and technological hazards could be ranked on a truly comparable basis, such a scheme would be extremely useful as an aid to policy and decision making. For example, it would greatly help those agencies that are responsible for emergency planning and response to make rational decisions about the allocation of scarce resources between different hazards. The purpose of this paper is to discuss some of the difficulties that arise in attempting to construct such a scheme. In particular, the discussion focuses on how to obtain the basic information on frequency and the associated range of consequences that is required for any quantitative scheme for ranking risks.

This report is based upon work that was carried out on behalf of the Federal Emergency Management Agency (Cheok, Kaiser and Parry, 1985). The list of hazards put forward by FEMA for consideration in this study was as follows:

- Avalanche
- Civil Disorder
- Dam Failure
- Drought
- Earthquake
- Fixed Nuclear Facility
- Flood
- Hazardous Materials Incident (fixed site, transport)
- Hurricane, Tropical Storm
- Landslide
- Nuclear/Conventional Attack
- Power Failure
- Resource Shortage (fuel, water etc.)
- Severe Winter Storm
- Tornado
- Transportation Accident (air, rail, marine)
- Tsunami
- Urban Fire
- Volcano
- Wild Fire

a. Present address: Science Applications International Corp., McLean, VA.

The subject of risk ranking has received a great deal of attention in recent years, and there are a number of schemes of varying degrees of subjectivity or objectivity. Examples include the straightforward comparison of complementary cumulative distribution functions (CCDFs — see below for further discussion) of fatalities or property damage such as appear in the pioneering Reactor Safety Study (USNRC, 1975); use of the areas under CCDFs, which are mathematically approximately equal to the average number of consequences per year; manipulation of the CCDFs to take account of people's perception of risks, such as by weighting high consequence events by the magnitude of the consequence before integrating (Wilson, 1975) or by applying subjective risk conversion factors that preferentially weight new technologies, hazards with catastrophic consequences and so forth (Rowe, 1977; Litai, Lanning and Rasmussen, 1983); and attempts to describe hazards by a number of descriptors that are related to such characteristics as annual human mortality, maximum human mortality in a single event, maximum spatial extent over which a single hazard has a significant impact, persistence in the environment, etc. These descriptors are assigned numerical values by a subjective process of discussion between experts (Hohensemer *et al.*, 1983). A good discussion of recent advances in the quantitative ranking of risks is to be found in the proceedings of a recent conference, "The Analysis of Actual versus Perceived Risk" (Covello *et al.*, 1983).

The risk ranking schemes briefly outlined above, and others to be found in the literature, are still controversial. The purpose of this report is not to try to narrow the range of controversy, but rather to address a question that is fundamental to the use of all risk ranking schemes, namely, is it possible to obtain the basic probabilistic information that is implicitly or explicitly used in any risk ranking scheme? This information consists of the range of consequences arising from the occurrence of a hazardous event, together with the associated range of frequencies of occurrence. It is convenient to express this probabilistic information concisely as a CCDF.

Figure 1 is an example of CCDF of the number of fatalities arising from the occurrence of tornadoes within the whole of the USA, based on historical databases that include records for about a hundred years. The best data in this case come from the National Severe Storms Forecasting Center. What the data do not do, however, is to allow any significant extrapolation beyond the recorded tornado of greatest consequence. In order to carry out this extrapolation, modeling is needed. It is when modeling becomes necessary that the expense of generating CCDFs begins to escalate.

In general, reliable US databases on various hazards, technological and man-made, do not exist for more than about a hundred years. Therefore, modeling extrapolation is necessary for events which occur with a frequency that is less than about one in a hundred per year. Statistical extrapolation of the data is generally not feasible because the shape of the probability density function at the high consequence end of the CCDF is not known until the modeling has been done.

The statistics given in the actuarial database upon which Fig. 1 is derived are represented in Table 1. It is apparent that, for most states, there is not enough data to support the development of an adequate CCDF. Overall, reducing the average size of the area for which CCDFs are being generated reduces the statistical soundness of the use of historical data. More modeling is required, and therefore more expense. Reducing the data base to the county level would mean that the 193 events recorded on Table 1 would have to be shared between 3100 counties, so that most counties would have no data at all. This is true for tornadoes, and also true for most other natural and technological hazards. Thus, we have established a point which may seem obvious but which has important implications; a risk ranking scheme at the county level cannot be established from historical data; modeling is required. These thoughts are summarized in Fig. 2.

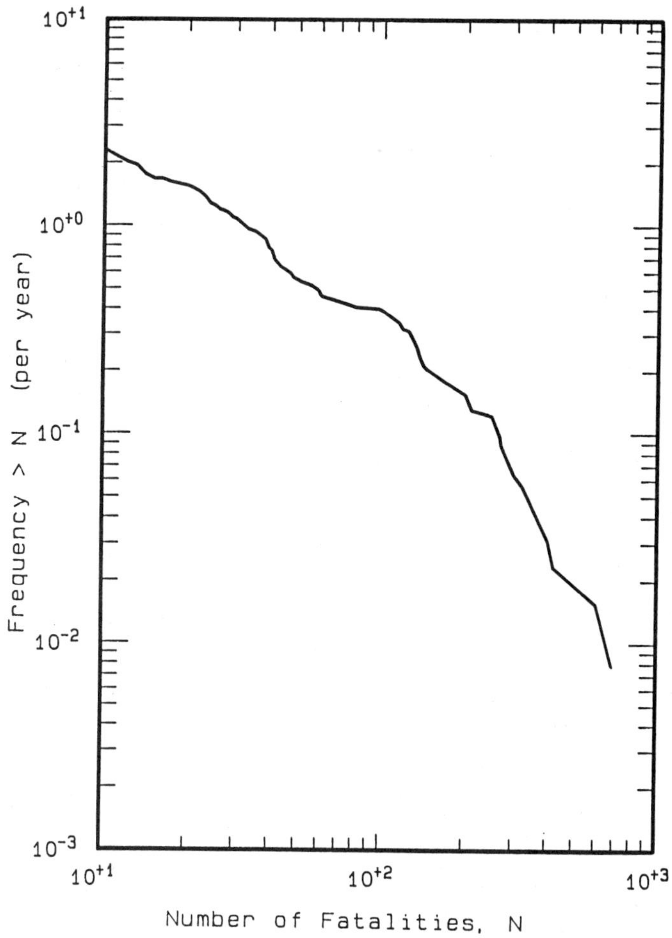

Fig. 1. Tornadoes — CCDF for fatalities.

SOME THOUGHTS ON MODELING

General Modeling Considerations

Many of the hazards can occur with a spectrum of severity. The severity can be characterized by one or more parameters which represent measurable quantities that are strictly associated with the physical manifestation of the event itself, but not with those factors such as population, type and numbers of buildings, etc., that are associated with the vulnerability of the community. Examples of such severity or intensity parameters are the peak ground acceleration of an earthquake, the height of a flood, the highest velocity wind speed, and the size and duration of release of a toxic chemical. In general the higher the severity, the lower the frequency of occurrence, so for a hazard whose severity can be characterized by a single parameter, a hazard curve (shown schematically in Fig. 3) can be established. The hazard curve gives the frequency $P(I)$ with which a severity rating I is equaled or exceeded.

Table 1. Number of Recorded Tornadoes Leading to Fatalities by State 1814-1981

STATE	NUMBER*
Arkansas	17
Texas	15
Michigan	14
Oklahoma	14
Missouri	13
Mississippi	13
Alabama	11
Illinois	11
Louisiana	11
Georgia	9
Tennessee	9
Kentucky	8
Ohio	7
Kansas	6
Iowa	5
Indiana	5
Minnesota	4
Wisconsin	4
Florida	2
Maryland	2
South Carolina	2
Virginia	2
Colorado	1
District of Columbia	1
Idaho	1
Massachusetts	1
Nebraska	1
North Dakota	1
Pennsylvania	1

*Some of the incidents recorded in the data base are assigned to more than one state.

Given the occurrence of an event, the amount of damage caused by that event is an increasing function of the severity of the event. This is displayed schematically in Fig. 4. This function $D(I)$ gives the expected value of the damage (in dollars or lives) given the occurrence of an event of severity rating I. This damage-intensity curve $D(I)$ is a function of the various factors which govern the vulnerability of the community, such as population density, strength of building, location relative to the event, etc. At any given intensity, for a structure of a given type, there may be a probabilistic spectrum of damage. However, in order to keep the discussion simple, a one-to-one relationship between intensity and damage has been assumed. In Fig. 4, $D(I)$ can be regarded as the mean or expected value of any probabilistic spectrum.

Both functions $P(I)$ and $D(I)$ are location dependent; earthquakes, for instance, are more likely in California than in the rest of the contiguous U.S.A., but buildings in California constructed since 1933 are built to a higher seismic standard than those elsewhere (Wiggins *et al.*, 1978). A detailed derivation of the functions $P(I)$ and $D(I)$ is a relatively complicated and time consuming exercise.

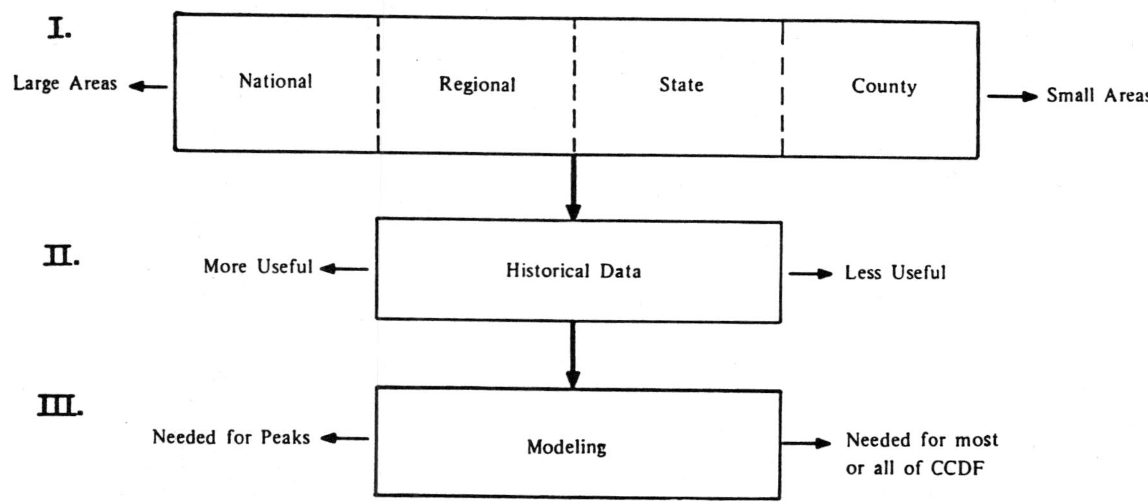

Fig. 2. Relationship between historical record, modeling, size of area and frequency of occurrence as inputs to the compilation of CCDFs.

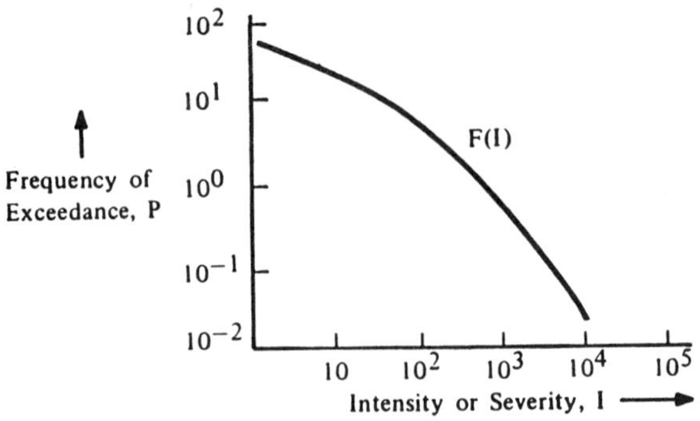

Fig. 3. Schematic of typical hazard cause.

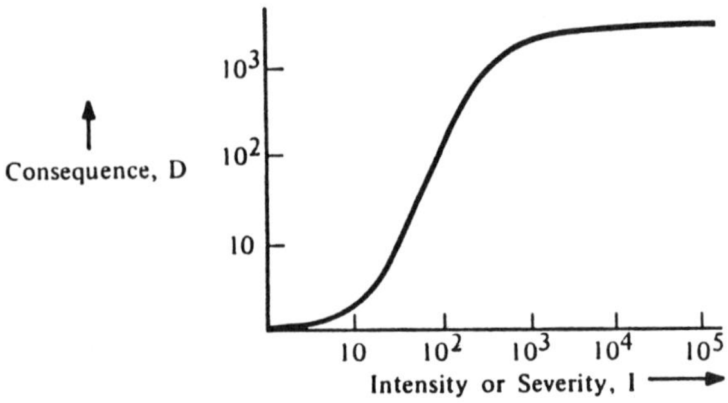

Fig. 4. Schematic of typical consequence—intensity curve.

A CCDF in which the relationship between frequency F and consequence C is given by a function $F(C)$ can be constructed from the functions $P(I)$ and $D(I)$ in the following way.

Let $P(I;R_i)$ be the location dependent hazard and damage intensity functions defined for a region R_i. For some hazards (e.g., chemical plant) the R_i can correspond to a point source. Then the CCDF, $F(C)$ giving the frequency of exceedance of consequence for a large region consisting of all the R_i in a given set, $i = 1, \ldots, N$ is given by

$$F(C) = \sum_i \int_I \frac{dP}{dI}(I, R_i)\, \theta(C - D(I, R_i))dI$$

$$\begin{aligned} \theta(C - D(I, R_i)) &= 0 \quad \text{if} \quad C > D(I, R_i) \\ &= 1 \quad \text{if} \quad C \leq D(I, R_i) \end{aligned} \tag{1}$$

Some care must be taken when the area over which the hazard is felt becomes on the order of the area of the regions R_i. Equation 1 is intended primarily as a symbolic equation for discussion purposes and would have to be looked at more carefully in constructing CCDFs for different hazards.

Localized or Non-Localized Hazards

The ease with which the CCDF can be modeled depends on whether the hazard is localized or non-localized. For the purpose of this discussion, a localized hazard is taken to be one for when the source can occur only at a point or in isolated regions. Examples of such hazards are dam failures, releases from chemical plants, avalanches and volcanoes. A non-localized hazard is one whose source can occur with similar characteristics anywhere over a large area at random, e.g., a tornado or in some cases an earthquake.

For non-localized hazards the source of hazard may occur in any location in the area and its position is therefore randomly distributed with respect to vulnerable areas of the community. In this sense the hazard function is spatially independent over a large area and is effectively decoupled from the damage intensity curve. So for every region R_i in a larger region R, $P(I, R_i)$ is independent of R_i and Eq. 1 can be rewritten as

$$F(C) = \int_I \frac{dP(I)}{dI} \sum_i A_i \, \theta(C - D(I, R_i))dI, \tag{2}$$

where $P(I)$ is now the hazard function per unit area and A_i is the area of region R_i.

For localized hazards the sources of the hazard are fixed. Thus, in an evaluation of CCDFs for localized hazards, the functions $P(I)$ and $D(I)$ are spatially correlated and the evaluation of $F(C)$ is performed as in Eq. (1).

Note that in both Eqs. (1) and (2) there is a sum over regions R_i; that is, for both localized and non-localized hazards the CCDFs in a given area may be regarded as being made up of the sum of CCDFs generated in smaller regions. This is an extremely important observation. It implies that CCDFs are generally constructed by building up from the bottom. The relevance to the present discussion is that it does not seem fruitful to try to derive CCDFs at the state level as an intermediate, cheaper step on the way to the county level CCDFs. On the contrary, many CCDFs are built up through the county to the state.

Scaling of CCDFs from National to Local Level

Modeling of the hazards, both their frequency of occurrence and their effects, can be relatively complicated and expensive, so it is of value to investigate whether some simple scaling of the national CCDFs can provide an approximate CCDF for some smaller area, such as a state or county. Here scaling means the application of simple multiplicative factors to the frequency and/or magnitude axis of the national CCDF.

Looking at Eqs. (1) and (2), it is possible to make some general statements about when scaling of national CCDFs is possible.

For non-localized hazards, the function $dP(I)/dI$ in Eq. (2) does not depend on the region R_i. Therefore, whether scaling is possible depends on the $D(I,R_i)$. For example, if the $D(I,R_i)$ do not depend on R_i — which is to say that quantities such as population density, mix of structure types and so forth do not vary between regions R_i — then $F_i(C)$ is simply proportional to A_i where $F_i(C)$ describes the CCDF in region R_i. If $D(I, R_j)$ in region R_j differs from $D(I, R_i)$ in region R_i by a fixed factor f (e.g. population density doubled, f=2), $D(I, R_i) = fD(I, R_j)$, then $F_i(fC) = F_j(C)$ and scaling between the regions R_i can be accomplished by multiplication factors on the consequences at fixed frequency.

If the differences in the $D(I, R_i)$ are more complex than simple factors, scaling between the $F_i(C)$'s is not possible. Similarly, a simple scaling relationship between the overall $F(C)$ and the individual $F_i(C)$'s is not generally to be expected. Even if the $D(I,R_i)$s differ only by constant factors, there is no simple relationship between $F(C)$ and the individual $F_i(C)$'s.

Thus, in general, simple scaling laws between CCDFs in large and small regions are not to be expected. This is not to rule out the possibility that reasonable approximations can be made which would allow such scaling. In general, however, the key to simple calculation of CCDFs in a region made up of a large number of subregions R_i is to try to combine the R_i into groups in which $D(I, R_i)$ are approximately the same. The number of groups can be reduced at the expense of accuracy, or, to put it another way, the required accuracy may dictate the number of groups.

For local hazards, the scaling must be demonstrated by considering both the functions $P(I, R_i)$ and $D(I, R_i)$ in conjunction. It is harder than for the non-local case to envisage cases where scaling from large areas is possible. The key to simple calculation of CCDFs is to combine the regions R_i into as few groups as possible in which both $P(I, R_i)$ and $D(I, R_i)$ are approximately the same.

The overall conclusion of this section is that, unless large areas exhibit a great deal of homogeneity in both $P(I, R_i)$ and $D(I, R_i)$, it is unlikely that simple multiplicative scaling factors will be found for converting CCDFs in large areas to CCDFs in small areas. This confirms the comment made previously, that it does not make sense to try to develop CCDFs for states as an intermediate stopping point on the way to compiling CCDFs at the county level.

CONCLUSION

The arguments of the foregoing sections lead to the conclusion that, in order to calculate risk at the county level, it is necessary to construct a model to estimate the occurrence frequencies and effects of the hazards. To make this process tractable and affordable some fairly broad generalizations and approximations need to be made. Such modeling has been performed by Wiggins *et al.* (1978) for nine natural hazards. The method is in principle the same for all hazards. Figure 5 shows the general features of the analysis for the evaluation of risk, with the measure of consequences being expected dollar loss due to property damage.

A hazard analysis results in a function which provides estimates of frequency of exceedance as a function of intensity. It is necessary also to have a detailed catalogue of structures. The structures are catalogued in various structural types, the different types being chosen to cover the spectrum of response to the hazard. For each structure type a fragility analysis is performed which evaluates the probable percentage damage as a function of intensity. When coupled with a function giving dollar loss versus probable percentage damage, a consequence curve for each structure type giving dollar loss as a function of intensity with the consequence curves for the different structural types produces a CCDF. It should be noted that Wiggins *et al.* did not generate CCDFs but generated the expected loss which is equivalent to the area under the CCDFs. The Wiggins study, and indeed any study of this type, relies heavily on engineering judgment to establish all elements of the analysis. The procedures are essentially the same for both localized and non-localized hazards in that the evaluation is performed by summing up contributions from discrete areas. However for non-localized hazards it may be possible to define larger areas over which hazard characteristics can be regarded as constant when compared with localized hazards.

The evaluation of CCDFs for certain localized hazards, for example, hazards arising from accidental releases from chemical plants, could be prohibitively expensive if each

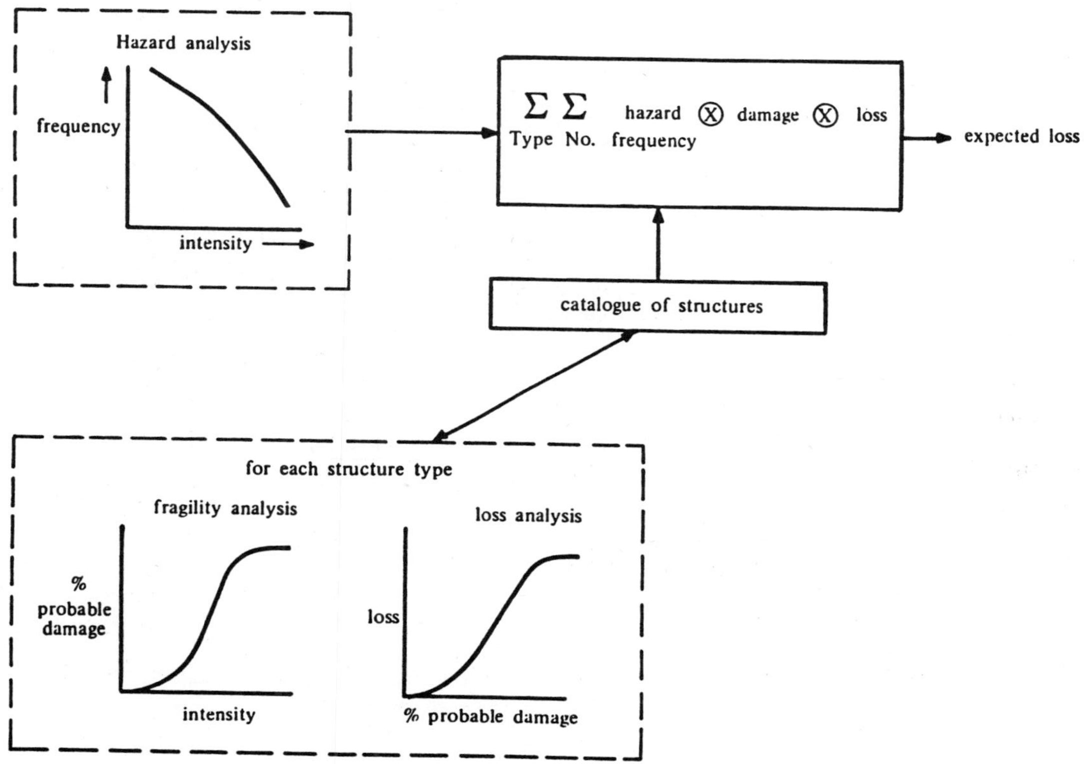

Fig. 5. Simplified schematic for risk analysis.

Table 2. Summary of the Feasibility of Calculating Risks for Hazards at the County Level

Feasible		Maybe (Development Work Needed)	Not Feasible
Localized	Non-Localized		
Dam Failure	Earthquakes	Drought (data based on the Palmer Drought Index could be used as a starting point)	Avalanche (practical model not available)
Hazardous Materials - Fixed Site	Floods (some development work needed to obtain CCDFs at county level)	Nuclear Attack (possibility that in-house FEMA methods for property damage could be adapted)	Civil Disorder (can only be dealt with qualitatively)
Fixed Nuclear Facility	Transportation of hazardous materials	Power Failure	Conventional Attack (eliminated on the grounds of low frequency)
Tornadoes	Hurricanes	Resource Shortage (possibility that FEMA adapt in-house methods)	Marine Disasters
Aviation Disasters	Landslides (but some development work needed)	Severe Winter Storm (further work needed to assess feasibility)	Volcanoes (practical model not available)
Tsunami	Railroad disaster		
	Urban fires		
	Wildfire		

plant were considered separately and some grouping were not done. An example of how such a grouping analysis might be carried out is given below.

1. Develop a grouping scheme, e.g., for chemical plants, in which each of the plants in the county or other areas are assigned to one of as few groups as possible.

2. Carry out one representative calculation for each group giving CCDFs for fatalities and economic costs with some assumed population density and property values.

3. Develop a simple scaling scheme for other plants based on population density and property values.

4. Sum over all plants in a given area.

This scheme really follows the suggestion made above that the only way to reduce the problem of calculating CCDFs for local hazards to tractable proportions is to make approximations in defining broad groups in which the $P(I, R_i)$ of Eq. (1) are set equal. Thus Step 1 above is equivalent to defining a limited number of $P(I)$'s; Step 2 is equivalent to defining one $D(I, R_i)$; Step 3 is equivalent to defining simple multiplicative relationship between $D(I, R_i)$'s so that scaling between $F_i(C)$'s is possible; and Step 4 is a simple sum. The procedure for non-localized hazards is not very different except that, by their very nature, there is no need for Step 1.

In the analysis that was performed in support of the work reported in this paper, the feasibility of the economical construction of CCDFs for each of the hazards listed in the introduction was considered. The results of the analysis are summarized in Table 2. As can be seen, it was judged that the development of county-level CCDFs is feasible for most of the hazards given existing techniques or modest adaptions thereof. It was thought that, with some development work, drought, nuclear attack, power failure, resource shortage and severe winter storms could be included. It was judged not to be feasible for only four hazards: avalanches, civil disorders, marine disasters and volcanoes.

For those models in the "feasible" column on Table 2, the simplified models for calculating CCDFs could be developed with modest effort—a few man years. The biggest expense would be in the collection of input on a county by county basis—transportation networks and what is carried on them, population distributions, characteristics of structures, flood plains, chemical plants and so forth. As a simple example, it would be extremely economical if the data could be collected, processed through the models and presented in a simple report format for an average of $10,000 per county. There are approximately 3,000 counties in the USA, leading to an estimated cost for obtaining local risk ranking data across the USA of $30,000,000.

REFERENCES

Cheok, M. C., Kaiser, G. D., and Parry, G. W., 1985, Development of a Methodology for Comprehensive Hazard Analysis — A Feasibility Study, Prepared for FEMA, NUS Report 4721.

Covello, V. T., Flann, W. G., Rodricks, J. V., and Tardiff, R. G., 1983, The Analysis of Actual Versus Perceived Risks, in *Advances in Risk Analysis*, Plenum Press, New York.

Hohenemser, C., *et al.*, 1983, Methods for Analyzing and Comparing Technology Hazards: Definitions and Factor Structures, A Report of the Clark University Hazard Assessment Group, Center for Technology, Environment and Development (CENTED), Clark University, Worcester, MA.

Litai, D., Lanning, D. D., and Rasmussen, N. C., 1983, The Public Perception of Risk, in *The Analysis of Actual Versus Perceived Risks*, Covello *et al.*, eds., pp. 213-224, Plenum Press, New York.

Rowe, W. D., 1977, *An Anatomy of Risk*, Wiley, New York.

USNRC (United States Nuclear Regulatory Commission), 1975, Reactor Safety Study, WASH-1400 (NUREG-74/014), Washington, D.C.

Wiggins, J. H., Slosson, J. E., and Krohn, J., 1978, Natural Hazards: Earthquake, Landslide, Expansive Soil Loss Models, Prepared for the National Science Foundation (NTIS No. PB 294 686/AS) by J. H. Wiggings Company, Redondo Beach, California.

Wilson, R., 1975, The Costs of Safety, *New Scientist* **68**:274-275.

Getting to Maybe: Implied Principles of Trust and Consent in Two Massachusetts Acts[a]

Mary R. English
University of Tennessee
Knoxville, TN

ABSTRACT

This paper compares two pieces of Massachusetts legislation — the Hazardous Waste Facility Siting Act passed in 1980 and the Low-Level Radioactive Waste Management Act submitted to the Massachusetts legislature in 1986. In particular, the paper looks at how these two acts differ radically in the means by which each implicitly seeks to develop trust in its risk management process and to gain consent to the process's outcome. As a prelude to this analysis, key points in the two acts are laid out. For the analysis itself, extensive use is made of a theoretical framework of culture and risk management presented by Steve Rayner at the 1984 meeting of the Society for Risk Analysis.

KEYWORDS: Risk management, government, trust, consent

INTRODUCTION

The 1980 Massachusetts Hazardous Waste Facility Siting Act (Massachusetts General Laws, Chapter 21D) was a bold departure from past approaches to siting locally unwanted but publicly necessary facilities. It was heralded as a deft combination of preemption and negotiation and was adopted as a model by several states. But less than five years after it was enacted, it was being cited as example of what *not* to do (see e.g., Kasperson, 1986).

Particularly interesting is the reaction of a Massachusetts legislative commission that had been established to develop an approach to the management of low-level radioactive wastes — wastes that, like toxic chemical wastes, provoke intense "not in my backyard" sentiments. This commission commented that "though it is inherently difficult to site any waste facility, the commission believes that serious flaws in both the language of Chapter 21D and its implementation have contributed to this state's inability to make any significant progress toward siting a hazardous waste facility" (Massachusetts Special Commission on Low-Level Radioactive Waste, 1986, p. 5).

---

a. A revised and expanded version of this paper has been published. See Mary R. English, "The Search for Political Authority in Massachusetts' Toxic Waste Management Law," *Boston College Environmental Affairs Law Review* 16(1):39-51 (1988).

New Risks
Edited by L. A. Cox, Jr., and P. F. Ricci
Plenum Press, New York, 1990

What caused this backlash, and how has it been addressed in subsequent legislation? This paper investigates an important root cause, using as a theoretical basis a framework of culture and risk management that was presented by Steve Rayner at the 1984 meeting of the Society for Risk Analysis (Rayner, 1986).

The Hazardous Waste Siting Act

Chapter 21D was adopted after a series of attempts by developers to site disposal or recycling facilities had failed, largely because of local opposition. (For a discussion of the events and conceptual work that led to Ch. 21D, see O'Hare, Bacow, and Sanderson, 1983.) As passed, it has three key elements:

1. It establishes a state Hazardous Waste Facility Site Safety Council which is responsible for making preliminary determinations of whether proposed projects are "feasible and deserving of state assistance" (M.G.L. Ch. 21D, sec. 7). The Council works with two existing state agencies: the Department of Environmental Management (DEM), which assesses the state's need for hazardous waste facilities, and the Department of Environmental Quality Engineering (DEQE), which licenses and regulates such facilities.

2. It gives a developer the right to construct a hazardous waste facility on land zoned for industry if the developer passes the "feasible and deserving" test, obtains the required state and local permits, and negotiates an agreement with the host community. Communities may not impose new permit requirements after the act's effective date.

3. It directs the community and the developer to negotiate an agreement on the facility's design and operation and on provisions for community compensation and impact mitigation. The Council is directed to provide technical assistance grants to the community to aid its negotiations. Binding arbitration may be used to break deadlocks.

To date, five projects have been attempted under Chapter 21D, but none has even made it to the negotiating table — all have been blocked by community protests and lawsuits in the early stages of the siting process. In particular, communities have sought to get the Council's affirmative "feasible and deserving" determinations reversed. While these efforts have not always succeeded in themselves, they have succeeded in their underlying intent: faced with intense local hostility and the prospect of long delays in court, each developer has withdrawn.

The Massachusetts experience is not unique; nationally, there has been at least a 50-percent failure rate in proposals made under state hazardous waste facility siting programs (Ryan, 1984). Throughout the United States, these facilities are perceived as threats to a community's physical and economic health — perceptions that are understandable in light of recent well-publicized disasters with hazardous materials and waste (e.g., Love Canal, Times Beach) and the prevalent image of hazardous waste facilities as "dumps."

Thus, the Massachusetts approach cannot be rejected out of hand. Perhaps none other would have been successful, especially in a politically contentious climate such as Massachusetts'. But it is useful, nevertheless, to consider what hasn't worked with the Massachusetts model.

A number of problems attended its inception: the lack of precedent because of the act's innovative approach; the fact that facility applications began to come in before rules could be promulgated; and the inexperience of all parties with this kind of negotiation, which contributed to a failure to begin talking before positions had hardened. But two problems in particular continued to plague Chapter 21D: a profound objection by proposed host communities to begin identified as targets for hazardous waste facilities, seemingly without a sufficient rationale; and an even more profound objection to the Council's early determination of a proposed project's "feasible and deserving" status — again, seemingly without a sufficient rationale.

Extensive revisions to Chapter 21D are being proposed (Senate Bill 1073, 1986 legislative session) which attempt to rectify these and other problems in the following ways:

1. by adding statewide hazardous waste management planning to the Council's responsibilities;

2. by making statutorily explicit the standards that the Council is to use in judging facilities;

3. by recommending a scale for the developer's direct incentive payments to the community and by providing that host communities may also apply to the state for compensatory grants;

4. by postponing and making less seemingly definitive the Council's "feasible and deserving" determination (now referred to as a determination of whether a project is "under review");

5. by giving the proposed host community more money initially in technical assistance grants and more time to negotiate before binding arbitration is initiated; and

6. by guaranteeing the host community access to the facility for environmental monitoring purposes.

The Low-Level Radioactive Waste Management Bill

Meanwhile, on a separate front, Massachusetts has been addressing the need to take responsibility for the disposal of low-level radioactive waste (LLW) generated within its borders, as mandated by the 1980 federal Low-Level Radioactive Waste Policy Act (42 U.S.C. 2021b et seq.). In response to this legislation, Massachusetts passed an act (Ch. 738, Acts of 1981) establishing a commission to study LLW issues.

The commission became staffed and operational in 1983. Over the next two years, it analyzed Massachusetts' current and projected LLW disposal needs and studied possible legislative approaches to siting a facility within the state, and by 1985 it had prepared a draft bill for a Low-Level Radioactive Waste Management Act. After circulation for public comment and a series of public hearings, a final draft was submitted to the legislature for consideration in 1986.

The bill (Senate Bill 1673) does the following:

1. It establishes a board with central responsibility for the state's LLW management.

2. It specifies a six-phase management process, including planning and rule promulgation, site selection, operator and technology selection, facility approval, facility operation and closure, and post-closure institutional control.

3. It provides that, following a preliminary screening process and the selection and detailed characterization of several candidate sites, the board shall choose a facility site by a 2/3 vote.

4. It allows the host community to select the facility operator and the disposal technology and to negotiate an impact mitigation and compensation agreement.

In so designing the bill, the commission noted that it sought to "devise a workable process that [would] not repeat the mistakes of Chapter 21D" (Special Legislative Commission on Low-Level Radioactive Waste, 1986, p. 5). In particular, it has attempted to distinguish its proposed process from that of Chapter 21D's by setting up a lead agency that would be independent of line agencies such as DEM and DEQE, which are regarded as having poor track records; by instituting a process that, through detailed planning and screening and extensive public involvement, would establish its credibility before any decisions were made; and by allowing local control on the issue of who would operate the facility and what that facility's technology would be.

Implied Principles of Trust and Consent

These two acts, and others like them, make silent but fervent appeals to the principles of trust and consent held by their constituencies, for without constituency support their chances of success are dim. Although the appeals to these principles go unstated, they can be elucidated with the help of a theory of culture and risk management that Steve Rayner has developed with his colleague, Robin Cantor. (For a current version of this theory, see Rayner and Cantor, 1987.)

Briefly stated, Rayner and Cantor's theory suggests that there are four ideal-typical kinds of social organization and that each type will have distinct cultural biases toward, among other things, issues of trust and consent:

1. *The competitive/market type.* Members of this type tend to trust successful individuals rather than organizations and to assume that consent can be given implicitly, by revealed preference.

2. *The bureaucratic/hierarchical type.* Members of this type tend to trust routine procedures and well-established organizations and to regard consent as something that can be assumed to have been given hypothetically, as part of a social contract that the citizen has entered into with the decision making institution.

3. *The egalitarian type.* Members of this type distrust both the market and the bureaucratic approaches — instead, they tend to trust decision making consensus and to require that consent be made explicit.

4. *The atomized-individual type.* Members of this type include those who are controlled by the system but who have little say in it and are simply struggling for survival; they tend to trust the forces of luck, God, or nature and have no special consent principles.

Using this analytical framework, it becomes clear that the Massachusetts Hazardous Waste Facility Siting Act (Ch. 21D) was dominated by an effort to serve two masters: the competitive/market type (the developers) and the egalitarian type (the host communities), with the bureaucratic type (the Site Safety Council) as a mediator between the two. As one on-the-scene lawyer commented:

> The idea of the [act] was that early notices of proposals, local briefing sessions, state grants for consultants and local review, mandatory environmental-economic impact reports, and negotiated settlements instead of local zoning vetoes could create a win-win situation for all parties. Companies would enjoy a streamlined approval process. Communities would be willing hosts. Citizens would appreciate negotiated benefits. Thus, the public would get needed

storage, treatment, and disposal capacity to serve Massachusetts business (McGregor, October 1983, p. 1).

Obviously, it didn't work. Instead as he went on to note:

> Some early actions by the Siting Council, compounded by some weaknesses in the Siting Act, disappointed business, cities and towns, and public interest groups. The result left citizens distrustful and galvanized in opposition to proposals. Local officials criticized the process as not fair. Companies despaired trying to get approval for new plants... (McGregor, p. 3).

The proposed revamping of Chapter 21D constitutes an attempt to turn failure into success by redirecting the act's appeals to underlying principles of trust and consent. By legitimatizing and routinizing the act's institutional procedures, its appeals to principles held by the hierarchical/bureaucratic type are strengthened, but by taking some of the pressure to negotiate off the prospective host community and by giving greater local control over monitoring, its appeals to egalitarian principles are also made stronger. Taken together, these revisions switch the emphasis of the act from a market/egalitarian approach with a relatively weak bureaucratic role to a bureaucratic/egalitarian approach with a relatively weak market role.

With these proposed amendments, Chapter 21D takes on some of the characteristics of the proposed Low-Level Radioactive Waste Management Act. That act, with its elaborate institutional procedure, emphasizes the development of trust in the government's LLW management process to enable, at key points in the process, the assumption of a tacit consent to decisions that are to be made centrally, at the state level. The Radioactive Waste Management Act thus makes a strong appeal to the principles held by the bureaucratic/hierarchical type. However, by providing for extensive public involvement in the process leading up to these decisions and by allowing some key decisions to be made at the local level, it also makes a strong appeal to egalitarian principles.

CONCLUSION

Like the proposed amendments to Chapter 21D, the Low-Level Radioactive Waste Management Act stresses bureaucratic and egalitarian values over market values, but it does so more extensively and forcefully. There is a good reason for this difference: hazardous waste disposal has traditionally been seen as a market-initiated, private responsibility, whereas LLW disposal has become by fiat a public responsibility. However, apart from this distinction in formal responsibilities (a distinction that will diminish with implementation of the October 1986 "Superfund Amendments," which require states to begin taking responsibility for hazardous waste generated within their borders), the public increasingly has expressed an interest in the safe disposal of chemically hazardous waste, and the states increasingly have assumed an initiatory rather than simply a regulatory role. Increasingly, too, with communities' objections to hosting hazardous waste facilities, states have had to intervene. Consequently, hazardous waste management is becoming more like radioactive waste management: a risk that is recognized as requiring public management.

Does risk management by a public institution necessarily mean a strong, if not an exclusive, appeal to bureaucratic/hierarchical principles? The two Massachusetts acts discussed here suggest that it doesn't but that there is a tendency to adopt this as a fall-back position if an appeal made primarily to the principles of the other group types doesn't work. This partial retrenchment also suggests that, by making the bureaucratic/hierarchical group more central to the siting process and by expanding that process to include related risk management concerns, an attempt is being made to reestablish (or establish for the first time) an elusive but crucial prerequisite for any risk management process: the *authority* to implement process.

Authority doesn't come automatically with either power or responsibility; to have it, people must be won over. These two acts represent two quite different approaches toward that aim. It remains to be seen whether either can work: neither the Chapter 21D amendments nor the Low-Level Radioactive Waste Management Act has yet been passed, so it is not clear whether they would fare better than the original Chapter 21D. So far, Chapter 21D has not been able to "get to yes" — all it has gotten to is "no." Perhaps the best that can be expected is "maybe."

REFERENCES

Kasperson, R. E., 1986, Hazardous Waste Facility Siting: Community, Firm, and Governmental Perspectives, in *Hazards: Technology and Fairness*, National Academy Press, Washington, DC.

Massachusetts Special Legislative Commission on Low-Level Radioactive Waste, January 1986, *Low-Level Radioactive Waste Management: A Siting Process for Massachusetts*, Boston.

McGregor, G. I., October 1983, A Formula for Hazardous Waste Facility Siting, unpublished paper.

O'Hare, M., Bacow, L., and Sanderson, D., 1983, *Facility Siting and Public Opposition*, Van Nostrand Reinhold, NY.

Rayner, S., 1986, Learning from the Blind Men and the Elephant, or Seeing Things Whole in Risk Management, in *Uncertainty in Risk Assessment, Risk Management, and Decision Making*, V. Covello *et al.*, eds., Plenum Press, NY.

Rayner, S., and Cantor, R., January 1986, How Fair is Safe Enough? The Cultural Approach to Societal Technology Choice, *Risk Analysis* **7(1)** (1987).

Ryan, A. S., September 1984, Approaches to Hazardous Waste Facility Siting in the United States, unpublished paper available from the Massachusetts Hazardous Waste Facility Site Safety Council.

Coping with Scientific and Technological Uncertainty: Federal Policy Analysis for Nonroutine Decisions

Frances M. Lynn, Curtis McLaughlin and Richard N. L. Andrews
University of North Carolina at Chapel Hill
Chapel Hill, NC

ABSTRACT

In the last 20 years, faced with an increasing number of decisions about complex scientific and technological issues, government has created new units and staff positions for "policy analysts." These analysts are frequently assumed to be trained in microeconomics, organization theory and quantitative decision analysis. In fact, a majority of them are often scientists and engineers who have been thrust into the role of policy analysts; few find quantitative decision analysis procedures helpful. In the face of scientific uncertainty and short timeframes, these analysts use rich and interactive methods to search and analyze information. However, there is an absorption of uncertainty and a narrowing in the richness of communication channels as information makes its way up the hierarchy.

KEYWORDS: Uncertainty, information, policy analysts, decisions

INTRODUCTION

Federal officials must frequently make policy decisions that involve important scientific and technical considerations, yet which are imperfectly understood and require action too quickly to permit systematic investigation. We refer to these as nonroutine decisions; examples include such issues as Toxic Shock Syndrome, groundwater contamination by hazardous wastes, the identification of high levels of toxic chemicals in foodstuffs, and the decision to permit the release of genetically altered microorganisms into the environment. They have in common scientific or technological uncertainty and the pressure to make a decision in a relatively short timeframe.

This research is a study of how policy analysts in the federal government gather, analyze and present scientific and technical information under conditions of uncertainty and when time is short. In such situations an analyst is not likely to have enough time to generate new or definitive scientific information. Moreover, the data that do exist can be analyzed using different assumptions and result in widely varying scientific conclusions and policy recommendations. For this study we have defined a nonroutine decision as having a timeframe of less than six months. These nonroutine, high-uncertainty situations provide

New Risks
Edited by L. A. Cox, Jr., and P. F. Ricci
Plenum Press, New York, 1990

the greatest potential for success or failure because they require both the analyst and the decisionmaker to go beyond existing formal structures and information sources.

METHODOLOGY

Our primary instrument was a set of one-hour structured interviews with 77 federal policy analysts. A study of 40 corporate staff personnel with analogous responsibilities is currently underway. The study is also supported by two case studies of nonroutine decisionmaking involving both the federal government and private industry: (1) Toxic Shock Syndrome and (2) field-testing of genetically-engineered microbes for the reduction of frost damage to plants.

Using the Federal Yellow Book Pages published by Barron's, we identified agency units that dealt with science and technology matters (other than intramural research), and we classified and stratified these agencies to include those whose purpose involved administration, regulation and oversight. Units included in our sample, for example, had responsibilities for recommendations about the toxicity of chemicals, the safety of foods and drugs, the economic impact of natural resource leasing arrangements, and policies concerning flood warnings and weather forecasting. Within those units we identified individuals with titles such as "program analyst" and "policy analyst," selected 77 of these individuals randomly, and called them to schedule interviews.

Each interviewee was asked to focus on a recent nonroutine issue which he had analyzed in a relatively short time period (3 to 6 months or less) and for which definitive scientific and technological information was not available. We believed that we would get a much richer description of the policy analysis process by asking each respondent to focus on a specific case associated with a recent set of events. The interview contained a combination of fixed-choice and open-ended questions. Respondents were assured of personal anonymity and anonymity regarding the specific case they elected to describe.

This study was conducted by a multidisciplinary team from the University of North Carolina Graduate School of Business Administration and the Institute for Environmental Studies, UNC School for Public Health. As a team, we drew heavily on research on information search in the R&D management literature, on the literature of organizational behavior and organizational development, on studies of policy analysis and analysts, and on the growing literature on risk assessment, management and communication.

WHO ARE THE POLICY ANALYSTS?

In selecting interviewees who carried the title "policy analyst" or "program analyst," we were expecting to encounter professionals with training in some of the skills which form the core of the curriculum in graduate schools of public policy — a mix of microeconomics, organizational theory, decision analysis and political science. Instead, we found ourselves interviewing scientists and engineers who had been thrust into the policy analyst role. As Fig. 1 indicates, of the 77 government policy analysts interviewed, 60% had a science or engineering background. The majority of the staff with whom they worked had similar (science or engineering) backgrounds. In addition,

- 74% were GS-14 or -15;
- 85% considered themselves career civil servants;
- 49% had a Ph.D., and 29% a master's degree as their highest degree;
- 60% saw themselves as supporting primarily a senior decisionmaker who was a secretary, administrator, deputy secretary, deputy administrator, or assistant secretary.

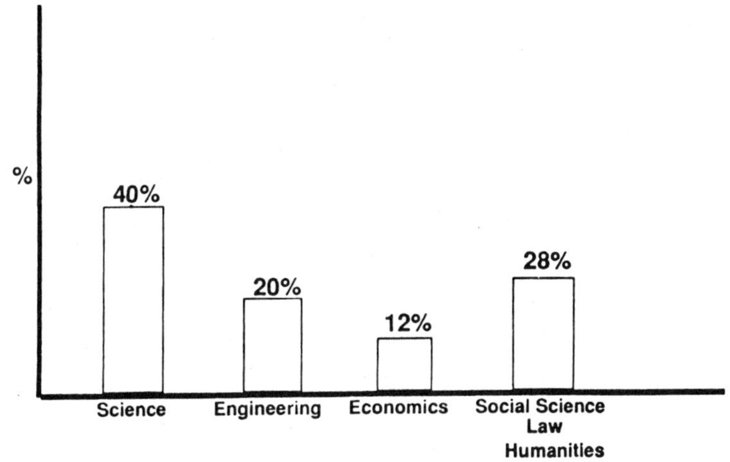

Fig. 1. Analysts' educational background.

We interviewed no graduates of formal programs in science, engineering and public policy, and only one who had a degree from a formal public policy program. One explanation for this lack of representation is the relative newness of these programs, most having been established in the last 20 years. Another explanation, however, is a preference, reported elsewhere and supported by the scientists and engineers we interviewed, for hiring people with specialized technical degrees.[1] The analysts we interviewed tended to select an MBA or law degree to complement their technical training.

ISSUE EMERGENCE

We asked our sample what the stimulus was for deciding to tackle the particular issue — that is, how did the issue come to be perceived as warranting analysis, and what were the main forces that brought it to decisionmakers' attention? One-third named Congress and 23% named interest groups as the source of stimulus (multiple stimuli were reported). Over half of the analyses, however, were initiated from within the agency. Some of these were preemptive strikes. One respondent noted, "We did it ahead of Congress to hold our ground." In other cases the analyst provided the stimulus, seeing it as an opportunity to move the organization in a specific direction. These were the "policy entrepreneurs" described by Arnold Meltsner, who seemed undaunted by layers of bureaucracy and had developed the skills necessary to push their ideas to higher levels for attention and possible actions.[2] One analyst described his strategy of "deciding the constituency" and "generating the forum" for his idea.

Very few issues described by our sample were totally new, supporting John Kingdon's observation that

> ...public ideas, proposals or issues are like perennials: flowering in one season, then lying dormant, only to flower anew...they may not come back in the same form; rather they are recast, combined with something else or attached to a problem different from the one they started with...a new recombination, a new twist — yes — but not wholly new material.[3]

Table 1. Sources of Information Chosen

Source	First n = 75	Second n = 70	Third n = 58	Fourth n = 39	Fifth n = 18
Individual	62.7%	50.7%	58.1%	58.1%	50.0%
face-to-face	36.0%	22.2%	32.3%	32.6%	31.8%
telephone	24.0%	22.2%	21.0%	16.3%	13.6%
Personal Files	6.7%	7.0%	0	2.3%	0
Documents	14.7%	36.6%	29.0%	25.6%	27.3%
Databases	16.0%	2.8%	6.5%	4.7%	4.5%
			Reasons for Selecting Source		
Reason					
Accessibility	32.9%	16.2%	1.8%	7.5%	4.8%
Credibility	12.9%	30.9%	21.4%	10.0%	9.5%
Expertise	31.4%	33.8%	44.6%	50.0%	28.6%
Representation	22.9%	17.6%	25.0%	22.5%	38.1%

This also fits the behavioral theory of the firm in which decisionmaking is the result of problems in search of solutions and solutions in search of problems, which combine at choice opportunities.[4] Kingdon suggests that successful analysts are like "surfers waiting for the big wave...out there...ready to paddle."[5] Indeed, one of our respondents described how an "issue percolated up. It didn't emerge from the background mode for a while, until pressure from the corporations and Congress increased." The press was mentioned as a stimulus for an issue in only 7.8% of the cases. This is less participation than many might expect. But this is consistent with Kingdon's theory that the press mostly reports what is happening, not what is not happening, and, when independent and responsive, does facilitate communication and magnify a movement which is underway.[6]

THE SEARCH PROCESS

Table 1 shows the sources, in order of preference, that our respondents turned to for information. Table 1 also lists the reasons that were given for selecting the sources. The S&T policy analysts chose face-to-face and telephone sources more than 50% of the time. At first, they turned to accessible individuals; they then turned to documents. One can see a pattern in their criteria for choice of sources: first, accessibility as well as expertise, then even more expertise and added credibility, followed by representation of key actors. Note that this implies a mixture of technical and organizational information. We heard statements such as:

"I talked to the key players to tap the institutional memory."

"We met face-to-face with the key decision maker to figure out expectations and concerns."

"Eye-to-eye contact provides more information than by telephone."

The methods of search used by these S&T analysts fit the patterns reported for scientists and managers elsewhere. The literature shows a strong correlation between uncertainty and the use of rich information channels. Richard Daft and Robert Lengel[7] have outlined a hierarchy of media richness with includes:

Rich and personal media:	group meetings integrator (coordinator) direct contact face-to-face telephone subjective written material planning special reports formal information systems
Less rich, impersonal media:	rules and regulations

William D. Bodensteiner's research with scientists found (1) that the more uncertain the individual is about the needed scientific information, the greater the pressure to seek additional information, and (2) that while seeking that information, the individual will choose channels/sources with increased information-carrying capacity and/or increased opportunity for personal support.[8] A survey of 384 engineers in a large government R&D organization showed a strong correlation between technical uncertainty and a decisionmaker's selection of face-to-face sources of information. Engineers saw such sources as richer, having the ability to communicate inflections, pauses, gestures and expressions.[9] William Fischer has noted the importance in nonroutine decisions of emphasizing informal sources which allow personal confrontation, interaction and reaction.[10]

The reliance on rich channels was not due to a general lack of knowledge of the issue involved. Roughly one-third rated themselves as not knowledgeable or slightly so; one-third rated themselves knowledgeable, and roughly one-third rated themselves as very knowledgeable or expert. They also had information on hand when the issue arose. One-third already had some personal files on the issue, 55% had relevant internal documents, 37% had documents from other agencies, a third had nongovernmental documents, and 21% had a database.

This suggests that technical uncertainty — lack of sufficient data to answer a specific question, is not the only form of uncertainty driving people toward richer channels for information. There is also organizational uncertainty or equivocality, defined as ambiguity, the existence of multiple conflicting interpretations about an organizational situation. High equivocality means confusion and lack of understanding... asking a yes-no question is not feasible. Participants are not certain about what questions to ask, and if questions are posed, the situation is ill-defined to the point where a clear answer will not be forthcoming.[11]

Table 2. Nature of Uncertainty

Technical Uncertainty (n=70; 91% — multiple answers allowed	
Different assumptions	62.2%
Conflicting data	34.7%
Different models	39.7%
Conflicting interpretation of same data	68.5%
Organizational Uncertainty (n=46; 60% — multiple answers allowed)	
Conflicting values	45.9%
Who would make key decisions?	29.5%
Unwillingness of key actors to participate	44.3%
Lack of precedent	36.1%
Ambiguity in mission/goals	31.1%

Most of the searches that we observed contained a combination of technical uncertainty and equivocality. When asked, "At the start of the study, what types of information seemed to have the most potential for being controversial," they responded:

61.6% scientific or technical
31.9% economic
16.7% legal
19.4% political (turf, cooperation, etc.)

Table 2 outlines the nature of the uncertainty reportedly associated with each search. Again, it is clear that, while technical uncertainty applies to each search, the majority also have organizational uncertainty associated with them. The S&T analysts were aware of these equivocalities which influenced decisions about the types of information to seek and the channels to use to acquire them. During the interviews we heard statements such as, "We had to be concerned with who should be involved," and "The most important information is the 'behind-the scene' information." A lesser influence was the presence of statutory or legal requirements that had to be met in acquiring information. Thirty-five percent reported such requirements.

In keeping with the nonroutine nature of the issues, the searches reported generally served a higher-than-usual decisionmaker (49.2%), or even one outside the chain of command (13.7%). The senior decisionmaker was not involved in the search in about 70% of the cases, and only monitored the search in another 20%. In only 10% of the cases did the senior decisionmaker work as a team member, and in only one case provided tips on

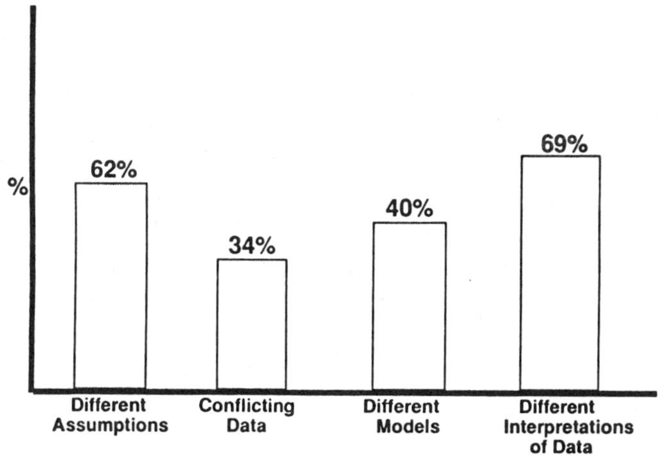

Fig. 2. Sources of uncertainty.

where to go for information. However, 40% of the respondents replied "yes" when asked if the educational background or personal characteristics of the persons who would be making the decision influenced where they went for information. Forty-two percent replied in the affirmative when asked whether or not political pressure influenced the gathering of the information.

ANALYSIS

We asked the analysts to focus on those situations in which definitive scientific and technological information was not available, but on which they were expected to make a recommendation. Such situations have become more common as the federal government has increased its responsibilities for protecting the environment, and health and safety in general. We wanted to get a sense of the nature of the scientific and technological uncertainties they faced; the extent to which formal analytic techniques were used; how the analysts dealt with conflicting data or alternative interpretations of the same data; and how they resolved (or didn't resolve) differences within their group.

In only 16% of the cases was the analysis a solo responsibility. More typically, it involved 3 to 5 people. In 44% of the cases it involved people from other units of government and, occasionally, individuals from other agencies (13.3%) or consultants (8.0%). Interest groups were represented in the analysis group in only one case.

The underlying issue that these groups faced was how to proceed in the face of conflicting scientific opinion. As Fig. 2 indicates, in two-thirds of the cases the uncertainty was not so much about conflicting empirical data as about which assumption, models or interpretations were appropriate. When asked how they went about resolving the technical uncertainty, their responses were:

14.5%	collected new data
19.7%	reinterpreted political/legal assumptions
33.8%	reinterpreted technical assumptions
7.0%	asked unbiased experts
18.3%	did not resolve them
5.6%	other

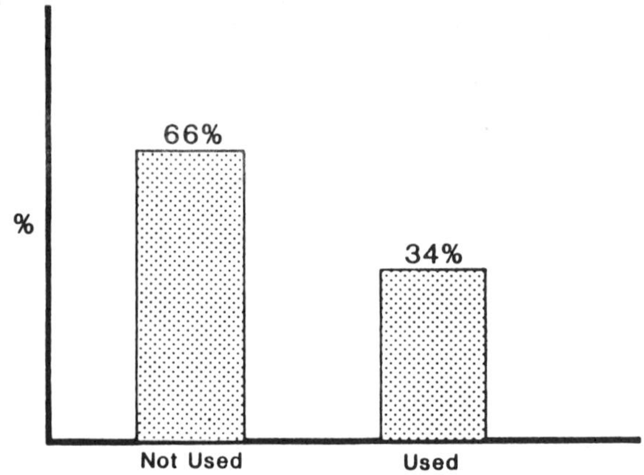

Fig. 3. Use of formal decision analytic techniques.

When asked what the process was for resolving intra-group differences, one-half reported using consensus; one-quarter, responding to a dominant individual; and 17%, referral to a supervisor. We heard numerous reports of groups conducting analyses using different assumptions and then sitting around the table arguing over what was most plausible.

It is this interaction, this richness in dialogue and debate, that perhaps best explains why, when we asked whether or not they used formal decision analytic techniques, only one-third of the respondents replied positively (Fig. 3). Most of these were econometric models. The analysts interviewed said that the techniques were not appropriate or applicable. Moreover, decisionmakers did not understand them. When policy analysis came into existence as a field almost 20 years ago, it borrowed heavily from the quantitative skills, techniques and definitions of rationality and decisionmaking dominant in economics. Analytic procedures were to replace the messiness of human beings arguing, learning from each other and changing their minds. Policy analysts were to provide decisionmakers with objective, value-free answers. Decisionmakers, however, turned out to be uncomfortable with these techniques' yes-no, right-wrong answers, and many of the problems they were hoped to solve did not resemble the easily structured, routine, inventory-like problems for which the techniques were originally developed.[12,13,14]

Communication of Analysis

We asked our interviewees a series of questions aimed at exploring how they communicated their deliberations to senior decisionmakers. Was there a standard reporting format? Had they found any one method of presentation more successful than others? How were elements of uncertainty presented? Did the senior decisionmaker want a detailed presentation of different options or a single recommendation?

Table 3 shows how the material in these nonroutine situations was presented to the senior decisionmaker. In this environment of relative urgency and uncertainty, the senior decisionmakers relied less on formal reports and more on issue papers, and also chose the richer medium of oral briefings. However, these seemed to be well-rehearsed presentations, almost to the point of being manipulative: "...best of all is a slide show because it keeps questions limited...you're the authority." Another mentioned distributing the briefing package at the end so that it wouldn't be rifled through during the briefing and also

Table 3. Method of Presentation to Senior Decisionmaker

Method	Nonroutine Case Cited	Standard Approach
Detailed report	28.9%	32.9%
Issue paper	55.8%	50.7%
Oral briefing	75.0%	53.4%
(Multiple answers allowed)		

"holding questions" until the end. The briefing papers were short, often in bulleted, outline form.

We asked the analysts if they presented decision options to the senior decisionmaker. As Fig. 4 indicates, in half of the cases the decisionmaker only received the consensus recommendation of the staff. We investigated two possible explanations for this. One explanation was that the analysts were successful in resolving the technical and organizational uncertainties in about 50% of the cases and therefore, in those cases, the presentation of consensus views was the only plausible approach. Consequently, in those other cases where there was considerable uncertainty remaining, the group doing the analyses would present alternative interpretations and minority views. Alternately, it could have been that the process of presentation itself was what narrowed down the diversity of information presented to the senior decisionmaker. Upon analysis of the data we identified an intervening consideration — namely, the way that the analysts perceived that the senior decisionmaker wanted the material presented. If the analyst perceived that the decisionmaker wanted consensus, it was supplied, and when the decisionmaker wanted to choose, options were given. We also asked respondents if differing opinions of groups not included in the team were represented in the presentation of the analysis results. When there were intergroup differences involved (43%), this was talked through only 21% of the time. The unresolved intergroup difference appeared as comments in the formal report 38% of the time or in a separate paper (24%).

In *Inside Bureaucracy*, Anthony Downs suggested leaders "encourage a variety of viewpoints...by structuring the interests of subordinates involved so that they are in direct conflict."[15] Mason and Mitroff in their book, *Challenging Strategic Planning Assumptions*, press for a dialectical approach to the presentation of arguments with "structured debate" replacing well-staffed papers or traditional briefings, especially when multiple interpretations of data and uncertainty prevail.[16]

THE ANALYSTS AND TRAINING

Although our analysts came from a wide variety of backgrounds, by and large they are upper-middle class, middle-aged, male technocrats with professional training in fields other than policy analysis. They arrived at the policy analysis jobs after a number of experiences in their agency and elsewhere. As we mentioned earlier, over 60% of the analysts had science or engineering backgrounds; 12% had economics backgrounds, and

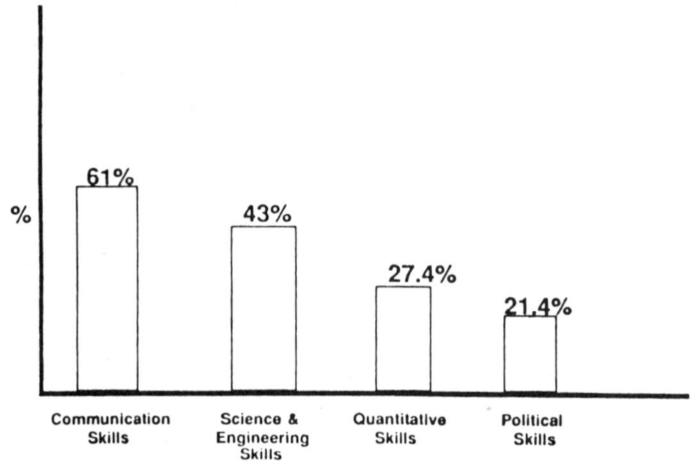

Fig. 4. Recommended skills for entering policy analysts.

27% had backgrounds in law or social science. Nearly 40% had doctorates (including M.D. and J.D.), while 29% had master's degrees. Their length of time in their current job was quite varied. Eighty-seven percent had prior experience with a different job in the same agency; 21% in academic institutions, 10% in state government, 43% in another federal agency, 12% in consulting and 13% in industry. A large majority (84%) saw themselves as career civil servants.

We asked the analysts to reflect on the structure of their job responsibilities and to estimate how they divided their time.

The heaviest use of time was in communication, then in administration, and then in analysis (see Table 4). The analysts often had other duties, such as preparing senior officials for important ceremonial meetings, including testimony before Congress. Their experiences seemed to fit the patterns observed by Mintzberg, in that productive output could be measured primarily in terms of verbally transmitted information and that negotiation is a significant managerial role, especially where interorganizational linkages are involved.[17] This supports Richard Elmore's recommendation that we need to focus training on the "skills of influence, coalition-building, motivating and understanding complex institutional and interpersonal relations."[18]

Scientists and engineers are not necessarily known for their communication abilities. Karl Weick recalls the admittedly stereotypical description of engineers as "smart people who don't talk." In fact, when we asked those we interviewed about the skills that they would advise a new, young science and technology policy analyst to acquire, communication skills and technical background were both at the top of the list (Fig. 4).

SUMMARY AND CONCLUSIONS

The purpose of this research was to examine how policy analysts use scientific and technological information to make decisions under conditions of uncertainty. We asked our sample, mainly scientists and engineers operating in the policy analyst role, to focus on a specific nonroutine decision. In the process of a one-hour interview, we examined the way issue emerged, the methods used to gather information, the processes used to combine or

Table 4. How Analysts Spend Their Time

%Time Spent	Administration	Research	Analysis	Communication
0%	17.1	56.0	18.2	6.5
1-25%	28.9	34.7	33.8	29.6
25-50%	39.5	8.0	35.1	45.5
51-75%	9.2	1.3	7.8	15.6
76-100%	5.3	-0-	5.2	3.9

synthesize information, and the techniques used to present findings to senior decisionmakers.

We found that, while the issues that our sample picked were indeed nonroutine — that is, they had to be resolved in a relatively short timeframe — many were not totally new. The individual or agency often had a variation of that topic on their agenda for some time. In some cases, the analysis or report or flap that others saw was the planned boiling-over of a pot that had been cooking for a long time, with a carefully orchestrated buildup of the heat. The analysts interviewed tended to use informal (face-to-face meetings or telephone conversations) means of searching for information. These findings support a growing recognition of the reliability and richness of face-to-face information for uncertain and complex information and for nonroutine decisionmaking, including crisis. This form of gathering information provides quick feedback, multiple cues, and avoids the sanitization of uncertainties and problems with data found in written documents. The analysts searched for answers to questions involving technical uncertainty, but at the same time, they were consciously probing for information which would allow them to reduce organizational and political uncertainties.

The analysts faced the not uncommon situation of posing policy questions concerning issues of science and technology that even the best information could not answer. In over two-thirds of our cases, analysts debated internally which assumptions or models to use to interpret the information. Only one-third of the respondents used formal decision analytic procedures, and most of these were econometric models. Some interviewees were unaware of their existence. Others were dubious of the ability of these highly structured models to capture the complexity of the environment in which the decision was being made and were also concerned about these models' acceptability to decisionmakers.

The method of presenting information to senior decisionmakers was uniform. Throughout both the bureaucracies and the analysts sampled, the modus operandi included some form of oral briefing, sometimes accompanied by graphics, by one member of the analysis group or their immediate superior. Accompanying the briefing was a short, bulleted memo. During this stage of the policy analysis process, we started to observe a well-established phenomenon which, given the uncertain nature of the data, we found unsettling. In half of the cases reported, even amidst technical and organizational uncertainties, decisionmakers were presented with only one consensus recommendation.

What we observed, in sum, was the absorption of information about risk at lower levels in the organization, and an accompanying narrowing in the richness of the communication channels as information made its way on up the hierarchy. In light of recent technological disasters and near-disasters, we may need to rethink the common wisdom that persists in government that, in a situation involving uncertainty, it is better to narrow choices facing the decisionmaker in order to avoid cognitive overload and paralysis. Studies of executives and managers in industry suggest that high-level decisionmakers are not uncomfortable with multiple and dissonant sources of information. Some scholars contend, in fact, that when under pressure to make unanticipated changes, decisionmakers are more likely to want to hear from new and diverse sources.

As research scholars and practitioners, we need to consider mechanisms and methods that ensure that all positions are aired in a manner which can be understood by the nontechnical decisionmaker and enable her or him to be an active participant in science policy dialogue. Specifically, we need research and experiments which will help us

1. develop approaches that will facilitate the use of expertise but at the same time ensure that competing perspectives and interpretations are adequately considered;

2. devise methods of presentation which successfully convey the nature of uncertainties to the nonexpert, and in a manner that does not sacrifice comprehension, nuance or emphasis for readability; and

3. develop mechanisms which do not overload already busy decisionmakers, but do involve them more in the final stage of choosing among plausible policy options.

The challenge, in sum, to those interested in decisionmaking under conditions of scientific and technological uncertainty, is to develop communication models which place analysts and decisionmakers in creative, face-to-face dialogue and deliberation. Conceptually, this is a two-stage process.

In the first stage, the analysts use their knowledge of a field to identify a full range of competing data, models and scientific perspectives on the issue. The goal of this first stage is not to reach a scientific consensus among the analysts; consensus at this stage masks too many uncertainties. It subtly involves the analyst in making choices which often have social, economic and ethical implications. The hardest part of the analyst's job at this stage may be identifying the implicit social and political values associated with selecting one set of data or one interpretation of data over another.[19]

This dissection of scientific arguments and underpinnings, is receiving attention from others interested in decisionmaking under conditions of uncertainty. Dale Hattis and David Kennedy recommend that analysts:

> ...communicate their findings so that policy makers and the public are able to comprehend the important assumptions, data gaps and choices almost as if they themselves had gone through the process.[20]

Sheldon Reaven's[21] observations about improving the role of the public in scientific and technological decisions are relevant. Reaven contends that it is not enough just to present scientific information, including gaps and uncertainties. He suggests that nonscientists also need to be prepared to (1) distinguish styles of argumentation, (2) differentiate parties' views toward how definitive the science is, and (3) appreciate the philosophical or conceptual underpinnings of scientific issues.

The second stage in this communication model involves reformulating the sender-receiver paradigm that pervades the communication of scientific, technological and risk information. This includes using what Barbara Ferman[22] labels "interactive problem-solving devices." But it could also entail creating opportunities for decisionmakers, should they so want, to have more direct contact with proponents of different perspectives. This is a change in the traditional role of the policy analyst from being mainly a conveyor of information to acting as a facilitator of informal meetings and seminars between the decisionmaker and primary sources of information.

This role has some of the attributes that a negotiator plays in the mediation of scientifically-intensive disputes, including sharing information and participating in joint fact-finding.[23] This role also shares characteristics of the boundary spanner and gatekeeper identified many years ago in organizational literature. The role extends beyond that of the boundary spanner and gatekeeper because of the need for strong interpersonal skills and a sensitivity and receptivity to ensuring that different value perspectives and points of view are heard at all stages in the policy process.

REFERENCES

1. Albert H. Teich, "Graduate Education and Career Directions in Science, Engineering and Public Policy," Washington, D.C.: American Association for the Advancement of Science (1986).
2. Arnold J. Meltsner, *Policy Analysis in the Bureaucracy*, University of California Press, Berkeley, CA (1976).
3. John W. Kingdon, *Agendas, Alternatives and Public Policies*, Little, Brown & Co., p. 149, Boston (1984).
4. M. D. Cohen, The Garbage Can Theory of Administration, *Administrative Science Quarterly* **21**:1-19 (March 1976).
5. Kingdon, Ref. 3, p. 173.
6. Kingdon, Ref. 3, p. 173.
7. Richard L. Daft and Robert H. Lengel, Organizational Information Requirements, Media Richness and Structural Design, *Management Science* **32(5)**:554-571 (May 1986).
8. W. D. Bodensteiner, Information Channel Utilization under Varying Research and Development Project Conditions: An Aspect of Interorganizational Communications Channel Usage, Doctoral dissertation, University of Texas, Graduate School of Business Administration, Austin, TX (1970).
9. W. E. Holland, B. A. Stead, and R. C. Leibrock, Information Channel/Source Selection as a Correlate of Technical Uncertainty in a Research and Development Organization, *IEEE Transactions on Engineering Management* **EM-23**:163-167 (1976).
10. William A. Fischer, The Acquisition of Technical Information by R&D Managers for Problem Solving in Nonroutine Contingency Situations, *IEEE Transactions on Engineering Management* **EM-26**:8-14 (1979).
11. Arnold J. Meltsner, Ref. 4, pp. 556-557.
12. Richard F. Elmore, Graduate Education in Public Management: Working at the Seams of Government, *Journal of Policy Analysis and Management* **6(1)**:81 (1986).
13. Research Briefing Panel on Decision Making and Problem Solving, *Scientific Futures: Selected Areas of Opportunity*, Committee on Science, Engineering and Public Policy, NAS/NAE/IOM, Washington, D.C., National Academy Press (1986).
14. Richard O. Mason and Ian Mitroff, *Challenging Strategic Planning Assumptions*, John Wiley & Sons, p. 10, New York (1981).
15. Anthony Downs, *Inside Bureaucracy*, Little, Brown & Co., p. 119, Boston, (1966).
16. Richard Mason and Ian Mitroff, Ref. 15.
17. Henry Mintzberg, *The Nature of Managerial Work*, Harper & Row, New York (1973).
18. Richard F. Elmore, Ref. 13, p. 77.

19. Frances M. Lynn, The Interplay of Science and Values in Assessing and Regulating Environmental Risk, *Science, Technology and Human Values,* **11(2)**:40-50 (Spring 1986).
20. Dale Hattis and David Kennedy, Assuming Risks from Health Hazards: An Imperfect Science, *Technology Review* **71**:60-67 (May-June 1986).
21. Sheldon Reaven, The Reliability of Individual and Pooled Expert Judgment, in *Risk Assessment and Risk Management*, L. Lave and V. Covello, Eds., Reidel Press, New York (1986).
22. Barbara Ferman, Bridging the Gap: Policy Analysis and the Governmental Policy Process, in *Bureaucratic and Governmental Reform*, pp. 157-170, JAI Press (1986).
23. Gail Bingham, *Resolving Environmental Disputes: A Decade of Experience*, The Conservation Foundation, Washington, DC (1986); Gerald W. Cormick and Alana Knaster, Mediation and Scientific Issues, *Environment* **28(10)**:6-16.; Connie P. Ozawa, Mediating Science-Intensive Disputes, *Journal of Policy Analysis and Management* **5(1)**:23-29 (1985).

Signal Detection and Perceived Risks: Homeowner Responses to Radon Risk Information

F. Reed Johnson
U.S. Naval Academy
Annapolis, MD

Economic Analysis Branch
U.S. Environmental Protection Agency
Washington, DC

ABSTRACT

Policies for reducing such risks as natural hazards, AIDS, household use of pesticides, and indoor air pollution often rely on dissemination of risk information to promote risk-reducing behavior. Economists often use state-dependent utility constructs to model this kind of problem. While this approach assumes a high degree of competence on the part of the decision maker, psychological evidence indicates that people frequently use simple rules when faced with complex decisions. This study employs Heiner's model of rule-governed behavior to analyze radon mitigation. Application of the reliability model is illustrated with observed mitigation responses of a sample of Maine homeowners. Data now being collected from a sample of New York homeowners will provide a basis for testing formal hypotheses arising from the model.

KEYWORDS: Radon, perceived risk, risk information, reliability model

INTRODUCTION

Much of the recent literature on social regulation has been concerned with risk management. For example, policy decisions setting emission standards for point sources of pollution, mandating passive restraints in automobiles, or selecting disposal sites for hazardous or nuclear wastes all attempt to reduce risks to which various groups are exposed. Economists have devoted increased attention to the conceptual and empirical issues related to valuation of the benefits and costs of such policies. An important implicit assumption of most such studies is that people form risk perceptions in a rational manner. It should follow that observed decisions reflect people's relative preference for the events at risk, assuming they are given proper information about objective probabilities.

Psychologists have long disputed this description of risk-related behavior. These critiques have begun to appear in the economics literature (see, for example, Kahneman and Tversky [1979], Thaler [1980], and Schoemaker [1982]). In a recent summary of research on this subject, Slovic, Fischoff, and Lichtenstein [1985] suggest that

New Risks
Edited by L. A. Cox, Jr., and P. F. Ricci
Plenum Press, New York, 1990

> ... people often are misinformed, rely on suboptimal risk-assessment heuristics, and fail to understand the limits of their own knowledge (p. 245).

Much of the research in cognitive psychology has been devoted to the ways that people process and use information. The results of this research often indicate that common strategies for dealing with complex problems frequently lead to errors or suboptimal choices. For example, Einhorn and Hogarth (1985) recently analyzed inferential judgments and choices under ambiguity, which they define as limited knowledge of the process that generates [probabilistic] outcomes. They model such behavior as a mental stimulation process governed by the amount of and attitude toward ambiguity.

This approach is quite different from the standard optimizing models used in economics. Although the economic approach has proven to be a powerful analytical tool in many applications, some economists have begun to question its relevance for some kinds of risk-related behavior. Preliminary analysis of data now being collected on homeowner responses to new information on radon health risks indicates that the standard economic model may not prove useful in this context.

Tversky and Kahneman (1977, 1979) have identified several probability heuristics that violate the economic axiom of expected utility. Although they refer to these heuristics collectively as "prospect theory," their characterization of the problem is a taxonomy of observed phenomena rather than a behavioral model that yields interesting testable hypotheses. However, several recent contributions by Heiner (1986, 1984, 1985, 1983) and Kaen and Rosenman (1986) suggest an alternative departure from expected utility that yields predictable responses under true uncertainty. This paper summarizes Heiner's reliability model and illustrates its application with some recent empirical results on homeowner responses to new information on radon exposure.

SIGNAL DETECTION EXPERIMENTS

There is a large literature based on experiments in psychological scaling of responses to sensory stimuli. Although research has been applied to a wide range of perceptual skills, one simple form of this type of experiment resembles a hearing test. The subject is required to detect an intermittent signal imbedded in a frequency distribution of background noise. Successful detections are rewarded and errors are penalized according to specified rules. Experiments involve measuring the rate of successful detections as various parameters change, including personal characteristics of subjects, environmental factors, the frequency, intensity, and probability distribution of signals, and rules for gains and losses based on performance.

Figure 1 shows a typical empirical signal-detection curve. Each point on the curve represents a battery of detection trials. For each set of trials, the experimenter observes the number of detections when the signal is present and the number of detections when the signal is not present. The axes measure the likelihood of detecting the signal given that the signal is or is not present. Increasing the signal probability, with all other variables held constant, traces out a concave path from the upper right hand corner to the lower left hand corner. This locus of points consists of all the feasible combinations of probabilities of right and wrong detections. Increasing the loss-gain ratio with constant signal probability traces out a similar path. The concavity of the curve is invariant with respect to variations in detection skills, the probability distribution of the signal relative to the background noise, and environmental variables.

The difficulty of the detection problem depends on the degree to which the signal changes the mean and variance of the noise distribution. The greater the difference between the noise mean and the signal frequency, the greater the shift in mean. The greater the

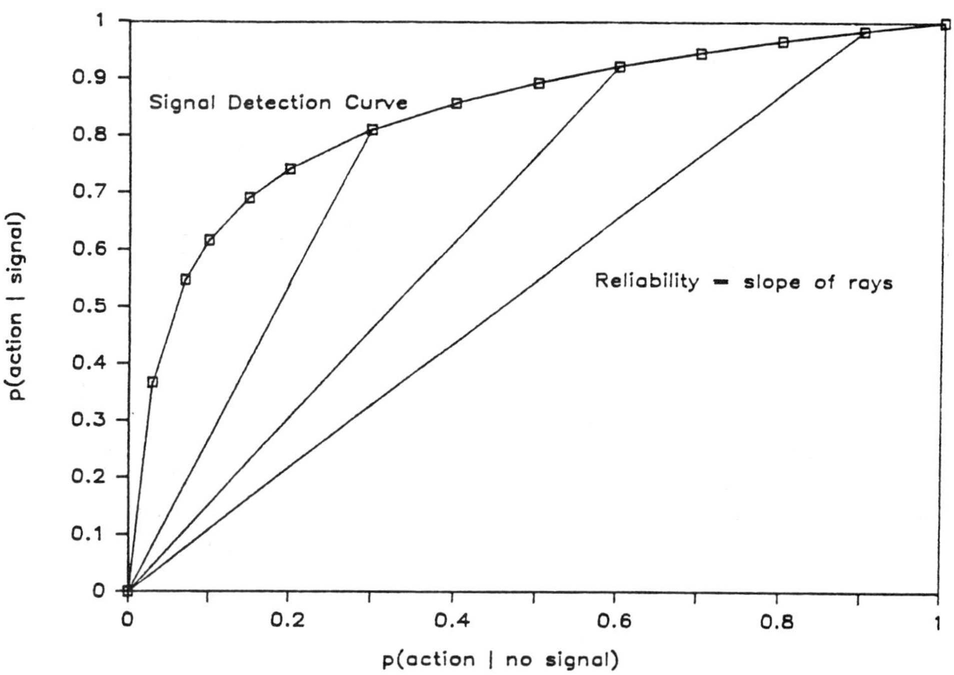

Fig. 1. Signal detection and reliability.

intensity of the signal, the smaller the variance. The ratio of the change in mean to the variance measures the degree of uncertainty in the problem and is called the signal-to-noise ratio. Increases in the signal-to-noise ratio (decreases in uncertainty) bend the curve toward the upper left hand corner, increasing the reliability of a given subject's detection performance. Heiner defines reliability as the ratio of the probability of correct detections to the probability of incorrect detections. The reliability ratio for three sets of trials is shown as the slope of rays emanating from the origin in Fig. 1.

Notice that movement to the left on a given curve increases reliability. Since the sum of the conditional probabilities of detection or action weighted by the probabilities of the signal being present or absent is equal to the unconditional probability of action, movement to the left requires reduction in the total number of actions for a set of trials, regardless of the probability of a signal being present.

The importance of this result is that predictable patterns of behavior are associated with true uncertainty. Economic models can accommodate risk, but generally require the decision maker to know, or assume, the relevant probability distribution. Signal detection experiments can be designed so that the subject cannot calculate the probability distribution of frequencies. This uncertainty results in greater reliability of responses rather than less. However, the agent achieves greater reliability by taking fewer actions, that is, fewer departures from the status quo.

The observed responses are conditioned on the interaction of the competence of the subject and the difficulty of the detection problem. The regularity of behavior arises out of some process, whether conscious or not, that affects the agent's alertness or sensitivity to information that might prompt selection of an action. The following section outlines the derivation of Heiner's condition for whether an agent will benefit from paying attention to such information.

THE RELIABILITY MODEL

The basic question is whether an agent can benefit from reacting to new information about whether to select a given action. This is equivalent to asking under what conditions average performance will increase by choosing action a relative to continuing the existing pattern of behavior.

$$\begin{aligned} dP &= p(R)[rV(a|R) + (1-r)V(z|R)] \\ &\quad + p(W)[wV(a|W) + (1-w)V(z|W)] \\ &\quad - p(R)V(z|R) - p(W)V(z|W) \\ &= Gp(R)r - Lp(W)w > 0 \end{aligned}$$

where dP is the change in average performance, $p(R)$ is the probability of right conditions R, r is the conditional probability of choosing a, given right conditions, V indicates the value of the indicated outcome, z is no change in behavior, $p(W)$ is the probability of wrong conditions, w is the conditional probability of choosing a, given wrong conditions W. G is defined as $V(a|R) - V(z|R)$ and L is defined as $V(z|W) - V(a|W)$. Rearranging gives Heiner's Reliability Condition:

$$r/w > T = (L/G)\,[p(W)/p(R)].$$

T is the "tolerance limit" or the minimum lower bound on reliability for which allowing flexibility to select an action will improve performance. Although this specification uses a standard expected-value formulation, the same result can be obtained by using recently proposed nonexpected utility formulations. (See Heiner, 1984.)

Figure 2 shows two signal detection curves. The lower curve indicates a higher degree of uncertainty (lower signal-to-noise ratio) than the upper curve. If the tolerance limit is T', then there is no condition under which performance can be improved by choosing a. By lowering the loss-gain ratio or increasing $p(R)$, T falls to T^*. Flexibility in choosing a will improve performance to the left of the intersection with the lower curve.

Holding the loss-gain ratio constant, Fig. 3 shows how the tolerance limit changes as $p(R)$ changes. If reliability is above the curve for any given $p(R)$, then the agent will respond to new information. He will ignore the information if reliability is below the curve.

We can then show the relations among the unconditional probability of selecting action a, reliability, and the tolerance limit in Fig. 4. The slopes of the two signal-detection curves from Fig. 2 are plotted as reliability curves in Fig. 4. The model thus indicates that increases in the probability of selecting an action are associated with either a decrease in uncertainty or a decrease in the tolerance limit arising from either a decrease in the loss-gain ratio or an increase in $p(R)$. A reliance on rules rather than a continual adaption to environmental changes follows as a predictable consequence of uncertainty. This result is intuitively attractive and does not rely on the strong assumptions about information and decision-making ability of the standard economic model.

There are four major implications of the reliability model:

1. The probability of selecting an action is inversely related to shifts in the tolerance limit.
2. The probability of selecting an action is directly related to shifts in the reliability ratio.

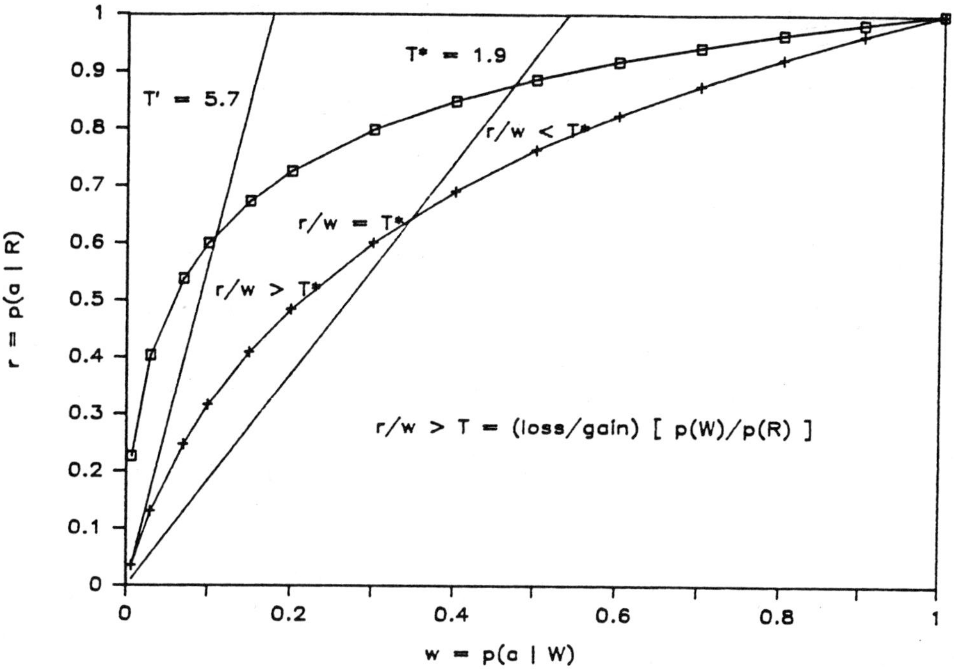

Fig. 2. The reliability condition.

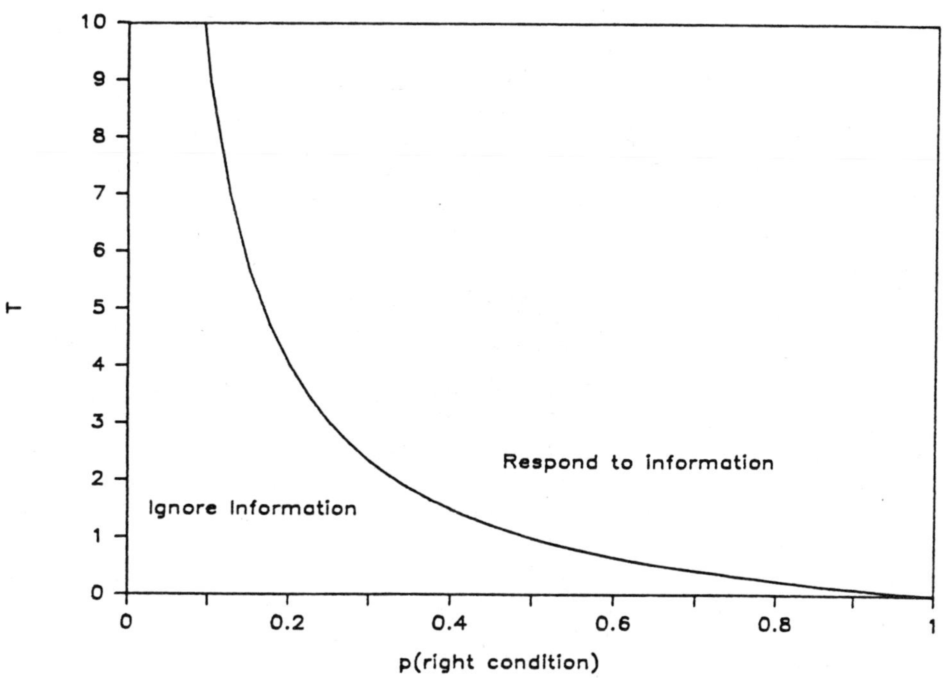

Fig. 3. The tolerance limit.

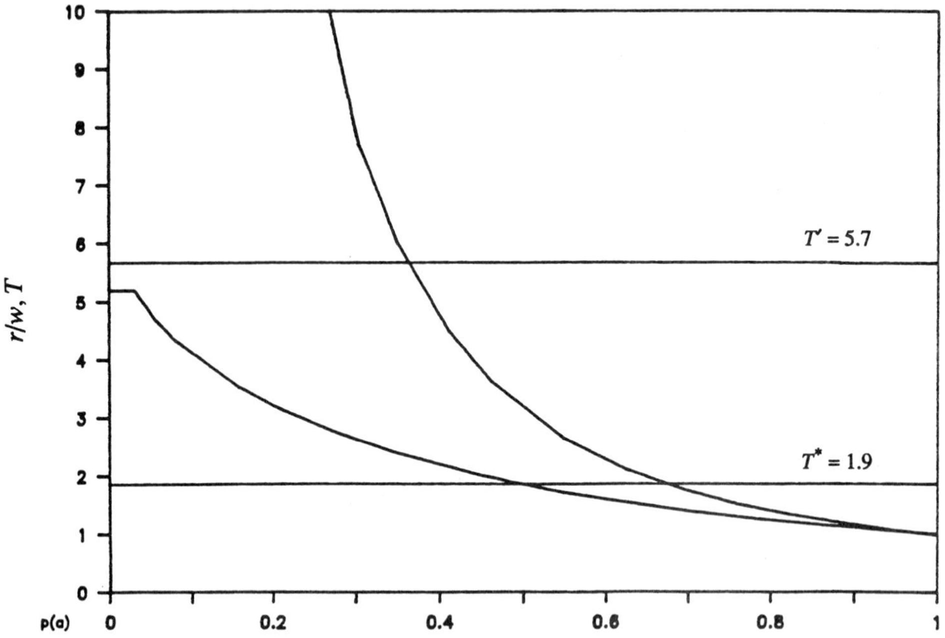

Fig. 4. Reliability curves.

3. Changes in the probability of selecting an action are greater when reliability is smaller.

4. Changes in the tolerance limit or reliability may not be large enough to cause the curves to intersect. Behavior is therefore often characterized by sudden switching when changes become large enough to favor selection of the action.

EVIDENCE ON RADON RISKS

Although most of the literature on risk perceptions relates to clinical experiments and general perceptual studies, our analysis relies on actual observed behavior of households who received new information on their exposures to significant health risks from indoor radon. The source of the health risk is a naturally occurring gas that enters buildings through water and soil gas infiltration. Indoor radon exposure results from the combination of residential location and structural characteristics of buildings. The trend toward reducing air exchanges between the indoor and outdoor air to conserve energy has greatly increased radon exposures in vulnerable locations. Because this problem was discovered only recently, there is an unknown but potentially very large number of individuals currently exposed to this hazard (U.S. General Accounting Office, 1986).

Indoor radon differs significantly from more familiar environmental problems. First, there is no identifiable polluter, since the source is naturally occurring. Second, health risk assessments for radon are based on human epidemiological studies of lung cancer incidence among uranium miners. Although there is some inevitable disagreement about dose-response coefficients, there appears to be less uncertainty than is common for risk assessments based on interspecies, low-dose extrapolations. Risk estimates are also extremely high, with lifetime risks of death from lung cancer 100 to 1000 times greater than most of the risks commonly regulated by federal agencies. Third, development of reliable techniques for reducing exposures is at a relatively early stage, while detection technology is readily available and inexpensive.

As part of an epidemiological study of the connection between radon exposure and lung cancer incidence in Maine households, the Maine Medical Center (MMC) tested a sample of homes and collected detailed medical histories. About half of the tested sample consisted of lung cancer and other cancer patients and the remainder consisted of randomly selected controls. MMC included an information pamphlet developed by the University of Maine at Orono when they sent test results to households having radon exposures of more than 20,000 pCi/l for water or more than three pCi/l for air.

The Radiological Health Program of the Maine Department of Human Services also has a radon testing program for homeowners requesting tests. All homeowners receive the University of Maine pamphlet with their test results, regardless of the exposure level. About 44% of the state tests exceeded three pCi/l. The combined MMC and State samples provided about 230 observations.

Figure 5 shows the distribution of perceived seriousness of the radon risk by actual exposure. Perceptions of low seriousness fall as actual risk increases, perceptions of high seriousness rise as actual risk increases, but the percentages reporting moderate seriousness at each risk level are approximately the same. Furthermore, large percentages of the respondents seriously misjudged the seriousness of their actual exposures. Table 1 indicates that over 58% of the respondents understated their objective risks by at least one order of magnitude after receiving the brochure designed to explain the relation between their test results and health risks. About 20% understated their exposure by two or three orders of magnitude.

An important question is whether homeowners invested appropriately in protective measures. Figure 6 shows mitigation responses by objective risk level. More than half of the whole sample did something to mitigate against radon exposure. However, many homeowners undertook such simple, low-cost measures as opening windows more frequently and avoiding basement areas.

Unfortunately, the results here are entirely inconclusive. A chi-square test indicates that the null hypothesis cannot be rejected, i.e. that mitigation is independent of exposure levels. Likewise, *t*-tests indicate that the differences in mean costs are not statistically significant. In other words, mitigating responses cannot be characterized simply in terms of exposure alone. The decision to mitigate or not to mitigate and, when mitigation is chosen, the decision of how much to spend involve complex interactions among multiple factors.

Johnson and Luken (1987) report the results of binomial probit and tobit analyses of the probability of mitigating and mitigation costs. Objective risk is statistically insignificant, while perceived seriousness of the risk is positive and statistically significant. Ease of understanding the pamphlet is positively and significantly related to the probability of mitigating. Smokers are significantly less likely to mitigate, presumably because smoking is a proxy for being less risk averse or less concerned about health generally. Respondents who have lived in their home for a long time are significantly less likely to mitigate, perhaps because they have difficulty believing that their house is a health hazard. The control group is significantly less likely to mitigate than households containing a member who has recently become a cancer patient or households who have requested testing. Variables expected to influence mitigating expenditures, such as income, education, and value of the house, were not significant in any of the models tested.

These results provide additional evidence that formation of risk perceptions and associated behavioral responses do not clearly follow the expectations of conventional

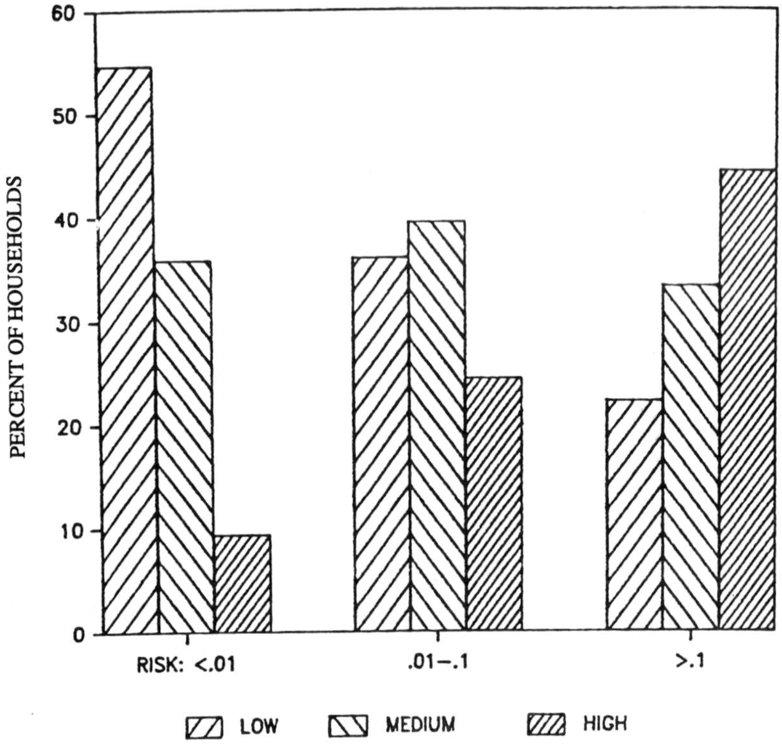

Fig. 5. Perceived seriousness of risk (percent of households at each exposure level).

Table 1. Difference Between Perceived Risk and Actual Risk (Percent of Observations)

Orders of Magnitude Difference	Understatement	Overstatement	Accurate
1	38.0	13.4	
2	19.3	1.1	
3	1.1	--	
Total	58.4	14.5	27.1

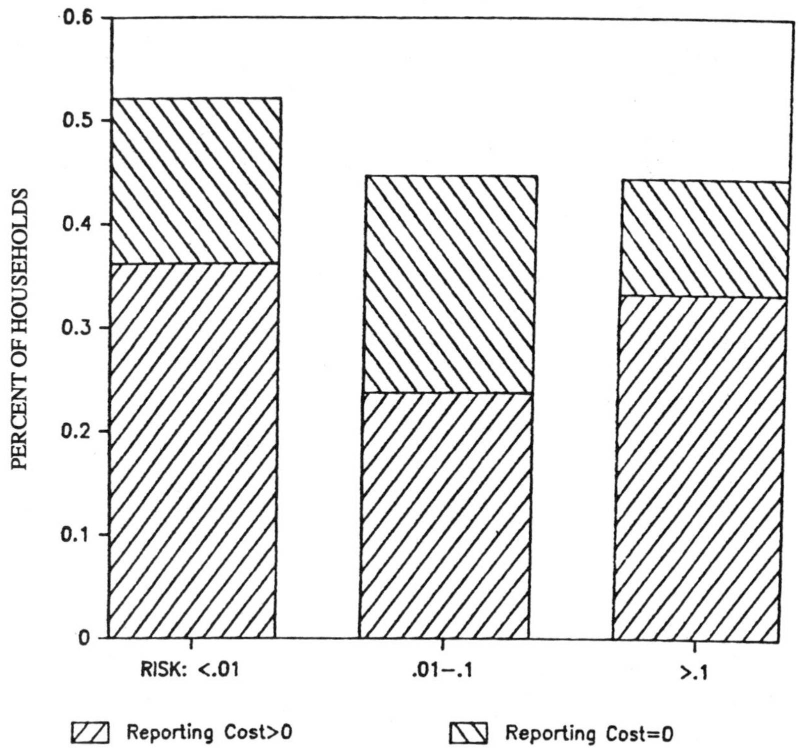

Fig. 6. Mitigation by risk level (percent of households at each exposure level).

economic models. The remainder of this paper suggests a way of interpreting these findings in light of the reliability model developed previously.

RADON MITIGATION AS A TWO-STAGE RELIABILITY PROBLEM

Radon mitigation decisions resemble complex signal detection problems. In the radon case the homeowner receives new information on radon-related health risks and mitigation alternatives against a background of existing beliefs and general "noise." In Heiner's terminology, homeowners' "competence-difficulty gap" may result in failure to respond to the message or in an inappropriate response. However, these responses are predictable consequences of uncertainty, potential losses and gains, the likelihood that the observed exposure warrants action, and personal characteristics of homeowners that affect their signal-detection competence.

Again reinterpreting Heiner (1986), suppose the right conditions R and the wrong conditions W for mitigating are correlated with the signal $s=s(e)$ contained in any information materials received, where e is the measured radon exposure. The signal is imbedded in the "noise" of background information and beliefs about risk exposures. This background information determines a probability distribution for the seriousness of risk exposures with a mean m_o. If the signal indicates exposures above the designated EPA action level, the mean shifts to m_1. Let m_o correspond to W and m_1 correspond to R in the reliability model.

We can expect that some homeowners will have difficulty detecting which value of s is actually observed. Denote the perceived message as $s'=s'\ (e,i,X)$ where i is the information on radon risks received by the household and X is a vector of household characteristics that are related to detection competence. Assume that the homeowner's decision rule dictates action only if $s \geq z$. The variable z incorporates the degree of acceptance of the designated action level, based on either acceptance of the criteria used to establish the standard or trust in the authority issuing the standard. Conditional probabilities of taking action under R and W can therefore be written as $r(z)$ and $w(z)$. Different values of z generate a signal-detection curve as in Fig. 1.

There may be circumstances where s' is below z when s is actually above z and vice versa. This additional uncertainty generates a two-stage detection problem. The first problem involves correct identification of the household's objective risks. Letting $r' = p(s' \geq z \mid s \geq z)$ and $w' = p(s' \geq z \mid s < z) = 1 - r'$, the first problem is subject to the reliability condition for r'/w'. The second problem requires choice of the decision switch point z that satisfies the reliability condition for $r(z)/w(z)$.

The combined reliability ratio over the two-stage problem is

$$\frac{r^*}{w^*} = \frac{r(z)[(r'/w') - 1] + 1}{w(z)[(r'/w') - 1] + 1}$$

The probabilities $r(z)$ and $w(z)$ both approach zero as z becomes larger. The concavity of the signal detection curve ensures that, as z increases for any given r', r^*/w^* will rise to a maximum after which r^*/w^* will decrease toward 1. Beyond this maximum the homeowner can no longer benefit from increasing z.

Figure 7 shows combined reliability curves for several values of r' (the chance of correctly detecting the objective risk). Figure 7 has two important properties: (1) the reliability curve is unbounded as h approaches zero only for the case of perfect detection of the objective risk ($r' = 1$); (2) second-stage reliability is severely constrained by the success of the first stage. Even for $r' = 0.9$, second stage reliability is bounded below 2 in this example. This restriction means that tolerance limits requiring $r(z)$ to be twice as large as $w(z)$ are not attainable for any positive probability of taking action.

The two-stage reliability model provides a useful framework for analyzing observed radon mitigation in response to new information. The reliability model yields a number of testable hypotheses related to the four implications listed previously. The first implication was that $p(a)$ is inversely related to $T = (L/G)\ [p(W)/p(R)]$. Variations in T via the loss-gain ratio are related to the perceived risk (–), cost of mitigation (+), number (–) and age (+) of household members, and household income (–), where the sign in parentheses indicates the direction of the effect on T. Whatever judgment the homeowner makes about the acceptability of the suggested action levels, z, we can presume that $p(R)$ increases with the difference between the measured exposure and the EPA action levels. Higher readings should therefore be associated with lower tolerance limits across the sample.

Heiner argues that uncertainty arising from the remoteness of the information from the homeowner's experience also affects reliability. Households with recent experiences of cancer may therefore be more responsive to information on cancer risks.

The third implication was that $p(a)$ is more responsive to changes in T at lower levels of reliability. This implication enables us to predict the relative magnitudes of coefficients in addition to the sign predictions of the previous discussions. The analysis will consist of disaggregating the sample by hypothesized reliability and testing the comparative sizes of coefficients on variables thought to shift T.

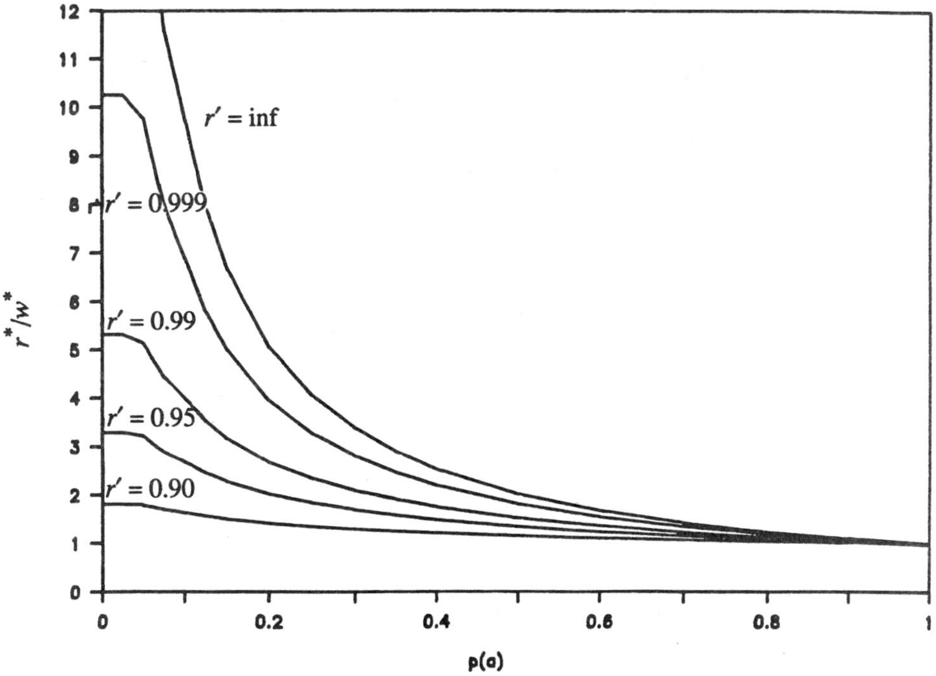

Fig. 7. Two-stage reliability curves.

The fourth implication was that abrupt switching (i.e., sudden changes from no mitigation activity to substantial mitigation activity, rather than gradual adoption of increasingly more costly measures) is likely to occur, but not necessarily only at the specified action levels. Various sources of uncertainty facing the homeowner may cause switching at levels higher or lower than the recommendations. Switching behavior may also take the form of searching for additional information on radon. At some point, shifts in T and r^*/w^* could cause the homeowner to shift abruptly from expending no effort to acquire additional information to expending substantial effort. Since the study design will allow us to observe responses at several points following receipt of test results and brochures, it will be possible to test such dynamic implications of the reliability model.

Unfortunately, the Maine data cannot provide a rigorous test of these hypotheses. Household responses were observed at a single point in time several months after receiving test results and the information brochure. Therefore there is no accurate measure of the relation between evolving risk perceptions and the decision to mitigate as signals change over time.

EPA is currently undertaking a more extensive study of 2300 households in New York state. Both the households being tested and a control sample were surveyed before any test results were sent in order to establish baseline knowledge, perceptions, and attitudes. Participants will be sent one of five different information packages. Each package presents risks and mitigation alternatives in a different way in order to generate differences in the information signal. Subsequent surveys of the participants will measure dynamic responses as perceptions evolve over time.

CONCLUSION

Problems in applying standard optimizing concepts to analyze anomalous risk-related behavior have led to such increasingly sophisticated models as state-dependent utility. These models implicitly require greater competence on the part of decision makers to solve complex optimizing problems. In contrast, extensive psychological evidence indicates that people tend to rely on simplifying heuristics or decision rules when faced with such complex decisions.

Radon mitigation is likely to manifest many of the analytical difficulties observed in other low-risk situations. Therefore it seems desirable to address the modeling of rule-governed behavior directly. The reliability model explains such behavior by explicitly incorporating uncertainty in an intuitively appealing way. It also generates a number of interesting hypotheses suitable for testing with data currently being gathered on signal responses of a large sample of New York homeowners.

REFERENCES

Einhorn, Hillel J., and Hogarth, Robin M., 1985, Ambiguity and Uncertainty in Probabilistic Inference, *Psychological Review* **92(4)**:433-461 (October).

Heiner, Ronald A., 1983, The Origin of Predictable Behavior, *American Economic Review* **73(4)**:560-95 (September).

Heiner, Ronald A., 1984, On Reinterpreting the Foundations of Risk and Utility Theory, Unpublished paper, Economics Department, Brigham Young University.

Heiner, Ronald A., 1986, Uncertainty, Signal-Detection Experiments, and Modeling Behavior, in *Economics as a Process: Essays in the New Institutional Economics*, Richard N. Langlois, ed., Cambridge University Press, New York.

Heiner, Ronald A., 1985, Rational Expectations When Agents Imperfectly Use Information, *Journal of Post Keynesian Economics* **8(2)**:201-207.

Heiner, Ronald A., The Necessity of Delaying Adjustment of Imperfect Decisions and Expectations, *American Economic Review* (in press).

Heiner, Ronald A., The Necessity of Imperfectly Using Information, *Journal of Economic Behavior and Organization* (in press).

Kaen, Fred R., and Rosenman, Robert E., 1986, Predictable Behavior in Financial Markets: Some Evidence in Support of Heiner's Hypothesis, *American Economic Review* **76(1)**:212-220 (March).

Johnson, F. Reed, and Luken, Ralph A., 1987, Radon Risk Information and Voluntary Protection: Evidence from a Natural Experiment, *Risk Analysis* **7(1)**:97-107.

Kahneman, Daniel, and Tversky, Amos, 1979, Prospect Theory: An analysis of Decision Under Risk, *Econometrica* **47**:263-91 (March).

Schoemaker, J. H., 1982, The Expected Utility Model: Its Variants, Purposes, Evidence, and Limitations, *Journal of Economic Literature* **XX**:529-563 (June).

Slovic, P., B. Fischoff, and S. Lichtenstein, 1985, Regulation of Risk: A Psychological Perspective, in *Regulatory Policy and the Social Sciences*, Roger Noll, ed., University of California Press, Berkeley.

Thaler, Richard, 1980, Toward a Positive Theory of Consumer Choice, *Journal of Economic Behavior and Organization* **1**:39-60.

Tversky, Amos, and Kahneman, Daniel, 1977, Causal Schema in Judgments Under Uncertainty, in *Progress in Social Psychology*, M. Fishbein, ed., Erlbaum, NJ.

U.S. General Accounting Office, 1986, Air Pollution: Hazards of Indoor Radon Could Pose a National Health Problem, GAO/RCED-86-170, Washington, D.C. (June).

Role of Publicity in Three Risk Management Decisions: Ethylene Dibromide as a Pesticide, Dioxin in Times Beach, and Screening Mammography

Barbara Berney
Independent Consultant
Washington, DC

Barbara Mandula
National Science Foundation
Washington, DC

Jack Needleman
Lewin/ICF
Washington, DC

ABSTRACT

To study the role of publicity in risk management decisions, we looked at national press coverage of three cases. Once the public and press became concerned about the risk issues in these cases, government and other groups were under pressure to act in ways that generated favorable press and that reassured the public that the risks were being controlled. The press reported the risk estimates accurately, but without questioning the validity of the estimates.

KEYWORDS: Ethylene dibromide, Times Beach, dioxin, mammography, publicity, risk management

INTRODUCTION

> "Since the early 1950s, much of public policy has been devoted to concern about the impact of science upon society—for example, dealing with the public's fears of environmental degradation and toxic hazards. At the federal level alone, 13 major laws are the basis for thousands of pages of regulation aimed at controlling how science will be developed and its fruits applied. . . . Science is a motivating force in our society. It goes out into the streets via the journalists, and it comes back in the form of myriad laws and regulations at all levels of society," (Smith, 1986).

New Risks
Edited by L. A. Cox, Jr., and P. F. Ricci
Plenum Press, New York, 1990

Decisions on occupational and environmental hazards are based on political as well as scientific considerations. The political considerations include media coverage of the hazard; public and government response to the hazard; and publicity about these responses.

We hypothesized that a risk assessment would engender responses of the public, press, and public interest groups, and that the vigor of these responses would, in turn, influence government action as much as did the results of the risk assessment itself. In examining this hypothesis, we have looked at three cases: occupational and environmental exposure to the pesticide ethylene dibromide (EDB); dioxin contamination of the soil in Times Beach, Missouri; and screening mammography. While these cases do not represent the universe of risk assessments, their analysis can suggest how risk assessment, publicity, and public pressure interact to affect risk management decisions. We did not look at unpublicized cases with comparable risks.

Risk management decisions take place within a larger political and historical context. This context was extremely important in the EDB and dioxin cases. These two cases received widespread publicity shortly before the federal government took definitive action to limit exposure to these substances. These government steps occurred at a time when the administration was under pressure on environmental issues in general. For example, the observation that the initial assessments and animal studies of EDB were done years before the press took an interest or the government took action suggests that the publicity itself strongly influenced at least the timing, if not the strength, of government action.

Table 1 shows characteristics of the three cases. Cancer was the endpoint of concern in all three cases; two cases involved primarily the Environmental Protection Agency (EPA), while the third was a concern of the Department of Health, Education, and Welfare (DHEW), now the Department of Health and Human Services (DHHS). The two EPA cases were part of a regulatory decision-making process, while the HEW case, mammography, was part of an advisory process that would lead to government guidelines. We asked the following questions about each case:

1. Why did the press begin to cover the case? What constituted news?
2. How accurately were risks and benefits described?
3. What level of concern (danger) was communicated?
4. What sources did the press use for information?
5. How did the press describe public response to the government risk assessments and to the press coverage?
6. How did publicity and public reaction influence risk management decisions?

Methodology

While television and community press play an important role in publicizing information about risks, access to archives for these media are limited. Consequently, the analyses presented here are generally based on review of major national newspapers. Table 2 shows the indexes we used for locating newspaper stories. Our systematic literature searches extended to mid-1986.

These searches identified approximately 200 newspaper stories about EDB as a pesticide and approximately 75 stories about Times Beach. Approximately 40 stories on mammography were found for 1976-1977; none were located for 1975.

The *Readers' Guide to Periodical Literature* was used to locate articles in magazines for the three cases.

Table 1. Characteristics of Cases

Case	Health Endpoint	Years of Publicity	Agency	Outcome or Status (1988)
EDB as pesticide	Cancer	1981-84	EPA, OSHA	EPA banned EDB from almost all pesticide uses
Dioxin in soil in Times Beach, Missouri	Cancer	1982-83	EPA, CDC	Times Beach buyout by EPA
Mammography screening for breast cancer	Breast cancer	1976-77; ongoing	NCI, NIH	Recommended for women above age 40; until 1987, NCI recommended screening below age 50 only for women at high risk for breast cancer

We compared print media reporting with technical risk assessments prepared by government agencies or by committees or groups associated with government agencies. Our conclusions are limited by the possibility that broadcast media and local papers might have had different coverage from the national print media that we examined. However, many of our conclusions agree with those of Greenberg *et al.* (1989).

ETHYLENE DIBROMIDE

Background

Ethylene dibromide (EDB) was used extensively as a grain, fruit, and soil fumigant for 30 years until 1983-84. At that time EPA suspended its use as a pesticide except for certain citrus intended for export.[a] This case study examines the efforts of EPA and the Occupational Safety and Health Administration (OSHA) between 1973 and 1985 to review their regulations on EDB and the publicity that attended these efforts. EDB was regulated under the OSHAct by OSHA; under the Federal Insecticide, Fungicide, and Rodenticide Act (FIFRA) by EPA; and by various states under legislation covering contamination of food and drinking water.

Risk assessments concentrated on cancer and reproductive harm. They were conducted by or for EPA, OSHA, NIOSH, and several states, including New York and California. Workers exposed at the allowable occupational exposure limit of 20 ppm were estimated to have an excess individual lifetime cancer risk of 0.2 to 1. EPA estimated the individual excess cancer risk from lifetime exposure to food contaminated with EDB at levels found in 1982-83 to be between 0.00004 and 0.0001. This estimate is at least 1000 times lower than the estimated risk permitted for workers.

a. This paper is not addressing EDB's primary use, as an antiknock agent in leaded gasoline. As leaded gasoline is phased out, this use of EDB is decreasing.

Table 2. Sources of Newspaper Articles

Newspaper	Time Period	Index Used	Cases
New York Times	1973-1979	Printed Index	EDB Times Beach Mammography
New York Times	1979-1986	National Newspaper Index, computerized (NNI)	EDB Times Beach Mammography
Washington Post	1973-1982	Printed Index	EDB Times Beach Mammography
Washington Post	1982-1986	NNI	EDB Times Beach Mammography
Wall Street Journal	1979-1986	NNI	EDB Times Beach Mammography
Christian Science Monitor	1979-1986	NNI	EDB Times Beach Mammography
Los Angeles Times	1982-1986	NNI	EDB Times Beach Mammography
St. Louis *Post-Dispatch*	1979-1986	Printed Index	Times Beach

Timing of Press Coverage

EDB was the center of a regulatory battle beginning in the mid-1970's, but gained little notice until it was used during the medfly crisis in California in 1981. During that period a few articles in national magazines and local newspapers covered California OSHA's emergency temporary standard for occupational exposure to EDB and the refusal of dockworkers in California and Japan to handle treated produce until the proposed California standard was put into effect.

EDB reappeared in the news in the fall of 1983 when Florida officials discovered it in groundwater and in wells in areas where it had been used extensively as a soil fumigant. Florida issued an emergency ban against its use as a soil fumigant. State officials also discovered EDB in baking mixes, flour and baked goods, and generated widespread publicity by pulling these items from grocery shelves. Further testing revealed EDB contamination of groundwater in California, Hawaii, and Georgia. California suspended its use in counties where it was detectable in the water supply. Publicity continued through

1984 while EPA suspended use of the pesticide, first as a soil fumigant, then as a grain and mill machinery fumigant, and finally on citrus intended for domestic consumption.

Rather than reporting on contamination *per se*, the press generally concentrated on the government reactions to the findings, such as the banning of the pesticide. Congressional hearings on the risks of EDB and on EPA's handling of the pesticide were extensively reported in the media.

Reporting of Risks

The press accurately reported the risk estimates found in government documents, but generally did not question the estimates. Most articles mentioned that EDB is a "potent animal carcinogen," and that it has reproductive and other adverse health effects in animals. However, the statements and quotes do little to help the reader determine the actual danger or the validity of the various conclusions.

Many articles explained the cancer risk in detail. One *New York Times* editorial said, "A lifetime of exposure to EDB-tainted food, the (EPA) now estimates, would cause cancer in 3 of every 1000 people." (NYT, 1/21/84) Another *New York Times* article (2/21/84) explained that EPA estimates that the average 30-year-old adult who eats EDB-contaminated food for the next two or three years would increase his chances of getting cancer by only 1 in one million or 1 in ten million. The article compared the risk of EDB with that of aflatoxin and quoted academics and government officials with contrasting judgments of the risk. It was one of the few articles reviewed that described sources of uncertainty in risk assessments.

Conflicts among findings or conclusions were reported but rarely analyzed or critiqued. In a September 1983 article, *The New York Times* (9/26/83) reported "Florida health officials consider amounts greater than 0.1 of a part per billion unsafe. Hawaii considers more than 0.02 of a part per billion unsafe and California considers more than one part per billion unsafe. Although the Federal Government has no safety standard for the chemical in drinking water, an advisory range will be issued next week..." Because the article makes no comment about the reasons for the differences in the assessments, the lay reader has no basis for understanding the significance of these differences.

In writing about the risk assessment, the press quoted various sources, such as industry or environmental groups, which have a clear point of view. Sources outside of government did not usually comment on the numerical estimates of the risk assessments, but did discuss whether a risk was present. For example, many stories quoted industry representatives saying that their products containing EDB are safe to eat, or that the pesticide is safe to use. *The New York Times* (9/26/83) quoted an industry spokesperson as saying that there is no evidence that EDB causes adverse health effects when used according to label instructions. Label instructions include keeping the chemical out of water supplies, a point that the article does not mention.

In a later story (NYT, 1/6/84), a spokesman from the Grocery Manufacturers of America pointed out that "People have been eating foods with EDB in them for 40 years and nobody has dropped dead." The same story includes statements by the Natural Resources Defense Council, pointing out that there is no safe level for a carcinogen in food.

The industry and public interest sources are balanced by more "neutral" sources, such as academics, government scientists, and regulators. EPA made an oft-quoted attempt to explain that its estimated risk referred to eating contaminated products consistently over a lifetime, rather than to an immediate or acute health threat. This explanation was transmitted quite well in relation to food containing traces of EDB, but less well regarding the hazard represented by drinking water containing EDB. William Ruckelshaus, then

Administrator of EPA, was quoted in many articles as saying that we are faced not with a public health emergency but with an effort to eliminate a long-term, chronic, and unacceptable health risk.

Level of Concern Communicated

Most articles quoted the risks in a neutral fashion, projecting a sense that most scientists and public officials believed that eating cake made from a tainted mix would not change a person's health status now or ever. Articles that discussed the occupational risk generally distinguished between exposure of workers and exposure of the general public. The risk to workers was portrayed as substantial and much greater than that of the general public.

There were exceptions to this reasonably neutral approach. The lead on a *Washington Post* (2/14/84) story read, "In April 1973, barely four months into the National Cancer Institute's laboratory test on ethylene dibromide (EDB), half of the rats in the high-dosage experimental group were dead." Although the story went on to say that the individual cancer risk from eating contaminated food is about one in a million, the reader was left with the feeling that EDB was extremely hazardous.

Television images also were sometimes extremely graphic and dramatic. Although the written press generally stated that the risk was long term, one set of TV images showed two workers being hosed down and taken to a hospital after they were accidentally exposed to large amounts of liquid EDB. The newscaster announced that the workers later died. Another TV report showed a grain elevator worker with a debilitating nervous disorder, supposedly related to EDB exposure. It is not clear if the TV stories attempted to distinguish between worker and community exposures in these two situations (Sharlin, 1986).

Publicity and Public Response

The public response, as discussed in the press, varied from disbelief to fear, anger and confusion. The press reported that people called health departments and EPA asking whether they should eat products listed as contaminated. In some areas, sales of cake mixes, cereals, and bread dropped.[b]

Both industry and government responded to publicity. One industry spokesman said that adverse publicity would require industry to stop using EDB regardless of government action. In fact, shortly after millions of viewers watched potentially contaminated grain products being removed from grocery shelves, the major grain millers and grain product manufacturers announced that they were voluntarily ceasing to use EDB.

Although EPA proposed suspending EDB in 1980, no further action was taken for three years. Under pressure from intense publicity and several congressional hearings, the agency carried out two emergency suspensions in less than six months; these were the second and third emergency suspensions that the agency had ever issued. OSHA also speeded its regulatory process while publicity was intense, taking less than a month to prepare a standard for publication in the Federal Register. However, the Office of Management and Budget did not approve publication, and, since OSHA has not issued a new standard for EDB as of mid 1988, the earlier standard of 20 ppm remains in effect. It is unclear to what extent OSHA failed to take further action because EDB fell from public view, or because bans by EPA largely mooted the issue.

---

b. Johnson (1988) discusses press coverage of the EDB contamination and the economic costs of the public's reaction.

DIOXIN AND TIMES BEACH

Background

The term dioxin in this report refers to the 2,3,7,8-isomer of tetrachloro-dibenzo-p-dioxin. It is a contaminant of herbicides and pesticides and a by-product of certain manufacturing processes. It is a potent carcinogen in animal tests.

In 1971, a waste oil dealer in Missouri acquired 29 kilograms of sludge wastes contaminated with dioxin, mixed it with waste oils, and sprayed it for dust control on dirt roads and riding arenas in the eastern part of Missouri, including Times Beach, a small blue collar town on the banks of the Meremac River. Immediately after the oil sludge spraying, hundreds of birds, several cats and dogs, and numerous rodents were found dead near a riding arena. At least one child became ill, and over the next year or two, dozens of horses sickened and died. In 1974, the Centers for Disease Control (CDC) established that the riding arenas were contaminated with dioxin. In 1982, EPA returned to Times Beach and found high levels of dioxin still present in the soil. In December of that year, Times Beach was flooded. CDC urged residents not to return to their homes because of the dioxin contamination, and early in 1983, EPA decided to buy out Times Beach.

This case looks at CDC's risk assessment of dioxin in the soil in Missouri, the publicity surrounding the discovery of dioxin in Times Beach, and the EPA buyout of the town.

Initial Coverage and Timing

Local (St. Louis *Post-Dispatch*) coverage of dioxin contamination in Missouri began in October 1982 with EPA's rediscovery of high dioxin levels in Jefferson City. Coverage in the national press began around the same time. Most stories were written in response to government statements, government release of reports, or other government actions.

Reporting of Risks

CDC's risk assessment concluded that dioxin in the soil presented a hazard to human health at 1 part per billion (ppb) or greater. The articles we reviewed reported this risk level accurately, generally with appropriate caveats. For example, many articles stated that dioxin was extremely toxic when tested in laboratory animals, although its effects on humans were unclear. Newspapers and more scientific publications, such as *Science* or *Science News*, used similar terms in reporting the risk.

However, publications differed in their emphasis. The *Wall Street Journal* tended to downplay the risks,[c] running, for example, an "Op Ed" piece by chemical company executives that said, "The Times Beach dioxin episode is an example of overreaction to a pollution problem without regard to the facts at hand" (WSJ, 6/17/83). Human interest publications tended to concentrate more on the potential dangers of the chemical: "Dioxin is one of the deadliest compounds known to man—ravaging every system in the body" (*Life*, 5/83).

Public officials and the media attempted to convey the appropriate level of concern and to make the risk meaningful. Gary Stein of CDC drew a parallel between dioxin exposure and smoking, noting that it was important in both cases to stop the exposure, i.e., leave Times Beach to avoid or minimize future health problems. Unlike the EDB case, where reporters queried many people for a description of the risk, almost all the sources for

---

c. The *Wall Street Journal* tended also to downplay the risks associated with EDB.

Times Beach were scientists or officials in EPA or CDC. The St. Louis *Post-Dispatch* also quoted state officials.

Publicity and Public Response

Times Beach residents were often quoted about their reactions to the risk. Reactions were mixed. Some residents expressed anger and fear that they had been living in a contaminated environment for 10 years. They blamed dioxin exposure for birth defects, cancer, and various other chronic health problems. Others felt that if the dioxin had not yet made them sick, it was not a great danger. In general, the residents wanted definitive answers about whether their town was safe—so they could return home. If unsafe, they wanted definitive government action.

There is little doubt that publicity affected both public response and government action. One resident complained, "We never get any information from the authorities, just what we see on TV, and that seems to change with each newscast" (NYT 1/10/83). According to Gough (1986), the investigation of dioxin in Missouri was slow and quiet until the Environmental Defense Fund released its list of sites and concentrations. Then "mounting public pressure got the sampling program under way relatively quickly" (Gough, 1986). The press also reported similar effects of publicity in accelerating and strengthening government actions.

The rediscovery of high levels of dioxin in Times Beach and the subsequent flooding and evacuation of the town coincided with increasing controversy about EPA's handling of Superfund in particular and toxic waste problems in general. Rita Lavelle, chief of EPA's toxic waste programs, was forced to resign in the midst of the Times Beach episode. A month after announcing the buyout of Times Beach, EPA Administrator Ann Burford had also resigned under pressure.

The press saw events in Times Beach being pushed by the overall problems of the agency. *Newsweek* (3/7/83) reported, "In a frantic effort to control damage from the spreading EPA controversy, EPA administrator Ann Burford ... flew to Times Beach, MO, where she informed residents that the government had decided to spend $33 million to buy out their toxics-poisoned town. Then the administrator abruptly announced the forced resignations of two high level EPA officials."

Town residents reportedly saw the buyout as a political deal and a publicity stunt (*Time*, 3/7/83). Various other stories also reported that the buyout was in large part a response to public pressure, to publicity, and to the agency's need for good press and definitive action (LA Times, 1/10/83; WP, 2/23/83; *US News*, 3/7/83). The same levels of contamination discovered—but not widely reported—in 1975 had not resulted in government action. This lack of earlier response may have been due partly to the mistaken notion that dioxin had a one- or two-year half-life in the soil, but it is also likely that lack of public pressure and press coverage played a role.

MAMMOGRAPHY

Background

Mammography refers to using x-rays to detect abnormalities in breast tissue, including tumors too small to be palpable. Small tumors can be more successfully treated than those found only when they are large enough to be felt. Screening mammography refers to using the technique in the absence of signs or symptoms of disease, in contrast to diagnostic mammography, which is used when signs or symptoms are already present. The value of diagnostic mammography has not been questioned and is not addressed here.

This case differs from the previous two in several aspects. Screening mammography involves questions of balancing health risks and health benefits. In addition, the decision to be exposed to radiation from mammography is an individual decision. Furthermore, health care professionals are an important source of information regarding the decision. The government role is advisory rather than regulatory.

In the United States, approximately 120,000 cases of breast cancer are diagnosed per year, about one-quarter of them in women below age 50 (ACS, 1985; USDHHS, 1985). Annual mortality from breast cancer is about 40,000. The risk from mammography concerns the possibility that the technique could induce tumors. The major controversy concerned the advisability of using mammography in screening women below age 50; above that age, there was reasonable agreement that benefits outweighed risks.

The technical basis for the question goes back to a controlled breast cancer screening program that the Health Insurance Plan (HIP) of New York carried out from 1962-1966 among 31,000 women offered screening and 31,000 unscreened women. A 30% decrease in breast cancer mortality occurred among the screened women, but this decrease was seen only in women older than age 50 (Breslow *et al.*, 1977).

Based on the HIP study, the National Cancer Institute (NCI) and the American Cancer Society jointly began a multi-year Breast Cancer Detection Demonstration Project (BCDDP) in 1973, involving 27 centers doing annual screening, including mammography, for 5 years on 270,000 women ages 35-74 (Baker, 1982).

During the project's early stages, Dr. John Bailar, then deputy associate director for cancer control at NCI, noted that the efficacy of screening mammography had not been shown for women below age 50 and that mammography might take as many lives as it saved (Bailar, 1976). NCI and the National Institutes of Health (NIH) set up several ad hoc committees during 1975 to 1977 to consider the risk/benefit issues. NCI's initial decision to study the problem seems to have occurred in the absence of press attention, but the possibility of adverse publicity presumably existed.

The major press coverage occurred in the summer and fall of 1976 and the fall of 1977. Reports and recommendations associated with the various committees provided the news angle for many of the stories. Thus, in July 1976, the press reported that a committee chaired by public health specialist Lester Breslow recommended against mammography screening of women below age 50 unless they were in high risk categories (NYT, 7/15/76). Soon after, NCI suggested that Breslow's recommendations be followed in the demonstration project (NYT, 8/23/76; *Newsweek*, 9/6/76). In May 1977, although little noted in the press, these guidelines became mandatory in the BCDDP (NYT, 5/12/77).

In September 1977, breast cancer mammography became the topic for the first NIH consensus conference, which was well publicized both for its open process and its conclusions. The consensus panel supported the guidelines issued a year earlier, recommending that women under age 50 avoid screening mammography unless they or a first degree relative had had breast cancer. The panel affirmed the advisability of screening women above age 50 (NYT, 9/17/77).

Reporting of Risks

Reporting in newspapers and in national magazines such as *Time* and *Newsweek* was accurate, with risks stated in terms similar to those used in the technical documents. In no case did the articles compare in detail the risks and benefits of screening below age 50, but they did state that the breast cancer incidence increased among older women. *The New York Times* (7/26/76) noted that NCI's ad hoc committee to assess mammography risks had estimated that one mammogram at one rad might increase a woman's lifetime chance of getting breast cancer from a background rate of 7% (the rate then) to 7.07%. This increase is somewhat greater than that later published by the committee (Upton *et al.*, 1977).

In describing benefits of early detection, one article noted that five-year survival rates are 84% if the disease has not spread beyond the breast, compared with 55% if the disease includes lymph nodes (NYT, 7/26/76). The article further notes that five-year survival rates of up to 95% have been reported for the very earliest tumors that are detectable only by mammography.

Stories presented several viewpoints. Reporters quoted spokespersons from the NCI and from the NCI ad hoc committees to study mammography issues, representatives of the American Cancer Society, and directors of demonstration project centers. The latter two groups were generally more enthusiastic about mammography for younger women than were NCI spokespersons. *Time* (8/2/76) quoted Dr. Lester Breslow, chair of an NCI committee to assess benefits of screening mammography, saying that mammography is "a striking example of a situation where the very disease may be caused by the technology." He urged prompt discontinuation of mammography in routine screening of women under age 50. *Time* also quoted NCI's deputy director, who said he would have no hesitation recommending mammography for any woman over 50. But Dr. Philip Strax, director of the New York detection center, stated, "The real risk is in not doing mammography."

Publicity and Public Response

The controversy clearly caused concern among women. The press reported a special meeting that the NCI director held to answer questions and get the views of NCI women employees (WP, 7/31/76). *The New York Times* (7/26/76) noted that the controversy had left women confused, since they had not been informed that mammography carried with it a small risk of causing breast cancer. Furthermore, some women with symptoms of possible breast cancer cancelled mammography appointments. The mammography issue caused enough concern that Dr. Theodore Cooper, then Assistant Secretary for Health, explained NCI's new guidelines (no screening mammography under age 50 except under special circumstances) on the television program, "Issues and Answers" on August 22, 1976. He emphasized that the technique was valuable for older women (WP, 8/23/76).

Although the stories were generally accurate, headlines sometimes gave a more serious view of the problem than did the full story. For example, two headline writers used the idea of "mammogram muddle" (*Time*, 8/2/76; NYT, 5/12/77) to indicate controversy and confusion.

Publicity in the national press decreased after 1977. Until late 1987, government guidelines on mammography screening remained the same as those issued by NCI in 1977, with mammography not recommended below age 50 except for women in high risk categories. On the other hand, the American Cancer Society has been encouraging women in their 40's to have regular mammograms, and revised ACS guidelines, issued in 1983, were reported on the front pages of *The Washington Post* and *The New York Times* (8/3/83). In late 1987, NCI (1987) issued "Working Guidelines for Early Cancer Detection," which included the following statement for breast cancer: "That beginning at the age of 40, a mammogram should be encouraged every 1 to 2 years until the age of 50 after which it should become annual." NCI's screening recommendations thus finally became similar to those of the other major professional groups concerned with breast cancer screening.

During the past decade, perhaps influenced by publicity about mammography risks, mammography technology has improved so the technique is more sensitive at lower radiation doses, thereby decreasing the risk from radiation exposure. In recent years, there has been increased publicity about the value of mammography and of regular screening for early detection of breast tumors. There is concern that relatively few women are undergoing screening mammography, and special concern for those above age 50, who are at increasing risk for breast cancer (WSJ, 7/30/85; NYT, 6/21/86, 8/31/88).

DISCUSSION AND CONCLUSIONS

In the three cases addressed here, we found that publicity about risks led to government actions. The government response appeared geared both to creating favorable press and to making the public feel that the hazard was being controlled. In the case of dioxin, EPA was undoubtedly trying also to compensate for critical press regarding its handling of other toxic substances.

The timing of risk management decisions in the three cases appeared to be tied to publicity, especially where risk assessments had been available for years before action was taken. While some information changed over time (e.g., learning about EDB in groundwater and estimating a longer half-life of dioxin in soil), it appears that the extensive publicity triggered risk management activities. In the case of screening mammography, NCI's consensus conference in 1977 took place after extensive publicity about mammography risks. It would be of interest to examine risk management activities in other cases where risk assessments indicated substantial risk, but there was little publicity.

In general, we found that reporting was clear, accurate and fair, with the actual risk numbers usually taken from government risk assessment documents. However, the stories rarely questioned the accuracy of the risk assessments or the methodology used to derive them. For example, while stories reported that the risk estimates for EDB and dioxin were obtained from animal studies, there was virtually no discussion of the problems with that methodology, or of human exposure scenarios. In the case of screening mammography, few of the stories discussed the risk-benefit issues.

Stories usually covered government action, such as state health departments pulling EDB-contaminated food from the shelves, or CDC's warning Times Beach residents not to return to their town. The results or publication of risk assessments *per se* were not considered newsworthy.

Reporters quoted people with different points of view—industry, labor, environmental groups, government officials, and academics—so that readers learned of various sides of the controversies surrounding risk assessments and consequent actions. However, stories and articles contained little independent analysis. Day-to-day coverage did not put the problem in a broader context or help the reader understand the risk assessment process. Coverage did, however, bring the existence of the problem and controversies to public attention.

REFERENCES

General

Smith, T. K., 8/1/86, Let Them Eat Cake Mix, EDB and other Risk Perspective Problems, Vital Speeches.

Greenberg, M. R., *et al.*, 1989, Network Evening News Coverage of Environmental Risk, *Risk Analysis* **9**:119-126.

Ethylene Dibromide

Johnson, F. R., 1988, Economic Costs of Misinforming About Risk: The EDB Scare and the Media, *Risk Analysis* **8**:261-269.

The New York Times, 9/26/83, Pesticide Inaction Prompts Inquiry.
The New York Times, 1/21/84, EDB: A Needless Cancer Crisis.
The New York Times, 1/6/84, Pesticide is Found in Food Samplings.
The New York Times, 2/21/84, Some Experts Say EDB Risk Is Small.
Sharlin, Harold Issadore, 1986, EDB: A Case Study in Communicating Risk, *Risk Analysis* **6(1)**:61-68.
Washington Post, 2/14/84, EDB Poses Unclear Human Cancer Risk.

Dioxin at Times Beach

Gough, Michael, 1986, *Dioxin, Agent Orange: The Facts*, Dioxin in Missouri, pp. 131-136, Plenum Press, New York.
Life, 5/83, The Ordeal of a Poisoned Town, pp. 58-65.
Los Angeles Times, 1/10/83, Superfund Not Rushing to the Rescue.
The New York Times, 1/10/83, In Dioxin-Tainted Town, No "Welcome" Signs.
Newsweek, 3/7/83, Storm Over the Environment, pp. 16-19.
Time, 3/7/83, Found: A Buyer (Times Beach, Missouri Houses), p. 39.
U.S. News and World Report, 3/7/83, Will EPA's Offensive Stem the Furor? p. 13.
Wall Street Journal, 6/17/83, Barrons, Keith C., Toxicity vs. Hazard: Dioxin and the Oleander.
Washington Post, 2/23/83, Russakoff, Dale, U.S. Offers To Buy Poisoned Homes of Times Beach.

Mammography

American Cancer Society, 1985, Cancer Facts and Figures, American Cancer Society, New York.
Bailar, J. C., III, 1976, Mammography: A Contrary View, *Ann. Int. Med.* **84**:77-84.
Baker, L. H., 1982, Breast Cancer Detection Demonstration Project: Five-year summary report, *CA* **32**:194-230.
Breslow, L., *et al.*, 1977, Report of the NCI Working Group on the Gross and Net Benefits of Mammography in Mass Screening for the Detection of Breast Cancer, *JNCI* **59**:474-478.
National Cancer Institute, 11/16/87, Working Guidelines for Early Cancer Detection.
Newsweek, 9/6/76, Cancer Warning, p. 34.
The New York Times, 7/15/76, Curb on Breast Cancer X-Rays Urged in Symptom-Free Cases.
The New York Times, 7/26/76, Mammography Puzzle: Benefits Versus Risks.
The New York Times, 8/23/76, Cancer Institute Proposes Limits on Breast X-Rays.
The New York Times, 5/12/77, Mammography Muddle (editorial).
The New York Times, 9/17/77, Breast Cancer Screening Is Urged.
The New York Times, 6/21/86, Early Detection is Key in Breast Cancer.
The New York Times, 8/3/83, Cancer Group Urges X-Ray Breast Tests in Younger Women.
The New York Times, 8/31/88, Breast Cancer, Caught Early (editorial).
Time, 8/2/76, Mammogram Muddle, p. 42.
Upton, A. C., *et al.*, 1977, Report of NCI ad hoc Working Group on the Risks Associated with Mammography in Mass Screening for the Detection of Breast Cancer, *JNCI* **59**:481-493.
U.S. Department of Health and Human Services, 1985, SEER Program: Cancer Incidence and Mortality in the United States 1973-1981, Biometry Branch, National Cancer Institute, NIH Publ. No. 85-1837.
Wall Street Journal, 7/30/85, X-Ray Screening for Breast Cancer Gains, but High Fees Hamper Widespread Use.
Washington Post, 7/31/76, Women Divided on Value of Breast Cancer X Rays.
Washington Post, 8/23/76, Breast Cancer Study to Urge X-Ray Curb.
Washington Post, 8/3/83, Cancer Group Prescribes Breast X-Rays After 40.

An Extension of the Coverage-Attitude Hypothesis: Coverage of Technological and Environmental Hazards in Dutch Newspapers and Readers' Reactions

Reinder J. Houwen, Henk Boer, and Jan M. Gutteling
University of Twente
The Netherlands

ABSTRACT

Most people depend on the mass media for information about technological and environmental hazards. In this paper we discuss the Coverage-Attitude Hypothesis which supposes a relationship between the amount of mass media coverage of technological and environmental hazards and the reactions of the public to these hazards.

Two studies are presented. In the first study an analysis was made of the content of three Dutch newspapers. The second study investigated the reactions of the readers of these newspapers.

The results show that major differences exist in the coverage of technological and environmental hazards in Dutch newspapers. Readers of the newspapers differed in their reactions to the risks. Readers who were exposed to more information about environmental hazards judged the risks more negatively, showed more feelings of insecurity, sought information more actively, and were less inclined to avoid the risks than readers who were exposed to less information.

KEYWORDS: Newspaper content analysis, risk perception, environmental and technological hazards, information seeking, feelings of insecurity, avoidance, dread, readers' reactions, mass media

INTRODUCTION

Risks to people and the environment arising from industrial production processes are a serious side-effect of economic and technological developments, as for example is the case with explosions at chemical plants (e.g., Mexico City, 1984), accidents during transportation or production of chemical substances (Los Alfaques, 1978; Bhopal, 1984), accidents at nuclear power plants (Harrisburg, 1979; Chernobyl, 1986) and serious soil pollution in residential areas (Love Canal, 1978; Lekkerkerk, 1980).

Environmental risks may lead to negative reactions from the public. It is known that living in the neighborhood of a hazardous industry can lead to stress (Stallen and Tomas,

New Risks
Edited by L. A. Cox, Jr., and P. F. Ricci
Plenum Press, New York, 1990

1984). Similar stress reactions also occur after confrontation with a technological disaster. This has been shown following the accident at Three Mile Island (Baum *et al.*, 1983).

However, only a small part of the population has first-hand experience of hazardous situations. Most people are informed about environmental hazards by the mass media. There are a number of different theories which seek to describe the influence of the mass media on the public. The hypodermic needle theory, for example, states that the mass media inject a passive public with information, causing a response in the short term of total agreement with the content of the message (Katz, 1963). On the basis of the idea that a single report can exert an influence in the short term, research has been conducted to investigate, among other things, the correctness and objectivity of the news reports (Scanlon *et al.*, 1978). Other theories state that the mass media only have an effect after a long period of time, and that indirect and cumulative influences are of great importance (Lowery and DeFleur, 1983). Possible effects of the coverage of risks in the mass media are feelings of fear and insecurity (Dierkes, 1980) and negative attitudes with respect to certain technologies (Mazur, 1981; 1984). The idea that the effect is determined by a continuous stream of reports has been elaborated by Mazur (1981) in the Coverage-Attitude Hypothesis:

> "The rise in reaction against a scientific technology appears to coincide with a rise in quantity of media coverage, suggesting that media attention tends to elicit a conservative public bias."

An essential feature of this hypothesis is that the quantity of media coverage of hazards is related to reactions to those hazards. Combs and Slovic (1979), for example, related the relatively great amount of news coverage of dramatic and sudden events in two newspapers to subjective estimates of numbers of victims of these events. This kind of research more or less assumes that the amount of coverage of certain types of hazards is about the same in all newspapers.

However, it is not unthinkable that in The Netherlands, with its pluriform press, newspapers could differ in the amount of attention they pay to certain types of hazards (e.g., environmental hazards). On the basis of the Coverage-Attitude Hypothesis it can be expected that readers of newspapers which pay a different amount of attention to environmental hazards will react differently to those hazards. We expected that this would not only be the case for the attitude towards environmental hazards, but also for perceptions of those hazards, feelings of insecurity associated with them, and the way people cope.

In order to test these assumptions, we conducted two separate studies: study 1 focused on the way in which three Dutch newspapers covered three types of hazards (Boer *et al.*, 1986) over an extended period of time, study 2 focused on the reactions to the hazards of readers of those newspapers (Houwen *et al.*, 1986).

STUDY 1

Subjects

Three newspapers — *De Volkskrant*, *De Telegraaf*, and *De Limburger* — were investigated. It was supposed that they differed in amount of coverage of technological and environmental hazards. A sample was drawn from editions which appeared between 1977 and 1984, by taking an edition every other day from each newspaper. In total, 1078 editions of each newspaper were analyzed. It was verified that the sample was representative for the editions in the stated period.

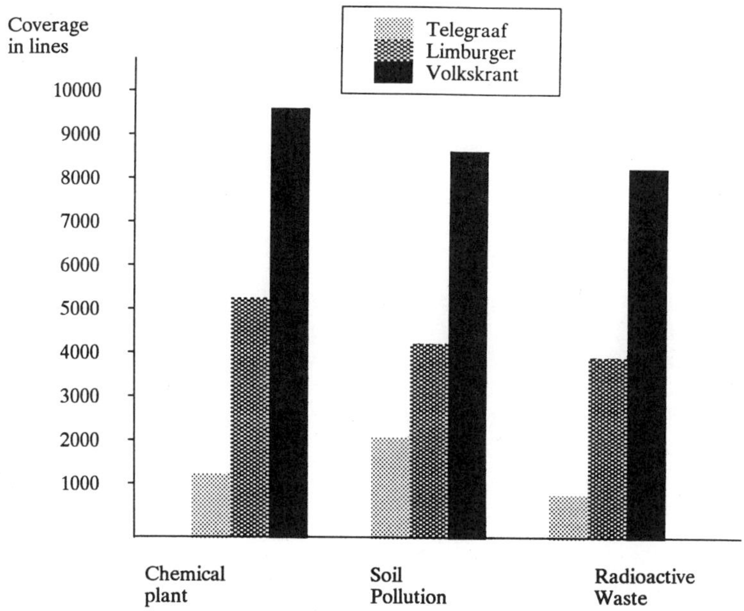

Fig. 1. Total coverage of three risks in the period 1-1-77 through 1-1-84.

Procedure

On the basis of a selection scheme, all articles concerning environmental and technological hazards were identified. This selection was found to be sufficiently reliable (86% concordance with two independent selectors). Articles were classified in a number of mutually exclusive categories, including (a) articles concerning hazards of chemical plants, (b) articles on soil pollution in residential areas, and (c) articles about radioactive waste. The classification of the articles in these categories was sufficiently reliable (90% concordance between two independent judges).

For the selected articles a number of characteristics were established, such as the number of lines in the article (see Budd, 1964). Homogeneity analysis for nominal data (HOMALS, see Gifi, 1981) showed that the number of lines of an article was the best indication of the amount of attention a newspaper paid to hazards.

Results

In all, more than 3000 newspaper articles were selected relating to environmental and technological hazards. The analyses showed that the newspapers differed systematically in their coverage of those hazards. Each year, *De Volkskrant* paid the most attention to the hazards, followed by *De Limburger*. *De Telegraaf* provided the least coverage. *De Volkskrant* devoted 96,000 lines to the hazards that were selected, *De Limburger* about 55,000 lines and *De Telegraaf* about 20,000 lines.

In Fig. 1 the coverage is shown for three specific hazards (the storage of radioactive waste, soil pollution in residential areas, and the hazards of chemical plants).

As Fig. 1 shows, there were also major differences between newspapers regarding the coverage of the three specific hazards. *De Volkskrant* paid the most attention to those hazards, followed by *De Limburger*. *De Telegraaf* provided the least coverage. *De Volkskrant* for example, paid about 2 times more attention to radioactive waste than *De Limburger* and 8 times more attention than *De Telegraaf*. The results for the other hazards were similar.

STUDY 2

Purpose

On the basis of the Coverage-Attitude Hypothesis, it can be expected that readers of *De Volkskrant*, *De Limburger*, and *De Telegraaf*, who are exposed respectively to a decreasing amount of information on radioactive waste, soil pollution in residential areas, and living in the neighborhood of a hazardous chemical plant, will differ in their reactions to those hazards. The purpose of study 2 was to test this assumption.

Subjects

In all, 148 readers of *De Volkskrant*, 166 readers of *De Limburger* and 324 readers of *De Telegraaf* participated in the study. These subjects were drawn at random from the Dutch population and filled in a questionnaire.

Procedure

In the questionnaire, subjects were confronted with three hazardous situations, namely, living in the neighborhood of a site for radioactive waste storage, living in a residential area with soil pollution, and living in the neighborhood of a hazardous chemical plant. Each situation dealt with a number of aspects of reactions to those hazards described in the literature: (a) attitude towards risks (see Mazur, 1981), (b) dread of risks (see Fischhoff *et al.*, 1978), (c) emotional reactions elicited by those risks (see Stallen and Tomas, 1984) and (d) the way people cope with them (see Baum *et al.*, 1983).

a. The subjects expressed their *attitude* towards risks on a 5-point scale on which they could indicate their degree of objection towards living in the neighborhood of the hazard, and the degree of danger they felt the hazard posed. These three items were found to form a reliable scale (Cronbach's alpha=0.75).

b. The *dread* of risks was measured on a 5-point scale on which the subjects expressed to what extent they perceived the possibility of an accident to be threatening, to what extent they considered the consequences of an accident could be controlled, and to what extent they considered accidents could be prevented by present safety measures. The 3 items formed a sufficient reliable scale (alpha=0.75).

c. The *feelings of insecurity* were expressed on a 4-point scale indicating the extent to which the thought of risks made the subject feel tense, nervous, restless, or fearful. The alpha of these items was 0.92.

d. We distinguished between two ways of coping with risks: problem-oriented coping (information seeking) and emotion-oriented coping (avoiding the problem).

   Information seeking was measured on a 5-point scale and defined by asking the subject to what extent he would consult anyone of 12 possible information sources such as the mass media, the authorities, or interest groups. The alpha of these 12 items was 0.88.

Avoidance was measured on a 5-point scale on which the subject could indicate the extent to which he did not worry about the risk, neglected the risk, resigned himself to the situation, or would try to think of something else. The alpha of these items was 0.77.

Some demographical variables such as age, sex, education and political opinion were also asked.

Analysis

The analysis was based on three groups of readers who were exposed respectively to a great amount (*De Volkskrant* readers), a medium amount (*De Limburger* readers) and a small amount (*De Telegraaf* readers) of news coverage relating to the three hazards. These groups are described in Figs. 2 to 6 as "High Exposure," "Medium Exposure," and "Low Exposure," respectively. The figures show the results for the three hazards, denoted by "Chemical Plant," "Soil Pollution," and "Radioactive Waste," respectively.

Analysis of Variance was used to test differences between the three groups of readers in their reactions towards the hazards. Because the three groups of readers differed on the demographical variables, these variables were taken as covariates in the analysis.

Results

The relationship between the amount of exposure to news coverage and the attitude towards the three hazards is presented in Fig. 2.

From Fig. 2 it can be seen that a significant relationship exists between the amount of exposure to the three hazards and attitude towards them (Chemical Plant: $F(2,524)=7.4$, $p=.001$; Soil Pollution: $F(2,523)=11.4$, $p<.0001$; Radioactive Waste: $F(2,520)=13.3$, $p<.0001$). These results are in agreement with the Coverage-Attitude Hypothesis of Mazur.

The relationship between the amount of exposure and the perceived dread of the three hazards is shown in Fig. 3.

As Fig. 3 shows, a significant relationship exists between the amount of exposure to the 3 hazards and the perceived dread of those hazards (Chemical Plant: $F(2,524)=13.5$, $p<.0001$; Soil Pollution: $F(2,523)=9.3$, $p<.0001$; Radioactive Waste: $F(2,520)=21.4$, $p<.0001$). These results indicate that the more one is exposed to information about hazards, the more threatening one perceives those hazards to be.

The relationship between the amount of exposure and feelings of insecurity of the subjects, in association with the 3 hazards, is presented in Fig. 4.

From Fig. 4 it can be seen that there is a significant relationship between the amount of exposure to the three specific hazards and feelings of insecurity of the subjects (Chemical Plant: $F(2,505)=8.8$, $p<.0001$; Soil Pollution: $F(2,501)=3.0$, $p=.05$; Radioactive Waste: $F(2,493)=11.4$, $p<.0001$). These results indicate that the more one is exposed to information about hazards, the greater the feelings of insecurity.

The relationship between amount of exposure to information about hazards and extent to which the subject copes with the hazards, in either a problem-oriented way or by avoidance, is shown in Figs. 5 and 6, respectively.

Figure 5 shows that a significant relationship exists between the amount of exposure to information about hazards and the extent to which subjects are inclined to seek information (Chemical Plant: $F(2,492)=8.3$, $p<.0001$; Soil Pollution: $F(2,502)=11.2$, $p<.0001$;

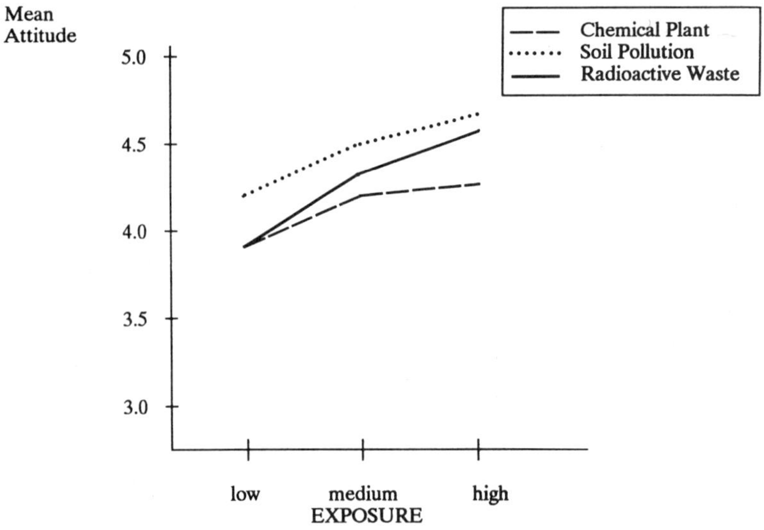

Fig. 2. The relationship between amount of exposure and the attitude towards the three risks.

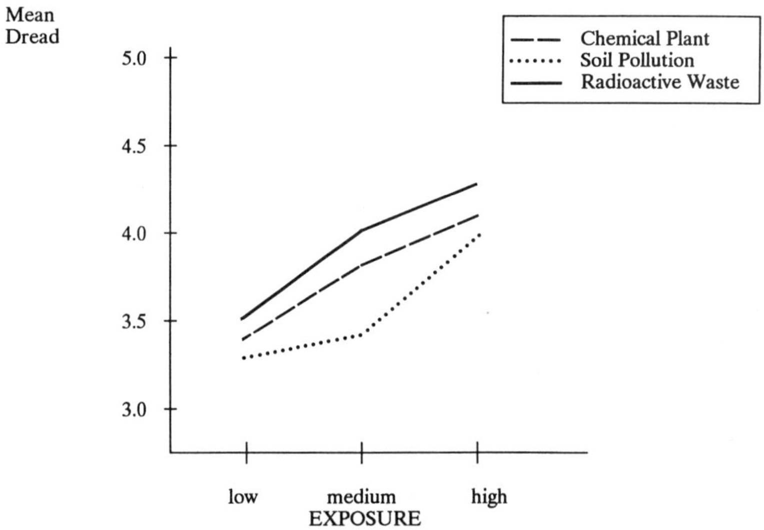

Fig. 3. The relationship between amount of exposure and the perceived dread of the risks.

Radioactive Waste: $F(2,495)=13.4$, $p<.0001$). There is a significant negative relationship between exposure and extent of avoidance (Chemical Plant: $F(2,492)= 11.2$, $p<.0001$; Soil Pollution: $(2,502)= 6.6$, $p=.002$; Radioactive Waste: $F(2,495)=14.8$, $p<.0001$).

DISCUSSION

The three Dutch newspapers studied differed strongly and systematically in the amount of attention they paid to environmental and technological hazards. The differences

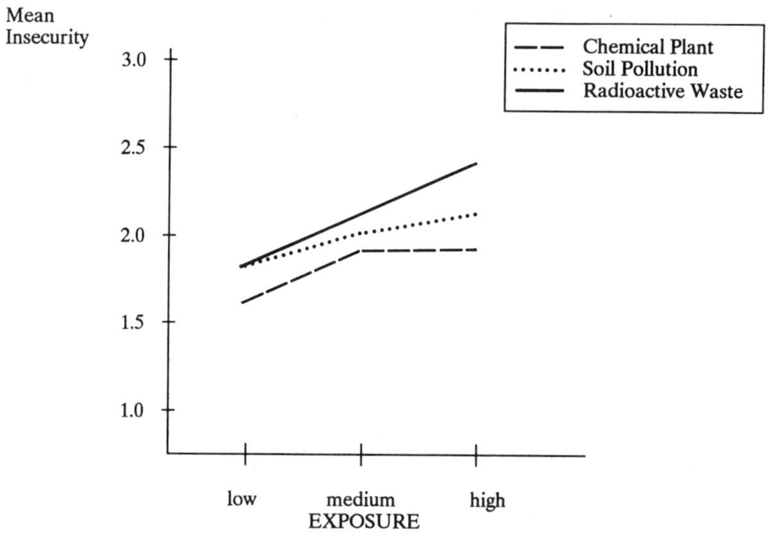

Fig. 4. The relationship between amount of exposure and the reported feelings of insecurity.

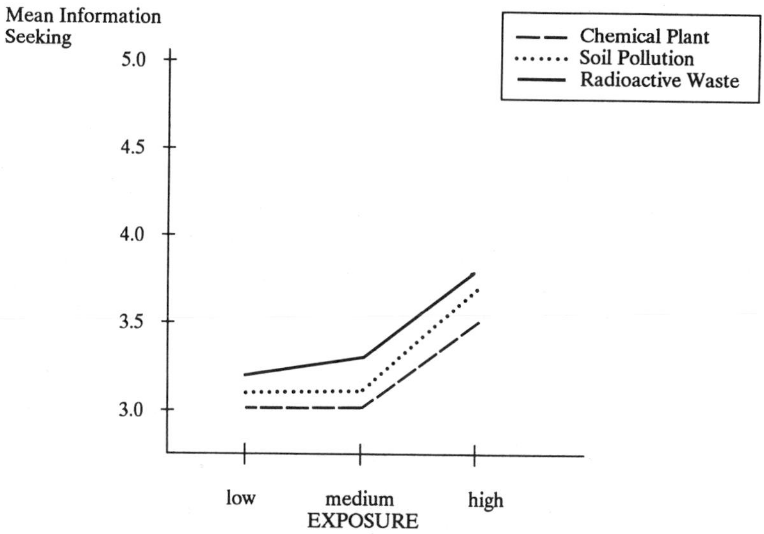

Fig. 5. The relationship between amount of exposure and the intention to seek information.

were even stronger for specific hazards such as radioactive waste, soil pollution in residential areas, and hazards of living in the neighborhood of chemical plants.

In this study it was shown that readers of those newspapers differed systematically in the way they reacted to the three hazards. The readers exposed to more information had a more negative attitude towards the hazards. These results are in agreement with the Coverage-Attitude Hypothesis of Mazur (1981). Further, it was shown that the readers exposed to more information perceived the risks to be more threatening, had more feelings

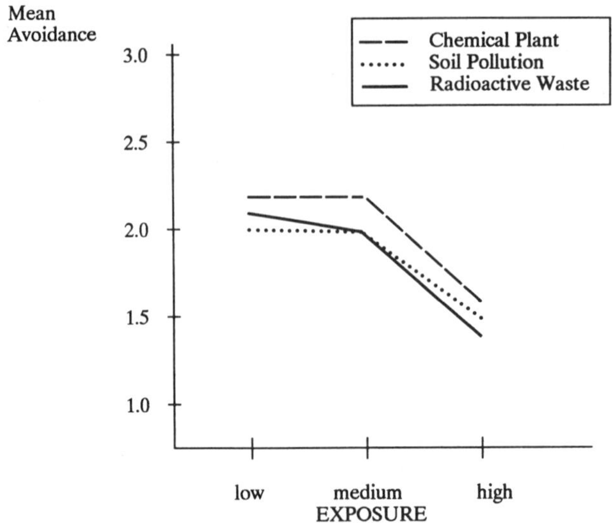

Fig. 6. The relationship between amount of exposure and the intention to avoid the risks.

of insecurity, and tended to cope with the risks. These results suggest that the Coverage-Attitude Hypothesis can be elaborated to include other aspects of the reaction to risks.

From the results of this study, implications can be drawn with regard to informing and educating the public about environmental and technological risks. Within the Dutch general public, distinct groups exist on account of exposure to varying amounts of risk information in the mass media. This segmentation of the general public is in accord with the marketing approach to risk communication (e.g., Earle, 1984). Prior to informing and educating the public about risks, it is necessary to gain insight in the reactions of the readers of different newspapers to the risk issue at hand. Further research on the generalizability of these findings to other countries is necessary.

The results are in agreement with the Coverage-Attitude Hypothesis. This strengthens the idea that increased coverage of technological and environmental hazards in the mass media leads to negative reactions from the public. In general, information on risks is aimed primarily at the negative consequences of a particular technology. Although newspapers may differ in their presentation of news, articles about technological risks are generally negatively biased. The differences in bias between and within newspapers can therefore be assumed to be small. Hence, the effect of the differences in message content will be negligible compared to the effect of differences in amount of coverage.

This is in agreement with Mazur (1984), who states with respect to the coverage of Love Canal and the Three Mile Island accident:

> "The bias in journalistic presentation is insufficient to account for the dramatic effects that the news organs had in shaping public conceptions of the accidents. Without the immense amount of attention, content alone—whatever its bias—would have had little impact on the audience."

REFERENCES

Baum, A., Gatchel, R. J., and Shaeffer, M. A., 1983, Emotional, Behavioral and Physiological Effects of Chronic Stress at Three Mile Island, *Journal of Consulting and Clinical Psychology* **51**:565-572.

Boer, H., Houwen, R. J., and Gutteling, J. M., 1986, *Analyse van de Berichtgeving in de Media Over Technologische en Milieurisico's*, Universiteit Twente, Enschede.

Budd, R. W., 1964, Attention Score: A Device for Measuring "Newsplay," *Journalism Quarterly* **41**:259-262.

Combs, B., and Slovic, P., 1979, Newspaper Coverage of Causes of Death, *Journalism Quarterly* **56**:837-843, 849.

Dierkes, M., 1980, Assessing Technological Risks and Benefits, in: *Technological Risk: Its Perception and Handling in the European Community*, M. Dierkes, S. Edwards, and R. Coppock, eds., Oelgeschlager Gunn & Hain Publishers, Cambridge.

Earle, T. C., 1984, Risk Communication: A Marketing Approach, NSF/EPA Workshop on Risk Perception and Risk Communication, Long Beach, California.

Fischhoff, B., Slovic, P., Lichtenstein, S., Read, S., and Combs, B., 1978, How Safe Is Safe Enough? A Psychometric Study of Attitudes Towards Technological Risks and Benefits, *Policy Sciences* **9**:127-152.

Gifi, A., 1981, Non-linear Multivariate Analysis, Leiden: Department of Data Theory.

Houwen, R. J., Boer, H., Gutteling, J. M., 1986, *Directe en Indirecte Confrontatie Met Milieurisico's, Reacties t.o.v. Milieurisico's en Het Gebruik Van Informatiebronnen*, Universiteit Twente, Enschede.

Katz, E., 1963, The Diffusion of New Ideas and Practices, in: *The Science of Human Communication*, W. Schramm, ed., Basic Books, New York.
New York: Basic Books, 1963

Lowery, S., and DeFleur, M. L., 1983, *Milestones in Mass Communication Research: Media Effects*, Longman, New York.

Mazur, A., 1981, Media Coverage and Public Opinion on Scientific Controversies, *Journal of Communication* **31**:106-115.

Mazur, A., 1984, Reporting about Love Canal and Three Mile Island, *Minerva* **22**:45-66.

Scanlon, J., Morton, G., and Luukko, R., 1978, Media Coverage of Crisis: Better Than Reported, Worse Than Necessary, *Journalism Quarterly* **55**:68-72.

Stallen, P. J. M., and Tomas, A., 1984, De Beleving van Industriële Veiligheid in Rijnmond, Studiecentrum voor Technologie en Beleid TNO, Apeldoorn.

Contrasting Risk Communication Tasks and Objectives for Superfund Sites, Underground Storage Tanks, and Biotechnology Field Tests

Bradley R. Brockbank[a]
ICF Incorporated
Washington, DC

ABSTRACT

Conventional "comparative" approaches to communicating risk issues can be used to improve the quality of the message to be communicated, but do not by themselves guarantee effective risk communication. This is particularly true for risk communication in the areas of biotechnology field releases and the regulation of underground storage tanks. As with many complex risk management challenges, the obstacles to effective risk communication involve limitations in the process by which information is communicated. Experience with the Superfund community relations program and other public involvement programs suggests that new channels of communication must be provided early in the risk management process to allow and encourage direct discussion between risk managers and the public.

KEYWORDS: Risk communication, biotechnology field tests, underground storage tanks, hazardous waste, Superfund

INTRODUCTION

The early 1980s witnessed the emergence of two new environmental risk management challenges: proposals for the deliberate release of genetically engineered microorganisms into the environment and the regulation of underground storage tanks containing petroleum or other hazardous substances. Each of these new activities brings with it a set of problems for risk managers who must communicate the nature of these risks to a public that, increasingly, demands a voice in decisions that affect local communities.

As it is used in this paper, the term "risk managers" refers to government officials, industry managers, and research scientists who must decide how to change or control an environmental situation in ways that create new risks or leave residual risks in communities. "Risk communication" — an increasingly prevalent "buzzword" in discussions about risk management — is used in this paper to refer to any activity intended to convey information about the nature and probability of a potentially hazardous

---

a. Now with Office of the Governor, Denver, CO. The assistance of Marion Cox and Robin Sandenburgh in the preparation of this paper is gratefully acknowledged.

environmental condition. Although decisions and actions taken by risk managers may convey risk information unintentionally, this paper focuses primarily upon risk communication that is undertaken deliberately.

This paper explores approaches to risk communication that may be useful to public and private decision makers who face difficult challenges in communicating the risks of biotechnology field tests and regulation of underground storage tanks to a concerned public. The growing interest in risk communication has prompted many individuals who work on hazardous waste issues — and, in particular, those of us who work closely with communities facing hazardous waste problems — to examine risk communication more closely. The community relations program established under the Superfund[1] hazardous waste cleanup program is not, and was not designed to be, a risk communication program. Rather, the community relations program was designed to encourage two-way communication between, on the one hand, agencies responding to a release of hazardous substances and, on the other, the local residents and public officials living in the vicinity of the site. In the discussion that follows, I have drawn on my experiences with the community relations program to identify potential tasks and objectives of risk communication in the areas of biotechnology field tests and underground storage tank regulation, two issues in which I have a long-standing interest and familiarity.

The first section of this paper examines a conventional "comparative risk" approach to risk communication and evaluates its applicability to the new risk communication challenges posed by biotechnology field tests and the regulation of underground storage tanks. The second section briefly discusses obstacles to effective risk communication in these two new areas that limit the usefulness of such a comparative approach. The third section presents relevant lessons drawn from ICF's experience with the Superfund community relations program and other public involvement programs that suggest more effective approaches to risk communication in general, and to the communication of biotechnology and underground storage tank risks in particular. The paper concludes by suggesting some possible risk communication tasks for these two new areas of environmental concern.

A COMPARATIVE APPROACH TO COMMUNICATING RISKS

One commonly used approach to communicating concepts and data about risks might be described as "comparative risk" communication. The objective of comparative risk communication is to put the risk in question into perspective by comparing it to risks with which the likely recipient of the information already may be familiar. For example, the risk of dying from cancer caused by benzene released from an underground storage tank might be compared with the more familiar risk of inhaling gasoline fumes at the gas station. The assumption underlying this approach is that the public will be better able to weigh the risk in question in the context of the more familiar risk.

This approach has shortcomings, however, because not all risks are alike in all respects. As Baruch Fischoff has pointed out, people tend to disagree more about the definition of "risk" than about the magnitude of a particular risk.[2] This tendency alone reduces the utility of the comparative approach, which seeks to communicate risk-related information by likening it to risks of similar magnitude and greater familiarity as a benchmark for people to put the new risk into perspective.

A potentially useful refinement of the "comparative risk" communication technique, then, might involve the development of a "profile" of the risk — a description of the newly-identified risk that takes into account many of the qualities shared by risks in general. Researchers in risk perception have identified a list of characteristics that influence people's judgments about risks.[3] In Table 1, some of these characteristics are listed in the left-hand

Table 1. Risk Perception Parameters and Risk Profiles for Biotechnology Field Tests and Leaking Underground Storage Tanks[a]

Parameter	Biotechnology	Tanks
Controllability	LOW	HIGH
Inspires Dread	MODERATE TO HIGH	LOW
Catastrophic Potential	MODERATE TO HIGH	MODERATE
Consequences Fatal	LOW TO MODERATE	MODERATE
Potential for Global Catastrophe	MODERATE	LOW
Risk to Future Generations	MODERATE TO HIGH	MODERATE
Ease of Reducing Risk	?	HIGH
Risk Decreasing	?	HIGH
Voluntary	LOW	LOW
Personalized (Affects Me)	MODERATE	MODERATE
Observable	LOW	LOW
Known to Those Exposed	LOW	LOW TO MODERATE
Effect Delayed	?	HIGH/LOW
New Risk	HIGH	MODERATE
Risks Unknown to Science	HIGH	LOW TO MODERATE

[a]Parameters adapted from various studies by Slovic, Fischhoff, and others. See, for example, P. Slovic, B. Fischhoff, and S. Lichtenstein, Characterizing Perceived Risk, in *Technological Hazard Management*, R. W. Kates and C. Hohenemser, eds., Oelgeschlager, Gunn, and Hain, Cambridge, MA, 1983.

column. The two right-hand columns of the table contain one set of possible assessments about how the risks of biotechnology field tests and leaks from underground storage tanks might be described with respect to each of these characteristics. The implication here is that, to communicate these risks more effectively, risk managers should choose familiar risks with similar "risk profiles."

Even this approach to risk communication has serious limitations, however. To reiterate Fischoff's observation, people are likely to disagree about the definition of the same "risky" situation. Thus, even if the risk manager had access to the information necessary to identify relatively objective risk profiles, he or she would still confront the reality of peoples' perceptions Thus, to frame the content of a risk communication "message," risk managers must apply creative methods of identifying the relevant audience and understanding the risk perceptions of that particular audience.

A more fundamental limitation of the "comparative risk" approach is that it addresses only the content of risk communication — i.e., at best, it improves only the substance of the message. An examination of the major risk communication problems related to biotechnology field tests and underground storage tank regulation, presented in the following section, suggests that the ability to communicate effectively about the risks of these activities depends not only on the content of the message, but also on the credibility of the sender of the message, the quality of the communication channel, and the attitudes of the receiver.

IDENTIFYING RISK COMMUNICATION PROBLEMS

Releasing Genetically Engineered Organisms Into the Environment

Genetic engineering has been subjected to intense public scrutiny since scientists first discovered gene splicing techniques in the early 1970s. Recently, the focus of public concern has shifted to proposals from industry and academia to release genetically modified microorganisms into the environment to combat or displace agricultural pests. In the past two years, for example, controversy has erupted over proposals to release, for test purposes, bacteria that have been genetically modified to protect crops from frost damage.[4] Expressions of concern and — at times — outright opposition to these proposals have come from environmental organizations, scientists, and local residents and political leaders living in the vicinity of the proposed test plots. The federal government has responded to these concerns by attempting to apply existing risk management laws and regulations to the new challenges.[5]

An equally important challenge in addressing public concerns about this new technology is to learn how to communicate risk-related issues effectively. Those who seek to reap the benefits of biotechnology confront several significant obstacles to effective communication about its potential risks, including:

- tremendous scientific uncertainties about microbial ecosystems and the behavior of specific organisms in the environment that hamper assessment of risks and ecological consequences of deliberate releases;[6]

- the tendency for people to seek certainty, which fosters an aversion to activities that pose unknown risks;[7]

- the polarization of the debate over biotechnology, which has impeded constructive discussion of legitimate questions raised by biotechnology critics and the public;[8]

- the objections of citizens living in the vicinity of proposed test sites, who do not want to bear the potential risks of these experiments in return for uncertain benefits;[9]

- the shallowness and sensational tendencies of the media as a channel for communicating complex technical information;[10] and

- the lack of established communication channels, other than the media, between risk managers and citizens concerned about the potential adverse effects of deliberate releases.[11]

Among these obstacles, only the first is directly related to the content of the "message" which is being communicated. The public's desire for certainty and its objections to bearing the risks are problems related to the reception of the risk communication message. Polarization is a problem stemming from the limitations of both senders and receivers of risk communication messages; officials and the lay public lack credibility in each other's eyes. The limitations of the media and the lack of alternative channels are "channel" problems.[12] The conventional comparative approach to risk communication described earlier is of limited usefulness because it only addresses the quality of the message. Efforts to improve communication about the risks of biotechnology field tests must address the source, channel, and receiver problems as well.

Leaking Underground Storage Tanks

In communities across the nation, leaks of petroleum and other hazardous substances from underground storage tanks pose risk management challenges to government officials at the local, state, and federal levels. In addition to posing immediate threats of fire or explosion, these leaking tanks may contaminate underground drinking water supplies, upon which half the nation's population depends.[13] In 1984, Congress amended the nation's hazardous waste management law, the Resource Conservation and Recovery Act (RCRA), to establish a program regulating the construction and management of underground storage tanks.[14] Congress also set aside $500 million when it reauthorized the Superfund law to pay for cleaning up sites where tanks are leaking.[15]

Although federal and state programs to regulate underground storage tanks are still in the early stages of development, government officials at all levels will face the challenge of communicating with citizens living in communities where leaks are occurring or have the potential to occur. The most significant obstacles to effective risk communication in this area include:

- the lack of reliable data needed to determine the probabilities of tank leaks in a given community and the effectiveness of leak prevention and detection measures;[16]
- the large number of people — over vast geographic areas — who are uninformed about the potential problems and need to know what is occurring;
- the reluctance of people who feel threatened by contaminated ground water to accept delays in government action created by resource limitations or by the necessity of diverting resources to higher priority sites;[17] and
- the limitations of the media as an effective communication channel and the lack of alternative methods for communicating with the affected communities.

Like the obstacles to effective risk communication about biotechnology field tests, the problems listed above go beyond the limitations of the information to be communicated. Only the first of these problems relates to the quality of the message — the lack of reliable data. The others are limitations of the parties to the communication or the channels through which such communication must take place.

NEW APPROACHES TO COMMUNICATING RISKS: INSIGHTS FROM THE SUPERFUND COMMUNITY RELATIONS PROGRAM

Effective risk communication requires attention to process — the factors that affect the transmission of the message from sender to receiver — not just to content. For this reason, the Superfund community relations program may be a useful source of insight for communicating risks effectively. It bears reemphasizing that the Superfund community relations program is not, and was never intended to be, a risk communication program *per se*. The purpose of the community relations program is to provide information to the public about the U.S. Environmental Protection Agency's investigations and planned activities at a Superfund site, and to create opportunities for concerned residents and officials to ask questions about and comment upon that information.

Nonetheless, two important features of the community relations program have potentially important implications for overcoming the obstacles to effective risk communication identified in the previous section of this paper. First, the community relations program is based on the principle of two-way communication. It is not simply a public information program. At several stages of the Superfund site investigation and cleanup process, citizens living in the vicinity of a site have an opportunity to communicate directly with the government officials who are responsible for managing and reducing the risks posed by the site. At many sites, EPA conducts public meetings before an in-depth investigation of the site is begun, after this investigation is complete, and after EPA has evaluated alternative remedies for the site. These meetings provide an opportunity for citizens to ask questions and to raise concerns about the Agency's planned activities. At the same time, EPA's technical staff has the opportunity to learn what concerns and perceptions the citizens may have about the site.

These meetings establish a channel for communication that otherwise would not exist. They serve as an important supplement to the limited and indirect channel provided by the news media. In some instances, these meetings have been a forum for explicit discussions about risks.

The second, related feature of the community relations program that provides an insight into improving risk communication is the stage in the process at which the public is involved. Before EPA develops its work plan for investigating potential problems at a site and identifying possible solutions, EPA community relations staff and/or their contractors visit the community in which the site is located and speak directly with concerned citizens and public officials. Based upon these community interviews, a "community relations plan" is developed that analyzes the public's concerns and integrates community relations activities into the technical activities planned for the site. When this process works effectively, EPA's credibility with the public can be significantly enhanced.

Building and preserving credibility is perhaps the most important objective common to both the Superfund community relations program and effective risk communication efforts. As Bud Ward, editor of *The Environmental Forum* has written:

> "Effective risk communication must have at its root a substantial reserve of public credibility in the public's mind. Those involved in the endeavor should treat credibility like an institutional asset, a resource to be protected and reserved for a 'rainy day.' They must, in effect, bank credibility and have it available for subsequent withdrawals."[18]

In summary, there are two important features of the Superfund community relations program that may provide important insights into how to communicate risk issues related to biotechnology field tests and underground storage tank regulation:

- communication must be two-way to be effective; and
- the public must be included early and consistently to enhance the credibility of those who make decisions about risks borne by the public.

The concluding section of this paper offers suggestions about how these principles may be applied in the context of biotechnology field tests and underground storage tank regulation.

TASKS AND OBJECTIVES FOR COMMUNICATING RISKS FROM BIOTECHNOLOGY FIELD TESTS AND LEAKING UNDERGROUND STORAGE TANKS

The first task in overcoming the obstacles to effective risk communication is to open up direct channels of communication with the public that encourage open debate about risks between risk managers and those potentially affected by their decisions. This task is both appropriate and important for risk managers concerned with either biotechnology field releases or the regulation of underground storage tanks.

The primary objective of this task is to avoid relying solely on the news media or upon public information pamphlets as communication channels. Although it is essential to cultivate and educate reporters on these issues, the media too often are limited by time, space, and the preferences of their readers or viewers to be a sufficient channel for communicating with an increasingly interested and active public. More importantly, the media may not represent accurately the concerns and interests of the public.

Similarly, fact sheets and other public information devices can be useful for conveying the content of risk communication messages, but they are worth little if the decision makers lack credibility. Moreover, fact sheets by themselves may be useless if they do not address the questions and concerns of their intended recipients. Only by communicating directly with the public can risk managers learn what kinds of risk information are useful and meaningful.

The second objective of opening up direct channels for communicating with the public is to encourage discussion of the potential risks involved in these new activities. Such encounters can, at times, be contentious, but risk managers will have great difficulty building the essential credibility if they are perceived as "shutting out" the public.

Biotechnology field tests are relatively well-suited to such site-specific communication techniques as public meetings and presence of on-site trailers staffed by people trained to convey technical information in terms understandable to the general public. Such techniques also may be useful for sites where leaking underground storage tanks are to be cleaned up. In general, however, communicating about the risks related to overall policies affecting the installation and management of underground storage tanks requires risk managers to reach a broader audience — for example, through the use of call-in radio shows or telephone "hotlines."

The second general task that may improve risk communication is to *involve the public early in the process, before key decisions are made* that could potentially affect people's well being. A third, corollary task is to *listen to the public's concerns and be willing to compromise* to address those concerns. The objective of these two tasks is to build credibility by demonstrating a responsiveness to those who bear the risks of the risk manager's decisions. If this credibility is established, the public will be more likely to listen to what risk managers have to say in future interactions. Moreover, by demonstrating a willingness to listen and respond, risk managers may also find the public more willing to extend its trust to the risk manager in his or her judgments about the safety of these new activities.

For biotechnology field tests, citizens must not be approached with the sole objective of "selling" them on the idea. It may be necessary to work with communities to understand what kinds of risks they are willing to accept and what kinds they are not. This personal interaction builds the credibility that is necessary for communicating the highly uncertain risks associated with biotechnology. It may also result in the development of a product or a technique that is best suited to those it is intended to serve.

At sites where underground storage tanks are leaking and no immediate threat to public safety is evident, government officials can follow procedures similar to those used by the Superfund community relations program. For each site, some assessment of site-specific community concerns ought to be made, based upon direct two-way communication with local residents and officials, before investigation and cleanup activities commence. These concerns should be factored into federal or state agency decisions about the appropriate actions needed to resolve the problems at the site.

For policy-level issues such as setting priorities for cleanup and establishing leak prevention and detection standards, it already may be too late for the public to feel invested in the federal decision process. Regional meetings conducted by state governments may be one way states can give the public an opportunity to express their preferences about the best ways that state regulations can address these risks.

Because these two environmental issues have only recently emerged as nationwide concerns, there is little historical experience to draw upon for support of the recommendations in this paper. In one illustrative case, however, researchers at the University of California at Berkeley conducted a public meeting in Tulelake, California, the proposed site of an experiment they sought to conduct using bacteria genetically engineered to inhibit frost formation on crops. According to one report, most of those attending the meeting listened to the scientists' presentations and had no substantive objections to the experiment.[19]

In contrast, a California company proposing essentially the same experiment on a strawberry patch in Monterrey, California, did nothing in advance to communicate with local residents and officials. The objections from local residents and political officials were much more strident and helped to block the experiment from taking place.[20] Although it may be too soon to draw definitive conclusions from these limited experiences, the outcomes in the two examples cited here are consistent with ICF's experience in the Superfund community relations program and other risk-related public involvement programs in which we have played a role.

CONCLUSION

Risk communication is a new area of inquiry that has attracted a great deal of attention in the past few years. One of the pressing tasks in this emerging field is to define with greater precision the term "risk communication" and to identify its usefulness both to existing risk management problems and to new challenges such as biotechnology and underground storage tank regulation. Limiting discussion of risk communication to the content of a risk communication "message" fails to address important obstacles that prevent effective risk communication and tends to reinforce the notion that risk communication is primarily a one-way process. This paper has explored the obstacles to effective risk communication in relation to biotechnology field tests and the regulation of underground storage tanks and emphasized the important "process obstacles" that can not be overcome simply by comparing new risks to more familiar ones.

Drawing on insights from the Superfund community relations program and other public involvement programs in which ICF has played a key role, it is our belief that risk managers may be able to communicate risks more effectively if they establish direct communication channels with the public and involve the public in the risk management process early on. By establishing a process of risk communication, risk managers may be able to establish the credibility needed to convey highly uncertain risk information in a way that fosters constructive debate about acceptable risk management decisions.

REFERENCES

1. The Comprehensive Environmental Response, Compensation, and Liability Act of 1980 (CERCLA; P.L. 96-510), more commonly known as "Superfund," gives the federal government authority to respond to releases or threatened releases of hazardous substances in the environment.
2. B. Fischhoff, Managing Risk Perceptions, *Issues in Science and Technology* (Fall 1985).
3. W. F. Allman, Staying Alive in the 20th Century, *Science* (October 1985).
4. P. J. Hilts, Microbe Test Hits New Snag, *Washington Post* (January 20, 1986).
5. 49 *Federal Register* 50856 (December 31, 1986).
6. J. Gillett, ed., Potential Impacts of Environmental Release of Biotechnology Products: Assessment, Regulation, and Research Needs, *Environmental Management* **10(4)** (July 1986); and Strauss, Harlee, *et al.*, Direct Release of Genetically Engineered Organisms: A Preliminary Framework for Risk Evaluation Under TSCA, MIT: Center for Technology, Policy, and Industrial Development (August, 1985).
7. Allman, op. cit.
8. A good example of this polarization is the lack of constructive dialogue between biotechnology supporters and outspoken critic Jeremy Rifkin.
9. Hilts, op. cit.
10. N. Pfund and L. Hofstadter, Biomedical Innovation and the Press, *Journal of Communication* **31**:2 (Spring 1981); also see V. T. Covello *et al.*, Risk Communication: An Assessment of the Literature on Communicating Information About Health, Safety, and Environmental Risks, paper presented at the 1986 SRA National Conference on Risk Communication and published by the Conservation Foundation (1987).
11. Concerns about trade secrets and the competition among researchers have created strong disincentives for public dialogue about innovative experiments.
12. An excellent discussion of risk communication problems can be found in the paper by Covello, op. cit.
13. V. D. Feliciano, Leaking Underground Storage Tanks: A Potential Environmental Problem, Washington D.C.: Congressional Research Service (January 11, 1984); cited in the Congressional Record S2028 (February 29, 1984).
14. The Hazardous and Solid Waste Amendments of 1984 (P.L. 98-616).
15. The Superfund Amendments and Reauthorization Act of 1986 (P.L. 99-499).
16. Based on EPA documents and conversations with staff in EPA's Office of Underground Storage Tanks.
17. This is a frequent phenomenon in the Superfund program, and may create problems for planned cleanups at leaking tank sites depending upon how EPA sets cleanup priorities.
18. B. Ward, Communicating On Environmental Risk, *The Environmental Forum* (January 1986).
19. Forum On Recombinant Bacteria Is Effort To Ease Way For Field Test, *Chemical And Engineering News* (May 12, 1986).
20. Hilts, op. cit.

Communicating Risk: Factors Affecting Community Acceptance of Information

Jenifer S. Heath
North Carolina Division of Health Services
Raleigh, NC

June Fessenden-Raden
Cornell University
Ithaca, NY

ABSTRACT

Local-level acceptance of risk information is crucial to successful risk management. We have studied the communication of health risk to local-level decision makers dealing with chemical contamination of groundwater that is used for drinking water. Several factors that affect local-level decision makers' acceptance of information and guidance provided by outside intervenors have been identified. These factors fall into two categories: those affecting local control and those affecting the credibility of the outside information provider.

A sense of local control is enhanced if the community is involved in identifying its own needs and if the community initiates interaction with the outside information provider. The credibility of the outside information provider is affected by his/her "localness"; by his/her history of interaction with the community; by the degree of compatibility between the information or guidance offered by the intervenor and any knowledge or expertise that may already exist locally; by indications that he/she shares values, goals and perceptions with the community or is sensitive to local politics and economic conditions; by the community's ease of access to the information provider; and by indications that the information provider respects local-level expertise.

KEYWORDS: Risk communication, risk management, community, drinking water

INTRODUCTION

The communication of risk-related information has received increased attention recently. The successful management of toxicologic risk is one area in which effective risk communication is essential.

This paper presents some insights about risk communication gained during research on toxicologic risk management at the community level. The research is part of a multidisciplinary project on individual and community management of risks posed by chemically contaminated groundwater used as drinking water. Concurrent projects have investigated hydrogeologic features of communities; the environmental fate of chemicals; the potential health risks; community members' knowledge about local water contamination problems; the cultural values and the risk management process itself. The research is guided by the tenet that enhanced risk management at the local level must include not only integration of the technical, scientific and legal aspects but must also mesh with the social and human aspects of the actual risk situation. This paper addresses one critical part of the risk management process — risk communication — focusing on what we have found thus far about factors affecting acceptance by community decision makers of risk information provided by outside intervenors.

Conventionally, risk management is conceptualized as an activity in which intervenors who possess specific skills and knowledge apply these skills to the resolution of the risk situation. The community within which the risk occurs tends to see risk management as something that is done to them. In reality, considerable interaction occurs (Heath and Fessenden-Raden, 1987). Fessenden-Raden *et al.* (1987) have proposed that successful management of toxicologic risk involves not only resolution of the risk situation itself, but also improved local-level ability to avoid similar risk situations in the future and enhanced local-level capacity to respond should a similar risk arise in the future. In order for risk management efforts to be successful on all three counts, local-level participants in risk management must possess and understand risk-related information.

Local-level decision makers do have access to information locally in some cases. For instance, in one study community some local-level participants educated themselves on the toxicology of some chemicals. In another community one of the village board members was a chemist who had experience in the evaluation of food safety. Also, one study community was able to recruit technical experts from local industry to serve in an advisory capacity to local decision makers. Yet even these communities lack the resources necessary to fully develop expertise in all of the many technical disciplines that play a role in the management of contaminated drinking water. This expertise is a critical input into the risk management process. Therefore, outside information providers (such as representatives of state and federal agencies) must convey to local-level participants in risk management an understanding of the risk, its remediation and the prevention of its recurrence.

An investigation of the provision of risk-related information by outside intervenors, as well as of the knowledge and understandings possessed by community members following communication attempts (Heath, 1987), provides the basis for this paper, which describes eight factors that seem to affect community members' evaluation and interpretation (and, ultimately, acceptance or rejection) of risk-related information provided by outside intervenors. These factors fall into two categories: those affecting the credibility of the information provider and those affecting the community's sense of local control. We argue that, to be effective, information providers must be as aware of these factors and the interactive nature of the risk communication process as they are of the facts they hope to communicate.

METHODS

Site selection for the study reported here was not random and no attempt was made to select "representative" communities. Five study communities were selected on the basis of the following criteria: (1) groundwater contributed to the public drinking water supply; (2) at least one contaminated public supply well was closed; (3) 1,1,2-trichloroethylene, 1,1,1-trichloroethane or tetrachloroethylene was the focus chemical; (4) all communities were non-metropolitan and in upstate New York.

Local risk management situations were followed through time. In each community we reconstructed a chronology of past events and observed ongoing events. Local, state, and federal level participants in risk management were interviewed. We analyzed local newspaper accounts of the events surrounding the risk management process and attended public meetings relevant to the risk situation. We also reviewed technical (engineering) reports and other documentation (such as written records of communication between participants or notes taken by participants at meetings) when available. Access to these information sources varied from one community to another and, within a single community, from one informant to another.

Purposive sampling was used to select informants. In each study community, local, state, and federal level individuals whose names regularly appeared on relevant documents or in newspaper articles or who were cited in other interviews were selected to be interviewed. These judgments were intended to err on the side of over-inclusiveness.

The interviews were open-ended and in-depth, yet semi-structured to allow the informant to determine the focus and direction of conversation; minimal direction was provided. Thus the informant's perceptions and understandings could be identified.

RESULTS

Toxicologic (and other) information and guidance provided by outside intervenors may or may not be accepted by local-level participants in the risk management process. The study of these five communities indicates that several factors affect the acceptance of information (see also Fessenden-Raden and Fitchen, 1986). Preliminary analysis suggests that these factors fall into one of two categories: those affecting the credibility of the information provider and those affecting local control. In general, community members' acceptance of information provided by intervenors seems to be enhanced when both the information provider's credibility and the community's sense of control over the interaction are high.

Credibility

Six factors were identified as affecting the credibility of the outside information provider.

1. The 'localness' of the intervenor affects community members' evaluation of information. Local-level participants in several study communities expressed the belief that risk communicators and other intervenors who are from a firm or agency located nearby are more concerned about the community than are less-local information providers (e.g., intervenors from Rochester or Syracuse, New York, are preferred over those from Atlanta, Georgia, or Nutley, New Jersey). Local-level participants believe that relatively local information providers both possess information about local conditions (e.g., weather) and treat community members with more respect.

2. Another factor affecting the evaluation of risk-related information is the history of interaction between the information provider and the community. For example, in one community the New York State Department of Environmental Conservation's wildlife and conservation representatives were perceived as interfering in local activities; as a result there was a hostile relationship between some members of the community and the agency. This negative reaction toward the wildlife and conservation workers of the Department of Environmental Conservation extended to those representing the hazardous waste branch, thus compromising all information provided by the representatives of the agency. Community members do not seem to distinguish effectively between the various branches of agencies such as the Department of Environmental Conservation and the Environmental Protection Agency (Heath, 1987). By contrast, in another study community the regional representative of the Department of Environmental Conservation had a long history of

positive interactions with community members. Risk-related messages funneled through this regional agency representative to the community were perceived as credible and were accepted.

3. The information provider's respect or lack of respect for expertise possessed by local-level participants often affects their response to the intervenor and their perceptions of the intervenor's credibility. For instance, we found that engineering consulting firms tend to act independently of the local knowledge of the risk situation. In one community, the consulting firm fell far behind schedule when drilling wells. The explanation offered was that they had expected the water table to be about fifty feet deep, and only upon drilling had learned that it was actually one hundred feet deep. Community members, who had been drilling wells in the area for decades and were aware of the depth of the water table, were appalled that the firm had neglected to seek local-level input on such an obviously local matter. This firm's credibility was all but destroyed by this episode.

4. Community members often respond to cues indicating that information providers are sensitive to local constraints and realities and that they share values and goals with the community. For instance, information providers who recognize that winter weather in upstate New York prohibits many outdoor activities (e.g., drilling) are considered more credible on other risk-related matters than are intervenors whose work schedules have water mains being put into place during January. Similarly, community members who view local industry as an integral and positive part of the community are unlikely to accept information provided by a person, for instance a representative of the Environmental Protection Agency, who advocates strong punitive measures against local industry.

5. The accessibility of the information providers also affects their credibility. Local-level participants often become suspicious of information providers who are difficult to contact. This problem extends to agencies which refer inquiries from one office to another and yet another. This runaround compromises the credibility of all information providers associated with that agency. Another accessibility problem exists when agency representation changes frequently. For instance, in one community the Environmental Protection Agency changed its project manager for a single Superfund site at least five times over the course of several years, causing community members to complain that they never knew where to go for information. The lack of continuity among project managers frustrated local level risk management participants, who were unable to determine even the identity of their contact. The credibility of all Environmental Protection Agency representatives suffered.

6. The compatibility between the risk information provided by intervenors and local knowledge or judgment also affects community members' evaluation of that risk-related information. In one study community, information providers focused on the carcinogenic risk from trichloroethylene in the drinking water, never mentioning the acute effects which could be expected at the levels actually found in the contaminated water. The credibility of these information providers suffered in the eyes of community members, who had heard of acute effects occurring locally but were not aware of any cases of cancer attributable to the contamination.

Sense of Control

Two factors were found to affect a sense of local control.

1. Local initiation of the interaction between local-level participants and the information-providing intervenor seems to enhance the sense of local control. While having to search and beg for information does not please community members, members of several study communities indicated a preference for information promptly provided upon their initial request. Unsolicited information from providers, who appeared to be

condescending by deciding for the community that risk information was required, seems to find less acceptance.

2. Local involvement in identifying the informational needs of the community is critical to a sense of local control. One study community established a citizens' committee which identified informational needs for intervenors, who provided the information that was requested plus other relevant risk information. Community members valued this information because it responded to their perceived informational needs.

CONCLUSIONS

Risk communicators must recognize that risk communication is not a one-way, top-down activity in which intervenors possessing unique information and skills educate a community about a specific risk as they, the communicators, perceive it. Rather, risk communication is a dynamic process and involves considerable interaction between information providers and the community members. Effective risk communication involves two-way dialogue in which the community is an active participant. Lack of sensitivity to this interactive relationship easily mars the process and decreases acceptance of risk-related information, no matter how accurate, by community members and so decreases the success of the entire risk management process.

We suggest that the interactive relationship between information providers and community members must be acknowledged and should be cultivated by all risk communicators. This approach to the provision of risk-related information recognizes that no two communities are the same — each has unique informational and risk management needs as well as unique perspectives on and reactions to risk management activities and risk-related information that is provided. In addition, different audiences within each community must be acknowledged and included in this interactive process of risk communication if risk management is to be successful.

We have identified eight factors that affect community members' acceptance and interpretation of risk-related information. Since community members must possess accurate risk-related information in sufficient quantity to participate effectively in risk management, it is critical that these factors be taken into consideration whenever information is to be communicated. Information providers must not only provide full, accurate information, but also provide that information under conditions that maximize the likelihood that it will be accepted. To be more effective, risk communicators must nurture the community's sense of control over the risk communication process. They also must make the effort and take the time to understand how the factors described in this paper affect their credibility and the acceptance of their information on risk.

REFERENCES

Fessenden-Raden, J., Bisogni, C. A., and Porter, K., 1987, Enhancing Risk Management by Focusing on the Local Level: An Integrated Approach, in *Risk Assessment and Management*, pp. 665-676, L. Lave, ed., Plenum Press, New York.

Fessenden-Raden, J., and Fitchen, J. M., 1986, Community Perspectives on Communication of Risk, paper presented at the Society for the Social Study of Science, October, 1986.

Gillespie, A. H., and Yarborough, P., 1984, A Model for Communicating Nutrition, *Journal of Nutrition Education* **16(4)**:168.

Heath, J. S., 1987, Toxicologic Risk Management at the Community Level: The Case of Chemically Contaminated Groundwater Used as Drinking Water, a thesis presented to the faculty of Cornell University, Ithaca, NY.

Heath, J. S., and Fessenden-Raden, J., 1987, The Relationship Between Risk Management Intervenors and the Community, in *Risk Assessment and Management*, pp. 687-695, L. Lave, ed., Plenum Press, New York.

AIDS: Risks and Public Policies

R. G. Mathias
University of British Columbia
Vancouver, B.C.

ABSTRACT

AIDS is a viral infection that has become epidemic in the USA and Canada since its introduction. Studies of the epidemiology of this infection have delineated many of the aspects of the agent, the host and the environment that are the determinants of its spread. A thorough understanding of these determinants will allow the establishment of appropriate public policy to limit the epidemic. Screening the blood supply will reduce risk to product recipients. The susceptibility of the host cannot be altered but exposure can be limited. The practical alternatives include the provision of sterile needles and syringes to drug users and the widespread use of barrier methods by promoting condom use to limit sexual transmission.

KEYWORDS: AIDS, condoms, sexually transmitted, percutaneous, policy

AIDS — the very name provokes a response in the public and in health professionals ranging from calm consideration to panic. The personal response is based on the perceived risks of this infection: risks to self, community or society, perception that is a meld of experience, knowledge, values and prejudices. The responses to the risk of this virus are deeply rooted in the societal context of the perception of these risks, and in that context, the risks and costs of infection are balanced against the need and effectiveness of control measures.[1]

Controlling communicable diseases is not a new exercise. Major religious texts deal with communicable disease control and other health threats. Behaviors are prescribed and proscribed with the purpose of reducing the risk of infection individually and the burden of disease on the community. The banning of the consumption of pork, using only the right hand to take food from the common dish, controls on persons identified as infected with leprosy, all stem from society's need to protect itself from the threat of infection.

The relative effectiveness of these measures can be gauged from the documentation of the epidemics that have spread across the world since records have been kept. Some infections, notably trichinosis from pork,were controlled in compliant groups, but other epidemics such as plague, smallpox, typhus, typhoid and cholera have affected the course of history of all of mankind.

New Risks
Edited by L. A. Cox, Jr., and P. F. Ricci
Plenum Press, New York, 1990

The causal web associated with these epidemics is complex and multifactorial. Plague does not spread without its vector, the flea. Epidemic typhus needs its vector, the louse. Typhoid and cholera need communal water supplies contaminated with feces containing the bacilli. Smallpox needs only the close contact of an infected person with a susceptible individual. All of these historical examples have the seeds of their control planted in knowledge of the modes of spread and the careful application of that knowledge to the measures taken by individuals and the community. The appropriate application of epidemiology and vaccine technology has resulted in the ultimate control of infection: the eradication of the transmission of smallpox worldwide.[2]

The spread of any infection requires that a sufficient dose of organism reach the appropriate portal of entry in a susceptible host. Organisms spread in two major methods: direct, either by contact such as sexual or percutaneous contact or by droplets, and indirect, via airborne, vehicle or vector spread. The portal of entry requires that cells of the appropriate type to allow attachment be present and that local defenses be ineffective. Once attachment has occurred, growth and multiplication either locally or after invasion occurs. Unless the human is to be a "dead-end" host, as in many arbovirus infections, the organism must also have a site of exit which will allow the organism to reach the environment and attack the next host.[3] The analysis of entry, growth, and multiplication and exit requirements of an agent allows the planning for its control.

How then does Human Immunodeficiency Virus (HIV) accomplish these virus determined objectives and how can we interrupt this chain? HIV is spread by direct contact. The rather hysterical pronouncements to the contrary do not change the epidemiologic observation that direct contact is necessary. Let us then rule out airborne, vehicle and vector as routes. There is no need to postulate any of these routes since there are virtually no cases where direct contact cannot be found. Logic would require that we need not invoke these routes.

Are there any direct routes that appear to be unlikely? The fact that a vehicle has not been found, in particular a food or water vehicle, indicates that oral spread may be either unusual or that it does not occur. Evidence from the studies of homosexual male cohorts that oral-genital contact, including the swallowing of semen, does not increase the risk of infection supports and strengthens the postulate that ingestion is not a risk.[4] Clearly direct contact with skin is not a risk as shown by family contact studies.

Sexual contact or percutaneous contact results in infection.[5] There can be no doubt that there is a risk to the receptive partner when the insertive partner is infected. The very high rates of infection after such contact indicate, but do not prove, that all individuals are susceptible to this virus. It seems also clear, however, that the risk of infection per contact is not 100% and indeed may be a good deal lower. The risk to the insertive partner is also not zero. However, it appears to be less than the risk to the receptive partner per exposure.[6] This is supported by the observation of the risk of infection in other sexually transmitted diseases such as gonorrhea, where the risk of the receptive partner is high (nearly 100%) while the insertive partner has less risk (about 20%).[7,8] Biologically this seems to make sense, as the ejaculate will carry the virus to the appropriate cells. Conversely the vaginal fluids must go retrogradely up the urethra to the cell(s) where attachment takes place. With few secretions in the rectum, the risk to the insertive partner may be quite low. An hypothesis that would fit these observations is that the cell where attachment takes place is a columnar epithelial cell and not squamous epithelium.

Intravenous drug abuse is associated with HIV infection. This appears to be most associated with the sharing of apparatus. The drawing back of blood to insure the intravenous position of the needle makes the transfer of virus infected cells relatively easy. This may explain why health care workers with needle puncture exposures have a very low infection rate.[9] The amount of blood may be too low to have the transmission of HIV infected cells occurring frequently or in sufficient quantity. In Canada and England, IV drug use does not seem to carry the same risks as have been noted in the U.S. The much lower risk in Canada may be due to the fact that needles and syringes are legally available in some jurisdictions while in the U.S. they are not.

How can we use this biological data to control or eradicate this infection, and how do we put this control into the individual and social context? The sine qua non of the prevention of infection is to prevent the agent from reaching a susceptible host. The intervention can be directed at the host, the agent or the environment. At this time we cannot change host susceptibility by the use of a vaccine, although this method has been spectacularly effective in eradicating smallpox. We have not used it to control a primarily sexually transmitted disease due to lack of knowledge of the immune response to these infections. The very property of being able to spread via sexual contact — that is being present in an infectious form in the individual long enough to maximize the chance of exposure to a new partner — indicates that the organism can escape the immune response. This property is also present in HIV and will make the development of a vaccine difficult. If a vaccine was developed, the testing and marketing of such a vaccine would also be difficult. I do not think that this is an approach that will have an immediate impact on this epidemic.

Another option is to prevent infected individuals from coming into contact in situations where spread may occur. This has been the approach taken with the blood supply where screening of blood reduces the risk of transmission to susceptibles.[10] This technique is susceptible to the uncertainties of diagnostic test procedures and predictably will fail occasionally but will result in a lessening of risk.

Is there another model that can be examined to determine if risk reduction is possible? I believe that syphilis control presents us with a model that is essentially the same as the model of HIV infection. Although we do not have a treatment option for HIV yet, the epidemiology of these infections is similar. Certainly screening the blood supply was done for syphilis and for similar reasons. Other control measures and perhaps the statutes used for syphilis could be adapted to try to protect susceptibles from HIV infection.[11] Would premarital serology be a viable test to re-institute? This was done in most states until recently with the object being to prevent the infection of the woman and hence transmission to her infant. Was this effective? With an estimate that 30% of pregnancies are started without the benefit of marriage, the amount of control that would be exercised seems to be relatively small. This may be even more true for high risk groups, but that leads to questions of the identification of high risk individuals. Should prenatal testing be implemented? The outcome measure for this intervention would need to be abortion, as treatment is not available. It would appear that prenatal testing also does not need to be done except in high risk populations. It would also be a program directed at the individual but would not control infection.[12]

How about more general measures? Isolation was a time honored procedure for the control of infections. Leper colonies were formed on the basis that this would protect the members of the community. There is no evidence to support this practice, mainly because the identification of infectious individuals was not possible. We are able to identify infected individuals, however, with EIA tests as used for the blood supply. It seems extremely unlikely that infected individuals can be identified before the infection has spread. Even in syphilis, where the infectivity of an individual can be abruptly terminated with treatment, contact tracing procedures have not resulted in disease eradication but at best are used for case finding. The screening of all individuals on admission to hospitals leads to treatment, but the evidence that this would control infection is lacking. To then attempt to carry out a similar operation for individuals who cannot be treated but who would need isolation for long periods would exact a social price that is too high to be considered seriously.

Is there then no effective preventive measure indicated by an analysis of the risks of infection? Education has been espoused as the mainstay of prevention. Unfortunately there is little evidence to show that sex education results in a behavior change, although learning may occur.[13] Education, while it will result in a behavior change in a portion of the population, cannot be seen as the control measure alone.[14,15] Smoking, seat belt use and drug abuse have all been approached with this technique. The drives that result in the spread of sexually transmitted diseases are at least as strong as the peer pressure that results in substance abuse and may be even more biologically-based. Physical barriers of time and space, as in isolation with its social disruption and lack of evidence of effectiveness, could also be considered in a more individual model, the use of the condom.[16] In Japan it is reported that 80% of couples use the condom. This rate of use in North America would reduce the risk of transmission considerably. Whether these rates could be achieved in the high risk populations is problematic. This appears to be the most effective method. Perhaps a widespread, well-funded program, supported by the development of an acceptable condom, will reduce risk. In making this suggestion I hear the laughter of generations of medical officers in the armed forces who have tried to persuade males to use condoms to prevent sexually transmitted diseases. Perhaps if women take the lead role in the use of condoms, at least they could lessen the risk.

As mentioned earlier, the availability of sterile needles and syringes in Canada may be one reason for the low rate of infection of IV drug users. If this group represents one of the channels which lead the virus from one risk group to another, then clearly closing the channel is necessary as soon as possible.

Experience with communicable disease control over many years indicates the powerful influence of socioeconomic conditions on disease control. I fear that without a procedure that can be applied once, such as immunization, or carried out without personal motivation, such as the pasteurization of milk, we will not be able to control this epidemic. Nevertheless we can inform the individuals who are at risk and make the means available to them to reduce their risk. We must make acceptable condoms available to anyone who wants to use them, and persuade them that this is acceptable, and we must make sterile needles and syringes available to IV drug users even as we have made the blood supply as safe as possible to individuals needing blood or blood products.

REFERENCES

1. G. D. Lundberg, The Age of AIDS: A Great Time for Defensive Living, (editorial) *JAMA* **253**:3440-1 (1985).
2. F. Fenner, Global Eradication of Smallpox, *Rev. Infect. Dis.* **4**:916-22 (1982).
3. A. S. Evans, ed., *Epidemiologic Concepts and Methods; in Viral Infections in Humans*, 2nd edition, pp. 3-42, Plenum Medical Book Co., New York (1982).
4. M. T. Shechter, W. J. Boyko, B. Douglas, *et al.*, Can HTLV-lll Be Transmitted Orally? *Lancet* **1**:379 (1986).
5. T. A. Peterman, and J. W. Curran, Sexual Transmission of Human Immunodeficiency Virus, *JAMA* **256**:2222-6 (1986).
6. E. Jeffries, B. Willoughby, W. J. Boyko, *et al.*, The Vancouver Lymphadenopathy-AIDS Study 2: Seroepidemiology of HTLV-lll Antibody, *Can. Med. Assoc. J.* **132**:1373-7 (1985).
7. K. K. Holmes, D. W. Johnson, and H. J. Trostle, An Estimate of the Risk of Men Acquiring Gonorrhea by Sexual Contact with Infected Females, *Am. J. Epid.* **91**:170-4 (1970).
8. R. R. Hooper, G. H. Reynolds, G. G. Jones, *et al.*, Cohort Study of Venereal Disease, 1: The Risk of Gonorrhea Transmission from Infected Women to Man, *Amer. J. Epid.* **108**:136-44 (1978).

9. E. McCray, Occupational Risk of the Acquired Immunodeficiency Syndrome Among Health Care Workers, *NEJM* **314**:1127-32 (1986).
10. Council Report, Status Report on the Acquired Immunodeficiency Syndrome, *JAMA* **254**:1342-45 (1985).
11. M. Mills, C. B. Wopsy, and J. Mills, The Acquired Immunodeficiency Syndrome: Infection Control and Public Health Law, *NEJM* **314**:931-6 (1986).
12. V. DeGruttola, K. Mayer, and W. Bennett, AIDS: Has the Problem Been Adequately Assessed? *Rev. Infect. Dis.* **8**:295-305 (1986).
13. M. D. Benson, C. Perlman, and J. J. Sciarra, Sex Education in the Inner City: Learning and Retention, *JAMA* **255**:43-7 (1986).
14. J. E. Osborn, The AIDS Epidemic: Multidisciplinary Trouble, *NEJM* **314**:779-82 (1986).
15. E. D. Acheson, AIDS: A Challenge for the Public Health, *Lancet*, pp. 662-6 (1986).
16. K. M. Stone, D. A. Grimes, and L. S. Magder, Primary Prevention of Sexually Transmitted Diseases: A Primer for Physicians, *JAMA* **255**:1763-6 (1986).

The Healthy Worker Effect: The Need to Reevaluate a Broad Spectrum of Occupational Risks

T. Sterling and J. Weinkam
Simon Fraser University
Burnaby, B.C., Canada

ABSTRACT

There is evidence that the practice of standardizing the mortality experience of occupational groups to that of the general population introduces a serious bias that systematically tends toward underestimating occupational health risk. This bias is called the Healthy Worker Effect (HWE) and is based on the observation that the mortality ratios for all diseases at all ages of employed individuals are approximately 25% less than that of the general population. The HWE introduces a constant bias which requires the modification of common epidemiological methods for detecting occupational hazards. Previous occupational health studies need to be reevaluated to assess whether or not their results are distorted because of HWE.

KEYWORDS: Risk, Standardized Mortality Ratio, Healthy Worker Effect, occupational health studies, occupational hazards, occupational risk

INTRODUCTION

The Standardized Mortality Ratio (SMR) is used as the basic index for risk determination in cohort follow-up mortality studies.[a] It is a summary statistic that compares an observed mortality experience (of a cohort of interest) with one that is deemed to be normal or typical. A large number of occupational health studies use as their norm the mortality experience of the general population of the country or of the regions where members of the study cohort reside (such as states, provinces, countries, or municipalities).

There is evidence that the practice of standardizing the mortality experience of occupational groups to that of the general population introduces a serious bias that systematically tends toward underestimation of occupational health risks. The existence of such a bias is of considerable concern.

1. Epidemiological methods used for detecting occupational hazards will have to be modified.

---

a. A similar index may be computed for morbidity in these few instances when morbidity rather than mortality is the endpoint of a follow-up procedure.

New Risks
Edited by L. A. Cox, Jr., and P. F. Ricci
Plenum Press, New York, 1990

2. A reevaluation may be required of many previous occupational health studies to assess whether or not their results had been distorted because they used what turned out to have been inappropriate standardizing procedures.

REVIEWING SOME BASICS

Follow-up studies determine mortality (or morbidity) rates for a particular population observed (followed) over a time interval. Each member of the cohort contributes to the time at risk until the termination of the study, death, or loss to follow-up.

The computation of the mortality rate is guided by factors known to influence mortality. The most common of these are:

1. age: because mortality increases sharply with age;

2. sex: because there are differences in mortality between the sexes for various causes;

3. race: because mortality differs by age and sex for different racial groups for various causes.

In most cohort studies, the major interest is centered on mortality from specific causes. Thus, results of cohort follow-up usually are given in terms of age, sex, race and disease specific mortality rates.

Epidemiologists are well aware of other factors besides age, sex and race that influence disease. But while it may be desirable to adjust disease-specific mortality rates for additional possible confounders, limitations are imposed by a dearth of data. Adjustment factors act in a multiplicative way. Adjusting a follow-up study for age in 5-year groups (say between ages 35 to 84) results in approximately ten distinct age groups; sex and race contribute two groupings each and, among causes of death, interest is usually focused on all causes, all cancers, all heart disease, all other diseases, and possibly also on one or more specific diseases such as lung cancer, resulting in at least five categories. Thus, the minimal adjustments for confounding require sufficient data for approximately 200 cells. As rates depend on death counts (in the numerator) and as one death per cell is neither a likely nor a desirable outcome, a cohort must be large enough and be followed sufficiently long enough to yield approximately 600 deaths, which, in turn, requires an original cohort of 9,000 to 10,000 workers who are followed for 15 to 20 years. If the investigation is especially concerned with causes of death with relatively low rates (such as cancer of the oral cavity) the cohort may need to be much larger than 9,000 to 10,000 individuals, or follow-up much longer than 15 to 20 years. As occupational cohorts of this magnitude are difficult to assemble and costly to follow, few studies are sufficiently large to adjust for more confounders than age, sex, and race or to analyze causes of death that normally occur with low rates.

Although the length of the follow-up period is important, the size of the cohort is the primary determinant of the power of investigation to detect an increase in risk. Relatively few occupational health studies have sufficient power to detect less than a doubling in risk.

In order to infer that the specific death rates observed for a cohort indicate an underlying increase in risk, two more steps are required.

1. A normal mortality rate has to be defined in order to judge whether the followed cohort's mortality exceeds expectation.

2. A summary index has to be computed that is suitable for measuring the magnitude of any increased or decreased risk.

The Standardizing Population

Inferences that the mortality observed in a follow-up cohort is greater or less than it ought to be are based on comparing the observed specific mortality rates to those of some population that may be deemed to have a normal mortality experience. The task is to ascertain whether the mortality observed in the cohort of interest is larger or smaller than normally would be expected. Suitable data must be available to permit the computation of specific rates for the standardizing population. Specifically there must be information on age, sex, race, and on any other relevant factors, and on the incidence of deaths from diseases of interest so that the same mortality rates can be computed. The standardizing population furnishes an expected number of deaths which then may be compared with the observed number of deaths.

The Summary Index, The Standardized Mortality Ratio

There are a number of ways in which a proper summary statistic may be computed. The most common is to utilize the age/sex/race/disease specific mortality rates of the occupationally exposed cohort to compute the expected number of deaths from specific diseases for comparable age, sex, race subgroups of the standardizing population. These expected numbers of deaths are then compared to the observed number of deaths. In this last step, an index is computed by taking the ratio of expected deaths to observed deaths, which is called the Standardized Mortality Ratio (SMR).

The SMR is obtained by adding the expected deaths for all ages and dividing that sum by the number of deaths observed to occur in the standardizing population. Another method is sometimes used, especially with smaller cohorts or for causes of death with low rates. This method is known as indirect standardization. It involves using the death rates of the standardizing population to compute the number of deaths which would be expected to occur in the cohort if it were subject to the same mortality experience as the standardizing population and comparing this to the number of deaths actually observed in the cohort by taking the ratio of observed deaths to expected deaths. The advantage of this method is that it is less subject to errors due to unstable estimates of age specific death rates. The disadvantage is that it depends on the age distribution of the cohort, making comparisons between subgroups more difficult.

The SMR would be 1.0 (or would not differ significantly from 1.0) if the same factors that determine mortality in the follow-up cohort also determine mortality in the standardizing population. The SMR serves as a convenient summarizing statistic. There are two advantages to the SMR.

1. Procedures for testing hypotheses based on expected variations around an expected value of 1.0 have been worked out.

2. The final ratio is less affected by empty cells among possible age, sex, race, and disease categories because it sums over a number of years (and possibly over other categories such as race and sex).

However, the SMR suffers from two serious drawbacks:

1. It implicitly relies on the assumption that the mortality of the standardizing population is the same as that of the followed-up cohort would be, without the (possible) effect of the occupational exposure.

2. It is not sensitive to compensating differences in mortality in different age ranges. For instance, if a cohort has a relatively low mortality rate at younger ages but an elevated mortality rate at older ages, the final SMR might not differ from 1.0 at all.

The Healthy Worker Effect

Persons who are employed come from a class of individuals who are sufficiently healthy to obtain and hold employment. Persons who are not employed include many who do not seek employment or have ceased working for reasons of health. Mortality is higher among the latter than among the former so that the mortality of the total or general population can be expected to be greater than that of a sample of employed individuals, that is, providing these employed individuals are not at some occupational risk (Fox and Collier, 1976; McMichael *et al.*, 1976; Ott *et al.*, 1976; Vinni and Hakama, 1980; Wang and Miettinen, 1982).

Studies of the Healthy Worker Effect have established the following (Seltzer and Jablon, 1974; Monson, 1986; Sterling and Weinkam, 1985, 1986):

1. Mortality rates for individuals who are sufficiently healthy to hold employment (or at least had been sufficiently healthy to hold employment when they were younger) are smaller by about 25 percent than those for the general population.

2. The difference in mortality rates between a healthy population and the general population persists over the usual human lifespan.[b]

3. The Healthy Worker Effects appears to hold for all causes of death — those that appear late in life as well as those that appear early.

Table 1 summarizes these properties of the Healthy Worker Effect. The table shows SMR's computed for a large cohort of U.S. Veterans who were holders of government life insurance policies. This U.S. Veterans study was initiated in 1954 and followed approximately 250,000 veterans of the armed services who held active government life insurance policies in December 1963 and who had served between 1917 and 1940 (Kahn, 1966). Veterans who were qualified to serve in the armed forces may be considered as having been healthy at the time of their induction. They are thus analogous to a population consisting of persons for whom there is a reasonable expectation that they are employable. The fact that the study population consists of paying holders of special insurance policies indicates that participating persons were employed (or had sufficient income to pay their insurance premiums). Also, few seriously disabled persons would be included, as support for service-connected disabilities is handled under special public laws. Follow-up of this large group of persons thus offered an opportunity to observe the magnitude, persistence and ubiquity of the Healthy Worker Effect.

Table 1 gives cause specific SMR's, using all U.S. white males as the standardizing population. (The veterans were almost exclusively white males.) SMR's for all causes and for all age ranges (35-84 years of age) are smaller than unity. (Because of the large number of veterans, all differences are due to other factors than chance.) While there appears to be some increase in SMR's from ages 35-54 to ages 55-84, the increases are small. The general population has a mortality from all causes that is roughly one-third higher than that of the veterans. That effect persists through the normal lifespan and holds for different disease groupings (for more detail see Sterling and Weinkam, 1986).

---

b. McMichael (1976) is usually quoted as a source for the belief that the Healthy Worker Effect disappears with age. However, McMichael does not present convincing data to support such a conclusion, and several available data sets support the conclusion that the HWE persists throughout the lifespan.

Table 1. Standardized Mortality Ratios for All Causes and for Specific Causes Observed in a Population of 250,000 U.S. Veterans, Holders of Government Life Insurance (extracted from Sterling and Weinkam, 1986)

Cause	Age Group (yr) 35-54	55-84	35-84
All causes	0.69	0.74	0.73
All cancer	0.68	0.77	0.76
Respiratory cancer	0.52	0.67	0.64
Lung cancer	0.55	0.67	0.65
Oral cancer	0.30	0.67	0.60
Upper digestive tract cancer	0.63	0.56	0.57
Colon and rectum cancer	0.79	0.86	0.85
Prostate cancer	1.15	0.84	0.85
Kidney cancer	0.93	0.97	0.97
Bladder cancer	0.66	0.81	0.80
Heart disease	0.67	0.75	0.74
Stroke	0.62	0.63	0.63
Accident	0.79	0.56	0.66

Similar findings were also obtained by Seltzer and Jablon (1974) on another group of veterans and by Monson (1986), who followed a large population of industrial workers.

Consequences of Using SMR's Based on the General Population

Do SMR's based on the general population consistently underestimate the effect of occupational hazards? One simple test is to review the outcome of studies in which cohorts exposed to known occupational hazards were followed and their mortality experience standardized by use of the general population. Table 2 summarizes a number of recent studies (all published since 1981) for known carcinogens.

Reports are grouped by type of primary exposure (as petrochemicals, metals, etc.). For each study, Table 2 gives the SMR's for All Causes and All Cancers. The list is remarkable. All SMR's for all causes and most SMR's for all cancers are smaller than 1.0 (and some of them are as low as 0.26), and while some SMR's for all cancers exceed 1.0, none of these studies have sufficient power to judge these increases as being significant. An inspection of Table 2 leaves the impression that occupational exposure to asbestos, chlorinated compounds and PCB's, metals, or petrochemicals protects against mortality. It is also remarkable that this apparent preservation of life is among populations that are known to be the heaviest smokers in the U.S. (Sterling and Weinkam, 1976; Weinkam and Sterling, 1987).

Duration of employment is an important precondition for the analysis of effects of occupational health hazards because many of the occupational hazards associated with chronic disease and disability make their effect felt only after considerable time has past. Table 3 summarizes the outcome of a study in which Standardized Mortality Ratios were computed for steel workers for ages "25-64," for "65 and up" and for all ages "25 and up" (Sterling *et al.*, 1962). We see that SMR's for ages 25-64 are uniformly lower than 1.0; that SMR's for ages 65 and older are uniformly larger than 1.0; and that the SMR for all ages, being a summary of the lifetime follow-up mortality, is slightly less than 1.0.

Table 2. SMR's for ALL CAUSES and ALL CANCERS From a Sample From Occupation Health Studies of Working Populations Exposed to Known Carcinogens

Exposure	Reference	SMR for All Causes	SMR for All Cancers
DIOXINS & FURANS	Brown & Jones (1981)	0.89	0.89
	Ott *et al.* (1980a)	0.54*	0.28*
	Riihimaki (1982)	0.78*	0.71*
	Zack & Suskind (1980)	0.69*	1.00
METALS	Axelson *et al.* (1980)	0.99	0.90
	Cragle *et al.* (1984)	0.86*	0.94
		0.89*	1.10
	Elinder *et al.* (1985)	0.92	1.15
	Enterline & Marsh (1982)	0.84	1.03
		0.82*	1.03
	Godbold & Tompkins (1979)	0.75	0.94
	Hayes *et al.* (1979)	0.92	1.02
	Lemen *et al.* (1976)	0.93	1.54*
	Sheffet *et al.* (1982)	0.68*	0.85
	Wong *et al.* (1984)	0.88*	1.02
ORGANIC CHEMICALS & SOLVENTS	Alderson & Rattan (1980)	0.83	0.91
	Hoar & Pell (1981)	0.82*	0.69
		0.47*	0.50*
	Katz & Jowett (1981)	0.80	1.00
	Ott *et al.* (1980b)	0.71	0.76
		0.98	0.85
	Pasternack *et al.* (1977)	0.66*	1.23
	Polednak (1980)	0.86	0.94
	Wang & MacMahon (1979)	0.72	0.82
POLYAROMATIC & MISCELLANEOUS HYDORCARBONS	Jarvholm *et al.* (1981)	0.82*	0.65*
	Olin & Ahlbom (1982)	0.89*	1.32
	Olin (1978)	0.93	1.89*
		0.88	1.63*
	Paganini-Hill *et al.* (1980)	0.92	0.98
	Rockette & Arena (1983)	0.99	0.91*
	Werner & Carter (1981)	0.90	1.10
	Wong *et al.* (1985)	0.81*	0.93*
PETROCHEMICAL	Alderson & Rushton (1982)	0.84*	0.89*
	Austin & Schnatter (1983)	0.80*	0.87
	Decoufle (1978)	0.80*	1.05
	Divine *et al.* (1985)	0.75	0.75
	Hanis *et al.* (1982)	0.92*	0.92
	Rushton *et al.* (1983)	0.84*	0.95
	Rushton & Alderson (1980)	0.84*	0.89*
	Schottenfeld *et al.* (1981)	0.55*	0.72*
	Theriault *et al.* (1979)	0.78*	0.89
	Waxweiler *et al.* (1983)	0.80*	0.81*
	Wen *et al.* (1981)	0.88	0.96
	Wen *et al.* (1985)	0.78*	0.88
PVC	Fox & Collier (1977)	0.75	0.90
	Heldaas *et al.* (1984)	0.84*	1.14*
	Tabershaw & Gaffey (1974)	0.75*	1.10
RADON	Beral *et al.* (1985)	0.76	0.79*
	Checkoway *et al.* (1985)	0.73*	0.78*
	Gilbert & Marks (1979)	0.75	0.85
	Hadjimichael *et al.* (1983)	0.82*	0.85
	Smith & Doll (1981)	0.83*	0.82*
	Stayner *et al.* (1985)	0.85*	0.76
RUBBER	Delzell *et al.* (1981)	0.89	0.96
	Delzell & Monson (1981)	0.87	0.96
	Delzell & Monson (1982)	0.88	1.04
	Fox & Collier (1976)	0.98*	1.16*
	Monson & Nakano (1976)	0.65*	0.73*
	McMichael *et al.* (1974)	0.99	1.03

*Indicates statistical significant, as reported, with p equal to or less than 0.05 at least.

Table 3. Standardized Mortality Ratios of Steelworkers for Ages 25-64, 65 and Older, and 25 and Older (extracted from Sterling *et al.*, 1962)

Causes	Age Range (yr) 25-64	65 and Older	25 and Older
Lung cancer	0.79	2.36	1.00
Other cancers	0.77	1.84	1.04
Coronary disease	0.90	1.27	1.00
All other causes	0.65	1.17	0.76
All causes	0.78	1.34	0.91

We know that in this population, exposed to high levels of carcinogens around coke ovens, the Healthy Worker Effect depresses mortality in young ages and that the effect of many years of exposure to toxic gases and particulates produces a sharply increased mortality in later years, especially from lung cancer. However, because of the Healthy Worker Effect, the SMR's computed over the lifetime of these workers are insensitive to changes due to many years of occupational exposures.

A Current Example: Does Formaldehyde Cause Cancer in Occupational Settings?

Formaldehyde has been suspected to be a possible carcinogen for the lungs and upper respiratory organs. Studies of some workers with potential exposure to formaldehyde, such as biologists, laboratory workers, pathologists and undertakers, have not found increases in lung cancers. However, these studies are of classes of individuals who are only occasionally exposed or of whom only a small number are actually exposed to formaldehyde, e.g., embalmers and morticians (Williams *et al.*, 1984) or are exposed more to solvents than formaldehydes, such as pathologists and laboratory workers. (The increase in risks for cancers of nervous system and lymph tissue found among pathologists and laboratory workers may reflect solvent and not formaldehyde exposure.)

There are a number of studies in the literature which followed industrial cohorts exposed to formaldehyde. A number of these studies fail to find significantly elevated risks for lung cancer when relying on SMR's with the general population as the referent, but do find significantly increased risk when analyzed by a method that controls for the Healthy Worker Effect. For instance, Wong (1983) fails to find an increase in all cancers judging by an SMR of 1.01, but shows a significant increase in lung cancer risk judging by the proportional mortality ratio (PMR), which is independent of the Healthy Worker Effect, (PMR=1.48, statistically different from 1.0 with $p<0.05$).

A study by Acheson *et al.* (1984) computes SMR's for lung cancer using British national and local mortality rates. Acheson *et al.* concluded that they had not demonstrated increased mortality from lung cancers in their populations. They report finding an SMR for lung cancer of 1.04.

But that report provides sufficient data to permit an internal comparison of employed individuals who differed with respect to exposure and so makes it possible to compute PMR's as partial control for the Healthy Worker Effect. Acheson *et al.* provide a detailed breakdown of the number of deaths in seven factories which either used or manufactured

Table 4. Mortality Among Men in British Industrial Plastics (BIP) and in All Other Factories and PMR's Corresponding to Different Causes of Deaths (extracted from Sterling and Arundel, 1985)

Cause of Death	BIP		Other Factories		PMR
	Number	%	Number	%	
Respiratory disease (not malignant)	145	17.2%	91	11.8%	146*
Malignant tumors (including cancer)	251	29.7%	204	26.3%	113
Lung Cancer	128	15.2%	77	9.9%	154*
All other	449	53.1%	479	61.9%	94
All Causes	845	100%	774	100%	--

*p<0.01.

formaldehyde and also of the number of workers in each factory who were exposed to low, moderate or high levels of formaldehyde. High exposure was almost entirely limited to employees of one factory (British Industrial Plastics), which contributed 90% of all high exposure. These data permit comparison of the mortality experience of the employees of BIP with that for employees for all other factories in which workers had, for the most part, low or negligible exposure. At the same time, workers in other factories should be comparable to those working for BIP for various indicators of social class and exposure to other risk factors. Using the Acheson data it is possible to compute crude proportional mortality ratios by comparing the percentage of deaths due to lung cancer and respiratory disease among employees at the high exposure factory against the percentage of death due to lung cancer for the employees of all other factories combined. PMR's for lung cancer and for nonmalignant respiratory disease are substantially and significantly elevated (see Table 4).

A third example is based on a study by the National Cancer Institute and the Formaldehyde Institute (NCI/FI) on the mortality of formaldehyde workers from all and specific causes. The investigators base their conclusions on SMR's, computed by using the general population as the referent. They conclude that the study provides little evidence of increased cancer risk associated with exposure to formaldehyde at levels experienced by these workers (Blair *et al.*, 1986). The data were reanalyzed, comparing workers with low exposure to workers with high exposure to formaldehyde. The high exposure group was compared to the low exposure group while simultaneously controlling for differences in length of employment, salaried or hourly job classification, and age. Substantially elevated lung cancer risks emerged (Sterling and Weinkam, 1988) (see Table 5).

Of interest also in Table 5 is the considerable age effect. It is clear that analysis of occupational data must take age into consideration.

DISCUSSION

Goldsmith (1975) suggested correcting SMR's from occupational cohort studies for the Healthy Worker Effect by using 0.9 as a base (rather than 1.0). However, studies such as Sterling and Weinkam (1986), Seltzer and Jablon (1974), and Monson (1986) indicate that such a correction (for a population of long term employees) should use 0.7 to 0.75 as a

Table 5. Comparison of Risks of Workers with High Exposure to Formaldehyde Obtained from the Same NCI/FI Data (1) Using the General Population as Referent (from Blair, 1986) or (2) Using Least Exposed Workers as Referent (based on Sterling and Weinkam, 1988)

Disease	General Population	Least Exposed Workers (Sterling,1988)
All Cancers	1.01	1.14
Lung Cancer	1.11	1.56

base. While having the merit of preventing a gross underestimate of occupational mortality, that kind of an adjustment could introduce a bias in the opposite direction in instances where selection of individuals for a working population does not limit itself to healthy candidates to the same extent as some other job selections do. Thus, the important factor to bear in mind in designing any historical cohort study is to ensure that it can compare individuals with low and high levels of exposure to toxic materials. And that comparison must not be based on differences in exposure between hourly (or blue-collar) workers and salaried (or white-collar) employees because the mortality of white-collar employees very likely differs from that of blue-collar workers for a variety of reasons, some of which are and some of which are not related to occupational exposures.

Computation of proportional mortality ratios has some advantages in that they are affected less by the Healthy Worker Effect. But, PMR's share one weakness of SMR's, namely that they are not sensitive to changes in relative mortality with age. If the incidence of cancers among young workers is low but there is a high level of cancer among older individuals, that change may be missed by computing proportional mortality. Similarly, a substantial increase in lung cancer rates at younger ages accounts for relatively few deaths which may be masked by the much larger number that occurs in both the exposed and unexposed groups at older ages. Taking care to compute proportional mortality ratios for different age ranges can adjust for age factors while still controlling for the Healthy Worker Effect.

It may well be best not to rely on historical occupational cohort studies, given the difficulties in interpreting mortality data from them. Case control studies are cheaper, yield results sooner, and are much easier to adjust for Healthy Worker Effects than historical cohort studies.

Our findings still leave unanswered the problem emerging from perusal of Table 2. There are today a large number of historical cohort studies which use the general population as a base to standardize mortality ratios. Many of these studies conclude that they find little or no evidence of occupational hazard. These studies need to be re-examined. The problems of reviewing conclusions from these types of studies indeed loom large.

REFERENCES

Acheson, E. D., Barnes, H. R., and Gardner, M. J., 1984, Formaldehyde in the British Chemical Industry, *Lancet* **i**:611-16.

Acheson, E. D., Barnes, H. R., and Gardner, M. J., 1984, Formaldehyde Process Workers and Lung Cancer, *Lancet* **i**:1066-67.

Alderson, M. R., and Rattan, N. S., 1980, Mortality of Workers on an Isopropyl Alcohol Plant and Two MEK Dewaxing Plants, *Brit. J. Ind. Med.* **7**:85-89.

Alderson, M., and Rushton, L., 1982, Mortality Patterns in Eight U.K. Oil Refineries, *Annals NY Academy of Sci.* **381**:139-145.

Austin, S. G., and Schnatter, A. R., 1983, A Cohort Mortality Study of Petrochemical Workers, *J. Occup. Med.* **25**:304-312.

Axelson, G., Rylander, R., and Schmidt, A., 1980, Mortality and Incidence of Tumors Among Ferrochromium Workers, *Brit. J. Ind. Med.* **37**:121-127.

Beral, V., Inskip, H., Fraser, P. *et al.*, 1985, Mortality of Employees of the United Kingdom Atomic Energy Authority, 1946-1976, *Brit. Med. J.* **291**:440-447.

Blair, A., Stewart, P., O'Berg, M. *et al.*, 1986, Mortality Among Industrial Workers Exposed to Formaldehyde, *J. Natl. Cancer Inst.*, **76**:1071-1084.

Brown, D. P., and Jones, M., 1981, Mortality and Industrial Hygiene Study of Workers Exposed to Polychlorinated Biphenyls, *Arch. Environ. Health* **36**:120-129.

Checkoway, H., Mathew, R. M., Shy, C. M. *et al.*, 1985, Radiation, Work Experience, and Cause Specific Mortality Among Workers at an Energy Research Laboratory, *Br. J. Ind. Med.* **42**:525-533.

Cragle, D. L., Hollis, D. R., Qualters, J. R. *et al.*, 1984, A Mortality Study of Men Exposed to Elemental Mercury, *J. Occup. Med.* **26**:817-821.

Decoufle, P., 1978, Further Analysis of Cancer Mortality Patterns Among Workers Exposed to Cutting Oil Mists, *J. Natl. Cancer Inst.* **61**:1025-1030.

Delzell, E., Louik, C., Lewis, J. *et al.*, 1981, Mortality and Cancer Morbidity Among Workers in the Rubber Tire Industry, *Amer. J. Ind. Med.* **2**:209-216.

Delzell, E., and Monson, R. R., 1982, Mortality Among Rubber Workers: V. Processing Workers, *J. Occup. Med.* **24**:539-545.

Delzell, E., and Monson, R. R., 1981, Mortality Among Rubber Workers: III. Cause-Specific Mortality, 1940-1978, *J. Occup. Med.* **23**:677-682.

Divine, B. J., Barron, V., and Kaplan, S. D., 1985, Texaco Mortality Study, *J. Occup. Med.* **27**:445-447.

Elinder, C-G., Kjellstrom, T., Hogstedt, C. *et al.*, 1985, Cancer Mortality of Cadmium Workers, *Brit. J. Ind. Med.* **42**:651-655.

Enterline, P. E., and Marsh, G. M., 1982, Mortality Among Workers in a Nickel Refinery and Alloy Manufacturing Plant in West Virginia, *J. Natl. Cancer Inst.* **68**:925-933.

Fox, A. J., and Collier, P. F., 1976, A Survey of Occupational Cancer in the Rubber and Cablemaking Industries: Analysis of Deaths Occurring in 1972-74, *Brit. J. Ind. Med.* **33**:249-264.

Fox, A. J., and Collier, P. F., 1976, Low Mortality Rates in Industrial Cohort Studies Due to Selection for Work and Survival in the Industry, *Br. J. Prev. Soc. Med.* **30**:225-230.

Fox, A. J., and Collier, P. F., 1977, Mortality Experience of Workers Exposed to Vinyl Chloride Monomer in the Manufacture of Polyvinyl Chloride in Great Britain, *Brit. J. Ind. Med.* **34**:1-10.

Gilbert, E. S., and Marks, S., 1979, An Analysis of the Mortality of Workers in a Nuclear Facility, *Radiation Res.* **70**:122-148.

Godbold, J. H., Jr., and Tompkins, E. A., 1979, A Long-Term Mortality Study of Workers Occupationally Exposed to Metallic Nickel at the Oak Ridge Gaseous Diffusion Plant, *J. Occup. Med.* **21**:799-806.

Goldsmith, J., 1975, What We Expect from an Occupational Cohort? *J. Occup. Med.* **17**:126-127.

Hadjimichael, O. C., Ostfeld, A. M., D'Atri, D. A. *et al.*, 1983, Mortality and Cancer Incidence Experience of Employees in a Nuclear Fuels Fabrication Plant, *J. Occup. Med. 25*:48-61.

Hanis, N. M., Holmes, T. M., Shallenberger, L. G. *et al.*, 1982, Epidemiologic Study of Refinery and Chemical Plant Workers, *J. Occup. Med.* **24**:203-212.

Hayes, R. B., Lilienfeld, A. M., and Snell, L. M., 1979, Mortality in Chromium Chemical Production Workers: A Prospective Study, *Intl. J. Epidem.* **8**:365-374.

Hoar, S. K., and Pell, S., 1981, A Retrospective Cohort Study of Mortality and Cancer Incidence Among Chemists, *J. Occup. Med.* **23**:485-501.

Jarvholm, B., Lillienberg, L., Sallsten, G. *et al.*, 1981, Cancer Morbidity Among Men Exposed to Oil Mist in the Metal Industry, *J. Occup. Med.* **23**:333-337.

Kahn, H., 1966, The Dorn Study of Smoking and Mortality Among U.S. Veterans: Report on Eight- and One-Half Years of Observations, in *Epidemiological Approaches to the Study of Cancer and Other Diseases*, W. M. Haenszel, ed., National Cancer Institute Monographs, **19**:1-25.

Katz, R. M., and Jowett, D., 1981, Female Laundry and Dry Cleaning Workers in Wisconsin: A Mortality Analysis, *Amer. J. Pub. Health* **71**:305-307.

Lemen, R. A., Lee, J. S., Wagoner, J. K. *et al.*, 1976, Cancer Mortality Among Cadmium Production Workers, *Annals NY Acad. Sci.* **271**:273-9.

Monson, R. R., 1986, Observations on the Healthy Worker Effect, *J. Occup. Med.* **28**:425-433.

Monson, R. R., and Nakano, K. K., 1976, Mortality Among Rubber Workers: II. Other Employees, *Amer. J. Epidem.* **103**:297-303.

McMichael, A. J., Spirtas, R., and Kupper, L. L., 1974, An Epidemiologic Study of Mortality within a Cohort of Rubber Workers, 1964-72, *J. Occup. Med.* **16**:458-464.

McMichael, A. J., 1976, Standardized Mortality Ratios and the 'Healthy Worker Effect': Scratching Beneath the Surface, *J. Occup. Med.* **18**:165-168.

Olin, G. R., and Ahlbom, A., 1982, Cancer Mortality Among Three Swedish Male Academic Cohorts: Chemists, Architects, and Mining Engineers/Metallurgists, *Annals NY Acad. Sci.* **381**:197-201.

Olin, G. R., 1978, The Hazards of a Chemical Laboratory Environment — A Study of the Mortality in Two Cohorts of Swedish Chemists, *Amer. Ind. Hygiene Assoc. J.* **7**:557-562.

Ott, M. G., Holder, B. B., and Langner, R. R., 1976, Determinants of Mortality in an Industrial Population, *J. Occup. Med.* **18**:171-177.

Ott, M. G., Holder, B. B., and Olson, R. D., 1980, A Mortality Analysis of Employees Engaged in the Manufacture of 2,4,5-Trichlorophenoxyacetic Acid, *J. Occup. Med.* **22**:47-50.

Ott, M. G., Kolesar, R. C., Scharnweber, H. C., *et al.*, 1980, A Mortality Survey of Employees Engaged in the Development or Manufacture of Styrene-Based Products, *J. Occup. Med.* **22**:445-460.

Paganini-Hill, A., Glazer, E., Henderson, B. E. *et al.*, 1980, Cause-Specific Mortality Among Newspaper Web Pressmen, *J. Occup. Med.* **22**:542-544.

Pasternack, B. S., Shore, R. E., and Albert, R. E., 1977, Occupational Exposure to Chloromethyl Ethers, *J. Occup. Med.* **19**:741-746.

Rockette, H. E., and Arena, V. C., 1983, Mortality Studies of Aluminum Reduction Plant Workers: Potroom and Carbon Department, *J. Occup. Med.* **25**:549-557.

Polednak, A. P., 1980, Mortality Among Men Occupationally Exposed to Phosgene in 1943-1945, *Environ. Res.* **22**:357-367.

Riihimaki, V., Asp, S., and Hernberg, S., 1982, Mortality of 2,4-Dichlorophenoxyacetic Acid and 2,4,5-Trichlorophenoxyacetic Acid Herbicide Applicators in Finland, *Scand. J. Work Environ. Health* **8**:37-42.

Rushton, L., and Alderson, M., 1980, The Influence of Occupation on Health — Some Results from a Study in the UK Oil Industry, *Carcinogenesis* **I**:739-743.

Rushton, L., Alderson, M. R., and Nagarajah, C. R., 1983, Epidemiological Survey of Maintenance Workers in London Transport Executive Bus Garage and Chiswick Works, *Brit. J. Ind. Med.* **40**:340-345.

Schottenfeld, D., Warshauer, M. E., Zauber, A. G. *et al.*, 1981, A Prospective Study of Morbidity and Mortality in Petroleum Industry Employees in the United States — A Preliminary Report, in *Banbury Report 9*, R. Peto and M. Schneiderman, eds., Cold Springs Harbor.

Seltzer, C. C., and Jablon, S., 1974, Effects of Selection on Mortality, *Am. J. Epidem.* **100**:367-372.

Sheffet, A., Thind, I., Miller, A. M. *et al.*, 1982, Cancer Mortality in a Pigment Plant Utilizing Lead and Zinc Chromates, *Arch. Environ. Health* **37**:44-52.

Smith, P. G., and Doll, R., 1981, Mortality from Cancer and All Causes Among British Radiologists, *Brit. J. Rad.* **54**:187-194.

Stayner, L. T., Meinhardt, T., Lemen, R. *et al.*, 1985, A Retrospective Cohort Mortality Study of a Phosphate Fertilizer Production Facility, *Arch. Environ. Health* **40**:133-138.

Sterling, T. D., Phair, J. J., and Rustagi, J., 1962, New Developments in Chronic Disease Epidemiology: Competing Risks and Eligibility, *Am. Ind. Hyg. Assoc. J.* **23**:433-446.

Sterling, T. D., and Arundel, A., 1985, Formaldehyde and Lung Cancer, *Lancet* **14**:1366 (December).

Sterling, T. D., and Weinkam, J. J., 1976, Smoking Characteristics by Type of Employment, *J. Occup. Med.* **18**:743-754.

Sterling, T. D., and Weinkam, J. J., 1985, The "Healthy Worker Effect" on Morbidity Rates, *J. Occup. Med.* **27**:477-82.

Sterling, T. D., and Weinkam, J. J., 1986, Extent, Persistence, and Constancy of the Healthy Worker or Healthy Person Effect by All and Selected Causes of Death, *J. Occup. Med.* **28**:348-353.

Sterling, T. D., and Weinkam, J. J., 1988, Reanalysis of a National Cancer Institute Study on Mortality Among Industrial Workers Exposed to Formaldehyde, *J. Occup. Med.* **30**:895-901.

Tabershaw, I. R., and Gaffey, W. R., 1974, Mortality Study of Workers in the Manufacture of Vinyl Chloride and Its Polymers, *J. Occup. Med.* **16**:509-518.

Theriault, G., and Goulet, L., 1979, A Mortality Study of Oil Refinery Workers, *J. Occup. Med.* **21**:367-370.

Vinni, K., and Hakama, M., 1980, Healthy Worker Effect in the Total Finnish Population, *Br. J. Ind. Med.* **37**:180-184.

Wang, H. H., and MacMahon, B., 1979, Mortality of Workers Employed in the Manufacture of Chlordane and Heptachlor, *J. Occup. Med.* **21**:745-748.

Wang, J., and Miettinen, O., 1982, Occupation Mortality Studies — Principles of Validity, *Scand. J. Work Environ. Health* **8**:153-158.

Waxweiler, R. J., Alexander, V., Leffingwell, S. S., *et al.*, 1983, Mortality from Brain Tumor and Other Causes in a Cohort of Petrochemical Workers, *J. Natl. Cancer Inst.* **70**:75-81.

Weinkam, J. J., and Sterling, T. D., 1987, Changes in Smoking Characteristics by Type of Employment from 1970 to 1979/80, *Am. J. Ind. Med.* **11**:539-561.

Wen, C. P., Tsai, S. P., Weiss, N. S. *et al.*, 1981, A Population-Based Cohort Study of Brain Tumor Mortality Among Oil Refinery Workers with a Discussion of Methodological Issues of SMR and PMR, in *Banbury Report 9*, R. Peto and M. Schneiderman, eds., Cold Springs Harbor.

Wen, C. P., Tsai, S. P., Weiss, N. S. *et al.*, 1985, Long-Term Mortality Study of Oil Refinery Workers, IV. Exposure to the Lubricating-Dewaxing Process, *J. Natl. Cancer Inst.* **74**:11-18.

Werner, J. B., and Carter, J. T., 1981, Mortality of United Kingdom Acrylonitrile Polymerisation Workers, *Brit. J. Ind. Med.* **38**:247-253.

Williams, T. M., Levine, R., and Blunden, P., 1984, Exposure of Embalmers to Formaldehyde and Other Chemicals, *Amer. Ind. Hyg. Assoc. J.* **45**:172-176.

Wong, O., Brocker, W., Davis, H. V. *et al.*, 1984, Mortality of Workers Potentially Exposed to Organic and Inorganic Brominated Chemicals, DBCP, TRIS, PBB, and DDT, *Brit. J. Ind. Med.* **41**:15-24.

Wong, O.,1983, An Epidemiologic Mortality Study of a Cohort of Chemical Workers Potentially Exposed to Formaldehyde with a Discussion on SMR and PMR, in *Formaldehyde Toxicity, Chemical Industry Institute of Toxicology Series*, pp. 256-272, J. E. Gibson, ed., Washington, D.C.

Zack, J. A., and Suskind, R. R., 1980, The Mortality Experience of Workers Exposed to Tetrachlorodibenzodioxin in a Trichlorophenol Process Accident, *J. Occup. Med.* **22**:11-14.

Cultural Values Affecting Risk Perception: Individualism and the Perception of Toxicological Risks[a]

Janet M. Fitchen
Ithaca College
Ithaca, NY

ABSTRACT

Cultural assumptions and values shape the way members of a society characteristically perceive and respond to risks. As new risks arise, they are perceived through existing cultural orientations. Individualism is often considered the quintessence of American culture; and given its lasting preeminence, it should influence the way Americans perceive such new risks as those posed by toxic chemicals in the environment.

Anthropological research in communities where groundwater contamination has affected local drinking water reveals that individualism does indeed shape public responses. This paper explores several interrelated facets of individualism expressed in these real-life toxic risk situations. Risk communicators and risk managers should be more aware of such cultural dimensions of risk perception.

KEYWORDS: Risk perception, individualism, groundwater contamination, cultural values

INTRODUCTION

Risk perception has important cultural as well as psychological dimensions, for basic cultural orientations shape the way members of a society characteristically perceive and respond to risks.[b] This paper reports some findings of anthropological research being conducted to identify assumptions and values of American culture that shape perception of and responses to environmental risks. To narrow the research focus, one specific environmental risk situation has been selected: chemical contamination of groundwater

a. A more extensive version of this article appeared in *Science, Technology, & Human Values* **12(2)**:1-12 (Spring 1987).

b. Anthropological attention to this field has been stimulated by such seminal works as Douglas and Wildavsky, 1985; and Rayner, 1984. Much of the work, however, comes from psychology and psychometrics: e.g., Slovic, Fischoff, and Lichtenstein, 1979, 1980, and 1984.

used for drinking water supply.[c] This particular environmental problem represents a new toxicological risk to which Americans are increasingly exposed. As the paper will demonstrate, Americans bring to this new risk some rather old cultural orientations that shape their interpretation and response.

The anthropological research reported on here is part of a project being carried out by a multi-disciplinary team that also includes environmental toxicologists, chemists, biochemists, and water resource specialists (Fitchen, 1987; Fitchen, Heath, and Fessenden-Raden, 1987). The group is conducting field studies in over a dozen communities in New York and Maine that have discovered chemical contaminants at concentrations exceeding state or federal standards for groundwater used as drinking water supply.[d] The social science part of this project involves open-ended interviews with local decision makers and other community leaders, and with personnel from state and federal agencies who are involved at the local level. It also includes informal discussions with local residents, attendance at public meetings, and review of local newspapers, public documents, and official correspondence. To explore in greater depth and breadth some of the cultural concepts and meanings identified in these contamination cases, supplementary research probes were conducted outside the study communities: questionnaires with diverse populations; scrutiny of newspaper articles about groundwater problems in different regions; interviews with water and groundwater specialists around the U.S.; and attendance at courses and workshops for water-related professionals.

With data drawn from these disparate sources, but primarily from the communities actually facing groundwater quality problems, one can begin to identify cultural patterning in the way people talk about and react to the degradation of their water by man-made chemicals. The patterns emerging from these data closely resemble the cultural orientations and values usually identified as characteristic of American culture by anthropologists, sociologists, and historians studying other bodies of data.[e] One cultural orientation has become particularly evident in this analysis: individualism. Research data will be presented here to elucidate the meaning of individualism with respect to this specific environmental problem, and the role individualism plays in shaping public perceptions of the risks posed by chemical pollution of groundwater.

FACETS OF INDIVIDUALISM REVEALED IN GROUNDWATER CONTAMINATION CASES

Although other national cultures, particularly in western Europe, exhibit some similarities in definitions of and importance accorded to the individual, individualism as a basic cultural orientation is generally thought to be preeminent in and distinctive of American culture. Tocqueville used this newly-coined term in 1835 to describe a

---

c. Half of all Americans, and the vast majority of rural and non-metropolitan residents, obtain their drinking water from groundwater—water stored in aquifers (permeable saturated zones of rock, sand, or gravel that underlie the water table in some areas). In recent years, this essential resource has increasingly been found to be contaminated by toxic chemicals from industrial disposal, agricultural use, accidental commercial spills, ordinary household usage, and leaking underground storage tanks. Public awareness of groundwater contamination has been growing, aided by informative articles in general-circulation magazines (e.g., *New York State's Conservationist*, 12/84; *Discover Magazine* 2/85; *U.S.A. Today*, 2/12/85; and *Blair and Ketcham's Country Journal*, 3/85), as well as numerous mass circulation magazine articles (*Time*, 10/14/85) and television documentaries on toxic waste problems of all sorts.

d. For our research project, "community" is defined operationally as a locality with a chemical contamination problem in the water supply. Therefore our sites need not be "true communities" in a geographic, political, or social-interaction sense. We are conducting research in communities, not studies of communities. For other case studies of contamination problems at the local level, see Edelstein, 1982 and 1988; Fowlkes and Miller, 1982; and Berry and Stoeckle. See p. 8.

e. See Spindler and Spindler, 1983, for a review of anthropological study of American culture.

phenomenon he saw as utterly new, utterly American (Vol. II, pp. 104-6). And some version of this dominant cultural theme has been identified by scholars ever since as the defining characteristic of American culture (Gorer, 1948; DuBois, 1955; Henry, 1963; Hsu, 1973; Inkeles, 1979; Spindler and Spindler, 1983; Varenne, 1977; Greenhouse, 1985; Bellah *et al.*, 1985). For purposes of analysis, we will disaggregate individualism into four related and overlapping facets, each of which seemed to affect perception of local problems of groundwater contamination.

Self-Reliance and Independence

The American cultural emphasis on independence shows up in the phrases people use, preferences they express, and actions they take with regard to drinking water. Even the categories for different kinds of water reflect ideas of independence: in addition to categories of use, (e.g., drinking water, bathing water), ordinary residents and water professionals both tend to classify domestic water by source as either "our own water" or "city water." Where both private and public water was available (as is often the case in smaller cities and villages), many people expressed a preference for having "our own well" or "our own water." Reasons stated for this preference included the obvious "better tasting water" (no chlorine) and lower costs (no hookup charges or regular bills), but also a common assertion that having one's own well is a condition or sign of "independence." Even where the choice is between a contaminated private well and a "clean" public water system, we found some reluctance to hook up to the public supply. Similar preference for private wells was found by Edelstein (1982, p. 120) in a new suburban community that discovered contamination. The first wave of suburban settlers there preferred to keep their contaminated private wells (retrofitted with filters), rather than hook up to the town water supply, in large part because they viewed private wells as a sign of independence. This same phenomenon has also been observed on a community-wide scale. The desire of a village to "maintain our independence" seemed to be a major factor in its refusal to hook up to a neighboring village's water system as a substitute for its own contaminated water (Lemley *et al.*, 1985).

The conceptual linking of private wells with independence was confirmed in a separate research probe conducted among three populations with widely different geographic and demographic characteristics. Respondents ranked 40 items, one of which was "Having my own well for home water supply," on a 7-point scale of "dependence" (1) to "independence" (7). The three samples were remarkably consistent in associating this phrase with "independence." Over 82% of the 214 respondents ranked "having my own well" towards the "independence" side, with just over one-third ranking it as number 6 (which was the median and mode).[f] About 10% ranked it in the middle (4), and less than 10% placed it anywhere on the "dependence" side. By contrast, when the phrase "being on a public water supply" was included for one sample (n=53), the modal answer was slightly towards the "dependence" side (3).

Self-reliance was indicated in people's discussion of their preference for private wells and in their responses to groundwater contamination. In one case, a man with contaminants in his well (and considerable mechanical experience in his background) made the following statement of self-reliance to explain his preference for continuing to use his own water (perhaps with a filter): "Even if something goes wrong, say if the pump breaks down in the middle of the night, it's my pump and my water supply. I can just go down in the basement and fix the problem. I don't have to run around trying to get somebody in city

f. This ranking of "having my own well..." compares to such other high "independence" items on the list as "maturity," "getting a driver's license," "homeownership," "a growing child should become," and "grown children, in relation to their parents, should be." The three samples were: 100 liberal arts students at an Eastern college, 61 agricultural and home economics Extension agents in Kansas, and 53 people attending a Northeastern university short course on groundwater.

hall to come to my house to fix it. I can take care of it myself." Self-reliance seemed more important here than cost savings.

The cultural value placed on self-reliance also shapes the way people interpret a given contamination situation. People place considerable confidence in their own senses as a reliable source of information about the world around them. If people begin to see, smell, or taste something "funny" about their water, or if they perceive an unusual pattern of illness in the community, then they may become quite worried that something is wrong. (If they also see "black ooze" or "orange goo" seeping onto the ground, then their anxiety over the drinking water is greatly exacerbated—probably for good reason.) They will seek answers and action from city, county, state, and federal officials. If individual perceptions and official pronouncements disagree, the public is apt to be suspicious of the latter, reflecting again a strong reliance on the evidence of their own senses. For example, if water tests come back "negative," people who have seen or tasted something unusual in their water are quick to question the validity of the tests, the procedures of the technicians who conducted them, or the honesty of the bureaucrats and politicians who, they suspect, are trying to hide the truth from the affected public.

The confidence people place in their own senses is also demonstrated in opposite cases, where the contaminant is undetectable by ordinary senses. In these cases, people do not see, taste, or smell anything unusual about their water, and may not perceive any unusual illness pattern (due to a long latency period and low-level probability of health effects from the chemicals). When told that "routine sampling" has shown their water to be contaminated, people may doubt the information and exhibit rather little concern about health issues because their individual sensors for judging reality give them no reason to worry. Some homeowners whose senses tell them "there's nothing wrong with my water" have declined to have their water tested, even though a neighbor's well has been found to contain toxic contaminants. Strong reliance on ones own senses, especially vision, is also revealed in the fact that a municipal water department will receive many complaints when rusty-looking but safe water is substituted for water with invisible toxic chemicals. People's own eyes seem to tell them when to worry.

Autonomy and Control Through Individual Choice

Both public discourse and private conversation in communities with contaminated groundwater revealed that people want "freedom" as individuals to decide what level of risk is acceptable to them and which risks they will accept, a finding that is consistent with the literature on risk perception.[g] Risks are more readily accepted if they are perceived as voluntarily assumed rather than involuntarily imposed. Where the individual has choice and control, a higher level of risk may be acceptable. This preference for voluntarily accepted risks explains the fact that, during public meetings and interviews, some of the individuals who speak out most strongly against allowing any detectable carcinogens in their drinking water may be smoking cigarettes, which may pose a much higher cancer risk than the chemicals in the water. People we have interviewed on this point have explained away the seeming contradiction by stating that "self-inflicted risks" are not comparable to risks imposed by others.

The desire to exercise choice or control in accepting risk often conflicts with the fact that standards for permissible levels of toxics are set by federal or state agencies far removed from individual citizens. In some of our study communities, residents wanted more stringent quality standards, and became angry that the government would not provide funding assistance or take action to clean up a situation simply because the official

---

g. Psychometric research has convincingly demonstrated that risks are more readily accepted and less feared if they are voluntarily assumed; Slovic, Fischoff & Lichtenstein, 1979, 1980. Rayner's work (1983) on siting nuclear power plants also indicates the importance of a sense of control.

contaminant standard or guideline had not been exceeded.[h] By contrast, in other communities people wanted discretion to accept a higher level of contaminants than state or federal guidelines allowed, partly because they feared that local jobs might be lost if a local factory were required to undertake expensive changes in procedures of handling toxic materials. In these latter cases, people's resentment over the perceived lack of individual or local choice concerning water quality standards seemed to exceed their concern about a small health risk in the distant future. Comments at public meetings and in interviews included: "I'm the one to decide if I want to drink this water or not," and "Who the hell are they to tell me what I can or can't drink?" Whether residents wants to set standards higher or lower than state/federal levels, their common argument is that local citizens ought to have a say in answering the question of "How safe is safe enough?" or "How much risk is acceptable?"

The desire for individual control is also evident in people's reactions to temporary measures instituted upon discovery of contaminants in water supplies. It has been reported elsewhere, for example, that people somewhat resent the free tanks of water delivered by a municipality to their home during a "water emergency" (Edelstein, 1988, pp. 34-36). In our study communities, some interviewees mentioned not only the inconvenience of delivered water, but also an unwelcome feeling of dependence on others and loss of control over their own lives. In contrast, a "boil water notice" commonly issued by a local health department when contamination is first discovered, is regarded as a nuisance but is not resented. People feel that they have some choice in whether or not to comply—and many choose not to. The longer-range solution of hooking up households with contaminated private wells to a safe municipal water system is not uniformly welcomed, in part because people feel they are making a tradeoff in which they lose some control over their own lives.[i]

The Problematic Relationship Between Individual and Collectivity

As Tocqueville forecast long ago, the American emphasis on individuals means that relationships between people may be underperceived, and that the whole is perceived as an aggregate of individuals more than as a social unit (Varenne, 1977; Greenhouse, 1185, p. 261). The heritage of John Locke is still recognizable in this society, where "The individual is prior to society, which comes into existence only through the voluntary contract of individuals trying to maximize their own self-interest" (Bellah *et al.*, 1985, p. 143).

The cultural deemphasis on the collectivity becomes more obvious by comparing the discourse from our sample communities with the discourse of American Indians in the same region who are also dealing with problems of contaminated groundwater.[j] When talking or writing about the issue, Indian people use the word "community" a great deal more frequently; and use it to refer to a collective entity that is both social and spiritual, to a set of relationships with long-term continuity. They see the community, thus defined, as being at risk from the chemicals. This greater emphasis on community and relative deemphasis on the individual shows up in other American sub-cultures as well, for example, among Mormons (Vogt and O'Dea, 1953), and the contrast makes the individualistic emphasis of the dominant culture even more apparent. In most mainstream talk, the perceived harm is

---

h. The strong preference for individual choice shows up where angry residents feel they don't have enough say in deciding NOT to accept an officially allowable risk. "These are our kids, but you (state Health Department) are making all the decisions. Who do you think you are?" — a Vermont resident in "Toxic Time Bomb," aired on HBO, 10/21/84, DBA Productions.

i. Individual choice and control are also at issue in disputes over fluoridation of drinking water. The real issue for fluoridation opponents seems to be retaining control over the chemicals they ingest; they are not swayed by the "scientific evidence" presented by the experts from the other side.

j. Discussions with members of the American Indian Program at Cornell University have emphasized this point, and perusal of their publications gives further substantiation.

to individuals and to their individually held tangible property. The word "community" is used much less frequently and seems to refer mostly to a specific geographic or municipal entity. In fact, American culture does not make it easy for people who do feel a stronger collective sense to verbalize this.[k] In relation to water, it might be noted that there is no term to denote the collectivity of consumers of a given public water supply. Correspondence from the water district to residents on the municipal system usually begins, "Dear water customer," never "Dear member of our water community."

Individualism affects American perceptions of the collectivity in another way also. Since each individual is perceived as unique, and since the community is a collectivity of individuals, then by definition each community must also be unique. This assumption of the uniqueness of one's own community (but not necessarily of other people's communities) is common in America,[l] and seems to make each community feel that because its water contamination problem is unique, therefore there is no need for communities to learn from each other about dealing with such problems. The uniqueness assumption may also underlie the often-voiced distrust of solutions that have worked elsewhere: for example, an aquifer protection plan developed for one community might not even be considered by a second community despite its merits, simply because it is perceived as "not designed for our particular community." The assumption of uniqueness may also give rise to the expressed belief that "outside agencies" (federal or state) who come in to supervise cleanup of groundwater contamination don't know the local community, aren't aware of its special features, and therefore will not be able to conduct their project satisfactorily.

Perception of Events, Including Risks, in Terms of Individuals

The health risk posed by a particular toxic chemical is always officially stated in terms of probabilities and aggregates, for example, as the number of excess cancers per million of population that would be expected from ingesting a certain amount of the chemical over a certain period of time. But our research shows that people interpret the risk as an individual issue, asking "What is my chance of getting cancer?" or "Should I drink the water?" This disjunction between the official presentation of macro-level aggregate risk information and the micro-risk question that people are actually concerned about was also seen in the case of the "EDB scare" of 1984 (Fischoff, 1985; Sharlin, 1985).

Since the health expert (toxicologist, local health officer, or physician) cannot answer the micro-risk question people are raising about their own health future, the listeners tend to translate the probability statements they are given into their own personal risk assessments. In cases where people perceive a cause-and-effect relationship between the contaminant and an existing health problem, they may overestimate the probability figures that are given, and their anxiety rises. On the other hand, where the health effects have low probability and long latency (the actual onset of illness from the chemical would not occur until long after exposure, if it occurs at all), people may underestimate the health risk of toxic chemicals. When groundwater contamination involves no clear and present danger, the line of reasoning for many people is more or less as follows: "What they're saying might not happen anyway... They could be wrong... And if it does happen to one person in a million, it probably won't happen in our city because we only have 25,000 people here... And if it does happen here, it probably won't happen to me...Besides, by then they'll probably have a cure for it... So why should I worry about the water?" This kind of personal arithmetic acts as a filter for the formal probability statements of risk that are supplied by government

---

k. In an historic court decision, Buffalo Creek, WV was awarded compensation for the loss of community, an intangible psycho-social loss resulting from collapse of a dam. See Erikson, 1976.

l. Observations of communities around the nation dealing with environmental problems support this "uniqueness" hypothesis. Vidich and Bensman (1958) present a small-town version of it.

authorities and technical experts. It resembles also the misperceptions of probability exhibited in the gambling behavior of elderly Americans (Jones, 1976).

This difficulty in understanding the experts' statements regarding health risk is not simply a failure in mathematical understanding, but a cultural phenomenon, or really the conjunction of several cultural phenomena, all of which are connected to individualism. We have found numerous instances in which the following culturally shaped perceptions interfere with perception of probability of harm: (a) a propensity to doubt experts, especially if the experts' opinions differ from one's own (see Fischoff *et al.*, 1983) or from each other's (Nelkin, 1982); (b) the difficulty in perceiving collective health risk apart from the risk to the individual; (c) a belief in personal immunity (Wolfenstein, 1957; Slovic, Fischoff, and Lichtenstein, 1980, p. 188), and, conversely, an alternative belief in personal susceptibility; and (d) a cultural preference for optimism and "doing something" about problems (Arensberg and Niehoff, 1971).

Perception of events in terms of individuals is also evident in media coverage of groundwater and other environmental pollution problems. Nearly every newspaper or television story of toxic chemicals in water supplies begins with one particular named individual (or a family) whose life has been affected by the contamination. This opener seems obligatory and is crucial to establishing the human interest value of the story. Most of these stories feature quotes and "eye-witness accounts" from "the man on the street"; and these lay interpretations of the contamination situation are usually given precedence over the opinions of experts. Citizens with no particular knowledge of the chemistry or toxicology involved are quoted more prominently than experts, presumably because their observations carry more credibility with the public.[m]

The individual-centered interpretation of events also shapes the way people perceive institutional and governmental response to contamination. In giving retrospective accounts of local contamination problems, both residents and government officials may claim that the course of events was shaped entirely by the nature of the particular individuals involved: "that town has such a forward-looking mayor," or "the county board is such a bunch of dinosaurs." Satisfactory outcomes are credited by local people to a particularly helpful person assigned by a government agency to handle the problem; unsatisfactory outcomes are blamed by local people on "that so-and-so from Washington." Rarely do people wonder why a particular community has the kind of political leaders it does, or why people in a government agency have the outlook they do. The concept of institutional world-views is not part of the way Americans explain behavior: we characteristically assume that individuals shape institutions and determine the course of history.

CONCLUSIONS

Groundwater contamination situations present an arena in which cultural understandings and values are enacted. Individualism is a theme running through the discussions as communities discover and grapple with problems of contaminated groundwater. Though the theme is not always at the conscious level, its centrality in American culture injects it into public perceptions and public dialogue in many of these cases. Self-reliance and independence, the desire to maximize autonomy through individual choice, the problematic relationship between individual and collectivity, and the perception of events in terms of individuals—all four facets of individualism shape the way American perceive the risks posed by contamination of groundwater. We would suggest that the same dominant cultural orientations that shape our perception of risks posed by groundwater

---

m. Compare to similar findings of The Media Institute in a content analysis of print and broadcast coverage of such cases as EDB residues in foodstuffs.

contamination will also shape our actions (or inactions) to protect ourselves from those risks and to protect groundwater from further degradation.

Since individualism appears to play such an important role in perception of toxic risks resulting from groundwater contamination, it seems likely that individualism will be a cultural lens through which Americans perceive other new environmental risks as well. From radon gas to radioactive wastes, people will perceive these high-tech risks on the basis of their enculturation in a society where individualism has long been both an unquestioned assumption and a cherished value. More effective communication with the public about these new risks will require that risk communicators, and risk managers as well, pay more attention to the cultural dimension of risk perception.

REFERENCES

Arensberg, Conrad M., and Niehoff, A. H., 1971, American Cultural Values, in *Introducing Social Change*, Arensberg and Niehoff, eds., Aldine-Atherton, Chicago.

Bellah, Robert, Madsen, Richard, Sullivan, William M., Swidler, Ann, and Tipton, Steven M., 1985, *Habits of the Heart: Individualism and Commitment in American Life*, University of California Press, Berkeley.

Berry, David, and Stoeckle, J. Andrew, Decentralization of Risk Management: The Case of Drinking Water, Abt Associates, Cambridge, MA (unpublished ms.).

Douglas, Mary, and Wildavsky, Aaron, 1985, *Risk and Culture: An Essay on the Selection of Technical and Environmental Dangers*, University of California Press, Berkley.

DuBois, Cora, 1955, The Dominant Value Profile of American Culture, *American Anthropologist* **57**:1232-39.

Edelstein, Michael R., 1982, The Social and Psychological Impacts of Groundwater Contamination in the Legler Section of Jackson, New Jersey (unpublished ms.).

Edelstein, Michael R., 1988, *Contaminated Communities: The Social and Psychological Impacts of Residential Toxic Exposure*, Westview Press, Boulder, CO.

Erikson, Kai, 1976, *Everything in Its Path*, Simon and Schuster, New York.

Fischoff, Baruch, 1985, Managing Risk Perceptions, *Issues in Science and Technology* **2(1)**:83-96.

Fischoff, Baruch, Slovic, P., and Lichtenstein, S., 1983, The 'Public' vs. the 'Experts': Perceived vs. Actual Disagreements About Risks of Nuclear Power, in *Analysis of Actual vs. Perceived Risks*, V. Covello, G. Flamm, J. Rodericks, and R. Tardiff, eds., Plenum Press, New York.

Fitchen, Janet M., 1987, The Importance of Community Context in Effective Risk Management, in *Risk Assessment and Management*, pp. 677-685, L. B. Lave, ed., Plenum Press, New York.

Fitchen, Janet M., Heath, J. S., and Fessenden-Raden, J. M., 1987, Risk Perception in Community Context, in *The Social Construction of Risk*, Branden B. Johnson and V. Covello, eds., pp. 31-54, D. Reidel Publishing Company, Dodrecht, Netherlands,

Fowlkes, Martha R., and Miller, Patricia Y., 1982, Love Canal: The Social Construction of Disaster, Washington, Federal Emergency Management Administration, unpublished ms.

Gorer, Geoffrey, 1948, *The American People: A Study in National Character*, Revised ed. 1964, W. W. Norton & Co., New York.

Greenhouse, Carol, 1985, Anthropology at Home: Whose Home? *Human Organization* **44(3)**:261-4.

Henry, Jules, 1963, *Culture Against Man*, Random House, New York.

Hsu, Frances L. K., 1973, American Core Values and National Character, Hsu, ed. in *Psychological Anthropology*, Hsu, ed., pp. 241-62, Schenkman Co., Cambridge, MA.

Inkeles, Alex, 1979, Continuity and Change in the American National Character, in *The Third Century: America as a Post-Industrial Societyl*, S. M. Lipset, ed., Univ. of Chicago, Chicago.

Jones, Rex L., 1976, Poker and the American Dream, in *The American Dimension*, Wm. Arens and Susan Montague, eds., pp. 170-180, Alfred Publishing Co., Sherman Oaks, CA.

Lemley, A. T., Fessenden-Raden, J., Bisogni, C. A., and Holway, J. M., 1985, Nitrate Contamination: Public Awareness, *Journal of the American Water Works Association* **77**:34-39.

Media Institute, The, 1985, *Chemical Risks: Fears, Facts, and the Media*, Washington, D.C.

Nelkin, Dorothy, 1982, The Role of the Expert at Three Mile Island, in *Accident at Three Mile Island: The Human Dimension*, Sills, *et al.*, eds., Westview Press, Boulder.

Rayner, Steve, 1983, Can Grid-Group Analysis Help Our Understanding of Social Process? paper presented at American Anthropological Association, Chicago.

Rayner, Steve, 1984, A Cultural Analysis of Occupational Risk Perception, *Royal Anthropological Institute Newsletter* **60**:10-12.

Sharlin, Harold I., 1985, *EDB: A Case Study in Communication of Health Risk*, HIS Associates, Washington.

Slovic, Paul, Fischoff, Baruch, and Lichtenstein, Sara, 1979, Rating the Risks, *Environment* **21**:14-39.

Slovic, Paul, Fischoff, Baruch, and Lichtenstein, Sara, 1980, Facts and Fears: Understanding Perceived Risk, *Social Risk Assessment: How Safe is Safe Enough*? R. C. Schwing and W. A. Albers, eds., pp. 181-214, Plenum Press, New York.

Slovic, Paul, Fischoff, Baruch, and Lichtenstein, Sara, 1984, Behavioral Decision Theory Perspectives on Risk and Safety, *Acta Psychologica* **56**:183-203.

Spindler, George D., and Spindler, Louise, 1983, Anthropologists View American Culture, *Annual Review of Anthropology* **12**:29-78.

Tocqueville, Alexis de, 1835, *Democracy in America*, 1945 transl., Henry Reeve, Vol. II, Vintage Books, New York.

Varenne, Herve, 1977, *Americans Together: Structured Diversity in a Midwestern Town*, Teachers' College Press, New York.

Vidich, Arthur J., and Bensman, J., 1958, *Small Town in Mass Society*, Anchor Books, Garden City.

Vogt, Evon Z., and O'Dea, Thomas F., 1953, Cultural Differences in Two Ecologically Similar Communities, *American Sociological Review* **18**:645-654.

Wolfenstein, Martha, 1957, *Disaster: A Psychological Essay*, Free Press, Glencoe.

Are Twenty-Fold Differences in "Lifesaving" Costs Justified?: A Psychometric Study of the Relative Value Placed on Preventing Deaths from Programs Addressing Different Hazards[a]

John Mendeloff and Robert M. Kaplan
University of California
San Deigo, CA

ABSTRACT

Government programs to reduce risks to health vary greatly in their cost per death prevented. This research asks whether the magnitude of these disparities is justified by citizens' preferences about the relative value of reducing different kinds of risks. Three groups were asked to rank the relative priority of preventing ten deaths in any of eight different programs and then to rate how large the differences in spending should be. The results indicate that, partly because individuals disagree about which projects deserve priority, aggregated preferences rarely show more than a two-fold difference in the amount that should be spent to prevent different types of premature deaths.

KEYWORDS: Valuation on risk reduction, psychometric studies

INTRODUCTION

Analyses of safety and public health programs have revealed great disparities in the cost per fatality prevented. For example, many health standards of the Occupational Safety and Health Administration (OSHA) and the Environmental Protection Agency (EPA) can be justified only by valuing the deaths prevented at millions and sometimes tens of millions of dollars each. In contrast, many state highway departments appear to use valuations of only several hundred thousand dollars per death. These differences have led many economists to argue that funds should be reallocated so that the marginal cost of death prevention in each program is equal, thus maximizing the number of deaths prevented for a given outlay.

The logic of this argument depends upon the view that the prevention of all deaths is equally worthwhile. While many would acknowledge that adjustments should be made for the years of life added and perhaps their quality, the more critical issue concerns the context in which the risk arises. Some assert that context makes little difference. Thus Bailey writes:

---

a. A version of this paper appeared in *Risk Analysis* **9**:3 (1989).

New Risks
Edited by L. A. Cox, Jr., and P. F. Ricci
Plenum Press, New York, 1990

> Although there is no rational case for spending a huge sum to avoid a death from one cause while refusing to spend a relatively small sum to avoid a death from another cause, it can be shown that rational, well-informed citizens do not equalize these incremental sums precisely in private choices. Hence a policy based on such choices will allow some, albeit minor, differences in these sums to remain.[1]

Graham and Vaupel present another view:

> ...policymakers, for numerous legitimate reasons, may explicitly decide to devote more resources to saving lives in some areas than in others. For example, some causes of death are particularly painful and anxiety-producing. To give another example, some causes of death may seem especially "unfair" since they result from largely involuntary exposure to, say, carcinogens in the air rather than from more voluntary factors such as cigarette smoking.[2]

Yet they acknowledge that the issue remains: "How much more is it reasonable to spend in some areas than in others? Do the huge disparities in lifesaving expenditures reflect defensible judgments?"

Many different research strategies have provided some insights into the answers to these questions. The most prominent have been studies of behavior regarding willingness to pay (WTP) for risk reduction[3] and studies of "acceptable risk" via expressed preferences.[4] Both have shortcomings for normative purposes. WTP studies are limited to the valuation of risks by those directly at risk; caring by third parties is difficult to include. Also, the conditions required for a study of WTP are met in only a limited set of risk contexts (e.g., labor markets, where most studies of WTP have been conducted). Thus the issue of whether WTP differs greatly in different contexts is difficult to study.

The literature examining what people judge to be "acceptable" levels of risk has generated many insights about how people think about risks, but it requires respondents to perform very difficult tasks which presume a high level of information. The judgment about how much a risk should be reduced should depend upon several factors, including the value placed on a given reduction in risks and the marginal costs of achieving the reduction. Yet there is no reason to believe that respondents have even sketchy information about the latter.

Thus there appears to be a need for supplementary approaches. Three choices guided the approach presented here. The first was to elicit information only about how people value a given benefit, not about the optimal budget for a program or the optimal level of risk. The reason for this choice was to minimize the cognitive effort required. The second was not to try to elicit dollar values. The reason was both the apparent shakiness of the figures derived in studies that have sought dollar values and our own experience with subjects' discomfort with this task. The third choice was to ask questions about programs that are real lifesaving efforts in order to enhance the usefulness of the results for policymakers.

METHODS

Three groups of subjects participated in the project. The first group included 190 undergraduates from San Diego State University (SDSU). In exchange for their participation in the project, the students received credit toward a requirement in an introductory psychology course. The second group consisted of 35 students enrolled in a course on health and safety policy taught by one of the authors (JM) at the University of California, San Diego (UCSD). Because of their participation in the class, the UCSD students were more familiar with issues in health resource allocation. The third group consisted of 18 retirees who were participating in the University of California's Institute of Continued Learning (ICL). Almost all of this group were over 60 years old.

The basic instrument we employed tells respondents:

> We would like you to play the role of a policy maker who is faced with difficult choices among competing programs. We are interested in your views about the priorities that should be given to different programs for preventing fatalities. Assume that we can prevent ten deaths in any of the programs listed in Table 1, but that we have only enough money to carry out some of the programs. Which of the programs, if any, do you think deserve priority over the others? Should we value the prevention of some deaths differently than others and thus be willing to spend more to prevent them? Or do you think that the prevention of all deaths should be valued equally and thus that the amounts we are willing to spend to prevent additional deaths should be the same for all programs?

Respondents were first asked to rank each of the programs from lowest to highest priority. Then they were asked to give a rating of 10 to the lowest ranked program and rate others in relation to that program; e.g., if they thought that a program merited spending twice as much per fatality prevented, it should get a rating of 20. We emphasized that they should not question the presumed effectiveness of each program and that they should assume that the costs of each program were the same. "The only issue we want you to think about is whether you value the prevention of ten deaths in one program more than the prevention of ten deaths in another."

Respondents were also asked to rate each program on three scales: whether they thought that people, by their own actions, can reduce the risk from the hazard; the extent to which they personally faced a risk from the hazard; and the extent to which family or friends faced a risk from the hazard. (The findings for these questions are not reported here.)

VARIATIONS

Several variations on the basic format were employed. Students in the SDSU sample were randomly assigned to five subgroups, the first of which followed the instructions above. The second group was first presented a five minute "Discussion of Issues," which reviewed the reasons why one might or might not choose to value the reduction of deaths from different programs by different amounts. The third group was asked to rate the highest ranked program as 100 and rate the others down from that level, rather than rating the lowest a 10 and moving up. The fourth group was also asked to use the 100-down approach, but was first presented the "Discussion of Issues." The fifth group heard the discussion of issues, used the 10-up approach, but was also given information on the average age of the victims and the latency period for the type of death potentially prevented by the program. The latency period was the length of time between the intervention and the benefit. Table 1 shows the eight programs with this information added.

The groups at UCSD and ICL were subjected to the same treatment as this fifth group at SDSU. Thus all comparisons among groups are based upon this treatment. The findings of the research are reported below.

Table 1. Eight Programs for Preventing Fatalities

	Program Description	Column A Rankings (A=lowest priority)	Column B Ratings (Assign 10 to program with "A" ranking)
1.	A fatality is prevented by enforcing standards to reduce workplace exposures to cancer-causing chemicals. The deaths prevented occur 30 years in the future. The average age of the victims is 65.		
2.	A fatality is prevented by creating bicycle lanes to separate and protect cyclists from cars. The deaths prevented occur this year. The average age of the victims is 20.		
3.	A fatality is prevented by enforcing regulations requiring that the slats on cribs be close enough together to keep a baby from getting its head caught and strangling. The deaths prevented occur this year. The average age of the victims is 1.		
4.	A fatality is prevented by removing roadside obstacles (e.g., trees, boulders) along dangerous curves in order to protect motorists who drive off the road. The deaths prevented occur this year. The average age of the victims is 27.		
5.	A fatality is prevented by increasing the number of paramedics and ambulances so that heart attack victims receive help more quickly. The deaths prevented occur this year. The average age of the victims is 65.		
6.	A fatality is prevented by enforcing standards to reduce air pollutants that especially affect people with lung diseases like emphysema and bronchitis. The deaths prevented occur in 10 years in the future. The average age of the victims is 75.		
7.	A fatality is prevented by building median barriers on roads to separate traffic and prevent head-on collisions. The deaths prevented occur this year. The average age of the victims is 27.		
8.	A fatality is prevented by enforcing standards to prevent falls by workers at construction sites. The deaths prevented occur this year. The average age of the victims is 40.		

VARIABILITY IN EACH INDIVIDUAL'S RATINGS

The median difference between the top and bottom ranked program ranged from five-fold to seven-fold in the three samples. Half of the individuals thought the maximum difference between their top and bottom programs was less than that amount; half thought it was larger. In both of our student samples, some valued their top ranked program 50 or more times higher than the lowest ranked; in the ICL sample, the maximum difference was ten-fold.

VARIABILITY AMONG AGGREGATED RATINGS

When we aggregate across individuals, the differences in ratings within a sample are smaller than five-fold or seven-fold. There were 28 ratio comparisons constructed by looking at the median individual's relative rating for each pair of programs; among the three samples, there are 84 ratios. In only five of these 84 comparisons did the median ratio show a two-fold difference or greater (i.e., greater than 2.0 or less than 0.5). The largest difference is 3.5-fold (in the UCSD sample, crib deaths over pollution deaths).

Table 2 illustrates the findings by showing the median ratio of the rating placed on preventing workplace cancer deaths compared to deaths in the seven other programs. Thus for the SDSU sample, the median ratio of the value of preventing cancer deaths to the value of preventing deaths by building bike lanes was 0.85. The median respondent thought we should be willing to spend only 85% as much to prevent cancer deaths as to prevent deaths to cyclists. In contrast, the median ratio comparing cancer deaths to deaths due to roadside obstacles was 1.38, implying that the median individual thought that the value of preventing the cancer deaths was 38% greater. The greatest difference in Table 2 is in the UCSD sample, where the median valued the prevention of cancer deaths only 0.46 as much as the prevention of crib deaths.

Table 2 also shows the cutoff points for the upper and lower quartiles. Thus for the bike lane comparison, one quarter of the SDSU respondents thought that the prevention of cancer deaths justified spending no more than half as much as bike deaths. However, another quarter thought that cancer deaths justified spending at least 50% more than bike deaths.

Table 3 indicates that a major reason why there is little variation among aggregated ratings is that individuals often disagree about which programs deserve the higher ranking. If 50% think that Program X ranks higher than Program Y and the other 50% think the opposite, then the aggregate judgment will tend to show them valued almost equally. Disagreement was always substantial, but it did vary greatly among the 3 samples. Table 3 shows that in the ICL sample, 60% agreement was achieved for only 4 of the 28 comparisons; in the SDSU sample, for 12 of the 28 comparisons; and for the UCSD sample, for 20 of the 28 comparisons. Eighty per cent agreement was achieved a total of only 5 times.

SUBSTANTIVE DIFFERENCES AMONG SAMPLES

The rank order of most programs was roughly similar across the three subject populations. The main exception was that the student groups, especially UCSD, gave higher relative rankings to bicycle paths and crib standards.

Table 2. The Median Ratio of the Value Placed on Preventing Ten Deaths from Workplace Cancers to the Value Placed on Preventing Ten Deaths in Other Programs

	1st Quartile	Median	4th Quartile
SDSU Sample, *N*=50			
Bike Lanes	0.50	0.85	1.50
Crib Slats	0.40	0.89	1.90
Roadside Obstacles	0.82	1.38	2.87
Paramedic Services	0.54	0.87	1.20
Pollution	0.92	1.13	1.54
Median Barriers	0.56	0.88	1.71
Construction Falls	0.70	1.20	0.55
ICL Sample, *N*=18			
Bike Lanes	0.63	0.86	1.75
Crib Slats	0.55	0.88	1.33
Roadside Obstacles	0.71	1.10	2.75
Paramedic Services	0.69	0.89	1.58
Pollution	0.64	1.00	1.50
Median Barriers	0.45	0.63	1.07
Construction Falls	0.60	1.00	2.00
UCSD Sample, *N*=35			
Bike Lanes	0.40	0.60	1.20
Crib Slats	0.20	0.46	0.94
Roadside Obstacles	0.40	1.00	2.00
Paramedic Services	0.50	0.86	1.25
Pollution	0.91	1.10	1.50
Median Barriers	0.20	0.67	1.00
Construction Falls	0.38	0.86	1.17

SUBSTANTIVE DIFFERENCES AMONG METHODS

When SDSU subjects were asked to rate programs from 100-down rather than from 10-up, the rank order remained very similar in the two groups. However, the ratings were more compressed in the 100-down responses. When the subjects were given the "Discussion of Issues," the median overall ratings for the eight programs ranged from 50 to 10 in the "10-up" group and from 95 to 53 in the "100-down" group. When the discussion was not presented, the medians ranged from 37 to 13 and from 90 to 55, respectively.

The provision of the "Discussion of Issues" was also associated with a reduction in variation in the "10-up" sample, but not in the "100-down" sample. The only program in which the ranking of the median shifted by more than one place was the crib program, which dropped from third highest to fifth highest when the "Discussion of Issues" was provided in the "10-up" sample.

Table 3. Percent Rating Column Program Over Row Program to the Percent Rating Row Program Over Column Program[a]

	Workplace Cancer	Bike Lanes	Crib Slats	Roadside Obstacles	Paramedics	Pollution	Median Barriers
SDSU Sample *N*=50							
Bike Lanes	39/55						
Crib Slats	38/60*	41/49					
Roadside Obstacles	62/36*	82/12*	62/30*				
Paramedics	38/54	45/49	44/46	22/68*			
Pollution	57/29	58/31	65/35*	43/49	73/25*		
Median Barriers	42/52	53/23	48/44	12/71*	50/44	33/63*	
Construction Falls	54/34	69/18*	66/24*	38/54	58/32	45/53	73/23*
ICL Sample *N*=18							
Bike Lanes	56/39						
Crib Slats	39/50	39/56					
Roadside Obstacles	50/44	33/33	67/28*				
Paramedics	39/56	44/33	56/39	39/50			
Pollution	44/39	44/50	44/39	44/44	56/33		
Median Barriers	28/61*	28/67*	33/44	6/83*	17/78*	33/61*	
Construction Falls	44/28	50/33	44/39	28/50	56/44	50/39	56/22
UCSD Sample *N*=35							
Bike Lanes	26/74*						
Crib Slats	14/77*	20/64*					
Roadside Obstacles	49/46	66/20*	80/11*				
Paramedics	31/63*	57/34	74/9*	34/60*			
Pollution	54/26	71/24*	80/6*	57/34	69/20*		
Median Barriers	23/69*	34/43	57/31	9/69*	29/63*	11/80*	
Construction Falls	29/60*	60/34*	74/23*	37/57	46/40	29/66*	66/14*

[a]When the two percentages do not add to 100%, the remaining comparisons were ties; an asterisk following the ratio indicates 60% or greater agreement.

Providing both the "Discussion of Issues" and information about the average age of victims in the different programs, as well as the latency period before the preventive effects would be felt, produced more dramatic changes. The median rating for workplace cancer fell from first place to fifth and from second place to fifth for median barriers. When age and latency information were provided, two programs moved up in the ratings: the crib standard (moving from fifth to first) and bicycle lanes (moving from sixth to second).

DISCUSSION

We are cautious in interpreting the findings for particular programs. Although there is some regularity in the responses from our three groups, there are differences as well. For example, the two student samples placed a relatively higher value on preventing bicycle deaths than did the elderly group. This response could reflect self-interest (people placing a higher value on preventing risks that they face). If self-interest plays a prominent role in responses to these questions, then the choice of samples may shift some ratings substantially. (Further analysis will see if individuals within each sample who face the risk more directly place a higher value on reducing it.) The results for the UCSD sample indicate that the experience of taking a class with a professor who argued for the worth of a goal of reducing "lifespan inequality" created more of a consensus about placing a high value on preventing deaths among the young. In addition, the apparent differences in the size of the ratios in the "10-up" and "100-down" studies raise a question about whether a true ratio scale can be constructed. Clearly, before policymakers look to this methodology for guidance about the relative rankings of particular programs, more extensive pretesting with more representative samples must be carried out.

The finding that should be emphasized, we believe, is that the typical differences in spending per death prevented that people feel are justified for these programs are not more than a few-fold. Other studies that have asked the same question have obtained similar results. In a 1979 study, Slovic, Lichtenstein, and Fischhoff departed from their usual approach and investigated whether respondents believed that some deaths should be valued less than others.[5] They were surprised to find that the differences were much too small to explain more than a small part of the variation in the levels of risk that people judged "acceptable" for each activity. For their student sample, the least valued of 34 causes of death (alcoholic beverages) was valued about 2.5 times less than the highest valued (nuclear power). For their League of Women Voters sample, the difference was 4.4 fold (smoking versus homicide).

Beggs asked another small League of Women Voters sample to rank programs, to assume that the most worthwhile prevented 10 deaths, and then to say how many additional deaths the less worthwhile programs would have to prevent to make them as worthwhile.[6] For 61 programs, the median ratios they generated varied by somewhat less than a factor of six. At the bottom were programs like deaths due to "drug overdoses by drug addicts." For programs similar to the ones used in this paper, the variability was less than two-fold.

Perhaps the most interesting and difficult issue is, if we assume that this finding of limited variation is indeed valid, what are the implications for policymakers?

It is important to remember that the question being answered with this method is different than the one posed by WTP studies. Those studies ask how much the individuals get paid to accept jobs with higher risks. These payments (or the amounts that people would be willing to pay for risk reductions) are surely an important component in deciding how large the benefits of risk reduction are. In contrast, the approach here asks how much more the members of society, not the individuals at risk, should be willing to pay, not are actually willing to pay, to prevent one kind of death versus another. The answers here depend, at least in part, on ethical, personally disinterested judgments. While WTP depends

Table 4. Dependence of the Policy Implications of the Findings of This Methodology on Two Key Factors (A&B)

A. Valuation of risk reduction by those directly at risk varies with the risk context.	B. Relative size of "third-party" valuation on risk reduction compared to the valuation of those directly at risk is: Small	Large
No	1. This methodology indicates the relative valuation on preventing deaths in different programs but the absolute differences in valuation are not large.	3. Societal assessments will be dominated by findings from this type of methodology.
Yes	2. Findings of this methodology will be swamped by variations in WTP of those at risk.	4. This methodology plays an important role in relative valuations, but so do the variations in WTP of those at risk. Potentially, this probably provides the basis for the largest variations among programs in spending per death prevented.

upon factors like risk aversion, wealth, and the initial level of risk,[7] the "ethical" rating expresses judgments about the value of the premature deaths prevented and the responsibility of the individual or society to reduce risks.

One possible way to view the methodology presented here is that it conveys an estimate of the relative amount of third party valuation on reducing particular risks. But our methodology gives no estimate of the absolute magnitude of this third-party valuation. As a conceptual exercise, we argue that two factors determine the policy significance of the ratings that our method generates: (1) how large the third-party valuation is compared to the valuation of those directly at risk, and (2) whether the WTP of those at risk varies substantially with the risk context. Table 4 shows four possible combinations and their policy implications.

First, third-party caring may be small in absolute magnitude and the WTP of those at risk may be invariant to the risk context, at least as long as the programs are basically preventive in nature (i.e., none of them are treatments for disease). It does seem plausible that the WTP of third parties is more context-dependent than the WTP of those directly at risk. Given these assumptions, our findings would potentially indicate the direction of the relatively small variations that third-party caring should generate around this basic individual WTP.

Second, third-party caring that is small in magnitude could be assumed to exist along with large variations in average WTP among programs. This assumption of large variation does not have to depend upon individuals behaving irrationally. The people facing one sort of risk are not necessarily the same people who face another. Although studies of revealed

WTP behavior have, in general, been too imprecise to allow firm conclusions about different risk contexts, Viscusi did find that individuals in the most risky quartile of jobs placed an implicit value on risk reduction 40% to 60% lower than people in less risky jobs.[8] The implications of our second set of assumptions would be that findings about differences in third-party caring could easily be swamped by differences in the WTP of those at risk.

The third possibility combines the assumption of third-party caring that is large in absolute magnitude with the assumption that there is no variance across risk contexts in the WTP of those at risk. In this case, the findings of our method could dominate in the calculation of societal WTP.

Finally, the fourth possibility combines substantial third-party caring with large variance in the WTP of those at risk. This case could give rise to the largest variations in societal WTP. The reason is that individual WTP depends significantly upon risk aversion, while third-party valuations depend upon judgments about the culpability of the victims; these two factors are likely to be negatively correlated. For example, Beggs's study found that respondents placed a higher value on reducing risks for workers in lower risk jobs, jobs presumably filled by more risk-averse workers. (However, a negative correlation could arise if third parties are willing to pay more to reduce risks faced by the poor, while the poor are likely to place a relatively low value on reducing risks to themselves.)

If we posit an additive relationship between individual and third-party WTP, then, as a crude exercise, we might add Vicusi's finding of a roughly two-fold difference with our finding of a maximum of a three-fold difference. But under most sets of assumptions, a difference this large would appear to be hard to justify for most programmatic comparisons. This is our most important conclusion.

The preliminary nature of our study should be emphasized again. One important caveat is that we have assumed that the responses to our methodology were disinterested. This may not be correct. Individuals may give higher ratings to the programs which reduce risks that they personally face or that people close to them face. To the extent that this is true, the disinterested quality of the judgments is diluted. The ratings would reflect the number of people facing each hazard as well as the ethical judgments; hazards which confront more people would get higher ratings, holding the ethical judgments constant. If this were true, use of our methodology would show that the more people who face a risk, the bigger the benefit from preventing a single premature death from that cause. A discussion of the evidence on the prevalence of disinterested responses or the implications that would follow if they were common is, however, beyond the scope of this paper.

REFERENCES

1. Martin J. Bailey, Reducing Risks to Life: Measurement of the Benefits, American Enterprise Institute, Washington, DC (1980).
2. John D. Graham and James Vaupel, The Value of a Life: What Difference Does It Make?, *Risk Analysis* **1**:1 (1981).
3. Robert S. Smith, Compensating Wage Differentials and Public Policy: A Review, *Industrial and Labor Relations Review*, 32 (1979).
4. Baruch Fischhoff, Paul Slovic, and Sarah Lichtenstein, How Safe Is Safe Enough? A Psychometric Study of Attitudes Toward Technical Risks and Benefits, *Policy Sciences* **9** (1978).
5. Paul Slovic, Sarah Lichtenstein, and Baruch Fischhoff, Images of Disaster: Perception and Acceptance of Risks From Nuclear Power, in *Energy Risk Management*, G. Goodman and W. Rowe, eds., Academic Press, London (1979).
6. Stephen D. Beggs, Diverse Risks and the Relative Worth of Government Health and Safety Programs: An Experimental Survey, U.S. Environmental Protection Agency,

Environmental Benefits Analysis Series, EPA Report No. 230-04-85-005 (June 1984).

7. W. Kip Viscusi, Labor Market Valuations of Life and Limb: Empirical Evidence and Policy Implications, *Public Policy* **26**:3 (1978).
8. W. Kip Viscusi, *Risk By Choice*, Harvard, Cambridge, MA (1983).

The Nature of PRA for the Chemical Industry

D. B. Aulenbach, R. W. Hockenbury, and
G. Pimentel-Romero
Rensselaer Polytechnic Institute
Troy, NY

ABSTRACT

The status of risk analysis and probabilistic risk assessment in the chemical industry has been examined and comparisons made to the nuclear power industry. Major events such as those at Flixborough, Mexico City, Toronto and Bhopal and many other minor accidents have been studied. Several types of chemical industries were identified as prominent contributors to the observed accidents due to the nature of their products.

Our comparisons to the nuclear industry have identified several differences, including

1. methodology—adequacy of fault tree, time scales, human error;
2. processes—any given chemical plant is usually engaged in making several different products with their associated hazards;
3. variety—the very large number of chemical plants and complexes to be considered;
4. the dynamic and competitive nature of the chemical industry, which complicates the risk analysis. The nature and types of processes are continually changing relative to those in the nuclear industry.

Procedures for identifying hazards and importance measures are needed for orderly allocation of funds and manpower. Some major industry-wide issues should be examined. For example, the magnitude of observed accidents: are they precursors or major events themselves? Suggested solutions to some of the problems in methodology and ranking will be presented. Attention is directed to the stage of consequence modeling by considering the effects of modes of release in the dispersion phase.

KEYWORDS: Risk analysis, probabilistic risk assessment, accident statistics, consequence modeling, jet, atmospheric dispersion

INTRODUCTION

Risk assessment techniques, born in the military-aerospace industry and widely practiced in the nuclear industry, are expected to be applicable to the chemical industry.

This study will (a) present reported incidents and an analysis of the statistics of industrial accidents; (b) review the PRA methodologies in both the nuclear and the chemical industries; (c) discuss the applicability and limitations of the practiced PRA

New Risks
Edited by L. A. Cox, Jr., and P. F. Ricci
Plenum Press, New York, 1990

techniques used in the nuclear industry; and (d) focus attention on the stage of consequences modeling.

STATISTICS ON ACCIDENTS IN THE CHEMICAL INDUSTRY

A list of accidents in the chemical industry, covering the period from 1921 to 1979, was compiled according to imposed damages, causes and hazards, products involved, location and date for identification and verification.[1] This list was then extended to 1986,[2,3] allowing the assessment of any significant trends in the last seven years.

This list does not include all worldwide industrial accidents, but could be considered representative of the major industrial accidents of the century. It was reported[4] that of the 6 million chemicals (100%) registered by the Chemical Abstract Service, 575,000 chemicals (10%) were in use at that date, 65,000 (1.0%) had *any* toxic information available (mostly on rats) 4,000 chemicals (0.1%) were reported to exhibit human health effects and 400 chemicals (0.01%) were regulated by OSHA. Figure 1 shows the overall accident incidences categorized according to the major hazards involved: explosions (E), detonations (D), fires (F), releases of toxic materials (T) and a set of accidents with no specified hazard (N). Explosions and the releases of toxic materials were the most frequent accidents-hazards. For the period 1921-1979, explosions were more than twice as frequent as the releases of toxics. The increase in the number of events for 1980-1986 is about the same for these two contributors, indicating the importance of the release of toxic materials as a major hazard since such increases were not proportional to the number of events accumulated during the previous period.

The damage as measured by loss of life and injuries incurred in these events is presented in Fig. 2. Here, a great increase is noted in impact per event in the last years due to toxic materials. Previous to 1979 the greatest contributor to damage was explosions. The increase in damage due to toxic releases and explosions is related to two major accidents — at Bhopal, India and San Juan Ixhuatepec, Mexico, respectively. Accidents involving ethylene and propane were the major contributors. There is also a continuous increase (period 1980-1986) in new chemicals involved in accidents, sometimes with devastating effects (MIC at Bhopal).

Ethylene was the chemical most frequently involved in explosions. Ammonia and chlorine were most frequently related to toxic releases. Pressurized liquefied gas was a major contributor to fires. There was a more or less equal participation of diverse chemicals to detonations and to nonspecified hazards. Methyl isocyanate (MIC), ammonia dicromate (NH4Cr2), and methylene chloride (CH2C12) made important contributions to the hazard statistics. The greatest impact of dead and injured was due to the accident at Bhopal by MIC followed by pressurized liquefied gas (PLG) on explosions and fires. The major contributor to events, ethylene, did not impose a high impact in human life. The greater impact in the last seven years indicates that as time passes we are exposed to increasing risks.

Figure 3 shows the number of events due to the different types of explosions: unconfined vapor cloud explosions (UVCE), confined, boiling liquid expanding vapor explosions (BLEVE), dust, and others reported as unknown. The unconfined explosions have the highest frequency, with ethylene and propane as the most frequently involved chemicals. Davenport's (1977) survey of vapor cloud incidents,[5] showed that of a total of 43 unconfined vapor cloud explosions, 8 were transport related and 32 were linked to chemical facilities (3 undetermined). The most prominent causes (or modes of release that led to incidents where overpressure was created) were valve, fitting and vessel failures.

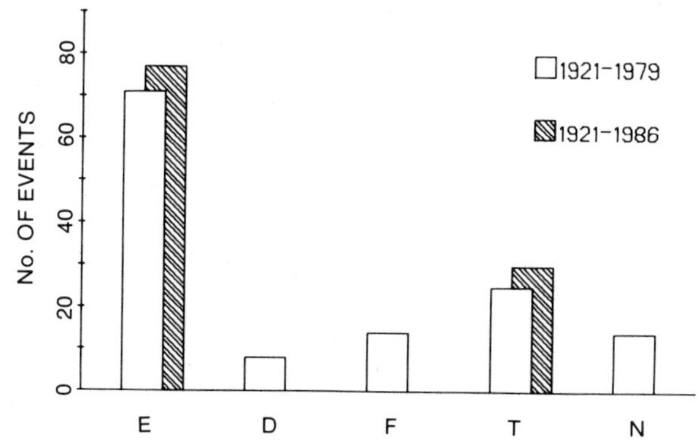

Fig. 1. Accident statics due to major hazards.

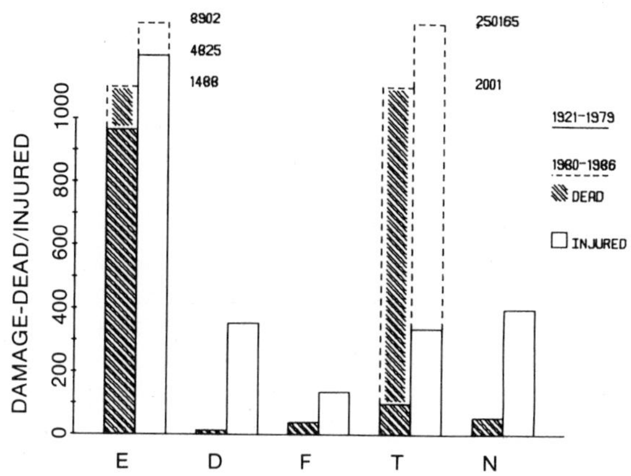

Fig. 2. Damage due to major hazards.

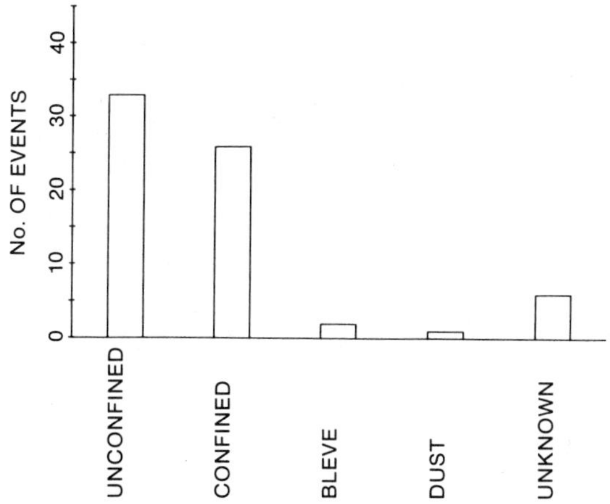

Fig. 3. Statistics on explosions 1921-1979.

The same initial cause may result in different accident sequences. Further, a chain of hazards often occurs in which the hazardous material is released, an explosion occurs, then a fire starts and finally toxics, including combustion products, are dispersed through the atmosphere or environment. For a proper consequence assessment, and depending on a postulated accident scenario (that is, type of chemical, presence of ignition source, chemical reactivity) these chains should be considered and ranked. Table 1 summarizes observations that relate property damage and different health effects to major industrial hazards.

PRA AND RA AS APPLIED TO THE CHEMICAL INDUSTRY

Risk analysis (RA) is the systematic process of examination of a complex to identify potentially serious hazardous occurrences and their possible consequences. Its principal purpose is to assist decision-makers in implementing risk avoidance or risk reduction measures. This is a broader context than probabilistic risk assessment (PRA) as applied in the nuclear industry. Even though PRA techniques have limitations, they are applicable to chemical complexes. Some characteristics of chemical plants include

1. multiproduct manufacturing,
2. high number of control loops and instrumentation (feed forward and feedback),
3. computer control,
4. delay times,
5. diversity in safety systems and devices,
6. strong human interaction, and
7. diversity of hazards.

All these characteristics are affected by (1) the dynamic nature of the chemical industry, (2) the competitive industry environment which makes safety records relatively inaccessible (due to privacy and secrecy) compared to the nuclear industry, and (3) the lack of legislation and regulations compared to the nuclear industry.

The chemical industry, following the nuclear philosophy, practices risk analysis in three steps:

1. hazard analysis,
2. consequence analysis, and
3. quantification of risk.

For step 1, identification of potentially hazardous situations and the minimization of plant and equipment, failure is commonly considered using the following techniques:

1. hazard indices,
2. hazard operability studies,
3. failure mode and effect analysis,
4. fault tree analysis,
5. event tree analysis, and
6. technique for human error prediction (THERP).

For step 2, an assumption of plant or equipment failure is made for the analysis of possible harmful consequences. The consequences are considered to have three components: physical, human, and environmental. The study of *the nature and magnitude* of these effects is carried out in the light of specific accident scenarios (chemicals, structures, environment, population etc.). Several risk analyses of different industries[6,7] give an idea of the complexity and the diversity of alternatives to carry out the consequence analysis step. From the study of these RA cases we conclude that:

Table 1. Damage and Industrial Major Hazards

Major Hazard	Property Damage	Fatalities	Long Term	Short Term
Explosions	High	High/Short range	No	No
Fires	High	Intermediate/ short range	No	No
Toxic releases	Recoverable	High/long range	Yes	Yes
Radioactive	Recoverable	Unknown potentially high/long range	Yes	Yes

1. With regard to the *nature of the effects*, an understanding of physical phenomena is required. Some of these levels are modes of release, atmospheric dispersion, pathways to humans, and toxicology.

2. With regard to the *magnitude of the effects*, one must have information on how much material may escape, and what may happen over time and space.

The final step is the calculation of risk, which then is compared against an acceptability criterion.

Comparisons and Limitations

Two of the main features of PRA in the nuclear industry are that it is applied to assess performance basically on safety systems (e.g., emergency safety features) and the principal hazard of concern is radioactive releases. The chemical industry requires modifications in these techniques since they do not consider

1. a comprehensive identification for diversity in hazards and accident scenarios;

2. means to incorporate in the system analysis stage, time delays caused by transport, and heat transfer processes;

3. the need to take into account multiplicity of stand-by systems (for process and safety purposes) requiring, for example, "demand gates" for fault tree analysis;

4. the non-repairability of systems and its implications on the calculation of the probability of occurrence of the top event;

5. the existence of feedback and mainly feedforward control loops; and

6. the need to assess the dynamics of different hazards or sequences of hazards where knowledge of the physical phenomena involved is imperative to avoid or mitigate the accident consequences.

Recognition of these facts has been reported[8,9] and efforts to solve some of the problems are being conducted by the scientific community and industry.[10,11]

CONSEQUENCE MODELING-MODES OF RELEASE AND ATMOSPHERIC TRANSPORT

The previous two sections have led us to focus attention on a problem of particular interest. At the level of consequence assessment, an initial stage in the ranking of hazards is to define *degree of consequence hazard* as the number of major industrial hazards occurring in an accident scenario. Then, the number of individual hazards or first degree consequence hazards that could occur for a given accident is

$$^{1}A_n = n \; . \tag{1}$$

For *n* major hazards involved, the number of possible sequences of *m* degree consequence hazards is then

$$^{m}A_n =^{m-1} A_n \left(n - m + 1\right) . \tag{2}$$

And the total number of hazard chains out of *n* hazards is given by

$$\sum_{m=1}^{n} =^{m} A_n \; . \tag{3}$$

For the nuclear industry, one first degree consequence hazard is considered (radioactive releases). Comparatively, the chemical industry is required to consider the possibility of three first order degree consequence hazards (E,F,T) six of second order (E-F,F-E,E-T,...) and six of third order (E-F-T,E-T-F,...).

This first stage should be followed by another level of ranking in which the likelihood of occurrence, as a function of time and space (distance), can be evaluated. Given the nature of system events or failures that have led to these major hazards, this second stage should be considered in light of the physical phenomena involved. This ranking scheme could lead to an assessment of the conditions that would avoid or reduce the consequences of industrial accidents.

Emission and Dispersion

The part of the accident scheme in which we will focus our attention is illustrated by Fig. 4.

The initial event common to the three major hazards in industrial accidents is emission. Issues of concern and uncertainty during emission are (1) quantity and duration of the emission, (2) energy and physical properties, and (3) exit geometry/configuration.

Assuming a continuous fluid released as vapor or gas in open air and above ground level, the two modes of release possible according to the momentum of the escaping fluid are low momentum release of gas or vapor forming a PLUME and high momentum release of gas or vapor giving place to a TURBULENT MOMENTUM JET.

Once the emission or release has occurred, the formation of a vapor cloud or the behavior of the released vapor follows three stages:

Stage I: The cloud is substantially unmixed with air and usually the vapors are heavier than air on release. This is so, because even vapors with a molecular weight of less than that of air have a mean temperature on release substantially below the ambient temperature.

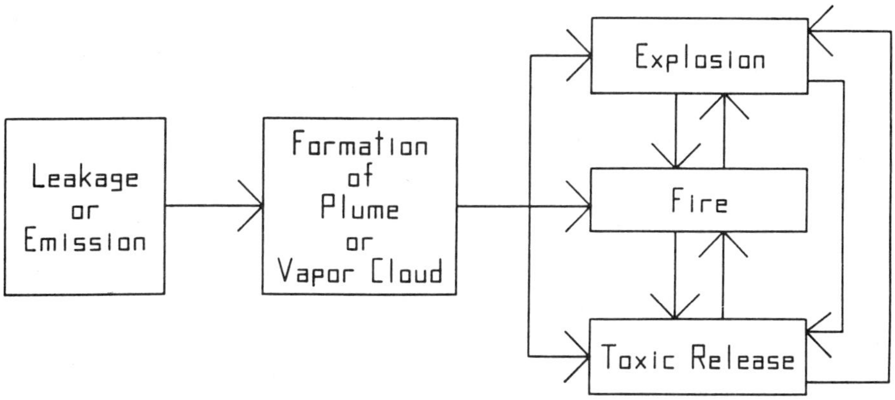

Fig. 4. Consequence accident scheme.

Stage II: The cloud is affected by the influence of gravity, and mixing with the ambient air takes place and the vapors start to interact with atmospheric currents.

Stage III: Mixing takes place to such extent that a difference no longer exists in density, leading to conditions of neutral buoyancy. Under these conditions, flammable vapor issuing from a high velocity jet may form a large vapor cloud explosion and similarly a toxic gas release could lead to a large gas cloud of lethal concentrations.

The properties of clouds containing flammable vapors and those containing toxic vapor are different. The lower limit of flammability lies in the region at which the density of the cloud differs markedly from that of the surrounding air (low for hydrocarbons). Thus, when diluted with the air, the gas has ceased to be dangerous long before the cloud has become neutrally buoyant. For toxic clouds, lethal concentrations could persist under conditions of neutral buoyancy and the ultimate interest is to know the point at which they have become so diluted that they do not pose danger or represent a hazard.

This leads us to consider the study of the transition from jet release conditions to plume dispersion. Some authors define a pointwise transition of these two phases as when the mean velocity of the jet decreases to the mean velocity of the atmosphere. The Gaussian atmospheric dispersion models used in the nuclear and chemical industries, either for routine releases or for accidents, also assume, among other conditions, a pointwise source and that they are applicable for distances beyond one hundred meters from the source. We are interested in knowing how the mode of emission can influence the dispersion phase.

Mode of Release-Jet

The importance of jet release conditions is directly related to development in the chemical industry. In the last decades operating conditions have become severe in terms of pressure and temperature, increasing the conditions of stored energy. An example of the ultimate consequences that this represents is the case of the Flixbrough accident[1,12] in which jet release conditions were reached.

Jet Structure

When fluid emerges from an exit (or typically a nozzle), this fluid interacts with the surrounding or environment (another fluid), forming a jet. As a result of the turbulence generated at the interface of both fluids, the jet fluid will undergo both lateral diffusion and

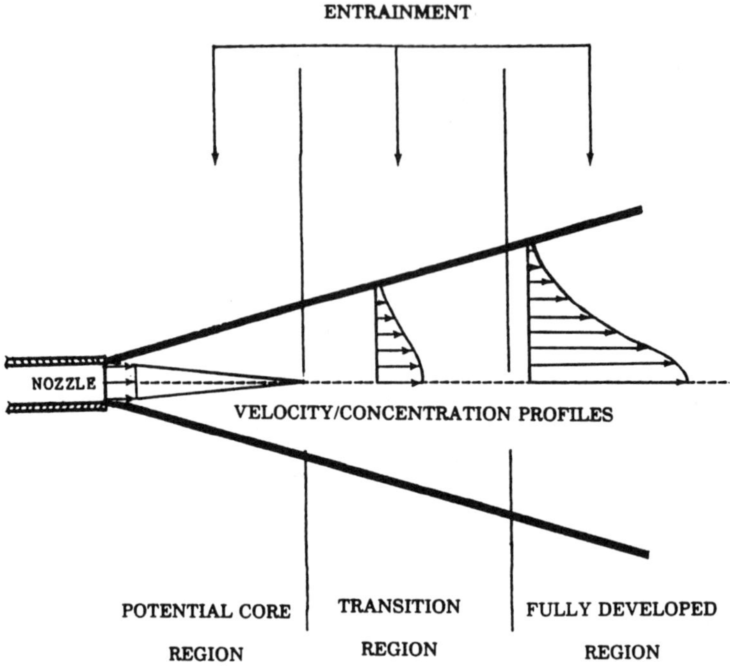

Fig. 5. Jet structure.

deceleration and the environment fluid will increase in motion. In this sense the jet is a source of momentum and energy.

In this process, the jet develops a characteristic structure shown in Fig. 5.

The potential core region of the jet includes the potential core in which the velocity and concentration profiles are constant, remaining unchanged with respect to their exit values. Then, due to the interaction of the jet at the outside free boundary, an entrainment process occurs in which momentum and mass transfer take place. Velocity and concentration profiles are no longer constant, giving place to the transition region at the end of which a similarity profile develops characterizing the fully developed region.

In general, the basic differential equations that describe a jet turbulent flow, relevant to industrial accident scenarios for chemical reacting components, are the conservation equations of mass momentum and enthalpy that include turbulent flux correlations. To provide closure to these systems of equations, use is made of Newton's constitutive stress-strain relation, Fourier's law and Fick's law. This approach ultimately led to develop or state a turbulence model for the turbulence viscosity to have closure.

Next we will consider the case of a steady state, horizontal configuration, axisymmetrical, submerged turbulent jet, assuming the following conditions:

- cylindrical coordinates,
- uniform static pressure,
- quiescent reservoir, one component (air),
- time independent,
- boundary-layer approximation,

- incompressible flow, and
- neglecting viscous effects,

which reduces our system of equations for the mass balance to

$$\frac{\partial}{\partial z}\overline{U}_z + \frac{1}{r}\frac{\partial}{\partial r}(r\overline{U}_r) = 0 \ , \tag{4}$$

and for momentum

$$\frac{\partial}{\partial z}\overline{U}_z^2 + \frac{1}{r}\frac{\partial}{\partial r}r\overline{U}_r + \overline{U}_z = \frac{1}{r}\frac{\partial}{\partial r}r(\overline{-u_r u_z}\ ) \ . \tag{5}$$

Similarity expressions for the velocities, $\overline{U}_z$, $\overline{U}_r$ and $\overline{u_r u_z}$, based on boundary-layer approximations and either Prandtl's mixing-length theory or Reichardt's inductive theory, eventually permit the demonstration of Gaussian distributions for $\overline{U}_z$[14] along the cross section of the jet, that is

$$\overline{U}_z = \overline{U}_{z\max} \exp\left\{-K_u\left(\frac{r}{z}\right)^2\right\} \tag{6}$$

and

$$\frac{U_o}{\overline{U}_{z\max}} = K_{1u}\frac{z}{d} + K_{2u} \ . \tag{7}$$

Similar treatment gives, for the axial distribution of concentration,

$$\frac{C_o}{\overline{C}_{z\max}} = K_{1c}\frac{z}{d} + K_{2c} \ , \tag{8}$$

and for the concentration profiles,

$$\overline{C}_z = \overline{C}_{z\max} \exp\left\{-K_c\left(\frac{r}{z}\right)^2\right\} \tag{9}$$

Where experimental evaluation for the *K*'s (constants) is required, Rodi reports[13] that experiments run from 1950 to 1979 yield mean values for these constants: Ku=87, Kc=63, $K2u$=$K2c$=0, $K1u$=0.16, $K1c$=0.194. These experiments provide excellent confirmation that the profiles of velocity and concentration are similar, fitting well the predicted Gaussian distributions.

Atmospheric Dispersion

The combination of effects due to atmospheric turbulent eddies causing diffusion and wind transportation is known as dispersion. Two approaches to the analysis of atmospheric turbulent diffusion are the Eulerian approach, which considers fluid properties of the fluid motion relative to a fixed coordinate system, and the Lagrangian approach, which studies the history of individual fluid motion particles. In both cases the turbulence closure problem or the lack of knowledge of turbulence properties remains. Approximations are required to solve the equation. Using the advection equation it is possible to show that the mean concentration has a Gaussian form. Assuming a continuous point source at the origin and steady state conditions it is possible to show that the slender plume approximation is given by

$$\overline{C} = \frac{q}{2\pi \overline{u} \sigma_y \sigma_z} \exp\left\{ -\frac{y^2}{2\sigma_y^2} - \frac{z^2}{2\sigma_z^2} \right\} , \tag{10}$$

where

$\overline{C}$ is the mean concentration (units/m^3, ppm),

q is the amount of material emitted (units/sec),

σ_y, σ_z are the vertical and horizontal standard deviation (m),

$\overline{u}$ is mean wind velocity (m/sec).

Comparison

The mathematical models for the behavior of the fully developed turbulent jet and the model for atmospheric dispersion will be compared assuming the following conditions: using the jet model for the fully developed turbulent region [Eqs. (6),(7),(8),(9)], Rodi's values for air at 60°F, nozzle exit of 0.5 m and an exit velocity of 1 m/sec, and for atmospheric dispersion the slender Gaussian model [Eq. (10)] for a wind velocity of 1 m/sec, dispersion parameters for stability class A,[15] and flow jet release conditions q=0.196 m^3/sec.

The selection of these parameters is based on the need to compare both models without violating their conditions of validity; that is, using air to have the submerged jet conditions and exit conditions which assure a turbulent jet, giving Re = 3.4×10^4, and for the Gaussian model, using air (low concentration of tracers) to exclude effects other than atmospheric transport and having unstable conditions to assure equal dispersion in all directions (at 1 m/sec). Finally, an extra consideration is that the initial concentration at the jet is normalized to 1 and the calculations for the virtual source to normalize the concentration for the Gaussian model are neglectable, giving a virtual distance of less than one nozzle diameter (0.195 m). Figure 6 shows the normalized axial concentration for the jet and the plume as a function of distance from the nozzle. This figure shows extrapolation results to regions where the models are not expected to be valid, therefore, observing the discrepancy and defining a region of transition from jet to plume.

Figure 7 shows the radial concentration at the region of interest (transition region) at two axial distances (Z=52.81 m, 152.81 m) from the source. The difference of concentrations confirms the need to develop proper models that can lead to precise evaluation in this region where explosions (UVCE's) and toxic releases can have a tremendous impact.

Transition Region and Future Research Considerations

The transition from the pure jet to pure plume deserves special attention to reach valid predictions for accidental modes of release (turbulent jet).

To appreciate the differences in the models (for jets and plumes), it is important to recall that for the jet discharge is into a still reservoir, and empirical data are required while the atmospheric dispersion model is based on homogeneity of turbulence, nonzero wind speed and empirically based parameters (σ_y,σ_z). Furthermore, real situations involve buoyancy effects, horizontal wind shears, diffusion along the downwind trajectory,

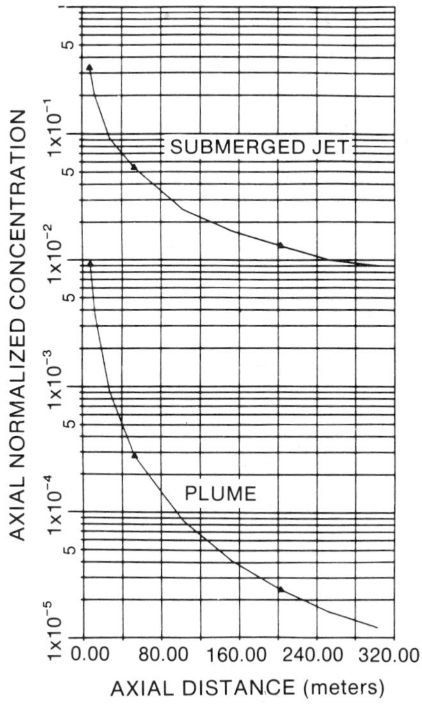

Fig. 6. Axial normalized concentration.

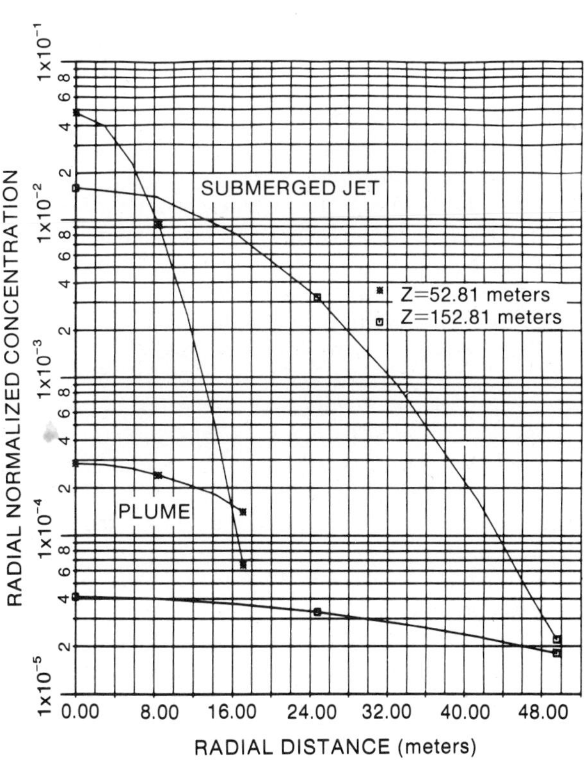

Fig. 7. Radial concentration profiles.

deposition and a temperature gradient which impose subsequent considerations with regard to the modeling of the transition region.

Future research will be concerned with modeling of the transition region. It is recognized that even the best turbulence models cannot predict plane and axisymmetrical turbulent jets in stagnant surroundings with the use of a single set of turbulence parameters. This is thought to be due to the fact that near the axis of the round jet, the gradient assumption is not valid. Works of Rotta and Spalding[16,17] will serve as grounds for future work in this area.

CONCLUSIONS

Statistics of industrial accidents show a trend with respect to their nature and scale by indicating the relative importance of major industrial hazards and the chemicals involved in accidents. A proposed a definition of "degree of consequence hazard" is expected to serve as an initial stage of ranking of hazards for actual accident scenarios, where a chain of hazards can take place. After characterization of chemical complexes in light of their environment and limitations, and a review of PRA/RA methodologies as practiced in the nuclear and chemical industries, attention is directed towards the consequence modeling stage. A preliminary study was conducted to assess the impact of the modes of release on dispersion which ultimately will define the dynamics of the consequences of the two major and most important hazards, UVCE's and toxic releases. This assessment pointed towards the identification of a region of transition between the fully developed turbulent region of the jet and the atmospheric current. At present, this region is being studied by the authors and it is expected that future result could permit assessment of these hazards at critical distances from the point of release.

REFERENCES

1. P. Lagadec, *Major Technological Risk*, Pergamon Press, Great Britain (1982).
2. Chemical Engineering News, *Amer. Chem. Soc.* (1980-1986).
3. *Chemical Week*, McGraw Hill (1980-1986).
4. New York Water Pollution Association Conference, Bear Mountain, April 1985.
5. J. A. Davenport, *Chemical Engineering Progress*, September 1977.
6. J. S. Arendt *et al.*, *Chemical Engineering Progress*, August 1984.
7. P. D. Moskowitz *et al.*, *Risk Analysis* **5(3)** (1985).
8. R. A. Freeman, *Plant/Operations Progress* **2(3)** (1983).
9. H. I. Joschek, *Plant/Operations Progress* **2(1)** (1983).
10. H. Kumamoto and E. J. Henley, *AIChE Journal* **25(1)** (1979).
11. M. Galluzo and P. K. Andow, *Reliability Engineering* **7** (1984).
12. T. A. Kletz, Loss Prevention N.9, *AIChE* (1975).
13. W. Rodi, *Heat and Mass Transfer (HMT-6)*, Vol. 6, Pergamon Press (1982).
14. J. O. Hinze, *Turbulence*, Second Edition, McGraw Hill (1975).
15. ASME, Recommended Guide for the Prediction of the Dispersion of Airborne Effluents (1973).
16. J. Rotta, Technical Notes TWF/TN/38,39, Imperial College Mech. Eng. Dept.
17. D. B. Spalding, Report N. EF/TN/A/16, Imperial College Mech. Eng. Dept.

Risk Uncertainties in Public Sector Safety Decisions: Assessment Methods and Management Implications

M. Elisabeth Paté-Cornell
Stanford University
Stanford, CA

ABSTRACT

In this paper, three levels of uncertainty analysis are examined for the different risk assessment methods used in public sector safety decisions. Level 0 relies on the worst-case approach. Level 1 is based on the estimation of a probability of event, failure, disease incidence, or death, and by a single probability distribution of the outcomes (a "risk curve"). Level 2 involves computation of a probability distribution of the probability of failure or death and leads to a distribution of risk curves. For each approach, we discuss the nature of the uncertainty analysis involved, the decision framework and criteria that are implicit in the information structure, and their effect on the management policy.

KEYWORDS: Risk analysis, risk uncertainty, decision theory, Bayesian probability, risk management, rationality, risk curves, uncertainty

INTRODUCTION

The type of information required for risk management decisions depends on its intended use. If the policy is simply to forbid an activity, a substance, or a facility whenever one can establish that some loss is possible (no matter how small or how unlikely), then it is sufficient to show the existence of the hazard. This is an easy task since it is often impossible to prove absolute safety. Most risk management decisions, however, involve trade-offs between residual risk, costs, benefits, risk displacement, and risk-benefit distribution. Considering these trade-offs explicitly permits informed and consistent safety decisions. When these trade-offs (such as safety vs costs) appear politically unpalatable, it is tempting to adopt publicly a zero-risk position; but in fact, other mechanisms are often found to make such a ruling economically practicable (Paté, 1983).

A balanced regulatory decision relies on an assessment of the effects of the regulation. There are large discrepancies, however, among the methods used by the different regulatory agencies for assessing and reporting risks. In particular, there are large variations in their levels of sophistication in the analysis of uncertainties about basic mechanisms, adequacy of models, and parameter values. For example, the Environmental Protection Agency (EPA), to assess the carcinogenicity of substances, uses a quasi-upper bound method that includes "conservatism" in the risk assessment phase without clear

New Risks
Edited by L. A. Cox, Jr., and P. F. Ricci
Plenum Press, New York, 1990

indication of the actual level of conservatism that is reached. By contrast, the Nuclear Regulatory Commission (NRC) requires that probabilistic risk assessment results for nuclear power plants not only include probability distributions of potential accidental losses, but also reflect the accumulation of analytical uncertainties beyond a single loss distribution. Each level of uncertainty analysis corresponds to a different rationale of decision making and to a particular spectrum of possible risk management criteria and safety goals. Some analytical results do not allow any economic efficiency considerations; some are suited to the classical decision analysis framework. Finally, some forms of risk information allow the analyst to accommodate, in addition to classical risk aversion, an aversion towards analytical uncertainties that seems apparent in public sector regulations.

It is proposed in this paper to examine three different levels of uncertainty analysis in risk assessment as currently practiced for public sector safety decisions. Level 0 relies on the worst-case approach (simple hazard identification, worst case, or quasi-worst-case analysis). Level 1 is based on the estimation of a probability of event, failure, disease incidence, or death, by a single number encompassing all uncertainties and leading to a single probability distribution of the outcomes (a "risk curve"). Level 2 involves computation of a probability distribution of the probability of failure or death and leads to a distribution of risk curves. For each approach, we discuss the nature of the uncertainty analysis involved, the decision framework and criteria that are implicit in the information structure, and their effect on the management policy (economic efficiency and the consistency of regulatory decisions).

LEVEL 0: WORST-CASE APPROACH AND VARIATIONS

Existence of a Hazard

This is the most basic level of risk identification: a loss is possible. No risk assessment is required. The hazard must simply be detected. There is, in particular, no attempt at exposure assessment. For example, a reservoir has been constructed and drowning is a possibility. This approach implies a zero-risk philosophy. In the case of the reservoir, swimming is simply forbidden. In spite of the intention, however, the goal of absolute safety is seldom reached in practice: someone might well disregard the interdiction. The question is then how high and how sophisticated should the fence around the reservoir be, unless one simply decides to eliminate the reservoir, which indeed would eliminate the risk of drowning but would entail other costs.

No trade-off is explicitly involved in the decision, such as the recreational benefits of swimming versus the risk of drowning. There is no consideration of the costs and benefits of mitigation of other risks or of the same risk by alternative methods, and therefore no consistency among regulations. There is no attention paid to risk displacement either. For instance, in the example above, swimmers might note the interdiction and choose a more hazardous place to swim. The costs — direct and indirect — of regulations based on the zero-risk concept might be substantial. The question that the funds might be spent more efficiently elsewhere is never raised. This philosophy, however, has the definite advantage of its unquestionable simplicity. It can be sufficient, for instance, when the costs of the regulation are low and the benefits of the activity are negligible.

Worst-Case Analysis

This approach relies on the notion of maximal loss in absolute terms. It can be applied when an absolute upper bound of the consequences of an accident can be found.The analysis is limited to an evaluation of the number of people exposed or to the property (public or private) at stake. In this perspective, it is assumed that all individuals exposed will become victims or that all property will be destroyed or suffer maximum damage. For example, the absolute upper bound of property damage in an earthquake in San Francisco is simply the total value of all the buildings in San Francisco, assuming that they are totally vulnerable to ground motions.

This method provides sufficient information only to the decision maker who chooses to minimize his maximum loss regardless of the costs, a position which in practice is often economically untenable. As with the zero-risk option, it does not allow the consideration of trade-offs and leads to global economic inefficiencies.

Quasi-Worst-Case Analysis

This approach represents an attempt either to assess an upper bound of the risk (losses or probability of death or accident) when there is some uncertainty as to what the worst case might be, or to obtain an evaluation of the worst possible conditions that can be "reasonably" expected.

The analysis results in a quasi-upper bound based on a "conservative" interpretation of existing statistics with some give and take in the risk assessment itself to support decisions that look economically and politically viable. A prime example is the EPA method of assessment of carcinogenicity by the Carcinogen Assessment Group (the "CAG numbers") (see for instance, Albert, 1980). Out of a spectrum of possible models and extrapolation assumptions (animals to humans and high dose to low dose), the method consists of choosing at each stage the option that will provide high risk results. Typically, this means choosing a multi-stage model of carcinogenicity resulting in a linear extrapolation of the dose-response relationship at low doses, and using the ninety-fifth upper confidence limit (UCL) of the animal data. Although by accumulation of unlikely models, assumptions, and parameter values, this method may yield a higher estimate than the actual effects, it is impossible to tell the probability of exceeding the computed figure and, therefore, where the results stand on a full distribution of the losses. There is no true analysis of the uncertainties and no consistency among substances as far as the "conservatism" of the assessment results is concerned. The EPA staff's rationale for this method is that, in their opinion, no established procedure exists to obtain a "best estimate." In particular, they refuse to aggregate experts' opinions about the probability of each of the possible models of carcinogenicity. Indeed, it may well be that the different models are not incompatible: they may represent different cases of a more general model and each of them may be true in different situations. Under the current state of knowledge, however, further analysis of uncertainty would allow more efficiency in the allocation of resources used for carcinogens regulation (Graham and Vaupel, 1981).

In hydrology, a similar philosophy underlies the popular concept of Probable Maximum Flood (PMF) based on the Probable Maximum Precipitation (PMP) and on an assumption of worst possible conditions such as an already saturated soil (National Research Council, 1985). The PMP itself is deduced from the worst historical record adjusted to reflect the worst possible conditions and it is unclear where it stands on the tail of the precipitation distribution. Consequently, the return period of the PMF probably varies from site to site (National Research Council, 1986). Using the PMF therefore does not allow any consistency in flood risk assessment from site to site, or any analytical coherence in analyses where a flood may be one of several types of initiating events of an accident (for example, floods and earthquakes can both be initiators of failure of dams.) Because using the PMF to judge the spillway adequacy of some dams seems to lead to an inefficient allocation of safety funds, some federal and state agencies use a fraction of the PMF, with the value of this fraction depending on the size of the dam (a surrogate for the consequences of failure). The goal is to provide a more reasonable criterion, but what actual level of safety is reached by this method is unclear. By contrast, the Institution of Civil Engineers in London uses probabilistic criteria such as the 10,000-, the 1,000-, and the 150-year floods (National Research Council, 1985).

The Maximum Credible Earthquake (MCE) used to judge the safety of some civil structures design relies on a similar concept (National Research Council, 1985). It is an attempt to reach an upper bound of the magnitude of earthquakes that can affect a given site. It is defined by the U.S. Army Corps of Engineers as "the earthquake that would cause the most severe vibratory ground motion or foundation dislocation capable of being produced at the site under the currently known tectonic framework." There is no place in this analysis, however, for the potential existence of other seismic mechanisms than those already known (e.g., as yet unidentified faults). Another method of quasi-upper bound computation is used to obtain the Safe Shutdown Earthquake (SSE), which is the peak ground acceleration used as a basis for the seismic design of nuclear power plants (National Research Council, 1985). It is the maximum earthquake intensity that can be generated by rupture of (nearly) the full length of the neighboring faults. Again the notion of absolute upper bound was softened to yield an economically reasonable figure but without indication of the actual safety level that is implied: the return period (or the annual probability) of the SSE is unknown, but probably varies from site to site.

In the quasi-worst-case approach, the criteria of risk management are imbedded in the risk assessment itself but in a manner that does not give the decision maker any information about the probability of exceeding the risk estimates. There is an obvious intention of risk aversion and a general desire to accommodate to some extent economic realities by ad hoc adjustments of the quasi-upper bound. It is impossible, however, to know the level of conservativeness of the final decision and to ensure the coherence of risk management decisions.

The major problem is therefore a serious lack of consistency and economic efficiency. Some minute environmental risks may be regulated at great expense, or some dams may be equipped with very large spillways, thus absorbing considerable efforts into marginal risk reductions when resources could be better used elsewhere. In the quasi-worst-case approach, although safety decisions might appear conservative case-by-case, the order of priorities among risk mitigation needs could well be reversed, leading to essentially unconservative overall effects. This reversal of priorities may occur within a given project (for instance, floods vs earthquakes as potential causes of dam failures), or project-to-project within a single agency such as the EPA, or area-to area (e.g., EPA versus NRC).

LEVEL 1: PROBABILISTIC APPROACH

This method is based on logic and probability. The probability of an event such as an accident, death, or a natural catastrophe is computed using tools that include event trees, fault trees, and stochastic processes. The Bayesian method involves the use of conditional probability and of the Bayes rule to obtain a distribution for the final outcomes as a description of the risk. Probability is defined as a degree of belief (de Finetti, 1972) and allows the use of experts' judgments in the risk assessment phase. Uncertainty about basic mechanisms, models, and parameter values is represented by probability distributions for the spectrum of models and for parameter values given each model. In the end, however, the estimation of these uncertainties is reduced to an expected value representing the final probability attached to each outcome, for example, the probability of exceeding a given level of losses in a given time period.

This method of risk analysis corresponds to the axioms of decision theory and to the maximization of expected utility (von Neuman and Morgenstern, 1947; Raiffa, 1968; Howard, 1984). The rationale of aggregation of uncertainties in a single probabilistic estimate regardless of the "softness" of the probabilities involved relies on the decomposition axiom that requires the aggregation of all uncertainties by virtue of the structure of preferences. Under this scheme, there is no distinction between risk and uncertainty. This axiom implies indifference between two lotteries represented by the same probability distribution for the outcomes, regardless of the amount of evidence that supports the probability assessment. Another key feature of this definition of rationality is the separation of risk assessment (models and probabilities) and risk management (utilities). The utility for an outcome is independent of the probability of this outcome.

In public sector safety decisions one encounters theoretical as well as practical difficulties in adapting the decision analysis methodology to essentially collective decisions. This requires collective aggregation of sometimes conflicting sources of information into one probabilistic estimate, and the use of collective preferences in the evaluation of the possible outcomes (Paté, 1986). One of the critical steps is to obtain an agreement about a probability distribution for a spectrum of possible models when basic mechanisms are poorly known. This problem arises, for example, in the analysis of carcinogenicity or seismicity.

In the nuclear power industry, a first study was conducted to estimate the potential losses in reactor accidents (USAEC, 1957). The PRA methodology was then developed (USNRC, 1975) to provide more information about the risks and to estimate the probability of accidental release of radioactive material and the probability distribution of accidental losses. Based on fault trees and event trees, the PRA method includes system reliability analysis, containment analysis, and consequence analysis. The results are used by the utility companies to assess potential measures of risk reduction, and by the United States Nuclear Regulatory Commission (USNRC) as a complement to more traditional decision supports. Recently the USNRC has adopted numerical guidelines that specify its risk management philosophy (comparison to risks of alternative methods of electrical production, comparison to background level of individual risk) along with more specific safety criteria, such as a maximum acceptable individual risk (Atomic Energy Clearing House, 1986).

PRA results allow the regulator to establish priorities among potential retrofit measures and, eventually, to check that the stated safety criteria are met. In the discussions that preceded the adoption of the criteria, the question was often asked: what should be the probability that the safety goals are actually reached? In other terms, what should be the relevance of the "softness" of the results in safety decisions? According to Bayesian decision theory, all uncertainties are properly accounted for in a single probability distribution of the outcomes, and there is no need for additional information about analytical uncertainties. Indeed, there now seems to be an agreement within the NRC that the mean of probabilities and future frequencies is to be used to check that the safety goals are satisfied. As we point out in Level 2, more information is needed if the desired structure of preferences does not fit directly the axioms of decision analysis.

In the field of environmental health risk assessment, the Bayesian approach to risk analysis is still controversial because it is felt that estimating risks by quasi-upper bounds leads to more conservative decisions (which, as we discussed before, may or may not be the case.) The Bayesian method for health risks assessment, however, has been illustrated in several places in the literature. For instance, Morgan *et al.* (1984) show how it can be applied to the problem of assessment of human exposure to sulfur air pollution. North (1983) shows how a probabilistic method can advantageously replace the EPA quasi-upper bound procedure of estimation of carcinogenicity. He proposes as an illustration the assessment of the risk of cancer from occupational exposure to perchloroethylene (Campbell *et al.*, 1982). A probability distribution is computed for the outcomes (number of deaths per year) in order to give the decision maker a more complete description of the analytical uncertainties than a single quasi-upper bound number. For many scientists, however, a major conceptual difficulty lies in the necessity to assign a probability distribution to a set of alternative models in order to express their uncertainty about which mechanism actually applies. They sometimes prefer to resort to sensitivity analysis and to a set of alternative results, but without further indication of the likelihood of each alternative model.

Yet the Bayesian method has important advantages over the worst-case approach or simple sensitivity analysis in the management of risks. Maximization of the expected value of the outcomes (or minimization of the expected value of losses) is the only criterion that leads to strict economic efficiency and to equity in protection from risks. Strict economic efficiency means that the maximum expected number of people are saved for a given level of investment. Equity in safety means that individuals' protection is independent of the potential number of victims in an accident. This is why this rule has often been recommended for public sector decisions. Risk aversion, when desired, can be introduced in the utility function independently from risk assessment. The decision maker can then be consistent in his preferences. Risk aversion, however, implies a risk premium, i.e., a willingness to make additional payment to avoid losses in large single events beyond what is spent to avoid the same losses in several smaller accidents. For this reason, it also leads to inequities among individuals because protection depends on the size of the potential accidents to which people are exposed.

LEVEL 2: RECOGNITION OF RISK UNCERTAINTY

The next level of uncertaintyanalysis involves a probability distribution for the probability (or future frequency) of failure, accident, death, or disease which leads to a distribution of risk curves for the description of the associated number of fatalities. In the nuclear power industry, it was judged from the beginning that a single estimate of a failure probability or a single probability distribution of the potential losses gave insufficient support to regulatory decisions. The need for a thorough treatment and communication of analytical uncertainties in PRA for nuclear power plants was recognized in the original version of the Reactor Safety Study (USNRC, 1975), and emphasized in the review of that work by the Lewis Committee (USNRC, 1978). The method of analysis is specified by the Nuclear Regulatory Commission in the PRA Procedures Guide (USNRC, 1983b). Uncertainties about future frequencies, probabilities, and models are assigned for the input variables, propagated in the analysis, and displayed in the results (Kaplan *et al.* 1981; Vesely and Rasmuson, 1984).

One interesting example of this method is the treatment of uncertainties in seismic PRA, which requires analysis of uncertainties about seismic hazard as well as system fragility (Kennedy *et al.*, 1980, USNRC, 1984). In the United States, the information regarding seismicity is substantial in some western parts of the country but much more limited in the East, where the seismic mechanisms themselves are often poorly known. The Bayesian method consists of identifying all the possible underlying seismic mechanisms and assessing a probability distribution to this set of possible models and parameter values (Electric Power Research Institute, 1986). The seismic PRA for the Limmerick power station provides an interesting illustration of this method. In this study, six different combinations of seismic mechanisms and parameter values are identified, leading to a set of six probability distributions for the annual maximum effective peak ground acceleration and their six probabilities, i.e., a degree of belief based on experts' opinions (USNRC, 1983a). The next step is to compute the effect of the uncertainty in seismic hazard and in fragility analyses on the future frequency of core melt. The future frequencies of core melts caused by earthquakes, fires, or other initiating events are represented and compared through their probability distributions. Obviously, internal consistency in the analysis of possible models and parameters values is needed to make meaningful comparisons about the relative contribution of each type of initiating event to the overall probability of core melt.

Finally, the question is to assess the effect of these uncertainties on the probability distribution of losses. Instead of a single cumulative distribution function, the results are presented as a family of cumulative distributions representing the range of analytical uncertainties. For each number of fatalities, these curves give a partial representation (e.g., by chosen fractiles) of the marginal probability distribution of the frequency of exceedence per reactor-year. An important use of these curves is to compute the effects on the potential losses of possible risk mitigation measures, i.e., the probability distribution of the reduction of the frequency of exceedence of a given casualty level per reactor-year. This information can be a major input in safety decisions, and this representation allows the analyst to communicate analytical uncertainties as part of the results.

In part because of large risk uncertainties, PRA is currently used by the NRC only "as a supplement to the regulatory process" and not as sole basis for decisions (Bernero, 1984). For problems such as nuclear safety, the decision analysis paradigm seems insufficient to capture the preferences expressed within the regulatory agencies, where there seems to be a definite preference for "hard" probability estimates (e.g., low uncertainty in core melt probability or loss distributions). The information given by the level 1 described above seems too densely aggregated. It is thus important to examine how risk uncertainties affect regulatory decisions in order to identify the level of analysis that provides the corresponding results.

Uncertainty in decisions involving risk can be separated into two types: the analytical uncertainty, which can be further refined between modelling and parametric uncertainties, and the observational uncertainty that would always remain even if the characteristics of the system were perfectly known. In other fields, the same distinction is sometimes made under the names of "ambiguity" and "risk" (Yates and Zukowski, 1976). The spectrum of uncertainties is continuous and is sometimes characterized (for lack of a better term) by the degree of "softness" of the probabilities involved. Analytical uncertainties thus reflect the limitations of the information available in many parts of the analysis ranging, for example, from small sample sizes upon which to base parameter estimates to current limitations in the study of human errors. To assess risk uncertainties, it is necessary to represent under one form or another a probability of a probability.

This notion of secondary probability is anchored in the definition of Bayesian probability in which a probability distribution can be attributed to the parameters (e.g., mean or standard deviation) of a probability distribution representing observational uncertainty. In Bayesian decision theory, however, the axioms of rationality imply that it is the mean value of this distribution of the probability itself that actually represents "the" probability of an event for decision purposes (Savage, 1954; Raiffa, 1968). One of these axioms implies that the rational decision maker is indifferent between two lotteries, one of which involves probabilities that are perfectly known (e.g., one chance in a million of drawing six straight times the same specified number out of ten perfectly equally likely numbers in six independent trials) and one of which is based on the occurrence of an event whose probability of occurrence is uncertain (e.g., lognormally distributed with mean 10-6). The axioms result in the maximization of expected utility and the mean is sufficient because the decision is the same regardless of the uncertainty in the probability of the event.

There are several possible interpretations of the apparent desire of the regulators and other interested parties to gain information about risk uncertainty beyond the expected values of probability. First, they may want to understand the results better. Second, they may want to be able to anticipate future changes in the risk estimates. Indeed, in the field of nuclear safety it might well be that, in addition to earthquakes themselves, the outcomes of interest in the future include new estimates of seismic risk that may prompt additional plant retrofits. Third, the uneasiness of the decision makers about using means of "soft distributions" and, therefore, about the von Neuman definition of rationality, suggests that they may be unwilling to follow literally the classical axioms of decision theory. In addition to an attitude towards risk such as risk aversion, they seem to have an aversion towards analytical uncertainties. This phenomenon, known as the Ellsberg paradox (Ellsberg, 1961), implies that they do not necessarily value equally two lotteries in which the distributions of the final outcomes are identical. They may indeed be willing to pay an additional premium to reduce the analytical uncertainty itself. This concept escapes the classical notion of value of information (Howard, 1966), which relies on the anticipated change in the distribution of the outcomes (based on mean probabilities) and, therefore, on the reduction of analytical uncertainties only to the extent that it results in a change in this outcome distribution and affects the expected utility. Although ambiguity avoidance has been documented in the literature (see, for example, Curley, Yates, and Abrams, 1986), invoking the Ellsberg paradox raises a host of uncomfortable issues, such as the existence of multiple levels of uncertainties and the dependence of preferences on the characteristics of each particular decision problem. By contrast, decision analysis seems simple and elegant but overlooks a critical element that needs to be explored further from an axiomatic and normative point of view: the existence of an attitude towards analytical uncertainty (or ambiguity) in addition to the attitude towards risk.

One of the reasons for this uncertainty attitude may be the collective nature of regulatory decisions even though an administrator in the end makes the safety decision. Bayesian probability as well as utility theory are based on individual notions of preferences and degree of belief that are difficult to translate into elements of collective decisions. Reaching an agreement about a collective degree of belief in a probability and making a collective value judgment about the outcomes may require a higher degree of confidence in the results than individual decision making, and, therefore, imply a greater willingness to pay for analytical information.

CONCLUSION

Uncertainties in risk analysis can be analyzed at three levels. Level 0 (the worst-case approach) does not account for uncertainties beyond upper bound or quasi-upper bound notions and mixes risk assessment and value judgments. It has the advantage of its simplicity but may lead to inconsistent and costly decisions and to reversals of management priorities. Level 1 is based on a single probability distribution of annual losses. The probabilities involved are based on the expected value of future frequency (or probability) of the considered event. This is the level of uncertainty analysis that is required by the axioms of decision theory and allows introduction of the desired risk attitude in the utility for the outcomes. Level 2 results in a family of risk curves or cumulative probability distributions for the potential annual losses. This formulation of the results allows the decision maker to judge the "softness" of the information and, if he wants to do it, to introduce in his preference structure an element of "uncertainty aversion" that goes beyond the classical risk aversion of utility theory.

There is a gap at this point between the level of sophistication of the uncertainty analysis performed in some sectors, such as the nuclear power industry, and the theoretical understanding of the actual management implications of the "softness" of the numbers. On one hand, the decision makers are told that "rationality" implies that the softness of the results does not matter. They are urged to focus on probability based on expected future frequencies (or classical Bayesian probabilities) because it contains all relevant description of uncertainties. On the other hand, they express the need for additional information to reduce analytical uncertainties. In fact, probabilistic results are sometimes either used or discarded according to their "softness." In some sectors, there is great reluctance even to change from worst-case concept to the first level of uncertainty analysis for fear of going further than what the scientific knowledge permits. In these sectors, developing methods that involve not only a level 1 but a level 2 of uncertainty analysis might alleviate the concern that probabilistic risk analysis can give a false impression of certainty.

REFERENCES

Albert, R. E. *et al.*, 1980, The Carcinogen Assessment Group's Method for Determining the Unit Risk Estimate for Air Pollutants, U.S. EPA, Washington DC.

Atomic Energy Clearing House, 1986, The Nuclear Regulatory Commission voted to adopt a policy statement on nuclear power plant safety goals, *Atomic Energy Clearing House* **32(26)**:23-31.

Bernero, R. M., 1984, Probabilistic Risk Analysis: NRC Programs and Perspectives, *Risk Analysis* **4(4)**:287-297.

Campbell, G. L., Cohan, D., and North, D. W., 1982, The Application of Decision Analysis to Toxic Substances: Proposed Methodology and Case Studies, Report to the EPA, Decision Focus Inc., Mountain View, California.

Curley, S. P., Yates, J. F., and Abrams, R. A., 1986, Psychological Sources of Ambiguity Avoidance, *Organizational Behavior and Human Decision Processes* **38**:230-256.

de Finetti, B., 1972, *Probability, Induction, and Statistics: The Art of Guessing*, Wiley, New York.

Electric Power Research Institute, 1986, Seismic Hazard Methodology for the Central and Eastern United States, Vol. 1: Methodology, EPRI NT-4726 (July).

Ellsberg, D., 1961, Risk, Ambiguity, and the Savage Axioms, *Quarterly Journal of Economics* **71**:643-669, 1961.

Graham, J. D., and Vaupel, J. W., 1981, Value of a Life: What difference Does It Make? *Risk Analysis* **1(1)**:89-95 (March).

Howard, R. A., 1966, Information Value Theory, *IEEE Transactions on Systems, Man, and Cybernetics* **SSC-2(1)**:22-26.

Howard, R. A., 1984, Risk Preference in Principles and Applications of Decision Analysis, Vol. 2, pp. 627-664, Strategic Decisions Group, Menlo Park, California.

Kaplan, S., Apostolakis, G., Garrick, B. J., Bley, D. C., and Woodard, K., 1981, Methodology for Probabilistic Risk Assessment of Nuclear Power Plants, Pickard, Lowe and Garrick, Inc., Irvine, CA (June).

Kennedy, R. P., Cornell, C. A., Campbell, R. D., Kaplan, S., and Perla, H. F., 1980, Probabilistic Seismic Safety Study of an Existing Nuclear Power Plant, *Nuclear Engineering and Design* **59(2)**:315-338.

Morgan, M. G., Morris, S. C., Henrion, M., Amaral, D. A., and Rish, W. R., 1984, Technical Uncertainty in Quantitative Policy Analysis — A Sulfur Air Pollution Example, *Risk Analysis* **4(3)**:201-216.

National Research Council, 1985, *Safety of Dams: Flood and Earthquake Criteria*, Committee on Safety Criteria for Dams, National Academy Press, Washington, DC.

National Research Council, 1986, Feasibility of Assigning a Probability Coordination, Washington DC.

North, D. W., 1983, Quantitative Analysis as a Basis for Decision Under TSCA, G. W. Ingle, ed., ASC Symposium Series, No 213: TSCA's Impact on Society and Chemical Industry.

Paté, M. E., 1983, Acceptable Decision Processes and Acceptable Risks in Public Sector Regulations, *IEEE Transactions on Systems, Man, and Cybernetics* **SMC-13(3)**:113-124 (March-April).

Paté, M. E., 1986, Probability and Uncertainty in Nuclear Safety Decisions, *Nuclear Engineering and Design* **93**:319-327.

Raiffa, H., 1968, *Decision Analysis*, Addison-Wesley, Reading, MA.

Savage, L. J., 1954, *The Foundations of Statistics*, Wiley, New York.

U. S. Atomic Energy Commission, 1957, Theoretical Possibilities and Consequences of Major Accidents in Large Nuclear Power Plants, WASH-740, Washington, DC (March).

U. S. Nuclear Regulatory Commission, 1975, Reactor Safety Study, WASH-1400 (NUREG-75/014), Washington, DC (October).

U. S. Nuclear Regulatory Commission, 1978, Risk Assessment Review, Group Report to the U. S. Nuclear Regulatory Commission. NUREG/CR-0400, Washington, DC.

U. S. Nuclear Regulatory Commission, 1983a, Probabilistic Risk Assessment, Limmerick Generating Station, Docket 50-352, Washington, DC.

U. S. Nuclear Regulatory Commission, 1983b, PRA Procedures Guide, NUREG/CR-2300, Washington, DC.

U.S. Nuclear Regulatory Commission, 1984, Probabilistic Risk Assessment Reference Document, NUREG-1050, Washington, DC.

Vesely, W. E., and Rasmuson, D. M., 1984, Uncertainties in Nuclear Probabilistic Risk Analysis, *Risk Analysis* **4(4)**:313-322.

von Neuman, J., and Morgenstern, O., 1947, Theory of Games and Economic Behavior, Princeton Univ., Princeton, NJ.

Yates, J. F., and Zukowski, L. G., 1976, Characterization of Ambiguity in Decision Making, *Behavioral Science* **21**:19-25.

Assessment of Risks from Acute Hazards at Monsanto

Raymond A. Freeman, Jerry M. Schroy, and James D. Wilson
Monsanto Co.
St. Louis, MO

ABSTRACT

Handling acutely hazardous materials safely is a skill that the modern chemical industry developed as it grew. The low rate of injury from exposure to reactive and dangerous substances such as chlorine, phosphorus, sulfuric acid, hydrogen cyanide, and ammonia testifies to that skill. Recently the changing structure of the industry, and especially the Bhopal disaster, have caused us to focus anew on risks from such materials. This paper describes Monsanto's approach.

At Monsanto, risk assessment is used to inform decision makers. Frequently the decisions involve choices to be made during process or plant design (e.g., which equipment to use or where to site a facility). Many of these choices do not require a sophisticated quantitative analysis. If the consequences of a credible accident or failure are unacceptable, an alternative is chosen even if the probability of occurrence is low. Thus, Monsanto would not erect a large volume vessel for high hazard liquid storage in a densely populated area.

Following Bhopal a senior management team reviewed Monsanto's manufacturing facilities using only conservative consequence analyses to identify situations possibly hazardous enough to warrant either action or more detailed study. A standard consequence analysis protocol was developed and used for all Monsanto sites. This paper outlines the Monsanto approach for the analysis of high hazard material handling and storage problems.

KEYWORDS: Risk assessment, screening, consequence analysis, dispersion analysis, credible accident

INTRODUCTION

Since 1974 Monsanto has operated under the following formal corporate policy: "Control pollutant discharges from Monsanto operations so as not to endanger health and the environment." Obviously, this language is very broad and covers both normal (planned) and accidental releases from Monsanto's plants. Several different Monsanto programs to control emissions, effluents, and accidental releases have been in place for a decade and are reviewed annually. The implementation of the program for identifying, analyzing, and controlling risk from accidental releases has been lodged with Monsanto's engineering personnel.

New Risks
Edited by L. A. Cox, Jr., and P. F. Ricci
Plenum Press, New York, 1990

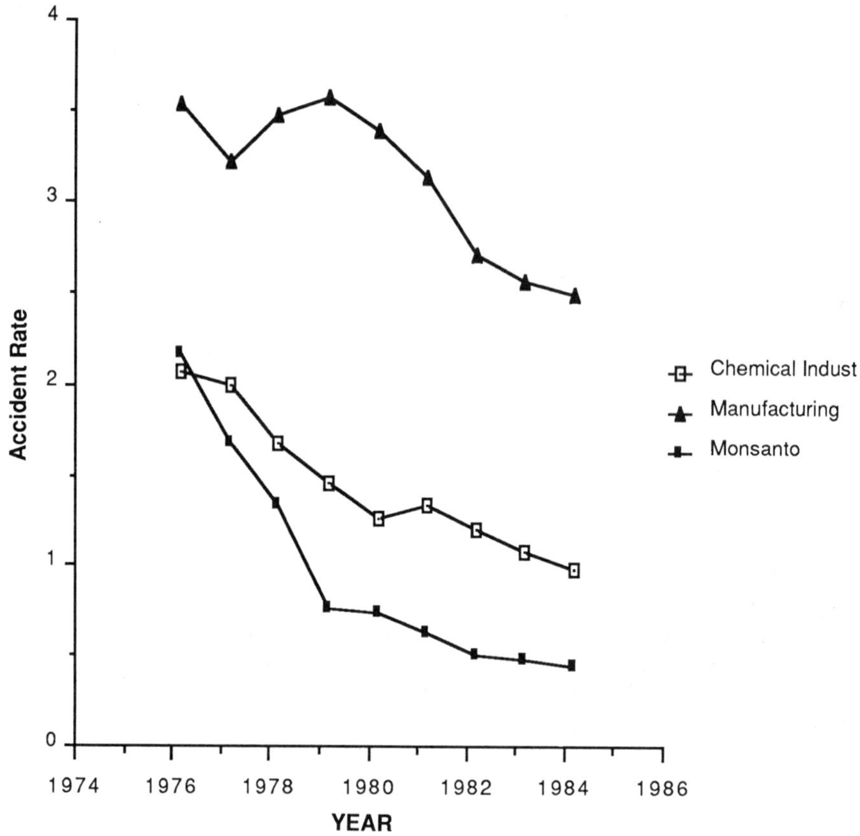

Fig. 1. Monsanto accident history.

Historically, Monsanto and the rest of the chemical industry have experienced a very low level of accidents. Figure 1 presents the Total Lost Workday Accident Rate for the Manufacturing Industries, for the Chemical Industry, and for Monsanto. As can be seen, the accident rates have been dropping for all three groups. Monsanto's accident rate has been lower than the average for the Chemical Industry in general. Monsanto's low accident rate is a result of the daily attention to safe operation in the operating plants.

Two years ago, two major disasters abroad shook the American chemical industry to its foundation. The Mexico City LPG explosion of November 19, 1984 killed approximately 500 people in the surrounding community. Two weeks later the Union Carbide plant in Bhopal, India, accidentally released a toxic material, methyl isocyanate, which killed approximately 2,000 people in the surrounding community. These two disasters emphasized the point that safety can not be taken for granted. Decades of safe operation will not ensure that an accident can not happen today. As a result of these accidents, the major American chemical companies again reviewed their policies and procedures for handling highly hazardous materials.

Ten days after the Bhopal disaster, Monsanto's President and CEO, Richard Mahoney, appointed a senior management task force to review Monsanto's safety practices. As part of this review process, a Consequence Screening Protocol was used at all Monsanto locations to define the impacts of potential accidents involving highly hazardous materials.

This paper outlines the Consequence Screening Protocol and describes its relation to standard risk assessments.

RISK ASSESSMENT PROCESS

The methods available for determining the consequences of hypothetical equipment failures that result in fires, explosions, or toxic vapor releases have been described in other literature and in summary documents.[1] Figures 2 and 3 are taken from the AIChE Guidelines for Hazard Evaluation[1] and describe the overall process of doing risk assessments. The methods involve definition of the hazards, prediction of the consequences, and definition of the frequencies of the events defined in the hazard analysis. This risk triplet (Scenario, Frequency, and Consequences) represents the impact of a facilities operations on the surrounding area. Figure 2 presents the standard sequence of steps for completing a risk assessment. Note that the estimation of a potential accident's frequency of occurrence is completed at the same time that the consequences of the accident are being estimated. Many times the consequences will be found to be of little importance and the effort spent on the frequency estimation will have been wasted.

A more efficient method for completion of a risk assessment is presented in Fig. 3. The consequences of various potential accidents are determined first. The initial consequence analysis is considered a form of risk assessment where the probability of the event is assumed to be one (i.e., the accident will occur). The effort to define the frequencies of each event is avoided for those incidents that can be eliminated or for the events where the consequences are acceptable. If the consequences are unacceptable and nothing can be done to reduce them, the analysis proceeds to estimation of the accident frequency.

Monsanto has built their protocol for risk assessment based on the model shown in Fig. 3. Full probabilistic risk assessments are done only when the consequence analyses suggests an unacceptable consequence can not be avoided if the accident is assumed to occur. The protocol for completing the initial consequence analysis is outlined in Table 1.[2] Table 1 summarizes the topics that must be addressed during the initial consequence analysis.

The various steps in the process may involve several different people with very different skills. The release scenarios depend on knowledge of the process chemistry and the operating characteristics of the equipment used in processing raw materials, intermediates, products and co-products. The characterization of the release rate and form involve the same types of skills but with a different focus. The vaporization and downwind dispersion involves knowledge of meteorology, chemical transport, and environmental fate.

CONSEQUENCE SCREENING PROTOCOL

The determination of consequences following an accident is a simple process. The problem facing engineers that work in the chemical process industry is how to identify potential consequences for each design option so serious accidents never take place. The same problem faces the engineer that works for companies that use chemicals in one form or another. However, the engineer often does not realize his role in dealing with potential fire, explosions and toxic vapor releases.

A key to doing consequence analyses without defining the probability of an event is the elimination of events which are not considered possible. This subjective step could be a major flaw for many evaluations. To avoid this concern, a common definition of a *credible event* is a must. The definition of credible event which is used at Monsanto is:

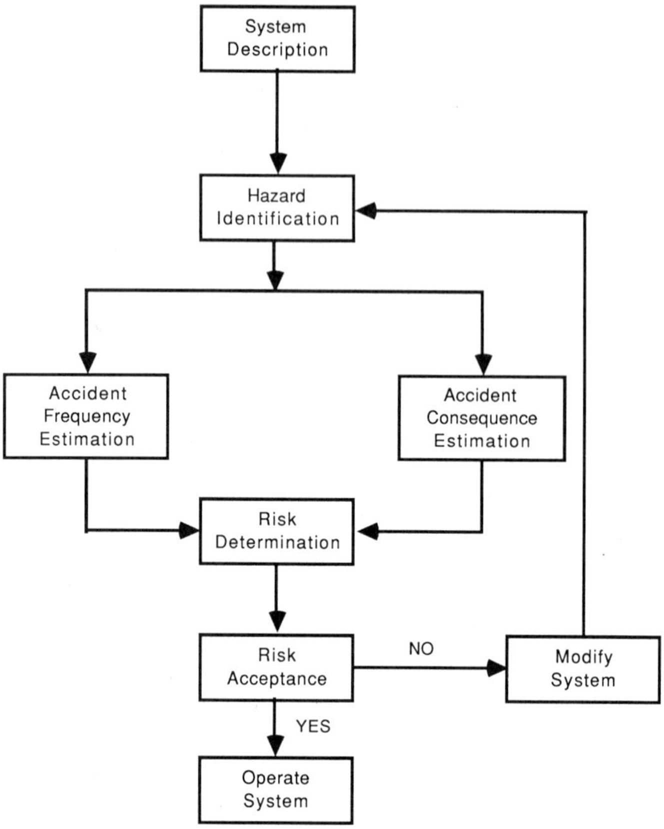

Fig. 2. Predictive risk assessment. (Reproduced by permission of the American Institute of Chemical Engineers.)

"A *credible event* is any accident for which a method can be defined which avoids the release of the chemical of concern to the atmosphere."

For example, the failure of an 8 foot by 8 foot panel on an ammonia sphere due to stress corrosion cracking of the welds is a credible accident because the cracking could have been detected by frequent inspection. However, an 8 inch hole magically appearing in the wall of a chlorine tank is not credible because the engineer is unable to protect against magic.

This simple definition requires the evaluation of many incidents which might be considered by some to be minor. To determine if they are, in fact, minor, the consequence analysis must be completed. When those "minor" events yield unacceptable consequences, they can no longer be considered minor. This approach is not compatible with only doing worst case analyses since many non-worst cases could lead to unacceptable consequences for very toxic chemical.

The *first step* in the risk analysis is to define the hazardous events and the types of releases from a facility that would result. The definition of a credible accident, given above, is used to define the hazardous events of concern. The Consequence Screening Protocol given in Table 1 is used to estimate impact of a credible accident on the area surrounding a facility.

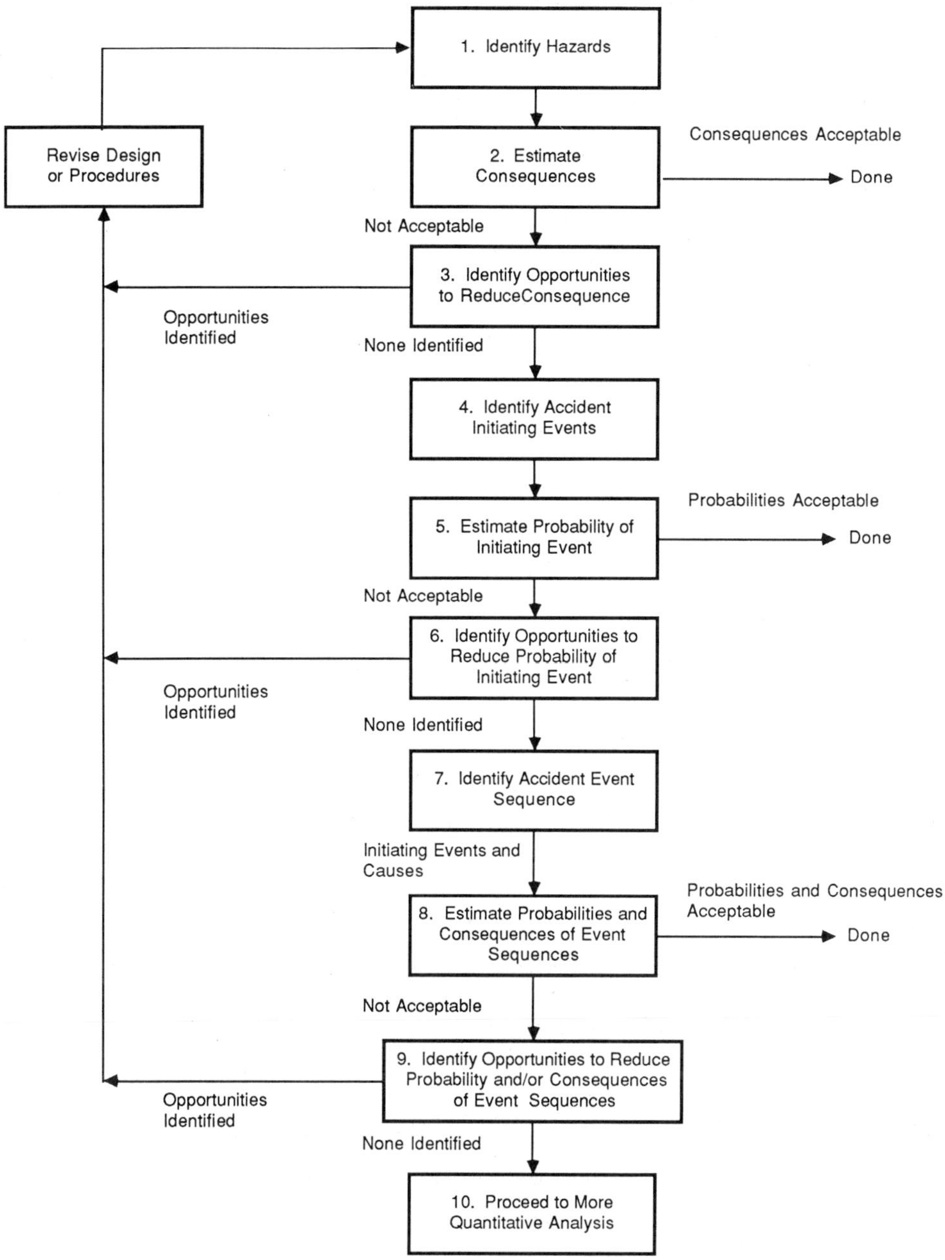

Fig. 3. Steps in predictive hazard evaluation. (Reproduced by permission of the American Institute of Chemical Engineers.)

In the first step the source type is defined. Is the release from a point source (vent line) or from a spill? The second step is to define the source characteristics. If the release is from a spill, what is the temperature? Is the spill area limited by a dike? A very precise definition of the problem is needed. The third step is to conduct a downwind dispersion analysis to estimate concentrations in the area surrounding the facility. From the downwind receptor locations found by use of the PLPLU/PTMAX model,[3] Monsanto uses the ISCST model[4] to determine the downwind impacts of all meteorological conditions.

Table 1. Consequence Screening Protocol*

1. Credible accident source type
   a. Point source (vent line)
      - Runaway reaction
      - Boilout during a fire
   b. Spill from a tank (pure or binary)
      - Onto land
      - Onto water
   c. Spill from a pipeline
      - Onto land
      - Onto water
2. Define the source characteristics
   a. Point source characteristics
      - Height
      - Temperature of plume
      - Temperature of air
      - Velocity of discharge
      - Emission rate
   b. Spill pool characteristics
      - Ambient temperature
      - Spill source, size
      - Limits of pool size (dike area)
      - Emission rate
   c. Plume characteristics
      - Short term puff
      - Buoyant
      - Dense cloud
3. Dispersion analysis
   a. PTPLU / PTMAX for receptor identification
   b. ISCST for 1 hour isopleths
4. Interpretation of results
   a. Target concentration
   b. Single breath / average exposure
   c. Health based population impacts

*Reproduced by permission of the American Institute of Chemical Engineers.

The fourth and final step in the Consequence Screening Protocol is to interpret the data. The predicted downwind concentrations from a credible accident are compared with an appropriate health based criterion. If the predicted downwind concentration is below the health based criterion, the credible accident is classed as of low concern. The above procedure is repeated for all identified credible accidents. Any credible accident which could adversely impact the area surrounding the facility is classed as of high concern.

A major step in any analysis of this type is the modification of the process to avoid the forecast event and its downwind impacts. If the raw materials can be replaced with less volatile materials or the processes or products can be modified to reduce the volatility of the material, the potential consequences may be reduced so they become acceptable.

Engineering efforts are directed at those high concern accidents to reduce the downwind impact.

To date, approximately 1000 credible accidents have been analyzed using the above protocol. Major changes in the methods of operation, equipment, and chemicals used have resulted from these studies. By these changes, the estimated downwind concentrations and the resulting risks from an accident at Monsanto facilities have been reduced. In five cases, the Consequence Screening Protocol has resulted in a more detailed analysis of the credible accident.

PROBABILISTIC RISK ASSESSMENT ISSUES

The Consequence Screening Protocol is structured to avoid the issue of acceptable risk. Steps 1 thru 3 of Fig. 3 can be completed using standard engineering procedures to reduce or often eliminate the consequences of a credible accident of concern. The problem becomes much more sticky when the frequency of an accident is estimated in step 5. Since there is no general agreement on what constitutes an acceptable level of risk, the use of Probabilistic Risk Assessment (PRA) to "prove that the system is safe" is a meaningless task. How does the risk analyst know when to stop reducing the frequency of a credible accident?

Monsanto has adopted a concept that we call "relative acceptable risk." The reduction in risks by design modifications or system changes continues until a "relative acceptable risk" is achieved. For example, the frequency of a major accident in a storage system may be computed as 1 event in 1,000,000 years.

The same accident can be initiated by a fire in an adjacent operating unit (expected frequency of 1 event in 1000 years). It makes little sense to improve the storage system until something is done about the adjacent operating unit's fire problem. Since the emphasis has switched to another portion of the plant, the analyst accepts the risk from the storage system. This "relative acceptable risk" concept forces the continued process of risk reduction by continued improvements to the entire facility. Thus the PRA is used to set priorities for improving the safety of a facility. The PRA is used to rationalize the expenditure of scarce resources to maximize the resulting safety and environmental improvements.

The use of the "relative acceptable risk" concept avoids the issue of what society will consider as an acceptable risk. The use of the "relative acceptable risk" concept is open ended and will continue to reduce the risks. Eventually, the expenditures on risk reduction will make the plant uneconomical to operate. Of critical importance for the future health of American industry is the determination of what constitutes an "absolute level of acceptable risk."

Related to the problem of acceptable risk are the uncertainties associated with any risk assessment. Freeman[5,6] has previously outlined some of the technical problems involved in the completion of a risk assessment for a typical chemical facility. In many cases the use of fault trees or event trees to represent occurrence of the credible accident of concern will yield incorrect estimates of the frequency of the accident. The errors will occur because of:

1. Use of generic data to represent a plant specific problem;

2. Long time delays that allow partial system failure without total system failure; and

3. Statistical dependence among the various basic events that are required to cause the credible accident.

These problems increase the uncertainty of the risk calculations and greatly complicate the communication of the results of the analysis to the public.

CONCLUSIONS

1. The Consequence Screening Protocol outlined in Table 1 can be used to systematically reduce the risks from a process facility without calculating the frequency of every credible accident.

2. Less than one percent of the cases analyzed by the Consequence Screening Protocol will require the estimation of frequency of the accident of concern.

3. The concept of "relative acceptable risk" can be used to set priorities based on the results of a probabilistic risk assessment.

4. There is a need for the various segments of American society to arrive at a mutually agreed upon definition of absolute acceptable risk.

REFERENCES

1. American Institute of Chemical Engineers, *Guidelines for Hazard Evaluation Procedures* (1985).
2. S. Kaplan, M. Kazarians, and J. M. Schroy, Risk Assessment in the Chemical Industry, *AICHE Today Series*, American Institute of Chemical Engineers (1986).
3. HMM Associates Inc., UNAMAP — 5, Available from HMM Associates Inc., 336 Baker Ave., Concord, MA 01742; PTPLU — Version 81036; ISCST — Version 82340.
4. H. E. Cramer Co. Inc., Industrial Source Complex (ISC) Dispersion Model Users Guide (December 1979); Volume I: NTIS PB80-133044, Volume II: NTIS PB80-133051.
5. R. A. Freeman, Problems With Risk Analysis in The Chemical Industry, *Plant/Operations Progress* **2**:185-190 (July 1983).
6. R. A. Freeman, The Use of Risk Assessment in the Chemical Industries, *Plant/Operations Progress* **4**:85-89 (April 1985).

Differences Between Industries in the Definition of Acceptable Risk

John C. Chicken
J. C. Consultancy Ltd.
United Kingdom

S. A. Harbison
H. M. Nuclear Installations Inspectorate
United Kingdom

ABSTRACT

This paper aims to identify and assess the differences that exist in the way acceptable risks are defined and proved in different industries. The paper is divided into five main parts: (1) the definition of acceptable risk in the nuclear industry, (2) the definition of acceptable risk in other industries, (3) reasons why certain levels of risk appear acceptable, (4) ways in which acceptability requirements are satisfied and (5) the cost implications of reducing risk. The final part of the paper speculates on possible future developments.

KEYWORDS: Acceptable risks, nuclear, comparison of industries

INTRODUCTION

This paper was written in the aftermath of the Chernobyl disaster, which underlined the differences between the safety standards applied to modern high technology installations in different countries. Other accidents like Three Mile Island, Seveso and Bhopal also have influenced thinking about the acceptability of risks.[1-3] For example, the Seveso disaster, which resulted in the release of dioxin, caused the EEC Directive on Major Hazards to be issued.[4,5] The Directive requires that, when the acceptability of an installation is being considered, a hazard report and an emergency plan must be prepared. This paper examines the differences between industries in the definition of acceptable risk in five steps: (1) the definition of acceptable risk in the nuclear industry, (2) the definition of acceptable risk in other industries, (3) reasons why certain levels of risk appear acceptable, (4) ways in which acceptability requirements are satisfied and (5) the cost implications of reducing risk.

DEFINITION OF ACCEPTABLE RISK IN THE NUCLEAR INDUSTRY

Examination of the way an acceptable level of risk is defined in the nuclear industries of Germany, Britain and France showed that the three countries apply quantitative dose/frequency criteria in superficially similar ways.[6] However, distinct differences exist

New Risks
Edited by L. A. Cox, Jr., and P. F. Ricci
Plenum Press, New York, 1990

in the way dose/frequency targets are applied and in the weight that is given to them in regulatory decisions.

In Britain, the Nuclear Installations Inspectorate (NII) assesses each safety case submitted to it against its Safety Assessment Principles.[7] These principles identify such requirements as ensuring redundancy and diversity in the reactor's protective systems; taking into account possible common mode failures; and stating explicitly the relevance, accuracy and reliability of the data used. The principles place an upper limit on the level of reliability that can be claimed for a single line of protective equipment in the range corresponding to one failure per 10^3 to 10^5 demands.[7]

The long public inquiry into the proposal to site a 1000-MWe Pressurized Water Reactor at Sizewell provided a forum for the Nuclear Installations Inspectorate to explain publicly how it attempted to define acceptability of risk. Two points made at the Sizewell Inquiry are particularly relevant to this paper:

(1) John Locke, once Director General of the Health and Safety Executive, explained HSE's view on the limit of risk acceptability to be the following: If the chances of a large-scale accident were estimated to be above one chance in 10,000 per year of operation, the HSE would regard this as unacceptable and the NII would refuse to allow such a plant to be built.[8] It was added that no such design has ever been submitted to NII; therefore it has not been necessary to define exactly the level at which a proposal would be refused.

(2) To illustrate the order of magnitude of the risk presented by a reactor designed just to satisfy safety assessment principles, NII presented an analysis of two notional cases, referred to as case A and case B.[9,10] In case A the product of frequency and dose was maintained constant down to the whole-body dose band 10^4 to 10^5 rem. It was then decreased to a cutoff value of 10^{-7} per year beyond this dose band. In case B a step reduction in frequency was introduced at a whole body dose of 10 rem [Emergency Reference Level (ERL)], and the same cutoff frequency applied in Case A was used for doses above 10^3 rem. Case B is considered to be a better representation of NII principles.

The estimated societal risk in terms of fatal cancers for cases A and B is shown in Fig. 1. Uncertainty in the analysis is indicated by the width of the shaded area. The conclusion that was drawn from the analysis was "that a reactor designed, constructed and operated in conformity with the NII safety assessment principles will result in a low level of risk to the most exposed individual and to the population as a whole. Moreover, because of a number of cautious assumptions adopted in the analysis, the predicted risks are likely to exceed those which might occur in practice."[9]

The Nuclear Installations Inspectorate's assumes that some upper level of risk exists above which a station should not be given a license and below which the Inspectorate's safety assessment principles should be applied. The overall objective is to ensure that the risks are as low as reasonably practicable. Preliminary work indicates that, provided a reactor conforms to the Nuclear Installations Inspectorate's safety principles, the level of risk to exposed members of the public and society as a whole should be acceptably low.

The licensing procedures and requirements in France lay down the following requirements:

> In general, a plant that goes to make up a nuclear reactor unit of the pressurized water type should be so designed that the overall probability of the unit giving rise to unacceptable consequences does not exceed 10^{-6} per year. Therefore, when a probabilistic approach is to be used for assessing whether a family of events should be taken into consideration in the design of a unit of this kind, that family of events should effectively be taken into consideration if the

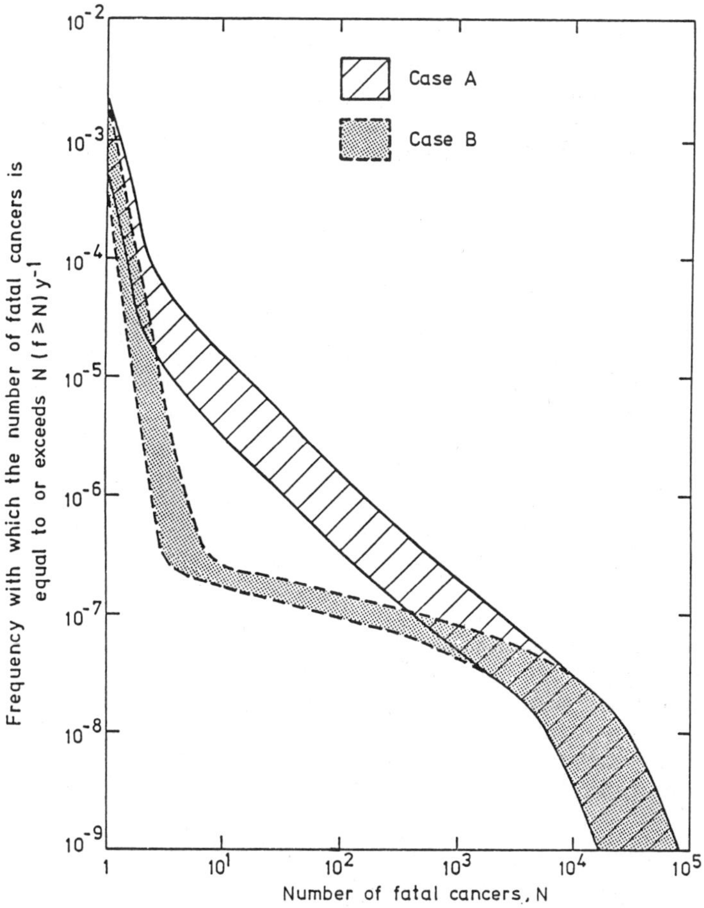

Fig. 1. Societal risk consequent upon a reactor sited at Sizewell and designed just to meet the NII principles for accidents (and their assumed extension).

probability of it giving rise to unacceptable consequences exceeds 10^{-7} per year.

Nevertheless, this figure should be considered more as a guide than as a strict mandatory limit: the probability figures should, of course, be considered as giving orders of magnitude for the overall objective and the various families of events.[11]

The present situation appears to be that regulatory authorities are not yet using probabilistic safety goals as a formal part of the licensing process; rather they are using them in an indicative fashion to supplement the traditional deterministic approach.

DEFINITION OF ACCEPTABLE RISK IN OTHER INDUSTRIES

In other industries there is more diversity in the way acceptable risks are defined and justified. To assess this diversity, the ways acceptable risks are defined in German, British, French and Dutch practices were examined.

In Germany a potentially hazardous installation has to satisfy a set of detailed requirements. These requirements call for a justification of the adequacy of the design, construction and operating procedures to be presented in the form of safety analysis.[12] No attempt is made to identify the level of risk that would be acceptable, even though the methods of analysis specified could give an assessment of the level of risk in probability terms. This suggests that the German Regulatory Authorities regard the quantitative assessment of risk as simply supplementing the Regulatory Authorities' professional judgment.

The British definition of what is an acceptable industrial risk is equally vague. It is, however, possible to construct a view of what is acceptable from official statements. The Advisory Committee on Major Hazards expressed a view that if for a particular plant a serious accident was unlikely to occur more often than once in 10,000 years, it might be regarded as just on the borderline of acceptability.[13] The committee admitted that they hoped to refine their thinking on this subject. Two other statements that indicate official thinking are related to the St. Fergus to Moss Morran LNG pipeline and the Canvey Island development. The Executive's assessment showed that the chance that leakage from the pipeline would put people in the area at risk fell in the range 1 to 4×10^{-6} per year and this was not considered to be a reason for stopping the project.[14] The risks around Canvey Island were put at 0.35×10^{-4} per year, and this was considered to be so high that there should be no further development.[15]

In France the Feyzin incident in 1966 was instrumental in initiating a general revision of the French safety system and the introduction of a new law in 1976. As a result of the 1976 act, a decree was made in 1977 requiring that installations in which there are more than certain threshold quantities of hazardous materials have to be justified by a technical evaluation of the hazards and the safety precautions. This procedure was first applied to the extensions to the Rhone-Poulenc ammonia plant near Rouen. In 1981 the Minister for the Environment announced his decision to apply these in-depth safety studies to 20 installations per year.[16]

Dutch industrial risk acceptability decision procedures were also modified by the Seveso Directive.[17] The Dutch procedure now requires a safety report to be prepared if the amount of hazardous material involved exceeds a "threshold value." The threshold value is defined as the quantity of material that would, in the event of an accident, seriously endanger human life beyond 100 m from the accident. The safety report should include an assessment of the significance of fire, explosion and toxic hazards, the criteria for staff selection, and the adequacy of arrangements for dealing with emergencies. Also the Dutch take into account the views of the local safety committee. The acceptability of risks from an accident has been defined quantitatively in the following strict way.[17] The maximum allowable level of risk to people living in the area of a potentially hazardous installation is an accident causing 10 deaths at 10^{-4} per year and the negligible level of risk is 10^{-6} per year. It is also suggested that if the consequence is n times greater, the chance must be n^2 smaller.

WHY CERTAIN LEVELS OF RISK APPEAR ACCEPTABLE

It has been shown that determining what is an acceptable level of risk requires assessment of the significance and interactions of a complex range of factors.[18] These factors can be broadly categorized as technical, economic and socio-political. The technical factors include design, construction, inspection, life estimation and operational aspects of the installation concerned. Provided the required quantitative data are available, the significance of these technical factors can be discussed in quantitative terms. Lord Ashby showed how the acceptability of risks could be ranked in quantified terms, and in Table 1 his ranking is amplified.[19] Risk acceptability economic limits may also be identified. The

Table 1. Ranking Technical Acceptability

Risk Rank	Risk of Death per Year	Acceptability Ashby Criteria	Justification of Rank
1	Close to greater than 1 in 1,000.	Unacceptable.	Public pressure to have risk reduced or eliminated.
2	Of the order of 1 in 10,000.	Acceptable only under special circumstances.	Locke mentioned that a plant with a risk of a large-scale accident occurring once in 10,000 years of operation would be unacceptable.
3	Of the order of 1 in 100,000.	Would require detailed evaluation to justify acceptability.	Risk recognized warnings given.
4	Close to or less than 1 in 1,000,000.	Acceptable without restriction.	Risk of natural disasters.

problem of defining the economic implications of risk is a challenge that has been successfully met by insurance underwriters. For them to remain solvent, they have to accurately assess the nature and frequency of occurrence of risks.

It is difficult to decide whether or not a risk is acceptable when little or no data are available. To say that no development should take place until all the required data is available would effectively stifle development. To overcome this problem the solution frequently adopted is to derive justification from a carefully structured development program. This solution is used in the nuclear, aircraft, motorcar and drug industries. But it does involve regulatory authorities monitoring the program.

HOW ACCEPTABILITY REQUIREMENTS ARE SATISFIED

The question of how the acceptability requirements of a project are satisfied has to be considered from two aspects. First, the risk characteristics and their accuracy must be established in scientific terms. Second, the evidence has to be tested to determine the degree of acceptability the project justifies. The nature of the appropriate and realistically available evidence is different at each stage of a project and is summarized in Table 2.

A variety of analytical techniques is required to build up the required evidence. The pattern of techniques varies from project to project. As a whole, the assessment procedure is an iterative process refining the acceptability of the risks as the project develops. It is usually the regulatory body that is formally responsible for determining acceptability on behalf of the public. The views of the public may be determined by allowing them to present evidence at a public inquiry, or to participate in discussion of acceptability by a safety committee, or by surveying their opinions. It is possible that such discussions will be helped by using the risk ranking techniques described in Ref. 20 or some variations of the technique.

Table 2. Technical Risk Acceptability Evidence Appropriate to Each Stage of a Project

Stage	Assessment Required	Detail Expected
Conceptual	To identify the nature of the main risks	Establish if worst risk is acceptable.
Design	Detailed assessment of relia-ability and comprehensive quantification of the risks.	Assessment of all possible fault conditions and their consequences.
Regulatory	To determine if the safety case submitted satisfies regulatory requirements and if the risks associated with the project are acceptable to the public.	This requires justification of the design. It includes assessment of the accuracy of the evidence. Also, assessment of acceptability to the public might be required.
Operational	To monitor operational reliability and effectiveness of management throughout the life of the plant.	The detail must be sufficient to demonstrate that regulatory requirements are satisfied.
Review	Throughout the life of the plant the acceptability arguments must be kept under review.	Monitoring the whole field of knowledge associated with the safety case.

THE COST IMPLICATIONS OF REDUCING RISK

The Chernobyl disaster showed just how extensive the cost implications of a major accident can be and how there can be international claims for compensation associated with such accidents. There have been claims in the U.S. that the capital cost of nuclear power has over the last 15 years increased by approximately 20% per year or roughly twice the rate of inflation.[21] Although some of these increases have been attributed to increased material requirements and persistent decreases in craft labor productivity, some is undoubtedly due to additional safety requirements. One suggestion has been that the total cost of regulatory delays to 13 nuclear reactors in the USA could be between 2.7 and 3.6 billion U.S. dollars.[22] In Britain the Monopolies and Mergers Commission found that finding satisfactory safety solutions to some of the novel features of the Advanced Gas Reactors was one of the reasons they cost more than originally estimated and were late in being completed.[23]

Financial problems can also be caused by inept management, as was the case in the 2.3 billion U.S. dollar municipal bond default by the Washington Public Power Supply System (WPPSS) in 1983.[24]

Escalation in project cost is not confined to the nuclear industry. Quite unrelated projects like the Sydney Opera House and Concorde escalated to about ten times the original estimate.[25] It has been suggested that for major civil projects the average cost escalation is over 50% and that for projects involving novel technology the escalation is always high.[25]

CONCLUSIONS

Several conclusions appear to be justified.

1. The differences between industries in the Western world in the quantitative definition of acceptable risk are probably not more than one or two orders of magnitude.

2. The techniques and methodology for establishing the nature and acceptability of risks are known. However, when a project is novel or insufficient data are available, problems may arise in making an objective assessment.

3. The convention of justifying acceptability with a very detailed risk assessment is more universally adopted in the nuclear industry than in some other industries, although potentially hazardous chemical complexes are now tending to be subjected to equally searching analysis.

4. There are indications that there may be merit in adopting universally a comprehensive and consistent way of ranking the acceptability of risks.

REFERENCES

1. Deadly Meltdown, *Time*, pp. 7-15, May 12, 1986.
2. The Bhopal Poison Gas Disaster, *The Financial Times*, p. 2, December 2, 1985.
3. No Easy Answer, India Still Faces Risk of Industrial Disasters a Year After Bhopal, *The Wall Street Journal*, p. 1 and 22, December 3, 1985.
4. P. Lagadec, *Major Technological Risk*, pp. 35-75, Pergamon Press, Oxford, (1982).
5. European Community Directive on the Major Accident Hazards of Certain Industrial Activities, *Official Journal of the European Community*, 82/501/EECd, L230 (1982).
6. J. C. Chicken, *Risk Assessment for Hazardous Installations*, published for the Commission of the European Communities, p. 29, Pergamon Press Oxford, (1986).
7. Nuclear Safety H. M. Nuclear Installations Inspectorate, Safety Assessment Principles for Nuclear Power Reactors, Fourth Impression 1983, published by Her Majesty's Stationery Office, London.
8. The Range of Risks from a PWR at Sizewell — An Overview, Evidence presented by H. M. NII to the Sizewell 'B' Power Station Inquiry, NII/S/92(SAF) (July 1984).
9. S. A. Harbison, G. N. Kelly, and C. R. Hemming, Procedures to Relate the NII Safety Assessment Principles for Nuclear Reactors to Risk, NRPB — R180, published by the National Radiological Protection Board, Didcot, Oxon (April 1985).
10. S. A. Harbison and G. N. Kelly, An Interpretation of the Nuclear Inspectorate's Safety Assessment Principles for Accidental Releases, paper presented at the IAEA seminar on implications of probabilistic risk assessment, Blackpool (March 1985).
11. Status Report on Safety Goals/Objectives, Commission of the European Communities, Brussels, XII/476/83 (May 1983).
12. J. C. Chicken, Risk Assessment for Hazardous Installations, Op Cit, pp. 33-41.
13. First Report of the Advisory Committee on Major Hazards, Her Majesty's Stationery Office, London, Second Impression, p. 12 (1976).
14. Health and Safety Executive Report, A Reappraisal of the HSE Safety Evaluation of the Proposed St. Fergus to Moss Morran LNG Pipeline, (September 1980).
15. Canvey, A Second Report, A Review of Potential Hazards from Operations in the Canvey Island/Thurrock Area Three Years After the Publication of the Canvey Report, published by Her Majesty's Stationery Office, pp. 14-15, London (1981).
16. J. C. Chicken, Risk Assessment for Hazardous Installations, Op Cit., pp. 45-48.

17. Appendix 2, "Dealing With Risks," of the Environmental Program of the Netherlands 1986-1990 established by three Ministries and edited by the Central Department for Information and International Relations.
18. J. C. Chicken, Risk Assessment for Hazardous Installations, Op Cit., pp. 58-100.
19. E. Ashby, *Reconciling Man with the Environment*, pp. 69-72, Oxford University Press, London.
20. J. C. Chicken and M. R. Hayns, Development of the Non-Dimensional Method of Ranking Risks, in *Risk Assessment and Management*, pp. 481-490, Lester B. Lave, ed., Plenum Press, New York (1987).
21. J. H. Crowley and D. Griffith, U.S. Construction Cost Rise Threatens Nuclear Option, *Nuclear Engineering International*, pp. 25-28 (June 1982).
22. The Impact of the Sholly versus NRC Decision, Hearings before the Sub-Committee on Nuclear Regulation of the Committee on Environment and Public Works of the United States Senate, Ninety Seventh Congress, First Session, March 1981, U.S. Government Printing Office, Washington (1981).
23. The Monopolies and Mergers Commission Report on the CEGB and the Reasons for the Delays and Cost Increases in the Construction of the AGR Power Stations, HG 315 Her Majesty's Stationery Office, London (1981).
24. Whoops: How It Happened, *Duns Business Month*, pp. 48-58 (October 1983).
25. P. Hall, *Great Planning Disasters*, p. 7, Penguin Books (1980).

Risk Education and Hazard Management in the Workplace: The Role of Workers' and Executives' Social Relations

Branden B. Johnson
Michigan Technological University
Houghton, MI

ABSTRACT

The recent emphasis on education, training and other risk communication programs as a means of improving workplace health and safety is likely to have limited success. This is partly because health and safety are also affected by availability of technology and capital. But worker response will be mediated by their general relations with corporate and union officials, views of work and alternate job opportunities, and perceptions of alternative corporate hazard management options. Executives' verbal and non-verbal communications are a function of the multiple pressures upon them, of which a mandate for risk communication is only one. Thus these programs can be unwittingly sabotaged from above as well as from below. These social constraints must be taken into account in a realistic approach to occupational hazard management programs.

KEYWORDS: Occupational safety and health, workplace relations, risk communication, risk education

INTRODUCTION

Management of hazards in the workplace requires that corporations have the technology available to control them and the economic resources to implement these technologies. Increasingly, corporations have argued that these resources are too limited for occupational safety and health to rely solely on their deployment. It is suggested that technology is unavailable or too expensive, that controls would unduly reduce production, or that human error is more important as a cause of safety and health problems than unsafe equipment, raw materials, or products. Simultaneously, unions have been successful at getting governments to enact "right-to-know" requirements. As a consequence there has been greater emphasis on education and training programs, warning signs and labels, and other means for communicating information about risks and appropriate coping behavior.

Some view the corporate arguments given above as self-serving for firms. By shifting responsibility for hazards to workers, industry reduces its non-productive capital costs while keeping lost workdays and workers' compensation costs down to the extent that education is effective. Worker hazard programs may also provide the benefit of making employees more amenable to managerial control generally (e.g., Noble, 1977).

Even if we dismiss these arguments as false or overblown, there still remains the question of whether worker-focused programs can be effective. Certainly business executives have also criticized government-mandated programs as costly, misdirected, or counter-productive. Critical literature in this area has tended to emphasize cognitive-visual constraints on the readability and comprehensibility of labels (e.g., Hadden, 1986).

It is the argument of this paper that constraints on the efficacy of workplace risk communication extend beyond those of political economy or cognition. Although these can be significant factors, they are by no means the only ones. A program which focuses exclusively on one or the other—they are hardly compatible in most ideologies—will be incomplete and ineffective. The constraints discussed here concern social relations of workers and executives in the workplace.

WORKERS' SOCIAL CONTEXT

The employee's daily environment makes multiple demands upon him or her. Among others, these include job performance sufficient to avoid censure, if not to merit rewards, and reasonably smooth relations with one's co-workers. Sometimes work demands reinforce each other; more often they conflict. In such an environment it is understandable that risk communication messages may be heard quite clearly but ignored. If the worker holds even one of these other demands to be more important than maintaining safety, risk education programs will lose effectiveness at the very least.

For example, workers can be ridiculed by fellow workers and supervisors for such safety precautions as wearing gloves and masks around dangerous chemicals. To acknowledge possible health threats would disrupt workers' relations with superiors and comrades, violating norms of job or worker loyalty—denial of risk often seems the best choice (Brown, 1987). Workers may fail to perform necessary safety routines to meet production quotas or because safety would require working overtime against company rules.

This clash of job demands suggests that accurately received risk messages can be overwhelmed by other important messages, explicit and otherwise, in the workplace. Thus behavior becomes distorted from the safety optimum, whether that optimum is set by corporate financial resources, production needs, or avoidance of penalties for non-compliance with government regulations.

The discussion so far has assumed that workers believe the risk messages they receive; the problem is that they will not act on them because other demands are more important. But workers may well not believe safety program messages. This lack of belief raises the issue of trust (Kasperson, 1986:277).

> If the communicator is viewed as having a compromised mandate or a lack of competence, credence in information provided tends to be weakened accordingly. Or if the particular risk has been mismanaged or neglected in the past, skepticism and distrust may greet attempts to communicate risks.

To be trusted, communicators must be seen as competent, unbiased, caring about those whom they "serve," and users of due process (Kasperson, 1986:277-278). Finally, effective participation may require (Kasperson, 1986:280)

> (a) the provision, or development, of an indigenous technical capability by which expertise under the control of risk bearers can be brought to bear on complex problems, and (b) institutional means by which the public can act on the enlarged information.

At stake is the issue of who controls decisions about risk management.

Is this model, based on public participation, applicable to workplace relations? The past relations of these groups certainly affect current relations; the parallel for conflict resolution is less clear. It can be argued that, by taking employment in a firm, workers have entered an implicit contract in which they trade risk autonomy for wages. Thus managers, unlike government officials, have no need to relinquish control over safety to workers, and caring and trust are not issues. On the other hand, it can be argued that money does not reimburse one for threats to health. One takes a job on the premise that it will be reasonably safe, unless one has no alternative, and thus worker rights to be involved in risk decisions are not surrendered (e.g., Graham and Shakow, 1981).

It is not possible to resolve what is, after all, a normative debate. But preliminary analysis of the 1977 national Quality of Employment Survey (QES) data provides some pertinent data. Brown (1987) suggested that workers who deny occupational risks do so to avoid anxiety and feelings of helplessness, and may reinforce this with the belief that their employers can be trusted to protect them. QES-polled workers who were more trusting of management's willingness to look after their safety were less likely to believe themselves exposed to numerous hazards (r=0.31), to see these hazards as serious (r=0.32), and to report the perceived hazard to superiors, unions, government agencies, or others (r=0.28). At first glance distrust appears to have been a function of union membership, which is associated with high hazard perception (r=0.23) and perception of hazards as serious (r=0.20). But those who saw their exposure as serious were less likely to be satisfied with their union's performance (r=0.14). It is possible that greater concern about hazards is associated with loss of perceived support from both business and labor institutions. Interestingly, 76% of the workers interviewed felt that they should have "complete" or "a lot of" say about safety equipment and practices; no other category of worker participation in decision-making was supported by more than 41% of the sample.

In addition to the deniers of risk, Brown (1987) suggested that a second group of workers would "accept" occupational risks because the job's benefits—wages, a rare local income opportunity, or the job's intrinsic satisfactions—compensate for the risks. The wage portion of the hypothesis has not yet been tested with QES data. Of four measures of perceptions that no alternative income sources exist, two were weakly (r=0.07 and 0.10) but significantly related to seeing few hazards at work; the others gave no significant correlation. Job satisfaction was negatively correlated with perceived hazards (r=–0.21), seriousness of hazards (r=–0.26), and hazard reporting (r=–0.22). Intrinsic satisfactions of the job—e.g., freedom to decide how to do one's job and to develop one's abilities—were somewhat better linked to low hazard perceptions (r=0.10) and low perceived seriousness (r=0.15) than were extrinsic satisfactions like pay, job security, and fringe benefits (r=0.07 and 0.10). These findings provide weak support for factors in worker acceptance of risks.

These perception findings may seem insignificant to the manager who is concerned about reducing actual workplace risks to a level commensurate with maintaining corporate viability, whether to keep production costs down or to comply with government mandates. One rebuttal of this viewpoint is that if one's workers over- or underestimate the actual risks they face, this misperception is likely to obstruct the utility of the firm's workplace safety program. Overestimation of risk will lead to avoidance of work areas or activities, reducing production; it may also increase hazards where safety is inconsistent with hesitation. Underestimation of risks leads to higher rates of occupational injuries and illnesses, greater production costs through lost workdays and halted production lines, and political costs because external actors will tend to blame the corporation for these dangers.

But it can also be said that workers' risk estimates are not entirely erroneous, based on preliminary analysis of QES data. A crude measure of hazard perception was the number of hazards out of some fourteen (e.g., dangerous chemicals and radiation, air pollution, noise) to which workers reported being exposed. Yet there was a correlation of 0.31 between this unsophisticated measure of perceived hazards and 1977 safety data per industry (Bureau of Labor Statistics, 1979) on numbers of injuries and illnesses per 100 workers and number of lost workdays. This correlation declined to 0.16 when the perception measure was the number of these exposures deemed serious by workers. Even with these primitive measures, however, workers' risk perceptions were still somewhat correlated with risk statistics.

These preliminary quantitative tests of Brown's qualitative findings suggest that social issues of trust and job satisfaction, among others, are indeed important in worker hazard perceptions. I suggest that these perceptions will play some role in worker safety behavior as well, if only to shape worker responses to technical and economic constraints on their behavior.

EXECUTIVES' SOCIAL CONTEXT

As noted above, human error has been discovered as a major cause of occupational safety and health problems and as a rationale for reduced engineering regulation and increased worker education. "Human error," however, can also be a euphemism for organizational or managerial problems, as Perrow (1984) suggested. Perrow's major interest was the problem of how bureaucracies deal with complex technologies. But the social context in which business executives work can affect their management regardless of technical complexity, and human error can be a worthy excuse for problems. A study of occupational risks in the British offshore oil industry illustrated this nicely.

Wright (1986) noted that accidents are assumed to be abnormal, unplanned or unexpected events. Thus the emphasis is on control over the work environment through appropriate response technology and worker safety training and incentives. Offshore oil accidents occurred when someone was in the wrong place or doing the wrong thing, as defined by the work rules. But rather than this indicating that safety equals conformity to routine, Wright (1986:279) concluded that "solutions to normal problems for the workforce had not been incorporated into formal work routine." Management dealt with the unexpected by treating it as usual. They assumed that operators' length of service, exclusive of any skills possessed, was sufficient evidence of experience coping with this problem (Wright, 1986:280-281). Length of service was

> believed to guarantee competence in any task even one such as this in which one cannot be experienced because of its infrequency. Through this device those responsible for ensuring that work was done properly were able to create the sense that they were dealing with a routine operation, one in which experience is relevant.

The workers would also identify an unusual task with a more familiar one, and it was the difference between them which was the critical accident generator. But Wright argued that the main responsibility for this failure to prepare for the unexpected in the workplace was management's. Blaming human error for accidents displaced blame from executives, deliberately or not.

Multiple work demands affect managers' behaviors just as they do those of workers. For example, factors which exacerbated offshore oil safety problems included the economic need for speedy work, overwhelming the economic incentive for safety. Also important were high levels of subcontracting which enhanced decision-making uncertainty, ignorance of work conditions, and confusion over safety procedures. In this case too many, rather than too few, channels for risk communication were the problem (Wright, 1986). Work rules are increasingly important as a focus of labor-management bargaining. Among the changes sought by executives are rights to subcontract and to shift workers from one task to another with which they may be less familiar (Serrin, 1986). The implications of these work rule changes for the effectiveness of corporate safety programs should not be overlooked.

Managers may sometimes use human error as a scapegoat to divert attention from problems that might otherwise be laid at their door. Authority is embodied for business executives "in their personal relationships with their immediate bosses and in their perceptions of similar links between other managers up and down the hierarchy" (Jackall, 1984:177). These shifting alliances are necessary to protect individual executives against unexpected failures; the price of such protection is a demand for team play and support for the way things are done.

For example, an upper-middle chemical firm executive was privy to the discovery that a certain process employed for years inadvertently exposed workers to a potent carcinogen. Other executives, responsible for this process, immediately sent workers home and the process was changed to prevent further exposure. But workers were not told that the reason for the change was health-related, nor that they had been exposed to a carcinogen. The executive justified his silence on several grounds: (1) the damage was already done and irreversible; (2) he would risk his career by exposing the responsible executives—who, after all, had been ignorant of the health effects until then—and violating the corporate code of trust; and (3) in a pinch one must cover for oneself and the firm (Jackall, 1985).

It is of particular interest that this chemical company had instituted an award-winning occupational safety and health program before the incident. The program was effective to the extent that production was immediately halted to revise the process to prevent further carcinogenic exposure. Given strong CEO support for the program, executives' alliances probably enhanced the formal incentives for its implementation. However, the workers' right to know of their exposure was given short shrift, because it could open both the company and the executives to high liability costs. This safety program was eliminated a few years later through the confluence of a change in the ruling clique of the firm, the relaxation of regulation under the new Reagan administration, and a revolt of line managers against staff oversight on the grounds that safety had been sufficiently integrated into normal operations (Jackall, 1984). Jackall drew the conclusion that executives' social networks are more permanent and more influential than any merely formal program.

Even when executives explicitly promote safety, these informal communication networks—conveying the "real" message—may undercut the program's effectiveness. Information in the firm is evaluated more in terms of the credibility, rather than the expertise, of its source (O'Reilly, 1983:119), and that credibility is to a large extent a function of how well one supports one's personal network. These executive networks are often the source of the safety-contradicting messages received by workers. Furthermore, "rational" elements of decision-making relevant to workplace safety—e.g., cost-benefit, financial analyses, and risk assessments—are used by organizational politicians to expand or protect gains, mask political motives, and legitimize self-interest. Appearing to meet norms of rationality validates one's position (e.g., Franz and Robey, 1984). Bureaucratic risk analyses may be slanted to ensure that the organization is not threatened, and they may appear only after the pool of competing definitions of acceptable risk has been sharply reduced (Clarke, 1985:12-13, 19). In this environment it is little wonder that maintaining the trust of workers and fellow executives may be difficult.

This discussion of executives' behavior is not meant to paint them as venal or only self-interested. These behaviors are often unconscious—corporate officials are merely human, not evil—and rational responses to the demands made upon them by the work environment. But such unconsciousness only enhances potential distortion of safety programs, since factors of which their implementors are unaware are unlikely to be accounted for in program design.

HAZARD MANAGEMENT IMPLICATIONS

What conclusions can be drawn from these comments about the social context of hazard management in the workplace? There are financial, legal and social limits to improvements that could be made in corporate safety programs. However, a few suggestions can be offered:

1. Emphasize other salient workplace objectives which safety observance can serve, such as avoidance of production losses and protection of fellow workers. The same social context occurs for product and environmental safety as for occupational safety; behaviors which serve more than one of these goals should be highlighted as well. This approach should reduce the incidence of contradictory messages and multiply worker incentives for compliance with the official corporate safety program. This depends, of course, on accurate executive identification of non-safety objectives salient to workers and on worker trust that risk education is sincerely promoted by managers.

2. Clearly reward successful safety actions and punish failure. It is vital that punishment be wielded impartially at all levels of the organization, even in the face of executives' protective alliances. If this is not done, workers are likely to see themselves as getting the stick and management getting the carrot, a view not calculated to promote trust.

3. Be sure that safety is truly important to the firm, even if only to comply with regulatory strictures. False claims will undermine corporate credibility without enhancing—and perhaps even undermining—safety.

4. Recruit other sources of messages to workers—e.g., unions— into the safety program so as to reduce contradictory demands. This may require that union demands be met directly or indirectly. For example, a firm unwilling to provide demanded wage increases might offer smaller wage increases and hefty bonuses for meeting safety targets, or it might sacrifice some desired work rule changes.

5. Give as much autonomy to workers as possible, which will make them feel like partners in both safety and production. The devolution of safety responsibility to workers will encourage them to be more concerned about safety and may have the added benefit in executives' eyes of leading to shared liability for any problems that may arise. However, sharing power directly contradicts the purpose of managerial networks, and in some cases—such as in occupations where machismo is customary—worker autonomy may not promote safe behavior. These constraints can perhaps be stretched beyond their immediately apparent limits, but narrow grants of true authority are superior to broader but insincere or counter-productive partnerships with workers.

6. Beware of unconsciously contradictory messages from management about work requirements. For example, safety violations may occur because there is a rule that overtime must be avoided at all costs. Ideas volunteered by workers may be derided even though their suggestions were solicited by executives. Risk education programs which are not designed to be consistent will fail.

The political economy of capitalism and the cognitive biases of workers may be important constraints on the effectiveness of occupational safety and health programs. It is the argument of this paper that workplace relations should also be taken into account in designing and understanding the limits of such programs.

REFERENCES

Brown, Michael S., 1987, Communicating Information About Workplace Hazards: Effects on Worker Attitudes Toward Risks, in *The Social and Cultural Construction of Risk: Essays on Risk Selection and Perception*, pp. 251-274, Branden B. Johnson and Vincent T. Covello, eds., Reidel, Dordrecht, Holland.

Bureau of Labor Statistics, 1979, Occupational Injuries and Illnesses in 1977: Summary (Report 561), Washington, D.C.: U.S. Department of Labor.

Clarke, Lee, 1985, Individual and Organizational Perspectives on Risk, American Sociological Association.

Franz, Charles R., and Daniel Robey, 1984, An Investigation of User-Led System Design: Rational and Political Perspectives, *Communications of the ACM* **27(12)**:1202-1209 (December).

Graham, Julie, and Don Shakow, 1981, Risk and Reward: Hazard Pay for Workers, *Environment* **23(8)**:14-20ff (October)

Hadden, Susan, 1986, *Read the Label: Reducing Risk by Providing Information*, Westview Press, Boulder, CO.

Jackall, Robert, 1985, The Bureaucratic Ethos and Dissent, *IEEE Technology and Society* **4(2)**:21-30 (June).

Jackall, Robert, 1984, The Moral Ethos of Bureaucracy, *State, Culture and Society* **1(1)**:176-200 (Fall).

Kasperson, Roger E., 1986, Six Propositions on Public Participation and Their Relevance for Risk Communication, *Risk Analysis* **6(3)**:275-281 (September).

Noble, David F., 1977, *America by Design: Science, Technology, and the Rise of Corporate Capitalism*, Oxford University Press, New York.

O'Reilly, Charles A., III, 1983, The Use of Information in Organizational Decision Making: A Model and Some Propositions, in *Research in Organizational Behavior*, Vol. 5., pp. 103-139, L.L. Cummings and Barry M. Staw, eds., JAI Press, Greenwich, CT.

Perrow, Charles, 1984, *Normal Accidents: Living with High-Risk Technologies* Basic Books, NY.

Serrin, William, 1986, Work Rules Overtaking Pay as Key Labor Issue, *New York Times*, p. Y-17 (October 26).

Wright, Chris, 1986, Routine Deaths: Fatal Accidents in the Oil Industry, *The Sociological Review*, **34(2)**:265-289 (May).

Consumer Response to Food Risk Information: Demographic Differences

Carol S. Kramer
Resources for the Future
Washington, DC

Karen P. Penner
Kansas State University
Manhattan, KS

ABSTRACT

Consumers were surveyed regarding their knowledge, attitudes and practices related to food safety and food safety information and governmental roles in providing food safety services. They were asked about their use of various information sources with respect to different food safety risks, their confidence in the sources and the information provided.

Generally, consumers evidenced concern about food safety; identified governmental involvement in food safety assurance approximately as important as environmental safety and more important than occupational or consumer product safety; stated concerns about the safety of meat and livestock products; stated preferences for safety information or guarantees on meat labels; indicated a modal willingness to pay of 2-4 cents per pound of meat for the label information desired. Significant differences in some food safety preferences and perceptions were found by sex, age, and educational level.

KEYWORDS: Food safety, consumer survey, food safety information, demographics

INTRODUCTION

Three fundamental safety policy questions applicable to the food market are: (1) What minimum levels of safety are tolerable in the market? (2) How should informed consumer choice be optimized? and (3) How should undesirable external effects be minimized? (Kramer, 1986). Policy decisions are, of course, a synthesis of consumer, producer and "public" interests. Yet from a policy perspective it is important to realize that consumers are diverse and that their stated preferences, safety concerns and information requirements differ. This paper presents consumer responses to food safety questions, then highlights some of the demographic differences that emerged during analysis.

New Risks
Edited by L. A. Cox, Jr., and P. F. Ricci
Plenum Press, New York, 1990

SURVEY OF KANSAS CONSUMERS

A sample of consumers chosen to be representative of the adult, non-institutional Kansas consuming public in important stratifications was surveyed by mail questionnaire and phone follow-up in summer 1983 (Appendix 1). The survey sought information about consumer attitudes toward food safety, governmental roles in food safety assurance, and various forms of food safety information. Expressions of willingness to pay for desired safety information on meat labels were solicited.

Socioeconomic information included gender; age groupings, 18-28, 29-37, 38-55, 56 or older; income range; employment status; community size; education; food buying and preparation responsibility (see Penner, Kramer, Frantz, 1985).

GENERAL GOVERNMENT ROLE

Consumers were questioned about the proper role of government in food safety assurance. Survey question 6 asked for degrees of agreement or disagreement with five assertions regarding proper governmental roles in food safety assurance (Table 1). Summarizing, 89.8 percent agreed or strongly agreed that keeping food safe should be a high government priority; 63.3 percent disagreed or strongly disagreed with the negatively worded statement, "The government should not try to eliminate all health risk from the food supply;" 70.5 percent agreed or strongly agreed that the government should not permit any cancer-causing chemicals in foods; 80.6 percent agreed or strongly agreed that "the government should require health risk information on food labels and let consumers choose;" while only 19.9 percent stated, "I think food producers and manufacturers can be relied on for safe food products." This simultaneous support of a governmental attempt to eliminate all risk and all carcinogens from the food supply and a requirement of health risk information on labels to let consumers choose is interesting and perhaps indicates a desire for choice, but among fundamentally safe products.

Among the socioeconomic differences in responses to these general attitudinal questions are the following: (1) the youngest age group (19-28) was more likely than expected to support the government attempt to eliminate all health risk from the food supply while the reverse was true of those 38 and over; (2) those employed full-time were more likely to be moderate or not sure in assessing whether all risk should be eliminated, while those unemployed were likely to be more extreme ("strongly" agree or disagree) than predicted; (3) food buyers agreed more strongly than non-buyers that the government should try to eliminate all risk. Comparing government involvement in food safety with other safety involvement including environmental safety, auto safety, consumer product safety, and occupational safety (Table 2), consumers tended as a whole to evaluate food safety and environmental safety efforts relatively equally and slightly higher than consumer product safety and occupational safety. Men were more likely than women to rate governmental involvement in consumer product safety, job safety, and auto safety as less important than food safety (Penner, Kramer, Frantz, 1985).

RANKING OF FOOD SAFETY RISKS

The Food and Drug Administration (FDA) ranked food safety hazards in descending order of public health importance: (1) microbiological contaminants; (2) malnutrition; (3) environmental contaminants; (4) toxic natural constituents of foods; (5) pesticides; and (6) food additives (Wodicka, 1985). Presented with the same selection, excepting malnutrition and toxic natural constituents of foods but including animal drug residues, consumers indicated highest concern about environmental contaminants ("such as mercury, PCBs, and dioxin") followed by "disease-causing organisms and parasites" and pesticides ("chemicals

Table 1a. Responses to Question 3, Parts a-e
(How much do you agree or disagree, if at all, with the following statements about the Federal government's role in keeping the food supply safe?)

Response	a. Keeping food safe should be a high government priority		b. The government should not try to eliminate all health risk from the food supply		c. The government should not permit any cancer-causing chemicals in foods.		d. The government should require health risk information on food labels and let consumers choose		e. I think food producers and manufacturers can be relied on for safe food products	
	No.	%	No.	%	No.	%	No.	%	No.	%
Strongly disagree	6	1.5	112	28.2	11	2.8	12	3.0	81	20.0
Disagree	12	3.0	139	35.0	46	11.5	30	7.5	143	36.0
Not sure	23	5.7	38	9.6	61	15.2	36	9.0	94	23.7
Agree	195	48.5	89	22.4	120	30.0	176	43.9	62	15.6
Strongly agree	166	41.3	19	4.8	162	40.5	147	36.7	17	4.3
Total	402	100.0	397	100.0	400	100.0	401	100.0	397	100.0

Table 1b. Responses to Question 3 by Age, Parts b and c
(How much do you agree or disagree, if at all, with the following statements about the Federal government's role in keeping the food supply safe?)

Part	Level of Agreement	Age of respondents*								Chi-Square	p
		19-28		29-37		38-55		56 or older			
		No.	%	No.	%	No.	%	No.	%		
b. "The government should not try to eliminate all health risk from the food supply	strongly disagree	6	12.5	23	28.7	41	32.0	41	29.5	26.4	.0094
	disagree	24	50.0	31	38.8	44	34.4	39	28.1		
	not sure	6	12.5	8	10.0	15	11.7	9	6.5		
	agree	11	22.9	16	20.0	26	20.3	36	25.9		
	strongly agree	1	2.1	2	2.5	2	1.6	14	10.1		
	Total	48	100.0	80	100.0	128	100.0	139	100.1		
c. "The government shall not permit any cancer-causing chemicals in foods"	strongly disagree	2	4.2	2	2.5	4	3.1	3	2.1	23.0	.0277
	disagree	11	22.9	10	12.3	13	10.2	12	8.4		
	not sure	11	22.9	17	21.0	20	15.8	13	9.2		
	agree	15	31.3	22	27.2	36	28.4	46	32.4		
	strongly agree	9	18.7	30	37.0	54	42.5	68	47.9		
	Total	48	100.0	81	100.0	127	100.0	142	100.0		

Table 2a. Response to Question 7 by Sex, Parts a,* b,** and d†

(How would you compare the importance of government involvement in other areas of safety in comparison to food safety?)

Importance	Survey Total		Sex of respondent: Male		Female	
	no.	%	no.	%	no.	%
a. Auto safety is						
Less important	94	23.5	56	34.4	38	16.2
Just as important	278	69.5	94	57.7	182	77.8
More important	28	7.0	13	8.0	14	6.0
Total	400	100.0	163	100.1	234	100.0
b. Consumer product safety is						
Less important	98	24.4	57	35.0	41	17.5
Just as important	273	68.1	91	55.8	181	77.0
More important	30	7.5	15	9.2	13	5.5
Total	401	100.0	163	100.0	235	100.0
d. Occupational safety is						
Less important	110	27.5	59	36.2	51	21.8
Just as important	258	64.5	91	55.8	165	70.5
More important	32	8.0	13	8.0	18	7.7
Total	400	100.0	163	100.0	234	100.0

*Chi-square = 19.5, (d.f. = 2); $p = 0.0001$.

**Chi-square = 20.2, (d.f. = 2); $p = 0.0001$.

†Chi-square = 10.4, (d.f. = 2); $p = 0.0055$.

used to kill insects, other pests"). Microbiological contaminants top the experts' list. Experts ranked pesticides next to last in terms of food safety concern. Kansas consumers ranked "residues of animal drugs or hormones in meat," and "food additives" fourth and fifth.

Consumers ranked food groups in order of food safety concern, led by perishable protein products (red meats, poultry, and dairy). Fruits and vegetables and fats and oils were third and fourth. Of least concern were dried staples such as flour, rice, beans, and pasta, and sugars, syrups and sweeteners.

Table 2b. Response to Question 7 by Age, Parts c* and d**

(How would you compare the importance of government involvement in other areas of safety in comparison to food safety?)

c. Environmental safety is

Importance	Survey Total		Age of respondent* 19-28		29-37		38-55		56 or older	
	no.	%	no.	%	no.	%	no.	%	no.	%
Less important	51	12.7	6	12.5	3	3.7	21	16.4	29	14.1
Just as important	284	70.8	32	66.7	58	71.6	93	72.7	100	70.4
More important	66	16.5	10	20.8	20	24.7	14	10.9	22	15.5
Total	401	100.0	48	100.0	181	100.0	128	100.0	142	100.0

d. Occupational safety is

Importance	Survey Total		19-28		29-37		38-55		56 or older	
Less important	110	27.5	15	31.3	24	29.6	43	33.6	28	19.9
Just as important	258	64.5	28	58.3	55	67.9	79	61.7	94	66.7
More important	32	8.0	5	10.4	2	2.5	6	4.7	19	13.5
Total	400	100.0	48	100.0	81	100.0	128	100.0	141	100.1

*Chi-square = 13.3, (d.f. = 6); $p = 0.0379$.

**Chi-square = 16.2, (d.f. = 6); $p = 0.0128$.

Consumers were asked to identify the stage in the meat products production and distribution chain where they felt food safety problems were most likely to occur. Food processing plants were ranked most problematic, followed by livestock feedlots. The feed manufacturing plant and warehouse were next, followed by grocery stores and restaurants. The farm and the home were ranked least problematic. These rankings highlight one obvious discrepancy between the low consumer rankings given the home, restaurant, and grocery store and experts' assessments of microbiological contamination as priority food safety public health hazards. Because many microbiological problems result or are exacerbated by time or temperature abuse of foods occurring in restaurants, homes and grocery stores, underestimation of risks is important. The final preparation site may be also the final line of protection against microbiological health problems. In many cases, organisms can be killed by proper cooking. In several questions related to home food handling under diverse conditions, respondents demonstrated lack of knowledge of safe handling practices (Penner, Kramer, Frantz, 1985). One third of respondents agreed or didn't know whether to agree with the statement: "Meats and poultry products bought at a grocery store do not contain bacteria or other microorganisms." In this case the youngest and oldest groups were more likely to agree erroneously than other age groups.

Over one-half of respondents answered they had not bought meat due to food safety concerns, including too much fat (66.8 percent of respondents), untidy meat departments (63.8 percent), colors or chemical ingredients (48 percent).

DESIRED LABEL INFORMATION ON MEAT

Asked to identify the kinds of information they would like to see on meat labels (Table 3), consumers most often indicated "guaranteed free of traces of animal drugs, hormones" (69.3 percent of respondents), followed by "guaranteed free of traces of other chemicals" (67.4 percent). "Additives" and "nutrient content" were third and fourth. "Cholesterol content" ranked next to lowest (sixth) as a label concern. Significant differences existed between response patterns of different age and gender groups. Nutrient content on labels was preferred by 56.9 percent of the youngest group (ages 19-28) but only 35.1 percent of those 56 or older. Women supported label guarantees of "drug-" and "chemical-free meat" more than men. In qualitative terms consumers indicated they would be willing to pay "slightly more" for desired food safety information on meat labels (Table 4). In quantitative terms, "from 1-3 cents per pound of beef" was the modal response (Table 5). Subsequent analysis using simple and multiple correlation techniques shows differences in willingness-to-pay by socio-economic characteristics (Penner, Kramer, Frantz, 1985; Pope, 1985).

CONSUMER ASSESSMENT OF GOVERNMENT ACTIVITIES AND INFORMATION

Consumers evaluated thirteen food safety activities the government now undertakes (Table 6). Inspection activities, particularly of meat, poultry and dairy products, were rated extremely important, including inspection at slaughter for animal drug and chemical residues, for disease, and for general safety of milk and dairy products. Respondents indicated less support for research, particularly university research. Label requirements for food safety information were supported by 85 percent, while government-supported education relating to food handling was supported by 75 percent, and more by women than men.

Consumers identified their most common information sources regarding different food safety perils (Table 7). The most frequent response concerning animal drugs or hormones and pesticide information sources was "I don't receive this information," followed by "Government." Out of seven listed possible sources, both university and the county extension service ranked extremely or relatively low as an identified source of information for food additives, animal drugs, and food exposed to environmental contaminants, while the government ranked high as an identified source. However, confidence by information source differed. Extension agents were rated higher than any other listed source (68 percent rated high or a lot of confidence), followed by university professors (38 percent) and "consumer spokesperson" (36 percent), "family or friend" (31.2 percent). Government ranked fifth (22 percent), followed by "popular media personality" (14 percent) and "food product spokesperson" (13 percent).

Highest media ratings by consumers for convenience in receiving food safety information were food labels, newspaper, and television. Intermediate were radio, newsletter, and magazines. Lowest ranks went to fact sheets or bulletins, consumer hotlines, and educational meetings. Lower educational levels were associated with higher convenience rankings for television.

POLICY RAMIFICATIONS, RESEARCH NEEDS

This consumer survey of food safety attitudes, preferences and knowledge indicates generally strong support for a high priority government commitment to food safety and

Table 3a. Survey Totals in Response to Question 27

(What food safety information, if any, would you like to see on meat labels?)

Information Preferred	Survey Total* No.	%
How to cook and handle	138	36.9
How to cook and freeze	116	31.0
Guaranteed free of traces of animal drugs, hormones	259	69.3
Guaranteed free of traces of other chemicals	252	67.4
Nutrient content	168	44.9
Cholesterol content	137	36.6
Additives	249	66.6

*n = 374 = 100%.

Table 3b. Response to Question 27 by Age

(What food safety information, if any, would you like to see on meat labels?)

Information Preferred	Age of respondents: 19-28 n = 44		29-37 n = 78		38-55 n = 119		56 or older n = 131				
	no.	%	no.	%	no.	%	no.	%	Chi-square	d.f.	*p*
Additives	30	68.2	59	75.6	86	72.3	74	56.5	10.7	3	.0135
Nutrient content	25	56.9	37	47.4	60	50.4	46	35.1	9.2	3	.0262
Guaranteed free of traces of animal drugs, hormones	23	52.3	58	74.4	89	74.8	89	67.9	8.8	3	.0325

Table 3c. Response to Question 27 by Sex

(What food safety information, if any, would you like to see on meat labels?)

Information Preferred	Sex of Respondent: Male (n = 151)		Female (n = 220)		Chi-square	d.f.	*p*
	no.	%	no.	%			
Guaranteed free of traces of animal drugs, hormones	91	60.3	166	75.5	9.7	1	.0018
Guaranteed free of traces of other chemicals	91	60.3	159	72.3	5.9	1	.0154

Table 4. Responses to Question 28

(As a consumer, would you be willing to pay more for meat items if there were safety information added to the label?)

Willingness to Pay	Survey Total	
	No.	%
I would pay slightly more	263	67.4
I would not pay more	105	26.9
I would pay considerably more	14	3.6
I do not buy meat items	8	2.1
Total	390	100.0

Table 5. Responses to Question 29

(In general, about how much extra per pound of beef would you be willing to pay for the types of food safety information you want?)

Price	Survey Total	
	No.	%
0¢ . . . I'm satisfied with current labels	70	18.9
0¢ . . . I think government (taxpayer) should pay	40	10.8
1-2¢ per pound	155	41.9
3-5¢ per pound	82	22.2
6-10¢ per pound	18	4.9
Over 10¢ per pound	5	1.3
Total	370	100.0

involvement in a variety of food safety activities. Meat, poultry and dairy inspection for disease and animal drug/chemical residues was strongly supported. Consumers thus indicate strong implicit support for enforcement of product standards setting safety levels.

Consumers also indicate strong support in the abstract for safety information on labels, although the specific content varied by socioeconomic characteristics in some cases. For women, meat label guarantees that meat is free of animal drug and other "chemical" residues were most important. Note that this desire is not for risk-benefit information per se, but for improved certification beyond current government levels. Consumers do not express strong confidence in government information, although it is the major source of many types of safety information (animal drugs, additives, environmental contaminants), when consumers receive information at all. Cooperative extension was rated highly in terms of confidence, but was not perceived as a source of most kinds of food safety

Table 6. Response to Question 6

(The government does many things to help keep our food supply safe. Please indicate how important, if at all, you think the following food-safety related activities are.)

Government activity to . . .		Importance					
		None	Slight	Some	Moderate	High	Total
a. provide money to universities for research on food safety	no.	11	28	106	127	116	388
	%	2.8	7.2	27.3	32.7	29.9	99.9
b. conduct research on food safety	no.	8	19	73	143	148	391
	%	2.0	4.9	18.7	36.6	37.8	100.0
c. require safety information on food labels	no.	4	8	46	82	249	389
	%	1.0	2.1	11.8	21.1	64.0	100.0
d. inspect livestock and poultry at slaughter for traces of chemicals or animal drugs	no.	2	5	24	77	289	397
	%	0.5	1.3	6.0	19.4	72.8	100.0
e. inspect livestock and poultry at slaughter for disease	no.	2	1	10	58	328	399
	%	0.5	0.3	2.5	14.5	82.2	100.0
f. inspect fruits and vegetables	no.	4	14	54	175	149	396
	%	1.0	3.5	13.6	44.2	37.6	99.9
g. inspect milk and dairy products	no.	2	2	32	113	240	389
	%	0.5	0.5	10.5	28.3	60.2	100.0
h. inspect grain and grain products	no.	3	17	69	168	139	396
	%	0.8	4.3	17.4	42.4	35.1	100.0
i. certify safety of food additives	no.	4	5	42	122	222	395
	%	1.0	1.3	10.6	30.9	56.2	100.0
j. certify safety of pesticides	no.	2	3	35	82	268	390
	%	0.5	0.8	9.0	21.0	68.7	100.0
k. certify safety of animal drugs	no.	3	7	40	104	243	397
	%	0.8	1.8	10.1	26.2	61.2	100.1
l. test for environmental contaminants in food	no.	2	4	38	106	247	397
	%	0.5	1.0	9.6	26.7	62.2	100.0
m. educate people to prepare and store food safely	no.	11	25	63	127	172	398
	%	2.8	6.3	15.8	31.9	43.2	100.0

Table 7. Response to Question 32

(Public information about food safety can come from many sources. For each of the types of food safety information listed, mark the one main source from which you are likely to receive information.)

Source	Type of food safety information																	
	Food Additives		animal drugs, or hormones used in livestock		pesticides used to kill insects or other pets on or near foods		food exposed to environmental contaminants (lead, mercury, dioxin)		Food preservation (canning, freezing, drying)		cooking and handling of meat and poultry		shelf storage of canned foods		refrigerator storage of meats and poultry products		federal and state agencies role in food safety	
	rank	%	rank	%	rank	%	rank	%	rank	%	rank	%	rank	%	rank	%	rank	%
Government	1	34.1	2	25.4	2	20.1	1	32.9	6	4.7	6	4.7	7	4.5	6	4.8	1	38.5
University	8	2.4	5	8.6	7	6.5	8	1.5	8	2.6	8	3.5	8	3.3	8	2.1	5	5.2
County extension agent or home economist	7	3.8	6	7.4	3	14.5	6	4.2	1	49.6	1	33.9	1	26.8	1	32.3	6	4.2
Food industry or company	2	19.4	8	3.6	8	2.9	7	3.3	4	6.4	4	8.6	3	15.7	5	10.1	9	0.0
Consumer group	4	11.8	4	8.9	5	9.2	4	9.5	7	3.3	7	4.1	6	6.5	7	3.9	6	4.2
Popular media personality	5	7.9	3	10.9	4	13.3	3	19.0	9	2.0	9	2.7	9	2.1	9	1.8	3	10.9
Family or friend	9	1.2	9	1.8	9	1.5	9	0.0	3	11.7	2	20.0	4	12.8	2	17.1	8	0.3
Source not listed above	6	6.5	7	7.1	6	7.7	5	7.1	5	5.9	4	8.6	5	6.8	4	10.8	4	5.5
I don't receive this information	3	12.9	1	26.3	1	24.3	2	22.5	2	13.8	3	13.9	2	21.4	2	17.1	2	31.2
Total	n=340	100.0	n=338	100.0	n=338	100.0	n=337	100.0	n=341	100.0	n=339	100.0	n=336	99.9	n=334	100.0	n=330	100.0

information. To the extent that extension can be a more visible supplier of various types of food safety information, consumers appear to accept its credibility.

Although labels are rated highly by consumers for convenience, much more research remains to be done on the appropriate uses of labels to transmit food safety information. Label information has been categorized in a number of ways: warnings, lists of ingredients, use instructions, risk-benefit information, and safety certification. Future research on consumer food choice, risk perceptions and preferences, and information processing should integrate consumer surveys with more empirical analysis of consumer market behavior. One policy issue potentially illuminable is the degree of correspondence of stated preferences for "safe" or healthful foods with actual purchase patterns.

BIBLIOGRAPHY

Cooper, Richard M., 1980, A Time To Warn and A Time To Ban, Product Labeling and Health Risks, Banbury Report 6, Cold Spring Harbor Laboratory.

Green, Sidney, 1985, Food and Health-Related Research Needs, *The American Food Supply: Are We At Risk?*, ICET Symposium II, Cornell University.

Hopper, Paul F., 1985, The Policy Issues of Risk Management, *The American Food Supply: Are We At Risk?*, ICET Symposium II, Cornell University.

Kramer, Carol S., 1982, An Economic Analysis of Food Safety Regulation: The Case of Sulfa and the Swine Subsector, unpublished Ph.D. dissertation, Michigan State University, East Lansing, Michigan.

Kramer, Carol S., 1986, Food Safety Policy Analysis: Analytical Methods, *Proceedings of the American Council on Consumer Interests*, St. Louis: American Council on Consumer Interest.

Lowrance, William W., 1976, *Of Acceptable Risk: Science and the Determination of Safety*, Wm. Kaufman, Los Altos, CA.

Mazis, Michael B., 1980, An Overview of Product Labeling and Health Risks, Product Labeling and Health Risks, Banbury Report 6, Cold Springs Harbor.

Penner, Karen P., Kramer, Carol S., and Frantz, Gary Lee, 1985, Consumer Food Safety Perceptions: A Survey of Consumers in Eight Kansas Counties, Kansas State University Cooperative Extension Service, Manhattan, Kansas.

Pope, Rulon, 1985, The Kansas Food Safety Survey: An Empirical Analysis, Draft report prepared for the Economic Research Service, U. S. Department of Agriculture (unpublished).

Roberts, Howard R., 1981, *Food Safety*, Wiley, New York.

Schell, Orville, 1984, *Modern Meat: Antibiotics, Hormones, and the Pharmaceutical Farm*, Random House, New York.

Schmidt, A. M., 1976, Food and Drug Laws of the United States, in *Food Quality and Safety: A Century of Progress*, Proceedings of the Symposium Celebrating the Centenary of the Sale of Food and Drugs Act 1875, Ministry of Agriculture, Fisheries and Foods, London.

Schultz, William B., 1980, Labels, Bans, and Consumer Preferences, Product Labeling and Health Risks, Banbury Report 6, Cold Springs Harbor.

Schwartz, Steven P., 1980, Consumer Attitudes toward Product Labeling, Product Labeling and Health Risks, Banbury Report 6, Cold Springs Harbor.

Staelin, Richard, 1980, Appliance Performance Labeling and Point of Purchase Information: The Results of Three Experiments, Product Labeling and Health Risks, Banbury Report 6, Cold Springs Harbor.

Stokes, Raymond C., 1980, Consumer Research On Food Label Information, Product Labeling and Health Risks, Banbury Report 6, Cold Springs Harbor.

Sporleder, Thomas L., Kramer, Carol S., Epp, Donald J., 1983, Food Safety, in *Federal Marketing Programs in Agriculture: Issues and Options*, Armbruster, Henderson and Knutson, eds., Interstate, Danville, IL.

U.S. Department of Agriculture, Economic Research Service, 1975, Consumer Satisfaction with Food Products and Marketing Services, Agricultural Economic Report No. 281, March.
U.S. Department of Agriculture, Economic Research Service, 1977, Food Safety: Homemakers' Attitudes and Practices, Agricultural Economic Report No. 360, January.
U.S. Department of Agriculture, Statistical Reporting Service, 1969, Homemakers' Opinions About Selected Meats: A Nationwide Survey, Marketing Research Report No. 854, July.
Wodicka, Virgil, 1985, Prioritizing Risks Associated with the Food Supply, in *The American Food Supply: Are We At Risk?* ICET Symposium II, Cornell University.
Wolf, Isabel, 1985,"Consumer Concerns and Expectations About Food Risks, *The American Food Supply: Are We At Risk?* ICET Symposium II, Cornell University.

Estimating Fatality Reductions from Increased Safety Belt Use[a]

Leonard Evans
General Motors Research Laboratories
Warren, MI

SUMMARY

This research was performed to calculate expected fatality reductions from increases in safety belt use which take into account that drivers who change from being nonusers to users have lower accident involvement rates than the remaining nonusers, a process referred to as "selective recruitment." The research relies on a recent quantitative determination of selective recruitment effects which found that drivers who do not wear belts had crash involvement rates about 50% higher than those who do. Other factors relevant to belt use and fatality reductions are also discussed in the present paper.

Analytical functions are derived which express expected fatality reductions in terms of changes in safety belt use rates from an initial rate. The function parameters are determined by requiring that computed average crash rates for nonusers be 1.53 times the rates for users, the recently determined empirical value. Typical results from the calculations are given below. By "benefit reduction" is meant that the estimated fatality reduction calculated by including selective recruiting is less than that estimated by a simple calculation which ignores selective recruiting.

	Safety belt use changes from	to	Fatality Reductions (%) Ignoring selective recruitment	Including selective recruitment	Difference
Largest possible benefit reduction	0%	58%	24.9	19.6	5.3
Largest possible benefit increase	70%	100%	18.5	23.8	−5.3
Small use increase from low base	14%	24%	4.6	3.2	1.4
Large use increase from low base	14%	96%	37.5	37.7	−0.2
Small use increase from high base	80%	90%	6.6	8.5	−1.9

a. Only the summary of this paper is published here, since the full paper has been published in *Risk Analysis* 7:49-57 (1987).

It is concluded that:

- Selective recruitment may *increase* or *decrease* expected fatality reductions.
- Effects are relatively small, in no case exceeding ±5.3%.

Are Special Trains Really Safer?

A. B. Boghani
Arthur D. Little, Inc.
Cambridge, MA

ABSTRACT

Certain hazardous materials can occasionally be transported by special trains composed of tank cars containing that material and buffer cars in front and rear. Such a train can be handled with special care by taking actions such as enforcing speed restrictions, establishing priority in passing, having a track inspection car in front and eliminating the need to go through classification yards. This paper compares risks involved in the shipment of hazardous material by special trains and by regular trains. The impact on comparative risks of changing the number of cars in a special train, the number in regular train, track quality, and puncture probability is also estimated.

KEYWORDS: Risk analysis, rail, hazardous material, special train

INTRODUCTION

Often, while arranging for the shipment of hazardous material by rail, the shipper is faced with two options:

- to allow the rail cars carrying hazardous material to go by the regular trains, or
- to arrange for the rail cars to be carried by a special train which incorporates several buffer cars in front and behind the "cut" (i.e., a group of cars) carrying hazardous material.

The second alternative is like putting your eggs in one basket and then guarding it. These special trains have certain privileges:

- they are not allowed to exceed a certain speed limit (say, 35 mph),
- they do not go through any yards or sidings, and
- a track inspection car checks the track prior to the special train's passage.

There are two contradictory effects at work when special trains are used to transport hazardous material instead of regular trains. The frequency of train accident will be reduced because of all the precautions taken. However, when an accident does happen, it

New Risks
Edited by L. A. Cox, Jr., and P. F. Ricci
Plenum Press, New York, 1990

will almost certainly involve the shipper's hazardous material cars (let us call them the "client" cars) because there are not many other cars in the train.

This study looked at the relative effects of these two factors to determine if the special trains truly reduce the incidence of hazardous material spill and if they do, by how much. The first part of the paper briefly describes the assumptions made in quantifying spill frequency for the two types of train. Next, the estimates of spill frequency and spill size for "base" case special and regular trains are shown. Finally, the effects of various parameters on the relative risks posed by the two train types are evaluated. In performing this study, a computer program, Risk.Pas, developed by Arthur D. Little, was used.

QUANTIFYING THE RISKS

The following factors need to be considered to determine the relative risks posed by the two train types:

- derailment and collision frequencies, fraction of cars derailing, probability of a client car derailing, puncture probability, and expected spill size.

Derailment and Collision Frequencies

The derailment and collision frequencies for yards and sidings were treated separately from those for mainline track. For yards the accident frequency is obtained in terms of accidents per visit of a car[1]; for sidings the measure is accidents per train visit; for the track, train-miles is assumed to be the exposure measure and the frequency is obtained in terms of accidents per train-miles.

Yard Accident Frequencies. In 1983, there were 19 million car loads carried by the railroads in the U.S., of which 6.5 million were of coal and grain. Since almost all coal and grain cars move in unit trains, only 12.5 million car loads must have gone through yards. Assuming six yard visits per car load (three loaded, three empty), we get

$$\text{Total yard visits} = 12.5 \times 106 \times 6 = 75 \text{ million.}$$

Now, in 1983 there were 1,230 derailments, 296 collisions, and 169 other accidents in yards.[2] Therefore, the frequencies of accidents in a yard are:

derailment = 1,230/75 = 16.4/million car visits;
collision = 296/75 = 4.0/million car visits;
other = 169/75 = 2.2/million car visits.

For the study a composite "derailment" rate of 20.4/million car visits is used.

Siding Accident Frequencies. According to a cooperating railroad, a train on a 750 mile run will on an average go on a siding six times. Assuming the same ratio for all mainline train-miles, the total number of visits to sidings are

$$\text{Total visits to sidings in 1983} = \frac{6}{750} \times 408.17 \times 10^6$$

$$= 3.26 \times 10^6$$

Here, the total train-miles in 1983 are assumed to be 408.17 million.[2] There were 25 collisions, 288 derailments and 31 other accidents on sidings, industry or unknown track type.[2] Assuming that one-third of these accidents took place on sidings, the frequencies of accidents on sidings will be

derailments	= 288/3/3.26	= 29.4/million train visits;
collision	= 25/3/3.26	= 2.6/million train visits;
other accidents	= 31/3/3.26	= 3.22/million train visits;

A composite rate of 32.0 "derailments" per million train visits has been used for the study.

Mainline Accident Frequencies. As mentioned earlier, the total number of mainline train-miles in 1983 was 408.17 million, while the number of derailments, collisions, and other accidents on the mainline track were 1486, 97, and 284, respectively. This means that the mainline accident rates were

derailment = 3.64/million train-miles
collision = 0.237/million train-miles
other = 0.695/million train-miles

The tank cars (used for most hazardous commodities) and boxcars (used as buffers for special trains) are as likely to derail as an average car; thus these rates can be assumed to be valid for a regular train as well as for a special train if both operate in the same manner.[3] However, a special train does not exceed 35 mph. Therefore, the accident rates have to be derived for two different speed requirements — above 35 mph and below that limit. This is done in Table 1.

As shown in Table 1, most of the accidents take place below a 35 mph speed. If we assume that a train would operate above 35 mph on a "good" track and below on a "fair" track, then most accidents take place on a fair track. The total exposure for calculating accident rates is 408.17 million train-miles. Next, it is assumed that 50 percent of this exposure takes place below 35 mph and 50 percent above.[a] Using this estimate, the accident rates on good track and fair track for the regular train can be calculated as shown in Table 1. A composite rate of 1.1 "derailments" per million train-miles on good track, and 6.6 "derailments" per million train-miles on fair track is used in the study.

For the special train, the accident rate caused by human factors is assumed to drop by 50 percent due to better training of the train operators. Since this train will not operate above 35 mph and the cars will be better maintained, the equipment caused accident rate is assumed to drop by 75 percent for good track, and by 25 percent for fair track. After analyzing Federal Railroad Administration (FRA) data,[2] it is estimated that having a track inspection car operating in front of the special train will reduce the track caused accident rate at most by 50 percent, but more likely by 30 percent. It will also reduce the accident rate due to other causes by about 25 percent. Distributing these overall accident rate reductions to the three categories of collision, derailment and "other" accident rates is a bit tricky. The relative proportions of these different types of accidents for each accident cause (track, equipment, human factors and others) was determined from FRA data.[2] This information was used to arrive at the accident rates for the special train shown in Table 1. The composite rates used in the study are 4.49 "derailments" on fair track and 0.315 "derailments" on good track per million train-miles.

---

a. Battelle had assumed a one-third, two-third split in their report on special shipment of radioactive material.[4] However, speeds have gone up substantially in the ten years since those data were obtained.

Table 1. Effect of Speed on Accident Rates

(a) Number of Mainline Accidents in 1983

Speed	Total Accidents	Number of Collisions	Number of Derailments	Number of Other Accidents	Track Caused	Equipment Caused	Human Factors Caused	Other Causes
Below 35 mph	1524	87	1272	165	635	315	299	275
Above 35 mph	343	10	214	119	65	145	29	104

(b) Accident Rates in Events/Million-Train Miles

Track Type	All Accidents	Collisions	Derailments	Others	Track Caused	Equipment Caused	Human Factors Caused	Other Causes
				Regular Train				
Fair	7.47	0.43	6.23	0.81	3.11	1.54	1.47	1.35
Good	1.68	0.05	1.05	0.58	0.32	0.71	0.14	0.51
				Special Train				
Fair	5.07	0.14	4.35	0.58	2.18	1.16	0.73	1.00
Good	0.85	0.005	0.31	0.53	0.22	0.18	0.07	0.38

One point to note is that the passing restrictions (i.e., the train being passed is completely stopped) which are applied to a special train will not affect the accident rates.[4] The same is assumed to be true in this analysis.

Fraction of Cars Derailing

According to a report published by the Federal Railroad Administration, the fraction of cars derailing depends on the speed of the train at the time of the accident and on the length of the train, among other factors.[5] From the data provided by this report, the fraction of cars derailing in an accident on a good track (speed more than 41 mph) is estimated to be 0.13, on a fair track (speed less than 40 mph) to be 0.11 and on the siding to be 0.07. For a special train, the effect of small size will offset the effect of speed (which is "average"). Thus, a fraction of 0.22 for both good and fair track is assumed, based on the data provided by the FRA report.

The data for collisions indicate that, on an average, a collision results in a fraction of cars derailing that is similar to that for a derailment.[5] Therefore, the collision and derailment accidents are treated identically.

Probability of a Client Car Derailing

Generally, the derailment will involve a string of cars. Thus all cars that derail will be sequential and located in one part of the train. Two other assumptions are made:

- all client cars are in a cut (they are together), and
- they are located in the middle of the train (which is a conservative assumption to make).

Then the following estimates hold true:

Total number of cars in the train = N_T
Number of client cars in the train = N_B
Total number of cars that derail = N_D (= fraction multiplied by N_T)
Number of ways the derailment can take place = $N_T - N_D + 1$
Maximum number of client cars that can derail = M = smaller of N_D or N_B

Conditional probability of one client car derailing = $P_q(1) = \dfrac{2}{N_T - N_D + 1}$

Conditional probability of two client cars derailing = $P_q(2) = \dfrac{2}{N_T - N_D + 1}$

Conditional probability of $M - 1$ client cars derailing = $P_q(M-1) = \dfrac{2}{N_T - N_D + 1}$

Conditional probability of M client cars derailing = $P_q(M) = \dfrac{|(N_D - N_B)| + 1}{N_T - N_D + 1}$

The above equations are valid only if there are as many as or more than $N_D - 1$ other (non client) cars in the front and rear of the train. If that is not so, then the minimum number of client cars that can derail will be larger than one. If there are N_F cars in the front and rear (the client cars are assumed to be in the middle) then the minimum number of client cars that can derail is equal to $N_D - N_F$, and the conditional probabilities of derailment of fewer cars than that equal zero. As an extreme example, if the train is a unit train carrying only the client's cars, then

conditional probability of M client cars derailing = 1.0, and
conditional probability of less than M client cars derailing = 0.0

One final deviation from the above occurs if the number of other cars (total number of cars in the train minus the number of client's cars) is odd and, therefore, the number of other cars in front does not equal those in the rear. If coupled in that situation, the number of cars in the front or rear (whichever is larger) is less than or equal to $N_D - 1$, and a correction has to be made in the conditional probability of the lowest number of client cars derailing. This conditional probability will, in that case, be equal to $1/(N_T - N_D + 1)$ instead of twice that amount.

Puncture Probability

In order to estimate the puncture probability for cars carrying hazardous materials, the accident history of 105 cars was obtained from Ref. 6 (these data were gathered before the cars were equipped with head shields or shelf couplers):

number of cars derailed in a specified period = 213;
number sustaining head puncture = 13;
number sustaining partial or total loss of lading = 39;
percent of cars losing lading = 18.3%.

The percent of cars losing lading will reduce if shelf couplers and head shields are incorporated. If we subtract the number of cars suffering head puncture (which will be almost eliminated by head shields/shelf couplers), the average conditional probability of a puncture of a car with head shields/shelf couplers, given that it has derailed, can be estimated to be

average conditional probability of puncture = (39–13)/213 = 0.12.

This is consistent with the 1983 data in which 11.4 percent of cars carrying hazardous material damaged in an accident released their contents.[2] Presumably many, if not most, of these cars were required to have head shields and shelf couplers.

The average probability of a puncture of cars carrying hazardous material is 0.167.[1] Thus the cars incorporating head shields/shelf couplers are 28 percent safer. (This is probably a conservative estimate; however, the conclusions of this paper will not be affected by changes in this estimate.)

Nayak *et al.* (1983) provide the relationship between speed of derailment and release probability from which the spill probability for various cases is estimated as shown in Table 2. Since the special train will operate below 35 mph, the puncture probabilities is assumed to be 0.25 for an average car in such a train and 0.18 for cars incorporating head shields/shelf couplers.

Now, a derailment can involve more than one client car. In this case, the calculations are as follows:

Define P_{fq} = Probability of f cars spilling out of q cars of client which have derailed.

Then $P_{fq} = {}^{q}C_{f}(P_p)^f(1-P_p)^{q-f}$

Table 2. Estimates of Release Probability

Derailment Location	Estimated Speed	Average Car Release Probability	Release Probability for Car Incorporating Head Shields/ Shelf Couplers
For Regular or Unit Train			
Good track	>35 mph	0.3	0.216
Fair track	≤35 mph	0.25	0.18
Yard, siding	<10 mph	0.1	0.072
For Special Train			
Good/fair track	≤35 mph	0.25	0.18

Where P_p = conditional probability of a derailed car releasing its contents (given in Table 2).

Adding up the values of P_{fq} for $f = 1$ to q will give the conditional probability of one or more client cars suffering a spill given q cars have derailed.

Another way of presenting the results is to derive the conditional probability of a spill, given that the *train* has derailed. This probability, P_s, can be calculated as:

$$P_s = \sum_{q=1}^{q=M} \left(P_q \sum_{f=1}^{f=q} P_{fq}\right)$$

where

P_q = conditional probability of derailment of q client's cars given that the train has derailed, and

M = maximum number of client's cars that can derail.

Expected Spill Size

If it is assumed that a client's tank car suffering a puncture will lose a fraction c of its contents, a spill involving f punctured cars will result in a spill of f multiplied by c tank car loads. When this is multiplied by the appropriate conditional probability values, we get

$$\text{Spill size} = \left(\sum_{q=1}^{q=M} \sum_{f=1}^{f=q} P_q . P_{fq} . f . c\right) / P_s$$

where P_s = conditional probability of a spill given that the train has derailed.

BASE CASE ANALYSIS

A computer program (called Risk.Pas) was developed to calculate estimated spill frequency and spill size for regular and special trains providing shipment to the same number of tank cars per year. The following base case values are used:

	Regular Train	Special Train
Total number of cars in train	100	17
Number of client cars in train	1	5
Number of trips per year	50	10
Miles of good track	750	750
Miles of fair track	250	250
Number of yards visited per trip	3	0
Number of sidings visited per trip	8	0

It is further assumed that the tank cars incorporate shelf couplers and head shields.

The results of the risk analysis performed using the computer program are:

	Regular Train	Special Train
Estimated number of client car spills per year	0.00342	0.00292
Average spill size (tank car)	1.0	1.20

The results show that, for the base case, the special train indeed reduces the likelihood of spill. But this 14.6% reduction in spill frequency is accompanied by a 20% increase in spill size. Also the reduction in spill frequency does not seem that significant.

The ultimate decision on whether or not to use the special train will depend on a trade-off study involving spill frequency, spill size and the cost differential between the two options.

The advantages of using a special train depend on a number of parameters as discussed next.

PARAMETRIC ANALYSIS

In order to determine the effects of various parameters on relative safety provided by special trains, several cases were run with changes in:

- Number of cars in special train,
- Percentage of good vs. fair track,
- Cars with or without head shields/shelf couplers (i.e., different conditional probabilities of puncture), and
- Number of cars in regular train.

While making these changes, the amount of hazardous material shipped was kept constant. Figure 1 shows the effects of the number of cars in special trains on relative safety. Two measures of relative safety are used:

$$\text{Spill frequency ratio} = \frac{\text{Spill frequency (per year) with special train}}{\text{Spill frequency with regular train}}$$

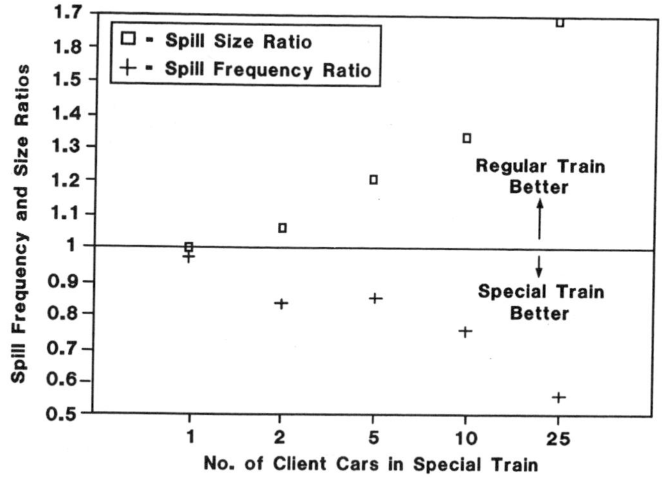

Fig. 1. Special train vs. regular train: effect of number of client cars in special train.

$$\text{Spill size ratio} = \frac{\text{Avg. spill size with special train}}{\text{Avg. spill size with regular train}}$$

As can be seen, the spill frequency ratio reduces if the number of cars in the special train is increased from one to two, but *increases* slightly if the number is increased from two to five. Beyond five, the ratio starts reducing again.

This type of behavior is explained by the fact that two things happen as the number of cars in a special train is increased:

- fewer trips are needed to ship the given amount of material (which reduces the spill frequency), and
- the likelihood of spill per train derailment increases (which increases the spill frequency).

The relative effects of these two changes determine the extent of improvement as the number of client cars in special trains is increased.

Figure 2 plots the spill frequency and spill size ratios for different percentages of good track. The other parameters for the regular and special trains are kept the same as in the base case (i.e., one client car in regular train vs. five in special train). The figure shows that the spill frequency ratio reduces as the percentage of good track increases, while the spill size ratio remains the same. The reason for this is that as the percentage of good track is increased, the safety effects of speed reduction for a special train become more and more important, since the regular train will now spend more and more time at speeds substantially higher than the special train. As can be expected, the spill size ratio is unaffected by these changes.

What is interesting is that for percentage of good track less than fifty, the spill frequency ratio is more than one, or the special trains are riskier than the regular train. This is because under those circumstances the effect of higher conditional probability of spill, given an accident, overpowers the effect of lower spill frequency due to slower speed.

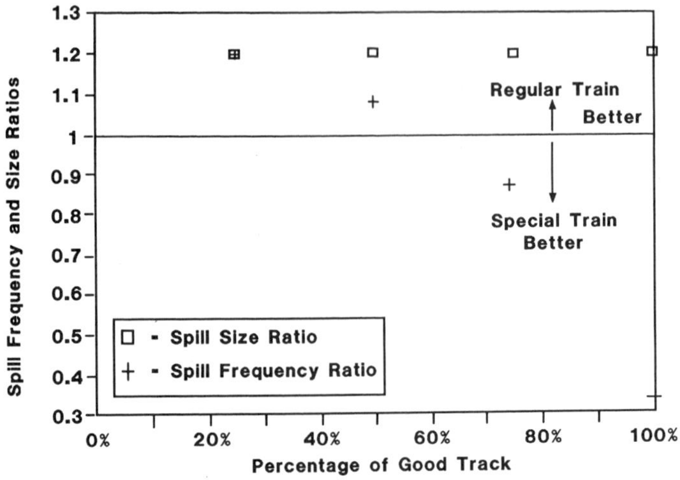

Fig. 2. Special train vs. regular train: effect of track quality.

Next, the effect of headshields/shelf couplers is considered. Note that in the base case the cars had these protective devices. If they are removed, the spill frequency ratio reduces and the spill size ratio increases as shown in Fig. 3. Note that in either case, the spill frequency ratio is smaller than one (i.e., the special train is safer) and the spill size ratio is greater than one (i.e., the spill when it occurs, will be larger if a special train is used). Thus the presence of protective devices does not affect the relative merits of special train.

Figure 4 shows the effects of increasing the number of client cars in a regular train, while keeping the number of cars in a special train the same (five). As can be seen, the spill frequency ratio increases as the number of cars in regular train is increased, while the spill size ratio reduces. One interesting point to note is when the number of cars in regular train equals that in the special train, the spill frequency (on a yearly basis) will be lower in the regular train than in the special train. This is because the effect of lower conditional probability of spill in the regular train (given that an accident has happened) overpowers the effect of increased frequency of accident. (The number of trips is the same for both.) The spill size, however, will be larger for the regular train, although it carries the same number of cars as the special train. This is because of higher speeds at which the accidents involving the regular train will occur.

CONCLUSIONS

The study reported in this paper shows that the use of a special train instead of a regular train to ship hazardous commodities may lead to a reduction in the spill frequency. However, such a reduction is accompanied by an increased spill size when the spill does occur.

The extent of reduction in spill frequency by using a special train is related to the number of cars in the special train. Generally, an increase in the number of cars in a special train leads to a reduction in the spill frequency (as related to a regular train); however, there are some exceptions. For the cases studied, an increase in the number of cars from two to five increases the spill frequency instead of reducing it, although it is still lower than that for the regular train.

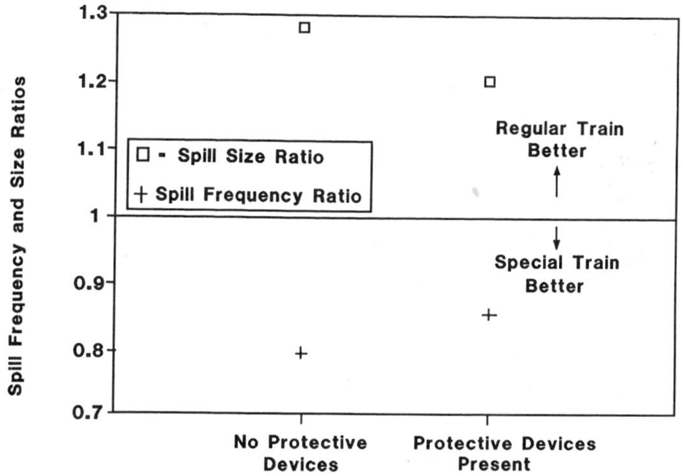

Fig. 3. Special train vs. regular train: effect of protective devices.

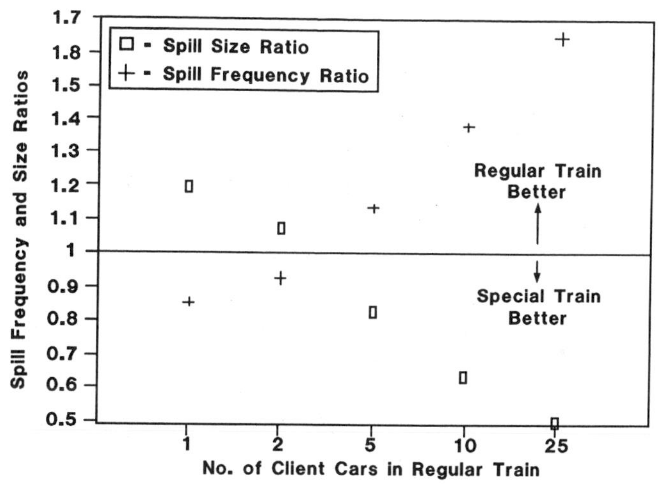

Fig. 4. Special train vs. regular train: effect of number of client cars in regular train.

The use of a special train becomes increasingly attractive when a larger and larger portion of track is good. However, the reverse is true for an increasingly worse track. On a truly bad track, it is likely that using a special train may lead to an increase in both spill frequency and spill size, relative to shipment by regular trains. The advantages of using a special train are greater when the cars do not have protective devices than when they do. However, this effect does not seem to be very significant.

Finally, the number of cars carried in a regular train also affects the relative advantages of using a special train. In fact, if the number of cars in a regular train equals that in a special train, the special train is worse than the regular train (i.e., the spill frequency is higher). However, the spill size in that case is smaller with a special train.

Special trains are not always better than regular trains in carrying hazardous materials. Each case should be studied and alternatives explored to ensure that the extra cost involved in shipping by a special train is worthwhile.

REFERENCES

1. P. R. Nayak, D. R. Rosenfield, and J. H. Hagopian, Event Probabilities and Impact Zones for Hazardous Materials Accidents on Railroads, prepared by Arthur D. Little, Inc., for the Federal Railroad Administration, DOT/FRA/ORD/-83/20, November 1983.
2. Accident/Incident Bulletin, No. 152, Calendar Year 1983, Federal Railroad Administration, June 1984.
3. Issues and Dimensions of Freight Car Size, U.S. Department of Transportation, November 1979.
4. R. E. Rhoads *et al.*, A Safety and Economic Study of Special Trains for Shipment of Spent Fuel, prepared by Battelle for U.S. DOE, Report No. BNWL-2263, December 1977.
5. P. R. Nayak, An Analysis of the Probability of Derailment with a Train Consist, prepared by Arthur D. Little, Inc., for the Transportation Systems Center, December 1979.
6. Anon., 105A Tank Car Study, prepared by Dynatrend, Inc., for the Transportation Systems Center, January 1980.

Comparison of Risks from Ocean-Based and Land-Based Incineration of Hazardous Wastes

Michael T. Huguenin
Industrial Economics, Inc.
Cambridge, MA

Jeffrey A. Kolb
Sobotka and Company, Inc.
Washington, DC

ABSTRACT

To evaluate the relative human health and environmental risks from ocean-based versus land-based incineration, we estimate the expected release of waste or hazardous by-products for each component of the two systems. We then evaluate the most important differences between the systems: human health risks resulting from continuous incinerator releases and the likelihood and consequences of waste spills during ocean transport. We find that human health risks from continuous incinerator releases are lower for the ocean-based system. However, the ocean-based system creates the remote possibility of a marine waste spill which could, under some circumstances, cause major environmental damage.

KEYWORDS: Hazardous waste, ocean incineration, environmental risk assessment, human health risk assessment

INTRODUCTION

This paper summarizes a number of analyses undertaken to evaluate the relative risks of incinerating hazardous wastes on land versus at sea. The work reported here is part of a larger effort undertaken by the U.S. Environmental Protection Agency (EPA) to assess incineration as a treatment option for hazardous wastes.[a]

ANALYTICAL APPROACH

A comparative risk assessment of ocean- and land-based incineration systems is complicated because the systems have different components, and releases of waste from

---

a. For a complete report of this work, see *Comparison of Risks from Land-Based And Ocean-Based Incineration* prepared for the Office of Policy Analysis, U.S. Environmental Protection Agency by Industrial Economics, Inc. in association with Applied Science Associates, Inc., Arthur D. Little, Inc., and Engineering Computer Optecnomics, Inc., 15 March 1985.

New Risks
Edited by L. A. Cox, Jr., and P. F. Ricci
Plenum Press, New York, 1990

each system affect different locations and ecosystems. The land- and ocean-based systems have three similar components:

1. transport of wastes by truck from the generator site to the incinerator or pier,

2. transfer and storage operations at the land-based incinerator or pier, and

3. incineration of the waste.

However, the ocean-based system has one additional component—transport of the wastes by ship from the pier to the burn zone—which has been the focus of much of the public's concern with the technology.

We develop estimates of releases of hazardous wastes and by-products for each of these components. Some releases are continuous, e.g., stack emissions, whereas others are episodic and occur with some probability, e.g., spills. We then develop estimates of human health effects from incinerator emissions, and the risk and consequences of spills during ocean transport of wastes to the burn zone. These are not only the most significant effects but also those that differ between the two incineration systems.

The land- and ocean-based incineration systems analyzed represent actual or proposed systems. The operation proposed by Chemical Waste Management, Inc., using the vessel Vulcanus II operating out of Mobile, Alabama, and burning wastes in the Gulf of Mexico burn zone, is used as the ocean-based incineration system. Two typical incinerators located in Deer Park, Texas and El Dorado, Arkansas are used to represent land-based systems.

We analyze two waste streams:

1. A waste containing 35 percent by weight of polychlorinated biphenyls (PCBs). The remaining 65 percent of the waste is assumed to be non-hazardous. Each system is assumed to burn 56,000 metric tons of this waste each year.

2. A waste containing 50 percent by weight of ethylene dichloride (EDC) and 50 percent non-hazardous substances. Each system is assumed to burn 68,400 metric tons of this waste each year.

In addition, both waste streams are assumed to include 100 ppm each of arsenic, cadmium, chromium, and nickel. Although all four metals have been found in a variety of actual waste streams, it is likely that this assumption overstates the average total concentration of carcinogenic metals in liquid incinerable wastes.

EXPECTED RELEASES

Tables 1 and 2 present the quantities of waste expected to be released from each system component for the PCB and EDC wastes, respectively. For releases due to accidents, spills, and other uncertain events, these estimates represent long-term average releases which include years with no release and years with one or more releasing events. Releases from the land transport and transfer and storage components of both systems are similar (between 3 and 4 metric tons per year). These releases depend on the average distance wastes are transported, the types of equipment and roads used, and so forth. We assumed a land transport distance of 250 miles over average roads in average weather, use of 5000 gallon tank trucks, a single handling step at the land-based incinerator or the pier, and temporary storage in floating roof tanks.

Table 1. Summary of Expected Quantities Released per Year, PCB Waste (metric tons per year)

Component	Ocean-based System	Land-based System
Land Transportation	2.1	2.1
Transfer and Storage	1.2	1.1
Ocean Transportation	0.6	--
Subtotal (Transport)	3.9	3.2
Incineration		
Undestroyed Wastes	0.1	0.1
PICs	0.0	0.0
Metals	22.4	4.5
Subtotal (Stack)	22.5	4.6
Scrubber Effluent		
Metals	--	17.9
Total Organics & Metals	26.4	25.7

Table 2. Summary of Expected Quantities Released per Year, EDC Waste (metric tons per year)

Component	Ocean-based System	Land-based System
Land Transportation	2.7	2.7
Transfer and Storage	1.2	1.1
Ocean Transportation	0.8	--
Subtotal	4.7	3.8
Incineration		
Undestroyed Wastes	6.8	6.8
PICs	20.6	0.6
Metals	27.4	5.5
Subtotal (Stack)	54.8	12.9
Scrubber Effluent		
Metals	--	21.9
Total Organics & Metals	59.5	38.6

Our estimates of releases for land transport are based on U.S. Department of Transportation (DOT) data on tank truck spill frequencies and sizes. These data likely overstate spill rates for hazardous waste carriers because of differences in tank truck design and safety procedures. Our estimates of releases from transfer and storage are based on additional DOT data and on an engineering analysis by Arthur D. Little, Inc. of spills and fugitive releases from storage facilities.

Ocean transportation releases are unique to the ocean-based system and are estimated to average 0.6 and 0.8 metric tons per year for the PCB and EDC wastes, respectively. While these releases are small on an expected value basis, they are comprised of low probability spills of significant quantities of waste which could have important consequences. The methods used to estimate these releases and their possible effects will be discussed later in this paper. We first turn our attention to releases from incineration and the potential human health effects resulting from these releases.

ANALYSIS OF INCINERATOR EMISSIONS

Incineration is the major cause of emissions in both systems. Undestroyed wastes are a significant release for the EDC waste but are small for the PCB waste. This is due to our assumed 99.9999 percent destruction of the PCB waste and 99.99 percent destruction for the EDC waste consistent with environmental permit requirements.

Metals account for the largest releases. Our estimates of metals emissions result from our assumptions about metals concentrations in the waste. Metals are not destroyed or transformed by either system, but are transferred to effluent in the land-based case by use of a scrubber which is assumed to remove from 50 to 90 percent of the four metals considered. Thus, stack emissions of metals in the land case are about 20 percent of those predicted for the ocean-based case.

Products of incomplete combustion (PICs) are not expected in significant quantities for the PCB waste based on EPA's trial burn data. However, the results of these trial burns are subject to large uncertainty because of the procedures used and the limited number of PIC compounds that were considered. PIC's could be generated in much larger quantities when the ocean-based system is burning a general organic waste like our EDC waste, but again, the PIC generation data are extremely uncertain. We do not know of any complete explanation for the lower level of PIC generation found for ocean-based PCB incineration and the higher level found for ocean incineration of general organic waste.

HUMAN HEALTH EFFECTS FROM INCINERATOR RELEASES

Our analysis of human health risks estimates the incremental risk of developing cancer for a hypothetical "most exposed individual" (MEI) who resides at the location of the highest air concentrations resulting from incinerator stack and transfer/storage fugitive releases. We chose to estimate risks to the most exposed individual in order to estimate the largest risks likely to be suffered by any person due to the releases considered. For the land-based system, the location of the MEI is based on census data, whereas for the ocean-based system the MEI is assumed to reside at that point on the coast where modelled concentrations are highest averaged over a year. All risk estimates assume 70 years of continuous exposure. In the ocean case different areas are affected by stack emissions (the coastline downwind from the burn zone) and transfer/storage fugitive releases (the area around the port), while for the land-based case the same area is affected.

Table 3 presents the incremental risk of developing cancer for the MEI due to land- and ocean-based incinerator releases. As shown, the incremental risk from land-based

Table 3. Incremental Cancer Risk to Most Exposed Individual from Incinerator Releases

	PCB Waste	EDC Waste
Ocean-based System		
Stack (coastline)	6.37E-7	1.06E-6
Fugitives (port)	2.02E-8	4.97E-10
Land-based System (average of two sites)		
Stack	2.74E-5	3.14E-5
Fugitives	7.05E-7	1.69E-8

releases is about three chances in one hundred thousand for the locations and wastes considered. Virtually all of the incremental risk is due to stack releases, with fugitive releases resulting in increased risks of less than one in a million. Incremental risks at the coastline for the ocean-based system range from one in one million to 6 in ten million. As shown, the risks from fugitives near the port facility are less than two per hundred million. Thus, the risks of the ocean-based system are smaller by one to two orders of magnitude than those from the land-based system.

The data and methods used to generate these incremental risk estimates are highly uncertain and tend to overestimate expected human health effects. Thus, the absolute risk levels indicated by these tables must be interpreted with caution. EPA has completed studies of incremental risks from other hazardous pollutant releases using similar methods with similar uncertainties and biases. For example, a recent study on toxic air pollutants found that, on average, individuals in the U.S. face incremental cancer risks of about 4 to 6 chances in ten thousand from toxic air pollutants.[b] Incremental risks to the persons most exposed by the incineration systems considered here would be one to three orders of magnitude lower.

SPILLS FROM INCINERATOR VESSELS

Incinerator ships have operated off the European coast in the North Sea since 1972. They have made over 320 voyages and have incinerated over 650,000 metric tons of hazardous wastes. No casualties such as collisions, groundings, rammings, or fires have occurred, nor have there been any spills from loading incinerator ships in port. Although these operations in Europe have established a good safety record, the number of voyages is too small to be used in directly estimating the probability of such ship casualties and spills. All that can be concluded is that there is no evidence that the operation of incineration ships or the physical incineration of hazardous wastes at sea constitutes a high risk activity.

In order to estimate the probability of spills for incineration ship operation, a much larger operating population and spill data base are required. We rely on a proprietary accident and spill data base entitled ECOTANK which is maintained by Engineering Computer Optecnomics, Inc. (ECO), of Annapolis, Maryland. This data base contains approximately 10,000 worldwide tank ship accidents and over 1,000 tank ship spill events

b. *The Magnitude and Nature of the Air Toxics Problem in the United States*, Draft Report, Offices of Air and Radiation and Policy, Planning and Evaluation, U.S. Environmental Protection Agency, 1984.

Table 4. Estimated Spill Rates per Voyage by Type of Casualty and Location

Location	Type of Casualty				
	Collisions	Groundings	Rammings	Non-Impact	All
Pier & Harbor	4.3E-6	2.2E-7	1.6E-7	1.8E-5	2.3E-5
Mobile Bay	1.9E-6	3.4E-6	1.4E-6	3.6E-7	7.1E-6
Coastal	6.4E-6	6.7E-6	2.4E-6	3.6E-6	1.9E-5
Burn Zone	4.5E-6	--	--	7.1E-6	1.2E-5
All Locations	1.7E-5	1.0E-5	4.0E-6	2.9E-5	6.0E-5

that occurred during the period from 1969 to 1982. ECO selected accidents and spill events for tank ships between 2,000 and 10,000 long tons. Ships in this category approximate the Vulcanus in size and also typically make short coastal voyages as opposed to long transoceanic voyages.

These data were used to estimate releasing vessel casualty or "spill" rates for four types of accidents at the following locations:

- Mobile Harbor and the pier,
- Mobile Bay,
- The Gulf coastal area, and
- The burn zone in the western Gulf of Mexico.

These spill rates were then adjusted for the design, location, and operating features for the Vulcanus that are expected to reduce the frequency of spills below that experienced worldwide by tank ships of similar size.

The results of ECO's calculations are summarized in Table 4. These data give spill rates (any size spill) for each location and type of casualty on a per voyage basis. The overall spill rate is quite low—about 6 per 100,000 voyages. This rate is about one order of magnitude less than the historic spill rate for tank ships worldwide.

ECO also estimated the probability that accidents resulting in a spill involve one, two, and three or more tanks of the vessel. Their estimates for impact type casualties (collisions, groundings, or rammings) are

- one tank — 80 percent,
- two tanks — 15 percent, and
- three or more tanks — 5 percent.

There is little information on which to base such estimates for non-impact casualties. Accordingly, the same distribution is assumed, but there obviously is uncertainty regarding this assumption. The data in Table 4, along with the assumptions on the percentage of

Table 5. Estimated Spill Rates per Voyage by Location and Number of Tanks Affected

	Number of Tanks Affected			
Location	One	Two	Three or More	Any Number
Pier & Harbor	1.8E-5	3.4E-6	1.1E-6	2.3E-5
Mobile Bay	5.7E-6	1.1E-6	3.6E-7	7.1E-6
Coastal	1.5E-5	2.9E-6	9.5E-7	1.9E-5
Burn Zone	9.3E-6	1.7E-6	5.8E-7	1.2E-5
Total	4.8E-5	9.0E-6	3.0E-6	6.0E-5

spills involving one, two, or three or more tanks, may be used to calculate the spill rates by location and number of tanks involved. Table 5 presents these estimates.

The Vulcanus would operate in a region of the country having substantial commerce in petroleum and hazardous materials. The volume of material transported by the Vulcanus would be a small fraction of shipments of such commodities in the Gulf—about 0.01 percent. Further, because the spill rate for the Vulcanus is less than for other vessels, the potential releases from the Vulcanus would be less than 0.002 percent of those from ongoing shipments of petroleum and hazardous materials in the Gulf.

POTENTIAL EFFECTS OF MARINE SPILLS

The effects of spills on the marine environment depend on the location and size of the spill, the type of material released, and the extent to which the spill is removed or contained. To develop an indication of the potential consequences of a spill, we asked Applied Science Associates, Inc. (ASA) to model the likely effects of spills in three locations—Mobile Bay, the continental shelf, and the burn zone. Spill consequences are summarized by changes in biomass levels and in bioconcentration levels for PCBs and EDC. For PCBs we consider both floating and sinking cases since this compound, although heavier than water, might float if entrained in lighter-than-water materials. For EDC we consider only a sinking case, which results in rapid diffusion since this compound is soluble in water. We modelled the effects of spills of one-half tank, two tanks and eight tanks. We assumed no actions were taken to contain and remove the spill.

ASA's analysis indicates that EDC spills would have relatively minor effects on the marine ecosystem. These small impacts are the result of EDC's rapid diffusion to low concentration levels and its relatively low toxicity to marine species. In addition, bioconcentration of EDC is not a significant phenomenon. The same results hold for all size spills.

In contrast, spills of PCBs are predicted to have major effects on the marine ecosystem. These effects are quite severe in Mobile Bay (substantial reduction in benthic species and large bioconcentration effects on fish and shrimp) but are less severe in the coastal and burn zone areas. Since PCBs are a persistent compound, such effects are expected to last for many years. Bioconcentration effects in commercial and recreational species would be of most concern in the Bay and coastal areas.

Estimating the effects of spills of persistent compounds such as PCBs in the marine environment is an imprecise science at best. Because of substantial uncertainties regarding the long term fate of PCBs in the marine environment and the biological mechanisms involved in the food web, the results of the modelling effort should be viewed as a general indication of potential effects rather than as a precise measure of those effects.

SUMMARY

The major differences in the risks posed by the ocean- and land-based systems are

- the human health risks from the ocean-based system are lower than the risks from land-based systems because of its greater distance from populated areas; but
- there is a small risk of a marine spill, unique to the ocean-based system, which could have minor or major effects depending on the type of waste and the location of the spill.

We conclude that risks are low for each system and that neither system is clearly preferable to the other.

Carcinogenesis Risk Assessment of Two-Carbon Alkylating Agents Using Dynamic Simulation of Absorption and Metabolism

Seth Tuler and Dale Hattis
Massachusetts Institute of Technology
Cambridge, MA

ABSTRACT

This paper describes a methodology used to incorporate specific mechanism-related information into carcinogenesis risk assessment. We are developing a coordinated series of pharmacokinetic models to aid in quantitative risk assessments of inhalation exposure to two-carbon alkylating agents. We have applied our system to perchloroethylene and ethylene oxide. Human, rat, and mouse multi-compartment pharmacokinetic models were developed on the Apple Macintosh using the STELLA modeling system. The interactive and easy to use STELLA provides a powerful tool in the study of dynamic behavior and underlying feedback mechanisms of a structure. The models allow the systematic exploration of the causal relationships between external ambient exposure and steps in the pathway to carcinogenesis, including tissue absorption and metabolism. Simulations run on the models were used to study the relative hazards of high and low doses under acute and chronic exposure conditions. In addition, the models were used to assess the sensitivity of risk of carcinogenesis to changes in physiological variables and different levels of activity.

KEYWORDS: Perchloroethylene, pharmacokinetics, metabolism, risk assessment, system dynamics

INTRODUCTION

Most current quantitative risk assessments for carcinogens are performed by applying a relatively simple statistical model to rodent chronic bioassay data. There are now opportunities to improve the state of the art of carcinogenesis risk assessment by making use of more detailed information on the biological processes involved.[1-3] The dynamics of absorption, metabolic processing, elimination, and DNA reaction and repair are all, at least in principle, readily characterizable with relatively short term experiments. The mathematical descriptions of these dynamic processes can then be integrated into pharmacokinetic models. In the context of risk assessment such models can help to

1. predict the biologically delivered dose of active metabolites to humans as a function of different external exposure levels, routes, and time patterns; and

New Risks
Edited by L. A. Cox, Jr., and P. F. Ricci
Plenum Press, New York, 1990

2. relate the dosage of active metabolites available to human cells to the dosage available in animal bioassays.

Our report for NIOSH is an initial step in constructing an integrated series of pharmacokinetic models for small molecular weight alkylating agents and their precursors.[4] It explores the implications of existing pharmacokinetic information for the human carcinogenic risk of perchloroethyleneperchloroethylene (PERC, or more formally, tetrachloroethene).

PERC is a relatively simple molecule whose uptake and distribution has been relatively well studied in both rodents and humans. It is a member of a family of compounds (including vinyl chloride) that are at least transiently converted to epoxides and other active intermediates which can then react with DNA and other cellular nucleophiles. PERC is used mostly as a solvent for dry cleaning. Exposure levels average somewhat over 30 ppm for machine operators and in the neighborhood of 6 — 10 ppm for other personnel, with wide variability with respect to time and establishment.[5,6]

A practical attempt to apply pharmacokinetic principles and data to a specific compound is a test of the potential helpfulness of this approach both for risk assessment and for identifying additional experimental and epidemiological information that can help reduce uncertainties in carcinogenic risk assessment. This paper will describe the methodology used in the construction of our pharmacokinetic models and an easy-to-use microcomputer software package used in their implementation. In addition, the paper will describe some of the features that distinguish our approach from previous efforts using pharmacokinetic analysis in quantitative risk assessment.

DEFINITION OF TISSUE COMPARTMENTS

Pharmacokinetic models are intended to represent quantitatively the dynamics of the uptake, transfer, metabolism, and excretion of chemicals in biological systems. Because the number of tissues and organs can be large, pharmacokinetic analyses are simplified by combining tissues with similar kinetic behavior into "compartments." The single most important characteristic used to define expected kinetic behavior for these purposes is the ratio of blood flow through the tissue (Q) to the tissue volume (V). This ratio defines the rate constant in an equation describing the equilibration of a compound's concentration in two media. Traditionally, compartments have been constructed using the heuristic that processes of first-order kinetics are experimentally distinguishable only if rate constants for different tissues differ by at least a factor of five. However, it is not necessary to place all tissues with similar rate constants into the same compartment, and, in general, tissues/organs important in metabolizing foreign substances are placed in distinct compartments.

For example, human tissue can be divided into five compartments based on the rate constants derived for human tissues/organs (Fig. 1):

1. Vessel Rich Group (VGR) containing the brain, kidney, coronary, adrenals, and thyroid tissues as well as additional small viscera;
2. Vessel Poor Group (VPG) containing the skeletal bone and connective tissue;
3. Fat Group (FG) containing the perirenal and subcutaneous fat and marrow;
4. Muscle Group (MG) containing the lean body tissue of muscle, skin, and tongue; and
5. Liver Group (LG) placed by many researchers in the VGR. In our models the liver is placed in a distinct compartment because of its key role in metabolism.

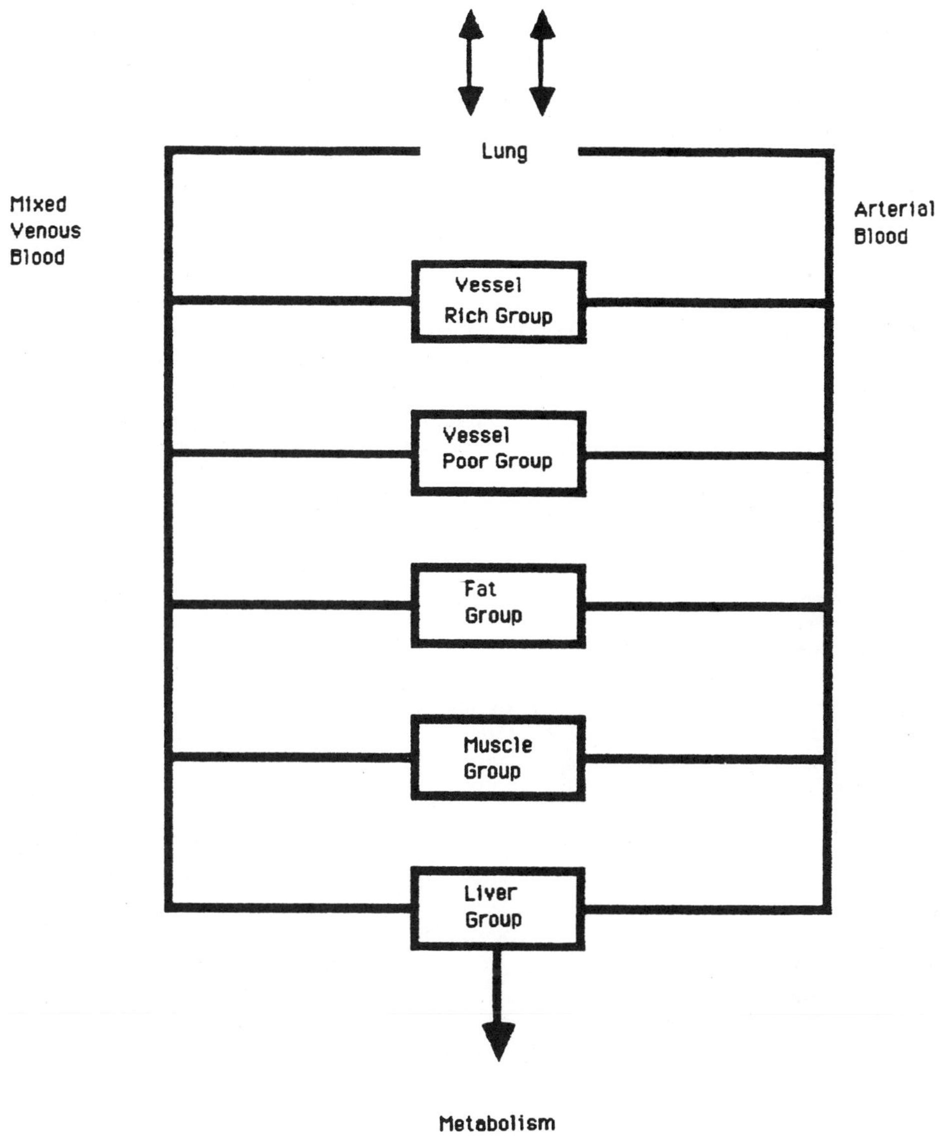

Fig. 1. Five compartment models of human.

DYNAMICS OF ABSORPTION OF THE TISSUE FROM THE BLOOD

We can demonstrate the mass conservation principles used in pharmacokinetic analyses by describing the dynamics of chemical exchange between tissue and blood. First, the partition coefficient, L_{t-b}, is defined as the equilibrium ratio of the concentration of a particular chemical in a specific tissue and the concentration in blood. If we assume rapid equilibration between the concentration, C, in the blood and the tissue, the output of the chemical from the tissue with respect to time is

$$dC/dt = C*Q/V*L_{t-b} \,. \tag{1}$$

Other things being equal, if the chemical has a higher affinity for the tissue than for blood, the blood flowing out will carry a smaller proportion of the chemical contained in the tissue. Effectively, the volume of the tissue relative to the blood flow is increased by the factor L_{t-b}. The relative solubility of the chemical in different media reduces the corresponding rate constant of tissue blood flow (Q) divided by tissue volume (V) by the factor L_{t-b} and the exponential equation derived by integration which describes the loss of chemical from the tissue is

$$C = C_o e^{-(Q/V*Lt-b)*t} \quad (C_o = \text{initial concentration}). \tag{2}$$

DYNAMICS OF ABSORPTION THROUGH THE LUNG

In the same way, a single equation describing the dynamics of absorption through the lungs, from the alveolar air to the arterial blood, can be derived. The flows of a chemical between the alveoli and the arterial blood can be described by the following equations:

$$\text{inflow} = Q*C_{\text{ven}} + C_{\text{exp}}*V_{\text{alv}}, \text{ and} \tag{3}$$
$$\text{outflow} = Q*C_{\text{art}} + C_{\text{alv}}*V_{\text{alv}}, \tag{4}$$

where C is concentration, Q is blood flow, V is the ventilation rate and the subscripts art, ven, exp, and alv correspond to arterial blood, and the mixed pool of venous blood, external air, and the alveoli, respectively.

To describe the relative affinity of the chemical for blood and air, a partition coefficient, L_{b-a}, is defined. Assuming rapid equilibration between arterial blood and alveolar air, the arterial concentration can be expressed as

$$C_{\text{art}} = C_{\text{alv}}*L_{b-a}, \tag{5}$$

and since the outflow from the lung must equal the total inflow to the lung,

$$C_{\text{art}} = (Q*C_{\text{ven}} + C_{\text{exp}}*V_{\text{alv}})/(Q + (V_{\text{alv}}/L_{b-a})) \tag{6}$$

or

$$C_{\text{art}} = L_{b-a}*(Q*C_{\text{ven}} + C_{\text{exp}}*V_{\text{alv}})/(Q*L_{b-a} + V_{\text{alv}}). \tag{7}$$

Equation (7) is directly incorporated into our models.

DYNAMICS OF METABOLISM

During exposure to a chemical, tissue compartment concentrations will begin to equilibrate with blood and the mixed venous concentration will rise. Eventually the net absorption through the lungs will decline to zero unless there are additional routes of elimination or the exposure level changes. However, in general, there is a route of permanent chemical change by metabolism in the liver, and in some cases additional routes exist. After an appreciable period of exposure, therefore, the rate of uptake of a chemical becomes a reasonable measure of internal rates of metabolism.

In our models we initially assumed that metabolism occurs solely in the liver group (LG) and is governed by basic Michaelis Menten enzyme kinetics:

$$\text{Metabolism rate} = V_{\text{max}}*C_{LG}/(K_m + C_{LG}), \tag{8}$$

where V_{max} is the maximum rate of metabolism and K_m is the liver concentration at which metabolism proceeds at half of V_{max}. This equation results from the fact that enzyme molecules have a maximum capacity to process their substrates. At high concentrations (i.e., high C_{LG} relative to K_m, a large proportion of the active sites are occupied and the reaction proceeds at close to a finite maximum rate, V_{max}. At low concentrations (i.e., low C_{LG} relative to K_m), metabolism is nearly directly proportional to C_{LG} with a rate constant equal to V_{max}/K_m. This form of metabolism results in the principal source of potential dose-response nonlinearity in our models.

COMPUTER IMPLEMENTATION

The human, rat, and mouse models used in our analysis were developed using STELLA, a "systems dynamics" package implemented on the Apple Macintosh.[7] "Systems dynamics" is a modeling methodology useful for exploring the behavior of complex systems. It has been shown that all dynamic systems can be represented using the concepts of stocks and flows: stocks represent accumulations through time; flows represent the movement of stock per unit time. An important premise of this methodology is that the cause and effect relationships in a system's behavior are primarily a function of the underlying feedback mechanisms and not of any external parameters.

The concept of feedback refers to closed sequences of causes and effects in a dynamic system. For example, the rate of exhalation of a substance during desorption depends on the concentration of the substance in the tissue. As the exhalation rate changes, the tissue concentration changes, thus creating a feedback loop as the new tissue concentration further modifies the exhalation rate.

Because STELLA makes extensive use of graphics, it can be easily used by people without extensive experience in programming. STELLA allows the construction of a model on the terminal screen as an explicit structural diagram defining the relationships between each model element. Figure 2 is an example of a STELLA diagram window. The program defines the relationship of each major element (accumulators, flow regulators, and converters) with other elements. STELLA facilitates the investigation of simulations through the animation of diagrams, the plotting of graphs, and the generation of tables of numerical data in real time.

Simulations in STELLA are run with user defined time steps and integration routines. Tissue compartment levels are calculated at the end of each time step; all inflows and outflows are integrated with respect to the time step, and the result is added to the previous amount contained in the accumulator. The maximum size of the time step required to obtain accurate computations is determined by the rate constants used in the model. The guiding heuristic for STELLA is that the time step must be no more than one fifth of the smallest rate constant in order to avoid rounding errors.

MODEL STRUCTURE

The STELLA implementation of the basic structure of flows in the human model (Fig. 1) is shown in Fig. 3. It incorporates the flows of blood and air between the external environment, the lungs, and the previously defined tissue compartments. The arterial blood concentration (ARTERIAL–CONC) and the mixed venous blood concentration) MIXED–VENOUS–CONC) are used to determine the rate of absorption/desorption through the lung. The rate of metabolism in the liver depends on the liver concentration and is calculated in LIVER–METABOLISM. The total amount of metabolites generated in a simulation are stored in TOTAL–METABOLISM.

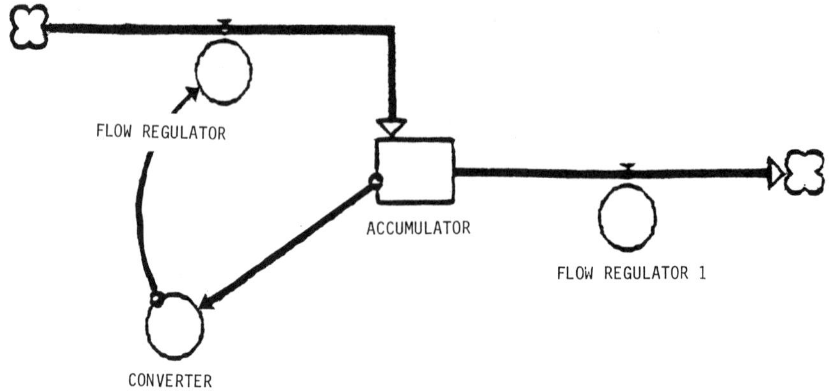

Accumulator: representation of levels of stock under analysis. They reflect conditions prevailing within the system at every point in time.

Flow Regulators: provide dynamic behavior to accumulators. Flow regulators use "decision rules" to compute flows from/to accumulators at each computation step given input from accumulators and/or converters.

Converter: allows information from other system components and constant parameters to be input into flow regulator decision rules.

Fig. 2. STELLA diagram.

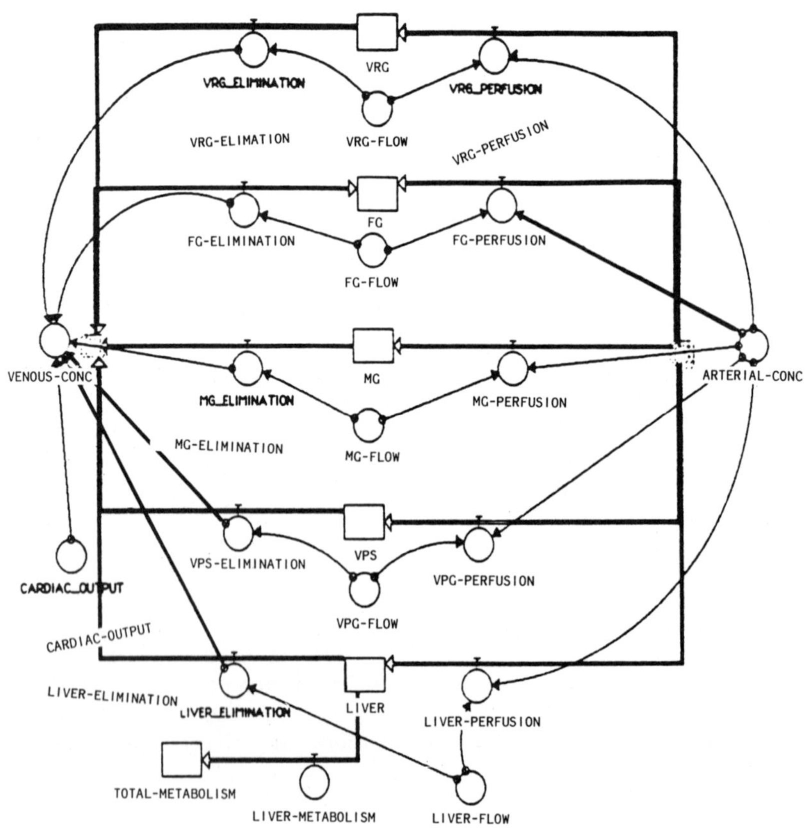

Fig. 3. STELLA implementation of human model.

Figures 4 and 5 show the relationships between parameters used to incorporate changes due to the time of day and the day of the week in the tissue blood flows and in the alveolar ventilation (V_ALV) and air exposure level (EXPOSURE), respectively. Using these relationships the models are able to accommodate dynamic changes in physiological parameters due to varying levels of activity during sleep, rest, and exercise.

MODEL RESULTS

Using the human, rat and mouse models, we simulated the dynamic behavior of various external dose regimens of PERC and calculated actual doses of metabolites delivered to the organisms. To construct our pharmacokinetic models, we began with physiological data on the tissues and blood flows of the organisms being modeled and the physical/chemical properties of PERC. The outputs generated by the models are the total active metabolites produced and the total amount of PERC exhaled unchanged. Literature data for metabolite production were used to calibrate the metabolism parameters in the human, rat and mouse models. No interspecies extrapolation of these parameters was necessary. Finally, the total amount of active metabolites produced was used in a multistage model for carcinogenic risk (GLOBAL82)[8] to predict lifetime risk of cancer. Both "best estimate" and 95% confidence level results were calculated.

An important aspect of our work was the ability to test the performance of our human model using extensive breath concentration data available from a previous human experimental study.[9] Modest but significant excesses of the breath concentrations predicted by the human models suggest the possibility of appreciable loss of PERC by a previously uncharacterized route. This could be a physical route of loss (e.g. through the skin) or a previously undocumented route of metabolism.

In our sensitivity analysis we found that a few important parameters cause most of the uncertainty in our models: different observations of human metabolite output for PERC reported in the literature result in five-fold differences in expected low dose metabolite output. Fat group perfusion has a significant effect on the amount of PERC retained during chronic exposure, and the ventilation rate has a significant effect on the delivered dose and rate of desorption.

CONCLUSION

Pharmacokinetic analysis provides an important tool to determine the differences between external exposure and delivered dose of a substance. It also enhances the ability to translate dose units between species. The user friendly systems dynamics modeling package STELLA provided a straightforward environment in which people with little experience in programming could construct complex models. Several features distinguish our modeling work from other efforts to incorporate pharmacokinetic analysis in quantitative risk assessments. In particular, we were able to

1. evaluate important dynamic effects resulting from administration of PERC via inhalation and gavage administration;
2. vary ventilation rates and blood flows to different tissues due to changing levels of activity;
3. vary exposure levels due to the time of day and days of the week; and
4. set metabolic parameters so that models reproduced actual human, rat and mouse data on metabolite output. Therefore, no interspecies projection of these parameters was necessary.

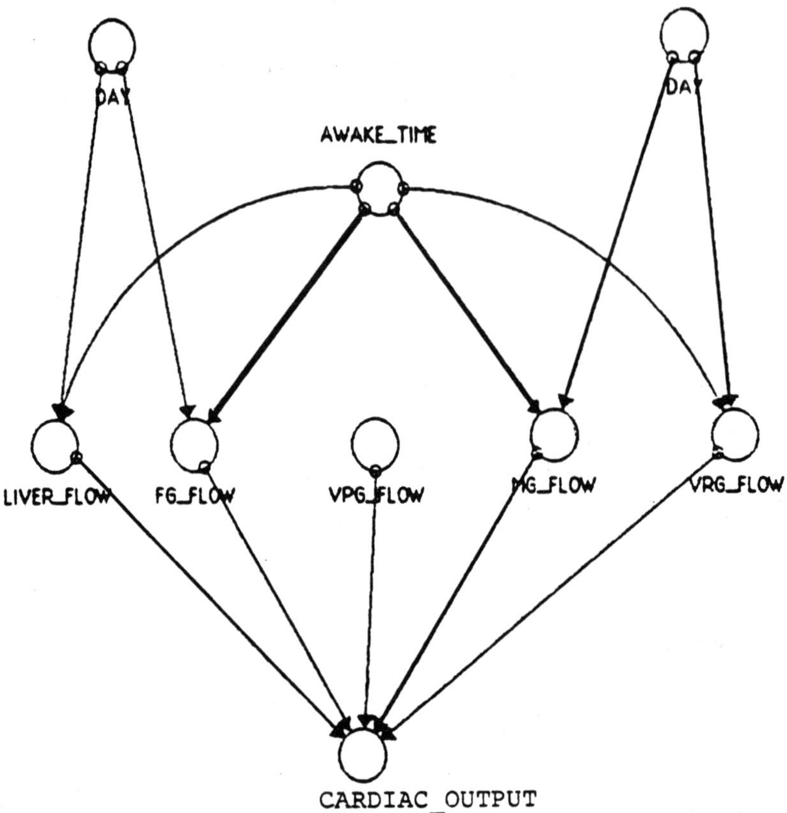

Fig. 4. STELLA implementation of blood flow variation.

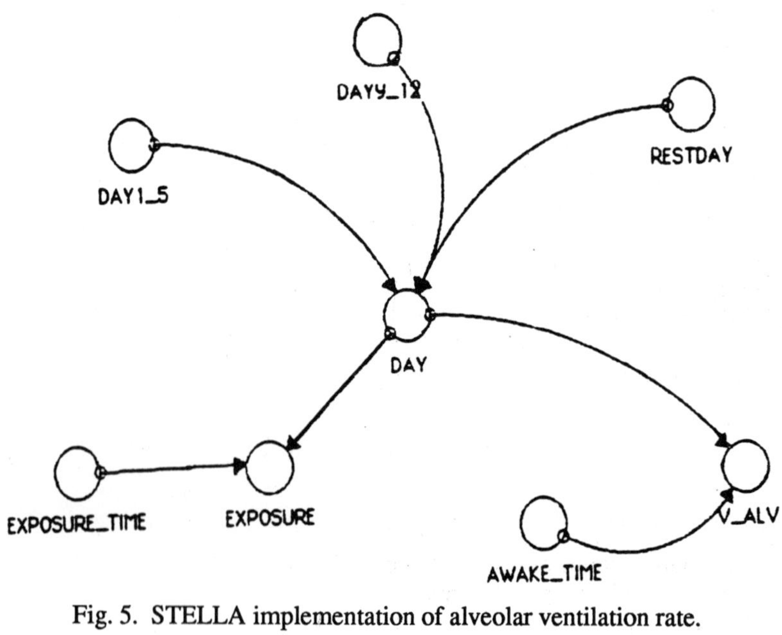

Fig. 5. STELLA implementation of alveolar ventilation rate.

REFERENCES

1. M. W. Anderson, D. G. Hoel, N. L. Kaplan, A General Scheme for the Incorporation of Pharmacokinetics in Low-Dose Risk Estimation for Chemical Carcinogenesis: Example — Vinyl Chloride, *Toxicol. Appl. Pharmacol.* **55**:154-161 (1980).
2. D. Hattis, The Promise of Molecular Epidemiology for Quantitative Risk Assessment, *Risk Analysis* **6(2)**:181-194 (1986).
3. P. J. Gehring, P. G. Watanabe, and C. N. Park, Resolution of Dose-Response Toxicity Data for Chemicals Requiring Metabolic Activation: Example — Vinyl Chloride, *Toxicol. Appl. Pharmacol.* **44**:581-591 (1978).
4. D. Hattis, S. Tuler, L. Finkelstein, and Z. Luo, A Pharmacokinetic/Mechanism-Based Analysis of the Carcinogenic Risk of Perchloroethylene, Report CTPID 86-7, Center for Technology Policy and Industrial Development, Massachusetts Institute of Technology.
5. H. R. Ludwig, M. V. Meister, D. R. Roberts, and C. Cox, Worker Exposure to Perchloroethylene in the Commercial Dry Cleaning Industry, *Am. Ind. Hyg. Assoc. Jour.* **44**:600-605 (1983).
6. B. L. Materna, Occupational Exposure to Perchloroethylene in the Dry Cleaning Industry, *Am. Ind. Hyg. Assoc. Jour.* **46(5)**:268-273 (1985).
7. Weigand, A STELLA Solution, *The MACazine*, 52-56 (1986). (STELLA is "Structured Thinking, Experimental Learning Laboratory with Animation.")
8. GLOBAL82 is a multistage carcinogenesis risk model developed by K. S. Crump and Company, Inc., Ruston, Louisiana. GLOBAL82 fits by maximum likelihood the multistage model to quantal dose-response data. It computes confidence limits by using a linear approximation valid at low doses and a more general procedure valid at high doses.
9. R. D. Stewart, C. L. Hake, H. V. Forster, A. J. Lebrun, J. E. Peterson, and A. Wu, Tetrachloroethylene: Development of Biologic Standard for the Industrial Worker by Breath Analysis, NIOSH, Report No. NIOSH-MCOW-ENVM-PCE-74-6, pg. 55.

Author Index

Subject Index

A

B

C

D

E

F

G

H

I

J

S

T

U

V

W